MW01331680

Conceptual Design of Distillation Systems

McGraw-Hill Chemical Engineering Series

Bailey and Ollis: *Biochemical Engineering Fundamentals*

Bennett and Myers: *Momentum, Heat and Mass Transfer*

Coughanowr: *Process Systems Analysis and Control*

deNevers: *Air Pollution Control Engineering*

deNevers: *Fluid Mechanics for Chemical Engineers*

Douglas: *Conceptual Design of Chemical Processes*

Edgar and Himmelblau: *Optimization of Chemical Process*

Gates, Katzer, and Schuit: *Chemistry of Catalytic Processes*

King: *Separation Processes*

Luyben: *Essentials of Process Control*

Luyben: *Process Modeling, Simulation, and Control for Chemical Engineers*

Marlin: *Process Control: Designing Processes and Control Systems for Dynamic Performance*

McCabe, Smith, and Harriott: *Unit Operations of Chemical Engineering*

Middleman and Hochberg: *Process Engineering Analysis in Semiconductor Device Fabrication*

Perry and Green: *Perry's Chemical Engineers' Handbook*

Peters and Timmerhaus: *Plant Design and Economics for Chemical Engineers*

Reid, Prausnitz, and Poling: *Properties of Gases and Liquids*

Smith, Van Ness, and Abbott: *Introduction to Chemical Engineering Thermodynamics*

Treybal: *Mass Transfer Operations*

Conceptual Design of Distillation Systems

M. F. Doherty and M. F. Malone

Department of Chemical Engineering
University of Massachusetts
Amherst

Boston Burr Ridge, IL Dubuque, IA Madison, WI New York
San Francisco St. Louis Bangkok Bogotá Caracas Kuala Lumpur
Lisbon London Madrid Mexico City Milan Montreal New Delhi
Santiago Seoul Singapore Sydney Taipei Toronto

McGraw-Hill Higher Education

A Division of The ***McGraw-Hill*** *Companies*

CONCEPTUAL DESIGN OF DISTILLATION SYSTEMS

Published by McGraw-Hill, a business unit of The McGraw-Hill Companies, Inc., 1221 Avenue of the Americas, New York, NY 10020.

This book is printed on acid-free paper.

1 2 3 4 5 6 7 8 9 0 DOC/DOC 9 3 2 1 0 9 8

ISBN 0-07-017423-7

Publisher: *Thomas E. Casson*
Executive editor: *Eric M. Munson*
Editorial coordinator: *Zuzanna Borciuch*
Senior marketing manager: *John Wannemacher*
Project manager: *Mary Lee Harms*
Media technology senior producer: *Phillip Meek*
Senior production supervisor: *Sandy Ludovissy*
Coordinator of freelance design: *Michelle D. Whitaker*
Freelance cover designer: *JoAnne Schopler*
Cover image: *PhotoDisc, Inc.*
Supplement producer: *Jodi K. Banowetz*
Compositor: *Techsetters, Inc.*
Typeface: *10.5/12 Times Roman*
Printer: *Quebecor World Fairfield, PA*

The credits section for this book begins on page 545 and is considered an extension of the copyright page.

Library of Congress Cataloging-in-Publication Data

Doherty, Michael F.
Conceptual design of distillation system / Michael F. Doherty, Michael F. Malone.—
1st ed.
p. cm.—(McGraw-Hill chemical engineering series.)
Includes index.
ISBN 0-07-017423-7
1. Distillation. I. Malone, Michael F., 1952– . II. Title. III. Series.
TP156.D5 D64 2001
600′.28425–dc21 00-063791
CIP

www.mhhe.com

To my mother and father,
to my wonderful wife, Margaret,
and to Sarah and Max
MFD

To the people who taught me something else:
Mom, Dad, and Mary,
Rick, Frank, and Kathy,
Christine,
Jim,
Pepper and the Chief,
Peter and the Ruggers,
and Suzanne
MFM

CONTENTS

PREFACE

One of the most important tasks in the chemical industry is the separation of multicomponent liquid mixtures into one or more high-purity products. Several technologies are feasible for this task, either alone or in combination, such as distillation, extraction, crystallization, etc. Among these, distillation is by far the most widespread and has a long history in chemical technology. However, until recently, there has been no *systematic* approach for understanding complex mixtures where azeotropes, multiple liquid phases, and simultaneous reactions may occur. This book describes such an approach including the basic data and models for understanding separation systems based on distillation. Some of the ideas and models are classical, but many are new and based on *geometric* methods of nonlinear analysis. This approach is made practical by modern methods for computer solution of these nonlinear models. The approach is suitable for simple as well as complex mixtures and can replace attempts to understand azeotropic, extractive, and reactive systems as separate and arcane topics. The ideas developed for distillation can be extended to other technologies such as liquid extraction and absorption.

The models developed here are the basis for software tools for *conceptual design* which are complementary to more detailed *simulation* methods developed in the last two decades. A major task in conceptual design is the generation and comparison of process alternatives, from which a promising subset is selected for more detailed study. Tools for computer-aided conceptual design of distillation systems have been developed and tested by several research groups, including our own at the University of Massachusetts. These or similar tools are available as part of at least two commercial products that are available for faculty and student use at reasonable cost. Although this book stands alone, our experience in teaching this material to both undergraduate and graduate students and to hundreds of industrial practitioners shows that these tools are valuable—probably more for the insight than for the numbers![1]

There are advocates for introducing more design concepts and problems throughout the engineering curriculum as an alternative to the traditional approach of a separate senior "capstone" course in design. It is our opinion that both are needed. The capstone course should provide a *systems approach* to whole plants, in which liquid separation systems are often a major component. In addition, this capstone course must reinforce the idea that students are expected to make and to justify decisions and must build on materials from earlier courses to offer a logical and manageable approach to such decision making. The systems approach is essential for this and,

[1]Paraphrasing Richard Hamming, who said "The purpose of computing is insight, not numbers."

we think, cannot be learned without a dedicated course. It makes more sense to us that elements of economics and optimization be introduced alongside the more fundamental material in earlier courses.

We have used the material in this book to teach a first undergraduate course in separations, which is followed by a second course on other separation technologies, many of them rate-controlled. The second course has more breadth than the first, but we do find that depth in the technology described in this book gives students confidence that they will be able to develop similar depth in other technologies on their own. The undergraduate course is preceded by a mass and energy balance course and two courses in thermodynamics; a heat transfer course is taken concurrently, but is probably not essential. One typical outline for an undergraduate course appears on page xii. A typical graduate course might cover almost all of the material, with approximately one hour of lecture on Chapter 1, two hours on Chapters 2, 3, and 6, and three hours each on the remaining chapters. For more information, or with questions, please e-mail to cdds@ecs.umass.edu.

Amherst, Massachusetts
July 14, 2000

ACKNOWLEDGMENTS

We are very grateful to the many people who helped us while we wrote this book.

First of all, it could not have been written without the creativity and hard work of our graduate students, postdoctoral fellows, and visiting scholars. We are particularly indebted to Domingos Barbosa, Christine Bernot, George Buzad, Glen Caldarola, Nitin Chadda, Madhura Chiplunkar, Fengrong Chen, Arthur Davydian, Marc DeGuiran, Dave Van Dongen, Tiziano Faravelli, Ted Fidkowski, Etienne Foucher, Sagar Gadewar, Kostas Glinos, Joel Grosser, Mooho Hong, Bob Huss, Vivek Julka, Jeff Knapp, Jennifer Knight, Leszek Krolikowski, Sandy Levy, Jason Manning, Mike Minotti, Ioannis Nikolaides, Matthew Okasinski, Raj Pai, Hoang Pham, Raymond Rooks, Maurizio Rovaglio, Peter Ryan, Wei Song, Ling Tao, Sophie Ung, Ganesh Venimadhavan, and Stan Wasylkiewicz. It was our pleasure to work with each of them, and we can only hope that they feel likewise.

Hundreds of undergraduates have suffered through early versions of this book without complaining about its faults. Their comments and suggestions were extremely helpful, and we thank them all for being such good sports. An equal or greater number of practicing engineers have attended a three-day short course based on these materials and offered numerous excellent suggestions.

Our academic and industrial colleagues, especially John Brierley, Richard Colberg, Jim Douglas, Dan Martin, Bill Parten, Jeff Siirola, Dave Smith, Jim Trainham, Don Vredeveld, and Clive Whitborn, gave constant advice and consistent support. We are especially grateful to Bruce Vrana, who probably read the manuscript as many times as we did! We are also grateful for excellent reviews and suggestions from Jeff Derby, Costas Maranas, Irv Rinard, Ross Taylor, and especially John Prausnitz.

We could not have completed this book without financial support from the American Chemical Society/Petroleum Research Fund, American Cyanamid, BASF AG, Dow-Corning Corp., E. I. DuPont Company, Eastman Chemical Company, GE Plastics, the GE Fund, Imperial Chemical Industries (UK), Mitsubishi Chemical Corporation (Japan), the National Science Foundation, Rohm and Haas Co., Rhone Poulenc, Shell International Chemicals BV, UOP, Unilever Research BV, Union Carbide Corporation, and the U.S. Department of Energy.

We are also grateful to the University of Massachusetts for support. Pamela Stephan, technical illustrator, created electronic versions of nearly all the figures and proved to be one of the most careful and patient people we have ever met. Thank you, Pam. Katie Crowley got more than she bargained for in preparing the final pages for publication. Sam Conti gave both advice and support on establishing our research programs that proved invaluable.

Finally, a special thank you to our families, who aren't sure what we do or why we do it but are willing to put up with us nevertheless.

COURSE OUTLINE

Class	Topic	Reference
Introduction and Fundamentals		
1	Introduction and Motivation	Chapter 1
2	Review of VLE Models and Data	2.1–2.4
3	Flash Separations, Sensitivity of Models	2.5
Basics: Binary Distillation and Economics of Design		
4	Balance Equations and Degrees of Freedom	3.1–3.2
5	Geometry: McCabe–Thiele Construction	3.3
6	Analysis: Fenske and Underwood Equations	3.4
7	Nonideal Mixtures and Examples	3.6
8	Computing: Performance Models	Handout
9	Exam 1	
10	Equipment Design	6.1
11	Cost Models and Economic Tradeoffs	6.2–6.3
Multicomponent Distillation		
12–13	Composition Profiles and Pinches	4.1–4.2
14	Minimum Stages and Flows	4.3
15	Fenske–Underwood–Gilliland Method	4.3
16	Column Sequencing for Ideal Mixtures	7.1–7.4
17	Feasibility and Product Distribution	4.4
18	General Approach for Nonideal Mixtures	4.4
19	Azeotropes, RCMs, and Distillation Boundaries	5.1–5.3
20	Exam 2	
21	Column Sequencing for Nonideal Mixtures	5.4–5.5
22	Sketching Residue Curves	5.3
Generalizations and Extensions		
23	Absorption and Stripping	Handout
24	Generalized Kremser Equation	Handout
25	Extraction: Liquid-Liquid Equilibrium	Handout
26	Extraction: Stage and Flow Requirements	Handout
27–28	Reactive Distillation	Chapter 10

LIST OF FIGURES

LIST OF TABLES

1

Introduction

1.1 MOTIVATION

Everyone has experience mixing liquids, and this is an important commercial activity, as in the blending of gasoline and the mixing of reactants in chemical reactors. It is less well known how to divide a liquid mixture into its separate components. This is surely due to the fact that mixing occurs spontaneously when two miscible liquids are put into contact (stirring and shaking act mostly to make the process occur faster), while separation does not normally occur without external intervention (e.g., heating or cooling, etc.). Because of this, separations have remained largely in the hands of experts from the days of alchemists to modern day engineers and scientists. The main purpose of this book is to pass along some traditional and some newer original material to educate beginners as well as experts in the design of separation systems.

Many separation devices exploit the fact that the equilibrium compositions of chemical species across coexisting phases are not equal. Thus, by repeatedly contacting the phases in a countercurrent fashion it is possible to isolate one or more of the components present in the feed mixture. Distillation, adsorption, absorption, liquid extraction, fractional crystallization, and other such devices all work on this principle. Although separation techniques have been used for thousands of years to make perfumes, alcoholic beverages, and medicines,[1] the formal physicochemical mathematical treatment of these devices is barely one hundred years old. In comparison with the study of mechanical systems, fluid motion, or heat transmission, separations science and engineering is still relatively young.

The first mathematical description of a countercurrent separation cascade was developed for distillation in a series of papers by Sorel (1889*a*, 1889*b*, 1889*c*, 1894). These results were greatly extended by Sorel (1893) in his book, *The Rectification of Alcohol,* and a few years later Schreinemakers (1901*a*, 1901*b*, 1903) laid the

[1]There are several fascinating accounts of the history and origins of chemical separations in the literature, e.g., Mohtadi (1982), Othmer (1982), Gentry (1995).

mathematical foundations of multicomponent batch distillation. A useful contribution on this subject for binary mixtures by Rayleigh (1902) also appeared about the same time.[2]

The advent of automobiles as a widespread consumer product in the first half of the twentieth century created a new and very important refinery industry for the manufacture of petroleum and petrochemicals. During this time much of the technology for separating multicomponent hydrocarbon mixtures was developed. These mixtures often contain many chemically similar species, and this allows the use of models that exploit the fact that these mixtures can often be treated as if they have ideal phase equilibrium behavior.

The invention of polyamides (e.g., nylons) by Caruthers in 1938 in the United States, and polyesters (e.g., Dacron) in England at about the same time led to the widespread use of polymeric materials that created a huge chemical industry for monomers, solvents, etc., in the second half of the twentieth century. The mixtures are often distinguished by the presence of very different functional groups, which leads to nonideal phase equilibrium behavior. Consequently, a great deal of effort in recent years has gone into developing the technology to separate nonideal mixtures into high-purity products.

Emerging technologies such as microelectronics, optical fibers, biotechnology, and high-performance materials have created a very active specialty chemicals industry. During their initial phases of development these industries create a demand for relatively small amounts of high-purity (and therefore expensive) raw materials. This has caused a resurgence of interest in batch processing and batch separations such as batch distillation. Moreover, the high molecular weight of many of these raw materials (most notably for the high-performance materials industry) means that they can be only partially separated by distillation. The final purification must be achieved by means of fractional crystallization, which is now entering a period of greater activity. The ultrahigh purities required for the microelectronics industry can only be achieved economically by final purification steps involving adsorption or reaction combined with separation. As industries mature, it is no longer economical to use batch separations. Thus, the latest generation of plants for making chlorosilanes (which are precursors for the microelectronics and optical fiber industries) all employ continuous distillation separations.

Separation methods such as distillation and extraction remain vitally important for the chemical industry and will do so for the foreseeable future. Driven by modern technological demands, the field has not stood still, and in the last twenty years many new techniques for understanding and carrying out these separations have been developed. Our purpose here is to try to present these techniques in a systematic and coherent way.

New separations based on differences in the rates of transport of the various species present are being developed for such industries as gas separations and

[2]It is interesting that the theory of Rayleigh was known to Schreinemakers, who had already developed the multicomponent version. This type of batch distillation and the equations that describe it carry Rayleigh's name in the English literature, but they are attributed to Schreinemakers in the Russian and German literature.

biotechnology. These separations include the use of membranes, hollow fibers, chromatography, and electrophoresis. Although these separation techniques are important, they are not nearly so well developed as the equilibrium fractionations, and probably will find important applications in a more narrow range of industries. Because the emphasis of this book is on the conceptual design of separation systems, we have chosen to concentrate on those separations which are reasonably well understood, for which there is a sufficient body of data available, and for which the applications are the most broad. As we will show, attention to a *systems* approach to the design of separation systems may lead to much more economical processes.

1.2 SEPARATIONS IN CHEMICAL PROCESSING

The Importance of a Systems Approach

Chemical process flowsheets can often be best understood if they are broken down into smaller interconnected subsystems (Siirola and Rudd, 1971; Siirola, 1996). For example, Douglas (1988) describes one decomposition in which the flowsheet is broken down into three main tasks: the reactor system, the liquid separation system, and the vapor recovery system, as shown in Fig. 1.1. These subsystems are connected together in the forward path by the downstream flow of the process stream, and in the backward path by the recycle flow of raw materials and intermediates. Each of these subsystems can be a quite intricate collection of units with its own internal or local recycle loops.

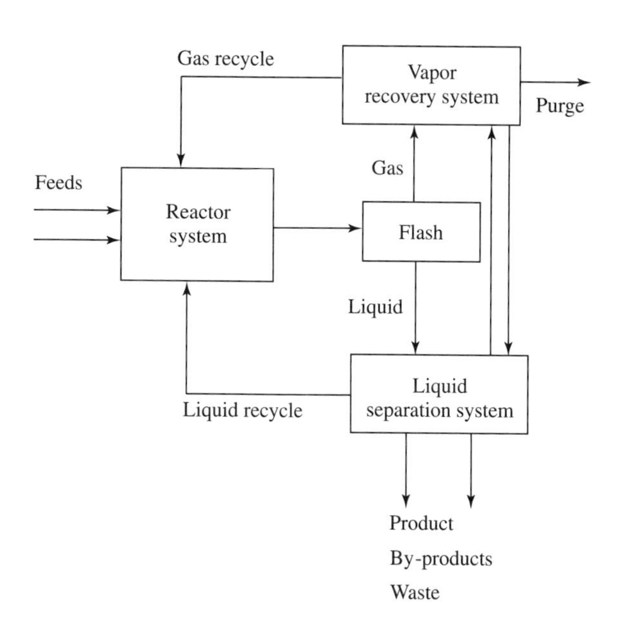

FIGURE 1.1
Douglas' decomposition of a process for the continuous production of a single product.

The main idea behind the decomposition in Fig. 1.1 is that the distribution of chemical products generated in the reaction dominates the conceptual design. The raw materials enter the reactor system, where they undergo chemical transformation into the main reaction products, and invariably into by-products. The foremost questions concern the *selectivity* s, which represents the efficiency with which raw materials are converted to valuable product. The selectivity is the moles of primary product produced divided by the moles of limiting reactant consumed. The selectivity varies from 0 to 1, depending on the stoichiometry, the type of reactor configuration, the catalyst, the reactor temperature(s), conversion(s), and molar feed ratio(s) of reactants. The extents for each individual reaction give additional information on the nature and amount of the by-products. Thus, the reactor product distribution and some mass balances determine the flowrates of all the species leaving the reactor system and entering the separation system.[3]

If the reactor products contain "noncondensable" components such as methane, hydrogen, argon, etc., then they are separated from the remaining reactor products by means of a flash separation. The process stream is normally cooled and/or depressurized before the flash separator. The vapor stream leaving the flash may be fed to a separation system in order to separate and purify the various components for recycle, product withdrawal, etc.; this is usually accomplished by absorption or membrane separations. Alternatively, the vapor stream may be used as fuel or, when valuable raw materials are present in the vapor, a gas recycle and purge stream may be used.

The liquid stream leaving the flash is fed to a liquid separation system where the components are separated into product, by-product, and recycle streams. These separations normally involve distillation and liquid-liquid extraction. Whenever absorption, adsorption, or liquid-liquid extraction is used in the separation system, further distillation is almost always required in order to separate the solute from the solvent. These separations therefore generate local recycles inside or between the vapor and liquid separation subsystems.

Each subsystem can be broken down further into the individual unit operations, which can be designed using the techniques described here and elsewhere. Once each of the subsystems has been designed, they must be connected in order to accommodate overall process interactions. This may entail, for example, adjusting the design of parts of the separation system in order to accommodate reactor system performance or constraints, or it may entail more significant flowsheet modifications such as the addition or removal of a separator, or the restructuring of the sequence of separators.

The normal yardstick used to measure the relative merits of one design over another is the process economics. The cost of each unit (together with any utilities required) in each subsystem can be estimated, e.g., see Chapter 6, and the total process cost can be assembled. The total process cost will depend on certain key design variables such as reactor conversion, recycle compositions, and purge compositions. What distinguishes the key design variables from the rest is that they influence not

---

[3]Douglas (1988) describes this procedure in detail with the assumptions and approximations that go into it.

only the cost of their own subsystem but also the cost of the other subsystems in the flowsheet. It is precisely this effect that gives rise to overall process interactions that cannot be ignored and it is information about these key design variables that is essential for a proper definition of the separation tasks.

In most cases, the choice of the key design variables requires information about economic trade-offs involving the raw materials and the reactor system and the separations subsystems. For example, in many cases, the selectivity *increases* as the conversion of the limiting reactant *decreases*. Increased selectivity means that less raw material is converted to by-product, thereby saving on raw material costs. However, as the reactor conversion is reduced, a larger amount of unreacted raw material must be separated from the products and by-products, and recycled to the reactor. The related separation costs increase quite steeply as the reactor conversion decreases toward zero. On the other hand, the reactor volume (and cost) increases sharply as the conversion approaches unity (or its equilibrium value, for the case of equilibrium-limited reactions). Therefore, an optimum value for the reactor conversion must exist which balances the reactor costs, the separation costs, and the raw material costs. This optimum affects the stream temperatures, pressures, and compositions, as well as the equipment sizes in both the reactor system and the separation system. The preferred conditions cannot be found by simple rules or by design and analysis of any one subsystem alone.

There are many other examples of how overall process interactions lead to modifications in the process subsystems that could not be discovered by analysis of the subsystems alone (Douglas, 1988; Siirola, 1996). In order to create an economical and competitive process, we need to explore the effect of the key design variables on the total processing cost for each of several promising flowsheet structures. To do this effectively, we need design methods that are equation-based, simple, and accurate. In this way, the design calculations can be performed repeatedly on a computer within a reasonable amount of time without introducing significant errors that could lead to a poor conceptual design of the process.[4] This latter point is actually quite important because poor initial designs are very costly to correct at the later design stages. A commonly accepted rule of thumb is that every dollar spent at the initial design stage to improve the conceptual design of the process costs $10 at the flowsheeting stage, $100 at the detailed design stage, $1,000 after the process is built, and $10,000 to clean up the mess after a failure Kletz (1989).

Synthesis vs. Analysis

Design requires both synthesis and analysis.[5] In design, we specify the feeds to a system together with targets on the output streams (e.g., desired purities of selected

[4]For these reasons, graphical techniques in this book are used primarily to explain or illustrate the equation-based methods.

[5]The latter is often carried out by *simulation,* for which there are many good and mature commercial software packages. In fact, simulation has been so successful in the last two decades that it has come to be the *only* sort of analysis or computing that is thought to be useful! We disagree strongly with this viewpoint and hope to show how other analysis tools are worthwhile in design.

components), and the goal is to find process flowsheets, equipment sizes and configurations, together with operating conditions that will do the job economically. The two major tasks in design are (1) to identify feasible flowsheets and ranges of operating conditions that are capable of meeting the desired process goals and (2) to rank the alternatives according to some economic measure in order to select a few candidates for more detailed studies.

Analysis is the complement of synthesis, where we specify the state of the feeds, the structure of the flowsheet, the equipment sizes, and values for all independent design variables (i.e., the operating conditions), and calculate the values of the remaining variables, including the state of the product streams.

Many separation systems have been invented by experience. However, this leaves open the questions of what potential there may be for improving existing separation systems as well as what methodology to use in efforts to devise new and improved systems. Furthermore, new technologies, along with environmental and economic conditions, present problems for which there is a relatively small base of experience to draw upon. The recent development of systematic design procedures, especially for nonideal mixtures, and startling decreases in the cost of computing for their implementation can be used to address these questions.

Some people may have the impression that the sole purpose of process synthesis is to provide the optimal specification of equipment based on known physical properties and specifications. We believe otherwise—that the major benefits from computing tools in conceptual design come when they are used in conjunction with experimental studies. For separations, this is partly because there remains a significant uncertainty in thermodynamic predictions made for very nonideal systems, despite the availability of large databases for vapor-liquid and liquid-liquid equilibrium. The data available for reacting mixtures is more sparse.

We would also like to keep in mind not only the difference between analysis and synthesis but also the fact that they are complementary activities. Synthesis relies on results from analysis, and much of the analysis is fueled by alternatives generated in the synthesis stage. Early work on process synthesis gives the impression that these two activities can be separated, but our experience suggests the contrary. Figure 1.2 portrays the idea that we need both synthesis and analysis for conceptual design.

A critical issue in any synthesis study is the number of alternative solutions. There may be many alternatives and there is no general rule to choose among them. Many may have comparable costs, or all but a few may be economically unattractive so some analysis for the ranking of alternatives is useful. Fortunately, this analysis need not be a rigorous design. When the differences among alternatives are small, high accuracy is required to identify the true optimum, but the choice among neighboring alternatives is unimportant because the differences are small. Conversely, larger differences among the alternatives mean that less accurate models will not lead to bad decisions. In other words, models need to be sufficiently accurate to discard poor alternatives and yield a smaller number of candidate designs worthy of further attention. A well-known example is the large number of alternative distillation sequences for ideal mixtures, but for very nonideal mixtures many of these alternative sequences are infeasible. This does not necessarily mean that there are no alternatives, but only that more work is required to find them.

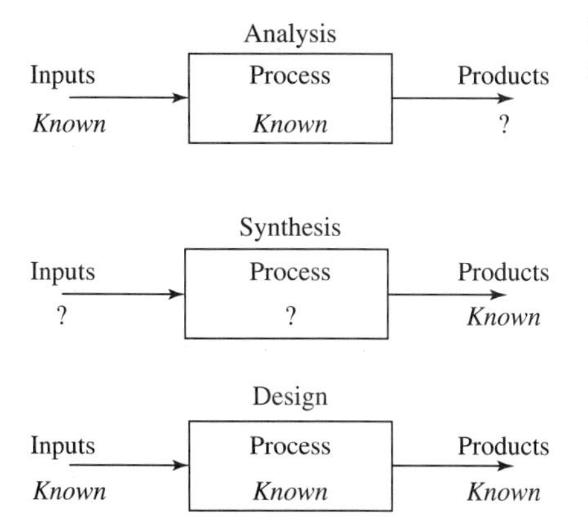

FIGURE 1.2
Synthesis + analysis = design.

In this book we describe an approach that combines heuristics and models for conceptual design. The models are typically nonlinear and make use of *geometric methods* to find solutions. The models are intended to preserve the essential nonlinearities and to address synthesis as well as analysis. On the other hand, for conceptual design, a complete first principles model is often too demanding of data, engineering time, and perhaps computational resources, so we will also use some heuristics to avoid expending resources to solve hard problems which either are not worth solving or typically have good solutions that are not very different from one another.

In order to illustrate the importance of separations in modern chemical technology, we now briefly describe two chemical processes that have been selected as prototypes for different sectors of the chemical industry. In these examples, we describe the complete process system and its chemistry, which may strike the reader as unusual in a book about separations. We hope to convince you that the design of separation systems is often coupled to effects on a higher level in the conceptual design and that the best designs recognize the intimate coupling between the reactor and separation subsystems.

1.3 EXAMPLES

Manufacture of Methyl Acetate

High-purity methyl acetate is used in large amounts as an intermediate in the manufacture of a variety of polyesters e.g., photographic film base, cellulose acetate, Tenite cellulosic plastics, Estron acetate for filter tow and textile yarns, etc., (see Mayfield and Agreda 1986; Agreda 1988; Agreda and Zoeller 1993). The manufacture of high-purity methyl acetate by the esterification reaction of acetic acid with methanol

$$CH_3COOH + CH_3OH \rightleftharpoons CH_3COOCH_3 + H_2O \tag{1.1}$$

is difficult because of reaction equilibrium limitations, the difficulty of separating acetic acid and water, and the presence of azeotropes between methyl acetate and methanol, and between methyl acetate and water. Since the thermodynamic equilibrium constant[6] is on the order of 20 and the reactor effluent contains significant amounts of all four components, conventional processes use one or more liquid-phase reactors with a large excess of one reactant in order to achieve high conversions of the other. One such conventional separation system described by Siirola (1996) requires 8 distillation columns, 1 liquid extractor, and 1 decanter[7] to break the azeotropes and tangent pinch present in the mixture. A schematic representation of the conventional flowsheet is shown in Fig. 1.3. This process requires a large capital investment; it has high energy costs and a large inventory of solvents. Therefore, there is an incentive to invent a better process in which some or all of these effects are reduced.

In 1975, the Tennessee Eastman Company[8] began a research program to develop a process to produce acetic anhydride, $(CH_3CO)_2O$, using coal instead of oil as the raw material source of carbon. Acetic anhydride is one of the company's key intermediates and it uses approximately one billion pounds per year in making the polyesters mentioned above. The polyester products are made by acetylating wood cellulose with acetic anhydride, giving acetic acid as a by-product which is used elsewhere in the process.

The overall chemicals-from-coal process consists of several interconnected plants in which the following (simplified) chemistry takes place.

Coal gasification plant

$$C + H_2O \rightarrow CO + H_2 \tag{1.2}$$

Synthesis gas (a mixture of CO and H_2) is generated from coal using the Texaco coal gasification process. An aqueous slurry of coal with oxygen is fed to the gasifier operating at high temperature and pressure. The product gas is mainly carbon monoxide and hydrogen, which is a good feedstock for the chemicals plants.

Methanol plant

$$CO + 2\,H_2 \rightarrow CH_3OH \tag{1.3}$$

Methanol is produced from carbon monoxide and hydrogen in a catalytic, gas-phase reactor operating at elevated temperature and pressure.

Methyl acetate plant

$$CH_3COOH + CH_3OH \rightleftharpoons CH_3COOCH_3 + H_2O \tag{1.4}$$

Methyl acetate is made by esterification reaction between recycled acetic acid from the polyester plant with methanol from the methanol plant.

---

[6]Defined as the product of the activity of each component raised to the power of the stoichiometric coefficient, with the convention that stoichiometric coefficients for reactants are negative and products are positive.

[7]The separation system consists of an extractive distillation subsystem, an azeotropic distillation subsystem, as well as individual columns that perform specific separations.

[8]Now the Eastman Chemical Company.

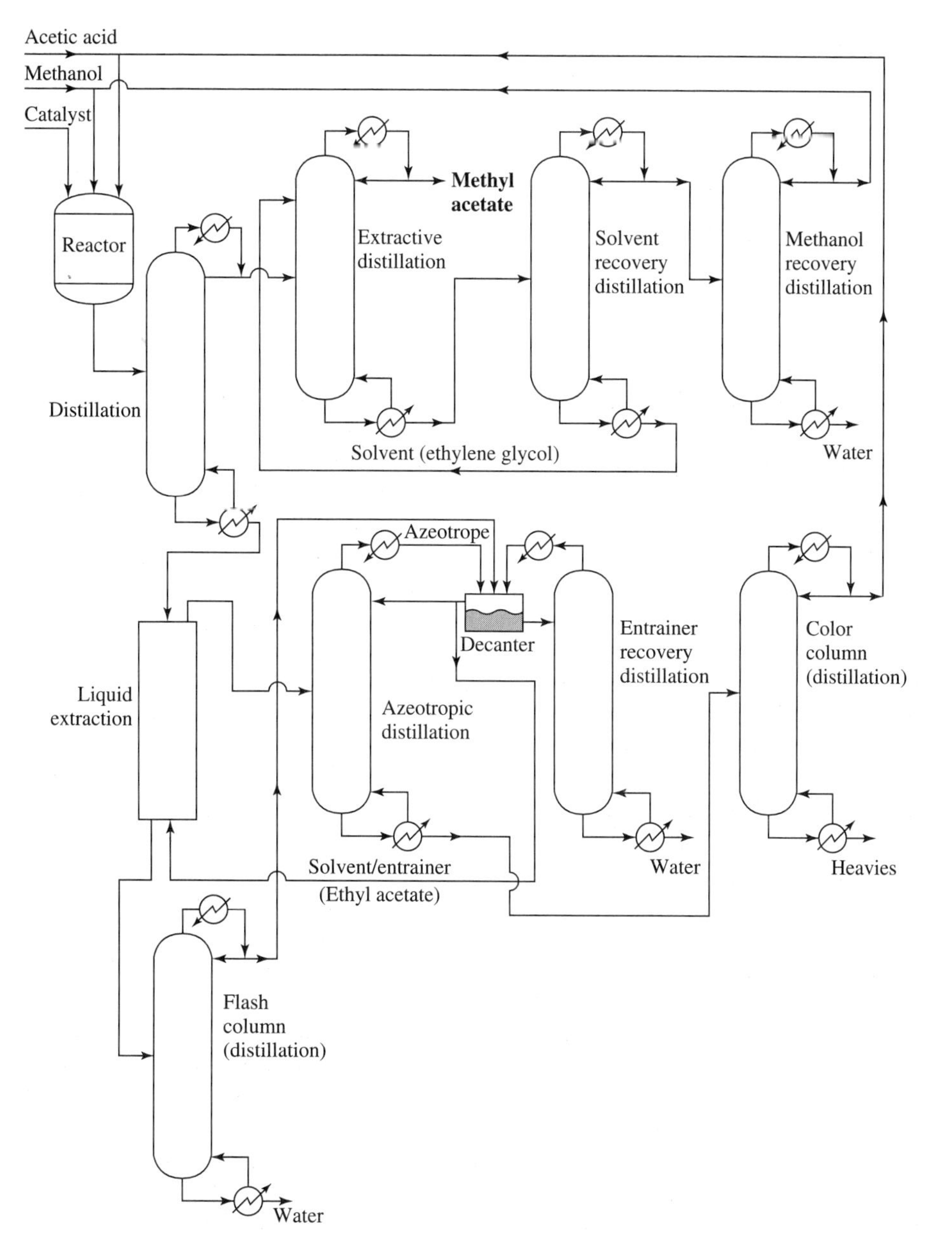

FIGURE 1.3
Flowsheet for a traditional process to produce methyl acetate. Siirola (1996).

Acetic anhydride plant

$$CH_3COOCH_3 + CO \rightarrow (CH_3CO)_2O \tag{1.5}$$

Acetic anhydride is made in a catalytic, gas-liquid carbonylation reactor by reacting purified CO with methyl acetate.

Polyester plant

$$(CH_3CO)_2O + \text{cellulose} \rightarrow \text{products} + CH_3COOH \quad (1.6)$$

The acetylation produces by-product acetic acid which is recycled from this plant to the methyl acetate plant.

Acetic acid plant

$$(CH_3CO)_2O + CH_3OH \rightarrow CH_3COOH + CH_3COOCH_3 \quad (1.7)$$

Additional acetic acid for use in this and other processes on the site is made by reacting acetic anhydride with methanol. The methyl acetate resulting from this reaction is sent back to the carbonylation reactor.

The entire plant complex was completed and successfully started up in 1983, and Eastman became the first manufacturer to produce a modern generation of industrial chemicals from coal. The process converts 900 tons of coal a day into 500 million pounds of acetic anhydride per year (and similar amounts of the intermediates). The plant is so economical and clean that it has been expanded in recent years.

One of the keys to the success of this project was finding efficient techniques for the many complex separations involved. The most challenging of these was the separation system for the methyl acetate plant, which was entirely replaced with a revolutionary reactive distillation unit (shown in Fig. 1.4) invented by Agreda and Partin (1984). In this single column, high-purity methyl acetate is made with no additional purification steps and with no unconverted reactant streams to be recovered. It also provides a method of driving the equilibrium-limited reaction to high conversions without a large excess of one of the reactants. It does this by allowing the liquid-phase reacting mixture to flash off methyl acetate, thereby increasing the conversion. The reactive column has stoichiometrically balanced feeds and is designed so that the light reactant, methanol, is fed to the bottom section and the heavier acetic acid is fed at the top. This provides good contact between the reactants in a countercurrent fashion. Most of the reaction takes place in the middle section of the column below the sulfuric acid catalyst feed point. As methyl acetate is formed, it produces a minimum-boiling azeotrope with methanol; the azeotrope is the lightest boiler in the mixture and it is the product from the top of the reactive zone. The

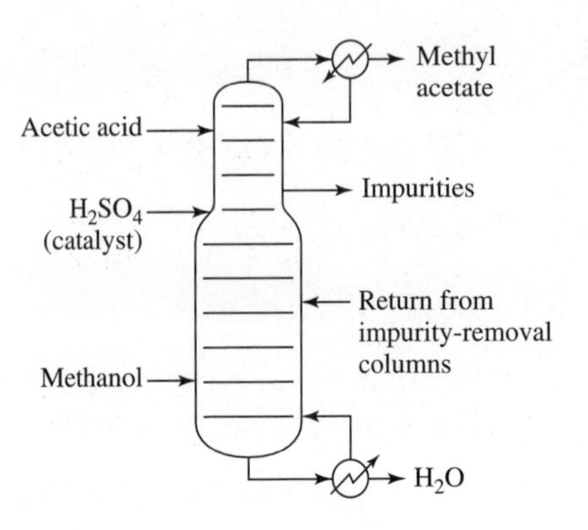

FIGURE 1.4
Reactive distillation column for production of methyl acetate. *[Adapted from Agreda and Partin (1984) and Agreda et al. (1990)].*

lower section of the column, below the methanol feed, performs a methanol-water separation to give high-purity water as bottoms product. The top section of the column, above the H_2SO_4 feed point, performs an extractive distillation using the acetic acid as entrainer to "break" the methyl acetate-methanol azeotrope, and high-purity methyl acetate is produced as the distillate. Composition and temperature profiles for the column are reported by Agreda et al. (1990) and are shown in Fig. 1.5.

The column is, in fact, an entire chemical plant in one unit, producing 400 million pounds of high-purity methyl acetate per year (Mayfield and Agreda 1986).

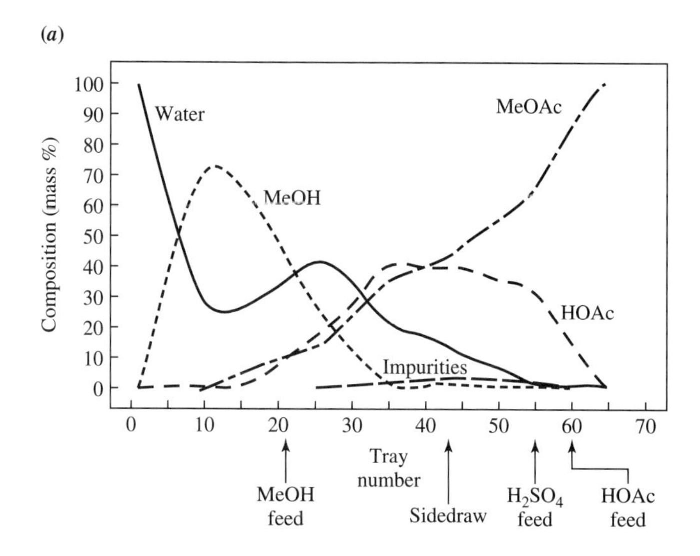

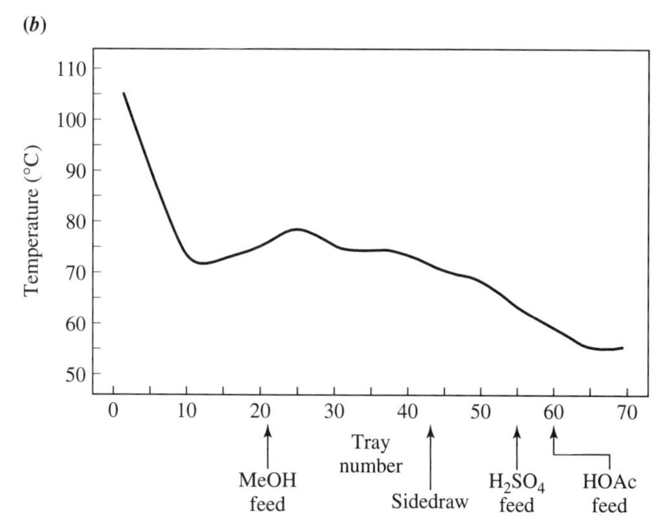

FIGURE 1.5
Composition and temperature profiles for reactive distillation.
Agreda et al., (1990).

This process is exceptionally economical and is reported to cost only *one-fifth of the capital investment of the conventional process and consumes only one-fifth of the energy* (Siirola 1996, p. 229). Since this plant came online, an improved second-generation reactive distillation unit has been built, and the two together produce nearly 1 billion pounds of methyl acetate per year. A photograph of one of these columns is shown in Fig. 1.6.

FIGURE 1.6
Methyl acetate column at Eastman Chemical Company, Kingsport, Tennessee.

Manufacture of Chlorosilanes for Microelectronics and Optical Fibers

The manufacture of solid-state electronic devices hinges on the semiconductor properties of certain materials that can be made into devices such as diodes and transistors and used in place of electron tubes in electrical circuits. In this way the circuits can be miniaturized, which leads to lower power consumption, higher device speed, greater durability, and of course much smaller device sizes. For example, the integrated circuits in a modern microprocessor chip contain more than 10 million transistors in a chip which measures only a few square centimeters in area.

Semiconducting materials include elements such as silicon, germanium, and diamond, and compounds such as silicon carbide (SiC) and gallium arsenide (GaAs). Although a variety of elemental and compound semiconductors are used for various research and specialty applications, silicon is by far the most widely used material.

In order to create the semiconducting devices in the integrated circuits on a silicon chip, it is necessary to deliberately introduce certain dopants. The dopant composition levels are carefully controlled, and are typically in the 0.5 to 5 parts per billion (ppb) range. Because of this, the silicon must be grown as a large and essentially perfect single crystal, so that when examined by conventional transmission electron microscopy, virtually no dislocations or physical defects can be detected. Moreover, the impurity levels must be less than 150 parts per trillion. Worldwide consumption of semiconductor-grade silicon is well above 10 million lb/year, which is small compared to conventional production in the chemical process industries. Nevertheless, the total value is quite substantial, since the selling price is hundreds of dollars per pound, or more depending on the scale.

High-purity silicon is manufactured by refining metallurgical-grade silicon, which in turn is produced by the carbon reduction of high-purity silica sand. The purified polycrystalline silicon, usually known as polysilicon, is used to grow large single crystals of silicon by continually pulling the crystal from the melt by a process known as Czochralski growth. These crystals are usually in the form of long cylindrical ingots 6 to 8 inches in diameter. The ingots are sliced to form thin wafers that are chemically and mechanically polished to produce perfect, damage-free surfaces. At this stage the device fabrication begins [for a summary of the remaining steps in the process see Larabee (1985), and Ravi (1981)].

The basic processing steps involved in manufacturing polysilicon are outlined below, and should be read in conjunction with the flowsheet shown in Fig. 1.7. This information was obtained from many sources, but most notably from Truitt et al. (1983).

Step 1. Metallurgical-grade silicon is made by reducing silicon dioxide (silica sand) with carbon (e.g., coal, coke, or wood chips) in a submerged electrode arc furnace at temperatures of 2000°C or higher. The primary reaction that takes place is

$$SiO_2 + 2C \longrightarrow Si + 2CO$$

which produces metallurgical-grade silicon that contains approximately 98.5 wt % pure silicon and 1.5 wt % impurities such as iron, aluminum, copper, boron, and phosphorus. Most polysilicon manufacturers buy metallurgical-grade silicon, so the flowsheet in Fig. 1.7 begins with this as a raw material.

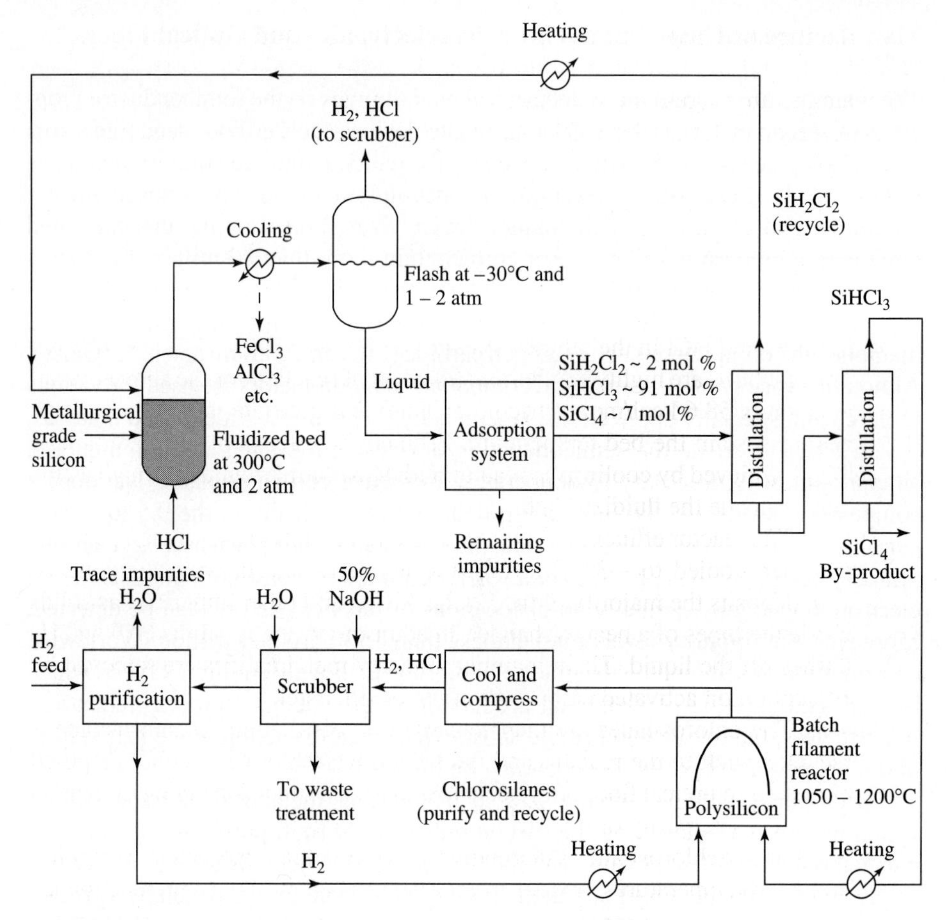

FIGURE 1.7
Flowsheet for manufacturing polysilicon.

Step 2. The metallurgical-grade silicon is dried with nitrogen and then reacted with anhydrous hydrogen chloride in a fluidized bed reactor at approximately 200°C and 2 atm pressure. The primary reactions that take place are

$$\mathrm{Si} + 2\mathrm{HCl} \longrightarrow \underbrace{\mathrm{SiH_2Cl_2}}_{\text{(dichlorosilane)}}$$

$$\mathrm{SiH_2Cl_2} + \mathrm{HCl} \longrightarrow \underbrace{\mathrm{SiHCl_3}}_{\text{(trichlorosilane)}} + \mathrm{H_2}$$

$$\mathrm{SiHCl_3} + \mathrm{HCl} \longrightarrow \underbrace{\mathrm{SiCl_4}}_{\text{(silicon tetrachloride)}} + \mathrm{H_2}$$

In addition, several metallic halides are formed such as $AlCl_3$, $FeCl_3$, BCl_3, etc. The chlorosilanes all boil at temperatures below 65°C, the $AlCl_3$ vaporizes at ~140°C, and $FeCl_3$ vaporizes at ~275°C. Therefore, the gas stream leaving the fluidized bed contains excess HCl, SiH_2Cl_2, $SiHCl_3$, $SiCl_4$, H_2, $AlCl_3$, $FeCl_3$ and the halides of other impurities. The conditions in the bed, e.g., temperature, pressure, residence time, and molar feed ratio of reactants, are designed to favor the production of trichlorosilane. The overall reaction in a typical commercial fluidized bed is

$$Si + 3.05HCl + \text{impurities} \longrightarrow$$
$$0.02SiH_2Cl_2 + 0.91SiHCl_3 + 0.07SiCl_4 + 1.05H_2 + \text{impurity halides}$$

It is typical to use an excess of HCl to metallurgical-grade silicon in a molar ratio of approximately 3.7:1. The excess HCl in the reactant stream appears as unreacted HCl in the effluent stream.

The reactions are highly exothermic (the heat of reaction for the above overall reaction is 58.6 kcal/mol of trichlorosilane) and generate more than enough heat to sustain the bed temperature. Efficient heat removal is essential, and this is achieved by cooling the reactor with Dowtherm running through a tube bundle inside the fluidized bed.

Step 3. The reactor effluent gas is passed through a filter to trap silicon particles and then cooled to −30°C in order to condense the heavier components. This deposits the majority of the $FeCl_3$, $AlCl_3$, and other impurities as solids inside the tubes of a heat exchanger. In addition, virtually all the HCl and H_2 flashes off the liquid. The remaining impurity metal halides are removed by adsorption on activated carbon, alumina, or silica gel.

Step 4. The chlorosilanes are then separated by distillation. Dichlorosilane is recycled back to the reactor, and the silicon tetrachloride is either disposed of or sold to optical fiber or fumed silica manufacturers depending on market demands. The purity of the trichlorosilane product is 99.9 mol %.

Step 5. The trichlorosilane is thermally reduced with a large excess of dry hydrogen at temperatures ranging from 1050 to 1200°C. The main reaction is

$$SiHCl_3(g) + H_2(g) \rightleftharpoons Si(s) + 3HCl(g)$$

Other chlorosilanes are produced by the side reactions

$$Si + 2HCl \rightleftharpoons SiH_2Cl_2$$
$$Si + 4HCl \rightleftharpoons SiCl_4 + 2H_2$$

The silicon is deposited as a solid on electrically heated silicon filaments of high purity.[9] In straight filament reactors polysilicon rods 100 to 120 cm long are grown to diameters of about 12 cm from 0.5-cm-diameter filaments. An alternative configuration is to use a hairpin filament.

The equilibrium reactions limit the amount of silicon deposited per pass through the reactor to about 25 mol % of that present in the feed.[10] The reactor effluent is compressed and cryogenically cooled to condense the chlorosilanes for purification and recycle. Hydrogen is stripped of HCl and residual

---

[9]The melting point of silicon is 1420°C.

[10]This low yield of silicon motivates the search for process alternatives. These include replacing the trichlorosilane by dichlorosilane or silane (SiH_4) which give higher silicon deposition yields; 37 and 56 mol %, respectively. These and other alternatives are described in Truitt et al. (1983); Bawa et al. (1980*b*); Bawa et al. (1980*a*) and Ravi (1981, pp. 3–5.)

chlorosilanes in a high-pressure scrubber and further purified and dried for recycle.

Step 6. The polysilicon is converted to a large single silicon crystal by either a Czochralski growth process or a zone flotation process Ravi (1981, pp. 5–17). This is the starting point for device fabrication.

The separation system in step 4 can be devised in one of several different ways. In Fig. 1.7 we have shown one particular configuration in which the lightest component is separated first, followed by separation of the two remaining components. This is usually called a "direct split" or a "direct sequence" of columns. Other possibilities include the "indirect" sequence in which the heaviest component is separated first, and complex sequences which involve special "nonsharp" separations. Systems issues of this sort are the main focus of Chap. 7.

A similar process is used to make silicon tetrachloride for use in the manufacture of optical fibers. In a survey article, Flamm et al. (1983) describe optical fibers in the following way.

> Lightguide fibers are thin glass filaments that can carry thousands of times more information than conventional copper wires. These fibers must be made of a special glass that can transmit a light signal over many kilometers without serious attenuation, that is, without diminishing amplitude of the signal, and without dispersion—the spreading of the light pulse. One has only to see the green tint when viewing the end of a piece of regular window glass to appreciate the problem of transmitting light through long lengths of glass. Even the world's best fused quartz is not up to the purity standards set for lightguide fibers.
>
> Recent improvements in long-wavelength lasers and high-purity glass have made it possible for lightguide technology to take advantage of a natural 'window' in the absorption spectrum of silica glass in the 1.2-1.6 micron, infrared-wavelength region. At these wavelengths, light-pulse scattering and ultraviolet and infrared absorptions are minimized, so signals can be transmitted over longer distances, permitting signal repeaters to be spaced further apart. Transmission in this range requires an essentially impurity-free glass.
>
> Lightguide fibers are pulled from the molten end of a glass preform, or solid rod, made by the modified chemical vapor deposition (MCVD) process. MCVD uses a high temperature, heat-driven reaction between silicon tetrachloride and oxygen to form SiO_2 by the reaction[11]
>
> $$SiCl_4 + O_2 \longrightarrow SiO_2 + 2Cl_2$$
>
> Small amounts of germanium tetrachloride, boron trichloride, fluorine and phosphorus oxychloride are used to modify the refractive index and fluid properties of the glass.
>
> The gaseous chemicals are heated to about 1800°C inside a silica tube, and the reaction products are fused to form a transparent film. As successive layers of glass particles are deposited within the tube, the relative proportions of the chemicals are adjusted to modify the refractive index. Finally, the tube is heated to more than 2000°C and collapsed to form the solid preform rod that is then heated to the softening point to draw the hair-thin fiber.

---

[11] This process starts with impure SiO_2 raw material and ends with ultrahigh purity SiO_2 product. The intermediate steps are mainly for the purposes of purification.

Since chemical reactions inside the MCVD process occur in the controlled environment inside the silica tube, the introduction of impurities is minimized to the point where nearly all impurities in the fiber come from the processing chemicals themselves.

The most significant impurity in commercial epitaxial grade silicon tetrachloride is trichlorosilane (about 1800 ppm), although hydrocarbons (40 ppm), hydrogen chloride (200 ppm), silanols (10 ppm) and iron (0.01 ppm) also have been found. *[The sum total of all the trace impurities is about an order of magnitude less that the* $SiHCl_3$ *present, and will therefore be ignored in calculations performed in this book.]* During the MCVD process, a fraction of hydrogen present in any of these chemicals is incorporated into the lightguide fiber as hydroxyl chemical groups. Bonds between these groups and the silicon atoms absorb light in the important 1.2 to 1.6 micron region of the infrared spectrum. Similar impurities are also found in the germanium tetrachloride.

The flowsheet for manufacturing optical fibers is similar to that shown in Fig. 1.7. Steps 1 and 2 remain the same except that the fluidized bed reactor is designed to give high yields of $SiCl_4$. A typical molar ratio of $SiCl_4$:$SiHCl_3$:SiH_2Cl_2 in the reactor effluent is 95:3:2. Step 3 is the same, and step 4 is essentially the same except that the primary product is now $SiCl_4$ with a final purity requirement of 99.998 mol % or better, i.e., optical-grade silicon tetrachloride. This is achieved with an intermediate step, 4*a*, which has no counterpart in the manufacture of electronic-grade silicon. In this step a series of photochemical chlorination reactors and distillation columns are used to purify the $SiCl_4$ to its final value. This procedure reduces the impurity levels from those values cited earlier to the new levels: trichlorosilane (10 ppm), hydrogen chloride (5 ppm), hydrocarbons (5 ppm), silanol (5 ppm), and iron (0.005 ppm, i.e., 5 parts per billion). Flamm et al. (1983) describe the relevant processing.

In step 5 the tetrachlorosilane is oxidized in a MCVD reactor to create the glass preform. This flowsheet provides motivation for ultrahigh-purity distillations that go far beyond even those required for making electronic-grade silicon.

1.4 STRUCTURE OF THIS BOOK

The chemical processes described in Sec. 1.3 demonstrate the continuing importance of separations technology in modern process systems. Multicomponent distillation remains the preferred technology when it is feasible. There is a trend toward higher-purity products that can be produced faster and cheaper, and this is driving the development of methods to find better processes where the structure of the separation system is determined by integration with the rest of the plant so as to minimize processing costs without sacrificing ease of operability.

The first designs of a distillation system (the conceptual designs) are almost always developed assuming that vapor-liquid equilibrium (VLE) is achieved on the stages in the column. Although this is not quite achieved in practice, it is much less demanding of data and detailed modeling, yet provides a rational basis for comparing alternative designs quickly. In Chap. 2 we review VLE models and the data that is needed to implement them. Chapter 3 introduces a classical analysis of binary distillation, especially the McCabe-Thiele method, which every chemical engineer should know. Most other binary separations (e.g., absorption and stripping, as well as some ternary separations with one degree of freedom, such as solvent extraction) can be represented and understood using an approach very similar to the McCabe-Thiele method. These extensions are described in many textbooks, but not here. Instead, we focus on multicomponent mixtures, including those with azeotropes and chemical reactions, which are important in many applications but hardly touched upon in existing textbooks.

The main ideas behind multicomponent distillation are developed in Chaps. 4 and 5. This is where we introduce the notions of node and saddle pinches, residue curve maps for azeotropic mixtures, and feasibility and product distributions in both ideal and nonideal mixtures, leading to system design for ternary azeotropic mixtures as well as to the important concept of identifying process alternatives. Alternatives are identified and then evaluated and compared on the basis of economics, so Chap. 6 introduces simplified models for equipment design and cost.

The material in Chaps. 4 and 5 is generalized for mixtures containing four or more components in Chap. 7. There, new approaches from recent process systems research are introduced: the state-task network and new methods for designing multieffect distillation. Chapters 8, 9, and 10 show how the basic concepts developed earlier in the book apply to mixtures with more than one liquid phase, to batch distillation, and to distillation with chemical reactions.

Similar approaches have also been successfully applied to liquid-extraction systems and to crystallization systems, but we have not attempted to include any of that literature in this book.

References

Agreda, V. H., "Acetic Anhydride from Coal," *ChemTech,* **18**(4), 250–253 (1988).

Agreda, V. H., and L. R. Partin, Reactive Distillation Process for the Production of Methyl Acetate (1984). U.S. Patent 4,435,595 assigned to Eastman Kodak Company.

Agreda, V. H., L. R. Partin, and W. H. Heise, "High-Purity Methyl Acetate via Reactive Distillation," *Chem. Eng. Progr.,* **86**(2), 40–46 (1990).

Agreda, V. H., and J. R. Zoeller, *Acetic Acid and Its Derivatives.* Marcel Dekker, New York (1993).

Bawa, M. S., R. Goodman, and J. K. Truitt, "Hydrogen Reduction of Chlorosilanes," *Semiconductor Engineering Journal (Texas Instruments, Inc.),* **1**(3), 42–45 (1980*a*).

Bawa, M. S., J. K. Truitt, and W. Haynes, "Polycrystalline Silicon Industry and Manufacturing Technology," *Semiconductor Engineering Journal (Texas Instruments, Inc.),* **1**(2), 47–51 (1980*b*).

Douglas, J. M., *The Conceptual Design of Chemical Processes.* McGraw–Hill, New York (1988).

Flamm, D. L., L. T. Manzione, J. W. Baumgart, and L. F. Thompson, "Continuous Production of Lightguide Chemicals," *Bell Labs Record,* pp. 17–21 (1983).

Gentry, J. W., "The 19th Century Legacy to Distillation from Kidd to Young," *Chem. Engng. Ed.,* **29**(4), 250–255 (1995).

Kletz, T., "Friendly Plants," *Chem. Engng. Prog.,* **85**(7), 18–26 (1989).

Larabee, G. B., "Microelectronics," *Chem. Eng.,* **92**(12), 51–59 (1985).

Mayfield, G. G., and V. H. Agreda, "The Eastman Chemical Process for Acetic Anhydride from Coal," *Energy Progr.,* **6,** 214–218 (1986).

Mohtadi, F., Pioneers in the Field of Diffusion, in Furter, W. F., editor, *A Century of Chemical Engineering.* Plenum Press, New York (1982).

Othmer, D. F., Distillation—Some Steps in Its Development, in Furter, W. F., editor, *A Century of Chemical Engineering.* Plenum Press, New York (1982).

Ravi, K. V., *Imperfections and Impurities in Semiconductor Silicon.* Wiley, New York (1981).

Rayleigh, L., "On the Distillation of Binary Mixtures," *Phil. Mag.,* **S.6, 4**(23), 521–537 (1902).

Schreinemakers, F. A. H., "Dampfdrucke Ternärer Gemische," *Zeitschrift f. Physik. Chemie,* **XXXVI,** 257–289 (1901*a*).

Schreinemakers, F. A. H., "Dampfdrucke Ternärer Gemische," *Zeitschrift f. Physik. Chemie,* **XXXVI,** 413–449 (1901*b*).

Schreinemakers, F. A. H., "Einige Bemerkungen der Dampfdrucke Ternärer Gemische," *Zeitschrift f. Physik. Chemie,* **XLIII,** 671 (1903).
Siirola J. J., "Industrial Applications of Chemical Process Synthesis," *Adv. Chem. Engng.,* **23**, 1–62 (1996).
Siirola, J. J., and D. F. Rudd, "Computer–Aided Synthesis of Chemical Process Designs," *Ind. Eng. Chem. Fundam.,* **10,** 353–362 (1971).
Sorel, E., "Sur la Rectification de l'Alcohol," *Comptes rendus,* **CVIII,** 1128–1131 (1889*a*).
Sorel, E., "Sur la Rectification de l'Alcohol," *Comptes rendus,* **CVIII,** 1204–1207 (1889*b*).
Sorel, E., "Sur la Rectification de l'Alcohol," *Comptes rendus,* **CVIII,** 1317–1320 (1889*c*).
Sorel, E., *La Rectification de l'Alcohol,* Gauthier–Villars, Paris (1893).
Sorel, E., "Sur la Rectification de l'Alcohol," *Comptes rendus,* **CXVIII,** 1213–1215 (1894).
Truitt, J. K., G. A. Carter, and M. S. Bawa, Chemical Alternatives in the Production of Electronic Grade Polysilicon (1983). Paper presented at the ACS Meeting, Seattle, WA.

2

Vapor-Liquid Equilibrium and Flash Separations

2.1 FUNDAMENTALS

Many separation devices exploit the fact that the equilibrium compositions of chemical species across coexisting phases are not equal. Before beginning the model development for these devices, we review the essential parts of phase equilibrium thermodynamics, since this plays a central role in the models.

The main achievement of phase equilibrium thermodynamics has been to provide a framework which makes it possible to calculate the (unknown) state of a phase, given the state of an adjacent equilibrium phase. For simplicity we will assume that the two phases are liquid and vapor, that they are each uniform with respect to temperature, pressure, and composition, that the interface between them has no mechanical rigidity, and that heat and all substances present can pass freely across the interface. We will also assume that no third phase is formed over the entire temperature, pressure and composition range of interest, and that chemical reactions do not occur. Many of these restrictions will be removed later in the text.

For each phase in such a system it is usually found that the specification of the chemical composition of the phase together with any two intensive variables will determine the values of the remaining variables (Denbigh, 1971, pp. 6–9). If there are c substances present, it is normal to characterize the state of each phase by the temperature T, pressure P, and $c - 1$ mole fractions, $x_1, x_2, \ldots, x_{c-1}$. It is not necessary to specify the last mole fraction x_c, since the mole fractions must sum to unity.

For nonequilibrium phases, the temperature, pressure, and composition of each phase may be set independently. When the phases are in equilibrium with each other, however, these variables are no longer independent but are related by the famous conditions derived by *J. Willard Gibbs* in 1875 (Gibbs, 1961, p. 65). Gibbs showed that for any two phases[1] in equilibrium, it is necessary and sufficient that the temperature, pressure, and chemical potential of each substance be uniform

[1] With the restrictions mentioned earlier.

throughout the system, i.e.,

$$T^L = T^V = T \tag{2.1}$$

$$P^L = P^V = P \tag{2.2}$$

$$\mu_i^L(T, P, x_1, x_2, \ldots, x_{c-1}) = \mu_i^V(T, P, y_1, y_2, \ldots, y_{c-1}) \tag{2.3}$$

These equations express the conditions of *thermal*, *mechanical*, and *chemical* equilibrium, respectively.

A degrees of freedom analysis on these equations leads to the *Gibbs phase rule*, e.g., Denbigh (1971, Chap. 5), and Smith and VanNess (1996, Chap. 12).

$$\mathcal{P} + \mathcal{F} = c + 2 \tag{2.4}$$

where $\mathcal{P}$ is the number of coexisting equilibrium phases, c is the number of substances present, and $\mathcal{F}$ is the number of degrees of freedom that must be specified in order to make a "square" set of equations in which there are as many variables as equations.[2] For two-phase mixtures, $\mathcal{F} = c$, which expresses the fact that if we specify the system pressure and the composition of the liquid phase, $x_1, x_2, \ldots, x_{c-1}$ (which is a total of c variables), the system temperature and the composition of the vapor phase may be calculated by solving the equilibrium equations. Such a calculation is called a *bubble-point calculation*. Another common situation is to specify $P, y_1, y_2, \ldots, y_{c-1}$ and calculate $T, x_1, x_2, \ldots, x_{c-1}$. This is called a *dew-point calculation*. In order to carry out these calculations we must first find suitable functional forms for the chemical potentials. This entails setting up *models* that are designed to represent the behavior of real systems under prescribed conditions. Creation of these models is *not* part of the program of classical thermodynamics. Models may be based on molecular considerations (this is the goal of statistical mechanics) or on empirical expressions that may be justified on the basis of available data. Some of the most useful models are derived using a combination of molecular and empirical arguments (Prausnitz et al., 1998). Classical thermodynamics places restrictions on the functional forms that these models can assume and provides relationships between certain variables that can be used for consistency checking of models and data, e.g., the Gibbs-Duhem equation, and for calculating dependent variables from the given values of the independent variables, e.g., Gibbs phase equilibrium equations.

2.2 MODELS AND DATA SOURCES

Ideal Mixtures

The simplest model for systems involving two or more components is the *ideal mixture* in which the chemical potential of every component is a linear function of

[2]This is a *necessary* but not sufficient condition for the existence of an isolated (or point) solution to the equilibrium equations. This condition does not imply that the equations actually have a solution, nor does it imply that if there is a solution it will be unique. There are well-known situations in which each of these outcomes occurs, e.g., in the retrograde condensation region of a phase diagram.

the logarithm of its mole fraction according to the relation[3]

$$\mu_i(T, P, \mathbf{x}) = \mu_i^0(T, P) + RT \ln x_i \tag{2.5}$$

where μ_i^0 is the chemical potential of pure component i at the temperature T and pressure P of the mixture. Notice that the mole fractions of the other components, x_j, $j \neq i$, do not enter into the expression for μ_i. For gas mixtures the term $\mu_i^0(T, P)$ may be simplified further by assuming that each component obeys the *perfect gas equation*, $Pv = RT$, at all temperatures, pressures, and compositions. That is, we assume the molar volume of each pure component is the same, and it is independent of composition and depends only on T and P according to the equation $v = RT/P$. Under these conditions the chemical potential of each component in the vapor mixture is given by (Denbigh 1971, Chap. 3)

$$\mu_i^V(T, P, \mathbf{y}) = \mu_i^{PG}(T, P^\dagger) + RT \ln \frac{P}{P^\dagger} + RT \ln y_i \tag{2.6}$$

where y_i represents the mole fraction of component i in the vapor and $\mu_i^{PG}(T, P^\dagger)$ is the chemical potential of pure i as a perfect gas at temperature T and pressure $P^\dagger$, and $P^\dagger$ is an arbitrary nonzero reference pressure. Such a mixture is called a perfect gas mixture.

The equilibrium relation between an ideal liquid mixture and a perfect gas mixture is found by inserting the model equations 2.5 and 2.6 into the Gibbs equilibrium condition (Eq. 2.3), giving

$$\mu_i^{0L} + RT \ln x_i = \mu_i^{PG} + RT \ln \frac{P}{P^\dagger} + RT \ln y_i \tag{2.7}$$

where μ_i^{0L} is the Gibbs energy per mole of pure liquid i at the temperature and pressure of the mixture, and μ_i^{PG} depends only on temperature.

Rearrangement leads to

$$\frac{Py_i}{x_i} = P^\dagger \exp \frac{\mu_i^{0L} - \mu_i^{PG}}{RT} \tag{2.8}$$

Generally, the exponential on the right-hand side is strongly dependent on temperature and weakly dependent on pressure [through the term $\mu_i^{0L}(T, P)$]. If we neglect the weak pressure dependence[4] and let $x_i, y_i \rightarrow 1$ at constant temperature, the pressure P must change and approach the saturated vapor pressure of pure component i, $P_i^{sat}(T)$, at temperature T. Therefore, at a given temperature the right hand side must equal $P_i^{sat}(T)$ *for all compositions*, which leads to the equilibrium relations

$$Py_i = P_i^{sat}(T)x_i \qquad i = 1, 2, \ldots, c \tag{2.9}$$

or *Raoult's law*.

---

[3]This may seem like a rather arbitrary or specialized choice, but Shapiro (1968) and Feinberg (1977) show that all the properties normally associated with ideal mixtures and perfect gases, including Eqs. 2.5 and 2.6, follow from some remarkably weak assumptions about the variables on which the chemical potential, the partial molar internal energy, and the partial molar volume depend.

[4]More precisely, we assume that the liquid is incompressible and that its molar volume is negligible compared to that of a perfect gas for the temperatures and pressures of interest, i.e., the Poynting correction is negligible. Whenever the major assumptions that went into the derivation of Eq. 2.7 are valid, i.e., a perfect gas mixture in equilibrium with an ideal liquid mixture, these additional assumptions are normally satisfied.

It is important to recognize that Raoult's law enables us to calculate the phase equilibrium behavior of certain *mixtures* using only *pure component* physical properties. The pure component saturated vapor pressures required in Eq. 2.9 are normally calculated from the Antoine equation, which has the form

$$\log P^{\text{sat}} = A - \frac{B}{T + C} \tag{2.10}$$

The Antoine coefficients A, B, and C are obtained by regressing experimental saturated vapor pressure data over a given range of temperature.[5] It would be convenient to have parameters available for each substance that were obtained from experimental data over the entire temperature range from the triple point to the critical point. This, however, is rarely the case, and it is important to know the temperature range over which the parameters can be safely used to predict vapor pressures. *Extrapolation outside this range must always be done cautiously.* Antoine coefficients are available for a large number of pure components. Useful sources for these parameters include the *Vapor–Liquid Equilibrium Data Collection* of the DECHEMA Chemistry Data Series;[6] (Reid et al., 1977, Appendix A; and Ohe, 1976). Ohe's data book provides quite extensive information for each pure component, including a graph comparing the Antoine equation to the original experimental data and the literature source of the data.

Values of the Antoine parameters for selected substances are given in Table 2.1, and a comparison of the Antoine equation with experimental data for acetone is shown in Fig. 2.1. The Antoine parameters used to plot this figure were fitted to experimental data in the temperature interval 57 to 205°C. Extrapolating the equation to lower temperatures produces systematic deviations from the experimental data, as can be seen in Fig. 2.1*b*. This can often be corrected by fitting the Antoine coefficients over a wider range of temperature, as done by Ohe (1976, p. 334), who reports the values $A = 7.29958$, $B = 1{,}312.253$, $C = 240.705$ for the Antoine equation in the same form and in the same units as that given in Table 2.1. Ohe's coefficients were fitted to data over the temperature range -13.99 to 234.45°C, and over this extended range of temperature the fit is very good, including at the lower temperatures. Another common approach is to regress separate sets of parameters over separate temperature intervals. For example, at lower temperatures, in the range -13 to 55°C, the DECHEMA Chemistry Data Series recommend using the following values for the Antoine coefficients: $A = 7.11714$, $B = 1{,}210.595$, $C = 229.664$. These values also give a fine fit of the experimental data in the lower-temperature range. Notice that Ohe's parameter values lie between the two sets reported in the DECHEMA Chemistry Data Series.

It is sometimes useful to estimate saturated vapor pressures from the Clausius-Clapeyron equation (Smith and VanNess, 1996, p. 198)

$$\frac{d \ln P^{\text{sat}}}{d(1/T)} = -\frac{\Delta h_{\text{vap}}}{R} \tag{2.11}$$

---

[5] See Press et al. (1986, Chap. 14) for a general introduction to regression. For a discussion of regression for VLE models, see any volume in the DECHEMA series and references there.

[6] An extensive compilation of phase equilibrium data has been published in the *Vapor-Liquid Equilibrium Data Collection* of the DECHEMA Chemistry Data Series.

TABLE 2.1
Antoine coefficients for selected substances

Substance	A	B	C	Temperature range, °C
Acetone	7.63132	1,566.69	273.419	57 to 205
Water	8.01767	1,715.70	234.268	100 to 265
Benzene	6.87987	1,196.76	219.161	8 to 80
Toluene	6.95087	1,342.31	219.187	−27 to 111
Ethylene glycol	8.09083	2,088.936	203.454	50 to 150
Hexane	6.91058	1,189.64	226.280	−30 to 170
p-Xylene	6.99053	1,453.430	215.310	27 to 166
Ethanol	8.11220	1,592.864	226.184	20 to 93
Acetic acid	8.02100	1,936.010	258.451	18 to 118
Acetaldehyde	7.20812	1,099.810	233.945	−82 to 20
Methanol	8.08097	1,582.271	239.726	15 to 84
Methyl ethyl ketone	7.06356	1,261.340	221.969	43 to 88
Chloroform	6.95465	1,170.966	226.232	−10 to 60
Ethylenediamine	8.09831	1,893.720	245.676	11 to 117
4-Methyl-2-pentanol	8.46706	2,174.869	257.780	25 to 133
Dichloromethane	7.08030	1,138.910	231.450	−44 to 59
1,3-Butadiene	6.85364	933.586	239.511	−75 to −2
Styrene	7.50233	1,819.810	248.662	−7 to 145
Ethyl acetate	7.10179	1,244.950	217.881	16 to 76
Vinylacetylene	7.02515	999.110	235.817	−93 to 5
Acetic anhydride	7.69301	1,781.29	230.395	2 to 140
Dichlorosilane[a]	7.18600	1,224.50	273.16	−45 to 121
Trichlorosilane	6.95524	1,102.900	238.865	−81 to 32
Silicon tetrachloride	7.02404	1,212.890	235.910	−63 to 57
Hydrogen chloride[b]	7.44899	868.358	274.904	−85 to 36

Source: The values were taken from the *Vapor-Liquid Equilibrium Data Collection* of the DECHEMA Chemistry Data Series, except where noted. The form of the equation is $\log_{10} P^{\text{sat}} = A - B/(T + C)$ with pressure in mmHg and temperature in °C.
[a]Bawa, 1988.
[b]Ohe, 1976.

where Δh_{vap} is the molar latent heat of vaporization and T is the absolute temperature. Although Δh_{vap} decreases with increasing temperature from the triple point to the critical point, where it is zero, experimental data of $\ln P^{\text{sat}}$ vs. $1/T$ for many substances show that such a plot produces a line that is nearly straight (see Fig. 2.1*b*). Therefore, integration of Eq. 2.11 with a constant value of Δh_{vap} often provides a quite accurate method for estimating saturated vapor pressures over a wide temperature range. Taking the lower limit of integration to be normal boiling conditions, i.e., $T = T_{nb}$ at $P^{\text{sat}} = P_{\text{atm}}$, we obtain

$$\ln \frac{P^{\text{sat}}}{P_{\text{atm}}} = \frac{\Delta h_{\text{vap}}}{R}\left(\frac{1}{T_{nb}} - \frac{1}{T}\right) \tag{2.12}$$

where P_{atm} is the atmospheric pressure, T is the absolute temperature, and Δh_{vap} and R are in a consistent set of units. It is usually convenient and accurate to set Δh_{vap} to its value at the normal boiling point. Tabulations of these values together with the normal boiling points of many substances are available in Reid et al. (1977, Appendix A). The saturated vapor pressure curve for acetone predicted by Eq. 2.12

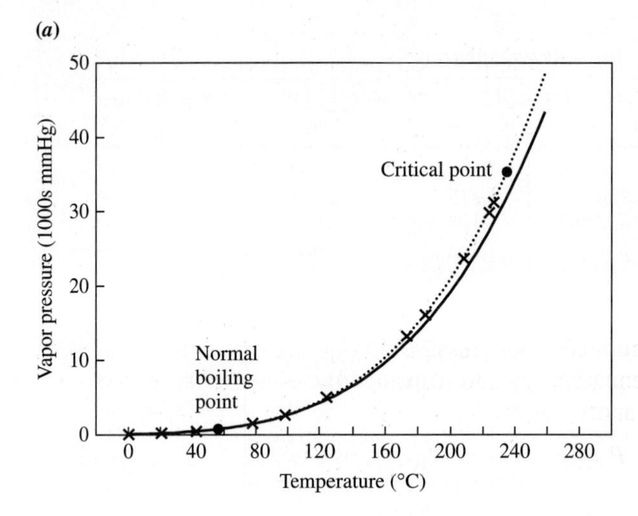

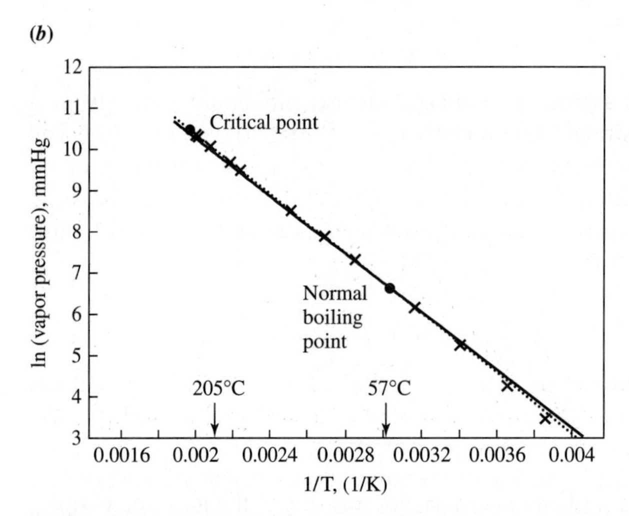

FIGURE 2.1
Vapor pressure of acetone correlated by the Antoine (dotted line) and Clausius-Clapeyron (solid line) equations. Values for the Antoine parameters are given in Table 2.1; values for the constants in the Clausius-Clapeyron equation are reported in the text. Experimental data (×) are reported by Ambrose et al. (1974).

is also shown in Fig. 2.1. The parameters were taken from Reid et al. (1977) and have the values $T_{nb} = 329.4$ K, $\Delta h_{\text{vap}} = 6{,}960$ cal/gmol, $R = 1.98721$ cal/gmol K. At pressures below 10 atm, the Clausius-Clapeyron equation reproduces the experimental data very closely and would be suitable for use in VLE calculations. At higher pressures, the predictions are not so good, although they are still quite respectable considering that the parameters in the model are not fitted to the experimental vapor pressure data. By comparison, the Antoine coefficients can only be obtained by regression of experimental vapor pressure data. If the constant Δh_{vap} in the Clausius-Clapeyron equation is treated as an adjustable parameter and fitted to the experimental vapor pressure data, a value of 7,175 cal/gmol produces a curve that fits the data as accurately as the Antoine equation.

Raoult's law imposes severe constraints on the shape of T, P, $\mathbf{x}$, $\mathbf{y}$ phase diagrams and their projections. This is most easily demonstrated for a binary mixture held at constant temperature. According to the phase rule, an equilibrium two-phase binary mixture has two degrees of freedom. Thus, a meaningful equilibrium problem would be to specify values for T and x_1 and then calculate the corresponding values of P and y_1 from Raoult's law[7]

$$Py_1 = P_1^{\text{sat}}(T)x_1 \tag{2.13}$$

$$P(1-y_1) = P_2^{\text{sat}}(T)(1-x_1) \tag{2.14}$$

The remaining compositions are obtained from the relations

$$x_2 = 1 - x_1 \qquad y_2 = 1 - y_1 \tag{2.15}$$

At fixed temperature, the saturated vapor pressures in Eqs. 2.13 and 2.14 are constant and known numbers. The equations can be rearranged to give explicit expressions for P and y_1 in terms of T and x_1, as follows:

$$P = P_1^{\text{sat}}x_1 + P_2^{\text{sat}}(1-x_1) \tag{2.16}$$

$$y_1 = \frac{P_1^{\text{sat}}x_1}{P} = \frac{\frac{P_1^{\text{sat}}}{P_2^{\text{sat}}}x_1}{1+\left(\frac{P_1^{\text{sat}}}{P_2^{\text{sat}}}-1\right)x_1} \tag{2.17}$$

The bubble-point curve on a pressure-composition phase diagram shows P as a function of x_1, which is a straight line for this type of mixture (see Fig. 2.2). The dew-point curve on a pressure-composition diagram shows P as a function of y_1, and can be obtained by solving Eqs. 2.16 and 2.17 and plotting the results as x_1 varies from 0 to 1. Alternatively, x_1 can be eliminated from these equations to give a direct relationship between P and y_1, as follows:

$$P = \frac{P_1^{\text{sat}}P_2^{\text{sat}}}{P_1^{\text{sat}} - y_1(P_1^{\text{sat}} - P_2^{\text{sat}})} \tag{2.18}$$

The dew-point curve for such a mixture is therefore part of a hyperbola and has positive slope and positive curvature for all values of y_1 in the interval 0 to 1; i.e., the curve is convex (see Fig. 2.2). Another useful phase diagram is the equilibrium y_1-x_1 diagram, obtained from Eq. 2.17. This curve has positive slope and negative curvature for all values of x_1, i.e., the curve is concave (see Fig. 2.2). It follows that such a mixture cannot exhibit azeotropes since it is not possible for x_1 and y_1 to be equal at any point other than at 0 and 1, nor is it possible for $(\partial P/\partial x_1)_T = (\partial P/\partial y_1)_T = 0$. Similarly, inflections in the y_1-x_1 diagram cannot occur in these mixtures.

Equations 2.13 and 2.14 are also valid for isobaric phase equilibria; e.g., given P and x_1, calculate T and y_1. Since the vapor pressures change with temperature, and the equilibrium temperature is unknown, it is not normally possible to obtain explicit expressions for T and y_1 in terms of P and x_1. Thus, the equilibrium equations must

---

[7] The initial set of equations is

$$Py_1 = P_1^{\text{sat}}x_1 \qquad Py_2 = P_2^{\text{sat}}x_2$$
$$x_1 + x_2 = 1 \qquad y_1 + y_2 = 1$$

The dependent mole fractions (which we take to be x_2 and y_2) can be eliminated from the equilibrium equations as shown in the text.

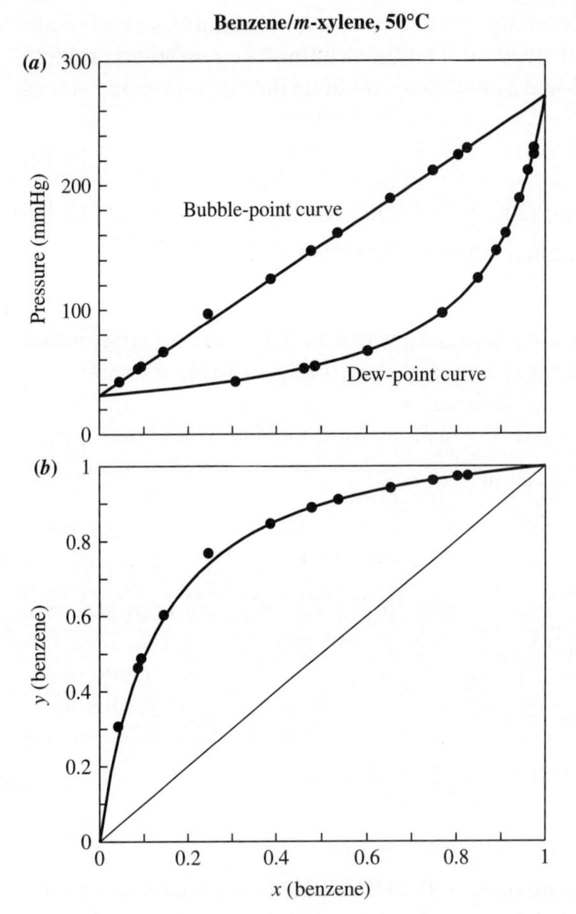

FIGURE 2.2
Isothermal phase diagrams for the mixture benzene and *m*-xylene at 50°C. (*a*) Pressure-composition diagram. (*b*) y_1-x_1 diagram. The solid lines are given by Eqs. 2.16 and 2.17; data are taken from Gmehling et al. (1980*a*, p. 309).

be solved iteratively, as discussed below. On the isobaric boiling-point diagram neither the dew-point nor the bubble-point curves are linear, even for ideal mixtures. For the special case of ideal mixtures with constant relative volatility it is possible to obtain closed-form solutions for T and y_1 in terms of P and x_1, showing that the equilibrium temperature varies inversely with $\ln x_1$, as discussed in Sec. 2.3.

Nonideal Mixtures

Few real mixtures are ideal, as even a cursory look through the DECHEMA Chemistry Data Series will reveal. In such cases it is often found that the chemical potential of each component depends on the composition in a more complex way than that given by the ideal mixture model. Current practice is to modify Eq. 2.5 and write the chemical potential of each component in a real mixture as

$$\mu_i^L(T, P, \mathbf{x}) = \mu_i^{0L}(T, P) + RT \ln x_i \gamma_i \tag{2.19}$$

where μ_i^{0L} is again the chemical potential of pure liquid i at the temperature and pressure of the mixture and γ_i is a correction factor, called the *activity coefficient* of component i, which depends on T, P, and $\mathbf{x}$. In order to complete the model for μ_i^L we need to create expressions for γ_i. Clearly, an ideal liquid mixture is one for which $\gamma_i = 1, i = 1, 2, \ldots, c$ for all values of $\mathbf{x}$. It must also be true that as $x_i \to 1$, the mole fraction of all other components tends to zero and the properties of the mixture must approach those of pure component i. That is, $\mu_i^L(T, P, x_i \to 1) \to \mu_i^{0L}(T, P)$. The only way that this can happen in Eq. 2.19 is if we impose boundary conditions[8] on each of the activity coefficients, so that $\gamma_i \to 1$ as $x_i \to 1, i = 1, 2, \ldots, c$.

The equilibrium relation for a nonideal liquid mixture in contact with a perfect gas mixture is found from Eqs. 2.3, 2.6, and 2.19. Following similar logic that led to Eq. 2.9, we find the equilibrium relation becomes

$$Py_i = P_i^{\text{sat}} \gamma_i x_i \qquad i = 1, 2, \ldots, c \tag{2.20}$$

Activity coefficients can be calculated from experimental data by methods described in Denbigh (1971, Chap. 9), and Smith and VanNess (1996, Chap. 11). If P, T, y_i, x_i data are available together with an accurate saturated vapor pressure equation for each component, then one way to calculate activity coefficients is to rearrange Eq. 2.20 explicitly for γ_i. These data are available for a large number of binary mixtures in the DECHEMA Chemistry Data Series. A plot of γ_1 and γ_2 versus x_1 for acetone (1)-water (2) mixtures at 1 atm pressure is shown in Fig. 2.3. The points are based on experimental data. It is convenient to represent these points with a model, and many models are available for this purpose. A widely used model is the Wilson equation, which has the following form for binary mixtures:

$$\ln \gamma_1 = -\ln(x_1 + \Lambda_{12} x_2) + x_2 \left(\frac{\Lambda_{12}}{x_1 + \Lambda_{12} x_2} - \frac{\Lambda_{21}}{\Lambda_{21} x_1 + x_2} \right) \tag{2.21}$$

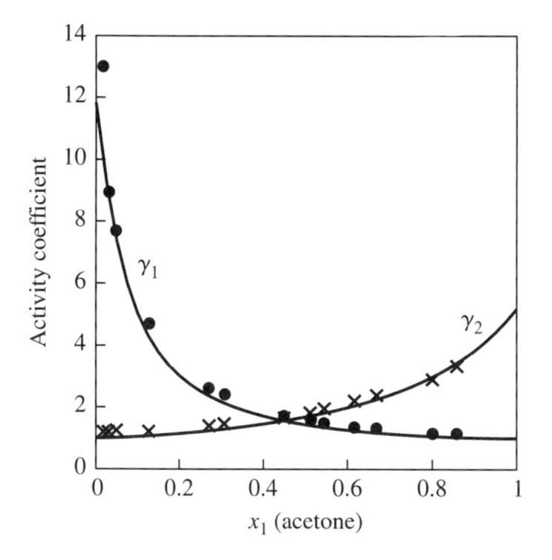

FIGURE 2.3
Plot of the activity coefficients in the acetone (1) and water (2) mixture at 1 atm pressure and under boiling conditions. The points are based on experimental data; the lines are calculated from the Wilson equation. All data and model parameters are taken from Gmehling and Onken (1977, p. 237). (•) γ_1; (×) γ_2.

[8]Other conventions are possible for Eq. 2.19 and the associated boundary conditions on the activity coefficients, but these are usually reserved for liquid mixtures in which some of the components are gases or solids at the same T and P as the mixture (Denbigh 1971, Chap. 9).

TABLE 2.2
Wilson equation parameters for selected binary mixtures under isobaric conditions at 1 atm pressure

Mixture		A_{12}	A_{21}
Component 1	**Component 2**	**(cal/gmol)**	**(cal/gmol)**
Acetone	Water	489.3727	1,422.849
Hexane	*p*-Xylene	−164.7038	390.7738
Benzene	Toluene	377.976	−354.9859
Ethanol	Ethylene glycol	−129.2043	1,539.4142
Ethanol	Water	464.2336	926.2759
Water	Ethylene glycol	−297.2946	16,553.5350
Methyl ethyl ketone	Water	892.6557	2,013.7175
Acetone	Methyl ethyl ketone	−232.9286	476.8721
Benzene	Ethylenediamine	490.0693	560.0207
Acetaldehyde	Water	285.5863	1,045.5669
Methanol	Water	19.2547	554.0494
Acetone	Chloroform	116.117	−506.8519
Chloroform	Ethanol	−268.7676	1,270.3897
Benzene	4-Methyl-2-pentanol	214.3044	363.9183
Dichloromethane	Ethyl acetate	−174.5487	−212.1071
1,3-Butadiene	Styrene	1,100.1231	−417.8319

Source: Taken from the *Vapor-Liquid Equilibrium Data Collection* of the DECHEMA Chemistry Data Series.

$$\ln \gamma_2 = -\ln(x_2 + \Lambda_{21}x_1) - x_1\left(\frac{\Lambda_{12}}{x_1 + \Lambda_{12}x_2} - \frac{\Lambda_{21}}{\Lambda_{21}x_1 + x_2}\right) \tag{2.22}$$

where

$$\Lambda_{12} = \frac{v_2^L}{v_1^L}\exp(-A_{12}/RT) \tag{2.23}$$

$$\Lambda_{21} = \frac{v_1^L}{v_2^L}\exp(-A_{21}/RT) \tag{2.24}$$

where v_i^L is the molar volume of pure liquid component i which is normally taken to be a constant, independent of T and P for the purposes of the activity coefficient model, and A_{12}, A_{21} are adjustable model parameters found by regressing experimental P, T, x, y data. Normally, v_i^L is in cm^3/gmol, T is in Kelvin, Λ_{12} and Λ_{21} are in cal/gmol, and $R = 1.98721$ cal/gmol K. Values for A_{12}, A_{21}, v_1^L, v_2^L are available for a large number of binary mixtures in the DECHEMA Chemistry Data Series. Values for selected binary mixtures are reproduced in Tables 2.2 and 2.3. Parameters for the following models are also available in the DECHEMA Chemistry Data Series: Margules, Van Laar, NRTL, and UNIQUAC. The forms of the equations are defined at the beginning of each volume in the series.

Equation 2.20 is solved iteratively for the most common types of phase equilibrium problems. For isobaric bubble-point problems where we specify P and $\mathbf{x}$ and calculate T and $\mathbf{y}$, the algorithm goes as follows:

1. Specify P and $\mathbf{x}$.
2. Guess T.
3. Calculate $P_i^{\text{sat}}(T),\ i = 1, \ldots, c$ from the Antoine equation.
4. Calculate $\gamma_i(T, \mathbf{x}),\ i = 1, \ldots, c$ from the Wilson equation (or another liquid solution model).
5. Calculate $y_i,\ i = 1, \ldots, c$ from Eq. 2.20.
6. If $\sum_{i=1}^{c} y_i$ is sufficiently close to unity, then stop. Otherwise, adjust T and go to step 3.

The algorithm is more complicated for isobaric dew-point problems where we specify P and $\mathbf{y}$ and calculate T and $\mathbf{x}$:

1. Specify P and $\mathbf{y}$.
2. Guess values for T and $x_1, \ldots, x_{c-1}$.
3. Calculate x_c from $\sum_{i=1}^{c} x_i = 1$.
4. Calculate $P_i^{\text{sat}}(T),\ i = 1, \ldots, c$ from the Antoine equation.
5. Calculate $\gamma_i(T, \mathbf{x}),\ i = 1, \ldots, c$ from the Wilson equation (or another liquid solution model).
6. Calculate the residuals f_i from Eq. 2.20 where $f_i = Py_i - P_i^{\text{sat}}\gamma_i x_i, i = 1, \ldots, c$.
7. If the residuals are sufficiently close to zero, then stop. Otherwise, adjust $T, x_1, \ldots, x_{c-1}$ and go to step 3.

TABLE 2.3
Pure component liquid molar volumes

Substance	v^L (cm^3/gmol)
Acetone	74.05
Water	18.07
Benzene	89.41
Toluene	106.85
Ethylene glycol	55.92
Ethylenediamine	67.84
Hexane	131.61
p-Xylene	123.93
Ethanol	58.68
Acetic acid	57.54
Acetaldehyde	56.62
Methanol	40.73
Methyl ethyl ketone	90.17
Chloroform	80.67
4-Methyl-2-pentanol	127.33
Dichloromethane	64.5
Ethyl acetate	98.49
1,3-Butadiene	87.09
Styrene	115.57

Source: Taken from the *Vapor-Liquid Equilibrium Data Collection* of the DECHEMA Chemistry Data Series. Note that these are taken as constants, independent of temperature for the models used there. Incorporating the temperature dependence requires a new regression for the parameters in the activity coefficient models.

There are many ways to implement this algorithm that range from simple successive substitution on each variable, e.g., VanNess and Abbott (1982, p. 355) to quasi-Newton methods where all variables are adjusted simultaneously.

The bubble-point algorithm was used to calculate the activity coefficients and phase diagrams shown in Figs. 2.3 and 2.4 for acetone and water. The activity coefficients for each component deviate significantly from unity. This is especially noticeable at infinite dilution of acetone where $\gamma_1^\infty \approx 13$. For this particular system the activity coefficients are always greater than or equal to unity and vary monotonically with composition. While this is the most common case, there are many systems in which the activity coefficients vary monotonically and are less than unity. Some mixtures are even capable of nonmonotone behavior, and others of exhibiting activity coefficients that are less than unity over a range of compositions and greater than unity elsewhere, e.g., benzene + hexafluorobenzene (VanNess and Abbott, 1982, Fig. 1.4–1*f*; Gaw and Swinton, 1968).

We know from experiment that phase diagrams for real mixtures can deviate appreciably from ideality. It is conventional to classify deviations from ideality in terms of *isothermal* positive or negative deviations from Raoult's law. Mixtures for which the P-x curve on an isothermal P-x-y diagram lies above the linear relation

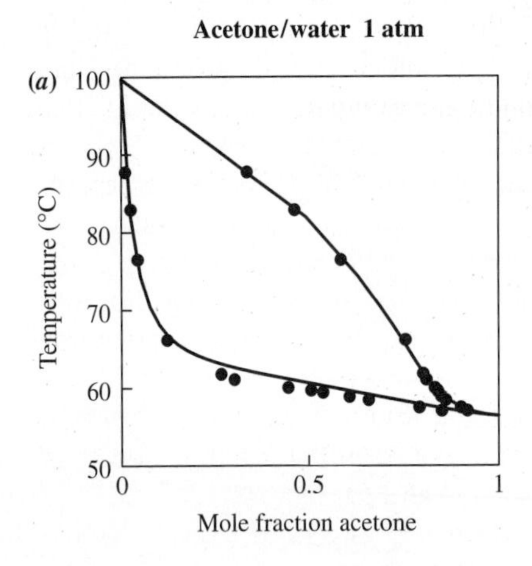

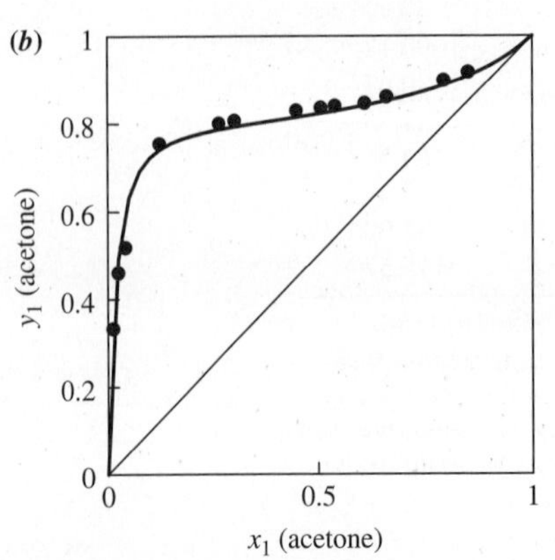

FIGURE 2.4
Isobaric phase diagrams for the acetone (1) and water (2) mixture at 1 atm pressure. (*a*) Temperature-composition diagram. (*b*) y_1-x_1 diagram. The points are experimental data; solid lines are calculated from the Wilson equation. All data and model parameters are taken from Gmehling and Onken (1977, p. 237). Also see Tables 2.1, 2.2, and 2.3.

provided by Raoult's law are said to exhibit positive deviations. Systems are said to show negative deviations when the isothermal P-x curve lies below the Raoult's-law line. A selection of such diagrams are given in Smith and VanNess (1996, Fig. 11.8); see also Exercise 1.

Since distillation processes are carried out at conditions which approximate constant pressure more closely than constant temperature, it is normal to use isobaric T-x-y diagrams. These diagrams are not so easy to classify, since the isobaric T-x line is not straight even for ideal mixtures. The isobaric activity coefficient diagrams at boiling conditions usually give useful information about the type of deviations displayed in the T-x-y diagram. Summing Eq. 2.20 leads to the following expression for the total pressure in binary systems:

$$P = P_1^{\text{sat}} \gamma_1 x_1 + P_2^{\text{sat}} \gamma_2 x_2 \tag{2.25}$$

Therefore, if γ_1 and γ_2 are ≥ 1 for all compositions under isothermal boiling, then the P-x line must lie above the straight line provided by Raoult's law, and the system exhibits positive deviations. If γ_1 and γ_2 are always ≤ 1, then the P-x line lies below the Raoult's-law line.[9] It normally follows that if γ_1 and γ_2 are ≥ 1 for all compositions under isobaric boiling, then the mixture exhibits positive deviations, and if γ_1 and γ_2 are always ≤ 1 the mixture exhibits negative deviations from Raoult's law behavior.

A selection of examples that exhibit deviations from ideality are shown in Fig. 2.5.[10] For each mixture we show the T-x-y, y-x, and $\ln \gamma_i$-x diagrams at 1 atm pressure. It is convenient to plot the logarithm of the activity coefficients rather than the activity coefficients themselves because values of γ_i in the interval $1 < \gamma_i < \infty$ have positive logarithms and values of γ_i in the interval $0 < \gamma_i < 1$ have negative logarithms, with zero on this logarithmic scale representing ideal behavior.

In Fig. 2.5*a* and *b* the logarithm of the activity coefficients is positive for all compositions while in Fig. 2.5*c* and *d* it is negative. This information alone, however, is not sufficient to fix the shape of the phase diagrams, for we see that azeotropes are present in Fig. 2.5*b* and *d* but not in Fig. 2.5*a* and *c*. Moreover, we have already seen an example in Figs. 2.3 and 2.4 where the logarithm of the activity coefficients is always positive and for which the T-x-y and y-x diagrams exhibit inflections, whereas no such inflections are present in Fig. 2.5*a* and *b*.

The azeotropes in Fig. 2.5*b* and *d* occur where $(\partial T/\partial x_1)_P = (\partial T/\partial y_1)_P = 0$, or equivalently where $x_1 = y_1$. Therefore, azeotropic mixtures produce y-x curves that cross the 45° line at the azeotropic composition; the slope of the curve at the crossing point is less than unity for minimum-boiling azeotropes and greater than unity for maximum-boiling mixtures, as shown in Fig. 2.5*b* and *d*, respectively.

---

[9] The converse is not true. For example, a mixture that exhibits positive deviations from Raoult's law may have values of γ_1 or γ_2 that are less than unity over a limited range of compositions provided the sum in Eq. 2.25 is always positive.

[10] The points in these figures represent experimental data. The solid lines are calculated from Eq. 2.20 using the Antoine and Wilson equations. All data and model parameters are taken from the DECHEMA Chemistry Data Series, as follows: Gmehling et al. (1982*b*, p. 554), Gmehling and Onken (1977, p. 285), Gmehling et al. (1982*a*, p. 449), and Gmehling et al. (1979, p. 92).

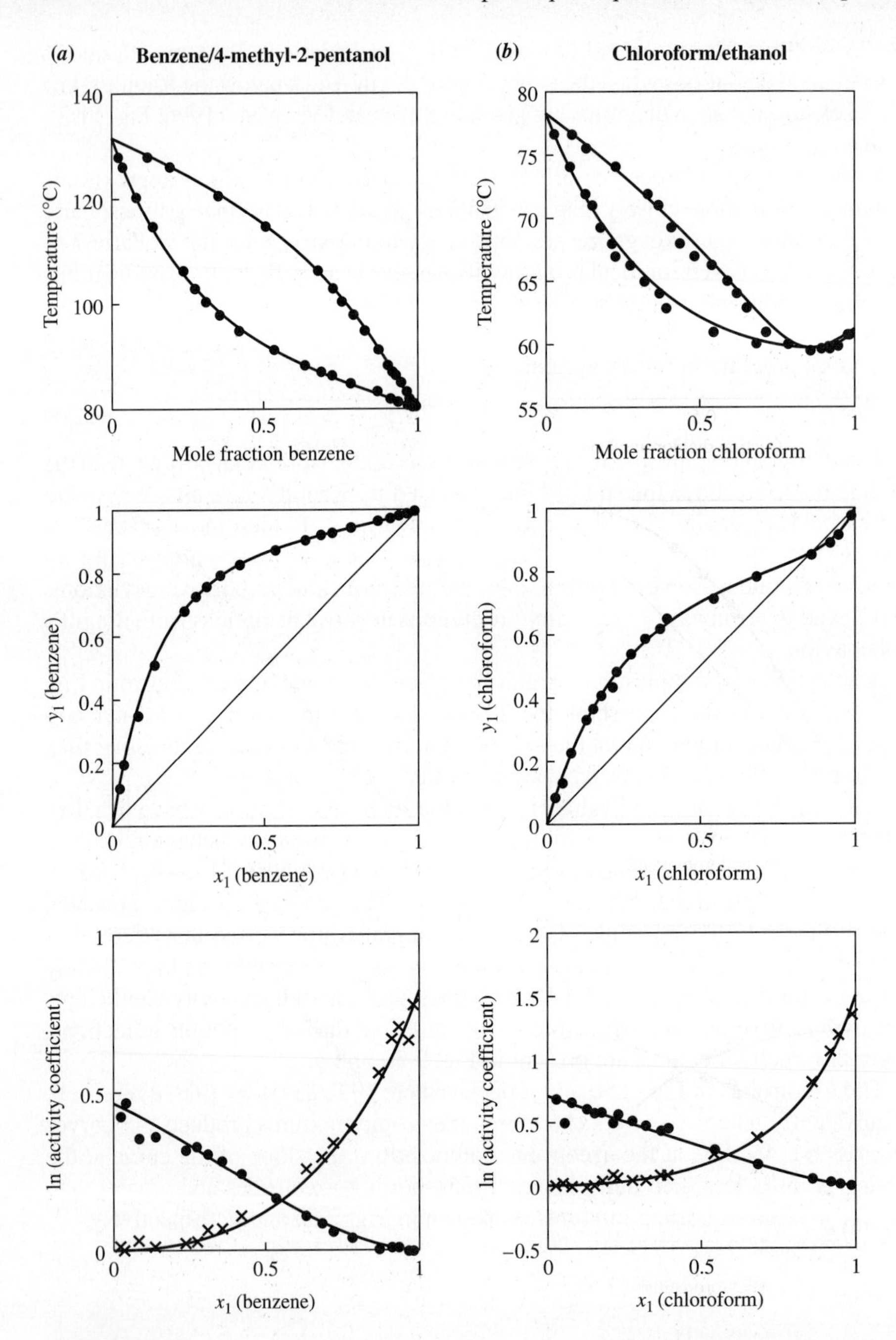

FIGURE 2.5
Selected examples of mixtures that deviate from Raoult's law. For each mixture we show the isobaric T-x-y diagram, the isobaric y-x diagram, and the isobaric $\ln \gamma_1$-$\ln \gamma_2$-x_1 diagram at boiling conditions. The pressure is held constant at 1 atm throughout. (*a*) Benzene (1) and 4-methyl-2-pentanol (2). (*b*) chloroform (1) and ethanol (2).

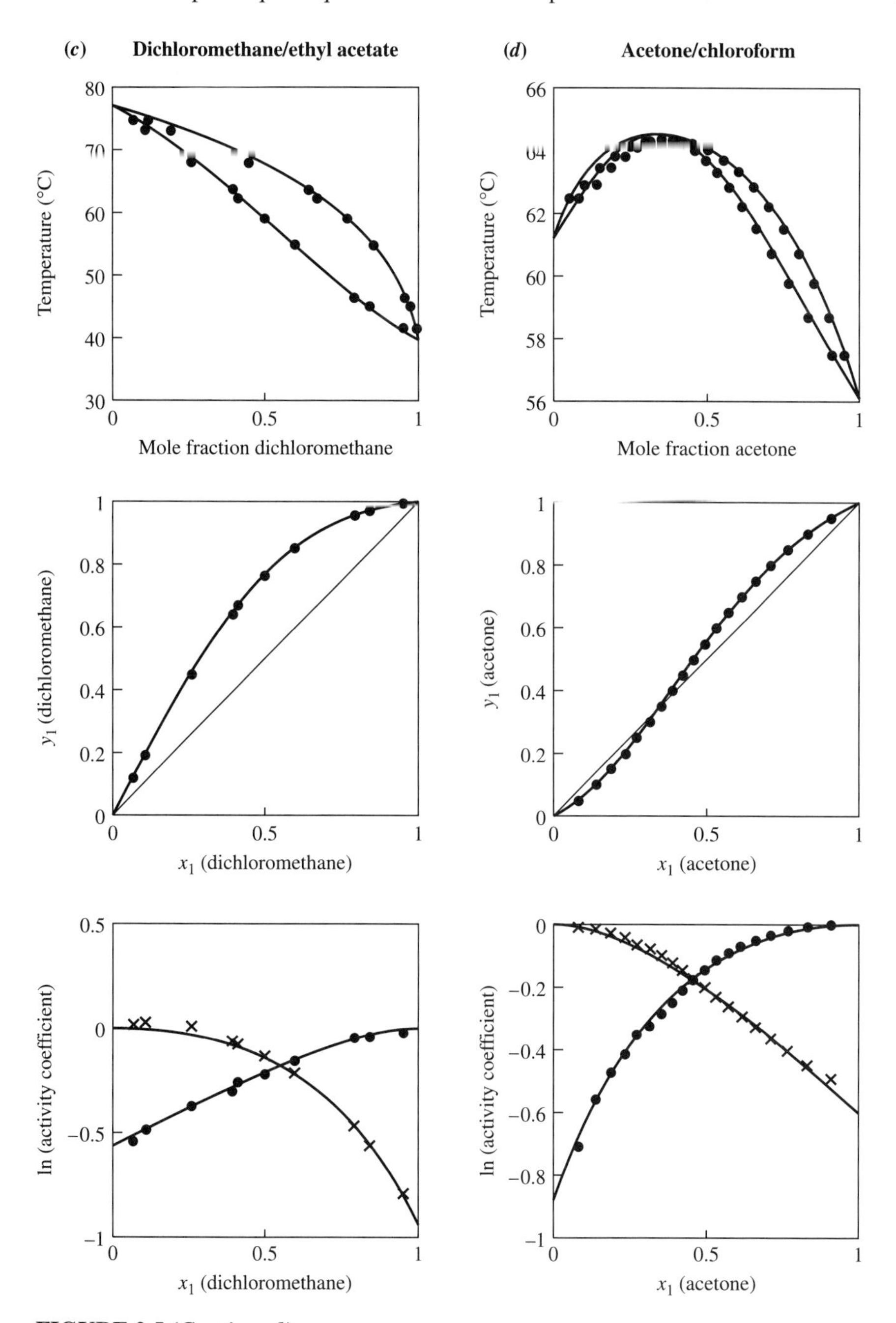

FIGURE 2.5 (*Continued*)
Selected examples of mixtures that deviate from Raoult's law. For each mixture we show the isobaric T-x-y diagram, the isobaric y-x diagram, and the isobaric $\ln \gamma_1$-$\ln \gamma_2$-x_1 diagram at boiling conditions. The pressure is held constant at 1 atm throughout. (*c*) Dichloromethane (1) and ethyl acetate (2). (*d*) Acetone (1) and chloroform (2).

A boiling liquid at the azeotropic composition produces a vapor of exactly the same composition, and the liquid therefore does not change its composition as it evaporates. Because of this, it is not possible to perform any separation of the components of a mixture at its azeotropic composition by means of distillation. Fortunately, it is often possible to separate such mixtures by distillation when a third component is added or when two columns are operated at different pressures. These topics are discussed later in the book.

Deviations from Raoult's law can be accounted for by the use of activity coefficients in the VLE relation. All of the activity coefficient models mentioned earlier, e.g., Wilson, Margules, Van Laar, NRTL, and UNIQUAC, are capable of representing phase diagrams similar to those shown in Figs. 2.4 and 2.5, although not always to the same level of accuracy. For this reason it is essential to plot the model predicted y-x diagram for each binary pair in the (multicomponent) mixture before blindly using the VLE model for design calculations. If the predicted diagrams do not give satisfactory agreement with experimental data, then other models should be explored. This initial step in the design process is frequently neglected, even by practicing engineers, and has been the cause of many difficulties and poor designs that could have been overcome with relatively little effort if proper attention had been paid to the VLE model, e.g., Roy and Hobson (1987).

Choosing a Model

There is no universal model for phase equilibrium in nonideal mixtures. However, some are better than others for different purposes and some are more demanding of good data for regression of parameters than others. Table 2.4 offers some guidelines for choosing a model.

2.3 SPECIAL LIMITING CASES

Constant Relative Volatility

For binary mixtures, the VLE relation (Eq. 2.20) becomes

$$P y_1 = P_1^{\text{sat}} \gamma_1 x_1 \tag{2.26}$$
$$P y_2 = P_2^{\text{sat}} \gamma_2 x_2 \tag{2.27}$$

Summing these equations and recognizing that the mole fractions in each phase sum to unity leads to the following equivalent pair of VLE relations:

$$P = P_1^{\text{sat}} \gamma_1 x_1 + P_2^{\text{sat}} \gamma_2 (1 - x_1) \tag{2.28}$$

$$y_1 = \frac{\alpha_{12} x_1}{1 + (\alpha_{12} - 1) x_1} \tag{2.29}$$

TABLE 2.4
Guidelines for choosing a VLE model for nonideal mixtures

1. Consider constant volatility primarily for mixtures of similar molecules, e.g., homologous series.
2. Raoult's law models and constant volatility models are often, though not always, close. The differences are usually largest for substances with very close or very different boiling points.
3. Avoid constant volatility models or Raoult's law for strongly nonideal mixtures; if they work, it's only by accident. Never use a constant volatility model or Raoult's law for an azeotropic mixture. Raoult's law or constant volatility are not useful for mixtures with azeotropes or tangent pinches. One exception is for very close boilers, where Raoult's law can predict an azeotrope if the vapor pressure curves for the pure components cross one another.
4. Mixture data (or parameters already estimated from mixture data) are needed for all models except Raoult's law (Raoult's law requires only pure component data). UNIFAC parameters are estimated for groups instead of molecules, but are nonetheless estimated from mixture data.
5. Be prepared to spend up to half the time solving a problem on the data and property models.
6. Engineering or "ϕ-γ" models are not accurate near the mixture critical point or to describe the behavior of very low boiling "noncondensables."
7. Engineering or "ϕ-γ" models that do well in predicting T-x-y behavior often make poor predictions of h-x, h^E, etc., for the same mixture at the same conditions (unless data on the desired quantity was used in the regression of parameters).
8. The Wilson model does not predict liquid-liquid phase behavior.
9. The NRTL or UNIQUAC models are generally preferred for liquid-liquid equilibria.
10. Most activity coefficient models with a single set of parameters will not make accurate predictions of both the liquid-liquid and vapor-liquid phase equilibrium. More recent models like the Kohler model (Talley et al., 1993) can do a better job at the expense of a larger data requirement.
11. Regressions for parameters that weight each point equally often produce model parameters that yield poor predictions in the dilute regions for mixtures. That is where accuracy often counts the most, because many stages are needed to reach high purities.
12. Extrapolation can be very inaccurate, even for models that are very accurate for interpolating data in a limited range. With a sufficiently large extrapolation, almost anything can happen (and probably will!).

where the *relative volatility* α_{12} is given by

$$\alpha_{12} = \frac{P_1^{\text{sat}}\gamma_1}{P_2^{\text{sat}}\gamma_2} \left(= \frac{1}{\alpha_{21}} \right) \tag{2.30}$$

and is a function of T and x_1, at fixed pressure.

The simplest description that is useful for the VLE in distillation is found for the special class of mixtures for which the relative volatility α_{12} can be treated as constant over the entire range of liquid compositions and equilibrium temperatures at a given system pressure. Such mixtures are often called "constant α" or "constant volatility" systems. Our convention is to refer to the lightest component (by boiling point at the system pressure) as component 1, in which case it is convenient to drop the subscripts and write the y-x relation as

$$y = \frac{\alpha x}{1 + (\alpha - 1)x} \tag{2.31}$$

which is an explicit relationship between y and x with a constant value of α.

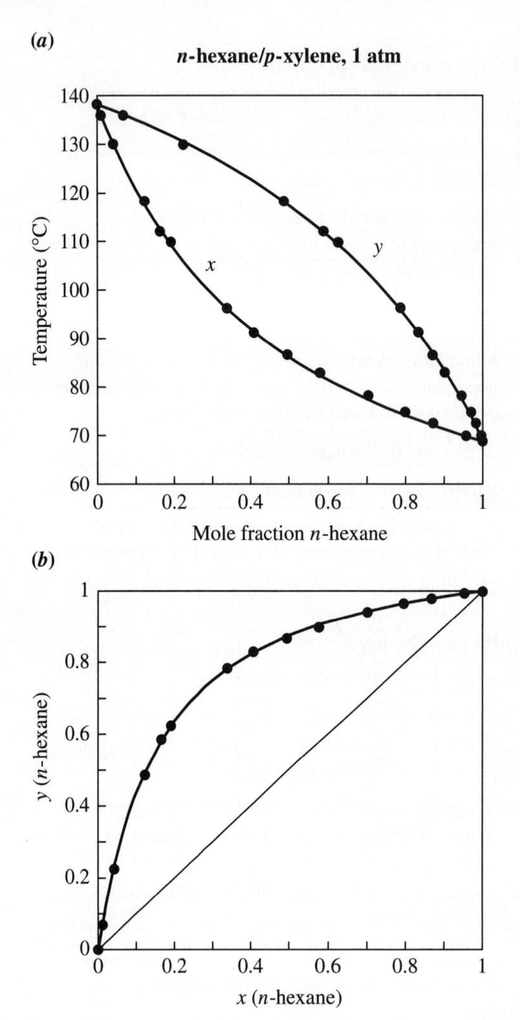

FIGURE 2.6
Phase equilibrium diagrams for the system hexane and *p*-xylene at 1 atm pressure. (*a*) T-x-y diagram. (*b*) y-x diagram.

Figure 2.6 shows diagrams of the temperature and composition behavior of a binary mixture of hexane and *p*-xylene, according to the data in Gmehling et al. (1980*b*, p. 608) The y-x diagram is particularly convenient for distillation calculations; it can be prepared by a cross-plot from the T-x-y plot, which contains more information about the phase equilibrium in the mixture. In this figure, we also show a comparison of the y-x data with Eq. 2.31. A constant value of $\alpha = 7.0$ accurately describes the y-x behavior even though the activity coefficients in the liquid phase change by as much as 50%, from 1.0 to 1.5. Selected examples of other mixtures whose VLE can be accurately represented with a constant value for α are given in Table 2.5.

For constant α mixtures, the y-x diagram has certain special features as noted in Fig. 2.7. These include:

1. The slope of the y_1-x_1 curve at $x_1 = 0$ is equal to $\alpha_{12} = 1/\alpha_{21}$.
2. The slope of the y-x curve at $x_1 = 1$ is equal to $1/\alpha_{12} = \alpha_{21}$.
3. The y_2-x_2 diagram is obtained from the y_1-x_1 diagram by turning the diagram upside down, i.e., rotation of 180°. This is a property common to all binary mixtures, whether they are constant α or not.

For each value of x, at a given pressure P, the equilibrium temperature is found from Eq. 2.28. This normally involves an iterative calculation. For the special case of ideal constant volatility mixtures, i.e., $\gamma_1 = \gamma_2 = 1$, the equilibrium temperature

TABLE 2.5
Selected examples of binary mixtures whose VLE can be accurately represented with a constant value for the relative volatility

Light component (normal bp, °C)	Heavy component (normal bp, °C)	α		
		Low	Optimal	High
Benzene (80.1)	Toluene (110.6)	2.27	2.34	2.40
Toluene (110.6)	*p*-Xylene (138.3)	2.21	2.31	2.44
Benzene (80.1)	*p*-Xylene (138.3)	4.25	4.82	5.75
m-Xylene (139.1)	*p*-Xylene (138.3)	1.0201	1.0202	1.0204
Pentane (36.0)	Hexane (68.7)	2.40	2.59	2.79
Hexane (68.7)	Heptane (98.5)	2.22	2.45	2.64
Hexane (68.7)	*p*-Xylene (138.3)	6.54	7.00	7.32
Ethanol (78.4)	Isopropanol (82.3)	1.14	1.17	1.19
Isopropanol (82.3)	*n*-Propanol (97.3)	1.71	1.78	1.84
Ethanol (78.4)	*n*-Propanol (97.3)	1.95	2.10	2.22
Methanol (64.6)	Ethanol (78.4)	1.46	1.56	1.71
Methanol (64.6)	Isopropanol (82.3)	2.20	2.26	2.33
Chloroform (61.2)	Acetic acid (118.1)	5.20	6.15	8.40
Acetone (56.2)	Propyl acetate (101.7)	4.50	4.74	5.05

Source: The values were determined using a minimum average absolute error in y from data in the DECHEMA *Vapor-Liquid Equilibrium Data Collection* at 1 atm pressure. The bounds correspond to one standard deviation of the absolute error.

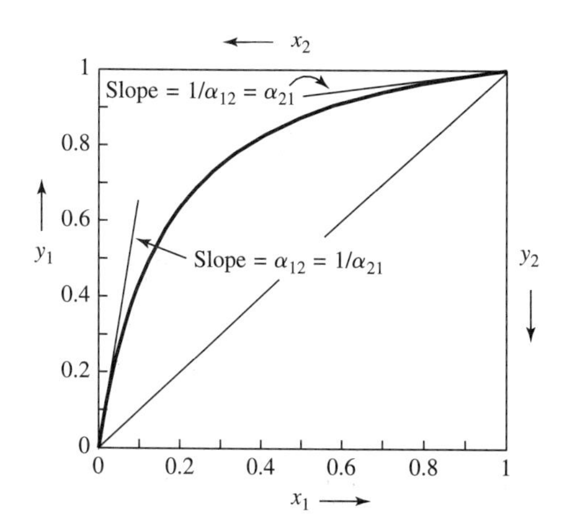

FIGURE 2.7
A binary y-x diagram showing some of its characteristic properties.

and the constant value of α can be obtained explicitly as

$$T = \frac{T_2}{(R/\lambda)T_2 \ln[1 + (\alpha_{12} - 1)x_1] + 1} \tag{2.32}$$

$$\alpha_{12} = \exp\left[\frac{\lambda}{R}\left(\frac{1}{T_1} - \frac{1}{T_2}\right)\right] \tag{2.33}$$

where T_1 and T_2 are the boiling points of pure components 1 and 2 at the system pressure, λ is the average heat of vaporization of the two pure components at the system pressure, and α_{12} is the relative volatility of the mixture at the system pressure. All of the temperatures are in Kelvin, and λ/R has units of Kelvin. The derivation of Eqs. 2.32 and 2.33 is set as a problem in Exercise 9. It can be shown that the value of α_{12} given by Eq. 2.33 guarantees that the T-x curve given by Eq. 2.32 satisfies the boundary conditions $T = T_1$ at $x_1 = 1$ and $T = T_2$ at $x_1 = 0$.

For example, in the hexane (1) and p-xylene (2) mixture at 1 atm pressure, $\lambda_1 = 28{,}850$ J/gmol, $\lambda_2 = 35{,}980$ J/gmol, $\lambda = 32{,}415$ J/gmol, $T_1 = 341.9$ K, $T_2 = 411.5$ K, $R = 8.314$ J/gmol K. The resulting value of α_{12} from Eq. 2.33 is 6.88. This is close to the optimum value of 7.0, found from regression of the y-x data, and is well within one standard deviation of the optimum; see Table 2.5. Figure 2.8 shows the T-x-y and y-x diagrams for this mixture predicted from Eqs. 2.31,

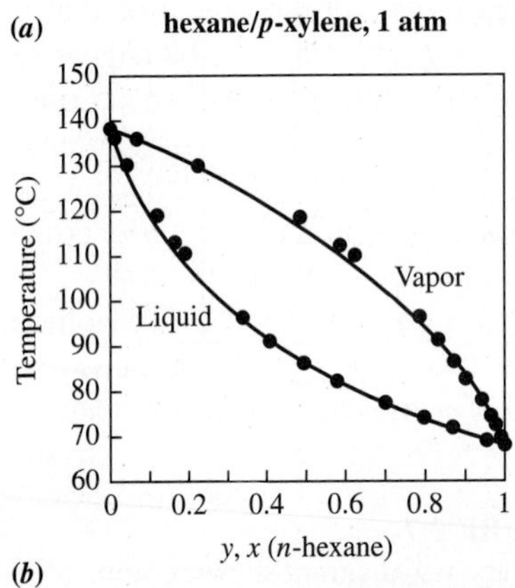

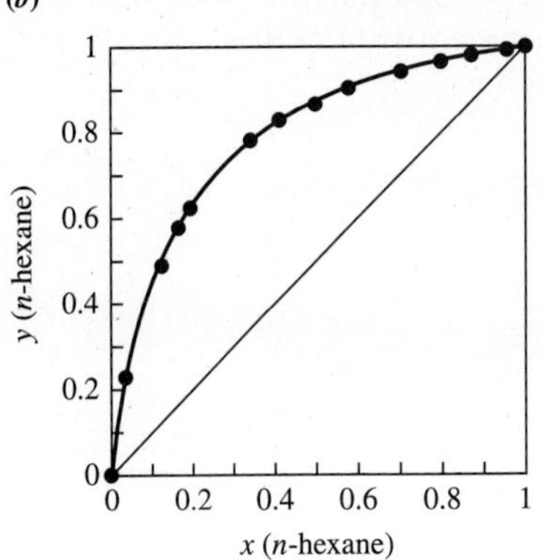

FIGURE 2.8
Binary VLE diagrams for hexane and p-xylene at 1 atm pressure. (a) T-x-y diagram. (b) y-x diagram. The solid lines are predicted from Eqs. 2.31, 2.32, and 2.33; experimental data are taken from Gmehling et al. (1980b, p. 608).

2.32, and 2.33. Note that only pure component properties have been used to compute these figures. Agreement of similar quality is obtained for other constant volatility ideal mixtures, e.g., benzene (1) and toluene (2) at 1 atm, for which $\lambda_1 = 30{,}760$ J/gmol, $\lambda_2 = 33{,}180$ J/gmol, $\lambda = 31{,}970$ J/gmol, $T_1 = 353.22$ K, $T_2 = 383.81$ K, giving $\alpha_{12} = 2.38$.

An Empirical Model for Nonideal Mixtures

Nonideal mixtures frequently display quite exotic phase behavior. This is reflected in the appearance of inflections and extrema in the phase diagrams.[11] As a consequence, the equilibrium y-x diagrams may show marked deviation from the uniform curves that are typical of constant α systems. In fact, thermodynamics places only one restriction on the shape of isobaric y-x diagrams for homogeneous, nonreactive, binary mixtures, namely, that the curve has strictly positive slope for all values of x, i.e.,[12]

$$\frac{dy}{dx} > 0 \tag{2.34}$$

e.g., Malesiński (1965, pp. 54–60). Inequality 2.34 is clearly satisfied for constant α systems, as differentiation of Eq. 2.31 will verify.

There are two common features of nonideal binary phase diagrams that are particularly troublesome for distillation column design. The first is the appearance of inflections in the isobaric T-x-y diagram, which in turn generate inflections in the corresponding y-x diagram. The second difficulty is associated with extrema in the phase diagram that give rise to homogeneous binary azeotropes. At such points, the equilibrium liquid and vapor compositions are equal and the y-x curve crosses the 45° line with *positive* slope.

It is quite convenient to have simple explicit y-x relations for selected nonideal binary mixtures so that certain types of distillation calculations can be demonstrated more easily for mixtures that exhibit tangent pinches and azeotropes. For example, data from Gmehling et al. (1980*a*, p. 170) show that the mixture benzene-ethylene-diamine at 1 atm pressure exhibits an inflection in its y-x diagram as shown in Fig. 2.9.

These y-x data are well correlated by the empirical expression

$$y = \frac{ax}{1 + (a-1)x} + bx(1-x) \tag{2.35}$$

with $a = 9$, $b = -0.6$. This expression was used to generate the solid curve through the experimental data in Fig. 2.9. Though empirical, this equation is quite versatile and can be used to correlate a variety of binary nonideal y-x data. Values of a and b for selected mixtures are given in Table 2.6.

---

[11] The practical consequences of such behavior are discussed more fully later.

[12] It is understood that pressure is held constant in the differential dy/dx.

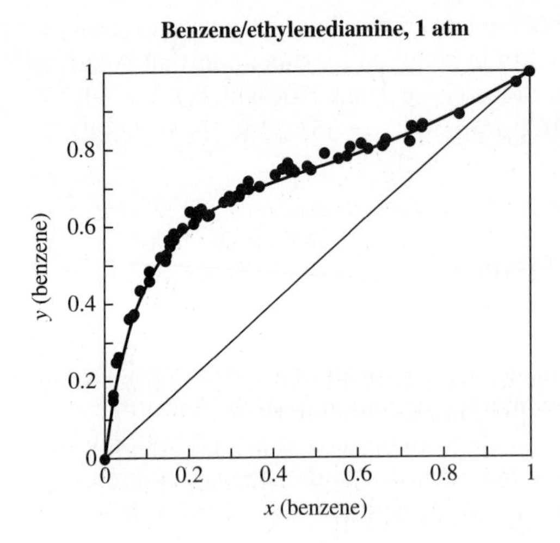

FIGURE 2.9
VLE data for the mixture benzene and ethylenediamine at 1 atm. The solid line is given by Eq. 2.35; experimental data are taken from Gmehling et al. (1980*a*, p. 170).

TABLE 2.6
Constants in the empirical VLE model (Eq. 2.35) for selected binary mixtures

Mixtures without inflections or azeotropes					
Light component	**Heavy component**	**(bp, °C)**		***a***	***b***
Acetaldehyde	Water	20.5	100.0	80.0	−0.10
Methanol	Water	64.6	100.0	7.15	−0.33
Acetaldehyde	Methanol	20.5	64.6	0.75	1.10
Water	Ethylene glycol	100.0	197.3	100.0	0.0
Mixtures with inflections but no azeotropes					
Light component	**Heavy component**	**(bp, °C)**		***a***	***b***
Benzene	Ethylenediamine	80.1	117.3	9.00	−0.60
Acetone	Water	56.1	100.0	32.0	−0.58
Mixtures with minimum boiling azeotropes					
Light component	**Heavy component**	**(bp, °C)**		***a***	***b***
Ethyl acetate	Ethanol	77.2	78.4	3.90	−1.13
Methanol	Ethyl acetate	64.6	77.2	5.50	−1.07
Ethanol	Methyl ethyl ketone (2-butanone)	78.4	79.6	3.20	−1.04
Tetrahydrofuran (235.5 mmHg)	Methanol (235.5 mmHg)	33.8	37.4	3.90	−1.05
Pentane (750 mmHg)	Dichloromethane (750 mmHg)	35.7	39.4	4.20	−1.20
Methyl acetate	Methanol	56.9	64.6	5.60	−1.15
Isopropanol	Water	82.3	100.0	15.0	−1.35
Acetone	Methanol	56.1	64.6	3.80	−0.88
Ethanol	Water	78.4	100.0	9.5	−1.00

Source: The experimental *y-x* data were taken from the DECHEMA *Vapor-Liquid Equilibrium Data Collection*. The pressure is 1 atm unless noted otherwise and the boiling points are given at the same pressure as the VLE data.

2.4 PRESSURE EFFECTS

Although most distillations are carried out at atmospheric pressure, it is not uncommon to distill at other pressures. Low-pressure distillation (typically in the range 30 to 300 mmHg) is frequently practiced for temperature-sensitive mixtures that polymerize or react at normal boiling conditions, e.g., acetic anhydride-acetic acid, diketene-acetic acid, or in cases where the boiling temperature of the bottoms stream is high enough to require the use of expensive heating methods such as high-pressure steam, electricity, heat transfer oils or furnaces, e.g., separation of *meta* and *para* di-isopropyl benzene. High-pressure distillation (typically 3 to 20 atm) usually occurs in thermally integrated processes or when the normal boiling point of the distillate is lower than the temperature of cooling water. In such cases, it is normally cheaper to pressurize the column in order to raise the boiling point of the distillate than to install a refrigeration system to condense the substance.

The main assumptions made in deriving the equilibrium relation (Eq. 2.20) are that the vapor phase can be treated as a mixture of perfect gases, and that the molar volume of each pure component liquid is independent of pressure and negligible compared to the molar volume of a perfect gas for all temperatures and pressures of interest. These assumptions improve as the pressure decreases. Thus, for low-pressure distillations, Eq. 2.20 can be used without modification.

At elevated pressures, the vapor phase deviates from the perfect gas law at the relatively low temperatures required for phase equilibrium, the molar volume of the liquid approaches that of the vapor (becoming equal at the critical point), which necessarily causes the liquid phase to become compressible, and the molar volume of the liquid may not be negligible. In view of these effects, modifications to Eq. 2.20 become necessary for high-pressure VLE calculations. In order to assess which modifications to make we must first establish the range of pressures of interest. The vast majority of distillations are carried out at pressures below 70% of the critical pressure, i.e., at reduced pressure, $P_r < 0.7$. At higher pressures either the column pressure or the associated boiling temperature is normally so high that distillation is not economical.[13]

Vapor phase deviations from the perfect gas mixture can be represented in terms of the virial expansion for the compressibility factor z $(= Pv/RT)$; thus

$$z = 1 + \frac{B}{v} + \text{higher-order terms} \tag{2.36}$$

where B is the second virial coefficient for the mixture, which depends only on temperature and composition [see VanNess and Abbott (1982, Chap. 4), and Rowlinson (1958, pp. 59 et seq.)]. The virial expansion can be written as a power series in

[13] Due to costs or constraints on temperatures, distillation columns are rarely operated at pressures above 20 atm or at temperatures above 250°C. Within these constraints, many substances can be distilled because their liquid and vapor phases will coexist and will be sufficiently far from the critical point to have significantly different properties. For instance, Hougen et al. (1964, Figs. 140*a* and 140*c*) show that when $P_r = 0.7$, the corresponding boiling temperature of many materials is $T_r \sim 0.95$. Some typical values are given in Table 2.7. For most of these substances, 20 atm is well below the critical pressure.

TABLE 2.7
Values of $0.7P_c$ and $0.95T_c$ for selected substances

Substance	$0.7P_c$ (atm)	$0.95T_c$ (°C)
Methane	31.8	−107.4
Propane	29.3	62.9
n-Octane	17.2	252.0
Propylene	31.9	58.4
Acetic acid	40.0	276.5
Benzene	33.8	245.6
Ethanol	44.1	202.0
Water	152.3	326.4

Source: Critical properties are taken from Reid et al. (1977, Appendix A).

pressure with the following form (VanNess and Abbott, 1982, Chap. 4):

$$z = 1 + \frac{BP}{RT} + \text{higher-order terms} \tag{2.37}$$

The second virial coefficient for the mixture is related to the virial coefficients for the like and unlike binary pairs in the following way:

$$B = \sum_{i=1}^{c} \sum_{j=1}^{c} B_{ij}(T)\, y_i y_j \tag{2.38}$$

where B_{ii} represents the virial coefficient for pure gas i and B_{ij} are the cross coefficients for molecules i and j. The coefficients B_{ij}, $i = 1, \ldots, c$; $j = 1, \ldots, c$ depend only on temperature since they are related to the intermolecular potentials between molecules i and j, which in turn depend only on temperature (McQuarrie, 1976). On physical grounds the cross coefficients are treated as symmetrical, i.e., $B_{ij} = B_{ji}$, although none of the properties of B depend on this assumption.

It is very convenient to rewrite Eq. 2.38 in the form [see VanNess and Abbott (1982, pp. 134 et seq.)]

$$B = \sum_{i=1}^{c} B_{ii}\, y_i + \sum_{i=1}^{c} \sum_{j=1}^{c} \delta B_{ij}\, y_i y_j \tag{2.39}$$

where the quantity δB_{ij} is

$$\delta B_{ij} = B_{ij} - \frac{1}{2}\left(B_{ii} + B_{jj}\right) \tag{2.40}$$

and clearly $\delta B_{ii} = 0$, $\delta B_{ij} = \delta B_{ji}$.

For pressures up to 70% of the critical pressure, the compressibility factor can be represented quite satisfactorily as a linear function of pressure at all temperatures of possible interest in distillation. This can be seen from Reid et al. (1977, Fig. 3.1), which shows this approximate linearity for all isotherms above and below the critical temperature for a wide variety of pure materials.

Rearranging Eq. 2.37 and neglecting higher-order terms leads to the following volume–explicit equation of state for the vapor mixture

$$v = \frac{RT}{P} + \sum_{i=1}^{c} B_{ii}\, y_i + \sum_{i=1}^{c}\sum_{j=1}^{c} \delta B_{ij}\, y_i y_j \tag{2.41}$$

The first two terms in Eq. 2.41 represent the molar volume of an *ideal mixture of imperfect gases*. The third term represents the excess molar volume or, equivalently, the volume of mixing. This is often a small contribution to the total molar volume of the gas mixture at reduced pressures up to 0.7. Therefore, a reasonable approximation is to set $\delta B_{ij} = 0$ for all values of i and j. This does not imply $B_{ij} = 0$ but rather $B_{ij} = \frac{1}{2}(B_{ii} + B_{jj})$; i.e., the cross coefficients are the arithmetic averages of the pure component coefficients. The net result is that we treat the gas phase as an ideal mixture of imperfect gases that obey the following equation of state:

$$v = \frac{RT}{P} + \sum_{i=1}^{c} B_{ii}\, y_i \tag{2.42}$$

Liquids do not normally exhibit appreciable compressibility until the pressure gets close to its critical value (as in the generalized P_r vs. T_r saturation envelope in Reid et al. (1977, Fig. 1–1). It therefore remains an acceptable assumption to treat the pure liquids as incompressible at reduced pressures up to $P_r = 0.7$, although they may not have negligible molar volume.

With these assumptions, the vapor-liquid equilibrium relation becomes

$$Py_i \exp\left(\frac{B_{ii}(P - P_i^{\text{sat}})}{RT}\right) = P_i^{\text{sat}}\gamma_i x_i \exp\left(\frac{v_i^{0L}(P - P_i^{\text{sat}})}{RT}\right) \tag{2.43}$$

The exponential term on the right-hand side is called the Poynting correction factor. The exponential term on the left-hand side is the fugacity coefficient for component i in the vapor phase, often denoted as ϕ_i. Combining the two correction factors in Eq. 2.43 leads to the form

$$Py_i = P_i^{\text{sat}}\gamma_i x_i \exp\left(\frac{(v_i^{0L} - B_{ii})(P - P_i^{\text{sat}})}{RT}\right) \tag{2.44}$$

The second virial coefficient is always negative for a saturated vapor,[14] and so $(v_i^{0L} - B_{ii})$ is positive. At low pressures, where we treat the vapor phase as a mixture of perfect gases ($B_{ii} = 0$), and where we neglect the volume of the pure liquids ($v_i^{0L} = 0$), Eqs. 2.43 and 2.44 reduce to the more familiar relation given by Eq. 2.20.

In order to solve Eq. 2.43 we need expressions for the variation of the second virial coefficients and the pure component liquid molar volumes.

The variation of pure component liquid molar volumes with temperature is represented quite accurately by the Rackett equation (Rackett, 1970)

$$v^{0L} = v_c z_c^{(1-T_r)^{0.2857}} \tag{2.45}$$

[14]That is, the slope of the isotherms on a z vs. P_r plot is always negative in the vicinity of the saturated vapor envelope, as can be seen from Reid et al. (1977, Fig. 3–1).

TABLE 2.8
Critical values and acentric factors for selected substances

Substance	T_c (K)	P_c (bar) [atm]	v_c (cm^3/gmol)	z_c	ω
Acetone	508.1	47.0 [46.4]	209.0	0.232	0.304
Water	647.3	221.2 [218.3]	57.1	0.235	0.344
Benzene	562.2	48.9 [48.3]	259.0	0.271	0.212
Toluene	591.8	41.0 [40.5]	316.0	0.263	0.263
Ethylenediamine	593.0	62.8 [62.0]	206.0	0.26	0.51
Hexane	507.5	30.1 [29.7]	370.0	0.264	0.299
p-Xylene	616.2	35.1 [34.6]	379.0	0.260	0.320
Ethanol	513.9	61.4 [60.6]	167.1	0.240	0.644
Methyl ethyl ketone	536.8	42.1 [41.5]	267.0	0.252	0.320
Acetaldehyde	461.0	55.7 [55.0]	154.0	0.220	0.303
Methanol	512.6	80.9 [79.8]	118.0	0.224	0.556

Source: Taken from Reid et al. (1987, Appendix A). To convert from bar to atm use the factor 1 atm = 1.013250 bar.

Values for the critical molar volume v_c, compressibility z_c, and temperature T_c are tabulated in Reid et al. (1987, Appendix A) and in Simmrock et al. (1986*a*, 1986*b*); values for selected substances are given in Table 2.8. Other methods for estimating liquid molar volumes are given in Reid et al. (1987, Chap. 3).

Second virial coefficients are often estimated from the Pitzer-Curl equations for nonpolar molecules (Smith and VanNess, 1966, pp. 86–92)

$$B = \frac{RT_c}{P_c}\left(B^{(0)} + \omega B^{(1)}\right) \tag{2.46}$$

with

$$B^{(0)} = 0.083 - \frac{0.422}{T_r^{1.6}} \tag{2.47}$$

$$B^{(1)} = 0.139 - \frac{0.172}{T_r^{4.2}} \tag{2.48}$$

where ω is the Pitzer acentric factor [tabulated in Reid et al. (1987, Appendix A)]; values for selected substances are given in Table 2.8. Using Eq. 2.46 with T_c in K, P_c in bar, and $R = 83.14$ (cm^3 bar)/(gmol K) gives values for B in units of cm^3/gmol. If T_c is in K, P_c in atm, and $R = 82.05$ (cm^3 atm)/(gmol K), Eq. 2.46 again gives values for B in units of cm^3/gmol. Methods for modifying Eq. 2.46 to estimate second virial coefficients for polar molecules are discussed by Reid et al. (1987, pp. 40–42) and by Perry and Green (1984, pp. 3–269).

Equation 2.43 is solved using an algorithm quite similar to that adopted earlier for low-pressure VLE calculations. For isobaric bubble-point problems where we specify P and $\mathbf{x}$ and calculate T and $\mathbf{y}$, the algorithm is

1. Specify P and $\mathbf{x}$.
2. Guess T.
3. Calculate $P_i^{\text{sat}}(T), i = 1, \ldots, c$ from the Antoine equation.
4. Calculate $v_i^{0L}(T),\ i = 1, \ldots, c$ from the Rackett equation 2.45. If the current estimate for T is greater than the critical temperature of any of the pure com-

ponents, do not use the Rackett equation for those components and instead set $v^{0L} = v_c$.

5. Calculate B_{ii}, $i = 1, \ldots, c$ from the Pitzer-Curl Eq. 2.46 or a modification.
6. Calculate $\gamma_i(T, \mathbf{x})$, $i = 1, \ldots, c$ from the Wilson equation or another liquid solution model.
7. Calculate y_i, $i = 1, \ldots, c$ from Eq. 2.43.
8. If $\sum_{i=1}^{c} y_i$ is sufficiently close to unity, then stop. Otherwise, adjust T and go to step 3.

In order to demonstrate the principles outlined in this section, we now present results of high-pressure VLE calculations for the mixture acetone (1) and water (2). For simplicity, we begin by setting the Poynting correction factor equal to unity and calculate isobaric phase diagrams at several pressures according to the algorithm above. Liquid-phase nonidealities are represented by the Wilson equation with a single set of parameters determined exclusively from atmospheric pressure measurements [see Gmehling and Onken (1977, p. 242) and Tables 2.2 and 2.3 for numerical values]. Antoine coefficients for the pure component vapor pressures are reported in Table 2.1; critical values and acentric factors for the pure components are given in Table 2.8.

Figure 2.10 reports the T-x-y and y-x diagrams at 1, 6.803, 17.01, and 34.01 atm pressure. The solid lines represent the calculated phase behavior with all of the second virial coefficients set equal to zero; i.e., they are the solutions to Eq. 2.20. Figure 2.11 reports the same diagrams at the same pressures but with second virial corrections included in the calculations according to Eqs. 2.43 and 2.46. As would be expected, these figures demonstrate that the predictions are improved by including second virial coefficient corrections.[15] An unexpected result, however, is that these corrections have a more significant effect on the T-x-y diagram than on the y-x diagram. It is not possible to anticipate this result from a casual inspection of Eq. 2.43. In fact, one of the biggest difficulties associated with Eq. 2.43 is that there is no method for determining a priori the relative importance of the various correction terms. This is in contrast to almost every other subfield of chemical engineering where scaling arguments and dimensional analysis provide guidelines for choosing an appropriate level of modeling detail a priori.

Equation 2.43 is not immune to these techniques, and we present here a perturbation solution that provides a framework for assessing the effects of various corrections. For simplicity, we neglect the Poynting correction, i.e., set $v_i^{0L} = 0$ in Eq. 2.43. Assume that P and $\mathbf{x}$ are known and that a solution to Eq. 2.43 has been calculated with $B_{ii} = 0$ for all i. Denote this solution by T_0 and $\mathbf{y}_0$. Then a first-order approximation to the solution of Eq. 2.43 with the second virial coefficients included is given by

$$T = T_0 \left(1 + \frac{\bar{\epsilon}}{\bar{\kappa}}\right) \tag{2.49}$$

[15] Addition of the Poynting correction as well as the correction to the Pitzer-Curl equations for polar molecules would improve the predictions a little further. However, the differences between the model and data in Fig. 2.11 can be attributed mostly to extrapolation of the liquid activity coefficient relation far from the temperature range over which it was correlated.

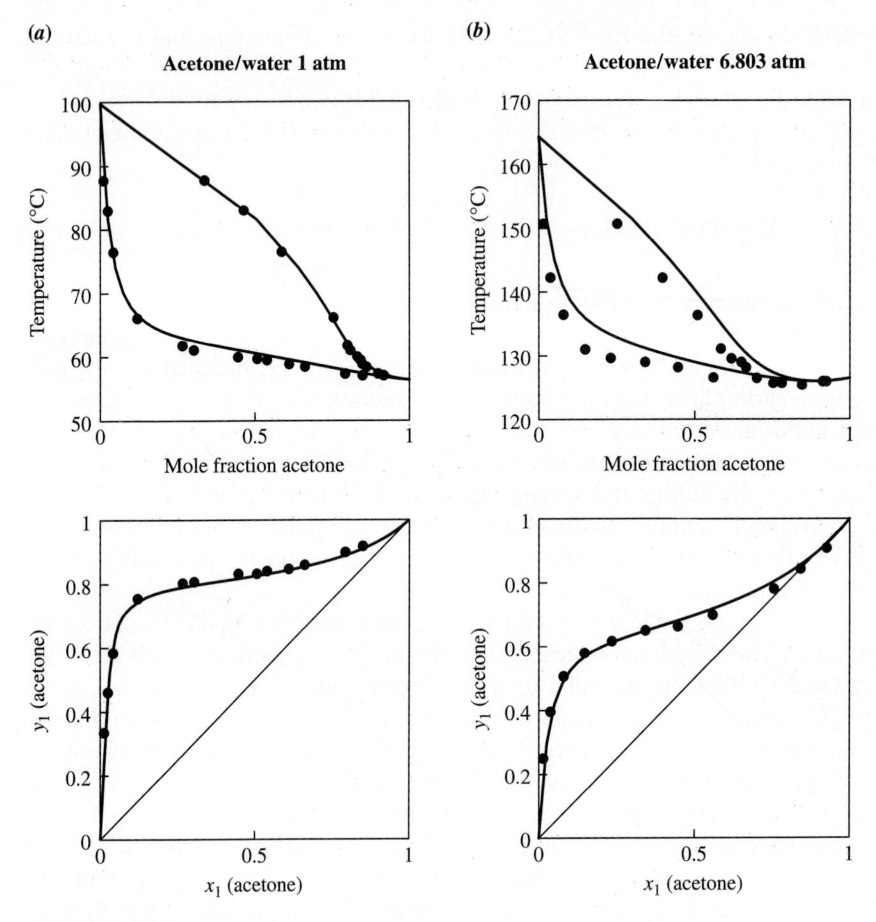

FIGURE 2.10
T-x-y (top) and corresponding x-y (bottom) diagrams for the mixture acetone (1) and water (2) at (*a*) 1 and (*b*) 6.803 atm. The solid curves are calculated from the equilibrium model treating the vapor as a mixture of perfect gases and neglecting the Poynting correction in the liquid (Eq. 2.20). The experimental data are taken from Othmer et al. (1952).

$$y_i = y_{i0}\left(1 + \kappa_i\left[\frac{\bar{\epsilon}}{\bar{\kappa}} - \frac{\epsilon_i}{\kappa_i}\right]\right) \tag{2.50}$$

where
$$\epsilon_i = \left[\frac{B_{ii}(P - P_i^{\mathrm{sat}})}{RT}\right]_{T=T_0} \tag{2.51}$$

$$\kappa_i = \frac{\Delta h_i}{RT_0} \tag{2.52}$$

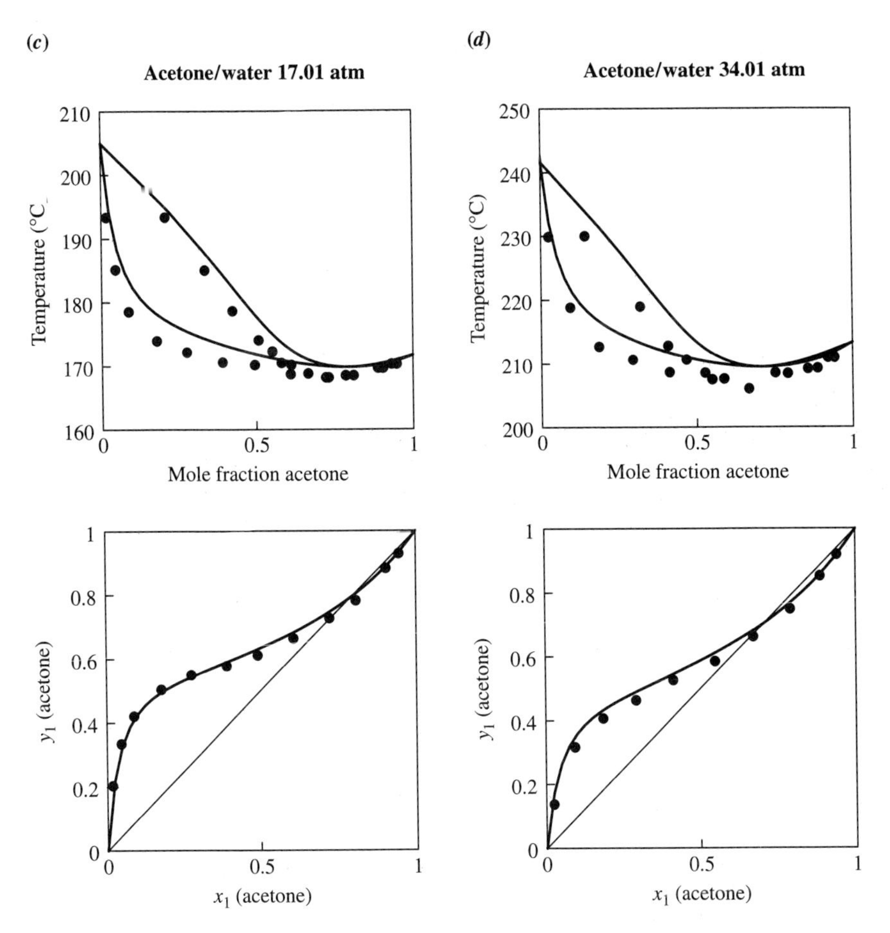

FIGURE 2.10 (*Continued*)
T-x-y (top) and corresponding x-y (bottom) diagrams for the mixture acetone (1) and water (2) at (*c*) 17.01 and (*d*) 34.01 atm. The solid curves are calculated from the equilibrium model treating the vapor as a mixture of perfect gases and neglecting the Poynting correction in the liquid (Eq. 2.20). The experimental data are taken from Othmer et al. (1952).

$$\bar{\epsilon} = \sum_{j=1}^{c} \epsilon_j y_{j,0} \tag{2.53}$$

$$\bar{\kappa} = \sum_{j=1}^{c} \kappa_j y_{j,0} \tag{2.54}$$

and all temperatures are in Kelvin and Δh_i is the latent heat of vaporization of pure component i at its normal boiling point; see Exercise 10.

Equation 2.49 shows that corrections to the temperature will be negligible whenever $|\bar{\epsilon}/\bar{\kappa}| < 0.001$. Equation 2.50 can be used to estimate corrections to the vapor-phase composition; these tend to cancel each other in the calculation of $\mathbf{y}$, in contrast

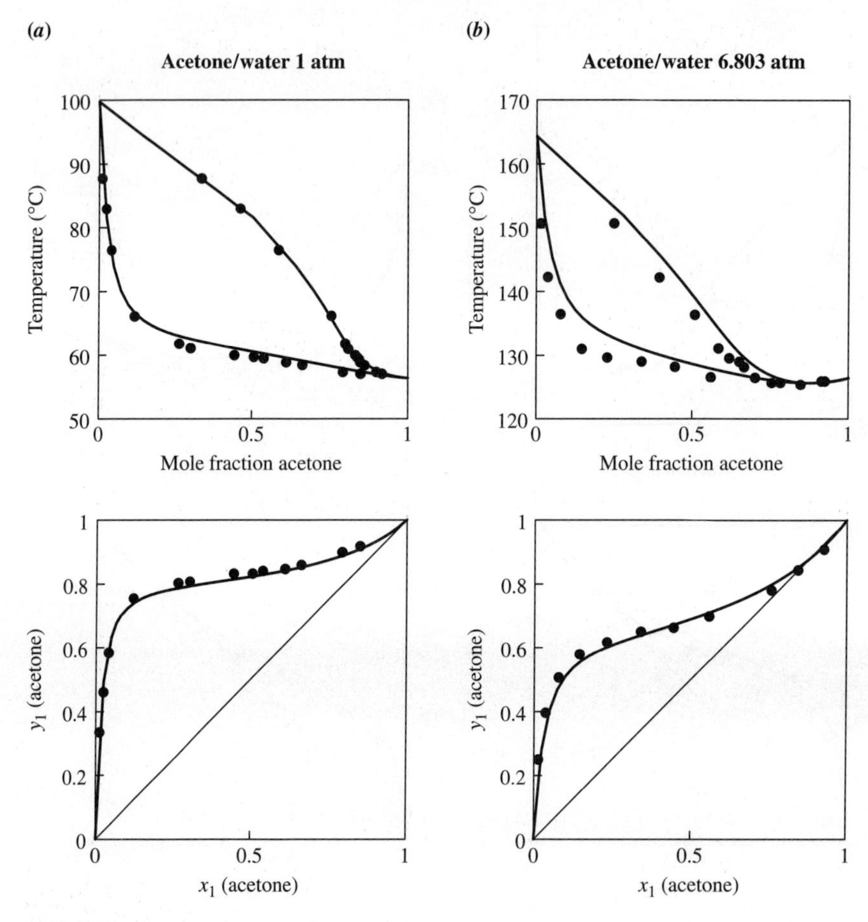

FIGURE 2.11
Phase diagrams for the mixture acetone (1) and water (2) at (*a*) 1 and (*b*) 6.803 atm. Solid curves are calculated from the equilibrium model treating the vapor as an ideal mixture of imperfect gases and neglecting the Poynting correction in the liquid (Eq. 2.43). Experimental data are taken from Othmer et al. (1952).

to the purely additive corrections to T. We can expect that the corrections will have a more significant effect on the shape of the T-x-y diagram than on the y-x behavior. Figure 2.12 compares the perturbation solution to the base case solution of Eq. 2.20 for the mixture acetone and water at a pressure of 34.01 atm. The Poynting correction is neglected in each case. For this example, the perturbation solution provides an accurate estimate of the full solution and a simple way of calculating phase diagrams. In addition, it is useful for assessing the relative magnitudes of the various terms contributing to the shape of the phase diagrams as the pressure is varied.

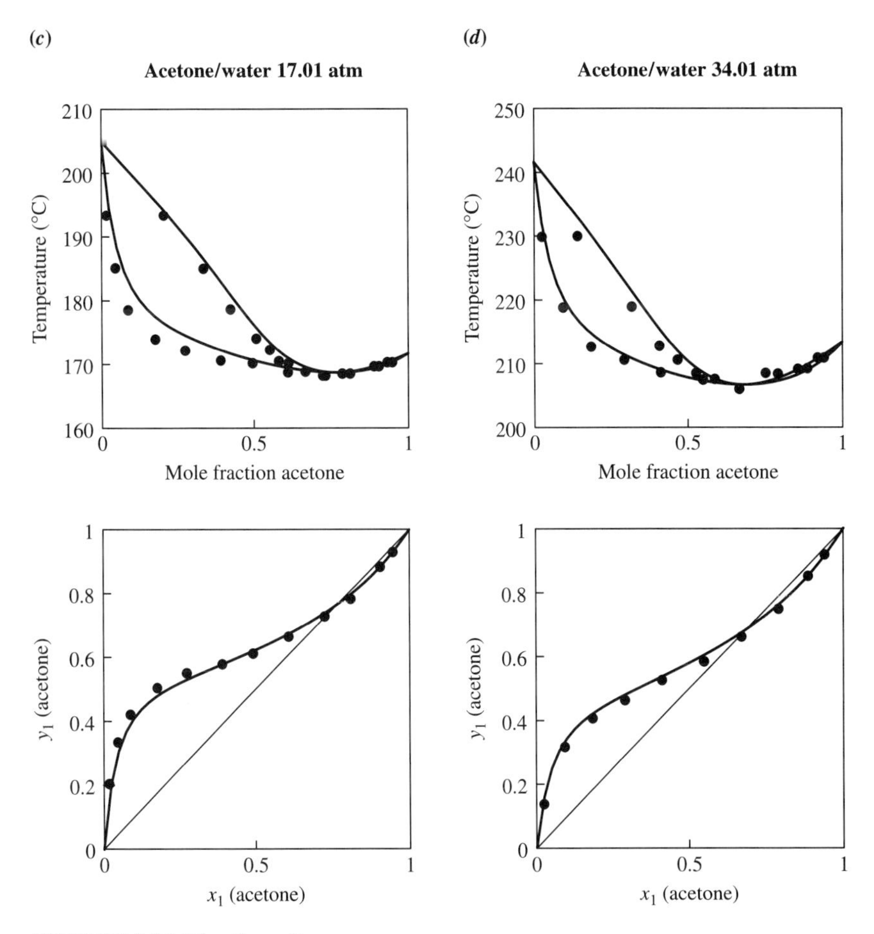

FIGURE 2.11 (*Continued*)
Phase diagrams for the mixture acetone (1) and water (2) at (*c*) 17.01 and at (*d*) 34.01 atm. Solid curves are calculated from the equilibrium model treating the vapor as an ideal mixture of imperfect gases and neglecting the Poynting correction in the liquid (Eq. 2.43). Experimental data are taken from Othmer et al. (1952).

2.5 FLASH SEPARATIONS

The simplest type of continuous equilibrium stage separation is the *two-phase flash*. A common arrangement is shown in Fig. 2.13. A gas-phase process stream is cooled to produce a two-phase mixture, which is passed into a vessel where the phases are separated. Vapor and liquid products are continuously withdrawn from the vessel and are close to equilibrium with each other. Alternatively, the process stream could be a liquid which undergoes heating and/or pressure reduction to produce a two-phase feed to the flash chamber. Flash separators are frequently used to separate light gases such as hydrogen, methane, or hydrogen chloride from liquid mixtures. Typical

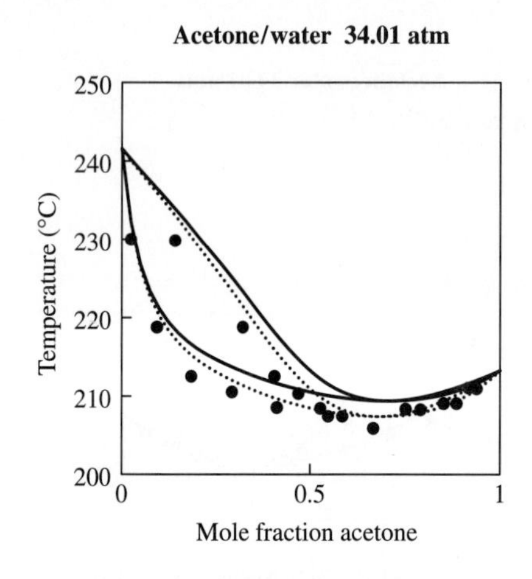

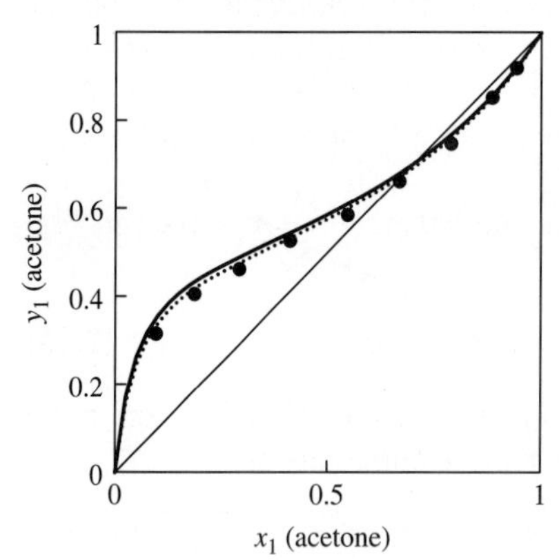

FIGURE 2.12
Phase diagrams for the mixture acetone (1) and water (2) at 34.01 atm. Dotted curves are calculated from the perturbation solution given in Eqs. 2.49 and 2.50; solid curves are calculated from the equilibrium model treating the vapor as an ideal mixture of perfect gases and neglecting the Poynting correction in the liquid (Eq. 2.20). Experimental data are taken from Othmer et al. (1952).

separations include hydrogen and methane from benzene, toluene, and diphenyl [as in the hydrodealkylation of toluene process (Douglas, 1998)], and HCl from a stream of mixed silanes (as in the manufacture of silicon, described in Chap. 1). If good separation cannot be obtained in one or two stages then a distillation column is normally used instead.

To obtain a mathematical model, a balance region is drawn around the vessel as shown in Fig. 2.13. A typical problem is to calculate the composition and flowrate of each product stream for given composition ($\mathbf{z}$), flowrate (F), temperature (T), and

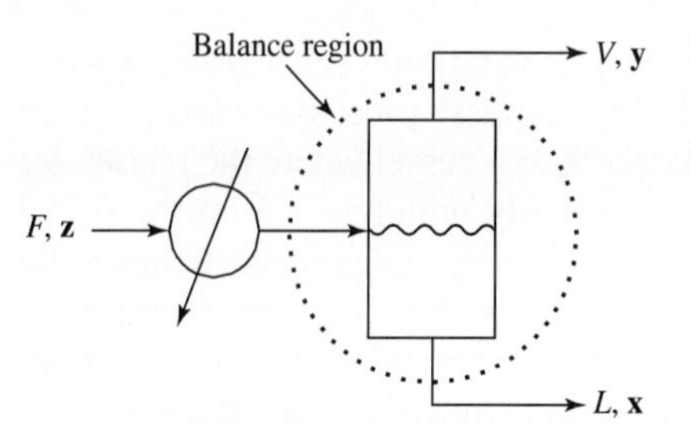

FIGURE 2.13
A simple flash.

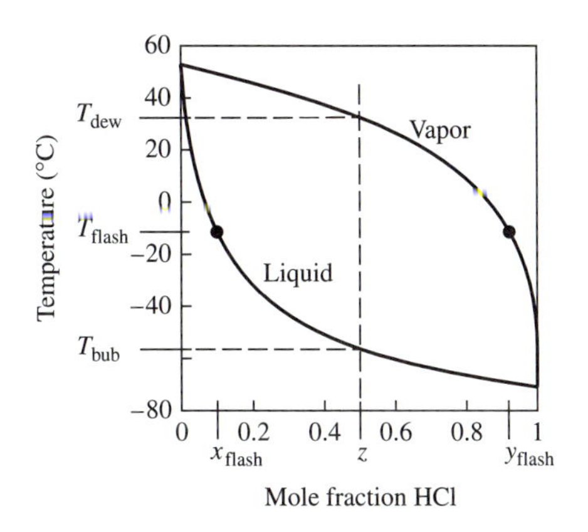

FIGURE 2.14
Flash separation represented on an isobaric T-x-y diagram for the mixture HCl and trichlorosilane at 2 atm pressure.

pressure (P) of the feed. This is called the *isothermal flash problem.* On a binary phase diagram, (Fig. 2.14), the isothermal flash problem corresponds to picking a flash temperature (T_{flash}) between the bubble point (T_{bub}) and the dew point (T_{dew}) temperatures at composition z. The corresponding phase compositions are denoted on Fig. 2.14 by x_{flash} and y_{flash}.

Material balances for each component over the balance region shown in Fig. 2.13 are

$$Fz_i = Vy_i + Lx_i \qquad i = 1, 2, \ldots, c \tag{2.55}$$

These equations are solved simultaneously with the phase equilibrium and mole fraction summation equations. For ideal mixtures these equations are

$$Py_i = P_i^{sat} x_i \qquad i = 1, 2, \ldots, c \tag{2.56}$$

$$\sum_{i=1}^{c} x_i = 1 \qquad \sum_{i=1}^{c} y_i = 1 \qquad \sum_{i=1}^{c} z_i = 1 \tag{2.57}$$

There are a total of $2c + 3$ equations and $3c + 5$ variables, T, P, V, L, F, z_i, x_i, and y_i, ($i = 1, 2, \ldots, c$), indicating that there are $c + 2$ degrees of freedom. Specifying values for T, P, F, and z_i ($i = 1, 2, \ldots, c - 1$) leads to a well-posed problem that is normally solved using the Rachford-Rice method (Rachford and Rice, 1952). For ideal binary mixtures the equations have an analytic solution which is useful for pedagogical purposes; see Exercise 15.

In the Rachford-Rice method, Eqs. 2.55, 2.56, and 2.57 are reformulated to give

$$z_i = \phi y_i + (1 - \phi)x_i \qquad i = 1, 2, \ldots, c \tag{2.58}$$

$$y_i = K_i x_i \qquad i = 1, 2, \ldots, c \tag{2.59}$$

$$\sum_{i=1}^{c} (y_i - x_i) = 0 \qquad V = \phi F \qquad L = (1 - \phi)F \tag{2.60}$$

where $\phi \equiv V/F$ represents the fraction of feed that is removed as vapor, $K_i = P_i^{sat}/P$ is the equilibrium K-value for component i, which is a constant for ideal mixtures at fixed T and P. Further rearrangement leads to

$$f(\phi) = \sum_{i=1}^{c} \frac{z_i(K_i - 1)}{1 + \phi(K_i - 1)} = 0 \tag{2.61}$$

$$x_i = \frac{z_i}{1 + \phi(K_i - 1)} \qquad i = 1, 2, \ldots, c \tag{2.62}$$

$$y_i = K_i x_i \qquad i = 1, 2, \ldots, c \tag{2.63}$$

$$V = \phi F \qquad L = (1 - \phi)F \tag{2.64}$$

Equation 2.61 is solved iteratively for ϕ; all other variables are then calculated explicitly from Eqs. 2.62 to 2.64.

The function $f(\phi)$ in Eq. 2.61 is well behaved, which is why the Rachford-Rice formulation is so popular. Specifically, it is simple to show that $f(\phi)$ has strictly negative slope for all values of ϕ, except in the trivial case where $K_1 = K_2 = \cdots = K_c = 1$. Therefore, $f(\phi)$ has a unique root in the interval $0 \leq \phi \leq 1$ if and only if $f(0) > 0$ *and* $f(1) < 0$. If either of these conditions is not satisfied, then no root exists. A little algebra shows that these conditions are equivalent to

$$\sum_{i=1}^{c} z_i K_i > 1 \tag{2.65}$$

and

$$\sum_{i=1}^{c} \frac{z_i}{K_i} > 1 \tag{2.66}$$

For mixtures obeying Raoult's law these conditions imply that the flash temperature lies between the bubble point T_{bub} and the dew point, T_{dew}. Therefore, they do not represent additional constraints on the process; see Exercise 22. Other formulations of the isothermal flash problem have been proposed, e.g., King (1980, pp. 71–80) and VanNess and Abbott (1982, Appendix G); however, the one presented here is generally preferred.

EXAMPLE 2.1. Mixtures of acetone and acetic anhydride occur in a process for making acetic anhydride from acetone and acetic acid. Consider a process stream of 30 kmol/h acetone (component 1) and 70 kmol/h acetic anhydride (component 2) at a temperature of 105°C and a pressure of 1 atm. The normal boiling points of the two pure components are 56.1 and 139.5°C, respectively. Since the boiling point difference is more than 80°C, we might expect a single-stage flash to give a good separation between the components. Does it?

- Vapor–liquid equilibrium for the mixture is represented quite well by Raoult's law. In fact, Raoult's law represents the *y*-*x* diagram as well as any of the nonideal solution models reported in Gmehling et al. (1979, p. 172). Using the Antoine coefficients reported in Table 2.1, we find $K_1 = 4.078$ and $K_2 = 0.317$.
- The feed composition to the flash vessel is $z_1 = 0.3$, $z_2 = 0.7$. We first check to see if two phases are present at the given conditions. The sums in Eqs. 2.65 and 2.66 are 1.45 and 2.28, respectively, indicating that two phases are indeed present and that a unique solution exists to the flash problem. This can be seen in Fig. 2.15*a*, which shows a plot of the Rachford-Rice function $f(\phi)$ for the example.

- Newton's method[16] is an effective way to solve Eq. 2.61. In this method successive iterates are calculated from the formula

$$\phi_{n+1} = \phi_n - \frac{f(\phi_n)}{f'(\phi_n)} \tag{2.67}$$

where $f'(\phi_n)$ is the derivative of f with respect to ϕ evaluated at the point ϕ_n. The method requires an initial condition, ϕ_0. Since the function is so well behaved, any initial value in the interval $0 \le \phi \le 1$ will do. The iterates are calculated until the *residual*, $|f(\phi)|$, is closer to zero than a specified tolerance ϵ. The tolerance will vary from one problem type to another depending on the scale of $f(x)$, and the desired accuracy for x. There is no foolproof method for setting tolerances, and they are normally selected initially by scaling arguments and then adjusted up and down until there is no further significant change in the answer. In flash calculations, $f(\phi)$ is typically no greater than +100 and no less than −100. This example was solved with $\phi_0 = 0.5$, and $\epsilon = 1 \times 10^{-4}$, which gives the answer $\phi = 0.212$.
- This value of ϕ is used in Eqs. 2.62 to 2.64 to calculate the phase compositions

	x_i	y_i
Acetone	0.182	0.741
Acetic anhydride	0.818	0.259

and flows, $L = 78.8$ kmol/h, $V = 21.2$ kmol/h.
- In this example the flash separator has certainly performed some separation since the liquid stream is approximately 80 mol % acetic anhydride and the vapor is roughly 75 mol % acetone. However, these streams are far from the high purities or high recoveries typically required in most processes, and further separation would certainly be necessary. Because of this we would choose not to flash this mixture, but instead to perform the separation in a distillation column.
- For this example, the graph in Fig. 2.15*a* is almost linear. This is typical of low-purity flash separations. However, for high-purity separations such as the one posed in the next example, the function is strongly nonlinear, with a sharp shoulder in the neighborhood of $\phi = 0.9$.

EXAMPLE 2.2. This example is based on the flowsheet in Chap. 1 for the production of electronic-grade polysilicon. The flash separation takes place directly after the fluidized bed reactor shown in Fig. 1.7. For simplicity we will treat the reactor effluent as a binary mixture of hydrogen chloride (component 1) and trichlorosilane (component 2). The more realistic multicomponent problem is set as Exercise 14. From the stoichiometry given in the description of the process flowsheet, we treat the mixture as approximately equimolar in the two components. The flash temperature and pressure are given as −30°C and 2 atm, respectively. If the feed flow to the flash chamber is 100 kmol/h, what is the flowrate and composition of the flash product streams?

[16]Newton's method alone is not generally recommended for solving one-dimensional problems, especially those defined on a bounded interval. The reasons for this are explained in Press et al. (1986, Chap. 9) together with recommended methods. We use Newton's method to solve the Rachford-Rice equation only because it is known to converge for these problems and because it is simple to describe.

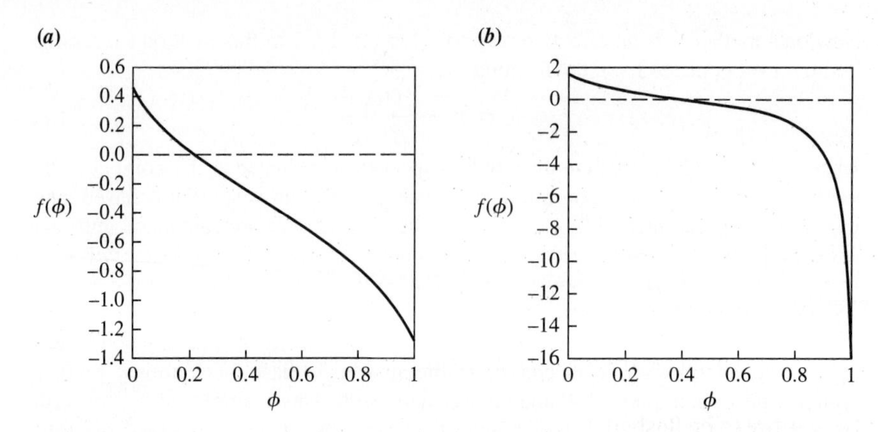

FIGURE 2.15
Rachford–Rice function for Examples 2.1 and 2.2.

- Phase equilibrium for the mixture is represented by Raoult's law, with Antoine coefficients taken from Table 2.1. The predicted T-x-y diagram is shown in Fig. 2.14. Since we are unable to find experimental phase equilibrium data for this mixture it is difficult to say how accurate the predictions are. The K-values are $K_1 = 5.270$ and $K_2 = 0.031$.
- Inequalities 2.65 and 2.66 are satisfied for the given feed composition and K-values, indicating that the mixture is two-phase. The graph of the Rachford-Rice function is strongly nonlinear, as shown in Fig. 2.15*b*.
- The flash equations are solved by the methods described in the previous example to give $\phi = 0.399$, $L = 60.1$ kmol/h, $V = 39.9$ kmol/h, and

	x_i	y_i
Hydrogen chloride	0.185	0.975
Trichlorosilane	0.815	0.025

- This example demonstrates the effective use of a flash separation. The goal of the front end of the flowsheet in Fig. 1.7 is to produce high-purity trichlorosilane as feed to the filament reactor, and to recycle unreacted HCl. The liquid leaving the flash vessel should contain almost all of the trichlorosilane, though not necessarily in high purity. This liquid is then separated further to recover the components in high purity, as shown in the flowsheet. The vapor, on the other hand, should be high-purity HCl, with very little trichlorosilane, so that it can be recycled without sending too much trichlorosilane back to the fluidized bed reactor. By calculating the component flows L_i and V_i in the liquid and vapor streams, respectively,

	$L_i = Lx_i$ **(kmol/h)**	$V_i = Vy_i$ **(kmol/h)**
Hydrogen chloride	11.1	38.9
Trichlorosilane	49.0	1.0

we see that the flash fulfills these objectives. The liquid contains 98 mol % of the entering trichlorosilane, even though it is not in very high purity (which means that appreciable amounts of HCl are also present in the liquid). The vapor is reasonably pure HCl, although the fractional recovery of HCl in the vapor is not high.

If the mixture to be flashed deviates from Raoult's law, the equilibrium equation must be modified in the Rachford-Rice method. For simplicity we neglect Poynting corrections and assume that Eq. 2.20 adequately represents the vapor–liquid equilibrium. Therefore, Eq. 2.56 is replaced by

$$Py_i = P_i^{\text{sat}} \gamma_i x_i \qquad i = 1, 2, \ldots, c \tag{2.68}$$

All other equations used to describe the isothermal flash problem remain the same. The Rachford–Rice formulation is again given by Eqs. 2.61 to 2.64 except now the K–values are equal to $K_i = P_i^{\text{sat}} \gamma_i / P$, which depend on the liquid composition. Since the K–values are not constant, Eqs. 2.61 and 2.62 must be solved simultaneously for ϕ and x_i, $i = 1, \ldots, c$. Once convergence is attained, the remaining variables are calculated explicitly from Eqs. 2.63 and 2.64. Methods for solving this problem are discussed in Boston and Britt (1978); see also Exercises 16 and 17.

The isothermal flash problem is sometimes formulated with P, ϕ, and z_i, $i = 1, \ldots, c - 1$, specified instead of P, T, and z_i. In this case the Rachford-Rice equation (Eq. 2.61) is solved for T; then all the other variables are calculated explicitly from Eqs. 2.62 to 2.64. For binary mixtures there is a simple graphical construction that illustrates how the material balance and phase-equilibrium relations are satisfied simultaneously. The material balance (Eq. 2.55) is written in the following way for the lightest component:

$$y = -\frac{L}{V} x + \frac{F}{V} z \tag{2.69}$$

Defining the quantity $\tilde{q} = L/F$ enables us to write the material balance in the form

$$y = -\frac{\tilde{q}}{1 - \tilde{q}} x + \frac{1}{1 - \tilde{q}} z \tag{2.70}$$

Note that $\tilde{q} = 1 - \phi$; therefore, specifying $\tilde{q}$ or ϕ amounts to the same thing. For fixed values of $\tilde{q}$ and z this equation is a straight line on y-x coordinates with slope $-\tilde{q}/(1 - \tilde{q})$ and intercept $z/\tilde{q}$ on the x axis. Therefore, the isothermal flash equations for binary mixtures with P, $\tilde{q}$, and z specified can be represented quite conveniently on a y-x diagram. The known value of P allows us to calculate the equilibrium y-x curve, and the known values for $\tilde{q}$ and z allow us to plot the material

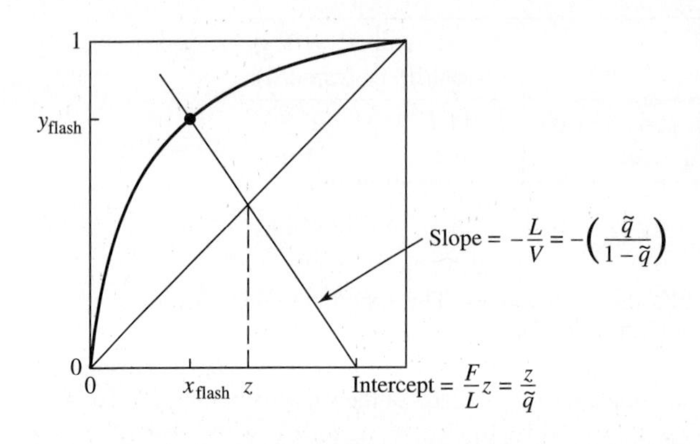

FIGURE 2.16
Graphical solution of the $P - \tilde{q}$ flash for a binary mixture.

balance line, as shown in Fig. 2.16. The intersection of these two curves is the point where the material balance and equilibrium equations are satisfied simultaneously, and is the solution to the problem; the corresponding temperature is read from the T-x-y diagram.

In many flash separations of practical interest, one or more of the components is above its critical temperature, typically hydrogen ($T_c = -240°C$), or methane ($T_c = -82.6°C$). In these cases, K-values cannot be estimated from Raoult's law. Instead they are estimated from one of the following methods:

1. *Experimental Data.* A good compilation of experimental data for mixtures containing low–boiling substances is now available through the DECHEMA Chemistry Data Series VI (Knapp et al., 1982). For each binary mixture the authors report:
   a. Experimental T-P-x-y data and the literature reference.
   b. Isotherms on a P-x-y phase diagram. Many of these diagrams show interesting supercritical behavior. Good discussions of phase diagrams for pure fluids and binary mixtures in the supercritical region are given by Smith and VanNess (1996, Chap. 12) and Rowlinson (1969, Chap. 6).
   c. A plot of $\log_{10} K$ vs. $\log_{10} P$ at various temperatures (Smith and VanNess, 1996, Chap. 13 gives a good interpretation of these figures).
   d. Pure component and binary interaction parameters for four equations of state.
2. *Equations of State.* In recent years nearly all the major chemical companies have placed greater emphasis on using equations of state to predict thermodynamic properties of mixtures. Phase equilibrium calculations using equations of state are much more complicated than the methods described in this book, and the reader is referred to more specialized texts for the particulars, e.g., VanNess and Abbott (1982, Chap. 4), and Prausnitz et al. (1998).
3. *Correlations.* For aliphatic hydrocarbon mixtures in the C_1-C_{10} range, approximate K-values can be determined from a set of nomographs prepared by DePriester in 1953. These charts are widely available and are reported in Perry's Handbook (Perry and Green, 1984, pp. 13–17 and 13–18). McWilliams (1973) regressed a correlation to the DePriester charts which is useful for computer calcu-

lations. Many other correlations have been proposed over the years for estimating K-values of hydrocarbon mixtures. Some of these are more accurate than the De-Priester charts; others apply to a wider variety of mixtures, and all are discussed briefly in Perry's Handbook.

Mixtures containing hydrogen are quite common in the chemical and petrochemical industries, and they are almost always flashed to remove the hydrogen. The most widely used technique for estimating K-values in such mixtures is the Chao–Seader method (Chao and Seader, 1961). The method can be quite accurate, as, for example, in the quaternary mixture of hydrogen-benzene-cyclohexane-n-hexane, where it predicts the experimentally measured K-values over a wide range of pressures and temperatures with an average deviation of slightly less than 5% (Brainard and Williams, 1967). Numerous variations and improvements to the Chao-Seader method have also been proposed over the years, and these too are discussed in Perry's Handbook.

2.6 STAGING FLASH SEPARATIONS

High purities and high fractional recoveries are often required in separations. Unfortunately, a single-stage flash rarely provides either of these, but certain arrangements of multiple flash operations or systems derived from these can. These can be implemented in more efficient ways described later, but it is instructive to consider them first in the simplest form and for binary mixtures. [Similar ideas are discussed by Furzer (1986), Chap. 4 and Wankat (1988).] An obvious but inefficient approach is to apply a series of flash operations, condensing part of the vapor products and boiling part of the liquid products from successive stages. Figure 2.17 shows a schematic.

The top portion of the cascade produces vapors from subsequent stages that are enriched in the lighter component, and the purity of the heavier component grows in the liquids in the bottom portion. If the feed flow F and its composition z are given, specifications of the pressure and temperature for each stage are sufficient to fix the other flows and compositions. Alternatively, the pressure and vapor fraction ϕ (or the liquid fraction $\tilde{q} = 1 - \phi$) can be specified for each stage. For example, if these are constant on each stage, then the flows are simply $V_{T,n} = F\phi^n$ and $L_{B,n} = F\tilde{q}^n = F(1-\phi)^n$ after n stages in the top or the bottom sections. The highest *purities* are found in the top section if flash stages $2, \ldots, N_T$ are near the bubble-point temperature, $\phi \to 0$. Conversely, stages in the bottom section, $2, \ldots, N_B$ should be near the dew point, $\tilde{q} \to 0$, for maximum increase of the purity for the heavy component in the liquid streams. This can be seen from the construction in Fig. 2.16.

In addition to the purities, the fractional recoveries of the components are important. Even for $\phi = 1/2$, the vapor flow in the top section is only 3.1% of the feed after 5 stages and 0.1% after 10 stages. Therefore, a basic deficiency in this simple arrangement is the fact that *high purities correspond to low factional recoveries and vice versa.* (There are also a large number of intermediate process streams with no

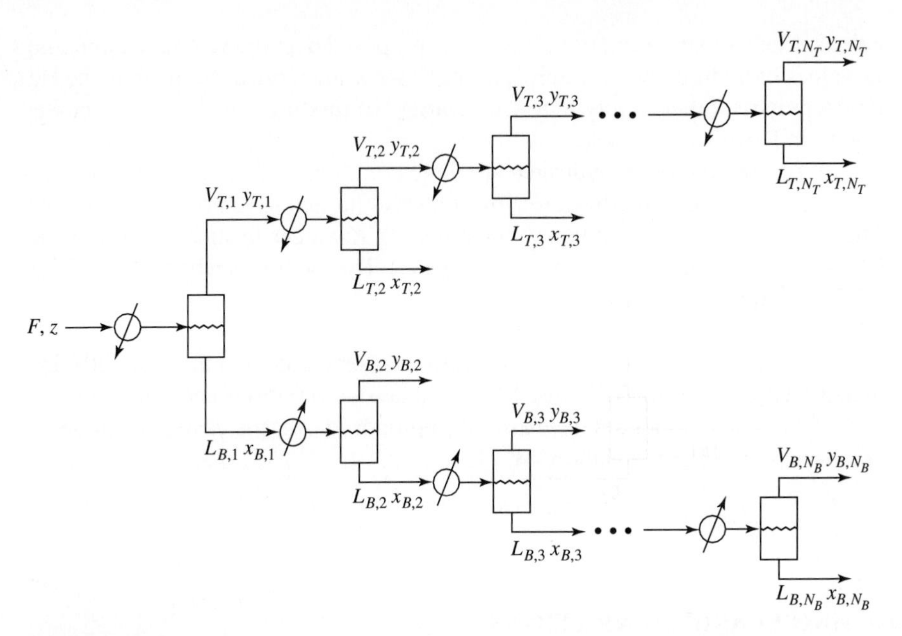

FIGURE 2.17
Series of flash separations.

destination, plus a large number of heaters and coolers with the associated costs and utility requirements!) Exercise 23 considers a simple analysis of this system and the choice of variables.

One alternative arrangement is to recycle all but the last vapor stream in the top and the last liquid stream in the bottom, e.g., sending stream $L_{T,4}$ to stage 3 in the top, stream $L_{T,3}$ to stage 2, etc., Fig. 2.18. Most of the intermediate coolers and heaters can also be eliminated by direct contact of the intermediate vapor streams with the recycled liquid streams, as follows.

Suppose that a single vapor product stream is taken from the top section, i.e., V_{N_T}. By material balance, each vapor stream in the top will have a flow larger than the liquid streams by exactly this product flowrate. Furthermore, the temperature of each vapor stream will be greater than the stage above, by an amount that depends on the choice of conditions, e.g., ϕ in each stage. Therefore, some of the intermediate vapor will condense if it is contacted directly with the liquid returning, and some of the liquid will boil. If the heat of vaporization is roughly constant and the system is adiabatic, the amounts of vapor condensed and liquid vaporized will be approximately the same, and there is sufficient energy at a suitable temperature in the vapor streams to boil the recycled liquids. Of course, on the last stage in the top section, a cooler is needed to condense some liquid.

A similar logic for the bottom section, which retains at least one heater, leads to the schematic in Fig. 2.18. This is a more effective system that offers the possibility for both high purities *and* high fractional recoveries. It is often economical to mix intermediate liquid and vapor streams at different temperatures and compositions,

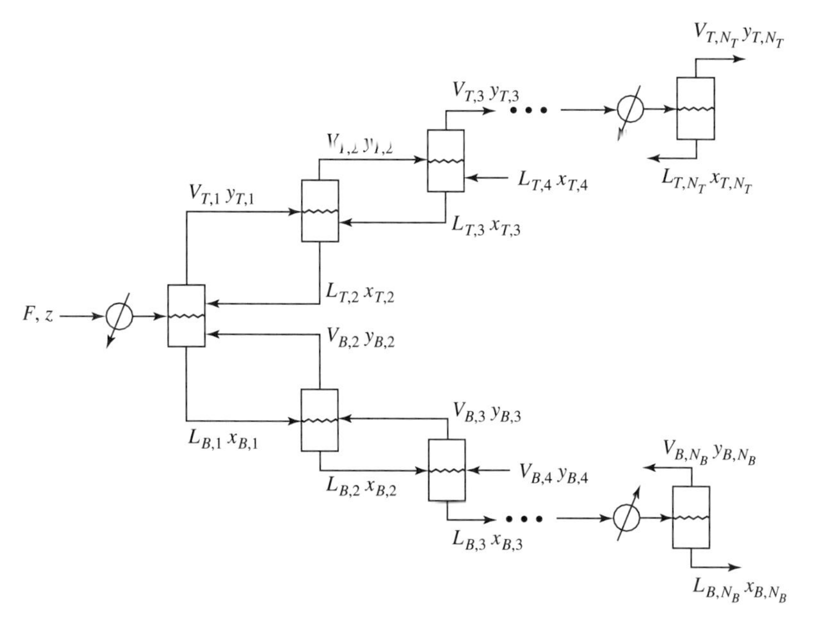

FIGURE 2.18
A more efficient series of flash separations with recycle of intermediate streams.

even though this introduces thermodynamic inefficiencies. This is often less expensive compared to the investment required for a large number of intermediate heat exchangers. However, in some cases one or two intermediate exchangers, the choice of intermediate streams as products, and other more complex configurations are better alternatives, especially for multicomponent mixtures. The configuration in Fig. 2.18 is analyzed in Chaps. 3 and 4; more complex configurations are discussed in Chap. 7.

2.7 EXERCISES

1. Calculate the isothermal P-x-y, y-x, and $\ln \gamma_1$-$\ln \gamma_2$-x_1 phase diagrams for the mixture shown in Fig. 2.5*b* at the average temperature of the pure component normal boiling points. Treat the system as a nonideal liquid in equilibrium with a mixture of perfect gases neglecting Poynting corrections, i.e., VLE equation 2.20. Use the Antoine and Wilson equations to represent the pure component saturated vapor pressures and liquid-phase activity coefficients, respectively. Parameters for these models are given in Tables 2.1, 2.2, and 2.3. Compare your isothermal phase diagrams with the corresponding isobaric diagrams shown in Fig. 2.5.
2. Using Raoult's law, calculate the phase diagram for benzene and toluene at 1 atm pressure. Plot your results in two ways, (*a*) as a T vs. x, y diagram and

TABLE 2.9
Data for Exercise 2 Gmehling et al. (1980*a*, p. 286).

T (°C)	x (benzene)	y (benzene)
108.70	0.0520	0.1100
107.20	0.0830	0.1720
105.90	0.1190	0.2380
105.10	0.1410	0.2740
103.90	0.1730	0.3220
102.80	0.1940	0.3600
101.50	0.2240	0.4020
100.60	0.2530	0.4340
99.00	0.2990	0.5010
98.30	0.3140	0.5230
96.60	0.3660	0.5810
94.70	0.4330	0.6430
92.70	0.5020	0.7060
90.70	0.5710	0.7610
89.10	0.6210	0.8010
87.10	0.7050	0.8530
85.30	0.7690	0.8890
84.20	0.8200	0.9190
83.10	0.8790	0.9460
81.80	0.9310	0.9680
81.30	0.9610	0.9830

Source: Gmehling et al., 1980*a*, p. 286.

(*b*) as a y vs. x diagram. Compare your results with the experimental data in Table 2.9.

3. Plot the y vs. x diagram for benzene and toluene at 1 atm pressure using the constant relative volatility model by adjusting α until the calculations give a best visual fit of the experimental data. On the same diagram plot the curve using $\alpha = P_1^{\text{sat}}/P_2^{\text{sat}}$ evaluated at the average boiling temperature.
4. Plot the y vs. x diagram for a constant α mixture for $\alpha = 1$, $\alpha = 5$, $\alpha = 10$, $\alpha = 100$.
5. Calculate the T-x-y and y-x phase diagrams for the mixture benzene and 4-methyl-2-pentanol at 1 atm pressure. Treat the system as a nonideal liquid in equilibrium with a mixture of perfect gases neglecting Poynting corrections, i.e., VLE equation 2.20. Use the Antoine vapor pressure equation and the Wilson activity coefficient model in your calculations, with parameter values taken from Tables 2.1, 2.2, and 2.3. Compare your phase diagrams with those shown in Fig. 2.5.
6. Calculate the T-x-y and y-x phase diagrams for the mixture acetone and water at pressures of 14.7, 100, and 500 psia, treating the system as a nonideal liquid in equilibrium with an ideal mixture of imperfect gases, i.e., VLE equation 2.43 and parameter values from Tables 2.1, 2.2, 2.3, and 2.8. Begin by neglecting Poynting corrections and compare your results to those given in Figs. 2.10 and 2.11. Repeat the calculations at 500 psia with Poynting

corrections included and observe what effect these corrections have on the calculated diagrams.

7. Calculate the T-x-y diagrams and y-x phase diagrams for the mixture acetone and methyl ethyl ketone at pressures of 14.7, 100, and 500 psia, treating the system as a nonideal liquid in equilibrium with an ideal mixture of imperfect gases, i.e., VLE equation 2.43 and parameter values from Tables 2.1, 2.2, 2.3, and 2.8. Compare your results with the experimental data of Othmer et al. (1952). Check the sensitivity of your results to the inclusion and exclusion of the following effects:
   a. Poynting correction factor.
   b. Vapor-phase deviations from the perfect gas law (use second virial coefficient corrections).

   Examine the predictions made by the perturbation solution (Eqs. 2.49 and 2.50), and compare with the above calculations and with the experimental data.
8. Calculate the T-x-y diagrams and y-x phase diagrams for the mixture methyl ethyl ketone and water at pressures of 14.7, 100, and 500 psia, treating the system as a nonideal liquid in equilibrium with an ideal mixture of imperfect gases, i.e., VLE equation 2.43 and parameter values from Tables 2.1, 2.2, 2.3, and 2.8. Neglect Poynting corrections and compare your results with the experimental data of Othmer et al. (1952). Repeat your calculations using the Margules activity coefficient model from Gmehling and Onken (1977, p. 364) in place of the Wilson equation. Interpret the meaning of the y-x diagrams at the various pressures.
9. Derive Eqs. 2.32 and 2.33. One way of doing this is to integrate the expression for the derivative of the equilibrium temperature T with respect to variations in liquid composition x_1 at constant pressure while maintaining equilibrium between the coexisting liquid and vapor phases. For ideal liquid and vapor mixtures this derivative is given by

$$\left(\frac{\partial T}{\partial x_1}\right)_P = \frac{(x_1 - y_1)RT^2}{x_1(1 - x_1)\lambda} \tag{2.71}$$

   where λ is the average heat of vaporization of the two pure components at the system pressure. This differential equation is a special case of a more general equation derived in Malesiński (1965, pp. 54–60). Most undergraduate textbooks on thermodynamics do not refer to this equation, preferring to study the derivative $(\partial P/\partial y_1)_T$ instead, e.g., Denbigh (1971, p. 236) or Smith and VanNess (1996, p. 507). There is obviously a family of such derivatives, each one suited to its own applications.

   If we assume that the phase equilibrium can be represented by Eq. 2.29 with a constant value of α, then this differential equation can be integrated in closed form to obtain the desired result.
10. Derive the perturbation solution in Eqs. 2.49 to 2.54 as follows.
   a. Begin by showing that the vapor-phase fugacity coefficients in Eqs. 2.43 can be written as $\phi_i \approx 1 + \epsilon_i$.

b. Using this approximation for ϕ_i and neglecting the Poynting correction, expand Eq. 2.43 using the following power series:

$$T = T_0 + \epsilon T_1 + \cdots$$
$$y = y_0 + \epsilon y_1 + \cdots$$

where $\epsilon = \max(\epsilon_i)$.

c. Find the approximate solution by equating terms of the same order in ϵ on both sides of Eq. 2.43.

Note: There is quite a lot of algebra in finding this solution, and you will discover that ϵ does not appear explicitly in the final solution. Also, don't forget that the activity coefficients must also be expanded, because they depend on temperature as well as the liquid-phase mole fractions.

11. Develop a perturbation solution for the VLE expression given by Eq. 2.43 in which Poynting corrections are included. Follow a procedure similar to that given in Exercise 10. Apply your solution to the acetone-water mixture at 34.01 atm pressure and compare your results with the data in Othmer et al. (1952) and with the results given in Fig. 2.12. Base case values for T_0 and y_0 as a function of x are shown in Table 2.10.
12. A binary mixture containing 40 mol % vinylacetate and 60 mol % toluene is to be flashed at 1 atm pressure and 60°C. The vapor–liquid equilibrium is adequately represented by Raoult's law, with the Antoine coefficients for vapor pressures given in Table 2.1.
    a. Solve the Rachford-Rice equations for the fraction of feed that is vapor ϕ, and for the composition of the vapor and liquid streams leaving the flash vessel.
    b. Repeat the calculations at the same temperature and pressure but with feed compositions of 30 mol % and 50 mol % vinylacetylene.
    c. Explain your results to parts *a* and *b* with the aid of a *T*-*x*-*y* phase diagram.
13. In this exercise we explore some process alternatives for the flash separation in the polysilicon flowsheet described in Chap. 1. As in Example 2.2, we treat the reactor effluent as an equimolar mixture of HCl (component 1) and

TABLE 2.10
Data for Exercise 11

x	T_0	y_0
0.0	241.63	0.000
0.1	221.12	0.358
0.2	216.15	0.442
0.3	213.59	0.494
0.4	211.85	0.539
0.5	210.58	0.587
0.6	209.76	0.640
0.7	209.44	0.702
0.8	209.75	0.777
0.9	210.93	0.873
1.0	213.30	1.000

trichlorosilane (component 2) with a total flow rate of 100 kmol/h. As discussed in the example, the objectives of this flash separation are to achieve high fractional recovery of trichlorosilane in the liquid stream, and to obtain high-purity HCl in the vapor. Solve the flash problem at the following conditions and assess whether these objectives are improved or not in each case.

a. Flash temperature = −40°C, pressure = 2 atm.
b. Flash temperature = −30°C, pressure = 1 atm.
c. Having explored some small perturbations, we now get serious and try a big perturbation $T = -80$°C, pressure = 2 atm.

14. Consider the process to manufacture polysilicon described in Chap. 1. Treat the feed to the flash as a mixture of the following components: hydrogen chloride, dichlorosilane, trichlorosilane, and silicon tetrachloride. Estimate the feed composition of each component from the information given in Chap. 1, and represent their VLE with Raoult's law. Antoine coefficients for each of the components are given in Table 2.1. If the flash temperature and pressure are −30°C and 2 atm, respectively, calculate the flow of each component in the exit streams. Remembering that the aim of this flash is to separate recycle gases from the main product (trichlorosilane), make an assessment of the unit's performance.
15. Show that the flash equations for ideal binary mixtures at constant T, P, and z have the following analytic solution:

$$x_1 = \frac{1 - K_2}{K_1 - K_2} \tag{2.72}$$

$$y_1 = K_1 x_1 \tag{2.73}$$

$$\phi = \frac{z_1}{1 - K_2} - \frac{z_2}{K_1 - 1} \tag{2.74}$$

Hint: The most convenient way of deriving these equations is to start from Eqs. 2.58 and 2.59 using Eq. 2.57 as necessary.
16. A stream containing 60 kmol/h water (component 1) and 40 kmol/h ethylene glycol (component 2) is to be flashed at 120°C and 603 mmHg pressure.
a. Solve the flash equations for the exit flows and compositions using Raoult's law to represent the VLE for the mixture. Antoine coefficients are given in Table 2.1. Raoult's law does a good job of representing the experimental VLE data for this mixture, as you can verify by calculating the T-x-y and y-x diagrams and comparing them with the data given in Gmehling et al. (1981, p. 172).
b. Solve the problem again, this time treating the system as a nonideal liquid in equilibrium with a mixture of perfect gases neglecting Poynting corrections, i.e., VLE equation 2.20. Use the Antoine vapor pressure equation and the Wilson activity coefficient model in your calculations, with parameter values taken from Tables 2.1, 2.2, and 2.3. You will notice a marked deterioration in the quality of your VLE predictions, and a corresponding loss of accuracy in your flash calculations. Why does this happen?

c. Repeat part *b*, this time using the Wilson parameters reported by Gmehling et al. (1981, p. 172) $A_{12} = 1{,}459.5128$, $A_{21} = -1{,}304.9609$. The quality of your predictions is now improved, although the T-x-y and y-x diagrams are no better than those predicted by Raoult's law.

Note that the method described in the text for solving the flash equations will not work for parts *b* and *c* of this exercise since the K-values are not constant but instead vary with composition. You will need to invent a modified algorithm for doing flash calculations with nonideal mixtures in order to solve this problem. Discuss the algorithm with your instructor.

17. A stream containing 80 kmol/h 1,3-butadiene (component 1) and 20 kmol/h styrene (component 2) is to be flashed at 20°C and 1 atm pressure.
    a. Solve the flash equations for the exit flows and compositions using Raoult's law to represent the VLE for the mixture. Antoine coefficients are given in Table 2.1. Raoult's law does not do a good job of representing the experimental VLE data for this mixture, as you can verify by calculating the T-x-y and y-x diagrams and comparing them with the data given in Gmehling et al. (1980*b*, p. 16).
    b. Solve the problem again; this time treat the system as a nonideal liquid in equilibrium with a mixture of perfect gases neglecting Poynting corrections, i.e., VLE equation 2.20. Use the Antoine vapor pressure equation and the Wilson activity coefficient model in your calculations, with parameter values taken from Tables 2.1, 2.2, and 2.3. You will notice a marked improvement in the quality of your VLE predictions, and a corresponding increase in accuracy in your flash calculations.

    Note that the method described in the text for solving the flash equations will not work for part *b* of this exercise since the K-values are not constant but instead vary with composition. You will need to invent a modified algorithm for doing flash calculations with nonideal mixtures in order to solve this problem. Discuss your algorithm with your instructor.
18. A stream containing 45 kmol/h hydrogen (component 1) and 70 kmol/h *n*-hexane (component 2) is flashed at 311 K and 40 bar. Find the flowrate and composition of the exit streams. Also calculate the component flows in the exit streams. Use the experimental VLE data given below (taken from Knapp et al., 1982, pp. 250–252) to estimate K-values for the mixture.

T (K)	**P** (bar)	x_1	y_1
310.93	34.473	0.0310	0.9860
310.93	68.947	0.0590	0.9920
310.93	103.420	0.0840	0.9940
310.93	137.894	0.1080	0.9950

19. A mixture containing 40 mol % methane (component 1) and 60 mol % benzene (component 2) flows at a rate of 250 kmol/h into a flash separator. The temperature and pressure of the flash vessel are 339 K and 35 bar, respectively.

What are the total flows, component flows, and compositions of the exit streams? Use the experimental VLE data given below (taken from Knapp et al., 1982, p. 465) to estimate K-values for the mixture.

T (K)	P (bar)	x_1	y_1
338.71	6.895	0.0140	0.9250
338.71	10.342	0.220	0.9470
338.71	13.789	0.0300	0.9570
338.71	27.579	0.0600	0.9770
338.71	41.368	0.0900	0.9800
338.71	55.158	0.1180	0.9800

20. A mixture of methane (component 1) and n-octane is flashed at 323 K and 12 bar. The feed is 30 mol % methane and has a total flowrate of 100 kmol/h. What is the composition and flowrate of the exit streams?
    a. Use the experimental VLE data given below (taken from Knapp et al., 1982, p. 477) to estimate K-values for the mixture.

T (K)	P (bar)	x_1	y_1
323.15	10.135	0.0430	0.9910
323.15	20.264	0.0830	0.9950
323.15	30.399	0.1220	0.9960

    b. Estimate K-values for the mixture using the DePriester charts (Perry and Green, 1984, pp. 13–18). Compare your answers with those obtained in part *a*. If the uncertainty in the K-values is $\pm$ 10%, what is the influence on your results?
21. A four-component mixture of hydrogen (1)-benzene (2)-cyclohexane (3)-n-hexane (4) is fed to a flash separator at 200°F and 520 psia. The feed contains 10 mol % hydrogen, 10 mol % benzene, 60 mol % cyclohexane, and 20 mol % n-hexane, at a total flowrate of 80 lb mol/h. At this temperature and pressure the experimentally measured K-values are $K_1 = 38.16$, $K_2 = 0.0684$, $K_3 = 0.0609$, and $K_4 = 0.0825$ (Brainard and Williams, 1967). Calculate the total flowrate, the component flowrates, and the composition of the exit streams.
22. For mixtures obeying Raoult's law that undergo an isothermal flash at temperature T_{flash}, pressure P_{flash}, and at overall composition z_i, $i = 1, \ldots, c$, show that

$$T_{\text{flash}} > T_{\text{bub}} \Leftrightarrow \sum_{i=1}^{c} K_i z_i > 1 \tag{2.75}$$

and
$$T_{\text{flash}} < T_{\text{dew}} \Leftrightarrow \sum_{i=1}^{c} \frac{z_i}{K_i} > 1 \tag{2.76}$$

where the K-values are evaluated at temperature T_{flash} and pressure P_{flash}. The temperature T_{bub} is the bubble point of the mixture at pressure P_{flash} and liquid composition z_i; temperature T_{dew} is the dew point at pressure P_{flash} and vapor composition z_i.

23. Consider a system of flashes like the one shown in Fig. 2.17. Suppose that the feed contains equal parts of hexane and p-xylene with $F = 100$ mol/h at $P = 1$ atm and $\tilde{q} = 1/2$. The entire system can be taken as isobaric at a pressure of 1 atm. Using the flash construction from Fig. 2.16, find the compositions, temperatures, and flows of each stream in the system for $N_T = N_B = 5$. *Hint:* Take $\phi = 1/2$ in every stage and plot the q-lines for each stage on the same y-x diagram. Use those results to find the temperatures from a T-x-y diagram. Also, find the maximum purities of the vapor leaving the last stage in the top and the liquid leaving the last stage in the bottom (adjust ϕ or $\tilde{q}$ for each stage).

References

Ambrose, D., C. H. S. Sprake, and R. Townsend, "Thermodynamic Properties of Organic Oxygen Compounds. XXXII. The Vapor Pressure of Acetone," *J. Chem. Thermodynamics,* **6,** 693–700 (1974).

Bawa, M. S., Private communication (1988). Texas Instruments.

Boston, J. F., and H. I. Britt, "A Radically Different Formulation and Solution of the Single-Stage Flash Problem," *Computers Chem. Engng.* **2,** 109–122 (1978).

Brainard, A. J., and G. B. Williams, "Vapor-Liquid Equilibrium for the System Hydrogen–Benzene–Cyclohexane–*n*-Hexane," *AIChE J.,* **13,** 60–69 (1967).

Chao, K. C., and J. D. Seader, "A General Correlation of Vapor-Liquid Equilibria in Hydrocarbon Mixtures," *AIChE J.,* **7,** 598–605 (1961).

Denbigh, K., *The Principles of Chemical Equilibrium.* Cambridge University Press, Cambridge, 3d ed. (1971).

Douglas, J. M., *The Conceptual Design of Chemical Processes.* McGraw-Hill, New York (1988).

Feinberg, M., "Constitutive Equations for Ideal Gas Mixtures and Ideal Solutions as Consequences of Simple Postulates," *Chem. Engng. Sci.,* **32,** 75–78 (1977).

Furzer, I. A., *Distillation for University Students.* University of Sydney, Sydney, N.S.W., Australia (1986).

Gaw, W. J., and F. L. Swinton, "Thermodynamic Properties of Binary Systems Containing Hexafluorobenzene. Part 4. Excess Gibbs Free Energies of the Three Systems Hexafluorobenzene + Benzene, Toluene, and *p*-Xylene," *Trans. Faraday Soc.,* **64,** 2023–2034 (1968).

Gibbs, J. W., *The Scientific Papers of J. Willard Gibbs.* vol. 1. *Thermodynamics.* Dover, New York (1961).

Gmehling, J., and U. Onken, *Vapor-Liquid Equilibrium Data Collection,* vol. 1/1, Aqueous-Organic Systems, of *Chemistry Data Series.* DECHEMA, Frankfurt/Main (1977).

J. Gmehling, U. Onken, and W. Arlt. *Vapour-Liquid Equilibrium Data Collection,* vol. 1/3+4, Aldehydes and Ketones, Ethers, of *Chemistry Data Series.* DECHEMA, Frankfurt/Main, 1979.

Gmehling, J. and U. Onken, *Vapor-Liquid Equilibrium Data Collection,* vol. 1/7, Aliphatic Hydrocarbons, of *Chemistry Data Series.* DECHEMA, Frankfurt/Main (1980*a*).

Gmehling, J., U. Onken, and W. Arlt, *Vapor-Liquid Equilibrium Data Collection,* vol. 1/6a, Aliphatic Hydrocarbons C_4-C_6, of *Chemistry Data Series.* DECHEMA, Frankfurt/Main (1980*b*).

Gmehling, J., U. Onken, and W. Arlt, *Vapor-Liquid Equilibrium Data Collection,* vol. 1/1a, Aqueous-Organic Systems, of *Chemistry Data Series.* DECHEMA, Frankfurt/Main (1981).

Gmehling, J., U. Onken, and P. Grenzheuser, *Vapor-Liquid Equilibrium Data Collection,* vol. 1/5, Carboxylic Acids, Anhydrides, Esters, of *Chemistry Data Series.* DECHEMA, Frankfurt/Main (1982*a*).

Gmehling, J., U. Onken, and U. Weidlich, *Vapor-Liquid Equilibrium Data Collection,* vol. 1/2d, Organic Hydroxy Compounds: Alcohols and Phenols, of *Chemistry Data Series.* DECHEMA, Frankfurt/Main (1982*b*).

Hougen, O. A., K. M. Watson, and R. A. Ragatz, *Chemical Process Principles Charts.* John Wiley, New York, 3d ed. (1964).

King, C. J., *Separation Processes.* McGraw-Hill, New York, 2d ed. (1980).

Knapp, H., R. Döring, L. Oellrich, U. Plöcker, and J. M. Prausnitz, *Vapor-Liquid Equilibria for Mixtures of Low Boiling Substances,* vol. VI. Chemistry Data Series VI, DECHEMA, Frankfurt/Main (1982).

Malesiński, W., *Azeotropy and Other Theoretical Problems of Vapour-Liquid Equilibrium.* Interscience, London (1965).

McQuarrie, D. A., *Statistical Mechanics.* Harper Collins, New York (1976).

McWilliams, M. L., "An Equation to Relate *K*-Factors to Pressure and Temperature," *Chem. Eng.,* **80**(25), 138 (1973).

Ohe, S., *Computer-Aided Data Book of Vapor Pressure.* Data Book Publishing Company, Tokoyo, Japan (1976).

Othmer, D. F., M. M. Chudgar, and S. L. Levy, "Binary and Ternary Systems of Acetone, Methyl Ethyl Ketone and Water," *Ind. Eng. Chem.,* **44,** 1872–1881 (1952).

Perry, R. H., and D. W. Green, *Perry's Chemical Engineers Handbook.* McGraw-Hill, New York, 6th ed. (1984).

Prausnitz, J. M., R. N. Lichtenthaler, and E. G. deAzvedo, *Molecular Thermodynamics of Fluid Phase Equilibria.* Prentice-Hall, Englewood Cliffs, NJ, 3d ed. (1998).

Press, W. H., B. P. Flannery, S. A. Teukolsky, and W. T. Vetterling, *Numerical Recipes.* Cambridge University Press, Cambridge (1986).

Rachford, H. H., Jr., and J. D. Rice, "Electronic Digital Computers in Calculating Flash Vaporization Hydrocarbon Equilibrium," *J. Petrol. Technol.,* **4**(10) (1952). Sec. 1, p. 19; Sec. 2, p. 3.

Rackett, H. G., "Equation of State for Saturated Liquids," *J. Chem. Eng. Data,* **15,** 514–517 (1970).

Reid, R. C., J. M. Prausnitz, and T. K. Sherwood, *The Properties of Gases and Liquids.* McGraw-Hill, New York, 4th ed. (1987).

Reid, R. C., J. M. Prausnitz, and B. E. Poling, *The Properties of Gases and Liquids.* McGraw-Hill, New York, 3d ed. (1977).

Rowlinson, J. S., Properties of Real Gases, in *Handbuch der Physik,* vol. XII, *Thermodynamics of Gases,* pp. 1–72. Springer-Verlag (1958).

Rowlinson, J. S., *Liquids and Liquid Mixtures.* Butterworths, London, 2d ed. (1969).

Roy, P., and G. K. Hobson, "The Selection and Use of *VLE* Methods and Data," I. *Chem. Eng. Symp. Ser.,* (104), A273–A290 (1987).

Shapiro, N., "Conditions for a Mixture to Be Ideal," *Chem. Engng. Sci.,* **23,** 1217–1218 (1968).

Simmrock, K. H., R. Janowsky, and A. Ohnsorge, *Critical Data of Pure Substances,* vol. 2 part 1, Ag-C_7. of *Chemistry Data Series* II. DECHEMA, Frankfurt/Main (1986*a*).

Simmrock, K. H., R. Janowsky, and A. Ohnsorge, *Critical Data of Pure Substances,* vol. 2 part 2 C_8-Zr. of *Chemistry Data Series* II. DECHEMA, Frankfurt/Main (1986*b*).

Smith, J. M., and H. C. Van Ness, *Introduction to Chemical Engineering Thermodynamics.* McGraw-Hill, New York, 5th ed. (1996).

Talley, P. K., J. Sangster, C. W. Bale, and A. D. Pelton, "Prediction of Vapor-Liquid and Liquid-Liquid Equilibria and Thermodynamic Properties of Multicomponent Organic Systems from Optimized Binary Data Using the Kohler Method," *Fluid Phase Equilibria,* **85**, 101–128 (1993).

Van Ness, H. C., and M. M. Abbott, *Classical Thermodynamics of Nonelectrolyte Solutions.* McGraw-Hill, New York (1982).

Wankat, P.C., *Equilibrium Staged Separations*. Prentice-Hall, Englewood Cliffs, NJ (1988).

3

Binary Distillation

3.1 INTRODUCTION

Distillation separates, or *fractionates,* chemically different species by exploiting the fact that the compositions of coexisting vapor and liquid phases are generally different. The liquid and vapor mixtures that coexist in distillation are at or very near the boiling temperature. In many mixtures the coexisting vapor and liquid phases rapidly approach equilibrium, and thermodynamics provides a framework for describing the composition differences.

Contact between the vapor and liquid phases can be achieved in many ways. One of the most common is the interconnection of successive *stages* in an arrangement like the one shown schematically in Fig. 3.1 for distillation. Vapor is generated in the *reboiler* and passes up through the stages of the column. The *bottoms* stream, enriched in one or more of the components, is removed from the reboiler. A *condenser* at the top of the column is used to provide the liquid that flows down through the stages. A second product stream called the *distillate* is removed at the top of the column.

The costs of distillation are determined primarily by the size of the column and heat exchange equipment (capital investment) along with the energy consumption (operating costs). The simplest case to analyze is the fractionation of a binary mixture, which is the topic of this chapter. We focus on the systematic analysis of the material flows, compositions, and temperatures in the column because the column size and the utility requirements can be estimated once these quantities are known. First, the basic model for binary distillation is developed. This is followed by the development of various techniques for solution of the model equations, beginning with the classic graphical technique on a *McCabe-Thiele* diagram. Because of the limited utility of graphical techniques, especially for mixtures containing more than two species, we also consider analytical and numerical solutions. Many of these are classical solutions that can be derived in several ways; we emphasize those that

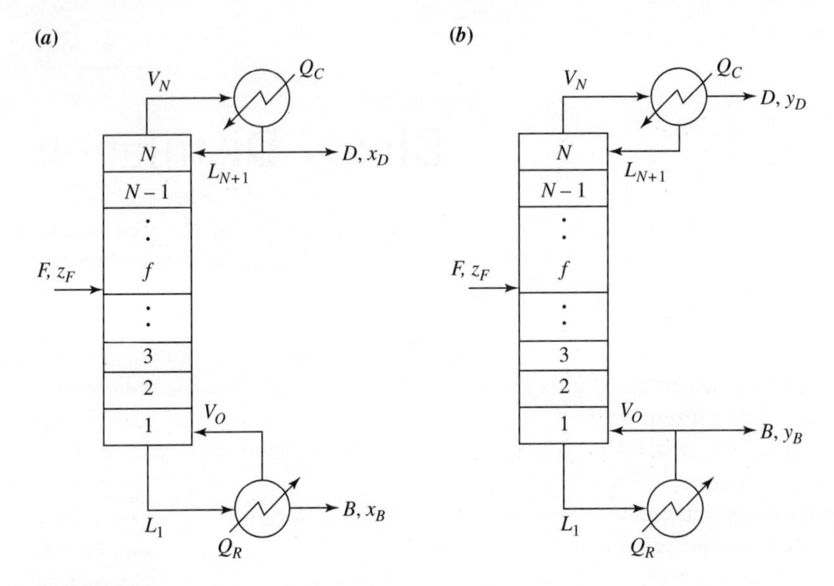

FIGURE 3.1
(*a*) A simple column for binary distillation producing saturated liquid products with a total condenser and a partial reboiler. Other combinations of condensers and reboilers are possible, and product coolers are also commonly used. (*b*) A simple column for binary distillation producing saturated vapor products with a partial condenser and a total reboiler.

unify ideas and that can be generalized. The more ad hoc solution techniques are sometimes useful and have been posed as analysis exercises at the end of the chapter.

The important special types of behavior that are peculiar to nonideal mixtures, including the notion of a *tangent pinch* and a treatment of *binary azeotropes* are discussed in Sec. 3.6. A more detailed description of heat effects can be found in Appendix A.

Simulation as Motivation

The rate of change in computer-based modeling of chemical processes in the last two decades is astounding. With these advances have come new databases for thermodynamics and new algorithms to solve the (sometimes complex) sets of equations to predict *performance*. There have also been major advances in *design* and *synthesis* tools. In fact, simulation is mistakenly thought by some to be simple, but new demands for high-fidelity models including mass transfer effects, chemical reactions, and process dynamics continue to challenge researchers.

Instead of a major activity for process engineers in model development, it is the use of models and a knowledge of their proper application and limitations that is a new challenge. Relatively little background is needed to make some simulations, although this can be dangerous for many reasons. It is vital to realize that an educated engineer who knows when and how to use the proper level of model, how to ask the next question, how to tell the correct converged answer, and when to stop modeling and do experiments is much less likely to be replaced by software.

Several excellent general-purpose simulators are available. The more widely used are Aspen Plus (Aspen Technology, 1999), Hysys (AEA Technology Enginering Software, 1997), and Pro-II (Simulation Sciences, Inc., 1999). All of these include a thermodynamic database, and capabilities for the simulation of complete plants, including distillation and many other unit operations. These tools also have the capability for adding special components or models—an activity often reserved for more experienced engineers. Once we have made sensible specifications, these software tools can assemble and solve the model equations (dozens to hundreds or thousands) and present the results; interpreting these is often another matter! Seider, Seader, and Lewin (Seider et al., 1999) give an excellent and comprehensive treatment of simulation.[1]

Suppose we are interested in an *existing* column. For example, we may have existing equipment and desire to use it for a new purpose. We may also have estimated a design using various approximations to get a rapid solution and want to examine its sensitivity to the assumptions by more rigorous simulations. Once a thermodynamic model and the associated parameters and data are selected and the feed composition, temperature, and pressure are specified, there are four degrees of freedom for a simple column (see Chap. 4 for the proof). For a *performance* simulation, these might be a specification for the number of stages, the location of the feed, and the heating and cooling rates. Actually, rather than the heating and cooling, it is more convenient to specify the ratio of the liquid returned to the top of the column to the distillate flow (reflux ratio) and the ratio of the vapor returned to the bottom of the column to the bottoms flow (reboil ratio).[2]

Figure 3.2 shows a schematic of a column that exists in our undergraduate unit operations laboratory. The column contains 12 stages, and a partial reboiler which is also modeled as an equilibrium stage. The location of the feed is adjustable from stage 3 to stage 7. The column is 8 inches in diameter, heated by low-pressure steam and cooled by cooling water. If we want to simulate the performance of the column as shown for a feed of acetone and water, the exercise takes only a few minutes. Two composition profiles are shown for different reflux ratios in Fig. 3.3. It is clear that neither case gives very high purity products at both ends. One strategy is to change the parameters and possibly the feed tray location in search of a better separation. Sometimes such a trial-and-error strategy works, but it is not efficient unless guided by principles and understanding of the behavior of the particular mixture at hand,

---

[1]The simulations intended here do not require advanced training in simulation. We recommend learning the tools as needed.

[2] It is a great advantage if students learn the use of a simulation tool at an introductory level sufficient to define and use performance models for distillation. We recommend brief training in this aspect to make maximum use of what follows.

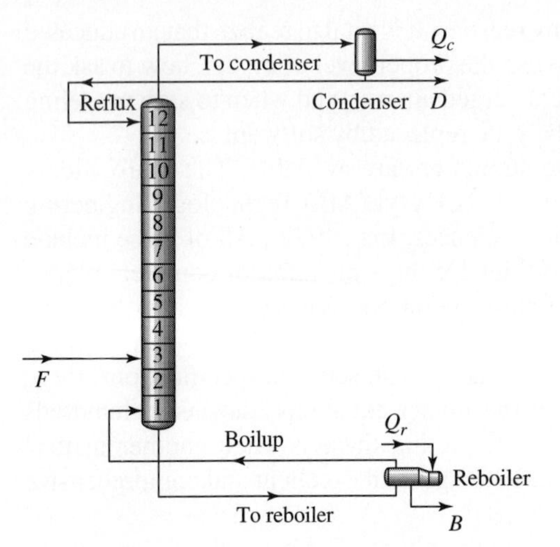

FIGURE 3.2
Schematic of 12-stage column with a partial reboiler. Simulations done with Hysys, version 1.2.

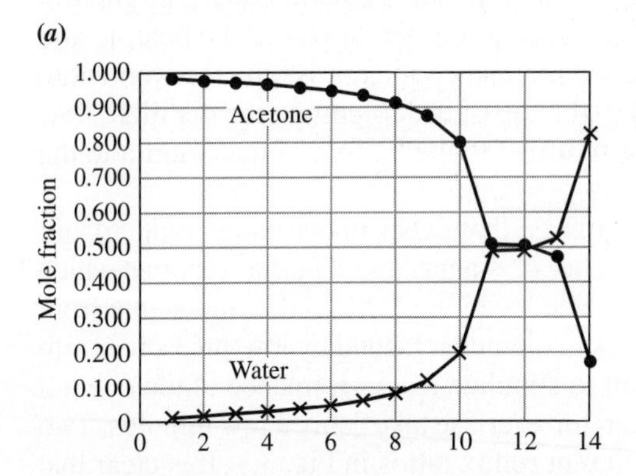

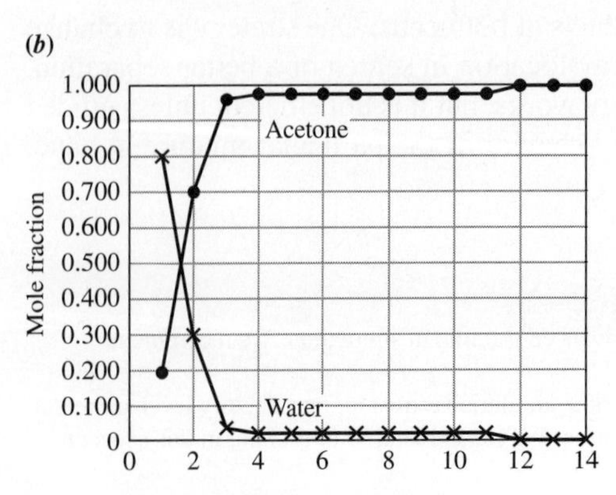

FIGURE 3.3
Composition profiles from simulation.
(*a*) Reflux ratio = 5, reboil ratio = 1.
(*b*) Reflux ratio = 1, reboil ratio = 1.

the fundamental column capabilities, etc. Some of the many questions one might want to answer in connection with this include:

- What is the highest-purity separation possible for this column?
- What is the minimum energy (reflux and reboil) needed for the desired separation?
- What is the minimum number of stages needed for the desired separation?
- How will changes in the column pressure, the feed composition, or the chemical components in the feed change the answers to these questions?

These principles and the development of intuition about sensitivities and capabilities is a major goal in this book. We begin with a treatment of binary mixtures, where a rather complete picture is readily available.

3.2 BASIC MODEL

Material Balances

In binary mixtures, the composition of each phase can be completely described by one variable, which we choose to be the mole fraction of the component with the lower boiling point, the *light* component. The mole fractions are denoted as x in the liquid phase and y in the vapor. The mole fraction of the second component is just $1 - x$ or $1 - y$. When the conditions of a stream can be either vapor or liquid, we denote the mole fraction as z. There are two independent material balances for the column shown in Fig. 3.1, an overall balance

$$F = D + B \tag{3.1}$$

and a balance for the light component[3]

$$Fz_F = Dz_D + Bz_B \tag{3.2}$$

From these material balances it is simple to find expressions for the fraction of the total feed that appears in the distillate and bottoms as

$$\frac{D}{F} = \frac{z_F - z_B}{z_D - z_B} \tag{3.3}$$

and

$$\frac{B}{F} = \frac{z_D - z_F}{z_D - z_B} \tag{3.4}$$

These simple relationships show that the specification of the feed composition and flowrate along with the product purities is sufficient to determine the flowrate of

---

[3]The mole fractions of streams are given the symbols x for a saturated liquid and y for a saturated vapor. When a stream may be either liquid or vapor or a mixture of both, we use the symbol z. The flowrates (normally moles/time) of streams are indicated by F for feed, D for distillate, and B for bottoms; these letters are also used as subscripts to refer to the respective streams.

each product stream. In fact, the ratio of the flowrates of the two product streams is completely determined by a *lever rule*

$$\frac{D}{B} = \frac{z_F - z_B}{z_D - z_F} \tag{3.5}$$

Although these results may seem intuitive, especially for binary mixtures, their consequences and importance in the analysis of the fractionation of mixtures containing more than two components should not be underestimated.

The use of *fractional recoveries* is an alternative to the compositions; the fraction of light component in the feed stream that is present in the distillate product is $f_l \equiv Dz_D/Fz_F$. Similarly, the fractional recovery of the heavy component is $f_h \equiv B(1-z_B)/F(1-z_F)$. The fractional recoveries can be related to the feed and product compositions by material balance. For example, $f_l = z_D(z_F - z_B)/z_F(z_D - z_B)$, and there is a similar expression for f_h. We will use compositions, fractional recoveries, or a combination of both as convenient.

The primary task in the *design* of a column is to relate the size and energy requirements to the specifications on the feed and product streams. The related *performance* calculation seeks to determine the flowrates and compositions of the product streams for a given feed, from information about the size and operating conditions of the column. In either case, a description of the internal flows and compositions is required.

Figure 3.4 shows a schematic of the column and a numbering scheme for the stages. On each stage, we assume that the compositions in the liquid and vapor phases

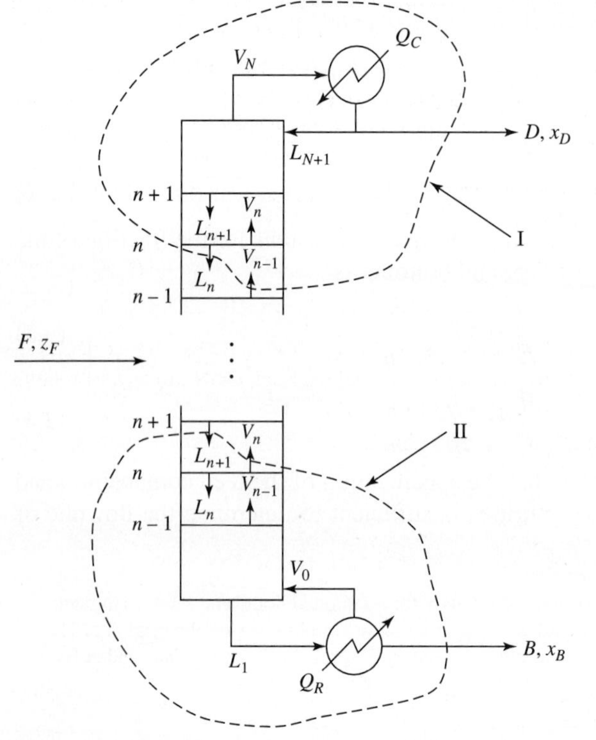

FIGURE 3.4
Stream labels and material balance envelopes for a simple column. Because the products are shown as liquids, their compositions are labeled with x.

leaving the stage numbered n are constant, with values given by the mole fractions x_n and y_n, respectively; i.e., the liquid and vapor phases are *perfectly mixed.* The steady-state material balances for the total molar flows in the top of the column are (see envelope I in Fig. 3.4)

$$V_{n-1} = L_n + D \tag{3.6}$$

and

$$V_{n-1} y_{n-1} = L_n x_n + D z_D \tag{3.7}$$

which gives

$$y_{n-1} = \frac{L_n}{V_{n-1}} x_n + \frac{D}{V_{n-1}} z_D \tag{3.8}$$

for $n = N+1, N, N-1, \ldots, f+1$ and f denotes the feed stage. In the bottom section, a similar balance on envelope II gives

$$L_{n+1} = V_n + B \tag{3.9}$$

and

$$L_{n+1} x_{n+1} = V_n y_n + B z_B \tag{3.10}$$

or

$$y_n = \frac{L_{n+1}}{V_n} x_{n+1} - \frac{B}{V_n} z_B \tag{3.11}$$

for $n = 0, 1, 2, \ldots, f-1$.

The material balances for the condenser and reboiler are also needed. A *partial condenser* provides the equivalent of one additional stage, because the vapors are taken as a product, while a *total condenser* provides a saturated or subcooled liquid product as the distillate. It is common practice to operate a partial condenser when one of the components has a low boiling point or when the product is desired as a vapor. However, the total condenser is a more common arrangement because it is less expensive to transport liquids than vapors and because a saturated vapor may partially condense in piping runs over any significant distance.[4] In either case, we number the condenser and its product stream as $N+1$. Figure 3.1 shows that part of the liquid condensate is returned to the column to provide liquid on the stages in the rectifying section; this stream has a molar flowrate L_{N+1} and is referred to as the *reflux*. The *external reflux ratio* r is

$$r \equiv \frac{L_{N+1}}{D} \tag{3.12}$$

This dimensionless number must be positive[5] and the minimum value needed depends on the difficulty of the desired separation. Columns must be designed so that r is somewhat above the minimum (the optimal value is discussed in Chap. 6). A material balance for a total condenser shows that the internal flows are related to the external reflux ratio by

$$\frac{L_{N+1}}{V_N} = \frac{r}{r+1} \tag{3.13}$$

[4]Total condensers are often vented to provide for the escape of trace amounts of very low boiling "noncondensable" components.

[5]A design calculation will yield a reflux which is zero (negative) if the distillate purity can be achieved (exceeded) in a single-stage flash. There is a similar result for the reboil ratio.

A specification of the external reflux ratio fixes the ratio of liquid to vapor flowrates on the top stage of the column. A partial condenser can be described with a similar approach.

The reboiler provides vapor flow at the bottom of the column. The bottoms product is often removed as a liquid but may be taken as a vapor when the purity requirements for the bottoms are high.[6] The flowrate of the vapor generated in the reboiler is related to the bottoms and the internal flows by the *external reboil ratio* s, defined as

$$s \equiv V_0/B \tag{3.14}$$

The relationship analogous to Eq. 3.13 is

$$\frac{L_1}{V_0} = \frac{s+1}{s} \tag{3.15}$$

There is also a minimum and an optimal value for s; r and s are related through an energy balance by the condition of the feed and the separation that is desired, as discussed in the next section.

The material balance for a total condenser can be combined with the definition of r to show that

$$V_N = (r+1)D \tag{3.16}$$

Similarly, for the reboiler,

$$L_1 = (s+1)B \tag{3.17}$$

Energy Balances

General relationships

The simplest energy balance for an adiabatic column is the overall balance

$$Fh_F + Q_R = Dh_D + Bh_B + Q_C \tag{3.18}$$

where h is the *specific molar enthalpy*, Q_R represents the heat supplied to the reboiler, and Q_C is the heat removed in the condenser.[7] The specific enthalpy of the feed h_F is evaluated at the pressure on the feed stage, excluding heating, cooling, or pressure changes before the column. We have ignored heat gain or loss to the surroundings other than through the condenser and reboiler.[8] We can relate the difference between the condenser and reboiler heat duties to the feed and product specifications by eliminating D and B from Eq. 3.18 with the mass balances in Eqs. 3.3 and 3.4 to find

$$Q_R - Q_C = \frac{(z_F - z_B)(h_D - h_F) + (z_D - z_F)(h_B - h_F)}{z_D - z_B} F \tag{3.19}$$

[6]Trace amounts of heavier components will then accumulate in the liquid contained in the reboiler and can be purged, perhaps only occasionally.

[7]Unless noted otherwise, the heat removed in the condenser Q_C is taken to be a positive quantity despite the usual thermodynamic convention.

[8]This assumption is better for larger and well-insulated columns.

This means that a specification of the feed rate, as well as the feed and product compositions, temperatures, and pressures, completely determines the difference between the condenser and reboiler heat duties. It is important to recognize this consequence of the overall balances in solving the model equations, e.g., care is needed in using software for design calculations when the product purities and heat duties are both specified, since these are not all independent.

The individual heat duties can be determined from energy balances around the appropriate units. For example, for a condenser, the energy balance can be combined with Eq. 3.16 to show the dependence of the condenser heat duty on the reflux ratio and the distillate flowrate.

$$Q_C = [r(h_N^V - h_{N+1}^L) + (h_N^V - h_D)]\ D \tag{3.20}$$

When the distillate is a saturated liquid, the enthalpy differences are equal to the heat of vaporization at the distillate conditions. The reboiler duty can be found by similar balances as

$$Q_R = [s(h_0^V - h_1^L) + (h_B - h_1^L)]\ B \tag{3.21}$$

The type of condenser or reboiler influences only the the terms h_B and h_D.

An energy balance around the column excluding the condenser and the reboiler can be combined with the overall material balance and Eqs. 3.5, 3.12, and 3.14 to relate the external reflux and reboil ratios to the product specifications and the enthalpy of the feed.

$$\frac{D}{B} = \frac{z_F - z_B}{z_D - z_F} = \frac{s(h_0^V - h_1^L) - (h_1^L - h_F)}{r(h_N^V - h_{N+1}^L) + (h_N^V - h_F)} \tag{3.22}$$

Equation 3.22 shows that a specification of the product compositions and the condition of the feed (temperature, pressure, and composition) completely determines the relationship between r and s.

The most detailed balances are, as always, at the level of the individual stages; in the rectifying section we have

$$V_{n-1}h_{n-1}^V = L_n h_n^L + Dh_D + Q_C \tag{3.23}$$

and for the stripping section,

$$L_{n+1}h_{n+1}^L = V_n h_n^V + Bh_B - Q_R \tag{3.24}$$

These energy balances must be solved simultaneously with the material balances and the phase equilibrium relationship to determine the variation of compositions and flowrates throughout the column. There are many cases, however, where a good approximation to the actual flow can be determined in a simpler manner; this approach is described next.

Constant molar overflow

It is useful (and convenient) to examine cases where the liquid and vapor rates within each column section are constant. This simplification is referred to as the condition of *constant molar overflow* (CMO). Intuitively, we might expect that a constant molar latent heat of vaporization for the mixture would lead to this condition in an adiabatic column, since identical amounts of energy are required to condense or vaporize a mole of the material. We make this argument more precise in Appendix A.

Accepting the CMO approximation for the moment, we denote the flowrates in the top or *rectifying* section as L_T and V_T, and in the bottom or *stripping* section as L_B and V_B. In this case, Eqs. 3.6 to 3.10 become

$$V_T = L_T + D \tag{3.25}$$

$$V_T y_{n-1} = L_T x_n + D z_D \tag{3.26}$$

for $n = N + 1, N, N - 1, \ldots, f + 1$ and

$$V_B = L_B - B \tag{3.27}$$

$$V_B y_n = L_B x_{n+1} - B z_B \tag{3.28}$$

for $n = 0, 1, 2, \ldots, f - 1$. It is convenient to rewrite these relationships in terms of the reflux and reboil ratios to find equations for the *top operating line*

$$y_{n-1} = \frac{r}{r+1} x_n + \frac{z_D}{r+1} \tag{3.29}$$

and the *bottom operating line*

$$y_n = \frac{s+1}{s} x_{n+1} - \frac{z_B}{s} \tag{3.30}$$

where $r = L_T/D$ and $s = V_B/B$. These relationships demonstrate that the vapor composition at any point in the rectifying or stripping section is a linear function of the liquid-phase composition on the stage above.

It is also convenient to rewrite the enthalpy of the feed as a linear combination of the enthalpies of a saturated vapor and a saturated liquid mixture, each at the overall composition of the feed.

$$h_F = q h_F^{L,\text{sat}} + (1 - q) h_F^{V,\text{sat}} \tag{3.31}$$

The *feed quality* q can be computed when the feed composition, temperature, and pressure are specified. For feeds that are mixtures of vapor and liquid, and for mixtures where the constant molar overflow assumption is accurate, q is simply the molar fraction of the feed that is liquid.[9]

Saturated liquid or vapor feeds correspond to q values of 1 and 0, respectively. When the feed is a superheated vapor or a subcooled liquid, q is < 0 or > 1, respectively. Equation 3.31 can be rewritten as

$$q \equiv \frac{h_F^{V,\text{sat}} - h_F}{h_F^{V,\text{sat}} - h_F^{L,\text{sat}}} = \frac{h_F^{V,\text{sat}} - h_F}{\lambda_F} \tag{3.32}$$

This *definition* of q is not restricted to constant molar overflow. However, if CMO is accurate and the enthalpy datum is selected as $h_F^{L,\text{sat}} = 0$, so that $h_F^{V,\text{sat}} = \lambda_F = \lambda$ so q is simply

$$q = \frac{\lambda - h_F}{\lambda} \tag{3.33}$$

[9]When CMO is not a good approximation, q is not precisely the fraction of the feed that is liquid, but this is usually a good approximation to its value for two-phase mixtures.

The single-stage, adiabatic flash was described in Chap. 2 using ϕ and $\tilde{q} = 1 - \phi$, the fractions of the total feed that exit as vapor and liquid, respectively. For the CMO case, and for two-phase feeds, $q = \tilde{q}$, so we need not distinguish these quantities when the CMO approximation is used.

The general relationships from the previous section are much simpler for constant molar overflow. For example, the heat duties become

$$Q_C = [\lambda(r+1) - h_D]D \qquad Q_R = [\lambda s + h_B]B \tag{3.34}$$

The relationship between the external reflux and reboil ratios in Eq. 3.22 becomes

$$\frac{D}{B} = \frac{z_F - z_B}{z_D - z_F} = \frac{s+1-q}{r+q} \tag{3.35}$$

Specification of the feed state and the compositions of the distillate and bottoms completely determines the relationship between r and s.

The behavior on the feed stage is slightly different from that on the other stages. At the feed stage, the balance of total mass is

$$F + V_B + L_T = V_T + L_B \tag{3.36}$$

and the energy balance is

$$F h_F + V_B h^V + L_T h^L = V_T h^V + L_B h^L \tag{3.37}$$

These two equations, along with the definition of q in Eq. 3.33, can be combined to relate the liquid flows in the two column sections

$$L_B - L_T = qF \tag{3.38}$$

as well as the two vapor flows

$$V_B - V_T = (q-1)F \tag{3.39}$$

The thermal condition of the feed determines the difference between the vapor and liquid flowrates in the rectifying and stripping sections of the column. For example, when the feed is a saturated liquid ($q = 1$) the vapor flowrate is constant throughout the column and the liquid flowrate in the bottom of the column is greater than that in the top by precisely the feed flowrate. The internal flowrate differences that result for other feed conditions are likewise easily found and in agreement with intuition.

The locus of points $\hat{x}$ and $\hat{y}$ where the top and bottom operating lines intersect is called the *q-line*. This can be found by subtracting the material balance for the light component in the top of the column (Eq. 3.26) from a similar balance in the bottom section (Eq. 3.28) and replacing the differences between the vapor rates and the liquid rates using Eqs. 3.38 and 3.39 to find

$$\hat{y} = \frac{q}{q-1}\hat{x} - \frac{z_F}{q-1} \tag{3.40}$$

EXAMPLE 3.1. A binary mixture containing 55 mol % hexane and 45 mol % p-xylene is to be distilled. The mixture consists of equimolar amounts of vapor and liquid at a pressure of 1 atm and a flowrate of 200 kmol/h. The desired purities are 95% hexane and 97% p-xylene. Find the flowrates of the product streams, the fractional recoveries of each component, the internal flows, and the heating and cooling requirements. Assume a value of 1.0 for the reflux ratio, an operating pressure of 1 atm, and saturated liquid products.

- The VLE data are available from Fig. 2.6*b*, and Table 2.5 shows that a constant value of $\alpha = 7.0$ can be used to describe the *y*-*x* curve. Additional data can be found in Reid et al. (1977, Appendix A). The normal boiling points are 68.7°C (341.9 K) for hexane and 138.3°C (411.5 K) for *p*-xylene; the light component is hexane. The heats of vaporization are $\lambda_l = 28{,}850$ J/mol and $\lambda_h = 35{,}980$ J/mol. Although there is a significant difference in the heats of vaporization, we will use the assumption of constant molar overflow using an average $\lambda = 32{,}420$ J/mol. The accuracy of the constant molar overflow assumption and design methods that account for heat effects are described in Appendix A.
- The distillate and bottoms flows can be computed from Eqs. 3.3 and 3.4 as $D = 113$ kmol/h and $B = 87$ kmol/h. The fractional recoveries are $f_l = 0.976$ for hexane and $f_h = 0.937$ for xylene.
- The vapor and liquid rates in the column can be determined once the constant molar overflow assumption is made, since the thermal condition of the feed is known, $q = 0.5$, and the reflux ratio has been specified. We find from Eq. 3.16 that $V_T = 226$ kmol/h. Equation 3.35 gives $s = 1.45$ and the liquid rate at the bottom of the column is given by Eq. 3.17, $L_B = 213$ kmol/h. The remaining liquid and vapor flows are determined by Eqs. 3.38 and 3.39 as $L_T = 113$ kmol/h and $V_B = V_T + (q - 1)F = 126$ kmol/h.
- The heating and cooling requirements can be estimated from the average heat of vaporization and Eqs. 3.34 ($h_D = h_B = 0$ for saturated liquid products) as $Q_C = 2{,}035$ kW and $Q_R = 1{,}136$ kW.

The value of r in the example was chosen to be larger than the minimum value required for the separation. The calculation of the minimum value, as well as the number of stages required in the column, generally requires a solution of the material and energy balances for the internal flowrates and compositions. In the constant molar overflow case, the internal flows are estimated as above, and only the difference equations for the material balances remain. Methods for solution of these equations are discussed in the next section.

3.3 GEOMETRY

When approximations such as constant molar overflow or constant α provide an accurate description of distillation, simple solutions of the balance equations and the VLE are possible. We describe some in this section.

The McCabe-Thiele Diagram

The first comprehensive solution of the design problem for binary distillation was described by McCabe and Thiele (1925). Their graphical solution of the material balances for the constant molar overflow model and the intuition it provides is a classic analysis in chemical engineering and is generally referred to as the McCabe-Thiele diagram.

Figure 3.5*a* reproduces the *y*-*x* diagram introduced in Fig. 2.6*b*, along with additional lines that represent the operating lines in each column section and their intersection on the q-line. Equation 3.29 is the top operating line, which intersects the line $y = x$ at the distillate composition, $x_D = 0.95$ in the example. At $x = 0$, the intercept of the line is at $y = x_D/(r+1)$. Thus, specifications of the reflux ratio and the distillate composition determine the location of this line. The bottom operating line (Eq. 3.30) intersects the line $y = x$ at the bottoms composition, $x_B = 0.03$, and has a slope $(s+1)/s$. The operating lines meet along the q-line which intersects the $y = x$ line at the feed composition, $z_F = 0.55$, and has a slope $q/(q-1)$; see Eq. 3.40.

The solution of Eqs. 3.29, 3.30, and 2.31 can be found by the following construction. The liquid composition x_N on stage N can be found from the intersection of the equilibrium curve with the horizontal line drawn through the point $y_N = x_{N+1} = x_D$ (assuming that the distillate is a saturated liquid). This is shown in Fig. 3.5*b* as the line segment AB. The vapor-phase composition y_{N-1} is related to x_N by the top operating line and the line segment BC reflects this fact graphically. This construction can be used to "step off" stages in the top of the column until an overlap with the identical construction using the bottom operating line and the equilibrium curve is found, as shown in Fig. 3.5*b*. Using this construction, all of the internal compositions can be found. It should be noticed in Fig. 3.5 that there is no strict requirement for switching from one operating line to the other. However, the smallest total number of stages will be found when this is done as soon as possible after the intersection of the two operating lines. This point corresponds to the *optimal feed stage location;* see Fig. 3.5*c*.[10] The composition on this stage must generally be determined from the thermal condition of the feed and the operating parameters of the column. For binary mixtures and saturated liquid feeds, the optimal feed stage composition is the same as the feed composition. However, for other feed conditions or for multicomponent mixtures, this is not generally true.

The required number of equilibrium or *theoretical* stages can be found by counting the number of steps in both sections. The reboiler is considered to be one stage, since vapor and liquid phases are both present. A partial condenser is treated likewise. Figure 3.5*b* shows that the constructions for the top and bottom sections of the column do not intersect at an integral number of stages; this means that with a reboiler plus three theoretical stages, and the feed entering at stage 2, the distillate purity would not be attained but if a fourth stage were added, the distillate purity would exceed 95%. Alternatively, the reflux ratio, the feed quality, or any of the specified compositions could be altered so that an integral number of stages would be found. Actually, this is not a matter for concern, since the actual stages will generally attain only some fraction of the equilibrium compositions and the design will contain some margin of safety to account for this (see Chap. 6).

The operating lines coincide with the $y = x$ line as the reflux (or reboil) ratio becomes very large. As shown in Fig. 3.6*a*, this yields the minimum number of

---

[10]Conceptually, a nonintegral number of stages means that the specified separation actually cannot be achieved with an integral number of stages. However, a higher or lower product purity would be found if the number of stages were rounded either up or down, respectively. The fact that a nonintegral number of theoretical stages is almost always found as a solution is not important for practical purposes.

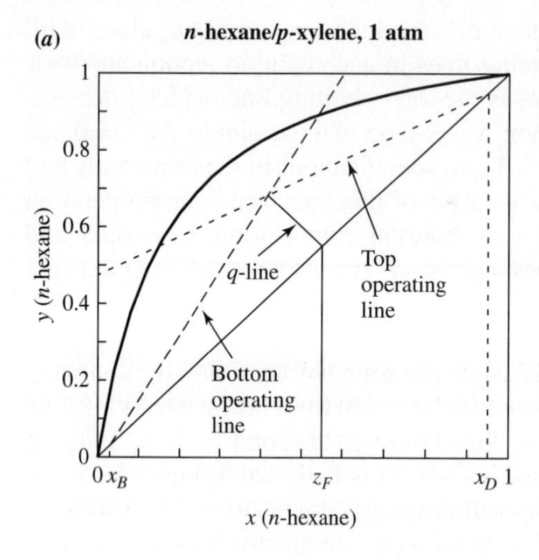

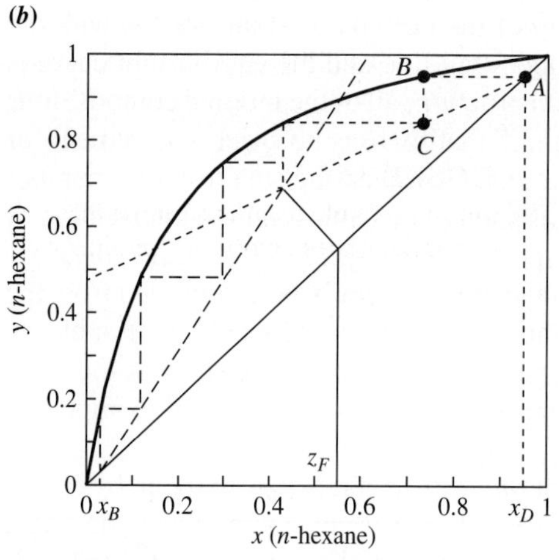

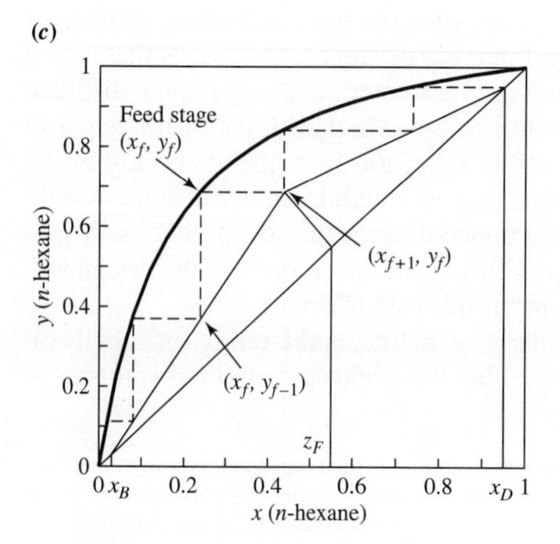

FIGURE 3.5
(*a*) The McCabe-Thiele diagram for hexane and *p*-xylene separation, showing the operating lines and the *q*-line. (*b*) Graphical calculation of the number of theoretical stages in each column section using the McCabe-Thiele diagram. This separation requires slightly less than five theoretical stages. The separate constructions for each section of the column begin at the desired product composition and end at the feed stage. This construction shows the optimum feed stage location, i.e., the one resulting in the minimum number of theoretical stages. (*c*) An alternative graphical calculation of the number of theoretical stages for the same example shown in part *b*.

stages $N_{\min}$, which is approximately 3.4 for the example. The number of theoretical stages N must be greater than this number in any column that is to have a chance of satisfying the conditions set in the example.

Another useful target is the *minimum reflux ratio,* which determines how much the liquid reflux to the column can be reduced; this also corresponds to the lower bound for the vapor rate and to the energy. Likewise, the minimum reboil ratio determines a lower bound on the vapor flowrate returning to the column from the reboiler and, indirectly, a lower bound on the liquid flow leaving the first stage. As the reflux ratio is decreased from the very large value in Fig. 3.6*a*, the intersection of the operating lines moves along the q-line toward the equilibrium curve. The number of stages increases and the compositions on successive stages are closer and closer together. Eventually, a value of r (and s) is reached where the number of stages required in each section is infinite. This is shown in Fig. 3.6*b*; this is termed a *pinch*, and the compositions at the intersection are called *pinch compositions* or *pinch points*. The reflux ratio at this point is the *minimum reflux* $r_{\min}$. In the example, $r_{\min} = 0.398$ can be calculated from the intercept or slope of the top operating line, and Eq. 3.35 gives $s_{\min} = 0.667$. The pinch point occurs at the intersection of the operating lines with the equilibrium curve, where the pinch compositions

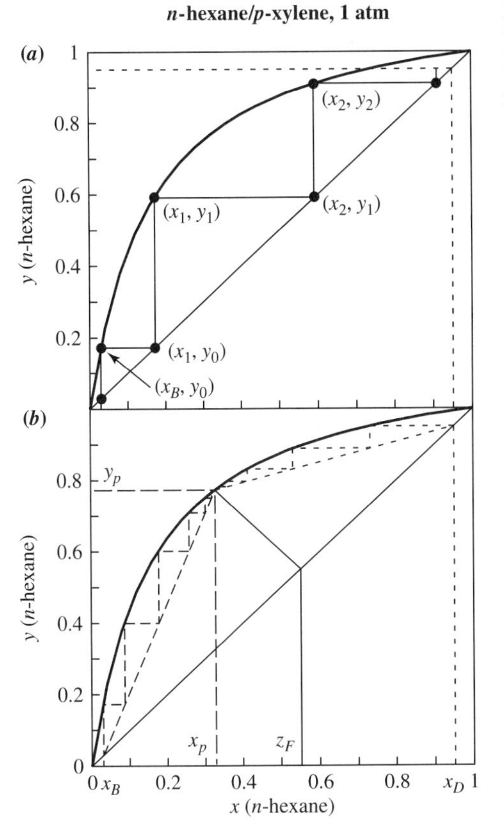

FIGURE 3.6
McCabe-Thiele diagrams for (*a*) the minimum stages and (*b*) the minimum reflux.

are $x = 0.33$ and $y = 0.77$. It is often convenient to find the minimum reflux or reboil ratio from the intersection of one of the operating lines, the q-line, and the equilibrium curve.

In constant α systems, there are always two pinches that occur simultaneously, one in each column section. In other words, there are an infinite number of stages in each column section on either side of the pinch point. Since the pinch points coincide at the intersection of the operating lines and the feed q-line, we shall call this a *feed pinch*; the corresponding McCabe-Thiele diagram will be called a *feed pinch construction*.

The McCabe-Thiele construction provides a tremendous amount of information, and a few exercises with these ideas can lead to the development of important intuition and understanding of binary distillation. However, the method is not very useful for analyzing the distillation of mixtures containing more than two components. It is also not convenient for studying preliminary process design or the behavior of interconnected systems of distillation columns where solutions are needed for a wide range of parameters. In the next several sections, we describe some analytical methods for the solution of the model equations.

Limiting Cases: Minimum Stages and Minimum Flows

One method for finding analytical solutions of the model equations for the constant molar overflow case is to recognize a pattern in the equations that are obtained by successive substitution or by finding the analytical forms for the graphical solutions shown in the previous section. This approach is particularly simple in the limiting cases of very large internal flows or a very large number of stages, as discussed in the two sections below.

3.4 ANALYSIS

Fenske's Equation for the Minimum Number of Stages

To find $N_{\min}$ we seek a solution for the hypothetical limiting case where the internal flows are much greater than the external flows; $V_T \gg D$ and $V_B \gg B$. In this case, r and $s \gg 1$ and the top operating line (Eq. 3.29) is

$$y_{n-1} = x_n \tag{3.41}$$

while the bottom operating line (Eq. 3.30) is

$$y_n = x_{n+1} \tag{3.42}$$

The q-line (Eq. 3.40) is unchanged.[11] This corresponds to conditions of *total reflux*

[11] The same results can be shown without the constant molar overflow used in deriving Eqs. 3.29 and 3.30.

and *total reboil* where the feed and product flows are essentially zero, but the column contains liquid and vapor on all the stages and a total condenser and partial reboiler are used. This is an interesting case because a lower bound on the number of stages is found.

We suppose that x_B is known and that α is a constant. The liquid-phase mole fraction leaving the reboiler is $x_0 = x_B$. The vapor-phase mole fraction leaving the reboiler is given by Eq. 2.31 as

$$y_0 = \frac{\alpha x_B}{1 + (\alpha - 1)x_B} \tag{3.43}$$

Equation 3.42 gives $x_1 = y_0$, and we can find

$$y_1 = \frac{\alpha x_1}{1 + (\alpha - 1)x_1} = \frac{\alpha^2 x_B}{1 + (\alpha^2 - 1)x_B} \tag{3.44}$$

This procedure can be followed for successive stages in the column, to find the vapor-phase mole fraction leaving stage n (the formula for a geometric series is helpful)

$$y_n = \frac{\alpha^{n+1} x_B}{1 + (\alpha^{n+1} - 1)x_B} \tag{3.45}$$

and a simple rearrangement gives the number of stages required to reach a composition of y_n, beginning at x_B

$$n + 1 = \frac{\ln \frac{y_n}{1-y_n} \frac{1-x_B}{x_B}}{\ln \alpha} \tag{3.46}$$

These expressions can be applied between any two stages to relate the compositions to the number of stages at total reflux. It can be seen that *complete* removal of either of the components or *any* separation for a volatility of unity would require an infinite number of stages and is therefore not feasible.

We introduce the *separation factor* S,

$$S \equiv \frac{x_D}{1 - x_D} \frac{1 - x_B}{x_B} \tag{3.47}$$

and the minimum number of stages can be found from a simple expression that results when $n = N$ so that $y_N = x_D$ in Eq. 3.46

$$N_{\min} + 1 = \frac{\ln S}{\ln \alpha} \tag{3.48}$$

Here, $N_{\min}$ is the minimum number of stages inside the column and $+1$ results from the equilibrium stage located in the partial reboiler. Equation 3.48 is usually referred to as the *Fenske* equation, although it appears to have been derived simultaneously by Fenske (1932) and by Underwood (1932).[12]

[12]The number of stages *inside* the column depends on the type of reboiler and condenser that is used because the partial condenser or partial reboiler generally provides one stage. For example, the minimum number of stages inside a column $N_{\min}$ that has a total condenser and a total reboiler is $N_{\min} = \ln S / \ln \alpha$.

The separation factor can be rewritten, using the definitions of the fractional recoveries and overall material balances for the column, as

$$S = \frac{f_l}{1 - f_l} \frac{f_h}{1 - f_h} \tag{3.49}$$

When either of the recoveries approaches 100%, S becomes very large, and the minimum number of stages grows in proportion to its logarithm. Thus, 100% recoveries are impossible, although very high purities can often be attained in a reasonable number of stages because of the relatively slow growth of the logarithm of S in Eq. 3.48.

For the example in the previous section, $S = (0.95/0.05)/(0.97/0.03) = 614$. The Fenske equation 3.48 gives $N_{\text{min}} + 1 = \ln 614 / \ln 7 = 3.30$, in good agreement with the graphical solution. (This means 2.3 stages inside the column plus a partial reboiler.)

When α cannot be considered constant the solution is more complicated, but results in an expression identical to Eq. 3.45 if α is replaced by

$$\tilde{\alpha} \equiv \left(\prod_{j=0}^{n} \alpha_j \right)^{\frac{1}{n+1}} \tag{3.50}$$

where α_j is the volatility at the composition on stage j.[13] The application of this procedure to find the minimum number of stages requires iteration. When the variation of α is not extreme, a geometric mean value of the volatility at the top and bottom, or at the top, bottom, and feed conditions provides a good estimate.

Fenske's equation can be generalized to treat mixtures of more than two components; see Chap. 4.

Underwood's Method for the Minimum Reflux and Reboil Ratios

An exact solution for the minimum internal flows can be found when the number of stages in both column sections is infinite. In binary mixtures with constant α and constant molar overflow, the pinch occurs when the intersection of the operating lines, which is always on the q–line, also falls on the equilibrium curve. When α is not constant, the pinch point can occur elsewhere; this *tangent pinch* is described in Sec. 3.6.

There are several ways to find expressions for the minimum reflux and reboil ratios. One of the more convenient is to first find the intersection of the q-line (Eq. 3.40) and the VLE curve (Eq. 2.31). This intersection takes place precisely at the pinch compositions y_p and x_p and, as seen in Fig. 3.6*b*, these are the same for both

[13]Each stage will have a volatility that depends on the liquid composition. Equations 3.44 become

$$y_1 = \frac{\alpha_1 x_1}{1 + (\alpha_1 - 1)x_1} = \frac{\alpha_1 \alpha_0 x_B}{1 + (\alpha_1 \alpha_0 - 1)x_B} \tag{3.51}$$

where α_1 is the volatility evaluated at the liquid composition on stage 1, x_1, etc.

sections of the column.[14] In some cases, it is simple to find these pinch compositions. For example, if the feed is a saturated liquid, $q = 1$, and

$$x_p = x_F \qquad y_p = \frac{\alpha x_F}{1 + (\alpha - 1)x_F} \tag{3.52}$$

For a saturated vapor, $q = 0$, and

$$x_p = \frac{y_F}{\alpha - (\alpha - 1)y_F} \qquad y_p = y_F \tag{3.53}$$

Other feed conditions require that the pinch compositions be found by equating the vapor-phase compositions calculated from the q-line with those from the VLE curve. This results in a simple quadratic to solve for the pinch composition $0 < x_p < 1$.

$$x_p = \frac{(\alpha - 1)(z_F + q) - \alpha \pm \sqrt{[(\alpha - 1)(z_F + q) - \alpha]^2 + 4z_F(\alpha - 1)q}}{2(\alpha - 1)q} \tag{3.54}$$

Equation 2.31 can then be used to find y_p.

The minimum reflux can be found from the top operating line (Eq. 3.29) once the pinch compositions are known[15]

$$r_{\min} = \frac{z_D - y_p}{y_p - x_p} \tag{3.55}$$

However, if $y_p \geq z_D$, then $r_{\min}$ can be taken as zero, since this means that the distillate purity can be met or exceeded in a single stage without reflux. The minimum reboil ratio can be found as

$$s_{\min} = \frac{x_p - z_B}{y_p - x_p} \tag{3.56}$$

There are important special cases where simple formulas for the minimum reflux and reboil ratios can be obtained. For a saturated liquid feed, constant volatility, and constant molar overflow

$$r_{\min} = \frac{1}{\alpha - 1}\left[\frac{z_D}{x_F} - \alpha\frac{1 - z_D}{1 - x_F}\right] \tag{3.57}$$

We see that $r_{\min}$ will generally increase as α approaches 1, and that it can also be very large when the feed composition is small. For high-purity distillates, the above expression is simply

$$r_{\min} = \frac{1}{(\alpha - 1)x_F} \tag{3.58}$$

Very dilute feeds or volatilities close to unity can result in extremely large reflux ratios.[16] However, for a given feed and volatility, the reflux ratio approaches a constant value as the purity of the separation is increased, unlike the number of stages

---

[14] At minimum reflux, the pinches for both column sections occur at the intersection of the q-line and the VLE if the volatility is constant. Section 3.6 shows why this is sometimes not the case for nonideal mixtures.

[15] Actually, Eqs. 3.55 and 3.56 are valid for any type of mixture once the controlling pinch has been found; see Sec. 3.6 for a discussion of tangent pinches.

[16] Dilute feeds ($x_F \ll 1$) do not require large vapor rates unless the volatility is also low. This is because the distillate flow is approximately Fx_F, so $V_{\min}/F \approx 1/(\alpha - 1)$.

which grows slowly but without bound. Similar conclusions can be made regarding the behavior of the reboil ratio; e.g., when the feed contains very little of the heavy component and a high-purity bottoms is desired, the reboil ratio can be very large.

Expressions such as Eq. 3.57 were found, apparently simultaneously, by Underwood (1932) and by Fenske (1932). Underwood's name is generally associated with the minimum reflux expressions, while Eq. 3.48 is called the Fenske equation.

In the hexane, p-xylene example already discussed, Eq. 3.54 gives $x_p = 0.327$, and from Eq. 2.31, $y_p = 0.773$. Equations 3.55 and 3.56 give $r_{\min} = 0.397$ and $s_{\min} = 0.667$. In this case, the graphical solution is in excellent agreement with the analytical results. However, in cases such as high-purity splits or small relative volatilities, the graphical construction can be quite tedious and the analytical results are valuable. Another use of the analysis is to examine the sensitivity of the solution to changes in parameters such as feed conditions and flowrate and the inevitable, but hopefully small, inaccuracies in the thermodynamic models.

3.5 DISTILLATION SEQUENCING EXAMPLE

Ethylbenzene (C_2H_5-substituted benzene ring), m-xylene (CH_3, CH_3 meta-substituted ring), p-xylene (CH_3, CH_3 para-substituted ring),

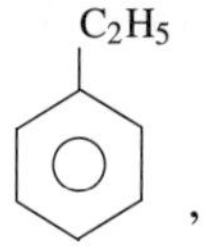

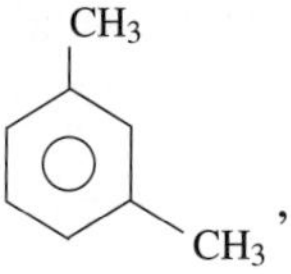

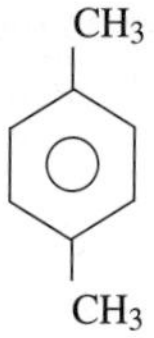

and o-xylene (CH_3, CH_3 ortho-substituted ring)

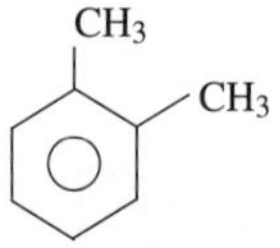

have the same molecular formula, C_8H_{10}, and are usually referred to as the "mixed C_8 aromatic isomers." These isomers are obtained from the mixed hydrocarbon stream produced in naphtha reforming units in oil refineries. The typical relative abundance of the four isomers, their normal boiling points and relative volatilities are given in Table 3.1, taken from deRosset et al. (1980). All of the C_8 aromatics are important intermediates in the petrochemical and polymer industries. The three xylene isomers are intermediates in the production of polyester fibers, films, and fabricated items, while the ethylbenzene is the precursor to styrene and polystyrene. The total U.S. production of the four C_8 aromatics as chemical intermediates is about 4 billion lb/year.

Orthoxylene is normally separated from the C_8 aromatics by conventional distillation. Some refiners also remove the ethylbenzene this way. However, meta- and paraxylene have such a low volatility relative to each other ($\alpha_{23} = 1.020$) that it is not practical to separate them by normal distillation.

There are two possible configurations of simple columns, i.e., columns with only a single feed and no side streams, to accomplish the separation task just described. These are shown in Fig. 3.7. Which configuration should we build? We can often

TABLE 3.1
Properties and typical composition of the mixed C_8 aromatic isomers. The relative volatilities are estimated from ratios of saturated vapor pressures at 138°C.

Component number	Component name	*n*-b.p. (°C)	Relative abundance (mol %)	Relative volatility at 1 atm pressure
1	Ethylbenzene	136.2	25	$\alpha_{14} = 1.235$
2	*p*-Xylene	138.3	14	$\alpha_{24} = 1.170$
3	*m*-Xylene	139.1	45	$\alpha_{34} = 1.147$
4	*o*-Xylene	144.4	16	$\alpha_{44} = 1.000$

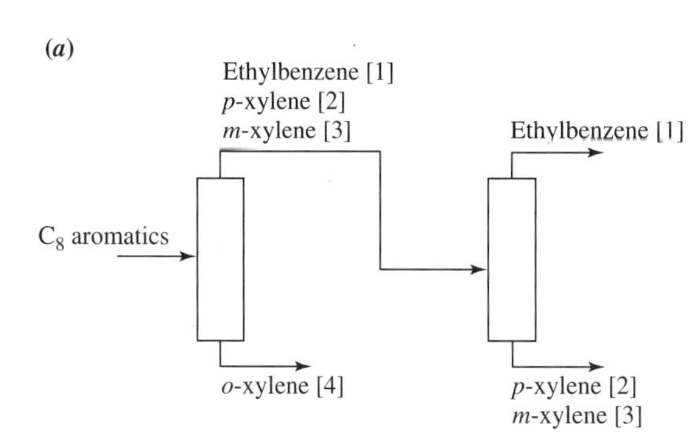

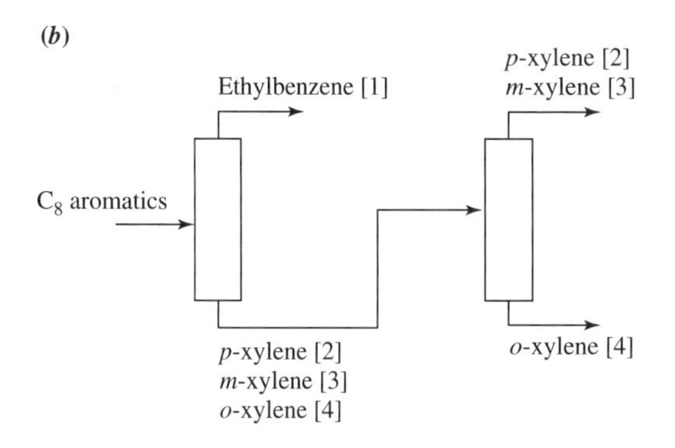

FIGURE 3.7
Separation sequences for the mixed C_8 aromatics; in *(a)* *o*-xylene is removed first in the *indirect* sequence or in (*b*) the lightest component, ethylbenzene, is removed first in the *direct* sequence.

discriminate between rival candidate sequences by estimating the total minimum vapor rate and the number of stages for each sequence. If this is not definitive, then more detailed evaluation and cost comparisons may be required (see Chaps. 6 and 7).

Suppose we consider the indirect sequence in Fig. 3.7*a* first. We estimate the design for the first column by treating the mixture as a pseudo-binary in which ethylbenzene, *p*- and *m*-xylene are lumped together as a single light component (represented by the symbol l). The relative volatility of the pseudo-binary mixture $\alpha_{l,4}$ is approximated as 1.147 (this is a conservative estimate which ensures that the product purity specification on *o*-xylene will be met). For this design, take the bottoms purity of *o*-xylene to be 95 mol % (a typical commercial value for this separation), its fractional recovery to be 99.9%, and assume that the feed, distillate, and bottoms are saturated liquids.

1. For the first column:

   a. The feed is 84% light and 16% heavy. We take a basis of $F = 1$ mole total feed flow. The relative volatility is $\alpha = 1.147$.
   b. For a purity of 95% *o*-xylene in the bottoms from the first column and a 99.9% recovery, mass balances show that $x_D = 0.9998$, $D/F = 0.8317$, and $B/F = 0.1683$.
   c. The minimum number of stages is 82.9 and the minimum reflux and vapor rate are $r_{\min} = 8.10$ and $V_{\min}/F = 7.57$.

2. For the second column:

   a. The feed is $0.8317F$ moles, containing 30.1% light and 69.9% heavy. The relative volatility is $\alpha_{12} = \alpha_{14}/\alpha_{24} = 1.056$.
   b. For a distillate purity of 99.7% and a fractional recovery of 90.3% for ethylbenzene, mass balances give $x_B = 0.040$, $D = 0.272(0.8317F) = 0.226F$ and $B = 0.728(0.8317F) = 0.605F$.
   c. The minimum number of stages is 174.7 and the minimum reflux and vapor rate are $r_{\min} = 72.0$ and $V_{\min}/F = 16.5$.

The reader should work out the corresponding solution for the direct sequence. Compare the alternatives according to the minimum total vapor and stages needed.

If no sequence is clearly better, a further comparison can be made based on choosing a reflux ratio (or several) above the minimum and finding the corresponding number of stages. Graphical methods are not very useful because of the low volatilities; see Exercise 8 for an equation-based method. Another more general equation-based method as well as a treatment of mixtures with more components are discussed in the next chapter.

3.6 NONIDEAL MIXTURES

Nonideal mixtures frequently display quite exotic phase behavior. This is reflected in the appearance of inflections and extrema in the phase diagrams. As a consequence, the equilibrium *y*-*x* diagrams may show marked deviation from the uniform curves that are typical of constant α systems. In fact, thermodynamics places only one restriction on the shape of isobaric *y*-*x* diagrams for homogeneous, nonreactive, binary mixtures, namely, that the curve has strictly positive slope for all values of x, i.e.,[17]

$$\frac{dy}{dx} > 0 \tag{3.59}$$

[17] It is understood that pressure is held constant in the differential dy/dx. Note that y and x refer to the most volatile component (1), although the inequality holds also for the less volatile component (2).

e.g., see Malesiński (1965, pp. 54–60). Inequality 3.59 is clearly satisfied for constant α systems, as differentiation of Eq. 2.31 will verify.

There are two common features of nonideal binary-phase diagrams that are particularly troublesome for distillation column design. The first is the appearance of inflections in the isobaric T-x-y diagram, which in turn generate inflections in the corresponding y-x diagram. The second difficulty is associated with extrema in the phase diagram that give rise to homogeneous binary azeotropes. At such points, the equilibrium liquid and vapor compositions are equal and the y-x curve crosses the 45° line with *positive* slope. Each of these cases will be treated in the remainder of this section.

Tangent Pinches

For given q, feed and distillate (bottoms) compositions, the minimum reflux (reboil) is the lowest possible liquid (vapor) rate in the rectifying (stripping) section that will make the separation feasible. Of course, an infinite number of stages are required under such conditions. Thus, the minimum reflux (reboil) is a theoretical concept which, rather surprisingly, has great practical value. Minimum reflux (reboil) occurs when the operating line for the rectifying (stripping) section touches the equilibrium curve. When the equilibrium y-x curve has no inflections, minimum reflux is determined by the feed pinch construction, as described above (see Fig. 3.6*b*). The characteristic feature of this situation is the requirement of an infinite number of stages in *each* section of the column, adjacent to the feed plate.

It is quite common for nonideal mixtures to exhibit inflections in their isobaric T-x-y phase diagrams, which in turn generate inflections in the corresponding equilibrium y-x diagrams. In such cases, it is possible for one of the operating lines to become *tangent* to the equilibrium curve, thereby causing a pinch in only one section of the column, away from the feed plate. A typical construction is shown in Fig. 3.8.

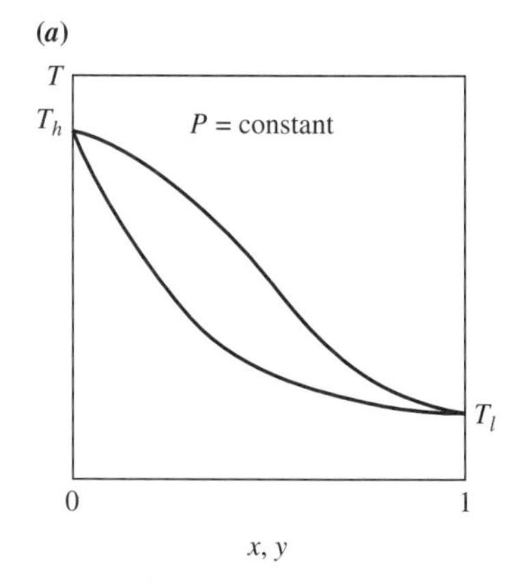

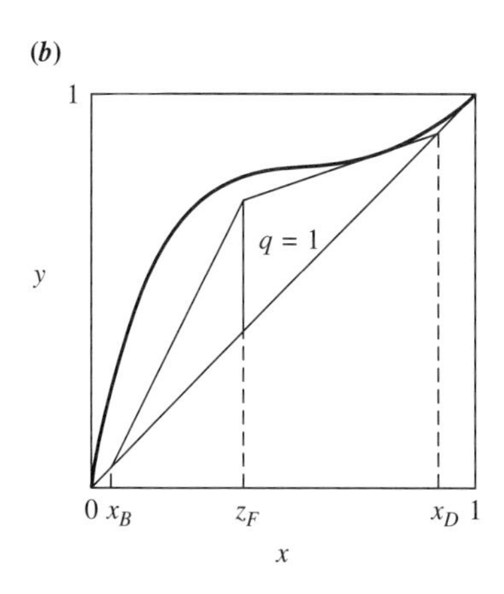

FIGURE 3.8
Schematic diagrams showing a tangent pinch. (*a*) T-x-y phase diagram. (*b*) Corresponding McCabe-Thiele diagram.

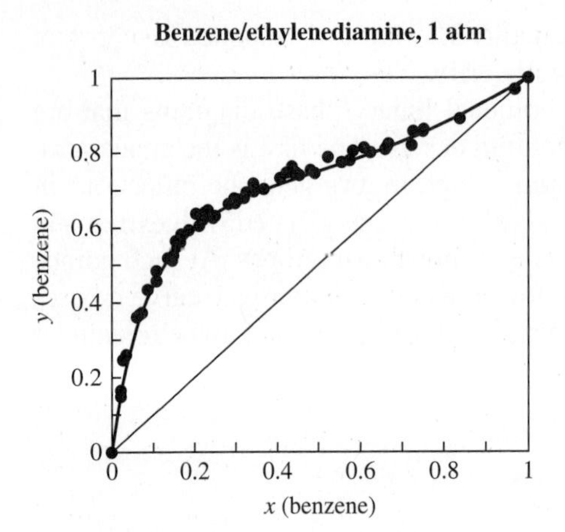

FIGURE 3.9
VLE data for the mixture benzene-ethylenediamine at 1 atm. The correlation is given by Eq. 3.62.

This sort of pinch is usually called a *tangent pinch* in order to distinguish it from a feed pinch.

For example, data from Gmehling et al. (180*a*, p. 170) show that the mixture benzene-ethylenediamine at 1 atm pressure exhibits an inflection in its y-x diagram as shown in Fig. 3.9. If a saturated liquid mixture containing 40 mol % benzene and 60 mol % ethylenediamine is fed to a distillation column in order to obtain a distillate that is 99.9 mol % benzene and a bottoms containing 1 mol % benzene, then the minimum reflux is determined by the tangent pinch shown in Fig. 3.10. The tangent pinch composition and the corresponding minimum reflux ratio can be found graphically by noting the composition at which the tangency occurs and by measuring the slope of the operating line. An algebraic procedure can also be devised by noting that any pinch, tangent or otherwise, on the top operating line has the property $x_n \to x_{n+1} \to x_{n+2} \to \cdots \to x_p$; $y_{n-1} \to y_n \to y_{n+1} \to \cdots \to y_p$ as $n \to \infty$. This means that the liquid and vapor compositions no longer change from stage to stage as the pinch is approached. Moreover, y_p is in phase equilibrium with x_p since the pinch point lies on the equilibrium y-x curve. Therefore, at a pinch point on the top operating line, Eq. 3.29 becomes

$$y_p - \frac{r}{r+1} x_p - \frac{z_D}{r+1} = 0 \tag{3.60}$$

For a given equilibrium y-x relation and for specified values of z_D and r, Eq. 3.60 will provide the value(s) of x_p at which the top operating line and the equilibrium curve intersect each other. At a tangent pinch there is the additional restriction that the slope of the equilibrium curve be equal to that of the top operating line, i.e.,

$$\left[\frac{dy}{dx}\right]_{x_p} = \frac{r}{r+1} \tag{3.61}$$

Equations 3.60 and 3.61 may be solved simultaneously for the liquid-phase mole fraction at the tangent pinch x_{tp} and for the corresponding minimum reflux ratio r_{tp}. If the tangent pinch controls minimum flows, the minimum reflux ratio is r_{tp}.

Exact expressions are available for the thermodynamic quantity dy/dx, e.g., Malesiński (1965, pp. 54–60). We discuss such results in Chap. 5, where the above

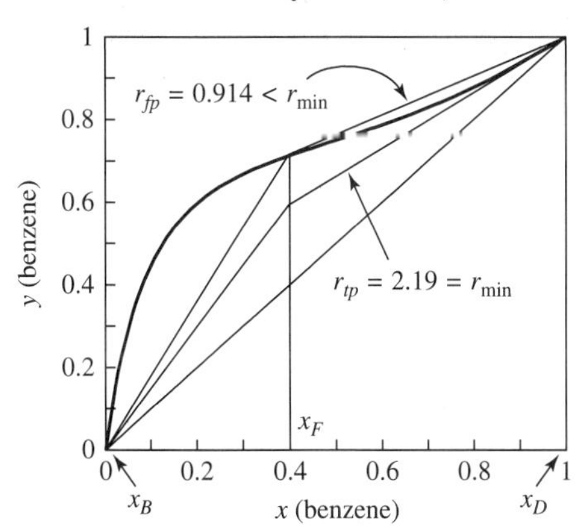

FIGURE 3.10
McCabe-Thiele diagram for the mixture benzene-ethylenediamine at 1 atm pressure. The tangent pinch controls the minimum reflux, despite the existence of a feed pinch at a lower value of the reflux ratio. You can test to see which pinch controls by asking if the separation can be done in a finite number of stages at a slightly higher reflux ratio.

method for calculating tangent pinch points is generalized to multicomponent mixtures. However, for the purpose of illustration, we note that the equilibrium y-x data shown in Fig. 3.9 can be correlated by the *empirical* expression

$$y = \frac{ax}{1+(a-1)x} + bx(1-x) \tag{3.62}$$

with $a = 9, b = -0.6$. This equation was used to generate the solid curve through the experimental data in the figure.[18]

Incorporating Eq. 3.62 into Eqs. 3.60 and 3.61 leads to the following pair of equations which can be solved for x_{tp} and r_{tp}:

$$\frac{ax_{tp}}{1+(a-1)x_{tp}} + bx_{tp}(1-x_{tp}) - \frac{r_{tp}}{r_{tp}+1}x_{tp} - \frac{z_D}{r_{tp}+1} = 0 \tag{3.63}$$

$$\frac{a}{\left(1+(a-1)\,x_{tp}\right)^2} + b(1-2x_{tp}) - \frac{r_{tp}}{r_{tp}+1} = 0 \tag{3.64}$$

For the example problem above, $a = 9, b = -0.6, z_D = 0.999$, and these equations can be solved to find $x_{tp} = 0.975, r_{tp} = 2.19$.[19] This value of r should be compared with the smaller value of $r_{fp} = 0.914$, which is found by the feed pinch construction. Both diagrams are shown in Fig. 3.10, where the geometrical interpretation of each construction may be compared. Clearly, the true value for $r_{\min}$ is the higher value given by the tangent pinch method, i.e., $r_{\min} = r_{tp}$, and we say that the tangent pinch controls the minimum flows.

We should note that this situation can sometimes be reversed. For example, if the saturated liquid feed contains only 20 mol % benzene, and the desired distillate composition is reduced to 90 mol % benzene, then the true minimum reflux ratio

---

[18]This empirical equation can be used to *correlate* a variety of binary nonideal y-x data. See Exercises 13, 14, and 17 and Table 2.6.

[19]When the tangent pinch occurs in the rectifying section, there is always a second lower root to these equations that occurs at smaller values of both x_{tp} and r_{tp}, e.g., for this example, the second root is $x_{tp} = 0.332, r_{tp} = 0.894$. The lower root does not correspond to any realizable design and must be avoided. This phenomenon is examined further in Exercise 13.

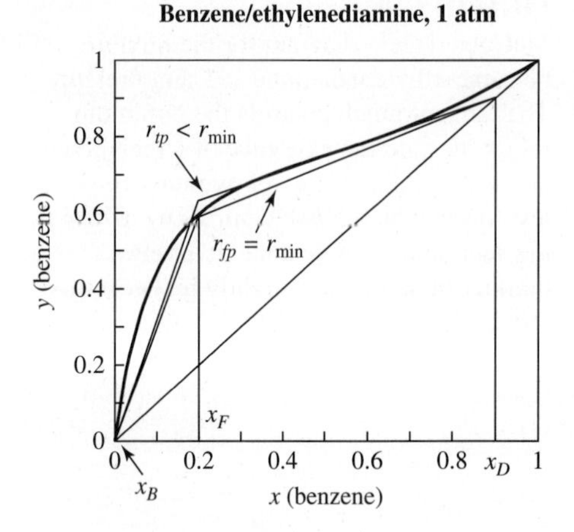

FIGURE 3.11
McCabe-Thiele diagram for the mixture benzene-ethylenediamine at 1 atm pressure. The feed pinch controls the minimum reflux.

is given by the feed pinch construction; i.e., the feed pinch controls, as shown in Fig. 3.11. In this case, the tangent pinch method underestimates r_{min}, as seen in the figure.

In order to understand these results in a systematic way, it is useful to ask first whether a tangent pinch exists for the specified distillate, bottoms, and feed compositions. If one does, then we ask whether it controls the minimum reflux or not.

Clearly, if the equilibrium y-x curve does not contain an inflection, i.e., a point where $d^2y/dx^2 = 0$, then a tangent pinch can never occur. If an inflection does occur, then a tangent pinch may or may not develop, depending on the value of the distillate composition.[20] If a tangent pinch does develop, then it may or may not control minimum reflux, depending on the value of the feed composition.

To find the critical value of the distillate composition z_D^{crit} below which it is not possible for a tangent pinch to occur, we refer to Fig. 3.12. From this figure, we see that the tangent at the inflection point on the equilibrium y-x curve intersects the $y = x$ line at a critical point which we will call z_D^{crit}. Distillate compositions below z_D^{crit} must be on top operating lines that cannot possibly become tangent to the equilibrium curve regardless of how we adjust their slope. For these distillate compositions, tangent pinches do not develop. However, distillate compositions greater than z_D^{crit} generate top operating lines that do become tangent to the equilibrium curve for suitable choices of the slope. The corresponding tangent pinches may or may not control the minimum reflux, depending on the value of the feed composition. This is seen most easily in Fig. 3.13.

In this figure, we have restricted attention to saturated liquid feeds, i.e., $q = 1$,[21] and fixed product purities such that $z_D > z_D^{crit}$. We now consider what happens as the feed composition falls in each of the three separate regions shown in Fig. 3.13. In region I, the feed pinch construction can be drawn in the normal fashion, and the

---

[20] Sometimes, an inflection occurs at the bottom of the y-x diagram, i.e., at low values of x and y. For these cases, it is the bottoms composition, not the distillate composition, that determines whether a tangent pinch develops. Such a situation occurs in binary mixtures of diethylamine and ethanol; see Gmehling and Onken (1977*a*, p. 381) and also Exercise 15.

[21] This restriction is removed in Exercise 16.

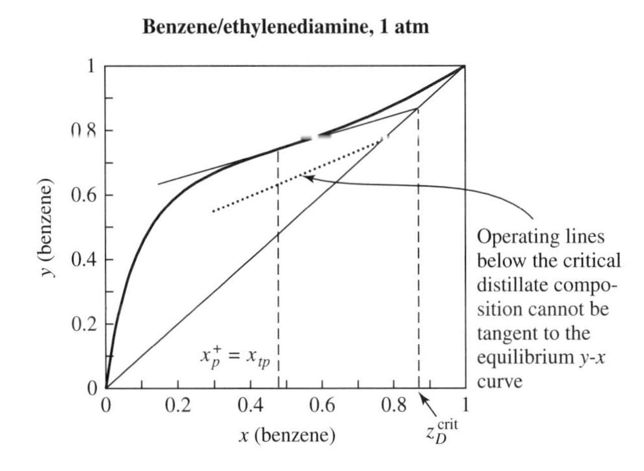

FIGURE 3.12
Equilibrium y-x diagram for the mixture benzene and ethylenediamine at 1 atm pressure showing the position of the critical distillate composition.

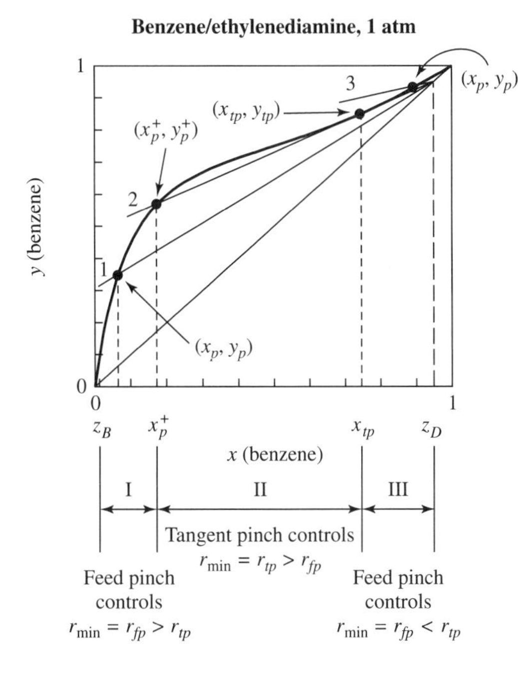

FIGURE 3.13
McCabe-Thiele diagram for the mixture benzene-ethylenediamine at 1 atm pressure showing the regions of feed pinch control (I and III) and tangent pinch control (II), when the feed is a saturated liquid.

feed pinch will control the minimum flows. For feed compositions in this region, it is also possible to draw a tangent pinch construction for which the slope of the top operating line is less than that in the corresponding feed pinch construction, i.e., $r_{fp} > r_{tp}$. Since the two operating lines intersect outside the equilibrium y-x curve for the tangent pinch construction, we conclude that such a construction is invalid for feed compositions in region I. Hence, in this region, $r_{\min} = r_{fp} > r_{tp}$. The constructions shown in Fig. 3.11 correspond to this case.

For feed compositions in region II, the tangent pinch controls the minimum flows. Within this region, the minimum reflux is independent of feed composition and the pinch zone with an infinite number of stages is no longer adjacent to the feed stage but instead is located in the "middle" of the rectifying section. In this region, it is still possible to draw feed pinch constructions for which the slope of the top operating line is less than the slope of the top operating line corresponding to the tangent pinch construction, i.e., $r_{tp} > r_{fp}$. However, these feed pinch constructions are not valid, since they generate top operating lines which lie partly outside the equilibrium *y*-*x* curve. Hence, for feed compositions in region II, $r_{\min} = r_{tp} > r_{fp}$, and a typical case is shown in Fig. 3.10.

Region III is similar to region I insofar as the feed pinch again controls the minimum flows. However, unlike region I, the tangent pinch is now a legitimate construction which generates a top operating line with a greater slope than the feed pinch construction. Therefore, in region III, $r_{\min} = r_{fp} < r_{tp}$.

We conclude that care must be taken to distinguish between correct, i.e., controlling, and spurious pinches in binary mixtures that contain inflections in their equilibrium *y*-*x* diagram. For graphical calculations it is visually obvious which pinch controls the minimum flows. Algebraic calculations must be performed more systematically by identifying the three regions noted on Fig. 3.13. The point x_{tp}, which divides region II from region III, is calculated by the methods described earlier in this section. The point x_p^+, which divides region I from region II, can be calculated from Eq. 3.60 once r_{tp} is known. With this information, it is easy to identify which region contains the given feed composition, and thus which pinch controls the minimum flows. As a final remark, we note that the relationship between Figs. 3.12 and 3.13 is simply that as z_D is reduced in Fig. 3.13, the points x_p^+ and x_{tp} get closer and closer until at $z_D = z_D^{\text{crit}}$ they become equal and the width of region II shrinks to zero.

The methods outlined in this section are developed further in Exercises 11 to 16.

Binary Azeotropes

Many binary mixtures exhibit maxima or minima in the equilibrium temperature and pressure with respect to composition. Often, this is due to the strength of the nonidealities in the liquid phase, but it can also happen for relatively ideal mixtures with close boiling points, depending on the temperature dependence of the vapor pressures. Horsley (1973) and the DECHEMA *Vapor-Liquid Equilibrium Data Collection* list several thousand examples of such behavior. One example is shown in Fig. 3.14 for a mixture of pentane and dichloromethane at 1 bar pressure (Gmehling et al., 1980*b*, p. 100). At the minimum equilibrium temperature, the dew- and bubble-point surfaces are tangent to one another and thus the phases have equal composition,[22] i.e., $y_i = x_i,\ i = 1, 2$. A liquid mixture at this special composition will boil with-

[22]This is more than mere observation, but a consequence of the Gibbs-Konavolov theorem (Malesiński, 1965, pp. 54–60) which guarantees that *stationary points* in the dew- and bubble-point surfaces occur simultaneously at points where $x_i = y_i,\ i = 1, 2, \ldots, C$, and vice versa. These conditions are valid only for nonreacting liquid-vapor mixtures with a single liquid phase. Azeotropes in more complex mixtures are characterized by different conditions, which will be discussed later in the book.

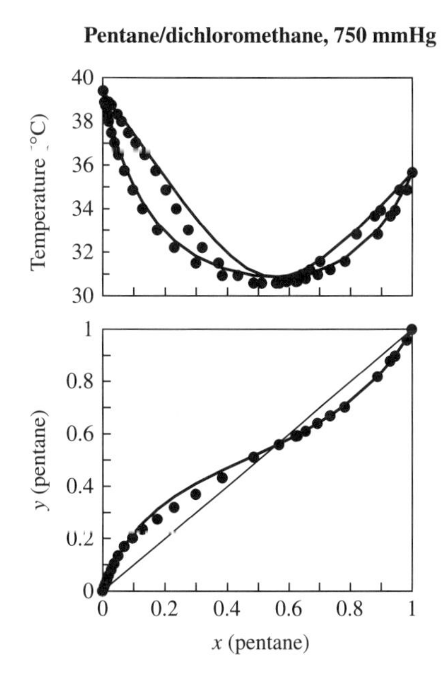

FIGURE 3.14
Equilibrium T-x-y diagram for the system pentane and dichloromethane at 1 bar pressure. The solid lines were calculated by the Wilson equation as reported in Gmehling et al. (1980*b*, p. 100).

out change of composition or temperature. For isothermal systems, the mixture will boil at constant composition and pressure. Such mixtures are called *azeotropes*. For binary mixtures, azeotropes are either minimum-boiling, as shown in Fig. 3.14, or maximum-boiling; the latter are much less common. In rare instances, both kinds of azeotrope are present in the same binary mixture; e.g., the system benzene-hexafluorobenzene exhibits two distinct azeotropes at the same pressure (Malesiński, 1965, Chap. 1; Gaw and Swinton, 1966; Doherty and Perkins, 1978).

As is clear from Fig. 3.14, azeotropes act as barriers to further separation by distillation. If the composition of the azeotrope changes significantly when the pressure is varied, a system of two columns operating at different pressures can be used to separate a binary azeotropic mixture into nearly pure components (see Exercise 20). In many cases, the azeotropic composition is insensitive to changes in the pressure, or the pressure changes are too costly or would lead to temperatures that cause degradation. In such cases, distillation-based separations involve the addition of a third component called an *entrainer* to facilitate the separation. The resulting separation systems often include a column to carry out a partial separation of the original binary azeotropic mixture, and we need to develop design tools for such problems.

Three design methods are available for binary azeotropic systems. At the simplest level, the McCabe-Thiele graphical method can be used. This method requires only that constant molar overflow be assumed; it is not limited to constant α or nonazeotropic y-x behavior. Although this method is sometimes useful for obtaining a quick estimate of the column design for a particular set of feed and product specifications, it suffers from the same limitations mentioned earlier. Namely, the procedure is not well suited to the repeated calculations involved in optimization, sensitivity, and operability studies where the feed and product compositions are varied. In addition, the graphical nature of the method makes it difficult to interface with computer-aided design tools that are useful for the design of the rest of the separation system, and for the remainder of the flowsheet. At the most complex level, a computer-aided design procedure can be employed whereby heat effects, in addition to the mixture nonidealities, can be taken into account. Appendix A gives a discussion of heat effects.

It is frequently possible to represent the binary azeotropic y-x diagram by dividing it into two separate figures, one on either side of the azeotrope.[23] We then introduce a set of transforms which stretch each of these figures over the interval 0 to 1 in the transformed variables; thus for $0 \leq x \leq x^{az}$ define $\mathcal{X} = x/x^{az}$ and for $x^{az} \leq x \leq 1$ define $\mathcal{X} = (x - x^{az})/(1 - x^{az})$. Similar transforms apply to y, recognizing that y^{az} has the same numerical value as x^{az}. The resulting $\mathcal{Y}$-$\mathcal{X}$ diagrams take on familiar shapes and can often be represented by an equation of the form

$$\mathcal{Y} = \frac{\alpha' \mathcal{X}}{1 + (\alpha' - 1)\mathcal{X}} \tag{3.65}$$

Inverting the transforms in this equation leads to the equilibrium models

$$y = \frac{\alpha_B x}{1 + (\alpha_B - 1)x/x^{az}} \tag{3.66}$$

for $0 \leq x \leq x^{az}$, with $\alpha' = \alpha_B$ and

$$y = x^{az} + \frac{\alpha_T (x - x^{az})}{1 + (\alpha_T - 1)(x - x^{az})/(1 - x^{az})} \tag{3.67}$$

for $x^{az} \leq x \leq 1$ with $\alpha' = \alpha_T$, where x is the mole fraction of the light component and α_T and α_B are constants determined from the experimental y-x data.[24] Since Eq. 3.65 has the same form as Eq. 2.31, all of the analytical design techniques developed in Secs. 3.2 and 3.4 apply to this class of problems provided that the transformed variables are used. This method is described in more detail by Anderson and Doherty (1984), and various aspects of these procedures are developed further in Exercise 18.

3.7 COMPLEX COLUMN CONFIGURATIONS

The methods and results discussed above are straightforward to extend beyond the "simple" column configuration, i.e., with a single feed, one overhead product, and one bottoms product.

For instance, multiple feeds introduce a new operating line and a new q-line. This can be advantageous when two binary feeds with different compositions are

---

[23] This procedure does not impose any artificial mathematical constraints on the design problem, since the azeotrope is itself an impassable barrier to any further separation by distillation.

[24] The parameters α_B and α_T are effectively relative volatilities between the azeotrope and the heavy component, and between the light component and the azeotrope, respectively. Thus, for minimum-boiling azeotropic mixtures like the example shown in Fig. 3.14, values for α_B are greater than unity (since this portion of the y-x diagram lies above the 45° line), and values for α_T are less than unity but greater than zero. See Table 3.2.

TABLE 3.2
Constants in the empirical VLE model for minimum-boiling binary azeotropic mixtures

Light component (b.p., °C)	Heavy component (b.p., °C)	α_B	α_T	x^{az}
Ethyl acetate (77.2)	Ethanol (78.4)	2.19	0.52	0.54
Chloroform (61.2)	Methanol (64.6)	3.16	0.19	0.66
Methyl acetate (56.9)	Methanol (64.6)	2.60	0.50	0.648
Methanol (64.6)	Ethyl acetate (77.2)	3.31	0.57	0.72
Ethanol (78.4)	Methyl ethyl ketone (2-butanone) (79.6)	1.59	0.58	0.51
Toluene (110.6)	1-Butanol (117.8)	3.28	0.37	0.67
Tetrahydrofuran (33.8 at 235.5 mmHg)	Methanol (37.4 at 235.5 mmHg)	2.28	0.50	0.63
Pentane (35.7 at 750 mmHg)	Dichloromethane (39.4 at 750 mmHg)	2.50	0.50	0.53

Source: The experimental VLE data were taken from the DECHEMA *Vapor-Liquid Equilibrium Data Collection*. The boiling points are given at the same pressure as the VLE data, which is 1 atm unless noted otherwise.

available; rather then mixing these, the composition difference already existing can be preserved which lowers the heating, cooling, and stage requirements. For example, see King (1980, pp. 223–225), and Wankat (1988, Chap. 6).

Intermediate heating or cooling can be useful to lower the stage requirements for tangent-pinched systems. In any system, intermediate reboilers and condensers can be used to adjust the temperatures at which heating and cooling must be done. This corresponds to a closer approach to reversibility, which can be especially important in cases where the boiling temperatures are extremely low or high. For more details, see Wankat (1988, Chap. 6), Agrawal and Herron (1997), Agrawal and Herron (1998*a*), and Agrawal and Herron (1998*b*).

3.8 EXERCISES

1. Repeat Example 3.1 for a feed composition of 25 mol% hexane, a saturated liquid feed, a reflux ratio of 2.5, and purities of 99.5 mol % in both the distillate and the bottoms. Draw the McCabe-Thiele diagram and find the number of stages required for the separation.
2. Consider the separation of a saturated liquid mixture of acetone (the light component) and propyl acetate, containing 50 mol % acetone. The distillate should contain 95% acetone, and the bottoms should contain 5% acetone. Assume that constant molar overflow applies and that the equilibrium diagram can be represented with a constant value of $\alpha = 4.74$.

Use the graphical McCabe-Thiele method to find

a. The minimum number of equilibrium stages required to achieve the separation.

b. The minimum reflux and reboil ratios.

c. The actual number of equilibrium stages required to achieve the separation at a reflux ratio of $1.5r_{\text{min}}$.

3. The following questions concern the distillation of benzene and *p*-xylene, available as a saturated vapor mixture at 1 atm pressure, containing 40 mol % benzene. The vapor-liquid equilibrium can be described with a constant relative volatility; values are given in Table 2.5.

For saturated liquid products with purities of 95 mol %, use the graphical McCabe-Thiele method to find

a. The minimum number of equilibrium stages.

b. The minimum reflux and reboil ratios.

c. The minimum vapor flows (relative to the feed flow) in each column section, and the corresponding energy requirements.

d. The number of equilibrium stages in each column section, and the energy requirements at a reflux ratio 60% larger than the minimum.

4. This problem concerns a mixture of chloroform[25] and acetic acid; see Table 2.5 for the α values.

a. Rework Example 3.1 for this mixture instead of the one from the text.

b. Draw the McCabe-Thiele diagram and estimate the minimum number of stages and the minimum reflux ratio. Compare your solutions to the Fenske and Underwood solutions for these quantities.

c. What is the uncertainty in your results on account of the uncertainty in the relative volatility?

d. Pick a sensible reflux ratio and explain why you chose it. Find the number of stages, the flows, the liquid and vapor rates, and the heat duties for this value of the reflux ratio.

5. The following questions concern the distillation of a mixture containing ethanol and *n*-propanol, available as a saturated liquid mixture at 1 atm pressure and containing 30 mol % ethanol. See Table 2.5 for the boiling points and relative volatility. It is desired to produce products with purities of 99.5 mol%.

a. Using the McCabe-Thiele diagram, estimate the minimum number of theoretical stages, and the minimum reflux and reboil ratios. Find the corresponding minimum vapor flows in each section of the column, as well as the heating and cooling requirements. Estimate the uncertainty in your results.

b. For a vapor rate 50% larger than the minimum, prepare a McCabe-Thiele diagram and find the number of stages in each section of the column.

c. Compare the results from question 5*a* to the solutions from the Fenske and Underwood equations.

---

[25] Check one of the data books cited in Chap. 2, or the database in a simulator for the heat of vaporization.

TABLE 3.3
VLE data for a mixture of methanol and 1,4-dioxane at 1 atm pressure

Temperature (°C)	x_{MeOH}	y_{MeOH}
101.10	0.0	0.0
94.9	0.05	0.212
89.6	0.10	0.365
82.2	0.20	0.542
76.9	0.30	0.657
73.1	0.40	0.735
70.7	0.50	0.779
68.8	0.60	0.818
67.5	0.70	0.85
66.2	0.80	0.892
65.1	0.90	0.947
64.7	0.95	0.974
64.5	1.00	1.000

6. Consider the distillation of a mixture containing methanol and 1,4-dioxane in a column operating at 1 atm pressure. The feed contains 3,000 lb/h of methanol and 7,000 lb/h of dioxane as a saturated liquid mixture, also at 1 atm pressure. The VLE data are available in Gmehling and Onken (1977*a*, p. 148) and are summarized in Table 3.3. It is desired to recover 98% of the methanol, which should have a purity of 99 mol%.
   a. Repeat Example 3.1 for this mixture but use a reflux ratio of 2.5 and an average heat of vaporization of 35,000 J/mol.
   b. How would the results in question 6*a* be changed if the distillate product were a saturated vapor?
   c. What is the minimum reflux ratio? What is the minimum number of stages?
   d. Draw a McCabe-Thiele diagram for the conditions described in question 6*a* and find the number of theoretical stages in each section of the column.
7. Imagine that you've been asked to work out a column design to separate methanol and water. The feed has about twice as much water as it has methanol and there are about 1,000 kg/h to process. You need the methanol to be fairly dry, e.g, less than 0.5% water, and are told you should recover "most" of it. Your first reaction is to ask for more information, but your supervisor has just left on an overseas trip and you've heard that she is really busy and expects all of the engineers in your group to be self-reliant. Unfortunately, none of your colleagues know much about the problem, and they can't give you more information.
   a. Monday morning: You find a fax from your boss, marked *Extremely Urgent,* asking for an immediate response to her question about the design. She is polite but insists on a response "ASAP." You decide to make a really quick estimate of the following based on Raoult's law VLE, plus the Fenske and Underwood expressions and whatever else you need.

(1) The number of theoretical stages
(2) The reflux and reboil ratios
(3) The total energy needed for heating and for cooling
(4) The production rate of methanol

b. Monday afternoon: Your boss called, but you were on your lunch break and drove off site to the bank. Your officemate, Lou, took the message. She received your answer to question 1, but told Lou that she "...could have worked that out herself..." and really needs a better estimate for a breakfast meeting tomorrow, which will be midnight in your office. According to Lou, she also says that the "...energy cost is way too high..." and that "...you had better get serious about this." Your first impulse is to phone her, but you realize that she is probably now asleep because of the time difference. After cancelling your plans for dinner, you work out a better solution. Submit a copy of the fax you would send to your boss.

c. (Optional) Tuesday morning: Your boss got the fax and replied that the solution is "...much better" and it would really be "...super if you had a simulation for your scheduled Thursday afternoon meeting." By the way, you had planned to ask her at the meeting to authorize some spending so that you can buy a nice new portable computer.

8. The Fenske and Underwood analytical solutions for total reflux and minimum reflux, respectively, are not useful to understand the relationship between stages and reflux. This can be done in many cases using the McCabe-Thiele diagram, but for high purities or small relative volatilities, an analytical solution is useful. For constant relative volatility and CMO, this can be developed by functional iteration (Smoker, 1938). The same equation can be derived using matrix techniques, as shown by Amundson (1946), which is an important paper because it was one of the earliest applications of matrix analysis in chemical engineering.

   a. Using a coordinate system with an origin at the pinch point for a stripping section, e.g., Fig. 3.15, show that the operating line can be written

$$Y_n = mX_{n+1} \tag{3.68}$$

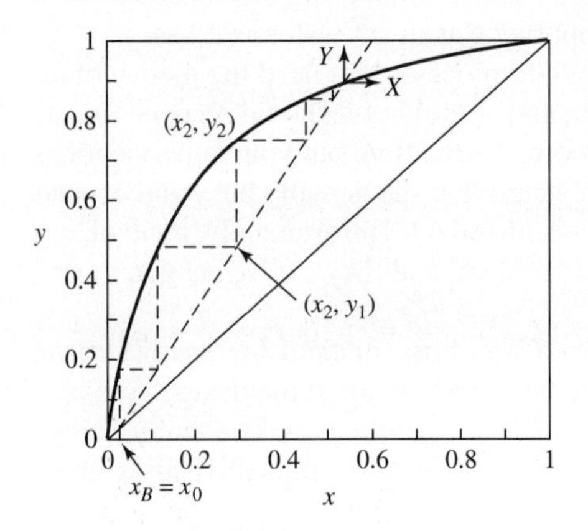

FIGURE 3.15
Coordinate system for Smoker's method.

and that the vapor-liquid equilibrium becomes

$$Y = \frac{\alpha X}{a^2 + a(\alpha - 1)X} \tag{3.69}$$

where m is the slope of the operating line, $a = 1 + (\alpha - 1)x_p$ and we have used the vapor-liquid equilibrium at the pinch point $y_p = \alpha x_p/(1 + (\alpha - 1)x_p)$.

b. Using an approach similar to the derivation of the Fenske equation, show that the vapor and liquid compositions after n stages are

$$Y_n = \frac{am\bar{\alpha}^{n+1}\,X_0}{a + (\bar{\alpha}^{n+1} - 1)\frac{\alpha-1}{\bar{\alpha}-1}\,X_0} \tag{3.70}$$

and

$$X_n = \frac{a\bar{\alpha}^{n}\,X_0}{a + (\bar{\alpha}^{n} - 1)\frac{\alpha-1}{\bar{\alpha}-1}\,X_0} \tag{3.71}$$

where $\bar{\alpha} = \alpha/(ma^2)$.

c. Next, show that the number of stages can be written

$$n = \frac{\ln \bar{S}}{\ln \bar{\alpha}} \tag{3.72}$$

where $\bar{S}$ is

$$\bar{S} \equiv \left[\frac{a - \left(\frac{\alpha-1}{\bar{\alpha}-1}\right) X_0}{X_0}\right]\left[\frac{X_n}{a - \left(\frac{\alpha-1}{\bar{\alpha}-1}\right) X_n}\right] \tag{3.73}$$

This is one form of *Smoker's equation.*

d. Finally, find the number of stages for Example 3.1 with Smoker's equation applied to the entire column. *Hint:* Treat the column sections separately, so that they reach the same feed composition, which should lie on the q-line.

9. A feed in the amount of 12,000 lb/h is a saturated liquid containing 70 mol % methanol and 30 mol % water. It is desired to separate this by distillation into a saturated liquid distillate containing 99 mol % methanol and a saturated liquid bottoms containing 99 mol % water.

   a. Find the minimum reflux and reboil ratios, the number of stages, the actual vapor rates and heating and cooling rates needed for the distillation of methanol and water.

   b. Repeat part *a* at a pressure of 5 atm.

   c. There is a way to reduce the net energy use for the system by heating the reboiler of the low-pressure column using one of the streams in one of the columns you just designed. Draw a flowsheet and explain how it works. Be certain to explain how you decide the amount of feed for each of the two columns.

   d. (Extra Credit) Simulate your flowsheet and check how it responds to small changes in one or two of the independent variables or parameters, e.g., flowrates, reflux ratio, etc.

10. For a binary mixture of acetone and propyl acetate, make a graph of number of stages vs. reflux ratio. Discuss the shape of the curve and the effects of the uncertainty in the relative volatility. *Notes:* Use Table 2.5 for the relative volatility. The feed contains 10 mol % propyl acetate and is available at 20°C and 1 atm pressure. The propyl acetate product should be 99.5 % pure, and 99% of the propyl acetate in the feed should be recovered in that product stream.
11. A saturated liquid mixture containing 40 mol % acetone and 60 mol % water is to be separated into saturated liquid products containing 95 mol % acetone and 5 mol % acetone, respectively. Determine the number of theoretical stages required to achieve the separation at a reflux ratio of $1.5r_{\text{min}}$. On which stage should the feed be located? Solve the problem graphically on the y-x diagram using the empirical VLE model in Eq. 2.35 and the numerical values for a and b from Table 2.6 to draw the equilibrium y-x curve.
12. Find a solution for Exercise 1 if the mixture is changed from hexane $+p$-xylene to benzene + ethylenediamine. How does the solution change if the purities of benzene change to 99.95% and 0.05% for the distillate and bottoms, respectively?
13. The general equations for finding tangent pinches in nonideal binary mixtures are Eqs. 3.60 and 3.61. These equations normally admit either zero or two solutions depending on the value of the distillate composition z_D. When there are two solutions, one of them will have larger values of both r_{tp} and x_{tp} than the second.
    a. Show the graphical position of these solutions on a McCabe-Thiele diagram and explain why the lower root is physically unacceptable.
    b. One way of finding the acceptable root to the tangent pinch problem is to calculate both roots and pick the larger of the two. Devise another test which identifies each root individually, without having to compare the roots with each other. If the VLE relation for the mixture is given by Eq. 3.62, write down the explicit form of your test in terms of a, b, and x_{tp}.
    c. Return to the benzene-ethylenediamine example described in Sec. 3.6 of the text. If the distillate purity is reduced to 99.0 mol % benzene, what is the new value of r_{min}.
14. For binary mixtures which exhibit inflections in their equilibrium y-x diagrams, there is a critical value of $z_D = z_D^{\text{crit}}$ below which tangent pinches no longer occur.
    a. What condition determines the critical value z_D^{crit} and what is the corresponding geometrical construction? If the VLE relation for the mixture is given by Eq. 3.62, write down the explicit form of the critical condition in terms of a, b, and x_{tp}. Explain how you would use this to calculate the corresponding critical values of z_D^{crit} and r_{tp}.
    b. Calculate the critical value z_D^{crit} for the benzene-ethylenediamine system described in Sec. 3.6. What are the corresponding values of r_{tp} and x_{tp}?
15. When tangent pinches occur, they often appear in the top of a continuous column for binary mixtures. However, it should not be assumed that this is always the case, as this exercise demonstrates.

    The VLE data for a mixture of diethylamine and ethanol are shown in Fig. 3.16. Estimate the minimum reboil ratio for the separation of a saturated liquid

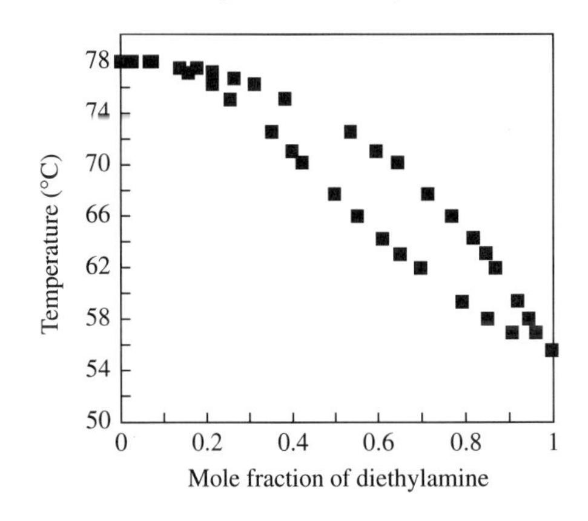

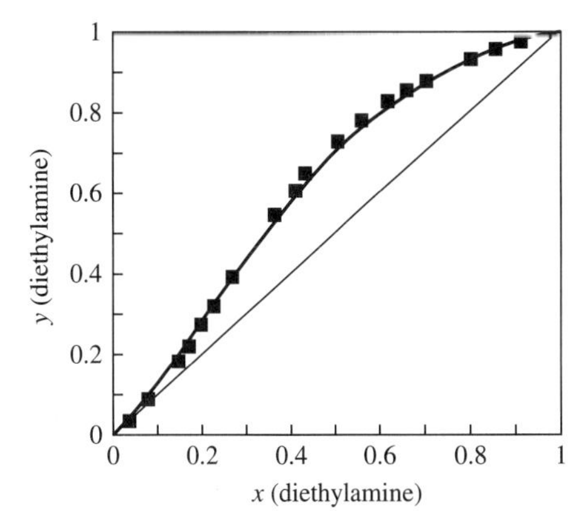

FIGURE 3.16
Vapor-liquid equilibrium data for the mixture diethylamine and ethanol at 1 atm pressure. *(Data from Gmehling and Onken, 1977a, p. 381.)*

feed containing 40 mol % diethylamine into products containing 95 mol % diethylamine and 98 mol % ethanol. Solve this problem graphically; the model in Eq. 3.62 will not describe this sort of behavior.

16. Binary mixtures which have inflections in their y-x diagrams can develop tangent pinches. Whether the tangent pinch controls the minimum flows depends on the value of the feed composition, as explained in Sec. 3.6 and as shown in Fig. 3.13. The discussion in the text is valid only for saturated liquid feeds; this exercise demonstrates the influence of variations in q on the controlling pinch.

    Consider a mixture of benzene and ethylenediamine at 1 atm pressure. The equilibrium y-x diagram is shown in Fig. 3.9 and can be represented by Eq. 3.62 with $a = 9$ and $b = -0.6$. A mixture containing 40 mol % benzene and 60 mol % ethylenediamine is fed to a distillation column in order to obtain a distillate that is 99.9 mol % benzene. Calculate the minimum reflux ratio as a function of q for values in the range $-1 \leq q \leq 2$. Draw a graph of $r_{\min}$ versus q and identify the intervals on the q axis over which the feed pinch controls and over which the tangent pinch controls.

17. Consider the separation of a saturated vapor mixture of methanol and ethyl acetate containing 30 mol % methanol into saturated liquid products. The

equilibrium y-x diagram for this mixture can be calculated using the model and parameters in Table 2.6. For a desired distillate purity of 65 mol % methanol and a bottoms purity of 95 mol % ethyl acetate, find

a. The minimum number of equilibrium stages.

b. The minimum reflux and reboil ratios.

c. The minimum vapor flow (relative to the feed flow) in the bottom section of the column.

d. The number of equilibrium stages at a reflux ratio of twice the minimum value.

With this feed, is it possible to get a distillate containing 99 mol % methanol?

18. Ethyl acetate (normal boiling point, 77.1°C) and ethanol (normal boiling point, 78.3°C) form a minimum-boiling binary azeotrope containing 54 mol % ethyl acetate at 1 atm pressure. A set of isobaric VLE data at this pressure, taken from Gmehling and Onken (1977*a*, p. 352) is given in Table 3.4; x and y are the mole fractions of ethyl acetate in the the liquid and vapor phases, respectively. These data can be adequately described by the empirical model given by Eqs. 3.66 and 3.67 with $\alpha_B = 2.19$, $\alpha_T = 0.52$, and $x^{az} = 0.54$. Draw a y-x diagram and compare the curves generated by these equations to the experimental data.

a. Using the transformed variables $\mathcal{X} = x/x^{az}$ and $\mathcal{Y} = y/y^{az}$, draw a $\mathcal{Y}$-$\mathcal{X}$ diagram and compare the curve generated by the transformed VLE model

$$\mathcal{Y} = \frac{\alpha_B \mathcal{X}}{1 + (\alpha_B - 1)\mathcal{X}}$$

to the transformed experimental data for the liquid compositions in the range $0 \leq x \leq x^{az}$.

b. Suppose a binary saturated liquid mixture containing 27 mol % ethyl acetate ($\mathcal{X}_F = 0.5$) is to be separated into saturated liquid products containing

TABLE 3.4
VLE data for a mixture of ethyl acetate and ethanol at 1 atm pressure

Temperature (°C)	x	y
78.3	0.000	0.000
76.6	0.050	0.102
75.5	0.100	0.187
73.9	0.200	0.305
72.8	0.300	0.389
72.1	0.400	0.457
71.8	0.500	0.516
71.8	0.540	0.540
71.9	0.600	0.576
72.2	0.700	0.644
73.0	0.800	0.726
74.7	0.900	0.837
76.0	0.950	0.914
77.1	1.000	1.000

53.73 mol % ($\mathcal{X}_D = 0.995$) and 0.27 mol % ($\mathcal{X}_B = 0.005$) ethyl acetate, respectively. Using the transformed Fenske equation, calculate the minimum number of theoretical stages required for the separation. Using the transformed Underwood equation, calculate the minimum reflux ratio. Optimally using Smoker's equation, calculate the number of theoretical stages in each column section for an operating reflux ratio of $r = 1.5r_{\text{min}}$.

c. Compare these results to those obtained using a graphical McCabe-Thiele analysis based on the experimental y-x data.

19. Methyl acetate (normal boiling point 56.9°C) and methanol (normal boiling point 64.6°C) form a minimum-boiling binary azeotrope containing 64.8 mol % methyl acetate at 1 atm pressure. The isobaric VLE data at this pressure are quite satisfactorily represented by Eqs. 3.66 and 3.67 with parameter values $\alpha_B = 2.6$, $\alpha_T = 0.5$, and $x^{az} = 0.648$.

    Suppose that a binary saturated liquid mixture containing 30 mol % methyl acetate is to be separated into saturated liquid products containing 64.48 mol % and 0.324 mol % methyl acetate, respectively. Calculate the minimum number of theoretical stages, the minimum reflux ratio, and the actual number of theoretical stages in each column section at an operating reflux 50% above the minimum. Solve the problem by using the VLE expression Eq. 3.65, taking $\alpha' = \alpha_B = 2.6$.

20. When the composition of an azeotrope changes significantly at different pressures, it is possible to separate a binary azeotropic mixture using a system of two columns operated at different pressures. This is sometimes called "pressure-shifting" or "pressure-swing" distillation. For example, this is used commercially to separate tetrahydrofuran from water, and several other systems have been proposed, though not all are economical (Knapp and Doherty, 1992).

    a. Calculate the x-y diagrams for methyl ethyl ketone (MEK) and water at pressures of 1 atm and 5 atm. These results should look like Fig. 3.17, which

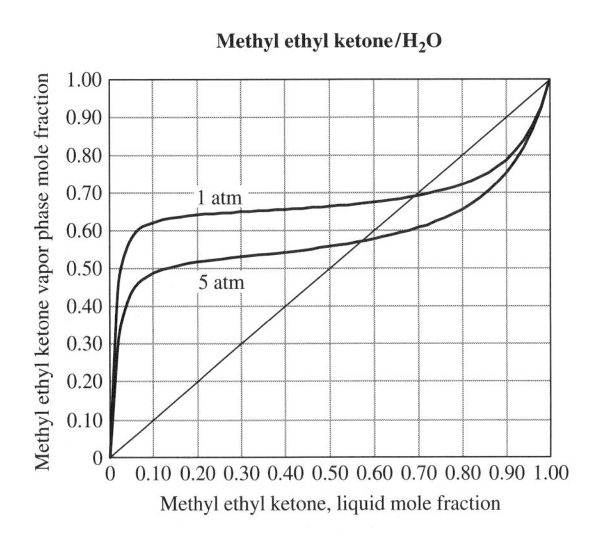

FIGURE 3.17
Effect of pressure on y-x diagram for methyl ethyl ketone and water. *(AEA Technology Engineering Software, 1999.)*

was calculated with the software tool Distil (AEA Technology Engineering Software, 1999) based on data from Gmehling and Onken (1977*b*, p. 372).

b. Draw an arrangement of two columns and their interconnections that can be used to separate a feed containing 20 mol % MEK into nearly pure MEK and nearly pure water. *Hint:* The azeotrope is minimum-boiling so both columns produce a composition close to the azeotrope as a distillate. The distillate from one column can be used as feed to the other.

c. Estimate the flows and compositions of all the streams if the MEK and water product streams are each 99.5 mol % pure. Especially discuss how you would choose the composition of the overhead streams.

References

AEA Technology Engineering Software, Hysys 1.2 (1997). www.software.aeat.com.

AEA Technology Engineering Software, Distil 3.1 (1999). www.software.aeat.com.

Agrawal, R., and D. M. Herron, "Optimal Thermodynamic Feed Conditions for Distillation of Ideal Binary Mixtures," *AIChE J.,* **43,** 2984–2996 (1997).

Agrawal, R., and D. M. Herron, "Efficient Use of an Intermediate Reboiler or Condenser in a Binary Distillation," *AIChE J.,* **44,** 1303–1315 (1998*a*).

Agrawal, R., and D. M. Herron, "Intermediate Reboiler and Condenser Arrangement for Binary Distillation Columns," *AIChE J.,* **44,** 1316–1324 (1998*b*).

Amundson, N. R., "Application of Matrices and Finite Difference Equations to Binary Distillation," *Trans. Amer. Inst. Chem. Engrs.,* **42,** 939 (1946).

Anderson N. J., and M. F. Doherty, "An Approximate Model for Binary Azeotropic Distillation Design," *Chem. Engng. Sci.,* **39,** 11 (1984).

Aspen Technology, Aspen Plus ver. 10 (1999). www.aspentech.com.

deRosset, A. J., R. W. Neuzil, D. G. Tajbl, and J. M. Braband, "Separation of Ethylbenzene from Mixed Xylenes by Continuous Adsorptive Processing," *Sepn. Sci. Tech.,* **15,** 637 (1980).

Doherty, M. F., and J. D. Perkins, On the Dynamics of Distillation Processes—II. The Simple Distillation of Model Solutions," *Chem. Engng. Sci.,* **33,** 569–578 (1978).

Fenske, M. R., "Fractionation of Straight-Run Pennsylvania Gasoline," *Ind. Eng. Chem.,* **24,** 482 (1932).

Gaw, W. J., and F. L. Swinton, "Occurrence of a Double Azeotrope in the Binary System Hexafluorobenzene-Benzene," *Nature*, **212,** 283 (1966).

Gmehling, J., and U. Onken, *Vapour-Liquid Equilibrium Data Collection*, vol. 1/2a, Organic Hydroxy Compounds: Alcohols, of *Chemistry Data Series*. DECHEMA, Frankfurt/Main, (1977*a*).

Gmehling, J., and U. Onken, *Vapour-Liquid Equilibrium Data Collection*, vol. 1/1, Aqueous-Organic Systems, of *Chemistry Data Series*. DECHEMA, Frankfurt/Main (1977*b*).

Gmehling, J., U. Onken, and W. Arlt, *Vapour-Liquid Equilibrium Data Collection*, vol. 1/7, Aliphatic Hydrocarbons, of *Chemistry Data Series*. DECHEMA, Frankfurt/Main (1980*a*).

Gmehling, J., U. Onken, and W. Arlt, *Vapour-Liquid Equilibrium Data Collection*, vol. 1/6a, Aliphatic Hydrocarbons C_4-C_6, of *Chemistry Data Series*. DECHEMA, Frankfurt/Main (1980*b*).

Horsley, L. H., *Azeotropic Data III*, number 116 in Advances in Chemistry Series. American Chemical Society, Washington, DC (1973).

King, C. J., *Separation Processes*. McGraw-Hill, New York, 2d ed. (1980).

Knapp, J. P., and M. F. Doherty, "A New Pressure-Swing Distillation Process for Separating Homogeneous Azeotropic Mixtures," *Ind. Eng. Chem. Research*, **31,** 346–357 (1992).

Malesiński, W., *Azeotropy and Other Theoretical Problems of Vapour-Liquid Equilibrium.* Interscience, London (1965).

McCabe, W. L., and E. W. Thiele, "Graphical Design of Fractionating Columns," *Ind. Eng. Chem.,* **17,** 605–611 (1925).

Reid, R. C., J. M. Prausnitz, and T. K. Sherwood, *The Properties of Gases and Liquids.* McGraw-Hill, New York, 3d ed. (1977).

Seider, W. D., J. D. Seader, and D. R. Lewin, *Process Design Principles.* Wiley, New York (1999).

Simulation Sciences, Inc. PRO/II ver. 5 (1999). www.simsci.com.

Smoker, E. H., "Analytic Determination of Plates in Fractionating Columns," *Trans. Inst. Chem. Engrs. (London)*, **34**(165) (1938).

Underwood, A. J. V., "Fractional Distillation of Multicomponent Mixtures—Calculation of Minimum Reflux," *Trans. Inst. Chem. Engrs. (London),* **16,** 112 (1932).

Wankat, P. C., *Equilibrium Staged Separations.* Prentice-Hall, Englewood Cliffs, NJ (1988).

4

Distillation of Multicomponent Mixtures without Azeotropes

Mixtures containing more than two components complicate the design problem somewhat. The specification of product purities is not as simple as in the binary case and the pinches are different in number and in character. In the first section of this chapter, we summarize the basic relationships, especially differences with binary mixtures. The next section illustrates the application of these models to ternary mixtures, followed by a separate section on mixtures with four and more components. Finally, we discuss tangent pinches in multicomponent mixtures.

4.1 BASIC RELATIONSHIPS

Phase Equilibrium

The generalization of Eq. 2.31 that describes the vapor-liquid equilibrium in a mixture containing c components is

$$y_i = \frac{\alpha_i x_i}{\sum \alpha_j x_j} \tag{4.1}$$

where i is the index of any component and α_i is the volatility of component i. Unless indicated otherwise, the summation is taken over all of the c components present in the mixture. The volatilities are given by $\alpha_i \equiv (\gamma_i P_i^{\text{sat}} \phi_k)/(\gamma_k P_k^{\text{sat}} \phi_i)$ relative to a reference component k. We will usually order the components by decreasing boiling point, with the highest-boiling component as the reference.[1]

These phase equilibrium relationships are nonlinear relations among all of the c mole fractions in each phase. However, we will usually need only $c - 1$ of the mole

[1]The choice of a reference component will sometimes be emphasized with the notation α_{ij}. Note that $\alpha_{ij} = \alpha_{ik}\alpha_{kj}$.

fractions in each phase with the understanding that the remaining mole fractions can be determined from the relations

$$\sum x_i = \sum y_i = 1 \tag{4.2}$$

Even when vapor-phase nonidealities can be neglected, the volatilities may not be constant because of the composition dependence of the activity coefficients or through the variation of the vapor pressures with temperature. In the important, but restrictive, special case where the volatilities can be treated as constants it is simple to find the composition of either phase given that of the other, i.e., to rewrite Eq. 4.1 as

$$x_i = \frac{y_i/\alpha_i}{\sum y_j/\alpha_j} \tag{4.3}$$

Material and Energy Balances

The overall balances for multicomponent separations are more informative than in the binary case. There are c independent balances that can be written

$$F = D + B \tag{4.4}$$

and

$$F z_{i,F} = D z_{i,D} + B z_{i,B} \tag{4.5}$$

where $i = 1, \ldots, c-1$. The analogs of Eqs. 3.3, 3.4, and 3.5 are

$$\frac{D}{F} = \frac{z_{i,F} - z_{i,B}}{z_{i,D} - z_{i,B}} \quad \text{and} \quad \frac{B}{F} = \frac{z_{i,D} - z_{i,F}}{z_{i,D} - z_{i,B}} \tag{4.6}$$

$$\frac{D}{B} = \frac{z_{i,F} - z_{i,B}}{z_{i,D} - z_{i,F}} \tag{4.7}$$

Thus, a specification of the product and feed compositions for any one component is sufficient to determine the ratio of the total product and feed flows and the ratio of the composition differences for every other component.[2]

The energy balances and the definitions of reflux ratio, reboil ratio, and feed quality are identical to the binary case when the composition of the light component in Eqs. 3.19, 3.22, and 3.35 is replaced by the composition of any component in

---

[2]It is sometimes convenient to relate the fractional recoveries to the mole fractions as follows:

$$f_{i,D} = \frac{z_{i,D}}{z_{i,F}} \frac{z_{j,F} - z_{j,B}}{z_{j,D} - z_{j,B}} \quad \text{and} \quad f_{i,B} = \frac{z_{i,B}}{z_{i,F}} \frac{z_{j,D} - z_{j,F}}{z_{j,D} - z_{j,B}} \tag{4.8}$$

where i and j are component labels and D and B indicate distillate and bottoms. If the fractional recoveries are known, the distillate and bottoms mole fractions are given by

$$z_{i,D} = \frac{f_{i,D} z_{i,F}}{\sum f_{j,D} z_{j,F}} \quad \text{and} \quad z_{i,B} = \frac{f_{i,B} z_{i,F}}{\sum f_{j,B} z_{j,F}} \tag{4.9}$$

With two product streams, $f_{i,D} + f_{i,B} = 1$ and only one of the fractional recoveries for a given component is independent.

the mixture. For example, the constant molar overflow approximation results in the following relationship between the reflux and the reboil ratios (cf. Eq. 3.35):

$$\frac{D}{B} = \frac{s + 1 - q}{r + q} \tag{4.10}$$

With the constant molar overflow assumption, the operating relationships for component i are

$$y_{i,n-1} = \frac{r}{r+1} x_{i,n} + \frac{z_{i,D}}{r+1} \tag{4.11}$$

in the rectifying section and

$$y_{i,n} = \frac{s+1}{s} x_{i,n+1} - \frac{z_{i,B}}{s} \tag{4.12}$$

in the stripping section. Stage-to-stage energy balances may also be needed to account for the variation of the liquid and vapor flows. It is straightforward to add these to the formulation if good models and data for the enthalpies as a function of composition and temperature are available.

4.2 COMPOSITION PROFILES AND PINCHES

The Boundary Value Design Method

It is useful to find a geometric method for mixtures with more than two components. With the addition of more components, the visualization is more difficult, but a complete picture is still possible for three components. This section considers the main geometric ideas for ternary ideal mixtures, and nonideal ternary mixtures are considered in the next section. The general case with four or more components is the subject of Sec. 4.5.

We will use a *triangular diagram* like the one shown in Fig. 4.1 to represent the compositions.[3] Each vertex of the triangle corresponds to a pure component and each edge corresponds to a binary mixture of the two pure components which lie at the vertices joined by that edge. The overall material balance in Eq. 4.7 can be used for any two components, say 1 and 2, to find

$$\frac{z_{2,B} - z_{2,F}}{z_{1,B} - z_{1,F}} = \frac{z_{2,D} - z_{2,F}}{z_{1,D} - z_{1,F}} \tag{4.13}$$

The term on the left is the slope of the line joining the bottoms and feed compositions and on the right is the slope of a line joining the distillate and feed compositions.

[3]Some prefer an equilateral triangle, or mass fractions instead of mole fractions. We use right triangles because rectangular coordinates seem easier to interpret and mole fractions because phase equilibrium models naturally use these variables. Modern computer-aided design tools make the coordinates easy to adjust.

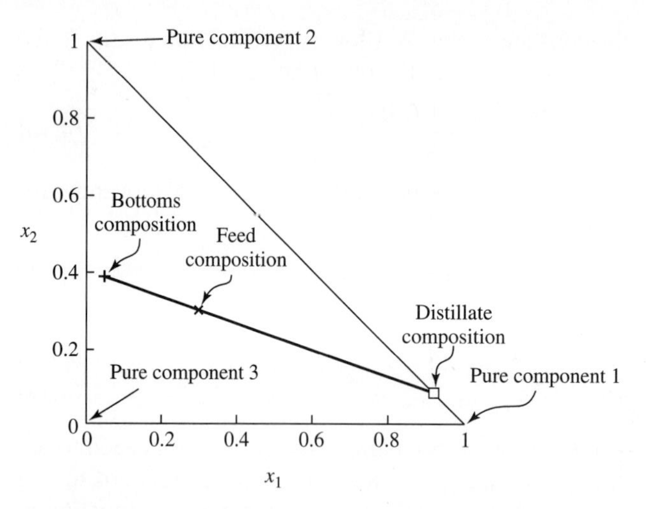

FIGURE 4.1
A composition diagram for ternary mixtures. The abscissa and ordinate represent the mole fractions of the components 1 and 2, respectively. Each vertex represents a pure component and each edge represents a binary mixture. The feed composition is $\mathbf{z}_F = (0.3, 0.3, 0.4)$ and the product compositions are $z_{1,B} = 0.05$, $z_{1,D} = 0.95$, $z_{2,D} = 0.049$. By mass balance $z_{2,B} = 0.397$, $D = 0.626F$, and $B = 0.374F$.

Thus, *on a triangular diagram, the distillate, bottoms, and feed compositions are collinear.*

Unlike the case for binary mixtures, it is necessary to consider the details of the column in order to determine the product distribution for mixtures with three or more components.

The liquid-phase mole fraction in each section of the column will vary from stage to stage according to the dictates of the material and energy balances and the vapor-liquid equilibrium. For ternary mixtures, the profiles of the liquid (and vapor) phase compositions can be determined by solving Eqs. 4.11 and 4.12 along with the appropriate VLE relationships for the two lightest components. Beginning with the distillate and bottoms compositions, the mole fractions on successive stages above the bottom and below the distillate can be computed. This procedure is completely analogous to the approach discussed in Sec. 3.3 for the treatment of binary mixtures. Beginning with some product compositions that satisfy the overall mass balances, we can examine the behavior of the composition profiles.

EXAMPLE 4.1. Figure 4.2 shows profiles of the liquid-phase mole fractions for a mixture of pentane, hexane, and heptane. The feed is a saturated liquid at atmospheric pressure with a composition $\mathbf{x}_F = (0.3, 0.3, 0.4)$; the volatilities are approximately constant at $\alpha_{13} = 6.35$ and $\alpha_{23} = 2.47$. The distillate contains 95% pentane and 4.9% hexane; the balance of 0.1% of the distillate is therefore heptane. If the bottoms is to contain 5% pentane, then the overall balances demand that this product stream also contains 39.7% hexane and 55.3% heptane. Each of the symbols represents the liquid mole fractions on an equilibrium stage; the stripping section requires $n_S = 5.0$ stages (one can be

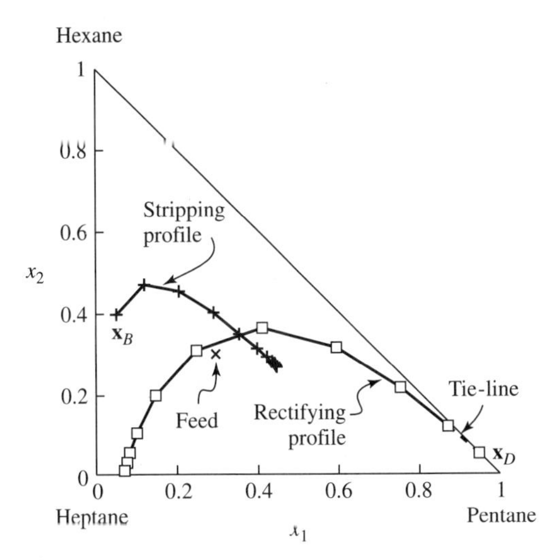

FIGURE 4.2
Composition profiles for a mixture of pentane (1), hexane (2), and heptane (3). The volatilities are taken as $\alpha_{13} = 6.35$, $\alpha_{23} = 2.47$; the pressure is 1 atm and the feed is a saturated liquid. The product specifications are for saturated liquid products with the distillate containing 95 mol % pentane, 4.9% hexane and 0.1% heptane and the bottoms 5% pentane. With a reflux ratio of 2.5, corresponding to a reboil ratio of 1.35, 8.3 theoretical stages are required. The feed is introduced five stages from the bottom as indicated by the intersection of the rectifying (□) and stripping (+) profiles. The dashed line is a vapor-liquid tie-line between the liquid composition on the top stage in the rectifying section and the distillate composition, $\mathbf{x}_D = \mathbf{y}_1$.

an equilibrium reboiler) and the rectifying section contains $n_R = 4.3$ stages. The total number of stages in the column is $n_R + n_S - 1 = 8.3$. (One stage is subtracted to avoid counting the feed stage twice, for both the stripping and rectifying sections.)

Because we specify the product compositions at the *ends* of the column, this is called a "boundary value" design procedure. The location of the feed is now a dependent variable, and it must be introduced at the intersection of the profiles.[4] If only two of the product compositions have specifications that *must* be met, the third composition can be adjusted to minimize the number of stages.

Figure 4.3 shows some profiles for cases where the mole fraction of pentane in the bottoms is reduced. A portion of the stripping profile approaches the ordinate in the lower stages, where the liquid contains little pentane. On successively higher stages, the mole fraction of hexane increases and the mole fraction of heptane drops. Eventually the mole fraction of hexane reaches a maximum, the composition of pentane increases rapidly, and the rate of decrease of the mole fraction of heptane is slowed abruptly. This point has mole fractions of approximately (0.0, 0.676, 0.324). This point is approached asymptotically by the stripping profile with an increasing

[4]Note that the composition of the liquid on the feed stage is generally *not* the same as that of the feed.

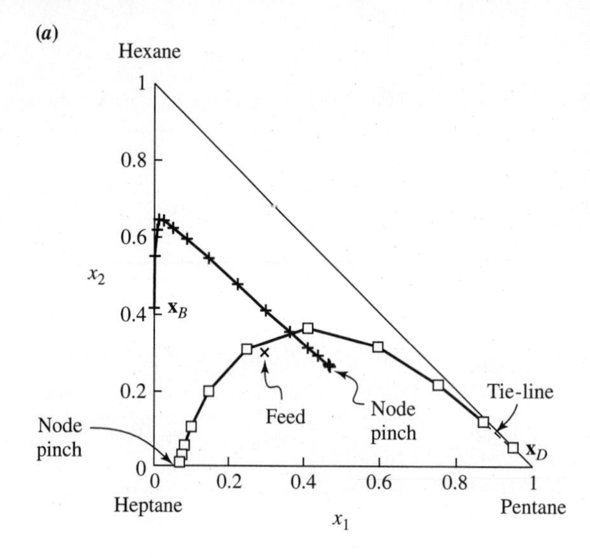

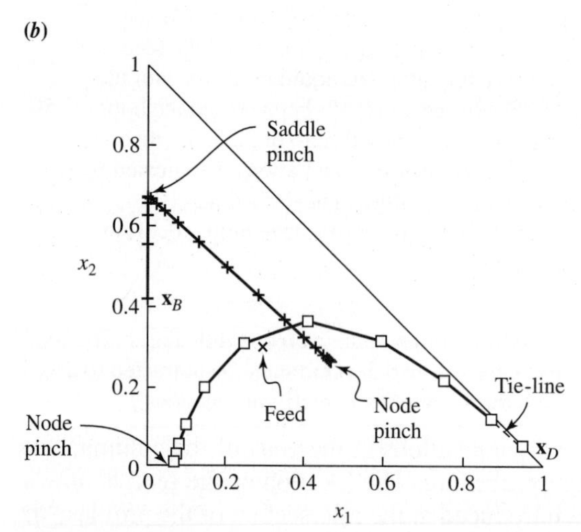

FIGURE 4.3
The effect of product purity on the liquid composition profiles for a mixture of the normal alkanes pentane, hexane, and heptane. The distillate contains 95% pentane, 4.9% hexane, and 0.1% heptane; the bottoms purity is (*a*) 0.1% pentane and (*b*) 0.0001% pentane. Both profiles show a *node* pinch. A *saddle pinch* in the stripping profile is apparent at the composition (0.0, 0.676, 0.324).

number of stages as $z_{1,B} \to 0$. That is, there is a large number of stages for which the composition is essentially constant; this is a *pinch.* The *asymptotic* approach to this point corresponds to a new behavior called a *saddle pinch* or simply a *saddle.* This has no analog in binary mixtures, but it is a common and extremely important feature in the separation of mixtures containing three or more components.[5] The pinches that appear at the end of the upper and the lower operating lines for binary

[5]Another pinch that does have a binary analog is the tangent pinch; this is found in nonideal mixtures, which are discussed in Sec. 4.6.

mixtures do have an analog in the ternary case; these *node pinches* or *nodes* are also shown in Figs. 4.2 and 4.3.[6]

As for the case of binary mixtures, the pinches can be used to find the minimum flows. However, for ternary mixtures there are several possibilities for the relative positions of the pinches. It is useful to classify these possibilities systematically once we consider the degrees of freedom.

We consider rectifying and stripping profiles which satisfy Eqs. 4.11 and 4.12 and cases where the feed composition, feed quality, and the pressure are known. The remaining variables are the distillate and bottoms compositions, along with the reflux and reboil ratios, for a total of $2(c-1)+2=2c$ variables.[7] The energy balance in Eq. 4.10 provides one relationship between r and s, and there are also $c-2$ independent versions of the mass balances Eq. 4.13 relating the product compositions. This gives $(2c-2)-(c-2)=c$ degrees of freedom in the product compositions and one degree of freedom between r and s for a total of $c+1$ degrees of freedom from the overall mass and energy balances. For a feasible column, we must also have an intersection of the profiles at the feed stage. If we denote the liquid-phase compositions in the stripping profile as $\mathbf{x}^s$ and in the rectifying profile as $\mathbf{x}^r$, then after n_T stages in the top of the column and n_B stages in the bottom we must have the $c-1$ additional constraints

$$\mathbf{x}^r_{n_T} = \mathbf{x}^s_{n_B} \tag{4.14}$$

and two additional variables n_T and n_B.

Thus, there are $(c+1)-(c-1)+2=4$ *degrees of freedom.* Common design specifications for binary mixtures consist of two product compositions (or fractional recoveries), a reflux *or* reboil ratio, and the condition that the feed stage location be optimal. In ternary mixtures, we might first choose the distillate and the bottoms composition for one of the components. Then, a specification of one more product composition and the reflux ratio will provide sufficient information to determine the composition profiles throughout the column, as well as the feed stage location. For a *performance* model, we might choose the number of stages in each column section and the reflux and reboil ratios. Then, the product compositions and flows are calculated. Iterative schemes can be devised to adjust these specifications with the goal of achieving certain desired products, but this is often inefficient, especially in nonideal mixtures.

Minimum Flows and Classification of Splits

In binary mixtures, a node pinch or a tangent pinch determines the minimum flows, but in multicomponent mixtures a saddle pinch may also be important. The node

---

[6]The terms "saddle" and "node" originate from a consideration of the composition diagram as a phase plane for Eqs. 4.11 and 4.12 and from further considerations for azeotropic mixtures, discussed in Chap. 5.

[7]We presume that Eqs. 4.2 can be trivially satisfied and therefore that there are $c-1$ independent compositions.

pinch is usually obvious, but the influence of the saddle pinch may not be immediately apparent in the profile. The position of the node usually depends most strongly on the ratio of vapor to liquid flows (reflux or reboil ratio), while the location of the saddle is more sensitive to the purity.

Figure 4.4 shows a few profiles for different values of the reflux ratio and product purities. Only the nodes are evident in Fig. 4.4*a*. Figure 4.4*b* shows a rectifying saddle

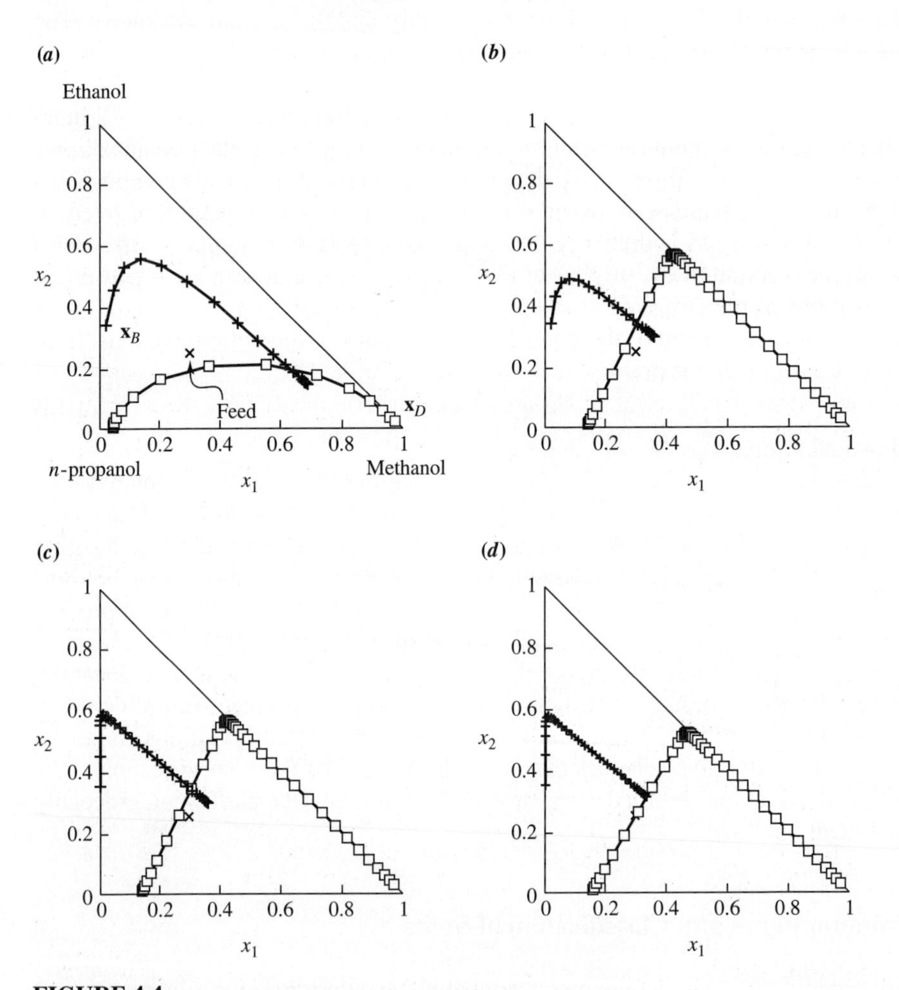

FIGURE 4.4
Column profiles and pinches for a ternary mixture of methanol, ethanol, and *n*-propanol at a pressure of 1 atm. The relative volatilities are 3.25, 1.90, and 1.00, the feed is a saturated liquid with a composition of (0.30, 0.25, 0.45), and the distillate contains 98% methanol. The reflux ratio and the mole fractions of methanol in the bottoms and propanol in the distillate are (*a*) 10, 0.02, 5×10^{-4}, (*b*) 3, 0.02, 5×10^{-11}, (*c*) 3, 5×10^{-4}, 5×10^{-11}, and (*d*) 2.7, 5×10^{-4}, 5×10^{-11}. The rectifying profile (□) and the stripping profile (+) each show a node pinch. The visibility of the saddle pinch depends primarily on the product purity.

and Fig. 4.4*c* shows both a node and a saddle in each profile. The saddle pinches in Fig. 4.4*c* correspond to a nearly infinite number of stages and a zone of nearly constant composition in each column section, so it may seem that this corresponds to the minimum reflux, but the reflux ratio can be further reduced, as shown in Fig. 4.4*d*. It is this case that corresponds to the minimum reflux (and flows). This is because the profiles do not intersect for lower values of the reflux ratio so that the product purities cannot be met even with an infinite number of stages. Also see Fig. 4.18*a*.

Valuable intuition is gained by generating these profiles yourself as the product compositions are systematically varied. Exercises 1, 2, and 3 are strongly recommended.

Figure 4.5 shows three possibilities for the minimum flow conditions, depending on the product specifications. When the node pinch in the stripping section controls,[8] the distillate may contain very little of the heaviest component, and this could be written symbolically as "AB/ABC." More specialized (but common) cases have a high purity and recovery of the lightest component in a *direct* or "A/BC" split; see Fig. 4.5*a* and Exercise 1. Conversely, when the node in the rectifying section controls, as in Fig. 4.5*b*, we produce a bottoms with a high purity and recovery of the heaviest component in an *indirect* or "AB/C" split. There is also a special case where the nodes from each section meet at the feed composition as in Fig. 4.5*c*; this is a *transition* or "AB/BC" split, since it marks the division between the direct and indirect geometries. Notice that the smallest value for $r_{\min}$ occurs for the transition split; see Exercise 3. This is sometimes used to develop novel systems of columns where the first column performs the AB/BC split, followed by other columns that split A from B and B from C. See Chap. 7 for more details.

In the direct split, the minimum reflux ratio is independent of the bottoms composition, although the *reboil* ratio will change for different values of the bottoms composition.[9] Consequently, the stripping profile will be somewhat different, although its node will still (just) meet the rectifying profile. Likewise, a variety of distillate purities could be specified for a separation like the one shown in Fig. 4.5*b* and the minimum reboil ratio along with the composition profile in the stripping section remain unchanged if the bottoms and feed compositions are constant.

This leads to a procedure for finding the minimum flows. The reflux or reboil ratio is adjusted until the liquid composition profiles meet in one of the three possible configurations. In fact, we can efficiently compute the minimum reflux (or reboil) knowing only the pinches. For instance, the pinches in the controlling profile, the node pinch in the other profile, and the feed composition are collinear for constant volatilities and high purities. When the volatilities are not constant, this collinearity is still a useful approximation, as discussed later. This method for finding the minimum flows will be discussed in more detail later in this chapter.

---

[8]"Control" by a pinch means that it is this pinch point which must align with the saddle in the opposite profile and the feed composition.

[9]The reflux and reboil ratios are related by Eq. 4.10.

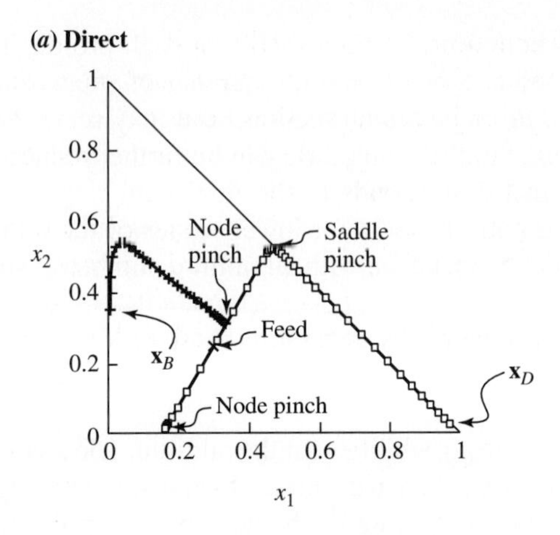

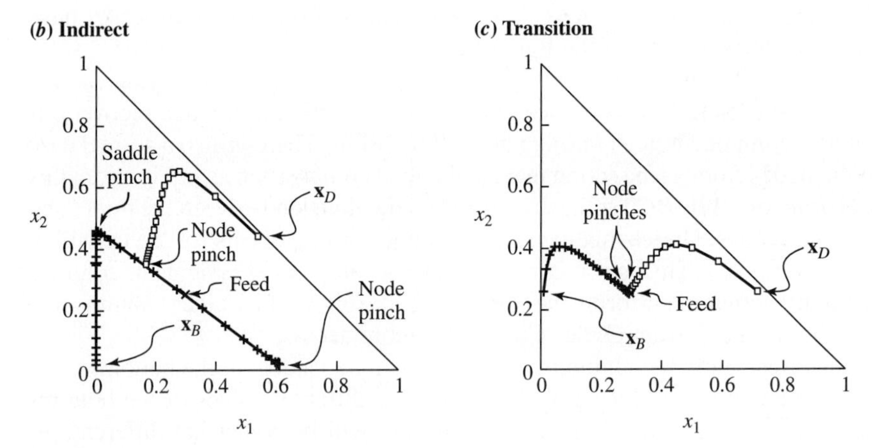

FIGURE 4.5
Composition profiles in (a) the direct, (b) the indirect, and (c) the transition splits. The feed is identical to the case described in Fig. 4.4, and we use $\alpha_{13} = 3.25$ $\alpha_{23} = 1.9$. In (a), the product compositions are $\mathbf{x}_D = (0.98, 0.02, 5 \times 10^{-11})$ and $\mathbf{x}_B = (0.005, 0.350, 0.645)$ while the reflux ratio is $r_{\min} = 2.7$. For case (b) these variables are (0.55, 0.44, 0.01), $(5 \times 10^{-11}, 0.022, 0.978)$ and 1.35. Case (c) is a transition or "AB/BC" split where the pinch in each section is adjacent to the feed stage; for this case, $\mathbf{x}_B = (0.01, 0.25, 0.74)$, $\mathbf{x}_D = (0.73, 0.25, 0.02)$, and $r = r_{\min} = 1.02$.

Minimum Stages

The minimum number of theoretical stages required for a separation can be found from a consideration of the limiting case of total reflux. Figure 3.6a shows an example for a binary mixture. For binary mixtures with a constant volatility, Fenske's equation 3.48 can also be used.

However, when the mixture contains more than two components, the minimum number of stages also corresponds to certain specific product compositions for some of the components. For total reflux and total reboil we have $r \to \infty$ and $s \to \infty$, respectively, so that the operating line in each section approaches $x_n = y_{n-1}$. In fact, the rectifying and stripping sections are indistinguishable. Two degrees of freedom remain; we can specify (only) two product compositions and the others must be computed, along with the number of stages.

Ternary mixtures can be understood from a consideration of the liquid-phase composition profiles as the reflux ratio is increased (Fidkowski et al., 1993, especially Fig. 2); also see Fig. 5.12. Figure 4.6 shows the composition profiles at $r = 1{,}000$ for the ternary mixture of normal alkanes discussed above. For each profile, the feed

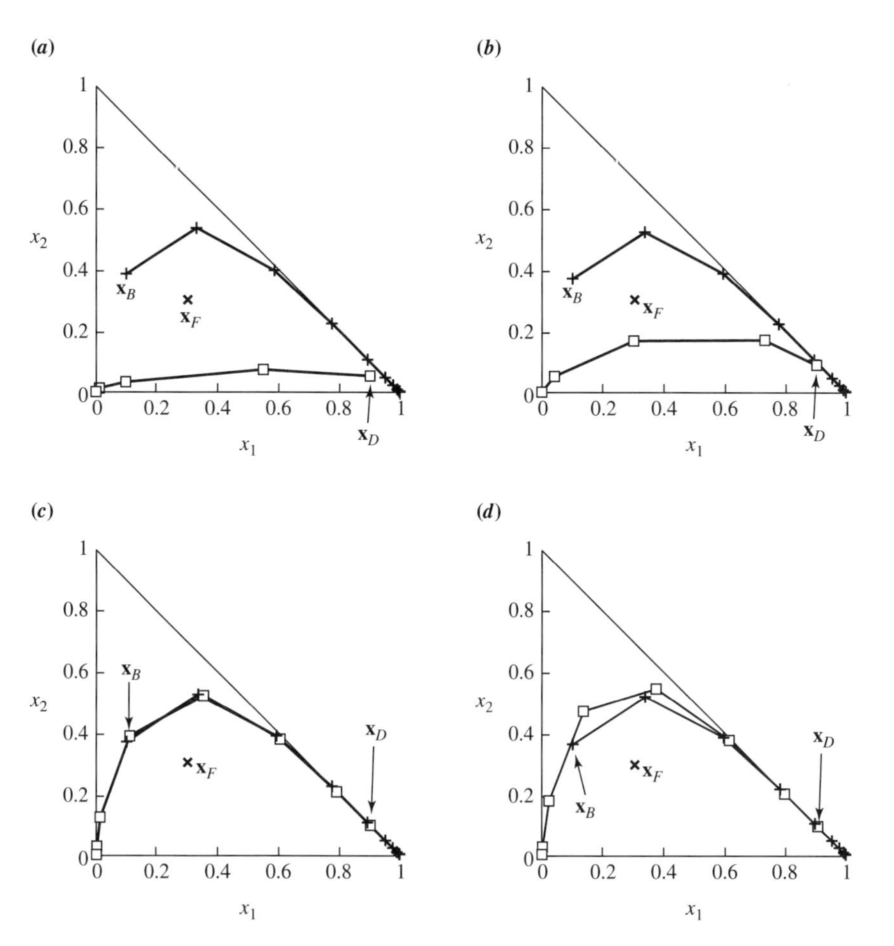

FIGURE 4.6
Composition profiles and the minimum number of stages for a ternary mixture. The feed is a mixture of hexane, heptane, and nonane with a composition of (0.3, 0.3, 0.4). The distillate contains 90% hexane and the bottoms contains 10%. The liquid-phase composition profiles in the rectifying (□) and stripping (+) sections are shown for distillates containing (*a*) 5.00%, (*b*) 1.00%, (*c*) 0.016%, and (*d*) 0.010% nonane. Case (*c*) corresponds to the minimum number of stages, which is slightly greater than 4.

composition is the same; the distillate and bottoms contain 90% and 10% of the lightest component, respectively. If the composition of another component in one of the product streams is specified, the composition and flows of both products can be determined. As the distillate composition of the heaviest component is reduced from an initial value of 5%, the profiles approach one another until they overlap. At this point, the distillate contains 0.016% of the heaviest component and the minimum number of stages is slightly more than 4.[10]

It is important to note that there is only one product distribution corresponding to the minimum number of stages. That is, despite the fact that the profiles shown in Fig. 4.6 were each calculated in the limiting case of large r and s, only one of these corresponds to a feasible solution, i.e., Fig. 4.6*c*. In other words, *it is not possible to specify each of the product compositions arbitrarily at total reflux.* Figure 4.6*a*, *b*, and *d* are *not* feasible. Once two product compositions are specified, the minimum number of stages *and* the other compositions can also be determined. This can be seen from the graphical construction or from Fenske's equation for mixtures with constant volatility, which we consider next.

4.3 ANALYTICAL RESULTS FOR CONSTANT VOLATILITY AND CONSTANT MOLAR FLOWS

The graphical approach described above is useful for developing intuition and because the basic ideas also apply to nonideal mixtures. However, analytical results are also useful, especially for the treatment of limiting cases in mixtures containing more components.[11]

Fenske's Equation for Minimum Stages

With a derivation that is essentially identical to that found in Chap. 3 (see Exercise 4), we can find that the minimum number of stages is given by[12]

$$N_{\min} = \frac{\ln[(z_i/z_j)_D(z_j/z_i)_B]}{\ln \alpha_{ij}} \tag{4.15}$$

which can also be written in terms of the fractional recoveries,

$$N_{\min} = \frac{\ln[(f_{i,D}/f_{i,B})(f_{j,B}/f_{j,D})]}{\ln \alpha_{ij}} \tag{4.16}$$

---

[10]The profiles do not overlap *exactly*, just as there is a difference between the compositions on each stage in the binary McCabe-Thiele construction, depending on whether the distillate or the bottoms is used as a starting point. The minimum number of stages is identical in either case.

[11]Unfortunately, these are available only for fairly simple mixtures such as those treated in this section.

[12]Fenske's equation 3.48 was written for the case of a total condenser and an equilibrium reboiler. For that case, one of the equilibrium stages is provided in the reboiler and one less than the minimum is needed in the column shell. In this chapter and subsequently, $N_{\min}$ is the total number, which can be provided inside or outside the column shell depending on the choices of condenser and reboiler. Note that this expression and the next assume constant volatilities and equilibrium stages at total reflux.

These expressions provide $c - 1$ independent equations because they must hold for any of the $c - 1$ pairs of components in the mixture (i and j). Therefore, if the fractional recoveries or the product purities of any two components are specified, then one of these $c - 1$ equations can be used to find the minimum number of stages. The remaining $c - 2$ equations, along with Eq. 4.2 and the overall balances, e.g., Eq. 4.6, can then be used to find the remaining product purities.

EXAMPLE 4.2. We consider the separation described in Fig. 4.6 (also see Exercise 5). Equation 4.15 can be written for the 1-2 and for the 2-3 pairs. We find $N_{\min} = 4.054$, $z_{2,D} = 0.09984$, $z_{3,D} = 0.00016$, $z_{2,B} = 0.3667$, and $z_{3,B} = 0.5333$. For higher-purity specifications of $z_{1,D} = 0.999$ and $z_{1,B} = 0.001$, we find $N_{\min} = 15.03$, $z_{2,D} = 0.001 - 1.54 \times 10^{-14}$, $z_{3,D} = 1.54 \times 10^{-14}$, $z_{2,B} = 0.4279$, and $z_{3,B} = 0.5711$. This case would be quite tedious to examine graphically.

It is a curious feature of multicomponent separations that some of the product compositions seem ridiculously small. This is partly because we are considering the total reflux limit, but also because distillation can be a very efficient separation method.[13] Cases with $r < \infty$ will be different numerically but may still be quite small. Although the product distribution at total reflux is not identical to the one found in the actual operation of the column, it provides a useful limiting case.

Underwood's General Method and the Minimum Flows

It's useful to have estimates of the minimum flows, the number of stages, and the product compositions which do not have to be determined graphically. The most important equation-based method was developed by Underwood (1946*a*, 1946*b*, 1946*c*) for the limiting case of constant volatilities and constant molar overflow. A very instructive geometrical interpretation of this approach is given by Forsyth and Franklin (1953) and developed further in Franklin (1988*a*) and Franklin (1988*b*). It is instructive first to consider the application to binary mixtures and then to generalize the results for cases with more components.

Binary mixtures

The material balances and equilibrium relationships in the rectifying section are

$$\frac{L_T}{V_T} x_{A,n+1} + \frac{D}{V_T} x_{A,D} = y_{A,n} = \frac{\alpha_A x_{A,n}}{\alpha_A x_{A,n} + \alpha_B x_{B,n}} \tag{4.17}$$

and

$$\frac{L_T}{V_T} x_{B,n+1} + \frac{D}{V_T} x_{B,D} = y_{B,n} = \frac{\alpha_B x_{B,n}}{\alpha_A x_{A,n} + \alpha_B x_{B,n}} \tag{4.18}$$

We have written the VLE relationship to agree with Eq. 4.1; the volatilities are usually defined so that α_B is unity. To solve these difference equations, Smoker recognized a recurrence relation after a transformation of variables as discussed in Exercise 3.8. Underwood was successful with a somewhat less obvious approach; the reader

[13] It is useful to pick a pair of components, usually with adjacent boiling points, and make purity specifications for these components. These are often termed the "key" components and the others are sometimes called "nonkey" components. It is the latter that may have very small compositions in one of the products.

familiar with integrating factor solution of differential or difference equations may be more comfortable with the following development.

First, multiply Eq. 4.17 by $\alpha_A/(\alpha_A - \phi)$, and Eq. 4.18 by $\alpha_B/(\alpha_B - \phi)$, then sum to find

$$\frac{L_T}{V_T}\left[\frac{\alpha_A x_{A,n+1}}{\alpha_A - \phi} + \frac{\alpha_B x_{B,n+1}}{\alpha_B - \phi}\right] + \frac{D}{V_T}\left[\frac{\alpha_A x_{A,D}}{\alpha_A - \phi} + \frac{\alpha_B x_{B,D}}{\alpha_B - \phi}\right] = \frac{\frac{\alpha_A^2}{\alpha_A-\phi}x_{A,n} + \frac{\alpha_B^2}{\alpha_B-\phi}x_{B,n}}{\alpha_A x_{A,n} + \alpha_B x_{B,n}} \tag{4.19}$$

Next, choose ϕ to simplify this expression by making the second term on the left-hand side exactly unity, i.e., *define* ϕ as any solution of

$$V_T = \frac{\alpha_A D x_{A,D}}{\alpha_A - \phi} + \frac{\alpha_B D x_{B,D}}{\alpha_B - \phi} \tag{4.20}$$

which is the same as

$$r + 1 = \frac{\alpha_A x_{A,D}}{\alpha_A - \phi} + \frac{\alpha_B x_{B,D}}{\alpha_B - \phi} \tag{4.21}$$

If the distillate purity is specified and r is known, ϕ can be found from Eq. 4.21.[14] Actually, there are two real solutions which satisfy $\alpha_A > \phi_1 > \alpha_B$ and $\alpha_B > \phi_2 > 0$. Equation 4.19 can be rearranged to find

$$\frac{L_T}{V_T}\left[\frac{\alpha_A}{\alpha_A - \phi}x_{A,n+1} + \frac{\alpha_B}{\alpha_B - \phi}x_{B,n+1}\right] = \phi\left[\frac{\frac{\alpha_A}{\alpha_A-\phi}x_{A,n} + \frac{\alpha_B}{\alpha_B-\phi}x_{B,n}}{\alpha_A x_{A,n} + \alpha_B x_{B,n}}\right] \tag{4.22}$$

which holds for both values of ϕ. It is useful to eliminate L/V by writing Eq. 4.22 once for each root to relate the liquid-phase mole fractions on adjacent stages.

$$\frac{\frac{\alpha_A}{\alpha_A-\phi_1}x_{A,n+1} + \frac{\alpha_B}{\alpha_B-\phi_1}x_{B,n+1}}{\frac{\alpha_A}{\alpha_A-\phi_2}x_{A,n+1} + \frac{\alpha_B}{\alpha_B-\phi_2}x_{B,n+1}} = \left(\frac{\phi_1}{\phi_2}\right)\left[\frac{\frac{\alpha_A}{\alpha_A-\phi_1}x_{A,n} + \frac{\alpha_B}{\alpha_B-\phi_1}x_{B,n}}{\frac{\alpha_A}{\alpha_A-\phi_2}x_{A,n} + \frac{\alpha_B}{\alpha_B-\phi_2}x_{B,n}}\right] \tag{4.23}$$

We can use Eq. 4.23 to relate the compositions on any tray in the rectifying section to those on the tray above, then to those two trays above, etc., from the feed stage up to the top of the column. At the top of the column, both the numerator and the denominator on the left-hand side are equal to $r + 1$ by Eq. 4.21. Therefore,

$$\left(\frac{\phi_1}{\phi_2}\right)^{n_T} = \frac{\frac{\alpha_A}{\alpha_A-\phi_2}x_{A,f} + \frac{\alpha_B}{\alpha_B-\phi_2}x_{B,f}}{\frac{\alpha_A}{\alpha_A-\phi_1}x_{A,f} + \frac{\alpha_B}{\alpha_B-\phi_1}x_{B,f}} \tag{4.24}$$

or

$$n_T = \ln\left(\frac{\frac{\alpha_A}{\alpha_A-\phi_2}x_{A,f} + \frac{\alpha_B}{\alpha_B-\phi_2}x_{B,f}}{\frac{\alpha_A}{\alpha_A-\phi_1}x_{A,f} + \frac{\alpha_B}{\alpha_B-\phi_1}x_{B,f}}\right) \Bigg/ \ln(\phi_1/\phi_2) \tag{4.25}$$

This is equivalent to Smoker's equation, 3.72, applied to the rectifying section. However, unlike Smoker's method, Underwood's approach can be extended to mixtures containing more than two components, as we shall see later in this chapter.

[14] Equation 4.20 is more convenient if the distillate flows of each component are known, e.g., from a specification of fractional recovery.

In the stripping section, a development analogous to the one above gives

$$-V_B = \frac{\alpha_A B x_{A,B}}{\alpha_A - \bar{\phi}} + \frac{\alpha_B B x_{B,B}}{\alpha_B - \bar{\phi}} \tag{4.26}$$

which is the same as

$$-s = \frac{\alpha_A x_{A,B}}{\alpha_A - \bar{\phi}} + \frac{\alpha_B x_{B,B}}{\alpha_B - \bar{\phi}} \tag{4.27}$$

If the compositions of the bottoms and the reboil ratio are known, either of these equations can be solved for the roots $\infty > \bar{\phi}_1 > \alpha_A$ and $\alpha_A > \bar{\phi}_2 > \alpha_B$. The number of stages in the bottom of the column is related to the feed stage composition by

$$n_B = \ln\left(\frac{\frac{\alpha_A}{\alpha_A - \bar{\phi}_1} x_{A,f} + \frac{\alpha_B}{\alpha_B - \bar{\phi}_1} x_{B,f}}{\frac{\alpha_A}{\alpha_A - \bar{\phi}_2} x_{A,f} + \frac{\alpha_B}{\alpha_B - \bar{\phi}_2} x_{B,f}}\right) \Big/ \ln\left(\bar{\phi}_1 / \bar{\phi}_2\right) \tag{4.28}$$

For binary mixtures, the liquid-phase mole fraction on the feed stage can be found from the q-line, but a more general alternative is to select the feed stage composition to minimize the total number of stages in the column, $n_B + n_T - 1$. The latter approach leads to a method that can be used for mixtures with more than two components.

EXAMPLE 4.3. The hexane-xylene separation considered in Chap. 3 can be treated as follows.

- We use $r = 1.2 r_{\min} = 0.477$ and $s = 0.770$ for a feed that is available at 760 mmHg pressure and that contains equal amounts of vapor and liquid. The products are saturated liquids containing 95% and 3% hexane. A constant volatility of $\alpha_A = 7.00$ is accurate at 760 mmHg pressure if we take $\alpha_B = 1.00$ as a reference. The feed contains 55 mol % hexane.
- In the top section, we need the roots of Eq. 4.21,

$$1.477 = \frac{7(0.95)}{7 - \phi} + \frac{1(0.05)}{1 - \phi}$$

which are $\qquad \phi_1 = 2.591 \qquad \phi_2 = 0.872$

The number of theoretical stages in the top of the column is related to the feed stage composition by Eq. 4.25.

$$n_T = \ln\left(\frac{\frac{7}{7-0.872} x_{A,f} + \frac{1}{1-0.872} x_{B,f}}{\frac{7}{7-2.591} x_{A,f} + \frac{1}{1-2.591} x_{B,f}}\right) \Big/ \ln(2.591/0.872)$$

If we assume that the feed is located at the intersection of the operating lines (the q-line), we can solve for $x_{A,f} = 0.345$ and $x_{B,f} = 0.655$. This gives $n_T = 3.40$.
- In the stripping section,

$$-0.770 = \frac{7(0.03)}{7 - \bar{\phi}} + \frac{1(0.97)}{1 - \bar{\phi}}$$

the roots are $\bar{\phi}_1 = 7.340 \quad \bar{\phi}_2 = 2.192$

and
$$n_B = \ln\left(\frac{\frac{7}{7-7.340}x_{A,f} + \frac{1}{1-7.340}x_{B,f}}{\frac{7}{7-2.192}x_{A,f} + \frac{1}{1-2.192}x_{B,f}}\right) \Big/ \ln(7.340/2.192)$$

The feed stage compositions found above give $n_B = 4.17$.

- Instead of finding the optimum feed stage composition from the intersection of the operating lines, we can plot the number of stages in each section and their sum against the feed stage composition, as shown in Fig. 4.7. The minimum is attained at $x_{A,f} = 0.345$ in agreement with the first analysis.

We can also see that the number of stages in the top of the column grows rapidly near $x_{A,f} = 0.29$ because of the pinch between the upper operating line and the y-x curve. Likewise, n_B grows unbounded as $x_{A,f}$ approaches 0.36 and there is a fairly small region of permissible feed stage compositions. As the reflux ratio is decreased toward the minimum, the two pinches merge into one.

The feed pinch always controls the minimum reflux for binary mixtures that can be described by Underwood's method. For such cases, $n_T \to \infty$ and $n_B \to \infty$ simultaneously, and it can be shown from Eqs. 4.25 and 4.28 that $\phi_1 \to \bar{\phi}_2$. Denoting this common root as θ, the addition of Eqs. 4.20 and 4.26 along with an overall mass balance on the column yields the expression

$$\frac{V_T - V_B}{F} = 1 - q = \frac{\alpha_A x_{A,F}}{\alpha_A - \theta} + \frac{\alpha_B x_{B,F}}{\alpha_B - \theta} \tag{4.29}$$

Thus, the value of θ depends only on the feed conditions and the volatilities. Once the value of θ is known, the minimum flows can be found from either Eq. 4.21 or Eq. 4.27, i.e.,

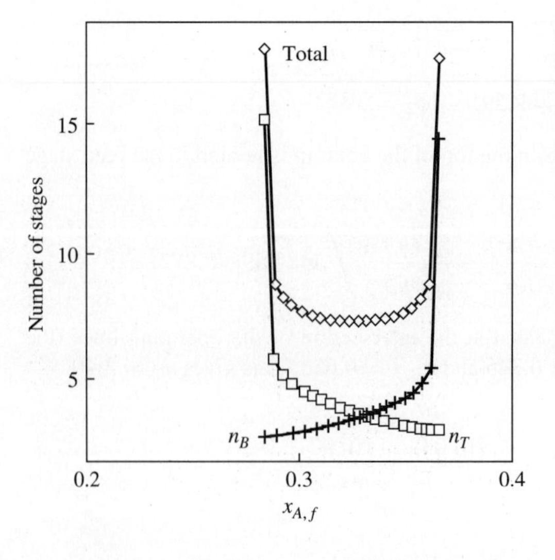

FIGURE 4.7
Optimum feed stage location in a binary mixture according to Underwood's method. The reflux ratio is 0.477.

$$r_{\min} + 1 = \frac{\alpha_A x_{A,D}}{\alpha_A - \theta} + \frac{\alpha_B x_{B,D}}{\alpha_B - \theta} \tag{4.30}$$

or

$$s_{\min} = \frac{\alpha_A x_{A,B}}{\theta - \alpha_A} + \frac{\alpha_B x_{B,B}}{\theta - \alpha_B} \tag{4.31}$$

This sort of analysis is more valuable for the case of multicomponent mixtures, as described below, because it gives an estimate of the product distribution as well as the minimum reflux. Exercises 6 and 7 provide some practice at using Underwood's approach for designing binary distillation columns.

Ternary mixtures

In the rectifying section, the analog of Eq. 4.20 is

$$V_T = \sum_{i=1}^{3} \frac{\alpha_i D x_{i,D}}{\alpha_i - \phi} \tag{4.32}$$

and instead of Eq. 4.21 we find

$$r + 1 = \sum_{i=1}^{3} \frac{\alpha_i x_{i,D}}{\alpha_i - \phi} \tag{4.33}$$

There are now three values for ϕ and for any pair of these, j and k, we can write the ratio as

$$\left(\frac{\phi_k}{\phi_j}\right)^{n_T} = \sum_{i=1}^{3} \frac{\alpha_i x_{i,f}}{\alpha_i - \phi_j} \Bigg/ \sum_{i=1}^{3} \frac{\alpha_i x_{i,f}}{\alpha_i - \phi_k} \tag{4.34}$$

In the stripping section,

$$-V_B = \sum_{i=1}^{3} \frac{\alpha_i B x_{i,B}}{\alpha_i - \bar{\phi}} \tag{4.35}$$

or

$$-s = \sum_{i=1}^{3} \frac{\alpha_i x_{i,B}}{\alpha_i - \bar{\phi}} \tag{4.36}$$

and

$$\left(\frac{\bar{\phi}_k}{\bar{\phi}_j}\right)^{n_B} = \sum_{i=1}^{3} \frac{\alpha_i x_{i,f}}{\alpha_i - \bar{\phi}_k} \Bigg/ \sum_{i=1}^{3} \frac{\alpha_i x_{i,f}}{\alpha_i - \bar{\phi}_j} \tag{4.37}$$

The three roots each for Eqs. 4.33 and 4.36 satisfy the inequalities

$$\alpha_A > \phi_1 > \alpha_B > \phi_2 > \alpha_C > \phi_3 > 0 \tag{4.38}$$

and

$$\infty > \bar{\phi}_1 > \alpha_A > \bar{\phi}_2 > \alpha_B > \bar{\phi}_3 > \alpha_C \tag{4.39}$$

EXAMPLE 4.4. We consider the separation of a ternary mixture of methanol (A), isopropanol (B), and n-propanol (C) where the relative volatilities are considered constant at $\alpha_A = 4.20$, $\alpha_B = 1.86$, and $\alpha_C = 1.00$.

- We use $r = 3.6$ for a saturated liquid feed that contains 20 mol % methanol, 30% isopropanol, and 50% n-propanol (C). The distillate should contain 96% methanol, and there should be 2% methanol in the bottoms. The feed rate is 200 kmol/h. Using the boundary value design method in Sec. 4.2, the minimum reflux is found to be approximately 2.4 for $x_{C,D} = 10^{-9}$ or lower.

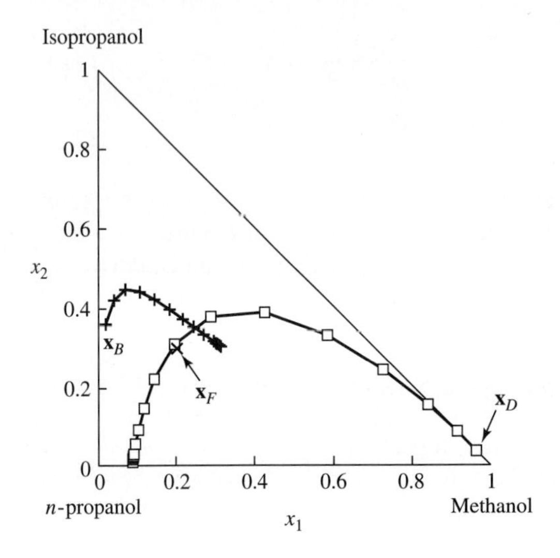

FIGURE 4.8
Composition profiles for methanol, isopropanol, and n-propanol at 1 atm. See the example in the text for the specifications and a comparison with the solution found via Underwood's method.

- It is a simple matter to find the distillate and bottoms flows from Eqs. 4.6; $D = 38.4$ and $B = 161.6$ kmol/h. The heats of vaporization are 35,250, 39,830, and 41,760 J/mol for the three components; the constant molar overflow approximation will be employed for this example. Equation 4.10 is used to find the reboil ratio $s = (r+1)(D/B) = 1.09$.
- In the top section, the roots of interest are those that satisfy Eq. 4.33:

$$4.60 = \frac{4.20(0.96)}{4.20 - \phi} + \frac{1.86 x_{B,D}}{1.86 - \phi} + \frac{x_{C,D}}{1 - \phi}$$

  The choice of r, $x_{A,D}$, and $x_{A,B}$ leaves one degree of freedom. In contrast to the binary case, it is not convenient to choose the location of the feed, and instead we specify a value for $x_{C,D}$. For example, if $x_{C,D} = 0.001$ we find $\phi_1 = 3.332$, $\phi_2 = 1.835$, and $\phi_3 = 0.99969$.[15]
- The composition of the bottoms stream is now completely specified with mole fractions of (0.020, 0.362, 0.618) and Eq. 4.36 can be solved to find the roots $\bar{\phi}_1 = 4.333$, $\bar{\phi}_2 = 2.716$, and $\bar{\phi}_3 = 1.230$.
- To complete the example, the number of stages in each column section and the composition of the liquid leaving the feed stage can be found by simultaneous solution of the four expressions obtained from Eqs. 4.34 and 4.37, along with Eq. 4.2. We find $n_T = 6.44$, $n_B = 6.97$, and a feed stage composition $\mathbf{x}_f = (0.244, 0.354, 0.402)$. These values should be compared to the graphical solution shown in Fig. 4.8.
- Of course, we could have chosen a different distillate mole fraction of n-propanol. Figure 4.9 shows solutions as this composition is varied. (This requires the repetitive solution of the example above, but that is not difficult to automate.) These results are for a fixed reflux ratio of 3.6, which is 50% larger than the smallest possible value of 2.4 that is found at very low compositions of heavy component in the distillate.

---

[15]The last root is near the volatility of n-propanol, and more significant figures are required than for the other roots. The sensitivity of this calculation can make it a challenging problem to find accurate solutions, especially when computing with a small number of significant figures. Of course, the same situation is encountered in computing composition profiles for high-purity products. The sensitivity is not introduced by the mathematical model but comes about because distillation can actually produce very small compositions for certain components relative to others.

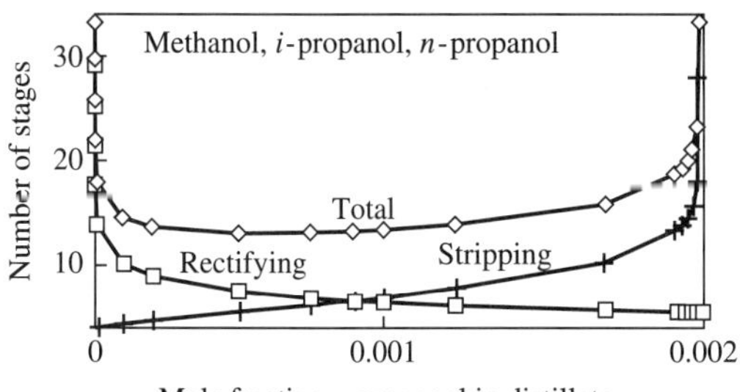

FIGURE 4.9
Effect of nonkey product purity on the stage requirements for a ternary mixture.

As the distillate mole fraction of n-propanol increases, the rectifying profile does not approach the saddle pinch closely so the number of stages in the top of the column drops sharply. When the distillate contains slightly more than 0.2% n-propanol, the number of stages in the bottom of the column increases sharply due to the node pinch in the stripping profile.[16]

Underwood's approach can be used to estimate the minimum reflux in multicomponent mixtures when the volatilities are approximately constant. At the minimum reflux or reboil conditions, the number of stages is indefinitely large for at least one of the compositions profiles. This is at the stripping node for the direct split or the rectifying node for the indirect split; there may also be saddle pinches before these nodes, as shown in Fig. 4.5*a* and *b* for a ternary mixture. Underwood proved that at least one root of Eq. 4.32 must be equal numerically to one root of Eq. 4.35 for conditions of minimum reflux. This fact can be used to develop a simple method to calculate the minimum flows as follows.

Although perhaps not obvious, it is simplest to assume that *all* of the roots for Eqs. 4.32 and 4.35 are equal and then later to discard those which are not. If we label these roots θ, then add Eqs. 4.32 and 4.35 and use the overall mass balance for each component, we find an expression for the roots that involves only the feed characteristics and the volatilities

$$\frac{V_T - V_B}{F} = 1 - q = \frac{\alpha_A z_{A,F}}{\alpha_A - \theta} + \frac{\alpha_B z_{B,F}}{\alpha_B - \theta} + \frac{\alpha_C z_{C,F}}{\alpha_C - \theta} \tag{4.40}$$

The roots of interest are the two between each pair of volatilities, i.e., $\alpha_A > \theta_1 > \alpha_B$ and $\alpha_B > \theta_2 > \alpha_C$. For a saturated liquid feed, these can be found by solving the quadratic equation that results from clearing the denominators on the right-hand side of Eq. 4.40 or, for general values of q, by solving the cubic equation. For any q (and for any number of components) this is a simple one-dimensional root finding problem to solve for the values of θ.

Once the roots are found, we seek to satisfy a system of equations obtained by writing Eq. 4.33 (or, alternatively, 4.36) once for each root, along with the summation equation for the mole fractions. For ternary mixtures, these are

$$r_{\min} + 1 = \frac{\alpha_A x_{A,D}}{\alpha_A - \theta_1} + \frac{\alpha_B x_{B,D}}{\alpha_B - \theta_1} + \frac{\alpha_C x_{C,D}}{\alpha_C - \theta_1} \tag{4.41}$$

[16]For $r = 3.6$, this node pinch lands on the rectifying profile. Solve Exercise 1 to see this effect in another mixture. Most mixtures behave this way.

$$r_{\min} + 1 = \frac{\alpha_A x_{A,D}}{\alpha_A - \theta_2} + \frac{\alpha_B x_{B,D}}{\alpha_B - \theta_2} + \frac{\alpha_C x_{C,D}}{\alpha_C - \theta_2} \tag{4.42}$$

$$1 = x_{A,D} + x_{B,D} + x_{C,D} \tag{4.43}$$

For example, if $x_{A,D}$ is specified we have three equations in three unknowns: $r_{\min}$, $x_{B,D}$, and $x_{C,D}$.

Finally, we need to check the assumption that the roots of Eqs. 4.32 and 4.35 are equal. This is accomplished by completing the mass balances for the bottoms compositions, then checking that all of the product compositions and fractional recoveries are feasible. Frequently, for high-purity splits, these conditions will not all be satisfied. For instance, if we specify a large value of $x_{A,D}$ ("large" depends on the mixture), then one or more of the heavier components may have an indefinitely small mole fraction in the distillate at the minimum reflux ratio—such components are termed "nondistributing."[17] For such a component, we set the distillate mole fraction equal to zero and discard the root between its volatility and the next lightest component. We also eliminate the equation corresponding to the discarded root, in this case Eq. 4.42 above. This leaves $c - 1$ instead of c equations to solve for the minimum reflux and the product compositions. For such a case, we have a "sharp" *direct* split, denoted as A/BC. The minimum reflux is given by the solution of Eqs. 4.41 and 4.43.

$$r_{\min} = \frac{\alpha_A x_{A,D}}{\alpha_A - \theta_1} + \frac{\alpha_B(1 - x_{A,D})}{\alpha_B - \theta_1} - 1 \tag{4.44}$$

For *indirect* splits, the composition of the heaviest component in the bottoms is large, and the lightest component may be "nondistributing." For indirect splits it is more convenient to write Eq. 4.36 instead of Eqs. 4.41, 4.42, and 4.43 to find the minimum reboil ratio.

$$-s_{\min} = \frac{\alpha_A x_{A,B}}{\alpha_A - \theta_1} + \frac{\alpha_B x_{B,B}}{\alpha_B - \theta_1} + \frac{\alpha_C x_{C,B}}{\alpha_C - \theta_1} \tag{4.45}$$

$$-s_{\min} = \frac{\alpha_A x_{A,B}}{\alpha_A - \theta_2} + \frac{\alpha_B x_{B,B}}{\alpha_B - \theta_2} + \frac{\alpha_C x_{C,B}}{\alpha_C - \theta_2} \tag{4.46}$$

$$1 = x_{A,B} + x_{B,B} + x_{C,B} \tag{4.47}$$

and we set $x_{A,B} = 0$ and drop Eq. 4.45 when A does not distribute. You may want to find an expression for $s_{\min}$ analogous to Eq. 4.44 for this case. (It's easy!)

EXAMPLE 4.5. Minimum reflux. We consider a ternary mixture of methanol (A), isopropanol (B), and n-propanol (C) for which the relative volatilities are approximately constant at $\alpha_A = 4.20$, $\alpha_B = 1.86$, and $\alpha_C = 1.00$. If the feed is a saturated liquid with mole fractions (0.3, 0.3, 0.4), Eq. 4.40 becomes

$$0 = \frac{4.20(0.3)}{4.20 - \theta} + \frac{1.86(0.3)}{1.86 - \theta} + \frac{1.00(0.4)}{1.00 - \theta}$$

[17] The minimum reflux and therefore the "nondistributing" components are a limiting case. While Underwood's method can find precisely zero product compositions, these cannot be depicted *exactly* in the composition profiles.

This equation has roots $\theta_1 = 2.74067$ and $\theta_2 = 1.28512$.[18] The system of Eqs. 4.41 to 4.43 becomes

$$r_{\min} + 1 = 2.88x_{A,D} - 2.11x_{B,D} - 0.575x_{C,D}$$
$$r_{\min} + 1 = 1.44x_{A,D} + 3.24x_{B,D} - 3.51x_{C,D}$$
$$1 = x_{A,D} + x_{B,D} + x_{C,D}$$

For $x_{A,D} = 0.99$, we find $r_{\min} = 1.58$, $x_{B,D} = 0.176$, and $x_{C,D} = -0.166$, which is infeasible. This means that n-propanol is nondistributing so $x_{C,D} = 0$. Therefore, $x_{B,D} = 0.01$ and Eq. 4.44 gives $r_{\min} = 1.83$, which compares quite closely with the results from composition profiles computed for very small but nonzero values of $x_{C,D}$. (Do it yourself and see.)

The bottoms compositions have not been calculated, but these are simple to find and will change only the reboil ratio and the distillate and bottoms flow rates, but not the minimum reflux ratio.

Four or more components

Underwood's method can be extended to mixtures with any number of components, provided that constant values of the volatilities are a reasonable approximation of the VLE and that the constant molar overflow approximation is adequate.

The rectifying equations are

$$V_T = \sum_{i=1}^{c} \frac{\alpha_i D x_{i,D}}{\alpha_i - \phi} \tag{4.48}$$

or

$$r + 1 = \sum_{i=1}^{c} \frac{\alpha_i x_{i,D}}{\alpha_i - \phi} \tag{4.49}$$

and

$$\left(\frac{\phi_k}{\phi_j}\right)^{n_T} = \sum_{i=1}^{c} \frac{\alpha_i x_{i,f}}{\alpha_i - \phi_j} \Bigg/ \sum_{i=1}^{c} \frac{\alpha_i x_{i,f}}{\alpha_i - \phi_k} \tag{4.50}$$

In the stripping section,

$$-V_B = \sum_{i=1}^{c} \frac{\alpha_i B x_{i,B}}{\alpha_i - \bar{\phi}} \tag{4.51}$$

or

$$-s = \sum_{i=1}^{c} \frac{\alpha_i x_{i,B}}{\alpha_i - \bar{\phi}} \tag{4.52}$$

and

$$\left(\frac{\bar{\phi}_k}{\bar{\phi}_j}\right)^{n_B} = \sum_{i=1}^{c} \frac{\alpha_i x_{i,f}}{\alpha_i - \bar{\phi}_j} \Bigg/ \sum_{i=1}^{c} \frac{\alpha_i x_{i,f}}{\alpha_i - \bar{\phi}_k} \tag{4.53}$$

There are c roots each for Eqs. 4.49 and 4.52, and these roots satisfy the inequalities

[18] *Mathematica* or a similar tool is **very** handy for solving these problems.

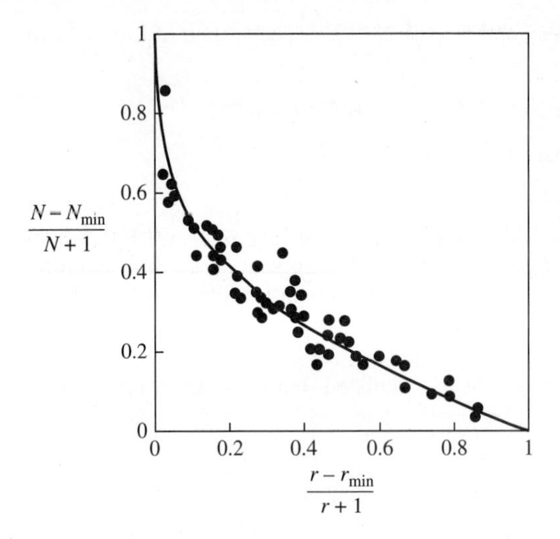

FIGURE 4.10
Gilliland correlation for the number of theoretical stages. Note that $r = L/D$, the subscript min has the obvious meaning, and N is the number of theoretical stages. The discrete points are the result of calculations made for 16 different cases and some variations on each case (Gilliland, 1940).

$$\alpha_A > \phi_1 > \alpha_B > \cdots > \alpha_c > \phi_c > 0 \tag{4.54}$$

and

$$\infty > \bar{\phi}_1 > \alpha_A > \bar{\phi}_2 > \alpha_B > \cdots > \bar{\phi}_c > \alpha_c \tag{4.55}$$

The implementation of the design procedure and minimum flow calculations is carried out in an analogous fashion to ternary mixtures.

Fenske-Underwood-Gilliland design method

After the minimum reflux ratio is estimated, a somewhat larger value can be chosen and the corresponding number of theoretical stages required for a design can be estimated. For ternary mixtures, it is possible to compute the profiles for the desired compositions and the specified reflux ratio. A more approximate method, which can also be used for mixtures containing more than three components, is based on the correlation reported by Gilliland and co-workers in 1940. (Robinson and Gilliland, 1950, pp. 347–348; Gilliland, 1940).

This correlation is based on the results of numerous calculations[19] for relatively ideal mixtures for which a relation between the number of theoretical stages and the quantities r, $r_{\min}$, and $N_{\min}$ was developed. Figure 4.10 shows the correlation and the data used in the original development. Several forms have been suggested for an equation to describe the Gilliland correlation, and one of the more common is due to Eduljee (1975).

$$\frac{N - N_{\min}}{N+1} = 0.75\left[1 - \left(\frac{r - r_{\min}}{r+1}\right)^{0.5688}\right] \tag{4.56}$$

There have also been other studies offering a marginal improvement over the original work of Gilliland, e.g., Erbar and Maddox (1961). However, the accuracy in all of these approaches is limited because other factors such as the feed quality, composi-

[19] These were quite tedious in the late 1930s!

tion, and the effects of variable volatility are not included. The data shown in Fig. 4.10 give an idea of the uncertainty involved in using the correlation.

The approximate number of theoretical stages can be found from Gilliland's correlation once values are known for N_{min}, r_{min}, and r. If the mixture can be described with constant volatilities, Fenske's equation (4.15 or 4.16) can be used to estimate N_{min}, Underwood's method gives r_{min}; as usual, r can be chosen as 1.2 to 1.5 times the minimum value or an optimization can be done. For obvious reasons, these are called the "Fenske-Underwood-Gilliland" or sometimes the "Fenske-Underwood-Erbar-Maddox" method.

Notes

1. This approach is limited to cases where the constant volatility and constant molar overflow approximations are accurate.
2. The feed stage location cannot be found from the method above but can be *estimated* using the approximation that the feed stage should be located approximately at the intersection of the operating lines on a binary diagram prepared for the "key" components, e.g., Treybal (1980, p. 482).[20]
3. In principle, Underwood's full set of equations as described above can be solved at the desired value of r to find an exact solution for the number of theoretical stages in each section of the column, and the product distributions. This would eliminate the need to use the Gilliland or the Erbar and Maddox correlations and would also give the location of the feed stage. However, this is quite complicated and frequently unnecessary because the Fenske-Underwood-Gilliland method gives a good approximation for the total number of stages and a performance calculation can be used to determine the final details of feed stage location and other variables. (In our opinion, this is not the best solution but is generally more practical than solving the full set of Underwood's equations because it is much less complex and gives sufficiently accurate results for most purposes.)
4. Approximate expressions for the minimum reflux were developed by Glinos and Malone (1984) for various types of splits as shown in Table 4.1. For mixtures containing more than four components, a "lumping" can be used to define a pseudo-four-component mixture with the rules (Glinos, 1984):

   a. Lump the components that are farthest from the keys, e.g., lump D with E for the split $A/BCDE$.

   b. Use the sum of the feed mole fractions of the lumped components for the feed mole fraction of the pseudo-component.

   c. Use the feed mole fraction average of the actual volatilities to find volatilities in the pseudo-mixture.

   d. When the pseudo-component includes the heaviest actual component, evaluate the other volatilities with respect to the new volatility for the heavy pseudo-component.

---

[20]The "key" components are A (light key) and B (heavy key) for the A/BC split or B (light key) and C (heavy key) for the AB/C split, etc.

TABLE 4.1
Approximate expressions for the minimum reflux ratio. A saturated liquid feed has been assumed and the volatilities are with reference to the heaviest component, and are taken to be constant. All mole fractions refer to the feed composition

Split	Minimum reflux ratio
A/BC^a	$\dfrac{\alpha_B(x_A+x_B)}{fx_A(\alpha_A-\alpha_B)}+\dfrac{x_C}{fx_A(\alpha_A-1)}$
AB/BC^b	$\dfrac{1}{\alpha_A x_A+\alpha_B x_B+x_C-1}$
AB/C	$\dfrac{(x_B+x_C)/(\alpha_B-1)+x_A/(\alpha_A-1)}{(x_A+x_B)(1+x_A x_C)}$
A/BCD	$\dfrac{\alpha_B(x_A+x_B)}{x_A(\alpha_A-\alpha_B)}+\dfrac{\alpha_C x_C}{x_A(\alpha_A-\alpha_C)}+\dfrac{x_D}{(\alpha_A-1)}$
AB/CD	$\dfrac{\alpha_C x_A/(\alpha_A-\alpha_C)+\alpha_C(x_B+x_C)/(\alpha_B-\alpha_C)}{(x_A+x_B)[1+x_A(x_C+x_D)]}+x_D\dfrac{x_A/(\alpha_A-1)+x_B/(\alpha_B-1)}{(x_A+x_B)^2}$
ABC/D	$\dfrac{x_A/(\alpha_A-1)+x_B/(\alpha_B-1)+(x_C+x_D)/(\alpha_C-1)}{(1-x_D)[x_D(x_A+x_B)]}$
AB/BCD	$\dfrac{\alpha_C(1-x_D)+x_D/\alpha_C}{\alpha_A x_A+\alpha_B x_B+\alpha_C(x_C+x_D)-\alpha_C}$
ABC/CD	$\dfrac{1-x_A(1-\alpha_B/\alpha_A)}{\alpha_B(x_A+x_B)+\alpha_C x_C+x_D-1+\alpha_B x_A/\alpha_A}$
ABC/BCD	$\dfrac{1}{\alpha_A x_A+\alpha_B x_B+\alpha_C x_C+x_D-1}$

[a] Here, $f=1+0.01x_B$.
[b] This is an exact result for the "sharp" *transition split,* and the distribution of the middle component satisfies $f_{B,D}/f_{B,B}=(\alpha_A-1)/(\alpha_B-1)$, where $f_{B,j}$ is the fractional recovery of component B in stream j.

4.4 GENERAL APPROACH FOR NONIDEAL TERNARY MIXTURES

Feasibility and Product Distribution

Some product purities cannot be attained in distillation. For instance, it is probably intuitive that the middle-boiling component cannot be isolated as a pure distillate or bottoms product from a two-product column. A proof of this for ideal mixtures hardly seems necessary, but the general idea is critical for azeotropic systems. However, because it is simpler to understand and to establish some points of comparison, we briefly describe the limits of distillation for nonazeotropic mixtures.

Along with overall mass balances, an important limit can be found from the behavior of the pinch points. We can use the composition profiles as described above, at least for ternary mixtures, to locate pinch points, though it is easy to miss the saddle pinch altogether. The pinches can also be determined as solutions of the operating relationships in Eqs. 4.11 and 4.12 along with the necessary material and energy balances. At a pinch, the compositions and temperatures are independent of the stage number and the following *pinch equations* can be found from Eqs. 4.11 and 4.12:

$$(r+1)y_{i,p} - rx_{i,p} - z_{i,D} = 0 \tag{4.57}$$

and

$$(s+1)x_{i,p} - sy_{i,p} - z_{i,B} = 0 \tag{4.58}$$

These balance equations are to be solved along with the stipulation that the vapor- and liquid-phase compositions at the pinch point(s) are in equilibrium, i.e., Eqs. 4.1 or 4.3, to find the pinch compositions $\mathbf{x}_p$ and $\mathbf{y}_p$. The pinches will be indicated by a subscript "p" or by a "^" as convenient.

In the general case, it is necessary to find the pinch compositions by a numerical solution for the roots of Eqs. 4.57 (or 4.58) due to the nonlinear VLE relationship. However, it is fairly simple to show that if there is none of the heaviest component in the distillate, i.e., $z_{3,D} = 0$, one of the pinch compositions will have $y_{3,p} = x_{3,p} = 0$. (See Exercise 8.) In this case, a pinch is located on the hypotenuse of the composition triangle at the same point as a binary 1-2 pinch, and the methods from Chap. 3 can be used to locate it. When the distillate contains even an infinitesimal amount of the heaviest component, the saddle pinch location is slightly off of the hypotenuse, but the difference from the binary value is small for high-purity splits.

At total reflux, rectifying profiles for an ideal mixture of hexane, heptane, and nonane are shown in Fig. 4.11. The unstable node is at the hexane vertex and the stable node is at nonane. The middle boiler, heptane, is a saddle. Figure 4.12 shows

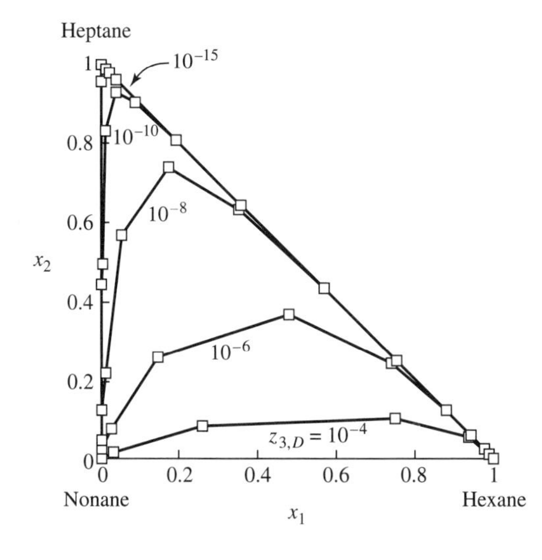

FIGURE 4.11
Rectifying profiles at total reflux. The components are the normal alkanes hexane, heptane, and nonane which have relative volatilities of 12.67, 5.35, and 1.00 at atmospheric pressure; the distillate composition is $z_{1,D} = 0.99$. Each vertex is a pinch point; the saddle is at pure heptane, while nodes are located at hexane (unstable) and nonane (stable).

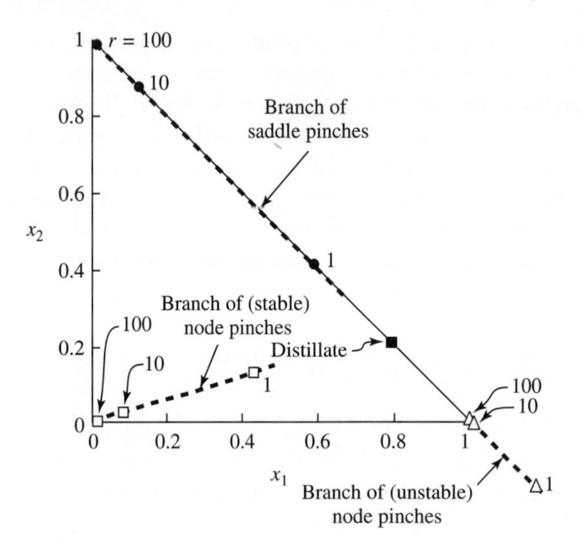

FIGURE 4.12
Pinch points in the liquid-phase rectifying profiles for a mixture of hexane, heptane, and nonane; we use $\alpha_{13} = 12.67$, $\alpha_{23} = 5.35$. At $r = 1{,}000$, there is a pinch point very close to each pure component vertex as shown in Fig. 4.11. Here, the movement of the pinch points away from the vertices is shown for $r = 100$, 10, and 1 as calculated from the pinch equations for a distillate composition of $(0.8, 0.2 - 1 \times 10^{-8}, 1 \times 10^{-8})$. The unstable node (△) gradually moves away from pure hexane and out of the composition triangle, while the stable node (□) is located inside, moving from the pure nonane vertex toward the distillate composition. The saddle pinch (•) remains very close to the hexane-heptane face.

the movement of the pinches with the reflux ratio for the mixture described in Fig. 4.11. For $r = 100$, the pinch compositions are close to the pure components. The saddle moves from pure heptane at total reflux toward pure hexane as the reflux is decreased, and it is located close to the line $x_3 = 0$. The (stable) node pinch at the nonane vertex moves inside the triangle and would eventually approach the distillate composition at low reflux ratios. The (unstable) node at the hexane vertex moves outside the composition triangle. The position of these pinches is important since they affect the shape of the rectifying profile.

A similar behavior is found in the stripping profiles. In that case, the saddle moves along the 2-3 face of the triangle, while the stability of the nodes that correspond to pure light and heavy components at total reflux is reversed.

To find the limits of distillation, we examine two cases (Wahnschafft et al., 1992; Fidkowski et al., 1993). The first is the locus of points where the vapor node pinch in the stripping profile coincides with the distillate composition (see Fig. 4.14), and the second is where the liquid node pinch in the rectifying profile coincides with the bottoms. In other words, the limiting cases of a single column section set limits on the product compositions. These limits are simple to find from the pinch equations 4.57 and 4.58, although it is more convenient to write them in a slightly different form.

In the rectifying section, Eqs. 4.57 can be solved for the reflux ratio to find

$$r+1=\frac{z_{i,D}-x_{i,p}}{y_{i,p}-x_{i,p}} \tag{4.59}$$

which must be true for every component in the mixture. Consequently, we also have

$$\frac{z_{2,D}-x_{2,p}}{z_{1,D}-x_{1,p}}=\frac{y_{2,p}-x_{2,p}}{y_{1,p}-x_{1,p}} \tag{4.60}$$

This means that the slope of a line joining the distillate composition and the liquid-phase pinch composition must be equal to the slope of a tie-line through the pinch compositions. In other words, the points $\mathbf{y}_p$, $\mathbf{x}_p$, and $\mathbf{z}_D$ are collinear, as shown in Fig. 4.13.

An analogous development holds for the stripping profile, and we find that the reboil ratio is given by

$$s=\frac{z_{i,B}-x_{i,p}}{x_{i,p}-y_{i,p}} \tag{4.61}$$

which must also be true for each component i. The bottoms composition must be collinear with both the liquid and vapor compositions at the pinch, i.e.,

$$\frac{z_{2,B}-x_{2,p}}{z_{1,B}-x_{1,p}}=\frac{y_{2,p}-x_{2,p}}{y_{1,p}-x_{1,p}} \tag{4.62}$$

In the limiting case where the vapor node pinch in the stripping profile coincides with the distillate composition, the following points must lie on the same straight line: $\mathbf{x}_D$, $\mathbf{y}_p$, $\mathbf{x}_p$, and $\mathbf{x}_B$. Therefore, $\mathbf{x}_F$ also lies on this line, by the lever rule (see Fig. 4.14). Equations 4.61, 4.62, and 4.13 can be used to find

$$s=\frac{x_{1,D}-x_{1,B}}{y_{1,D}-x_{1,D}} \tag{4.63}$$

and

$$\frac{y_{2,p}-x_{2,p}}{y_{1,p}-x_{1,p}}=\frac{x_{2,p}-x_{2,F}}{x_{1,p}-x_{1,F}} \tag{4.64}$$

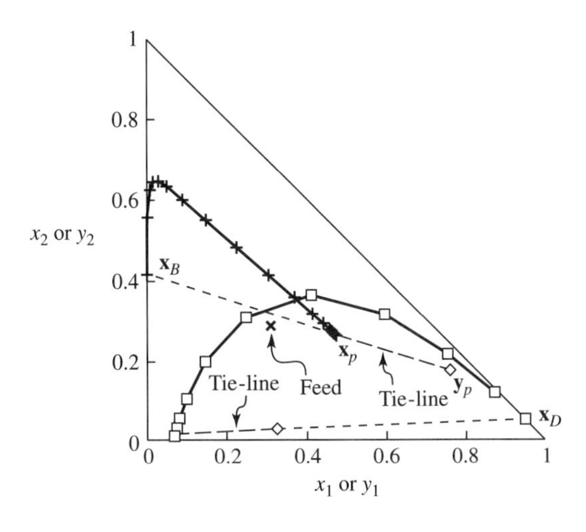

FIGURE 4.13
Collinearity of the pinch and product compositions. The distillate composition is collinear with a tie-line at the node pinch in the rectifying profile according to Eq. 4.60. A similar relation holds for the node pinch in the stripping section and the bottoms composition in agreement with Eq. 4.62. This is the same mixture of normal alkanes that is described in Fig. 4.3a. Note that the symbol ◇ marks the vapor composition.

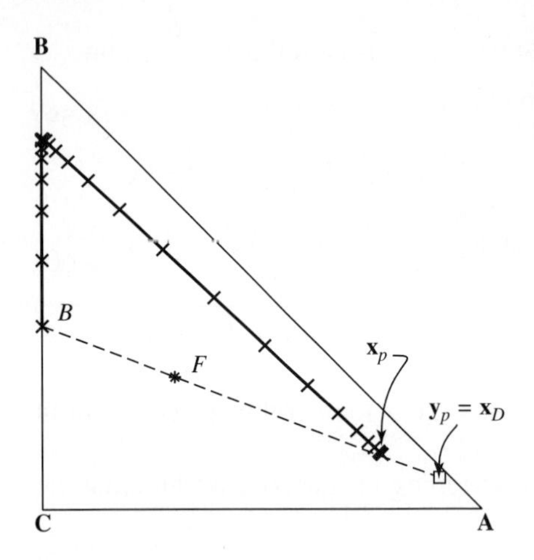

FIGURE 4.14
Stripping profile showing collinearity of the tie-line at a pinch with the product and feed compositions.

Equation 4.64 along with 4.1 and 4.2 is solved for $y_{1,p}$, $y_{2,p}$, and $x_{2,p}$ as a function of $x_{1,p}$. *The limiting distillate compositions are given by the locus of vapor pinch points. These values are independent of the bottoms composition, which affects only the reboil ratio in Eq. 4.63.* Similar expressions for the reflux ratio and the bottoms composition of the second component can be found from Eqs. 4.59, 4.60, 4.1, and 4.2 as

$$r + 1 = \frac{x_{1,D} - x_{1,B}}{y_{1,B} - x_{1,B}} \tag{4.65}$$

and

$$\frac{y_{2,p} - x_{2,p}}{y_{1,p} - x_{1,p}} = \frac{x_{2,F} - x_{2,p}}{x_{1,F} - x_{1,p}} \tag{4.66}$$

Instead of tracking all of the solutions of Eqs. 4.64 and 4.66 a trial-and-error approach using composition profiles can be used to find these limits for product compositions. This is discussed in Exercise 1; that approach is more practical when one of the desired product purities is known, and the limits for the other compositions are sought.

The curves defined by Eqs. 4.64 and 4.66 are shown in Fig. 4.15*a* for a mixture of normal alkanes. Product purities below these curves cannot be accomplished by distillation in a single feed column. For a saturated liquid feed, it can also be shown that the two curves meet at the feed composition $\mathbf{x}_F$, that the tangent to the liquid pinch curve at this point is a tie-line through the feed composition, i.e., through the point $\mathbf{y}_F$, which is the composition of a vapor in equilibrium with the liquid feed. This tie-line is simple to compute and is also shown in the figure. The construction of this tie-line and its extension to the binary edges identifies the product compositions for which the composition profiles in *each* section have a node pinch at the feed composition at the minimum reflux; cf. Fig. 4.5*c*. This is the special case of a transition split, and we will refer to the result of this construction as the *transition line*.

In Fig. 4.15*a*, the shaded regions correspond to compositions that are feasible according to the material balance constraints. Product compositions in either one of

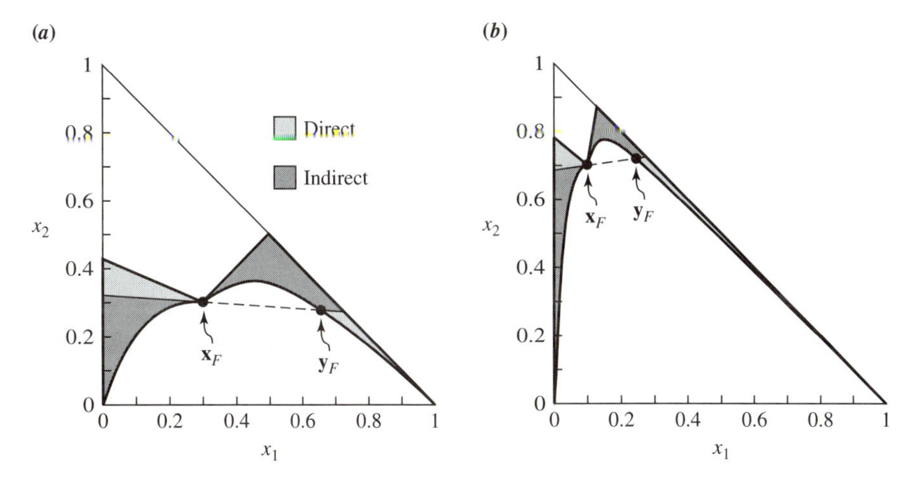

FIGURE 4.15
Feasible regions for product compositions in ternary mixtures. The lower curves are determined from the locus of node pinch compositions in the rectifying (left) and stripping (right) profiles. If the product compositions were chosen on this curve, a single column section could accomplish the separation at minimum flows. The tie-line through the feed composition is tangent to the boundary and marks the transition between the direct and the indirect splits. The volatilities and feed compositions are (*a*) (12.67, 5.35, 1.00) and (0.3, 0.3, 0.4), respectively and (*b*) (12.67, 5.35, 1.00) and (0.1, 0.7, 0.2), respectively.

the two similarly shaded regions in the figure correspond to the direct or indirect splits. Transition splits occur for product compositions on the transition line, with the "sharp" transition split, AB/BC, corresponding to the endpoints of the transition line. The vapor rates increase rapidly as the product composition moves away from the boundary defined by the transition line toward either the lower portion of the hypotenuse or the ordinate in the figure. The lowest possible minimum reflux (or minimum reboil) ratio for any given product compositions of the lightest component are found close to the transition line. Figure 4.15*b* shows the shapes of these feasible regions at another feed composition.

Equations 4.63 to 4.66 are independent of the VLE model, and the same approach can be applied for nonideal mixtures. However, when mixtures have tangent pinches or azeotropes, the pinch tracking is more complicated and distillation boundaries further complicate the picture. In Sec. 4.6 we discuss tangent pinches, and Chap. 5 gives a more detailed analysis of azeotropic mixtures.

Minimum Flows

A useful measure for comparing alternative distillation systems and for setting the vapor rate inside a column is the minimum value for the vapor-to-feed ratio $(V/F)_{\min}$, where V is the vapor rate at the end of the column with the most expensive utility costs. This is normally the vapor rate leaving the reboiler. However, in cryogenic and other low-temperature distillations that require refrigerated condensers V represents the vapor rate entering the condenser. As a rule, values of $(V/F)_{\text{min}}$ less than 0.5 are considered low (attractive) while values greater than 3 are high. The minimum value for V/F is normally calculated from the minimum reflux or minimum reboil ratio using the relations

$$\left(\frac{V_T}{F}\right)_{\text{min}} = (r_{\text{min}} + 1)\frac{D}{F} \tag{4.67}$$

$$\left(\frac{V_B}{F}\right)_{\text{min}} = s_{\text{min}}\frac{B}{F} \tag{4.68}$$

When the constant molar overflow assumption applies, the ratio $(V_B/F)_{\text{min}}$ is related to r_{min} as follows:

$$\left(\frac{V_B}{F}\right)_{\text{min}} = (r_{\text{min}} + q)\frac{D}{F} + (q - 1)\frac{B}{F} \tag{4.69}$$

The ratios D/F and B/F are known functions of the specified product compositions.

For binary mixtures, three of the four degrees of freedom for a design can be taken as $x_{D,1}$, $x_{B,1}$, and r. The composition changes can be represented on a McCabe-Thiele diagram, which is shown in Fig. 4.16 for the separation of hexane (1) and heptane (2).

The composition profiles are also shown in Fig. 4.16 by projection of the liquid mole fractions onto the x_1 axis. The profiles start at product compositions and end at the fixed points (pinches) where the operating lines intersect the equilibrium curve. There are at least two fixed points in each profile, one a stable node and the other an unstable node. For $r < r_{\text{min}}$, the profiles do not intersect. At $r = r_{\text{min}}$, the stable nodes in both profiles occur at the same point. For a saturated liquid feed, this happens exactly at the feed composition. For other values of q the common pinch point can be calculated easily. When $r > r_{\text{min}}$, the profiles overlap and we can choose an optimal feed stage location; this is the fourth degree of freedom.

If we define the *fixed point distance* between the node pinch in the rectifying section and the node pinch in the stripping section as e_1, it follows that on a graph of e_1 versus $r/(r+1)$ minimum reflux corresponds to the point where the curve goes through zero (see Fig. 4.17). For this example, $r_{\text{min}} = 0.968$, which gives $(V_B/F)_{\text{min}} = 1.084$ from Eq. 4.69.

The utility of this approach is that it generalizes to mixtures with more components. Liquid composition profiles for the separation of benzene from a mixture of benzene (1), toluene (2), and xylene (3) (a direct split) are shown in Fig. 4.18. The profiles start at the product compositions and end at the stable nodes; an additional (saddle) fixed point is present close to the binary edge in each profile. The saddles correspond to points where one of the components almost disappears from the column section, i.e., becomes present only in trace amounts. There is no analog to this in binary mixtures. For $r < r_{\text{min}}$ the profiles do not intersect and the separation cannot be achieved even with an infinite number of stages. For $r > r_{\text{min}}$ the profiles intersect transversely and the separation can be achieved with a finite number of stages. At $r = r_{\text{min}}$, the stripping profile ends (pinches) somewhere on the rectifying profile

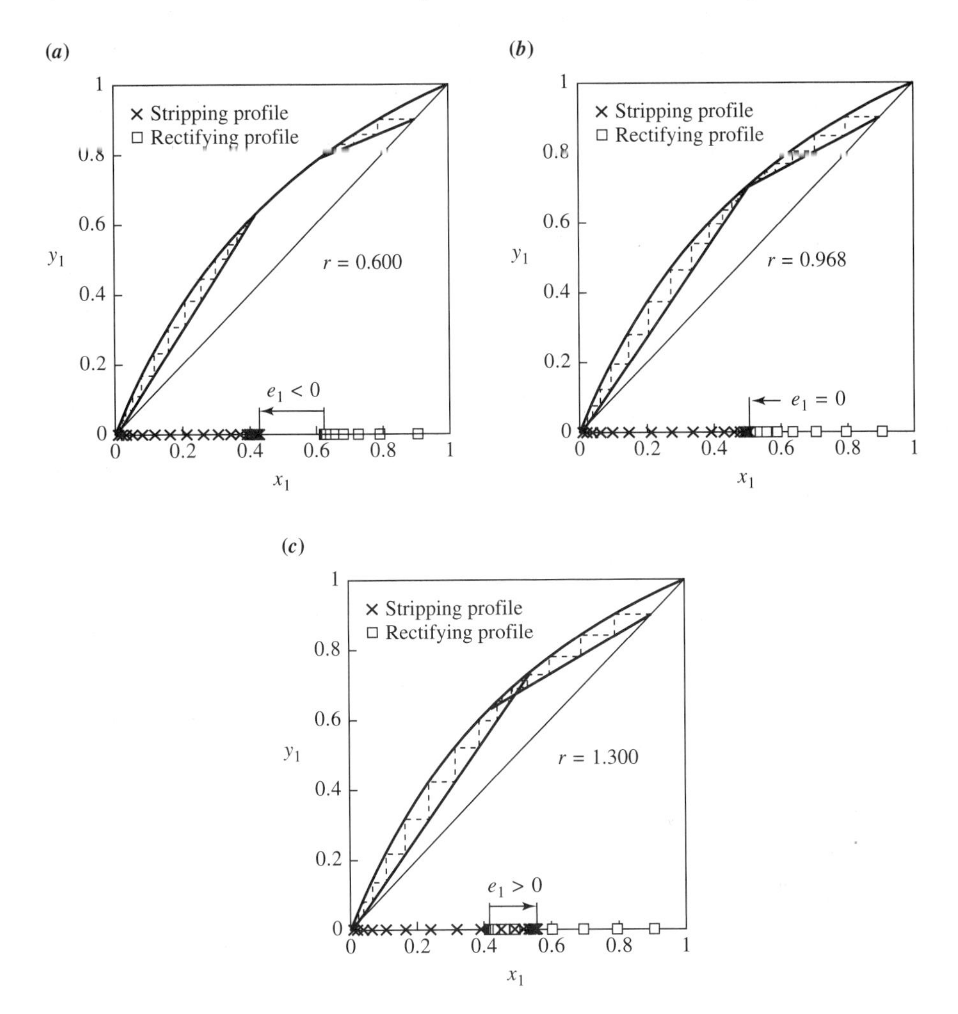

FIGURE 4.16
McCabe–Thiele diagram for a binary separation of hexane (1) and heptane (2) at 1 atm pressure with $x_{F,1} = 0.5$, $x_{D,1} = 0.9$, $x_{B,1} = 0.01$, and $q = 1$. Vapor-liquid equilibrium was calculated with a constant volatility of 2.37: (*a*) $r < r_{\min}$, (*b*) $r = r_{\min}$, (*c*) $r > r_{\min}$.

and the separation can just be accomplished with an infinite number of stages. *Two* pinch zones control the minimum reflux. For a direct split (Fig. 4.18*b*) one occurs in the stripping section below the feed stage (stripping node), and the other occurs in the rectifying section several stages above the feed stage (rectifying saddle).

For the distillation of constant volatility mixtures in columns with constant molar flows and a saturated liquid feed, the stripping node, the rectifying saddle, the rectifying node, and the feed composition (the points $\hat{\mathbf{x}}^{1,s}$, $\hat{\mathbf{x}}^{2,r}$, $\hat{\mathbf{x}}^{3,r}$ and $\mathbf{x}_F$) are *aligned* at minimum reflux for direct splits (Julka and Doherty, 1990). Since the pinches all move as the reflux ratio is varied, we can find minimum reflux by varying r until any three of these points lie on a straight line. The most convenient points to align

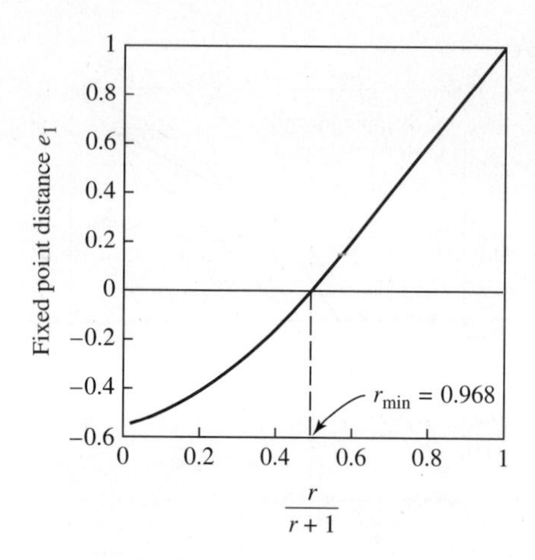

FIGURE 4.17
Fixed point distance as a function of $r/(r+1)$ for the example shown in Fig. 4.16.

are the stripping node, the rectifying saddle, and the feed composition.[21] Aligning these points is accomplished by constructing the following vectors originating from the rectifying saddle

$$\mathbf{e}_1 = \mathbf{x}_F - \hat{\mathbf{x}}^{2,r} \tag{4.70}$$

$$\mathbf{e}_2 = \hat{\mathbf{x}}^{1,s} - \hat{\mathbf{x}}^{2,r} \tag{4.71}$$

These vectors are shown in Fig. 4.18, where it can be seen that minimum reflux occurs when they lie on top of each other, i.e., the vectors are *linearly dependent.* Therefore, the minimum reflux condition is equivalent to

$$\det(\mathbf{e}_1, \mathbf{e}_2) = 0 \tag{4.72}$$

The value of r that solves this equation is the minimum reflux ratio. The absolute value of the determinant represents twice the area spanned between the vectors $\mathbf{e}_1$ and $\mathbf{e}_2$, and therefore the determinant is proportional to the *fixed point area* defined by these vectors. A graph of the fixed point area versus $r/(r+1)$ is shown in Fig. 4.19 for the benzene-toluene-xylene distillation described in Fig. 4.18. The zero in this graph gives a value for r_{min} of 1.518 and a corresponding value for $(V_B/F)_{min} = 0.763$. An efficient and robust method for performing these calculations is given by Fidkowski et al. (1991).

For nonsaturated liquid feeds, the point $\mathbf{x}_F$ should be replaced by $\tilde{\mathbf{x}}$ in the alignment condition (Eq. 4.72), where $\tilde{\mathbf{x}}$ is a linear combination of $\mathbf{z}_F$ and $\hat{\mathbf{x}}^{1,s}$ (Julka and Doherty, 1990).

$$\tilde{\mathbf{x}} = \mathbf{z}_F + (1-q)(\hat{\mathbf{x}}^{1,s} - \hat{\mathbf{y}}^{1,s}) \tag{4.73}$$

and $\hat{\mathbf{x}}^{1,s}$ and $\hat{\mathbf{y}}^{1,s}$ are in vapor–liquid equilibrium.

For nonideal mixtures the minimum reflux condition still requires that the stripping profile ends (pinches) on the rectifying profile for direct splits. However, now the stripping node, the rectifying saddle, and the feed composition lie on a *curve*

[21] We can also pick the rectifying saddle, the rectifying node, and the feed composition because the minimum reflux is *independent* of the bottom composition for direct splits. However, the minimum *vapor rate* will depend on the bottom composition, since the ratios D/F and B/F in Eq. 4.69 both vary with bottom composition.

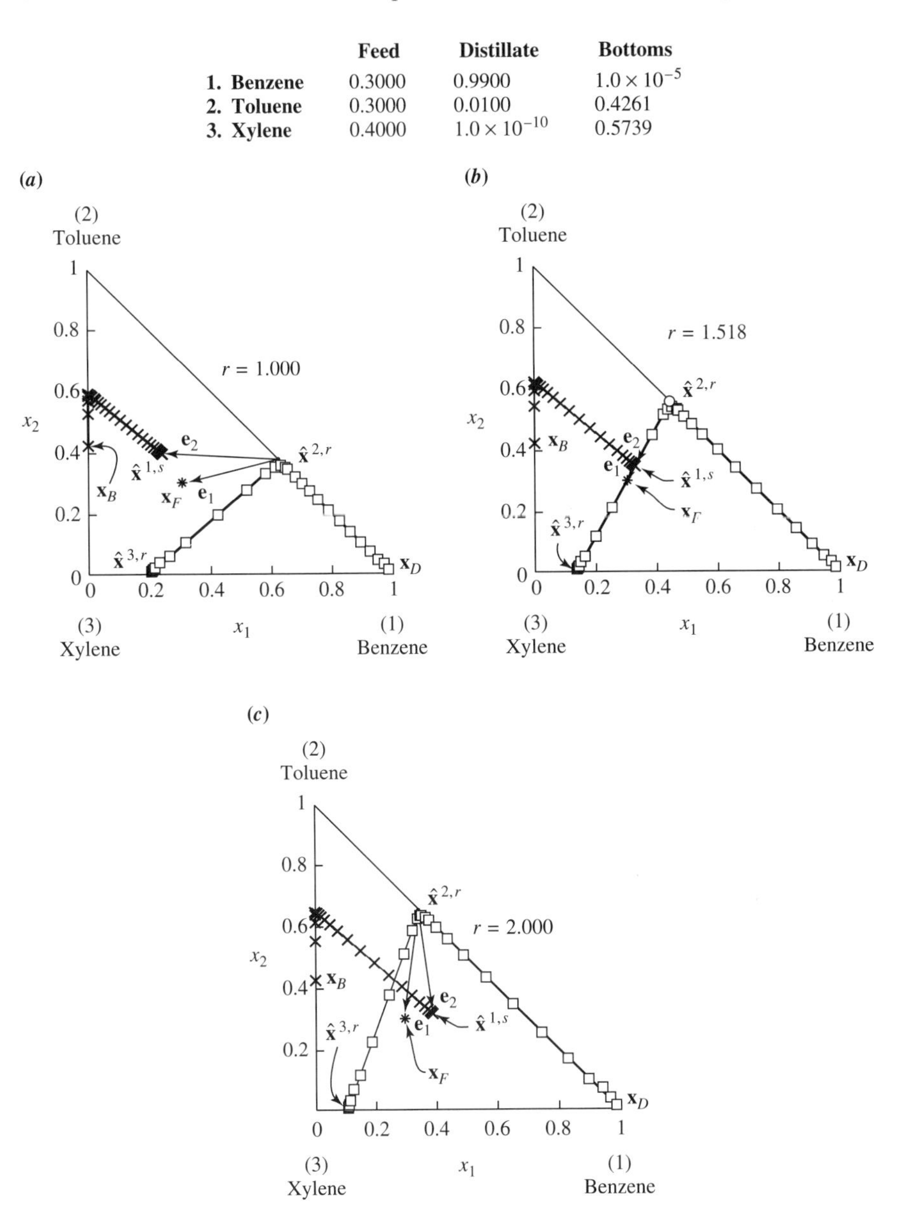

	Feed	Distillate	Bottoms
1. Benzene	0.3000	0.9900	1.0×10^{-5}
2. Toluene	0.3000	0.0100	0.4261
3. Xylene	0.4000	1.0×10^{-10}	0.5739

FIGURE 4.18
Liquid composition profiles for a ternary mixture of benzene (1), toluene (2), and xylene (3) at 1 atm pressure and $q = 1$. Vapor-liquid equilibrium was calculated using Raoult's law: (*a*) $r < r_{\min}$, (*b*) $r = r_{\min}$, (*c*) $r > r_{\min}$.

rather than a straight line. Nevertheless, the curve joining these three points is nearly linear even for highly nonideal mixtures (Levy et al., 1985), and the zero area method typically gives values for $r_{\min}$ that are within a few percent of the exact value. This is well within the combined error introduced by imperfect physical property models, and by the assumptions of a perfectly adiabatic column and of equilibrium stages.

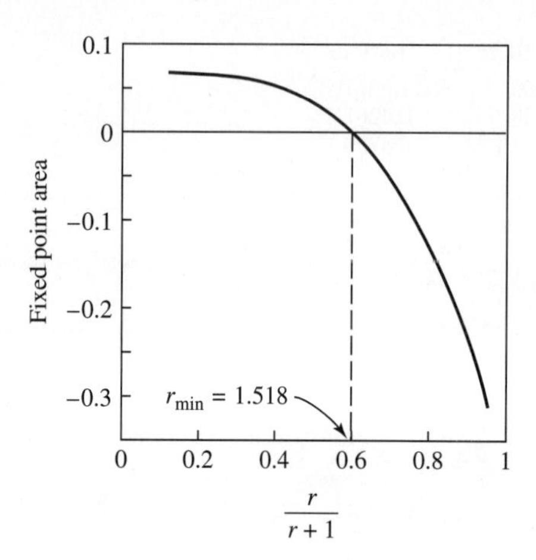

FIGURE 4.19
Fixed point area as a function of $r/(r+1)$ for the example shown in Fig. 4.18.

Therefore, the zero area method can be used safely to calculate minimum flows in both ideal and nonideal mixtures.

EXAMPLE 4.6. A saturated liquid mixture containing 30 mol % acetaldehyde, 30 mol % methanol, and 40 mol % water is to be distilled at 1 atm pressure. The desired products are saturated liquids with purities 99.9 mol % acetaldehyde and 10 ppm water in the distillate, and 1000 ppm acetaldehyde in the bottoms. Estimate the minimum reflux ratio and the minimum value of V_B/F for the separation. On the basis of your results, would you recommend continuing with the design of a distillation column to achieve the separation?

- The first step is to look for vapor-liquid equilibrium data and assess whether the mixture has any azeotropes or tangent pinches. A search through an available database reveals that a two-parameter Margules model exists for this mixture, which predicts the binary *y*-*x* diagrams shown in Fig. 4.20. The predicted diagrams should now be validated against experimental data, which is left as an exercise for the reader! The mixture is nonideal, as can be seen clearly from the asymmetric shapes in Fig. 4.20*a* and *b*. We also note that the mixture has no azeotropes, although there is expected to be a tangent pinch in the vicinity of pure methanol (see Fig. 4.20*a*). The normal boiling points are 20.4°C for acetaldehyde, 64.6°C for methanol, and 100°C for water (Reid et al., 1977, Appendix A). Therefore, the absence of azeotropes and the fact that acetaldehyde is the lightest component indicate that the desired separation is feasible. The remainder of this exercise addresses whether the separation is likely to be economical.
- Converting the problem specifications into mole fractions and calculating the remaining mole fraction $x_{B,2}$ from the overall material balance (Eq. 4.13) leads to the complete set of mole fractions for every stream; see the legend in Fig. 4.22.
- Although the pure components have different latent heats of vaporization, it is convenient (and simple) to assume constant molar flows in order to get an estimate of the minimum flows needed for the separation. A plot of the fixed point area versus $r/(r+1)$, calculated using the algorithm described in Fidkowski et al. (1991), is shown in Fig. 4.21. The minimum reflux is determined as $r_{\text{min}} = 0.559$, and from Eq. 4.69 we calculate $(V_B/F)_{\text{min}} = 0.467$. The column composition profiles are shown in

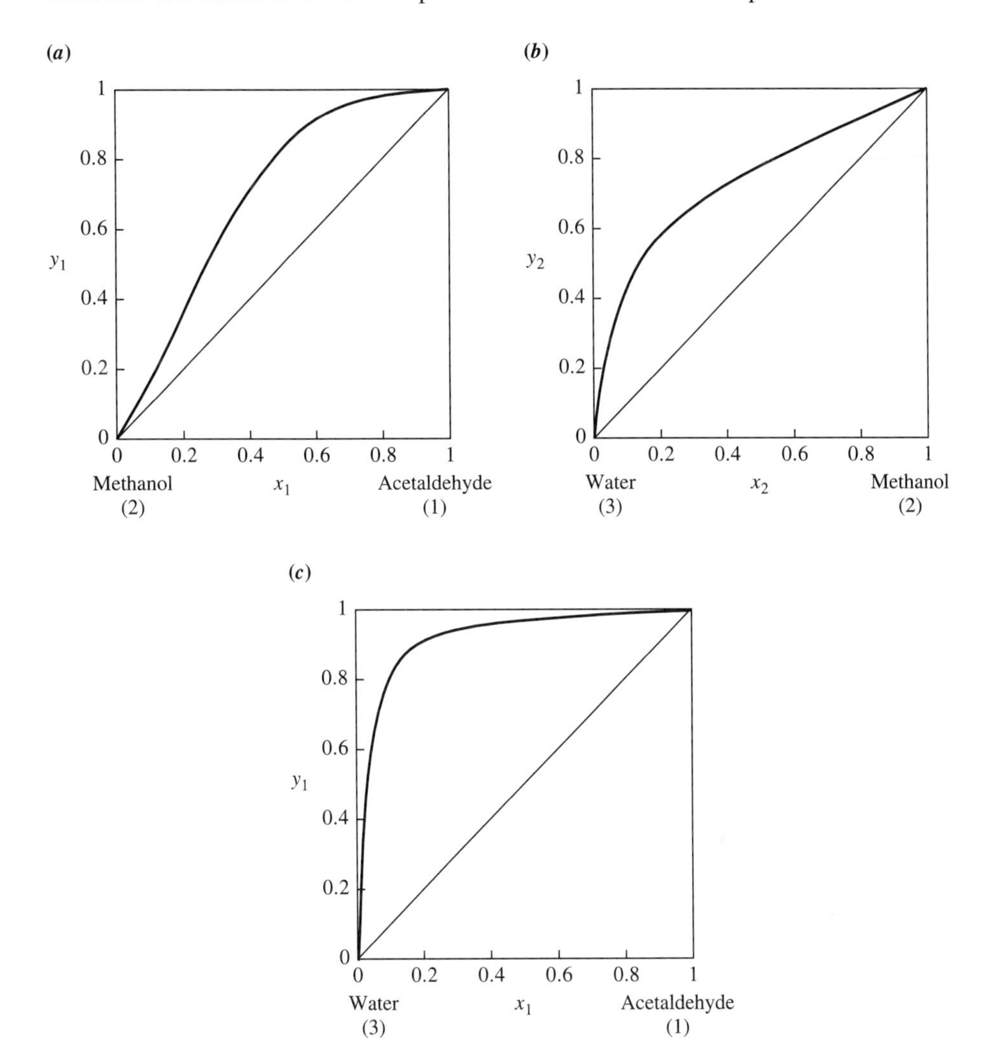

FIGURE 4.20
Binary y-x diagrams predicted by a two-parameter Margules model at 1 atm for the mixtures (a) acetaldehyde-methanol, (b) methanol-water, (c) acetaldehyde-water.

Fig. 4.22 at this value of r. Notice that the saddle fixed point in the rectifying section, the node fixed point in the stripping section, and the feed composition are visually collinear, in spite of the fact that this condition was not imposed on the calculated profiles.

- The minimum value of V_B/F is 0.467, which is an attractively low number. There is a strong incentive to pursue distillation as a means of achieving the separation. Therefore, the recommendation is to continue with the design of a distillation column.
- As a first step to continuing the design, we repeat the minimum flow calculations with a nonconstant molar flow model. This requires additional effort, which is justified by the results obtained above. The addition steps include:

1. Gathering and validating the addition physical property data required for calculating stream enthalpies, e.g., pure component latent heats of vaporization, parameters for pure component liquid and vapor heat capacity correlations as a function of temperature, and heat of mixing data for the binary and ternary mixtures.
2. Augmenting the mathematical model of the column with enthalpy balances for each stage as well as adding enthalpy balances to the fixed point equations (Knight and Doherty, 1986).
3. Developing the capability to calculate enthalpies of multicomponent liquids and vapors at specified temperatures and compositions.

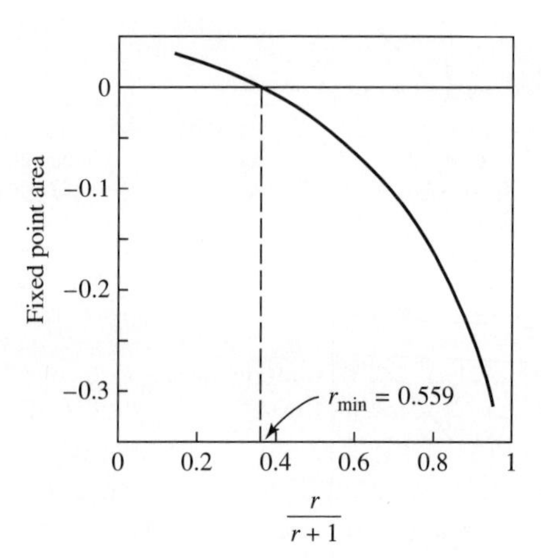

FIGURE 4.21
Fixed point area as a function of $r/(r+1)$ for the acetaldehyde-methanol-water example.

	Feed	Distillate	Bottoms
1. Acetaldehyde	0.3000	0.9900	0.0010
2. Methanol	0.3000	0.0010	0.4279
3. Water	0.4000	1.0×10^{-5}	0.5711

Methanol (2)
x_2
$r = 0.559$
$s = 0.667$
$\mathbf{x}_B$
$\mathbf{x}_F$
$\mathbf{x}_D$
x_1
Water (3)
Acetaldehyde (1)

FIGURE 4.22
Liquid composition profiles for a ternary mixture of acetaldehyde (1), methanol (2), and water (3) at 1 atm pressure and $q = 1$. Calculations were performed with a constant molar flow model and a nonideal vapor-liquid equilibrium relation.

With these additions to the column model, the minimum reflux is recalculated by aligning the (recalculated) pinches. We find $r_{\min} = 0.786$, with a corresponding value $(V_B/F)_{\min} = 0.535$, which is still very attractive. Column composition profiles for the nonconstant molar flow model at this value of r are shown in Fig. 4.23. Notice that the rectifying saddle, stripping node, and feed composition are still collinear to within a high degree of precision in spite of the fact that the profiles are calculated with heat effects and nonideal VLE.

The method just described can be modified to calculate minimum flows in other kinds of splits, e.g., *indirect* splits where the heaviest component is the desired pure component (in the bottom stream) from the separation, and *nonsharp* splits where the lightest and heaviest components do not distribute between the product streams but the middle-boiling component does. These and other kinds of splits are important cases to consider for distillation column sequencing, i.e., systems problems, which we discuss further in Chap. 7.

Design Procedure

We now describe a procedure to find the number of theoretical stages above and below the feed tray that are required to achieve given product purities at a specified value of the reflux (or reboil) ratio. We could, of course, solve this problem using the boundary value design method described in Sec. 4.2, but this approach requires that we specify the composition of both components in one product stream (thereby

	Feed	Distillate	Bottoms
1. Acetaldehyde	0.3000	0.9900	0.0010
2. Methanol	0.3000	0.0010	0.4279
3. Water	0.4000	1.0×10^{-5}	0.5711

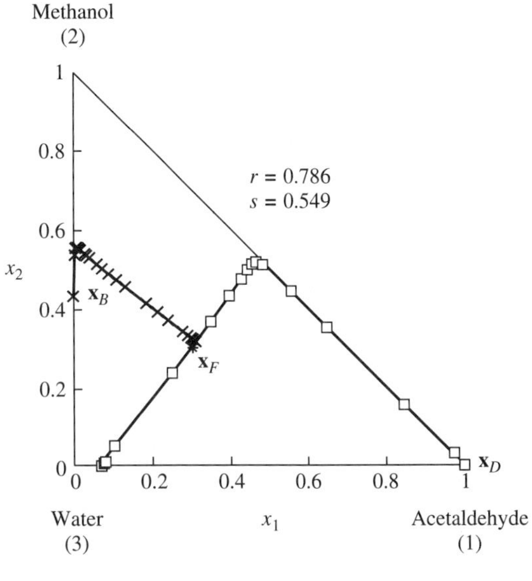

Methanol (2)

$r = 0.786$
$s = 0.549$

x_2 — $\mathbf{x}_B$ — $\mathbf{x}_F$ — $\mathbf{x}_D$

0 0.2 0.4 0.6 0.8 1

Water (3) x_1 Acetaldehyde (1)

FIGURE 4.23
Liquid composition profiles for a ternary mixture of acetaldehyde (1), methanol (2), and water (3) at 1 atm pressure and $q = 1$. Calculations were performed with a nonconstant molar flow model and a nonideal vapor-liquid equilibrium relation.

	Feed	Distillate	Bottoms
1. Benzene	0.3000	0.9900	1.0×10^{-5}
2. Toluene	0.3000		0.4261
3. Xylene	0.4000		0.5739

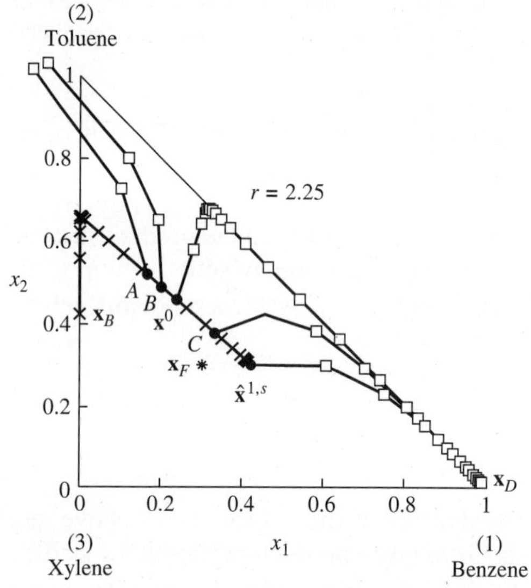

FIGURE 4.24
Liquid composition profiles for the benzene-toluene-xylene mixture. A single stripping profile begins at $\mathbf{x}_B$ and several rectifying profiles begin at various points along it.

fixing the composition of all three components in that stream). Frequently, we only wish to set the composition of one component in each product stream (e.g., 99.9% light component in the distillate stream and 0.01% light component remaining in the bottom stream; this is equivalent to setting the purity and fractional recovery of the light component). Moreover, the degrees of freedom do not allow us to specify the composition of all components in one product stream when the mixture contains four or more components,[22] so the boundary value design procedure does not extend to multicomponent mixtures. The procedure described below overcomes both these difficulties. This method is also based on the geometry of the composition profiles. We describe it for a direct split, although it works equally well for indirect splits after making changes to the formulation.

Consider the benzene-toluene-xylene mixture discussed earlier, but now we calculate the profiles (shown in Fig. 4.24) at a reflux ratio of $r = 2.25$, which is above the minimum value of $r_{\min} = 1.518$. The shape of the rectifying profile depends on the distillate purity. As the amount of heavy component in the distillate is decreased, the initial portion of the rectifying profile gets closer to the 1-2 edge of the composition triangle and travels farther along it before turning toward the interior of the triangle (see Fig. 4.24). This corner asymptotically approaches the rectifying saddle as $x_{D,3} \rightarrow 0$.

[22]Remember, there are only four degrees of freedom no matter how many components are present in the mixture being distilled. One of these is used to set the internal flows, e.g., by specifying a value for r that leaves only three left to specify product compositions.

Now suppose that we try to compute a rectifying profile starting from various points on the stripping profile (see Fig. 4.24). We order these points by their arc-length from the bottom composition (i.e., the distance from the bottom composition measured along the stripping profile). If we attempt to switch profiles too early (e.g., points A or B in Fig. 4.24), the rectifying profile will eventually turn away from the distillate composition. Such profiles are infeasible, because they cannot reach the distillate composition even with an infinite number of stages. The first point on the stripping profile from which a feasible rectifying profile can be drawn is denoted as $\mathbf{x}^0$. Notice that the profile beginning at $\mathbf{x}^0$ is asymptotically close to the rectifying saddle and is also asymptotically close to the 1-2 edge as it travels toward the desired distillate composition of 99% benzene.

Feasible rectifying profiles can also be calculated by starting anywhere between the points $\mathbf{x}^0$ and $\hat{\mathbf{x}}^{1,s}$ on the stripping profile (see Fig. 4.24). For any point $\mathbf{x}$ on this line segment, the fractional distance from $\mathbf{x}^0$ is denoted by ω and given by

$$\omega = \frac{l(\mathbf{x}) - l(\mathbf{x}^0)}{l(\hat{\mathbf{x}}^{1,s}) - l(\mathbf{x}^0)} \tag{4.74}$$

where $l(\mathbf{x})$ is the arc length of point $\mathbf{x}$ along the stripping profile.

The algorithm to determine the number of stages above and below the feed with the associated nonkey compositions in the product streams is:

1. Given $\mathbf{z}_F$, q and also r, $x_{D,1}$, $x_{B,1}$.
2. Calculate the reboil ratio from Eq. 4.10.
3. Approximate $x_{D,2} = 1 - x_{D,1}$, i.e., although we do not yet know the composition of the heavy component in the distillate we begin by assuming that it is small. This is a good assumption for all the designs shown in Fig. 4.24, and it is generally a good assumption for both ideal and nonideal mixtures. This merely states what we already know, namely, that distillation is a good way of removing heavy components from distillate streams. At the end of the procedure we will calculate a value for $x_{D,3}$ and see whether our assumption was good. It can be corrected at that point by repeating the algorithm with a better initial estimate for $x_{D,3}$. Normally such an iteration is not necessary.
4. Calculate the bottoms composition from the overall component balance (Eq. 4.13).
5. Calculate the stripping profile from $\mathbf{x}_B$ all the way to the stable node $\hat{\mathbf{x}}^{1,s}$.
6. Find the point $\mathbf{x}^0$ on the stripping profile. This can be done either by fixed point methods (Julka and Doherty, 1993) or by bisection, which is simpler and more accurate (Fidkowski et al., 1991).
7. For various values of ω ($0 < \omega < 1$) calculate the rectifying profile until it intersects the desired distillate composition of component 1. Count the number of stages in each column section and also note the distillate compositions of components 2 and 3.

The number of stages and the composition of the heavy nonkey component in the distillate for the benzene-toluene-xylene mixture are shown in Fig. 4.25.

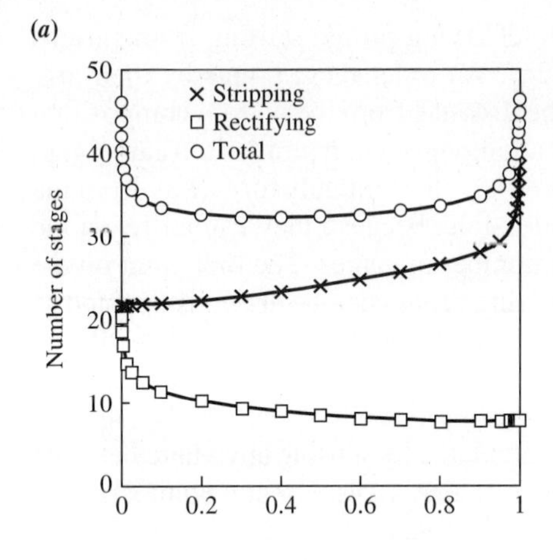

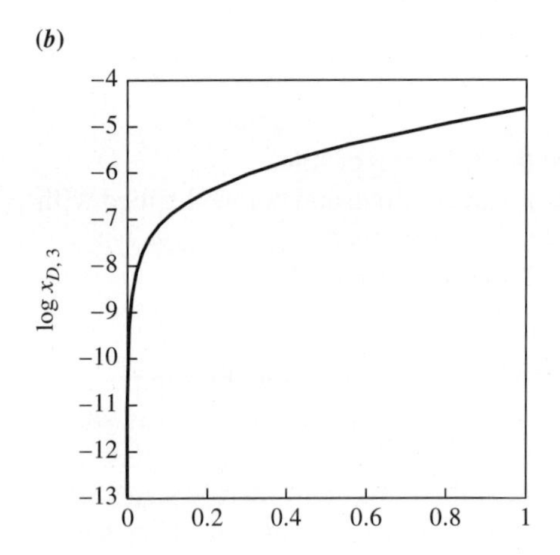

FIGURE 4.25
Spectrum of designs for the benzene-toluene-xylene mixture. (*a*) Number of stages in each column section, and (*b*) distillate mole fraction of xylene, the heavy nonkey component for the design specifications: $\mathbf{x}_F = (0.3, 0.3, 0.4)$, $q = 1$, $r = 2.25$, $x_{D,1} = 0.99$, $x_{B,1} = 1.0 \times 10^{-5}$.

Figure 4.25*a* is similar to Fig. 4.7 for binary mixtures. The total number of stages approaches infinity at either end of the range of feasible feed-stage compositions.[23] When $\omega = 0$ there is an infinite number of stages in the rectifying section due to a saddle pinch in the rectifying profile, and when $\omega = 1$ there is an infinite number of stages in the stripping section due to a node pinch in the stripping profile. For each of the designs represented in Fig. 4.25*a* there is a different amount of heavy component in the distillate stream, as seen in Fig. 4.25*b*. When $\omega = 0$, the large number of stages in the rectifying section has the effect of eliminating the heavy component from the distillate stream. As ω increases, the number of stages in the rectifying section decreases and the amount of heavy component in the distillate therefore increases. Column composition profiles at small, medium, and large values of ω are shown in Fig. 4.26.

[23]For multicomponent mixtures the feed-stage composition is a vector quantity, but can be conveniently represented by the scalar quantity ω.

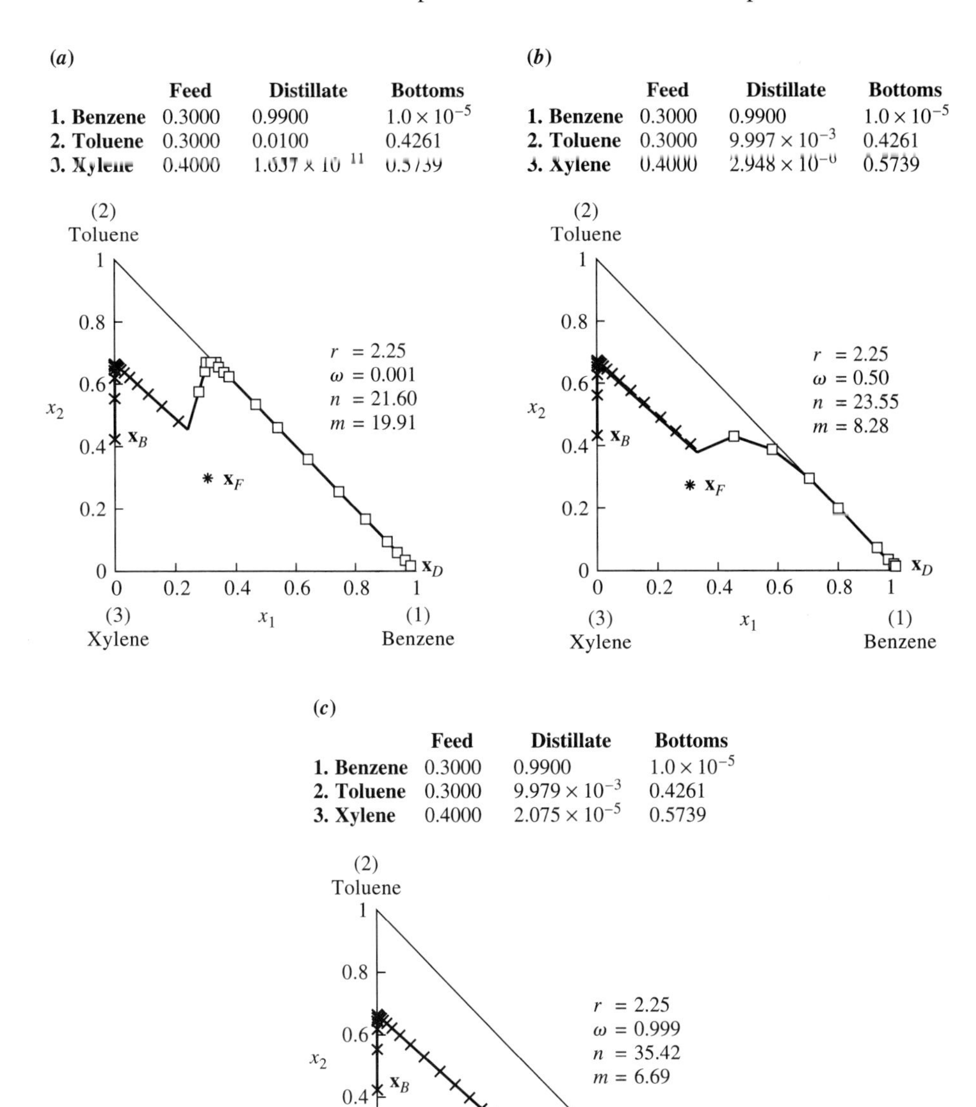

(*a*)

	Feed	Distillate	Bottoms
1. Benzene	0.3000	0.9900	1.0×10^{-5}
2. Toluene	0.3000	0.0100	0.4261
3. Xylene	0.4000	1.637×10^{-11}	0.5739

(*b*)

	Feed	Distillate	Bottoms
1. Benzene	0.3000	0.9900	1.0×10^{-5}
2. Toluene	0.3000	9.997×10^{-3}	0.4261
3. Xylene	0.4000	2.948×10^{-6}	0.5739

(*c*)

	Feed	Distillate	Bottoms
1. Benzene	0.3000	0.9900	1.0×10^{-5}
2. Toluene	0.3000	9.979×10^{-3}	0.4261
3. Xylene	0.4000	2.075×10^{-5}	0.5739

FIGURE 4.26

Liquid-phase composition profiles for the benzene-toluene-xylene mixture at various values of ω. (*a*) $\omega = 0.001$, (*b*) $\omega = 0.50$, (*c*) $\omega = 0.999$.

EXAMPLE 4.7. For the mixture from the previous example, consider an indirect split.

- We use the same feed composition and quality as before, but now the product compositions for water are $x_{B,3} = 0.9897$, $x_{D,3} = 0.001$.
- The reflux ratio is $r = 0.87$, which is approximately 50% larger than the minimum value (see Example 4.9 below).
- Figure 4.27 shows a spectrum of designs as a function of ω and Fig. 4.28 shows column profiles for three different values of ω.

Double–Feed Columns

In many applications it is desirable to have two separate feeds to the same distillation column, e.g., if we have two streams to separate containing the same components but with very different compositions. In this case the minimum flows are governed by pinches in either the rectifying and middle-section profiles or in the stripping

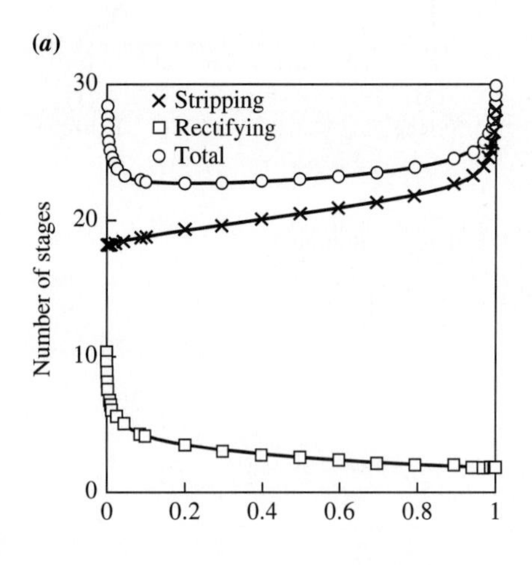

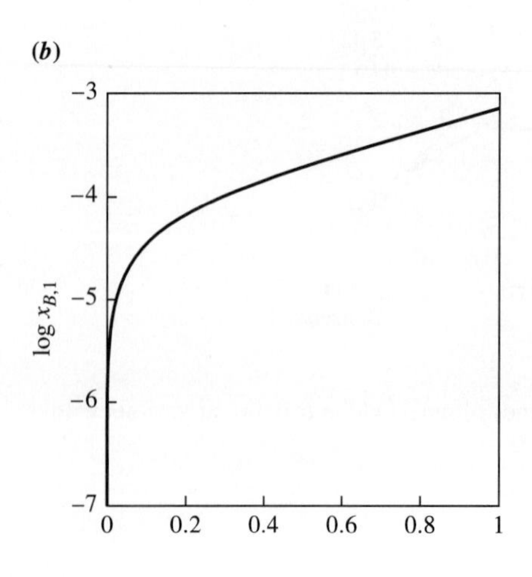

FIGURE 4.27
Spectrum of designs for an indirect split in the acetaldehyde-methanol-water mixture. (*a*) Number of stages in each column section, and (*b*) bottoms mole fraction of acetaldehyde, the light nonkey component for the design specifications: $\mathbf{x}_F = (0.3, 0.3, 0.4)$, $q = 1$, $r = 0.87$, $x_{B,3} = 0.9897$, $x_{D,3} = 0.001$.

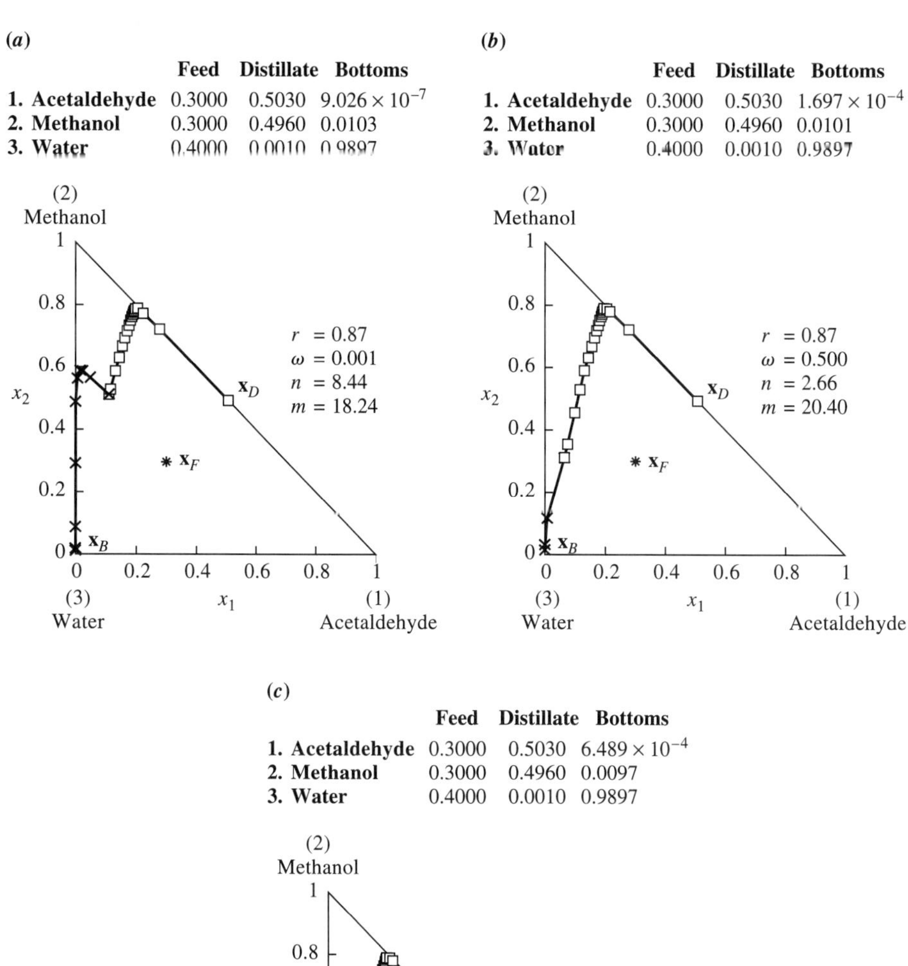

(a)

	Feed	Distillate	Bottoms
1. Acetaldehyde	0.3000	0.5030	9.026×10^{-7}
2. Methanol	0.3000	0.4960	0.0103
3. Water	0.4000	0.0010	0.9897

(b)

	Feed	Distillate	Bottoms
1. Acetaldehyde	0.3000	0.5030	1.697×10^{-4}
2. Methanol	0.3000	0.4960	0.0101
3. Water	0.4000	0.0010	0.9897

(c)

	Feed	Distillate	Bottoms
1. Acetaldehyde	0.3000	0.5030	6.489×10^{-4}
2. Methanol	0.3000	0.4960	0.0097
3. Water	0.4000	0.0010	0.9897

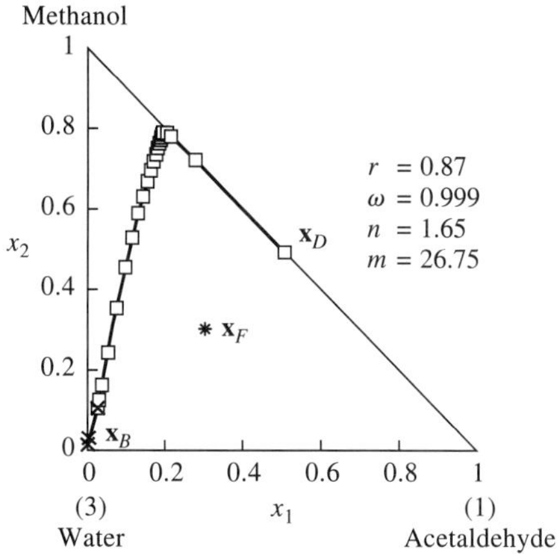

FIGURE 4.28
Column composition profiles for three values of ω in Example 4.7.

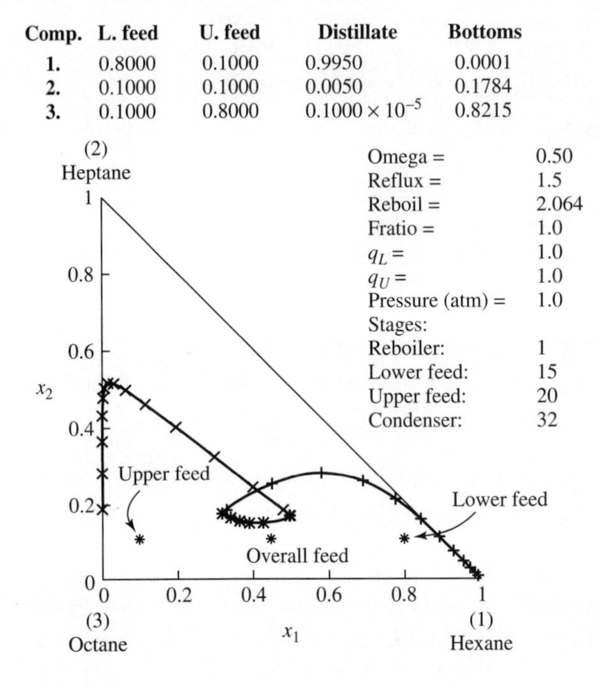

FIGURE 4.29
Column design for the mixture hexane-heptane-octane with incorrect feed positions.

and middle-section profiles. Algebraic methods for constant volatility mixtures are found in Nikolaides and Malone (1987). Geometric methods are found in Levy and Doherty (1986*a*), and Knapp and Doherty (1994).

One of the key design questions for these systems is to decide which stream should be the upper feed and which should be the lower feed. Intuitively, we would expect that the higher-boiling feed should be fed to the hotter section of the column, i.e., the lower feed stream. However, there are counterintuitive cases, even for ideal mixtures, so the problem deserves a brief mention. Levy and Doherty (1986*a*) study the mixture hexane-heptane-nonane; stream 1 contains 20 mol % hexane, 10 mol % heptane, and 70 mol % nonane, stream 2 contains 30 mol % hexane, 60 mol % heptane, and 10 mol % nonane. The bubble point of stream 1 is 30°C higher than that of stream 2. We would expect stream 1 to be the lower feed to the column. However, this arrangement requires more reflux (and in some instances more stages) than the case of mixing the feeds and using a single-feed column. If the feeds are reversed so that the higher-boiling stream is the upper feed, then the minimum reflux is reduced significantly.

Another case is shown in Fig. 4.29 for the mixture hexane-heptane-nonane. The lower feed consists of 80 mol % hexane, 10 mol % heptane, and 10 mol % octane (stream 1); the upper feed contains 10 mol % hexane, 10 mol % heptane, and 80 mol % octane (stream 2). As can be seen from the figure, this design requires more reflux and more stages than a single-feed column with a mixed feed.[24] In this case the best design is obtained by reversing the feeds and making the lower feed be

[24]This is because the stripping and rectifying profiles overlap and therefore the middle section serves no purpose.

the higher-boiling stream 2. Therefore, sometimes the higher-boiling stream should be the lower feed and other times it should be the upper feed. Failure to make the correct choice results in poor designs. Rules for deciding are given in Nikolaides and Malone (1987), and Levy and Doherty (1986*a*).

4.5 NONIDEAL MIXTURES WITH FOUR OR MORE COMPONENTS

The ideas described in the last section for ternary mixtures extend to multicomponent systems. Feasibility and product distribution for multicomponent mixtures are discussed in Chap. 7 in the context of distillation system synthesis. Extension of the methods for calculating minimum flows and column designs for mixtures with four or more components is described here briefly.

The pinch alignment method for calculating minimum flows in multicomponent mixtures is discussed in detail by Julka and Doherty (1990). The main ideas will be described for the direct split of a four-component mixture. For such a split, the rectifying profiles all approach a surface (called the *rectifying surface* or, in the case of constant volatility mixtures, the *rectifying plane*) as the compositions of the heavy components are varied.[25] The rectifying surface is suspended from three fixed points of the rectifying profile (a stable node $\hat{\mathbf{x}}^{1,r}$ and two saddles $\hat{\mathbf{x}}^{2,r}$, $\hat{\mathbf{x}}^{3,r}$); see Fig. 4.30*a*. Since each fixed point moves as the reflux ratio is changed, the position and

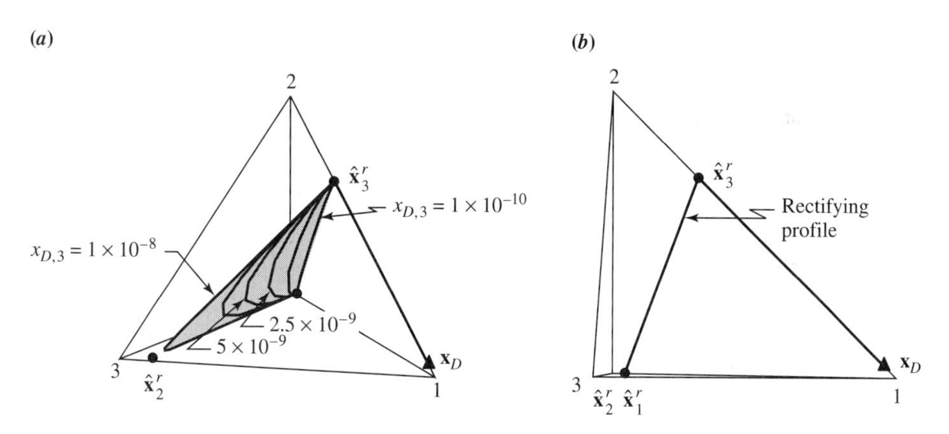

FIGURE 4.30
(*a*) The rectifying plane for a constant volatility mixture with $\alpha_{1,4} = 8$, $\alpha_{2,4} = 4$, $\alpha_{3,4} = 2$, $\alpha_{4,4} = 1$. The reflux ratio $r = 2.0$, $x_{D,1} = 0.95$, $x_{D,2} = (1.0 - x_{D,1} - x_{D,3} - x_{D,4}) \simeq 0.05$, $x_{D,4} = 5 \times 10^{-17}$. The value of $x_{D,3}$ varies as noted on the figure. (*b*) Natural projection illustrating that the rectifying surface is planar for this mixture.

[25] This means that we fix the value of $x_{D,1}$ at the desired product purity and treat the amount of heaviest component in the distillate as negligible by setting $x_{D,4}$ to a small number like 5×10^{-17}. The value of $x_{D,3}$ is varied and the final mole fraction $x_{D,2}$ is found by summing the mole fractions to unity.

size of the surface depends on r; see Fig. 4.31. For constant volatility mixtures these surfaces are planes (Julka and Doherty, 1990), as can be seen from the "natural" projection shown in Fig. 4.30*b*. In this projection the line of sight coincides with the straight line joining the fixed points $\hat{\mathbf{x}}^{2,r}$ and $\hat{\mathbf{x}}^{1,r}$. In this projection the geometric properties of the profiles appear to behave like those for a ternary mixture.

If we now specify a saturated liquid feed composition and one composition in the bottom stream, the remaining compositions in the bottom stream are determined by overall material balance. The stripping and rectifying profiles can be calculated for different values of r, but in general they will not intersect because we have *not* specified the correct value for $x_{D,3}$. Fortunately, we do not need to know this value in order to calculate the minimum flows! We know that for a fixed value of $x_{D,1}$ and a negligibly small value of $x_{D,4}$, the location and orientation of the rectifying surface is determined by the reflux ratio, while the value of $x_{D,3}$ determines the shape and location of the rectifying profile within this surface. Since the locus of the rectifying composition profiles sweeps out the entire rectifying surface as $x_{D,3}$ is varied, a special property of this surface is that for any point on the surface there exists a unique value of $x_{D,3}$ that generates a rectifying profile through that point. This implies that if, for a given value of r, the stripping profile intersects the rectifying surface, then it is always possible to find a unique value of $x_{D,3}$ that makes the column feasible, i.e., that makes the two composition profiles intersect. Therefore, to find the minimum reflux ratio it is only necessary to find the value of r that generates a stripping profile that just intersects the rectifying surface for an arbitrary value of $x_{D,3}$.

This concept is discussed next for a constant volatility mixture. If r is too small the stripping profile will not reach the rectifying plane; if r is too large it will intersect the rectifying plane and go beyond it, but at minimum flows the value of r is such that the stripping profile just reaches the rectifying plane, as shown in Fig. 4.32. There are an infinite number of stages in the rectifying section in the vicinity of the saddle pinch and an infinite number of stages in the stripping section below the feed stage. Notice that the geometry of the profiles shown in the natural projection in Fig. 4.32*b*

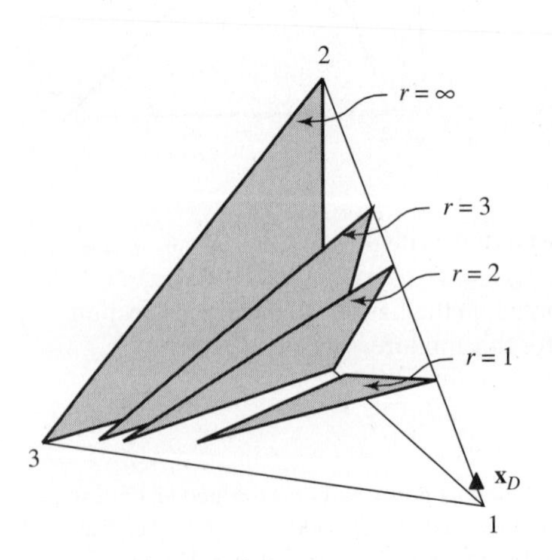

FIGURE 4.31
Typical rectifying planes at different reflux ratios for the constant volatility mixture described in Fig. 4.30. For each plane the distillate composition is $x_{D,1} = 0.95$, $x_{D,2} \simeq 0.05$, $x_{D,3} \simeq 0.0$, and $x_{D,4} \simeq 0.0$.

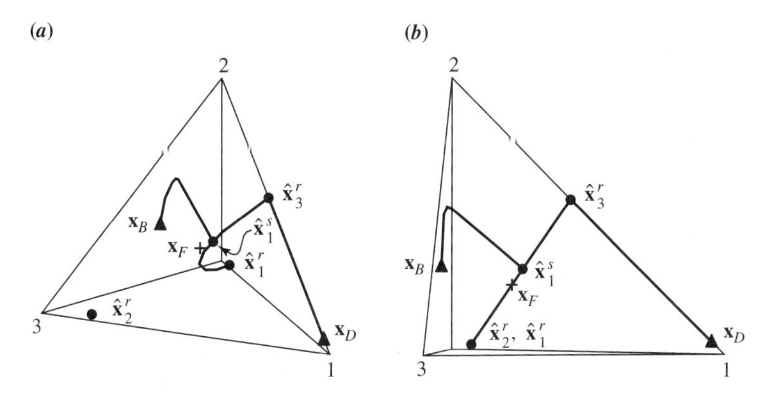

FIGURE 4.32
(*a*) Composition profiles for the constant volatility mixture described in Fig. 4.30 with $r = 2.015$ and $q = 1$ for an equimolar feed. (*b*) Natural projection illustrating that the stripping profile pinches on the rectifying plane, i.e., $r = r_{\min}$. Note that the three rectifying fixed points $\hat{\mathbf{x}}^{1,r}$, $\hat{\mathbf{x}}^{2,r}$, and $\hat{\mathbf{x}}^{3,r}$, the stripping node $\hat{\mathbf{x}}^{1,s}$, and the feed composition $\mathbf{x}_F$ are all collinear in this projection. In this figure the value of $x_{D,3}$ has been adjusted to the value $x_{D,3} = 3 \times 10^{-9}$ so that the rectifying profile and the stripping profile touch in the rectifying plane.

is similar to those for ternary mixtures. For nonsaturated liquid feeds the rectifying profile goes through the point $\tilde{\mathbf{x}}$ (defined in Eq. 4.73) instead of the point $\mathbf{x}_F$.

For the distillation of four-component constant volatility mixtures in columns with constant molar flows the three rectifying fixed points $\hat{\mathbf{x}}^{1,r}$, $\hat{\mathbf{x}}^{2,r}$, and $\hat{\mathbf{x}}^{3,r}$, the stripping node $\hat{\mathbf{x}}^{1,s}$, and the point $\tilde{\mathbf{x}}$ all lie in the same plane at minimum reflux. Choosing one of these points as a new origin (e.g., the rectifying saddle $\hat{\mathbf{x}}^{3,r}$) we define the following vectors:

$$\mathbf{e}_1 = \tilde{\mathbf{x}} - \hat{\mathbf{x}}^{3,r} \tag{4.75}$$

$$\mathbf{e}_2 = \hat{\mathbf{x}}^{1,s} - \hat{\mathbf{x}}^{3,r} \tag{4.76}$$

$$\mathbf{e}_3 = \hat{\mathbf{x}}^{2,r} - \hat{\mathbf{x}}^{3,r} \tag{4.77}$$

$$\mathbf{e}_4 = \hat{\mathbf{x}}^{1,r} - \hat{\mathbf{x}}^{3,r} \tag{4.78}$$

The minimum reflux ratio is determined by requiring that any three of these vectors be coplanar (i.e., linearly dependent). This is determined by finding the value of r that satisfies the equation

$$\det(\mathbf{e}_1, \mathbf{e}_2, \mathbf{e}_3) = 0 \tag{4.79}$$

Notes

1. The value of the determinant is related to the volume enclosed by the column vectors $\mathbf{e}_1, \mathbf{e}_2, \mathbf{e}_3$, and so this method for finding the minimum reflux has been called the *zero volume method.*
2. Selecting vectors $\mathbf{e}_1$, $\mathbf{e}_3$, and $\mathbf{e}_4$ in Eq. 4.79 shows that for saturated liquid feeds and constant volatility mixtures the only parameters to enter in the calculation are the distillate composition and the feed composition. Therefore, for such systems

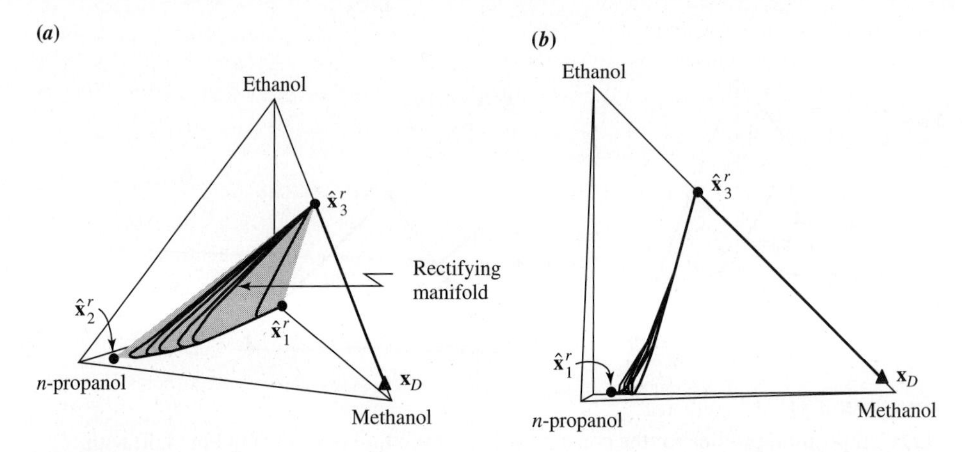

FIGURE 4.33
(*a*) Typical rectifying surface for the mixture methanol-ethanol-*n*-propanol-acetic acid, with $r = 2.50$, $x_{D,1} = 0.95$, $x_{D,2} = (1.0 - x_{D,1} - x_{D,3} - x_{D,4}) \simeq 0.05$, $x_{D,4} = 1 \times 10^{-17}$. The value of $x_{D,3}$ varies. (*b*) Natural projection illustrating that the rectifying surface is curved for this mixture.

the minimum reflux ratio is the same for all feasible bottom product compositions, although the reboil ratios are different.

3. An efficient and robust method for solving Eq. 4.79 is given by Fidkowski et al. (1991).
4. Minimum flows for indirect splits are found by similar methods.
5. The relationship between Underwood's method and the zero-volume method is discussed in Julka and Doherty (1990).
6. This approach extends to mixtures with more than four components by defining more vectors in Eqs. 4.75 through 4.79, also discussed in Julka and Doherty (1990).

For nonconstant volatility mixtures[26] the rectifying surface is no longer planar but exhibits curvature. Minimum flows are still determined by the value of r that just makes the stripping profile pinch on the rectifying surface, but this is not *exactly* determined by Eq. 4.79. However, it is possible to provide a good linear approximation to the rectifying surface in the vicinity of the stripping node using the vectors $\mathbf{e}_1, \mathbf{e}_2, \mathbf{e}_3$, and Eq. 4.79 provides a very good estimate of the true minimum reflux ratio (Julka and Doherty, 1990, Appendix E).

Consider, for example, the nonideal nonazeotropic mixture methanol-ethanol-*n*-propanol-acetic acid. A typical rectifying surface is shown in Fig. 4.33. While the bottom of the surface is curved, the top is nearly planar.

Now let's consider the separation of a saturated equimolar feed into a distillate product containing 95 mol % methanol and 5 mol % ethanol, and a bottoms product

[26]In addition to nonideal and azeotropic mixtures this class includes ideal mixtures with wide differences among the pure component boiling points, e.g., benzene-toluene-styrene-diphenyl. Such mixtures obey Raoult's law but do not have constant relative volatilities and therefore can only be treated by Underwood's method if average or conservative values for the volatilities are used.

containing 2 mol % methanol. A detailed examination of the corresponding rectifying and stripping composition profiles shows that they just touch each other on the rectifying surface when $r = 3.58$, and this is the exact value for $r_{\min}$. The estimate for $r_{\min}$ calculated using the zero-volume method is 3.552, which is in error by less than 1%. The resulting composition profiles are shown in Fig. 4.34.

The design method described earlier extends to mixtures with more components, Julka and Doherty (1993). Figure 4.35 shows the range of designs that are possible for

	Feed	Distillate	Bottoms
1. Methanol	0.25	0.95	0.02
2. Ethanol	0.25	0.05	0.316
3. *n*-propanol	0.25	1.0×10^{-10}	0.332
4. Acetic acid	0.25	1.0×10^{-17}	0.332

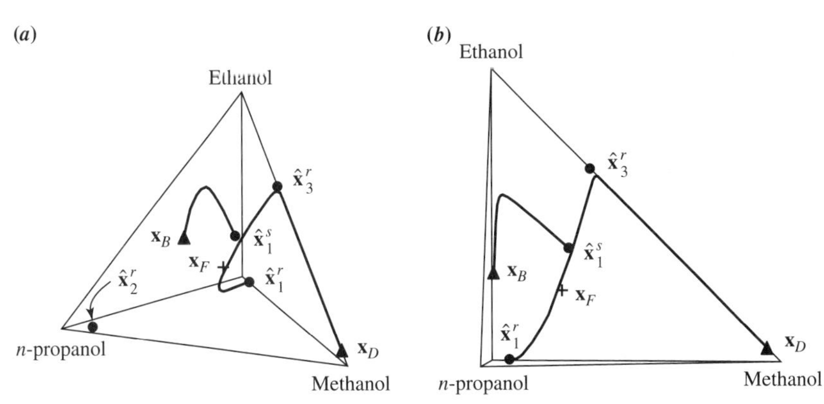

FIGURE 4.34
Methanol-ethanol-*n*-propanol-acetic acid mixture. (*a*) Rectifying and stripping composition profiles for $r = 3.552$ and $q = 1$. (*b*) Natural projection illustrating that the reflux ratio is slightly below $r_{\min}$.

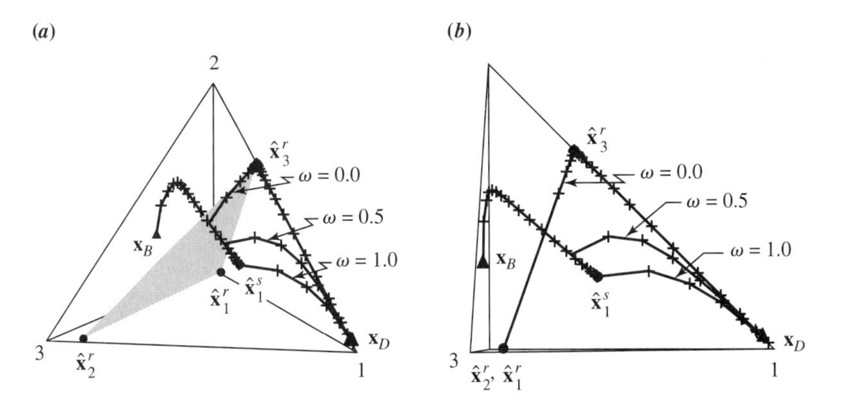

FIGURE 4.35
(*a*) Range of designs for the four component constant volatility mixture that we have been following in Figs. 4.30 and 4.32 at a reflux ratio of $r = 3.02$ ($= 1.5 \times r_{\min}$). (*b*) Natural projection.

the constant volatility mixture that we have been following. In the natural projection these designs take on the appearance of the designs shown earlier for ternary mixtures (see Fig. 4.24).

It is useful to plot the number of equilibrium stages and the distillate composition as a function of the feed–tray position (as expressed by the quantity ω), and this is done in Fig. 4.36.

Picking the fourth degree of freedom as $\omega = 0.4$ we see from this figure that the separation can be accomplished in 21 theoretical stages, which corresponds to 13 stages in the stripping section and 7 stages in the rectifying section. For such a design the distillate composition contains 100 ppm of component three and 1 ppm of component four (see Fig. 4.36(b)). The design does not change very much for a range of values of ω in the neighborhood of 0.4. Smaller amounts of components 3 and 4 in the distillate can be accomplished with a design at a smaller value of ω. Such designs will have more than 21 stages, the extra stages being located in the rectifying section.

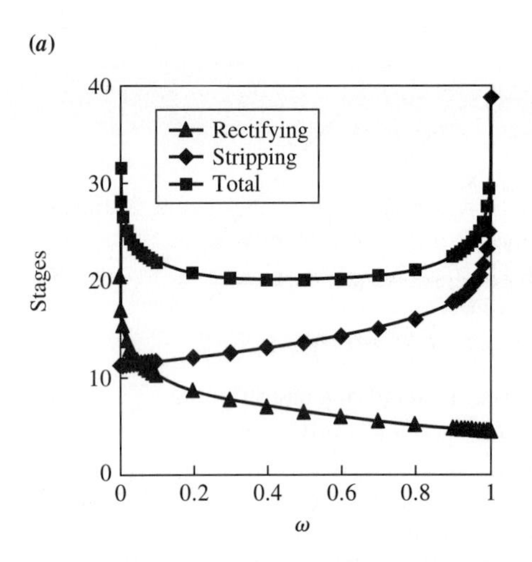

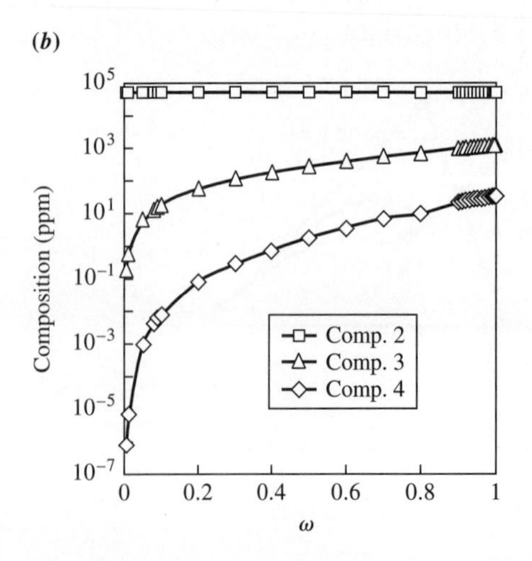

FIGURE 4.36
Spectrum of designs for the example shown in Fig. 4.35. The feed is an equimolar saturated liquid, and the values of the design variables are $r = 3.02$, $x_{D,1} = 0.95$, and $x_{B,1} = 0.002$. (*a*) Total number of stages, and number of stages in each column section as a function of ω. (*b*) Distillate composition (in parts per million molar) as a function of ω.

TABLE 4.2
Antoine coefficients for seven aromatic components. The form of the equation is $\ln P^{sat} = A - \frac{B}{T-C}$ with pressure in Pa and temperature in Kelvin

Substance	*A*	*B*	*C*
Benzene	20.793	2,788	52.4
Toluene	20.893	3,096	53.7
Ethylbenzene	20.893	3,279	60.0
Styrene	20.893	3,328	63.7
Diethylbenzene	20.993	3,657	71.2
Triethylbenzene	21.093	4,000	81.0
Diphenyl	21.593	4,602	70.4

EXAMPLE 4.8. Seven component mixture. In the production of styrene from toluene and hydrogen (Douglas, 1988) it is required to recover most of the benzene from a saturated liquid stream containing 5 mol % benzene, 20 mol % toluene, 5 mol % ethylbenzene, 20 mol % styrene, 15 mol % diethylbenzene, 5 mol % triethylbenzene, and 30 mol % diphenyl (the components are numbered from 1 to 7 in the order presented). Design a distillation column to separate the feed into a distillate stream containing 99.5 mol % benzene and a bottom stream containing 0.1 mol % benzene (this corresponds to a benzene fractional recovery of 98%).

- Since the mixture consists of similar aromatic materials, it is expected to obey Raoult's law. In order to calculate vapor-liquid equilibria we need to find a vapor pressure correlation for each of the seven components. Table 4.2 gives the Antoine coefficients for each component.
- Using the zero-volume method for the direct split of a seven component mixture we calculate $r_{min} = 5.18$ and $(V_B/F)_{min} = 0.30$.
- We take the operating reflux ratio 20% higher than the minimum,[27] giving $r = 6.22$ and $V_B/F = 0.36$.
- We now use the design procedure to find the number of stages and the product distributions that correspond to the design specifications: $r = 6.22$, $x_{D,1} = 0.995$, $x_{B,1} = 0.001$. The results of these calculations are given in Fig. 4.37.
- We pick the fourth degree of freedom as $\omega = 0.2$, which corresponds to a design with the smallest number of stages that meets our design specifications. This design has a total of 27 theoretical stages, 10 in the rectifying section and 17 in the stripping section.
- The volatilities for this mixture are not constant (check for yourself). Nevertheless, if we approximate the VLE for the mixture with the constant volatility model, using the geometric mean of the volatilities at the top and bottom of the column, the minimum reflux ratio for the separation is calculated to be 4.63,[28] which is 10.6% less than the value obtained using Raoult's law. Choosing an operating reflux ratio of $r = 5.57$ $(= 1.2 \times r_{min})$, the distillation column requires 27 stages (calculated using the above design procedure). The Fenske-Underwood-Gilliland estimate is 19 stages (rounded

[27]This is a little closer to r_{min} than we normally suggest. We justify this on the basis that the mixture is ideal and there is not much uncertainty in its physical properties. However, it is worth repeating the calculations with $r = 1.5 \times r_{min}$.

[28]This value is obtained from either Underwood's method or the zero-volume method.

(*a*)

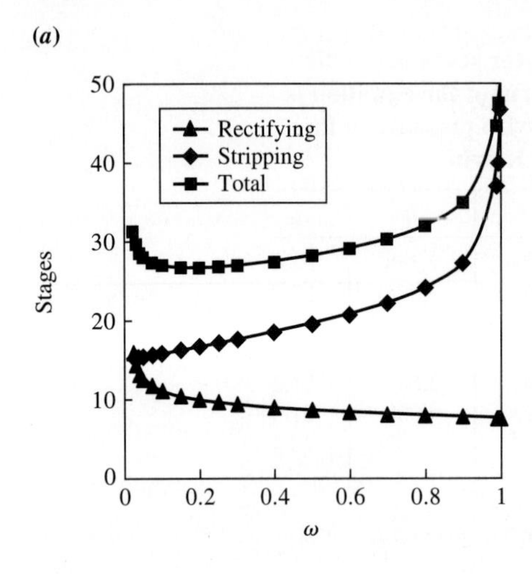

(*b*)

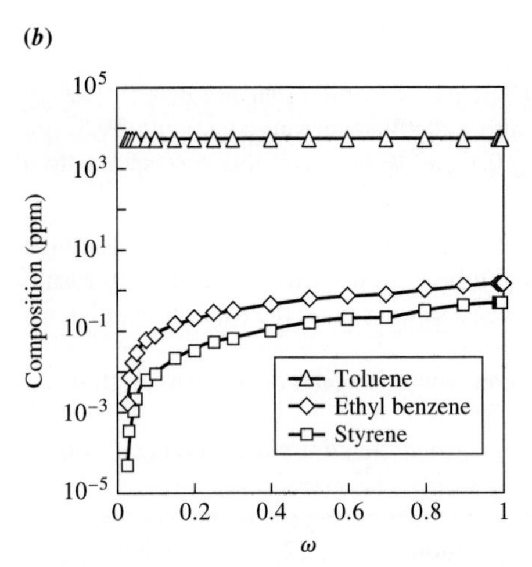

FIGURE 4.37
Spectrum of designs for the seven component mixture of benzene-toluene-ethylbenzene-styrene-diethylbenzene-triethylbenzene-diphenyl (the components are numbered from 1 to 7 in the order presented). The feed is a saturated liquid with a composition given by $\mathbf{x}_F = (0.05, 0.2, 0.05, 0.2, 0.15, 0.05, 0.3)$. The values of the design variables are $r = 6.22$, $x_{D,1} = 0.995$, and $x_{B,1} = 0.001$. (*a*) Total number of stages, and number of stages in each column section as a function of ω. (*b*) Distillate composition (in parts per million molar) as a function of ω. Note that the composition of diethylbenzene and heavier components in the distillate is vanishingly small (i.e., less than 10^{-5} ppm, which cannot be measured by modern instrumentation).

up from 18.4). However, at this operating reflux ratio the separation actually requires 36 theoretical stages when Raoult's law is used to represent the VLE. Therefore, the constant volatility approximation underestimates the number of theoretical stages in the column by 25 to 50% depending on the design method used.

4.6 TANGENT PINCHES

In Chap. 3 we showed that tangent pinches are commonly encountered in the design of distillation columns for the separation of nonideal binary mixtures. At a tangent pinch, one of the operating lines becomes tangent to the equilibrium curve on a McCabe-Thiele diagram. For binary mixtures it is easy to visualize these diagrams, and as a result it was relatively straightforward to develop the theory presented in Chap. 3.

In Sec. 3.5, the key figure that determines the controlling pinch as a function of feed composition is Fig. 3.13 (redrawn here as Fig. 4.38*a*). The same information can be represented on a plot of pinch composition x_p versus reflux ratio r, as shown in Fig. 4.38*b*. When $r = 0$ the top operating line is horizontal, i.e., line 3 in Fig. 4.38*a* is horizontal, and the pinch composition of the vapor y_p is equal to x_D. The pinch composition of the liquid is equal to the corresponding equilibrium liquid composition $x_p = x(y_p = x_D) < x_D$. As the reflux ratio increases, the value of x_p decreases smoothly (along the upper branch of pinches in Fig. 4.38*b*) until line 3 in

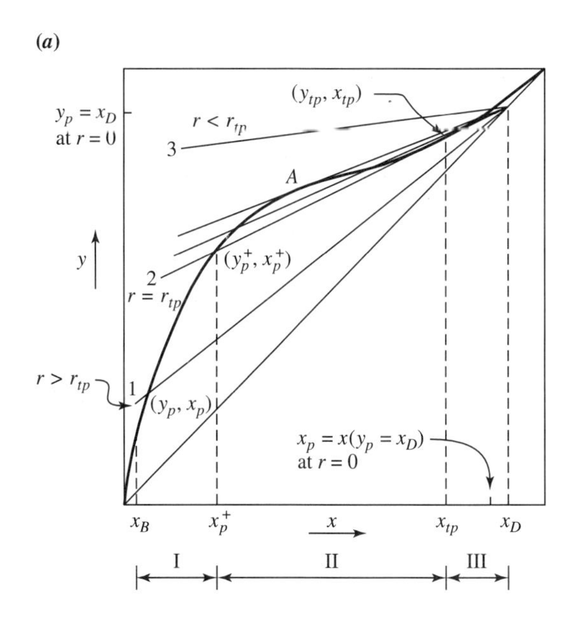

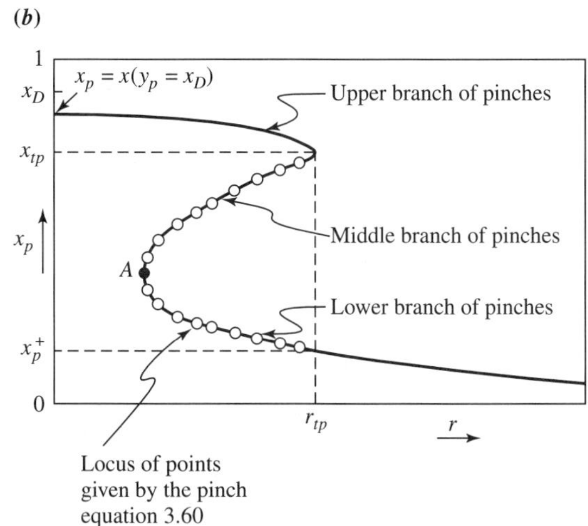

FIGURE 4.38
Two ways of representing tangent pinches (*a*) a McCabe-Thiele diagram and (*b*) a bifurcation diagram.

Fig. 4.38*a* becomes tangent to the equilibrium curve from above (denoted by point *A* in the figure). At this point a second solution to the pinch equation appears. The corresponding point *A* on Fig. 4.38*b* is called a *turning point* or a *bifurcation point*. Further increase in r causes three transverse intersections between the equilibrium curve and the top operating line. The two new branches of pinch points grow out from point *A*, as can be seen more clearly on Fig. 4.38*b*. Since these new branches of pinch points correspond to intersections between the equilibrium curve and the top operating line that have no physical meaning, we have labeled them with open circles in Fig. 4.38*b*. The three branches of pinches remain distinct until $r = r_{tp}$. At this point the equilibrium curve and the top operating line became tangent again (line 2 in Fig. 4.38*a*), and the middle branch in Fig. 4.38*b* coalesces with the upper branch at a second turning point. This is the tangent pinch point. Further increase in r results in a single intersection between the equilibrium curve and the top operating line; that is, only the lower branch in Fig. 4.38*b* survives.

In order to emphasize that the branches in Fig. 4.38*b* labeled with open circles are physically meaningless, we have omitted them in Fig. 4.39. This figure also shows the relationship between the feed composition x_F and the controlling pinch composition for saturated liquid feeds. The horizontal arrows in regions I and III signify that under feed-pinch control, the pinch composition is equal to the feed composition, and the corresponding value of $r_{\min}$ is read off the abscissa. In region II, the tangent pinch controls, and both $r_{\min}$ and x_p remain constant at r_{tp} and x_{tp}, respectively, for all values of x_F.

Now that we have introduced the *bifurcation diagram* in Fig. 4.38*b* and shown that it contains information equivalent to Fig. 4.38*a*, we abandon the McCabe-Thiele

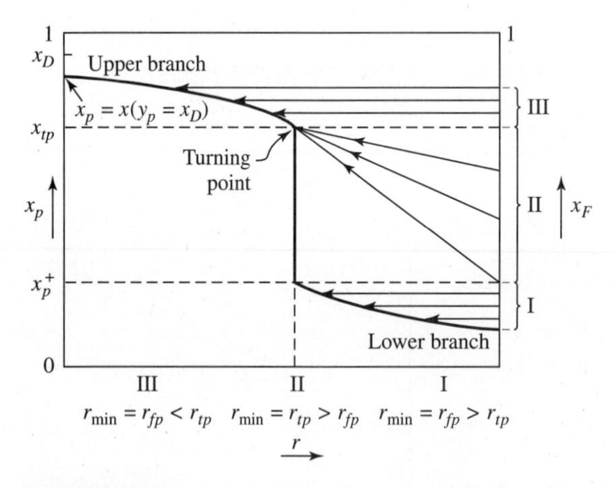

FIGURE 4.39
Bifurcation diagram of the pinch composition x_p vs. r showing the feed composition regions and the corresponding controlling pinches for saturated liquid feeds. In region I, the minimum reflux is controlled by the feed pinch while the tangent pinch controls in region II; in both cases, the minimum reflux is the larger of the two values. In region III, the feed pinch controls even though the corresponding reflux is *less* than the value determined by the tangent pinch.

diagram and develop an analysis for the new representation. This is the key step in developing a theory of multicomponent tangent pinches, since the McCabe-Thiele diagram in Fig. 4.38*a* does not readily extend to multicomponent systems while the bifurcation diagram in Fig. 4.38*b* does.

Each point on the curve in Fig. 4.38*b* is a pinch point and therefore satisfies the pinch equation (Eq. 3.60).

$$y_p - \frac{r}{r+1}x_p - \frac{x_D}{r+1} = 0 \tag{4.80}$$

This equation can be written in the more abstract functional form

$$g(x_p; r, x_D) = 0 \tag{4.81}$$

where r and x_D are treated as parameters. For a fixed value of x_D, Eq. 4.81 defines an *implicit function* between x_p and r. For each value of r there may be one or more values of x_p that satisfy Eq. 4.81. The graph of x_p versus r is called a bifurcation diagram, and the turning points in it are called bifurcation points. Whenever there is a range of values of r over which there are multiple solutions for x_p, then a tangent pinch occurs at the bifurcation point on the upper branch of solutions. Conversely, if for every value of r there is a unique value of x_p which satisfies Eq. 4.81, then a tangent pinch does not occur. Thus, the question of whether a tangent pinch occurs is reduced to the mathematical question of establishing the conditions under which an implicit function will give rise to unique solutions. This is addressed by one of the fundamental theorems of real analysis called the *implicit function theorem.*

An introduction to the statement and use of this theorem is given in Appendix B. Stripped of its technicalities, and expressed in terms of the problem formulated by Eq. 4.81, this theorem states that if the equation

$$g(x_p; r, x_D) = 0$$

has a solution at the point $(x_p^0; r^0, x_D)$ such that $g_{x_p}(x_p^0; r^0, x_D) \neq 0$,[29] then it will have a locally unique solution for x_p for all values of r near r^0. Thus, a necessary condition for a solution $(x_p^0; r^0, x_D)$ of $g(x_p; r, x_D) = 0$ to be a bifurcation point is that

$$g_{x_p}(x_p^0; r^0, x_D) = 0 \tag{4.82}$$

for otherwise we could solve uniquely for x_p as a smooth function of r. As shown in Appendix B, sufficient conditions for a turning point bifurcation to occur at a point $(x_p^0; r^0, x_D)$ are

$$g(x_p^0; r^0, x_D) = g_{x_p}(x_p^0; r^0, x_D) = 0$$

$$g_{x_p x_p}(x_p^0; r^0, x_D) \neq 0$$

$$g_r(x_p^0; r^0, x_D) \neq 0$$

[29] Throughout this analysis we keep x_D constant and draw families of bifurcation diagrams of x_p versus r parameterized by x_D, as discussed in Appendix B. The symbol $g_{x_p}(x_p^0; r^0, x_D)$ means $(\partial g/\partial x_p)_{r,x_D}$ evaluated at the point $(x_p^0; r^0, x_D)$. Similar interpretations apply to the second derivative $g_{x_p x_p}(x_p^0; r^0, x_D)$ and to the first derivative $g_r(x_p^0, r^0, x_D)$.

These conditions lead to

$$y_p^0 - \frac{r^0}{r^0+1}x_p^0 - \frac{x_D}{r^0+1} = 0 \tag{4.83}$$

$$\frac{dy}{dx}(x_p^0) = \frac{r^0}{r^0+1} \tag{4.84}$$

$$\frac{d^2y}{dx^2}(x_p^0) \neq 0 \tag{4.85}$$

$$\frac{x_D - x_p^0}{(r^0+1)^2} \neq 0 \tag{4.86}$$

Thus, for a fixed value of x_D, a point (x_p^0, r^0) that satisfies Eqs. 4.83 and 4.84 simultaneously, and also satisfies the inequalities 4.85 and 4.86, will be a turning point bifurcation. Notice that Eqs. 4.83 and 4.84 are the same as those obtained by the more traditional arguments in Sec. 3.6.

We expect Eqs. 4.83 and 4.84 to have two solutions, one for the turning point on the upper branch of the bifurcation diagram, i.e., the tangent pinch point, and one for the turning point on the lower branch, i.e., the nonphysical solution corresponding to point *A* in Fig. 4.38*b*. We can discriminate between these solutions, since the tangent pinch point is characterized on the bifurcation diagram by the condition $d^2r/dx_p{}^2 < 0$. From Appendix B, this leads to $-g_{x_px_p}/g_r < 0$ and since $g_r > 0$ (as is easily verified by checking the expression for g_r in Eq. 4.86) we find that $g_{x_px_p} > 0$ at a tangent pinch point. This in turn simplifies to $d^2y/dx^2(x_p) > 0$ at the tangent pinch, and may be used as a test to discriminate between the nonphysical solution and the tangent pinch solution to Eqs. 4.83 and 4.84.

A representative family of bifurcation diagrams, parameterized by x_D, for the mixture benzene-ethylenediamine is shown in Fig. 4.40.[30] Notice that as x_D decreases, the *S* shape in the curves gets shallower until it eventually disappears altogether at the critical value of $x_D = x_D^c$. Using the techniques described in Appendix B, sufficient conditions for a critical point to occur at a point $(x_p^c; r^c, x_D^c)$ are

$$y_p^c - \frac{r^c}{r^c+1}x_p^c - \frac{x_D^c}{r^c+1} = 0 \tag{4.87}$$

$$\frac{dy}{dx}(x_p^c) = \frac{r^c}{r^c+1} \tag{4.88}$$

$$\frac{d^2y}{dx^2}(x_p^c) = 0 \tag{4.89}$$

$$\frac{d^3y}{dx^3}(x_p^c) \neq 0 \tag{4.90}$$

$$\frac{x_D^c - x_p^c}{(r^c+1)^2} \neq 0 \tag{4.91}$$

[30]The locus of tangent pinch points in this figure is mathematically identical to the spinodal curve in the phase diagram for a pure fluid, e.g., Modell and Reid (1983).

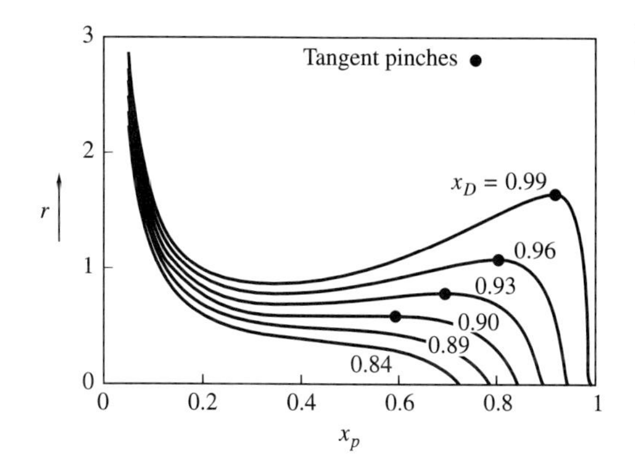

FIGURE 4.40
The bifurcation diagram for the mixture benzene-ethylenediamine at 1 atm pressure; the VLE relation is given by Eq. 3.62.

Since inequalities 4.90 and 4.91 are normally satisfied for all values of x_p, r and x_D, we may solve the three equations 4.87 to 4.89 for the three unknowns $(x_p^c; r^c, x_D^c)$. Again, these results are in agreement with those obtained by the more conventional treatment in Sec. 3.6.

These bifurcation concepts have been used by Fidkowski et al., (1991) to simplify the analysis of multicomponent mixtures with tangent pinches. They showed that turning points in a branch of pinches causes turning points in the *fixed-point volume* vs. r diagram. The S shape in this curve can be "cut off" at the turning point that occurs at larger r because this portion of the curve has no physical meaning. The minimum reflux ratio occurs where the modified *volume* curve has its zero. The result is shown in Fig. 4.41 for the benzene–ethylenediamine mixture.

Tangent pinches also occur in nonideal multicomponent mixtures, for which it is practically impossible to develop an analysis along the lines of Sec. 3.6, since it is so hard to visualize how the two operating lines are positioned in relation to the $c - 1$ dimensional equilibrium surface embedded in the $2c - 2$ dimensional $\mathbf{y}$-$\mathbf{x}$ space. Thus, in order to develop methods for treating multicomponent tangent pinches we need to exploit concepts which do not rely on our ability to interpret multidimensional $\mathbf{y}$-$\mathbf{x}$ diagrams. As we have demonstrated above, such a mathematical framework is provided by bifurcation analysis.

In order to keep the analysis as simple as possible, we again assume that the tangent pinch occurs in the rectifying section of the column, that constant molar overflow applies, and that the feed and product streams are saturated liquids. (Once this case has been mastered, it is possible to relax *all* these restrictions.)

For multicomponent mixtures, the pinch equations for the rectifying section are given by Eq. 4.57. In vector form they are written as

$$\mathbf{y}_p - \frac{r}{r+1}\mathbf{x}_p - \frac{\mathbf{x}_D}{r+1} = 0 \tag{4.92}$$

or

$$\mathbf{g}(\mathbf{x}_p; r, \mathbf{x}_D) = \mathbf{0}$$

where each composition vector contains the first $c - 1$ independent mole fractions, and $\mathbf{g}$ is a vector function in the variables $\mathbf{x}_p$ and the parameters r and $\mathbf{x}_D$. According to the implicit function theorem for systems of equations (see Appendix B), a

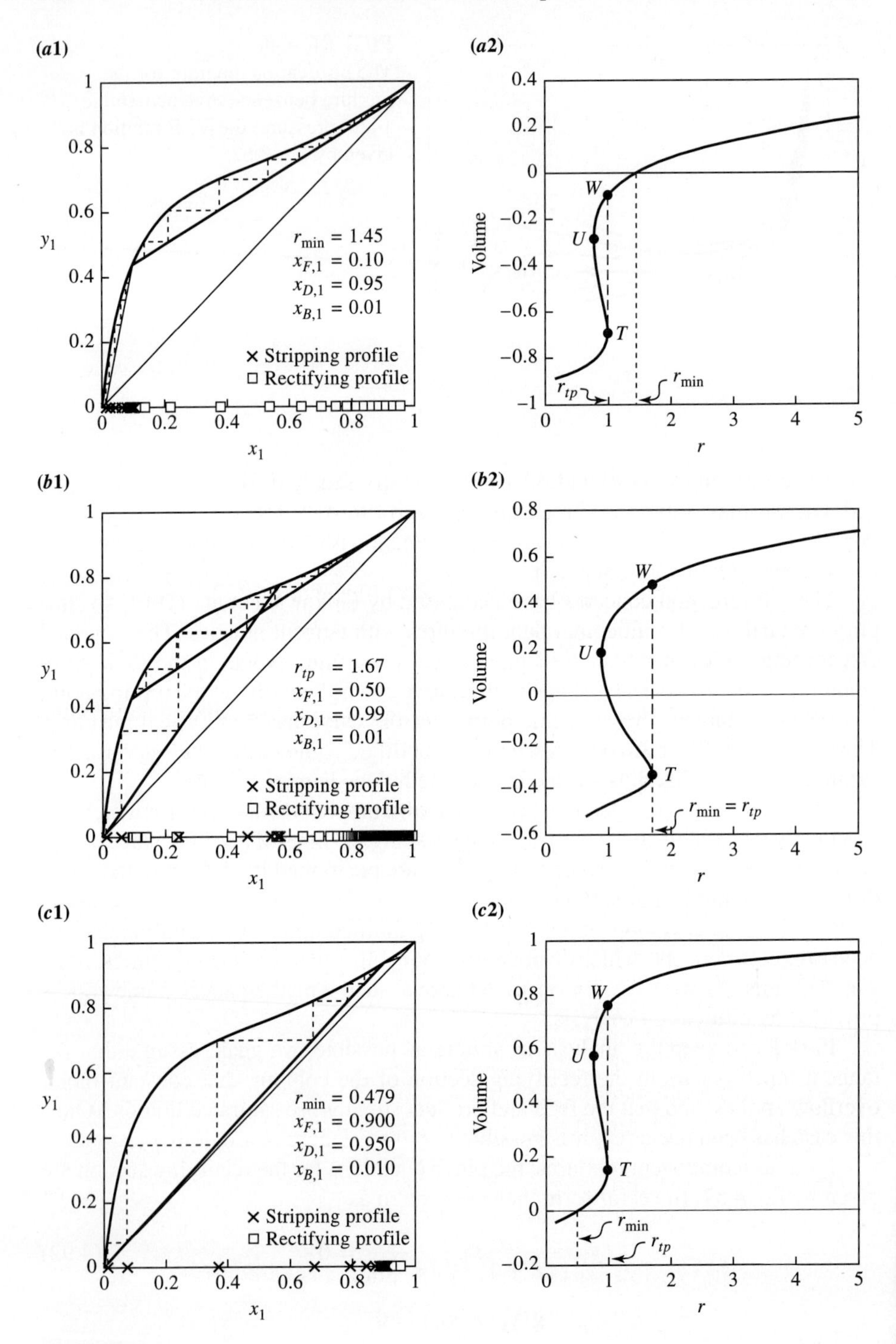

FIGURE 4.41
Three possible locations of a tangent pinch and corresponding *y*-*x* diagrams. (*a*) Tangent pinch at $r < r_{min}$. (*b*) Minimum reflux controlled by the tangent pinch. (*c*) Tangent pinch at $r > r_{min}$.

necessary condition for a solution $(\mathbf{x}_p^0; r^0, \mathbf{x}_D)$ of Eq. 4.92 to be a bifurcation point is $\det \mathbf{J}_{\mathbf{x}_p} = 0$, where

$$\mathbf{J}_{\mathbf{x}_p} = \begin{bmatrix} \frac{\partial y_1}{\partial x_1} - \frac{r}{r+1} & \frac{\partial y_1}{\partial x_2} & \cdots & \cdots & \frac{\partial y_1}{\partial x_{c-1}} \\ \frac{\partial y_2}{\partial x_1} & \frac{\partial y_2}{\partial x_2} - \frac{r}{r+1} & \frac{\partial y_2}{\partial x_3} & \cdots & \frac{\partial y_2}{\partial x_{c-1}} \\ \vdots & \vdots & \vdots & \vdots & \vdots \\ \frac{\partial y_{c-1}}{\partial x_1} & \cdots & \cdots & \cdots & \frac{\partial y_{c-1}}{\partial x_{c-1}} - \frac{r}{r+1} \end{bmatrix}_{x=\mathbf{x}_p^0, r=r^0}$$

or

$$\mathbf{J}_{\mathbf{x}_p} = \mathbf{Y} - \frac{r}{r+1}\mathbf{I}$$

where

$$\mathbf{Y} = \begin{bmatrix} & \vdots & \\ \cdots & \frac{\partial y_i}{\partial x_j} & \cdots \\ & \vdots & \end{bmatrix}$$

Thus, for a given value of $\mathbf{x}_D$, the necessary conditions for a bifurcation point are

$$\mathbf{y}_p - \frac{r}{r+1}\mathbf{x}_p - \frac{\mathbf{x}_D}{r+1} = 0 \tag{4.93}$$

$$\det(\mathbf{Y} - \frac{r}{r+1}\mathbf{I}) = 0 \tag{4.94}$$

This is a system of c equations in c unknowns, $\mathbf{x}_p$ and r. The equations will normally have multiple solutions, only one of which is the tangent pinch point.

By way of interpretation, we see from Eq. 4.94 that at the point of bifurcation, one of the eigenvalues of the matrix $\mathbf{Y}$ becomes equal to $r/(r+1)$. This is the natural extension of the binary tangent pinch condition, since for binary mixtures $\mathbf{Y}$ reduces to dy_1/dx_1. As an aside, it is worth noting that for multicomponent mixtures the eigenvalues of $\mathbf{Y}$ will generally be strictly positive and distinct. For multicomponent nonideal mixtures with constant latent heat, this property of $\mathbf{Y}$ is exact (Doherty, 1985). Therefore, at a tangent pinch, $\mathbf{J}_{\mathbf{x}_p}$ has a single zero eigenvalue, which is the simplest type of bifurcation problem to analyze and compute. Other ramifications of this result are discussed by Levy and Doherty (1986*b*), where this bifurcation theoretic approach was first introduced.

The simplest way to detect tangent pinches in multicomponent mixtures is to plot the fixed point volume vs. r bifurcation diagram to check for turning points, as demonstrated in Example 4.9.

EXAMPLE 4.9. In order to demonstrate this approach, we consider the ternary nonideal mixture acetaldehyde (20.5°C) + methanol (64.6°C) + water (100°C). The feed is a saturated liquid containing 30 mol % acetaldehyde and 30 mol % methanol. The targets for the

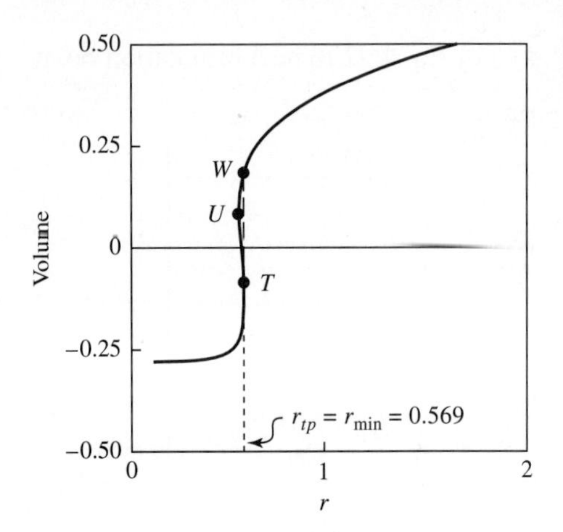

FIGURE 4.42
Fixed point volume as a function of reflux ratio for Example 4.9.

separation are saturated liquid products with the compositions $x_{1,B} = 1 \times 10^{-10}$, $x_{2,B} = 0.0103$, $x_{3,B} = 0.9897$, $x_{1,D} = 0.503$, $x_{2,D} = 0.496$, and $x_{3,D} = 0.001$. (The reader may verify that $\mathbf{x}_F$, $\mathbf{x}_B$, and $\mathbf{x}_D$ are collinear). The bottom stream is almost pure water and this is an indirect split. The phase equilibrium is represented with a two-parameter Margules model. We plot the fixed point volume vs. r in Fig. 4.42. The diagram has two turning points, indicating that a tangent pinch occurs. We cut off the S-shaped part at the higher value of r. Because the resulting vertical portion of the diagram crosses zero, a tangent pinch controls the minimum flows.[31]

To demonstrate that a tangent pinch occurs under these conditions, we plot the column composition profiles at values of r slightly below and slightly above r_{tp}, as shown in Fig. 4.43. At $r = 0.568 < r_{tp}$ the rectifying profile pinches before intersecting the stripping profile, while at $r = 0.570 > r_{tp}$, the rectifying profile not only intersects the stripping profile but extends well beyond it. This abrupt, discontinuous behavior is a fingerprint of tangent pinches and occurs in binary mixtures as well as multicomponent systems. Other examples of this behavior are given in Levy and Doherty (1986*b*) and Fidkowski et al. (1991).

4.7 EXERCISES

1. Consider a saturated liquid feed containing 30 mol % methanol (1), 25 mol % ethanol (2), and 45 mol % *n*-propanol (3) that is to be separated into saturated liquid products by distillation. The distillate and bottom streams contain 98 mol % and 2 mol % methanol, respectively. Use the geometric boundary value design method in Sec. 4.2 to find r_{min} and s_{min} as the mole fraction of *n*-propanol in the distillate is decreased. Try at least the following compositions for $x_{3,D}$: 0.01, 0.001, 1×10^{-5}, 1×10^{-8}. Show your calculated column composition profiles for each case on a separate triangular diagram and discuss your results.

[31] Solving Eqs. 4.93 and 4.94 leads to the following values for the tangent pinch: $r_{tp} = 0.569$, $x_{1,tp} = 0.156$, $x_{2,tp} = 0.617$.

	Feed	Distillate	Bottoms
1. Acetaldehyde	0.3000	0.5030	1.0×10^{-10}
2. Methanol	0.3000	0.4960	0.0103
3. Water	0.4000	0.0010	0.9897

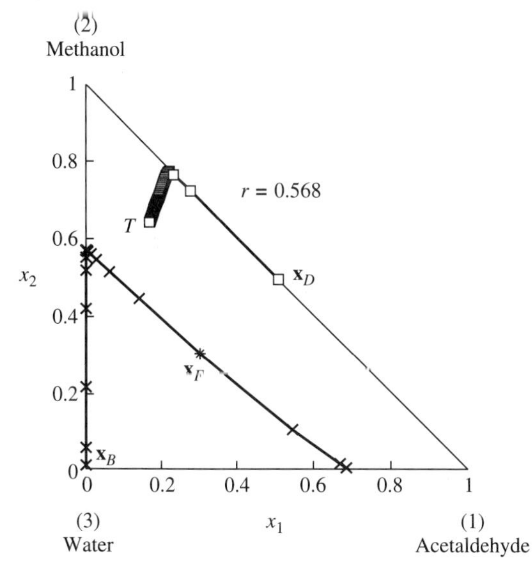

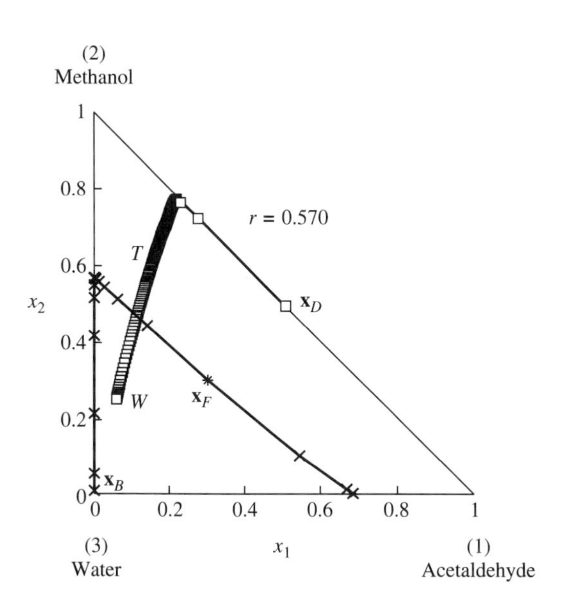

FIGURE 4.43
Composition profiles for a ternary mixture of acetaldehyde, methanol and water at (*a*) $r = 0.568$ and (*b*) 0.570. The abrupt change in the profile is indicative of a tangent pinch.

The VLE for this mixture can be represented either as an ideal solution (i.e., use Raoult's law) or with the following constant values for the relative volatilities: $\alpha_{13} = 3.25$, $\alpha_{23} = 1.90$, $\alpha_{33} = 1.00$.

2. Again consider the mixture methanol-ethanol-*n*-propanol with the same feed composition as in Exercise 1. This time keep the distillate composition fixed at

$x_{1,D} = 0.98$, $x_{3,D} = 1 \times 10^{-8}$. Using the geometric boundary value design method in Sec. 4.2, find the minimum reflux and reboil ratios at the following bottoms compositions: $x_{1,B} = 0.02, 0.01, 0.001, 1 \times 10^{-5}$. Assume the feed and product streams are saturated liquids. Show your calculated column composition profiles for each case on a separate triangular diagram and discuss your results.

3. A saturated liquid mixture containing 30 mol % pentane (1), 30 mol % hexane (2), and 40 mol % heptane (3) is fed to a distillation column with a partial reboiler and a total condenser. The products are withdrawn as saturated liquids.

   Plot a graph of the minimum reflux ratio versus the distillate composition of pentane. For selected distillate compositions of pentane show your computed column composition profiles on a triangular diagram. Be sure to include the profiles for the transition split. Keep the distillate composition of heptane constant at $x_{3,D} = 1 \times 10^{-11}$, and the bottom composition of pentane constant at $x_{1,B} = 0.0001$.

   Model the VLE either as an ideal solution (Raoult's law) or with constant relative volatilities of $\alpha_{13} = 6.35$, $\alpha_{23} = 2.47$, $\alpha_{33} = 1.00$.

4. Derive Fenske's equation 4.15 for multicomponent mixtures. The analogy with the binary case in Chap. 3 is helpful.

5. Consider the calculation of the minimum number of stages shown in Fig. 4.6 for a ternary mixture of the normal alkanes hexane, heptane, and nonane. The volatilities are $\alpha_{13} = 12.67$ and $\alpha_{23} = 5.345$ and we specify a feed composition $\mathbf{z}_F = (0.3, 0.3, 0.4)$ with $z_{1,D} = 0.9$ and $z_{1,B} = 0.1$. Use the material balance in Eq. 4.13, and Eq. 4.2 to find expressions for $z_{2,B}$, $z_{3,B}$ and $z_{3,D}$ in terms of the distillate mole fraction of the intermediate component $z_{2,D}$. Next, write Fenske's equation twice and solve for the four compositions above along with the minimum number of stages. Resolve for $z_{1,D} = 0.999$ and $z_{1,B} = 0.001$; why are the compositions of heavy components in the distillate so small?

6. This exercise refers to Sec. 3.5 and Table 3.1. Design the first column in Fig. 3.7*b*, treating the mixture as a pseudo-binary consisting of light component (l) ethylbenzene, and heavy component (h) which is taken to be a lumped component of *p*-xylene, *m*-xylene, and *o*-xylene. The relative volatility of the pseudo-binary mixture $\alpha_{1,h}$ is approximated as $\alpha_{1,2}$, i.e., 1.056 (this is a conservative estimate which ensures that the product purity specification on ethylbenzene will be met).

   If the distillate composition of ethylbenzene is 99.7 mol % and its fractional recovery is 90.3%, how many theoretical stages are required in each column section? Base your design on a saturated liquid feed, a total condenser, and a partial reboiler operating so as to provide saturated liquid products; use a reflux ratio of $1.5r_{\min}$. Solve this design problem using Underwood's method described in Sec. 4.3.

7. In contrast to the mixed C_8 aromatics, which are produced in large amounts, we now consider the separation of the isotope C^{13}, which is produced at the rate of about 20 kg/year in the United States. This isotope is used as a tracer in organic reactions and biological systems.

   In the late 1970s, a 700-ft-tall packed column was built below ground at the Los Alamos Scientific Laboratory to separate the rare but stable isotope

^{13}CO by distillation of the naturally occurring mixture of ^{13}CO and ^{12}CO (see Fig. 4.44). The natural abundance of ^{13}CO is 1.11 mol %. The normal boiling point of ^{13}CO is −191.46°C, while that of ^{12}CO is only a few tenths of a degree lower at approximately −192.0°C.

During startup, the column is cooled initially to liquid nitrogen temperature (the normal boiling point of N_2 is −195.8°C) as natural carbon monoxide gas at ambient conditions enters, rises to the condenser, and liquefies. The liquid flows downward and cools the column as it vaporizes. When liquid carbon monoxide collects at the bottom, it is vaporized by carefully regulated heat supplied to the reboiler. The column takes about

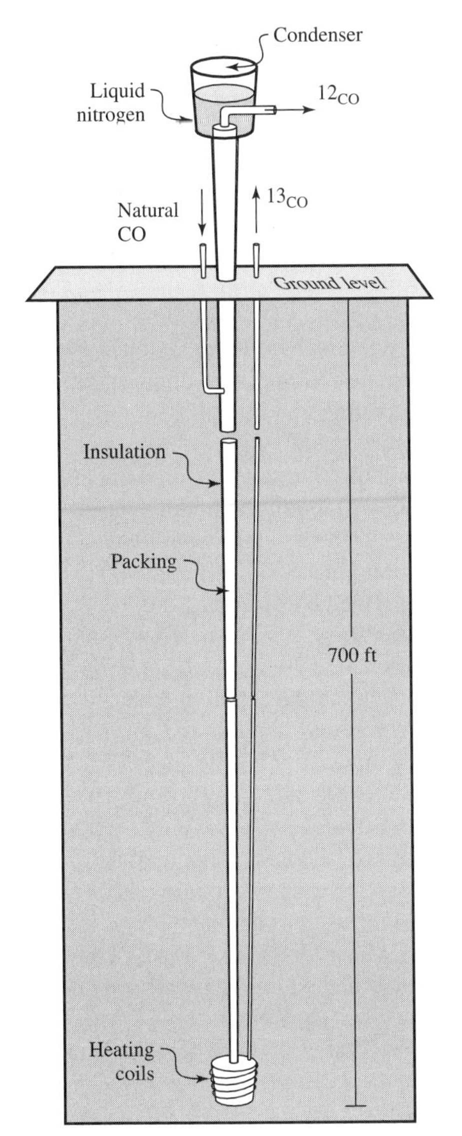

FIGURE 4.44
Isotope separation column. (From Matwiyoff et al., 1983.)

5 months to settle down to steady state, and during this startup period it changes length by about 2 ft.

At steady state, the column is fed with 302.5 mol/day of natural carbon monoxide gas at ambient conditions (note that $q < 0$). The bottoms stream is withdrawn at a composition of 82 mol % ^{13}CO, and the distillate contains 0.43 mol % ^{13}CO. The heat input to the reboiler is 3 kW.

a. Calculate the distillate and bottoms flow rates.

b. Treating the column as an adiabatic staged system and taking the products as saturated liquids, calculate the minimum reboil ratio required to achieve the given separation. What is the corresponding reflux ratio, and what are the minimum internal liquid and vapor flows in each column section? Calculate the minimum heat input to the reboiler. The thermodynamic and physical properties of the isotopes are as follows:

i. The VLE curve can be represented with a constant value for α of 1.005.

ii. The vapor-phase heat capacity of naturally occurring carbon monoxide is given by

$$c_p^V = 30.849 - 0.0128T + 0.27874 \times 10^{-4}T^2 - 0.12707 \times 10^{-7}T^3$$

where c_p^V is in J/mol K and T is in K.

iii. The latent heat of naturally occurring carbon monoxide at its normal boiling point is 6,040 J/mol.

c. The actual column at the Los Alamos Labs operates at a reboil ratio of $s = \beta s_{min}$. From your calculations, determine a value for the factor β. Now calculate the number of theoretical stages required in each column section to achieve the given separation at these operating conditions. Use Underwood's design equations for n_T and n_B.

d. In order to get some quantitative intuition about the stage requirements for a different purity ^{13}CO bottoms product, plot a graph of the bottoms purity of ^{13}CO (from 10 mol % to 99 mol %) on the ordinate vs. the minimum number of stages required for a fixed distillate purity of 0.43 mol % ^{13}CO.

This problem is based on the process described by McInteer and co-workers (McInteer, 1980; Matwiyoff et al., 1983). The heat capacity data were taken from Reid et al. (1977, Appendix A).

Additional Comments. In part *d*, if you calculate the actual number of stages in the column for a heat input of 3 kW you will find that the curve increases sharply at a bottoms composition of approximately 85 mol % ^{13}CO. A heat input of 4 kW, however, flattens out the curve and allows for much higher bottoms purities with a reduced number of stages.

8. Show that when the distillate mole fraction of the heaviest component is exactly zero, the pinch in the rectifying profile for a ternary mixture lies on face $x_3 = 0$. Also prove that this is identical to the location of a binary 1-2 pinch.
9. Repeat Example 4.4, but for an indirect split, where 99.5% of the n-propanol in the feed is recovered in the bottoms stream, which should have a purity of 98 mol % n-propanol.
10. Suppose that the feed in Example 4.4 contains ethanol in place of isopropanol. How do the results of the example change?

11. Repeat Example 4.5 but for an indirect split, where 99.5% of the *n*-propanol in the feed is recovered in the bottoms stream, which should have a purity of 98 mol % *n*-propanol.
12. Repeat Example 4.5, but for a feed that contains ethanol in place of isopropanol. Compare the result for minimum reflux to the value calculated using the approximate expression from Table 4.1.
13. Repeat Example 4.5, but for a feed that contains ethanol in addition to the original three components. Take the feed composition to be 30 mol % methanol, 15% ethanol, 15% isopropanol, and 40% *n*-propanol.
    a. Compare the results for r_{min} for the direct split to the value calculated using the the approximate expression in Table 4.1.
    b. Estimate r_{min} for the other splits that can be done as alternatives to the direct split.
    c. Find V_{min} for each of the splits and compare these.
14. Use the Fenske-Underwood-Gilliland method to answer the following.
    a. Estimate the number of equilibrium stages needed for Example 4.4. What is the uncertainty in your results on account of the uncertainty in the Gilliland correlation?
    b. Compare the Fenske-Underwood-Gilliland design with a more exact estimate using the geometric boundary value design method.
    c. Compare the results from both cases with a performance simulation.
15. Calculate the liquid and vapor pinch compositions for node pinches shown in Fig. 4.2. Show that these satisfy the collinearity condition given in Eq. 4.62. Compare the predictions of Eq. 4.61 for the reflex and reboil ratios with the values given in the figure caption.
16. Figure 4.13 shows the vapor and liquid pinch compositions for the rectifying and stripping profiles. Using the pinch points shown in the figure and equations 4.59 and 4.63, find the reflux and reboil ratios.
17. How would Fig. 4.15 change if (*a*) the mixture were the one in Example 4.4 at the same feed composition as shown in the figure and (*b*) the feed composition changes to 45% methanol, 10% isopropanol, and 45% *n*-propanol? Caution: This is potentially a long problem if you do not have a software tool to make the calculation. The ability to calculate profiles is sufficient.
18. To get an idea of the effect of the vapor-liquid equilibrium model on the accuracy of the model predictions, solve Example 4.8 using constant volatility models with values for the volatilities estimated as follows. Estimate the volatility from the vapor pressure ratios at temperatures corresponding to (*a*) your best estimate of the lowest temperature in the column, (*b*) your best estimate of the highest temperature in the column, and (*c*) geometric mean values of the volatilities calculated in (*a*) and (*b*).
19. Consider a saturated liquid mixture containing 40 mol % acetaldehyde, 20 mol % methanol, and 40 mol % water. We want to separate this mixture into pure saturated liquid products using distillation. The feed flowrate to the distillation system is 1,000 kmol/h. Compare the total vapor rate and the total number of stages required for the direct sequence with that for the indirect sequence if all the columns operate at $1.5r_{\text{min}}$ and use cooling water in the condenser.

References

Doherty, M. F., "Properties of Liquid–Vapor Composition Surfaces for Multicomponent Mixtures with Constant Latent Heat," *Chem. Engng. Sci.*, **40,** 1979–1980 (1985).

Douglas, J. M., *The Conceptual Design of Chemical Processes*. McGraw-Hill, New York (1988).

Eduljee, H. E., "Equations Replace Gilliland Plot," *Hydrocarb. Proc.*, **54,** 120 (1975).

Erbar, J. H., and R. N. Maddox, "Latest Score: Reflux vs. Trays," *Petrol. Refiner,* **40**(5), 183–188 (1961).

Fidkowski, Z. T., M. F. Doherty, and M. F. Malone, "Feasibility of Separations for Distillation of Nonideal Ternary Mixtures," *AIChE J.,* **39,** 1303–1321 (1993).

Fidkowski, Z. T., M. F. Malone, and M. F. Doherty, "Nonideal Multicomponent Distillation: Use of Bifurcation Theory for Design," *AIChE J.,* **37,** 1761–1779 (1991).

Forsyth, J. S. and N. L. Franklin, "The Interpretation of Minimum Reflux Conditions in Multicomponent Distillation," *Trans. Instn. Chem. Engrs.*, **31,** 363–388 (1953).

Franklin, N. L., "Counterflow Cascades: Part II," *Chem. Engng. Res. Design,* **66,** 47–64 (1988*a*).

Franklin, N. L., "The Theory of Multicomponent Countercurrent Cascades," *Chem. Engrg. Res. Design,* **66,** 65–74 (1988*b*).

Gilliland, E. R., "Estimation of the Number of Theoretical Plates as a Function of the Reflux Ratio," *Ind. Eng. Chem.,* **32**(9), 1220–1223 (1940).

Glinos, K., *A Global Approach to the Preliminary Design and Synthesis of Distillation Trains,* PhD thesis, University of Massachusetts, Amherst MA (1984).

Glinos, K., and M. F. Malone, "Minimum Reflux, Product Distribution and Lumping Rules for Multicomponent Distillation," *Ind. Eng. Chem. Process Design Dev.,* **23,** 764 (1984).

Julka, V., and M. F. Doherty, "Geometric Behavior and Minimum Flows for Nonideal Multicomponent Distillation," *Chem. Engng. Sci.,* **45,** 1801–1822 (1990).

Julka, V., and M. F. Doherty, "Geometric Nonlinear Analysis of Multicomponent Nonideal Distillation: A Simple Computer-Aided Design Procedure," *Chem. Engng. Sci.,* **48,** 1367–1391 (1993).

Knapp, J. P., and M. F. Doherty, "Minimum Entrainer Flows for Extractive Distillation. A Bifurcation Theoretic Approach," *AIChE J.,* **40,** 243–268 (1994).

Knight, J. R., and M. F. Doherty, "Design and Synthesis of Homogeneous Azoetropic Distillations. 5. Columns with Nonnegligible Heat Effects," *Ind. Eng. Chem. Fundam.*, **25,** 279–289 (1986).

Levy, S. G., and M. F. Doherty, "Design and Synthesis of Homogenous Azeotropic Distillations. 4. Minimum Reflux Calculations for Multiple Feed Columns," *Ind. Eng. Chem. Fundam.,* **25,** 269–279 (1986*a*).

Levy, S. G., and M. F. Doherty, "A Simple Exact Method for Calculating Tangent Pinch Points in Multicomponent Nonideal Mixtures by Bifurcation Theory," *Chem. Engng. Sci.*, **41,** 3155–3160 (1986*b*).

Levy, S. G., D. B. Van Dongen, and M. F. Doherty, "Design and Synthesis of Homogeneous Azeotropic Distillations; 2. Minimum Reflux Calculations for Nonideal and Azeotropic Columns," *Ind. Eng. Chem. Fundam.*, **24,** 463 (1985).

Matwiyoff, N. A., B. B. McInteer, and T. R. Mills, "Stable Isotope Production—a Distillation Process," *Los Alamos Science*, **65**(8) (1983).

McInteer, B. B., "Isotope Separation by Distillation: Design of a Carbon-13 Plant," *Sepn. Sci. Tech.*, **15**(491) (1980).

Modell, M., and R. C. Reid, *Thermodynamics and Its Applications*. Prentice-Hall, Englewood Cliffs, NJ, 2d ed. (1983).

Nikolaides, I. P., and M. F. Malone, "Approximate Design of Multiple-Feed/Sidestream Distillation Systems," *Ind. Eng. Chem. Research*, **26,** 1839–1845 (1987).

Reid, R. C., J. M. Prausnitz, and T. K. Sherwood, *The Properties of Gases and Liquids*. McGraw-Hill, New York, 3d ed. (1977).

Robinson, C. S., and E. R. Gilliland, *Elements of Fractional Distillation*. McGraw-Hill, New York, 4th ed. (1950).

Treybal, R. E., *Mass Transfer Operations*. McGraw-Hill, New York, 3d ed. (1980).

Underwood, A. J. V., "Fractional Distillation of Ternary Mixtures. Part I.," *J. Inst. Petrol.*, **31,** 111–118 (1946*a*).

Underwood, A. J. V., "Fractional Distillation of Ternary Mixtures. Part I.," *J. Inst. Petrol.*, **31,** 598–613 (1946*b*).

Underwood, A. J. V., "Fractional Distillation of Multicomponent Mixtures—Calculation of Minimum Reflux," *J. Inst. Petrol.*, **32,** 614–626 (1946*c*).

Wahnschafft, O. M., J. W. Koehler, E. Blass, and A. W. Westerberg, "The Product Composition Regions for Single-Feed Azeotropic Distillation Columns," *Ind. Eng. Chem. Research*, **31,** 2345–2362 (1992).

5

Homogeneous Azeotropic Distillation

Azeotropic and extractive distillations are important and widespread separation techniques in the chemical and biochemical industries. Many organic solvents, monomers, and fermentation products are purified by these methods.

The first successful distillation of an azeotropic mixture is credited to Young (1903), who received a patent for his batch distillation process for making anhydrous alcohol in 1903. This was later converted to a patented continuous process by Kubierschky (1915). Initially, these advances were largely ignored because there was little demand for anhydrous alcohol at the time. Nevertheless, Young's invention can be regarded as a landmark event in the history of the chemical processing industry.

Since that time, research and development on azeotropic distillation has received periodic attention, usually as a result of some crisis in world history. For example, World War I brought about a tremendous increase in demand for butanol which was met by the introduction of an azeotropic distillation process. Acetic acid demand also climbed, and engineers found that substantial savings in energy and increased production rates could be realized by converting from a nonazeotropic to an azeotropic distillation process. Then again, during World War II, the loss of the U.S. source of rubber from Southeast Asia was overcome by the development of a domestic synthetic rubber industry. This industry relied heavily on the use of azeotropic and extractive distillation for the recovery of high-grade butenes and butadiene. In recent decades the great increases in demand for high-grade chemicals, solvents, monomers, and biochemicals have transformed azeotropic separations from a specialty operation to one of the most important and commonplace separation techniques in the chemical industry.

To begin, it is worthwhile to describe the overall strategy in order to give some perspective to the individual steps which will be developed later in the chapter. When attempting to separate a binary azeotropic mixture[1] into its constituent pure components, the first thing to assess is whether or not the azeotrope is pressure-sensitive. If the azeotropic composition changes by a reasonable amount with moderate changes

[1]The methods developed in this chapter are not restricted to binary azeotropic mixtures. However, for clarity, our preliminary remarks discuss this case.

in pressure or, better still, if the azeotrope disappears altogether, then relatively straightforward distillation sequences can be devised to separate the original binary mixture into its pure components.[2] If the binary azeotrope is insensitive to pressure changes, or if pressure variations have been ruled out for processing reasons, then it is necessary to add a third component, which we will refer to as the *entrainer*,[3] in order to make the separation possible. Entrainers fall into at least four distinct categories, which may be identified by the way in which they make the separation possible:

1. Liquid entrainers that do not induce liquid-phase separation in the ternary mixture. Distillations of this type are called *homogeneous azeotropic distillations*, with classical extractive distillation being a special case.
2. Liquid entrainers that induce a liquid-phase separation in the ternary mixture. Distillations of this type are called *heterogeneous azeotropic distillations*.
3. Entrainers that react with one of the components in the original binary mixture, i.e., reactive entrainers.
4. Entrainers that ionically dissociate (e.g., inorganic salts) in the original binary mixture and move the composition of the binary azeotrope.

Within each of these categories not all entrainers will make the separation possible, i.e., not all entrainers will "break" the azeotrope. Thus, there is an incentive to develop methods that distinguish between feasible and infeasible entrainers. For feasible entrainers we identify the alternative feasible sequences and set targets on the distillate and bottom streams. Some of these targets are hard constraints that *must* be satisfied, e.g., product stream compositions, while others are more flexible and may be adjusted to reduce total processing costs, e.g., entrainer recycle flow and composition. This accomplished, each column in the sequence can be designed to meet its targets using the methods described in this and other chapters. Once a base-case design for each sequence is complete, alternatives can be optimized and ranked according to their economic potential. In order to do this properly, it is often necessary to incorporate each alternative into the entire process flowsheet and to *choose the alternative that optimizes the overall process economics.* Therefore, rapid identification of alternatives and estimates for the corresponding equipment sizes and energy use is very valuable for conceptual design.

Whether a given entrainer is feasible or not depends largely on the phase equilibrium behavior of the resulting ternary or multicomponent mixture. Thus, entrainer feasibility is an *intrinsic* thermodynamic property of the *mixture* and not of the individual components. In order to develop this point further, we first study the structure and properties of phase diagrams for azeotropic mixtures, a subject which will be considered in the next section and also in Appendix C. These diagrams lead to a more economical class of figures, called *residue curve maps*, which are the starting point of systematic methods for distinguishing between feasible and infeasible separations.

---

[2]A more quantitative treatment of this topic is given in Knapp and Doherty (1992) and Knapp (1991, Chap. II, A Bifurcation Theoretic Approach for Predicting the Effect of Pressure on Azeotropic Composition).

[3]Other names include *solvent, mass separating agent,* or simply *separating agent.*

In this chapter we concentrate on homogeneous entrainers, leaving heterogeneous and reactive systems until later in the book.

5.1 AZEOTROPY

In order to interpret the structure and properties of phase equilibrium surfaces for multicomponent azeotropic mixtures, it is useful to begin with a definition of what is meant by an azeotrope. Several definitions appear in the literature, only one of which is sufficiently general to apply under all circumstances. For homogeneous vapor-liquid mixtures, the various definitions all imply each other. However, this is not the case for more complex equilibria such as when multiple liquid phases or chemical reactions are present.

The formation of azeotropes was first reported by Dalton in 1802 (Dalton, 1802), who noticed that at the end of the distillation of aqueous solutions of hydrochloric and nitric acids both the boiling temperature and the composition of the mixture remained constant. However, the boiling temperature was always higher than that of the higher-boiling component. It was Wade and Merriman (1911) who first introduced the term "azeotrope" to designate mixtures that have a minimum (or maximum) boiling point. Malesiński (1965, pp. 1–9) gives a good historical account of the discovery and early scientific development of the phenomenon of azeotropy.

Over the years, two definitions of an azeotropic state have appeared widely in the literature. Wade and Merriman (1911) define an azeotropic state as a stationary point in the equilibrium T-$\mathbf{x}$, $\mathbf{y}$ or P-$\mathbf{x}$, $\mathbf{y}$ surface, which is commonly used as the working definition of an azeotrope, e.g., Malesiński (1965), Swietoslawski (1963). Alternatively, an azeotropic state has been defined as a state in which the composition of each component is the same in each of the coexisting phases. In Appendix C we show that these definitions are equivalent for nonreactive vapor-liquid mixtures with a single liquid phase. They cease to be equivalent, however, when reactions occur or when there are multiple liquid phases. In fact, for heterogeneous vapor-liquid-liquid systems *neither* definition is correct, since the derivative of the boiling temperature with respect to composition is not defined at the azeotropic point (Smith and VanNess, 1996, pp. 357–359) *and* the composition of each component is different in each of the coexisting phases.

In general *an azeotropic state is defined as a state in which mass transfer occurs between phases while the composition of each phase remains constant, but not necessarily equal* (see Prigogine and Defay, 1967; Rowlinson, 1969). With this as the fundamental definition of an azeotropic state, we can derive the necessary and sufficient conditions for an azeotropic transformation to occur in any kind of mixture (see also Appendix C and Chaps. 8 and 10). The "definitions" cited above may be shown to be special cases of this more comprehensive treatment.

In Appendix C we provide a more in-depth treatment of azeotropy and vapor-liquid phase diagrams for ternary mixtures.

Azeotropy and the resulting phase behavior have a profound effect on the feasibility and technology for distillation-based separations. An approach to understanding these effects that includes the minimum necessary information about the phase behavior, and that can also be closely connected to feasibility and design of systems is the residue curve map. These ideas are fundamentally displayed in the behavior of simple distillation, which we explore first, before making connections of the ideas with the behavior of continuous systems.

5.2 SIMPLE DISTILLATION RESIDUE CURVE MAPS

The least complicated of all distillation processes is the *simple distillation*, or open evaporation, of a mixture. The liquid is boiled and the vapors are removed from contact with the liquid as soon as they are formed (see Fig. 5.1). Thus, the composition of the liquid will change continuously with time, since the vapors are always richer in the more volatile components than the liquid from which they came. The trajectory of liquid compositions starting from some initial point is called a *simple distillation residue curve* or simply a *residue curve*. The collection of all such curves for a given mixture is called a *residue curve map*. These maps contain exactly the same information as the corresponding phase diagram for the mixture, but they represent it in a way that is much more useful for understanding and designing distillation systems. The concepts which we are about to develop for simple distillation serve as prototypes that can be extended to batch and continuous systems.

We begin by deriving the general conservation equations for multicomponent simple distillation. We then investigate their properties, first for binary mixtures and then for ternary and multicomponent systems.

General Equations

The pioneering work on simple distillation was published in the early 1900s by Schreinemakers (1901*a*, 1901*b*, 1903). He was the first to develop the general equations and analyze their properties, which was a remarkable achievement because he

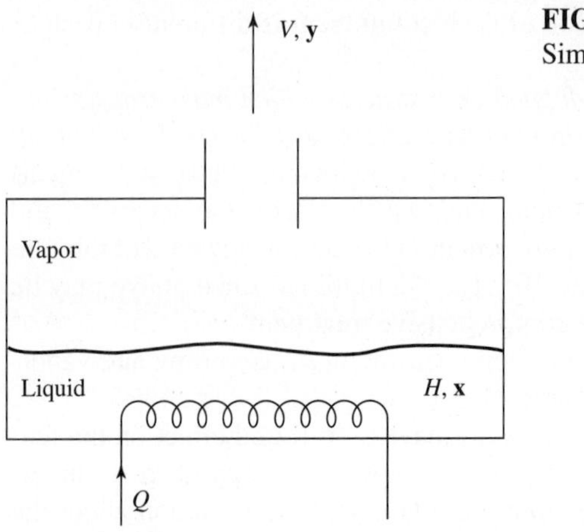

FIGURE 5.1
Simple distillation.

did this without the aid of the modern qualitative theory of nonlinear ordinary differential equations, which was being developed at about the same time by Poincaré and Liapounov.

We consider a simple distillation vessel in which the vapor is in phase equilibrium with the remaining well-mixed liquid (see Fig. 5.1). Overall material balance requires that the rate at which liquid is lost from the vessel exactly balances the rate at which vapor escapes. That is,

$$\frac{dH}{dt} = -V \tag{5.1}$$

where H and V represent the instantaneous total number of moles of liquid in the still, and the vapor rate (moles per unit time), respectively. The same must be true for each individual component; therefore, the $(c - 1)$ independent component balances are

$$\frac{d(Hx_i)}{dt} = -Vy_i \tag{5.2}$$

for $i = 1, 2, \ldots, c - 1$. Expanding the derivative in Eq. 5.2 and incorporating Eq. 5.1 leads to

$$H\frac{dx_i}{dt} - Vx_i = -Vy_i \tag{5.3}$$

for $i = 1, 2, \ldots, c - 1$ or, equivalently,

$$\frac{dx_i}{dt} = \frac{V}{H}(x_i - y_i) \tag{5.4}$$

for $i = 1, 2, \ldots, c - 1$, where H and V both vary with time as dictated by the energy balance. For binary mixtures, Eq. 5.4 is often called the *Rayleigh equation* (Rayleigh, 1902) and simple distillation is often called a Rayleigh distillation.

One way to proceed is to augment Eq. 5.4 with the overall energy and material balances, thereby having the facility to calculate the time dependence of V, H, and $\mathbf{x}$ once an energy input policy $Q(t)$ is prescribed. However, if we are interested only in the residue curve map; i.e., in knowing the x_i's relative to each other but not relative to time, then we can simplify the calculations by noting that the term V/H is a common factor in Eq. 5.4. Introducing a new warped time variable ξ, defined by a starting value, say, $\xi = 0$ at $t = 0$ and

$$d\xi = \frac{V}{H}dt \tag{5.5}$$

enables us to write Eq. 5.4 in the final form

$$\frac{d\mathbf{x}}{d\xi} = \mathbf{x} - \mathbf{y} \tag{5.6}$$

where $\mathbf{x}$ represents the state vector of $(c - 1)$ independent liquid phase mole fractions, and $\mathbf{y}$ the corresponding vector of equilibrium vapor phase mole fractions.

The new time variable ξ is a nonlinear transformation of the real time t. It has the advantage of being dimensionless and is defined on the interval 0 to $+\infty$ instead of 0 to t_{max}, where t_{max} is the time at which the still boils dry. These properties are shown by incorporating Eq. 5.1 into Eq. 5.5, which leads to

$$d\xi = -\frac{dH}{dt}\frac{1}{H}dt = -d\ln H \tag{5.7}$$

Integration of this equation with the initial condition $\xi = 0$ and $H = H_0$ at $t = 0$ gives

$$\xi(t) = \ln\left(\frac{H_0}{H(t)}\right) \tag{5.8}$$

Hence the above properties of $\xi(t)$ follow immediately.[4]

Equation 5.6 can be solved numerically using any of the usual integration schemes (e.g., Runge-Kutta, Gear's method, etc.; see Press et al., 1986, Chap. 15). The only subtlety is that at each function evaluation, a phase equilibrium calculation for $\mathbf{y}$ must be performed. If the distillation is isobaric, then an isobaric bubble-point calculation is required; if it is isothermal, then an isothermal bubble-point calculation is needed. We restrict our attention to isobaric systems, but the extension to isothermal processes is straightforward. In general, Eq. 5.6 poses no special numerical difficuties (i.e., they are not normally stiff or otherwise ill-conditioned) and can even be solved by Euler's method (see Greenberg, 1978, Chap. 24) with suitably small step size h. The algorithm for such a numerical integration scheme is then the simple, explicit recursive formula

$$\mathbf{x}_{n+1} = \mathbf{x}_n + h[\mathbf{x}_n - \mathbf{y}_n(\mathbf{x}_n, \mathbf{P})] \tag{5.9}$$

starting from the initial condition $\mathbf{x}_0$.[5]

Binary Mixtures

For binary mixtures with equilibrium behavior that can be represented by the empirical model

$$y = \frac{ax}{1 + (a-1)x} + bx(1-x) \tag{5.10}$$

Equation 5.6 takes the form

$$\frac{dx}{d\xi} = x - \frac{ax}{1 + (a-1)x} - bx(1-x) \tag{5.11}$$

[4] To a first approximation $H(t)$ varies linearly with time; thus Eq. 5.8 takes on the simple and explicit form $\xi(t) = \ln(H_0/(H_0 - (Q/\lambda)t))$, where Q is the heating rate to the still (assumed constant) and λ is the average latent heat of the mixture, also assumed constant (Doherty and Perkins, 1978*a*).

[5] We actually recommend integrating Eq. 5.6 with either Gear's method or a fourth-order Runge-Kutta method. However, Eq. 5.9 might be a useful place to begin for those readers who have no prior experience at solving ordinary differential equations numerically. A value for h of 0.01 will often give satisfactory results. However, this should be tested for each new problem by ensuring that the position of the trajectory does not change with further reduction of h.

with the initial condition $x(\xi = 0) = x_0$. This differential equation can be integrated using the method of partial fractions combined with separation of variables to obtain the solution

$$\xi = \frac{1}{a-1+b} \ln \frac{x_0(1-x)}{x(1-x_0)} + \frac{(a-1)^2}{(a-1+b)(a-1+ab)} \cdot \ln \frac{(1-x)\{a-1+b[1+(a-1)x_0]\}}{(1-x_0)\{a-1+b[1+(a-1)x]\}} \tag{5.12}$$

For constant volatility mixtures (i.e., $b = 0$, $a = \alpha$), this solution reduces to

$$\xi = \frac{1}{\alpha - 1} \ln \frac{x_0(1-x)}{x(1-x_0)} + \ln \frac{1-x}{1-x_0} \tag{5.13}$$

We now present two examples which illustrate the behavior of Eqs. 5.12 and 5.13. For the first example we consider the mixture ethanol-isopropanol, which exhibits phase behavior that can be well represented with a constant value for the relative volatility of 1.9. Figure 5.2 shows the transient response x vs. ξ for six different initial compositions in the still pot. These curves were generated by plotting Eq. 5.13 with $\alpha = 1.9$. Notice that in all cases, the ethanol mole fraction in the liquid decreases with time, eventually approaching zero as ξ gets large. This is in line with our intuition. Since the mixture does not form an azeotrope, we expect the lighter component (i.e., ethanol) to preferentially boil off, thereby making the liquid richer in the heavier component as time proceeds. By the same reasoning we would expect the temperature of the boiling liquid to continuously increase with time.

The behavior of the system can be represented in a more concise way by projecting the transient responses onto the composition axis. The resulting *trajectories* on the *phase line* show how the still composition varies with time; increasing values of ξ are indicated by the direction of the arrows. As argued above, this is coincident with the direction of increasing boiling temperature. For this example, all trajectories move away from the point $x = 1$ toward the point $x = 0$. Thus, we call these

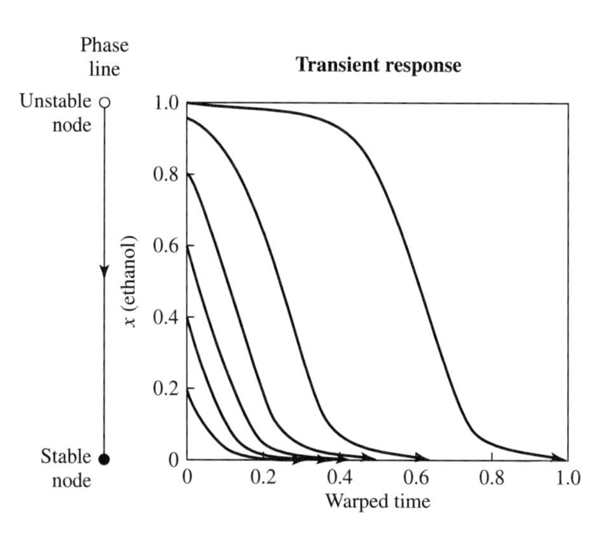

FIGURE 5.2
Simple distillation residue curves for a binary mixture of ethanol and isopropanol at 750 mmHg pressure.

points unstable and stable nodes, respectively. These terms are all borrowed from the field of nonlinear analysis, e.g., see Greenberg (1978, Chap. 23), Guckenheimer and Holmes (1983), and Hale and Koçak (1991) which provides a unified framework for studying nonlinear systems in one or more dimensions.

In the second example, we consider the mixture pentane-dichloromethane at 750 mmHg pressure. The VLE data for this mixture are shown in Fig. 5.3 (Gmehling et al., 1980, p. 100). The solid line in this figure represents the correlation Eq. 5.10 with $a = 4.2$ and $b = -1.2$. As can be seen from the data, this mixture has a minimum-boiling azeotrope at approximately 52 mole % pentane. Now we find that the outcome of a simple distillation depends on the initial composition in the still. Mixtures that are initially richer than the azeotropic compositions in pentane get further enriched with increasing time, while mixtures that are initially leaner than the azeotrope get progressively exhausted. This is shown in Fig. 5.4, which was generated by plotting Eq. 5.12 for six different initial conditions. Since the azeotrope is minimum-boiling (its boiling point is 30.6°C), the equilibrium temperature increases on either side (the boiling points of pentane and dichloromethane are 34.9°C, and 38.9°C, respectively). Thus, the trajectories move away from the azeotrope, making it an unstable node on the phase line. Correspondingly, the two pure components are stable nodes on the phase line.

The concepts introduced by way of these examples are not restricted to mixtures that obey the empirical equilibrium model (Eq. 5.10). When more detailed equilibrium relations are used, e.g., the Wilson equation or UNIQUAC equation, etc., it is no longer possible to obtain analytical solutions like Eq. 5.12, and the simple distillation equations (Eq. 5.6) must be solved numerically. However, the trajectories on the phase line still point in the direction of increasing temperature. Actually, if this is the only information required, then no calculations are necessary since the direction of increasing temperature is easy to establish once the boiling points of the two pure components and any azeotropes which may be present are known.

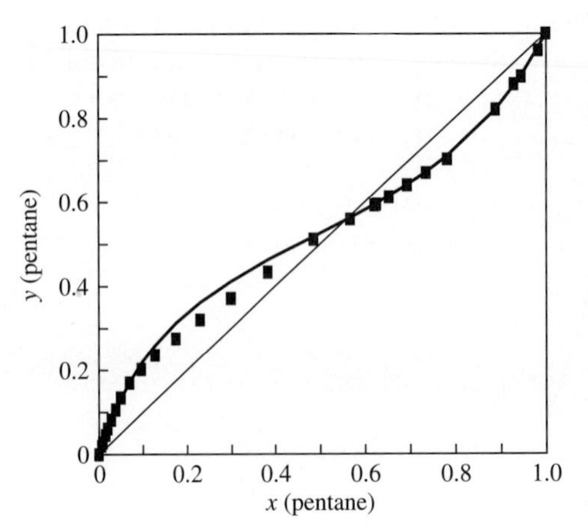

FIGURE 5.3
VLE data for a binary mixture of pentane and dichloromethane at 750 mmHg pressure. *(Taken from Gmehling et al., 1980, p. 100.)*

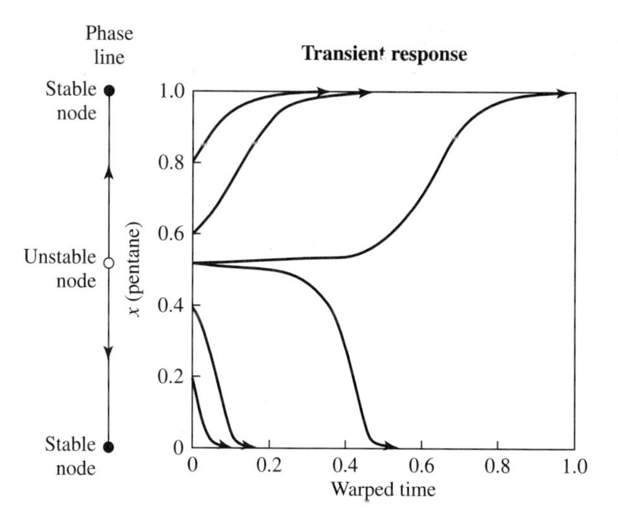

FIGURE 5.4
Simple distillation residue curves for a binary mixture of pentane and dichloromethane at 1 atm pressure.

5.3 RESIDUE CURVE MAPS FOR TERNARY AND MULTICOMPONENT MIXTURES

To begin, we discuss the residue curve map for an ideal ternary mixture of methanol–ethanol–n-propanol at 1 atm pressure. Integration of Eq. 5.6, using Raoult's law to represent the VLE, leads to the residue curve map shown in Fig. 5.5.[6] In this figure, all the residue curves begin at the pure methanol vertex (the lightest component) and end at the pure n-propanol vertex (the heaviest component). Along the path, the mixture gets initially richer in ethanol (the intermediate-boiling component), reaches a maximum, and then gets progressively leaner in ethanol. We would expect this to happen for an ideal mixture. At the beginning of the distillation the vapor is rich in the lightest component, which means that this component is preferentially removed from the liquid. Initially, therefore, the residue curve (which tracks the liquid composition) moves away from the methanol vertex as the liquid gets enriched in the two heavier components. After a while, most of the methanol has been removed from the liquid and ethanol starts to become the dominant component in the vapor. At this point, the liquid composition of ethanol starts to drop until the liquid consists of n-propanol with only trace amounts of the other two components.

Since all the residue curves begin at the methanol vertex, this point acts as a source of curves and is called an *unstable node*. n-Propanol is a sink for the curves and is called a *stable node*. Ethanol is an interesting vertex because it acts as a

[6]This map is generated by integrating the equations from an initial condition (i.e., a feed composition) and integrating forward in warped time until a stable node is reached. Then we go back to the starting point and integrate the equations backward in warped time until an unstable node is reached. In this way we produce the entire curve, from unstable node to stable node. We then repeat this procedure for several other feed compositions.

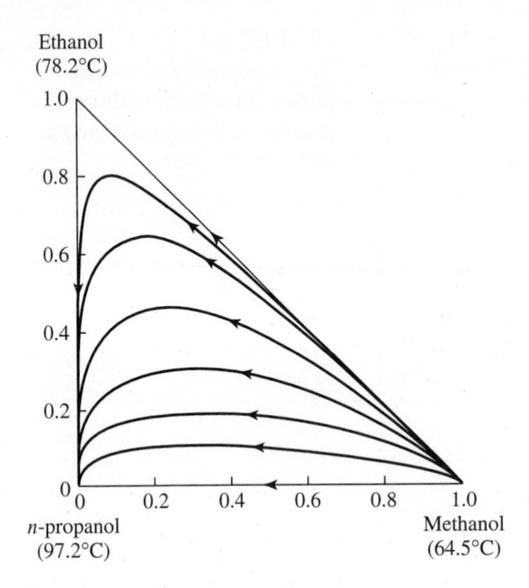

FIGURE 5.5
Residue curve map for a ternary mixture of methanol, ethanol, and *n*-propanol at 1 atm pressure. The arrows point in the direction of increasing time.

sink for a single curve on the binary edge leading from the methanol vertex[7] and as a source for a single curve on the binary edge leading to the *n*-propanol vertex; yet all the other curves generated from feed compositions inside the triangle pass by this vertex without stopping. Such a point is called a *saddle point*. Notice that the unstable node is the lightest component (lowest-boiling), the stable node is the heaviest component, and the saddle point is the intermediate-boiling component. Along each residue curve the boiling temperature continually *increases* as the liquid gets richer in the heavier components. You will also notice that the direction of the residue curves inside the triangle is consistent with the direction of the curves on the binary edges. These edge directions are easy to determine from the boiling temperatures of the individual pure components. It is clear, therefore, that every ternary mixture that obeys Raoult's law will have a residue curve map qualitatively similar to the one shown in Fig. 5.5. As azeotropes are introduced into nonideal mixtures, however, the residue curve maps get more complicated, as we shall see below. But first we summarize the general properties of residue curve maps in order to provide a framework for understanding nonideal mixtures.

The general properties of residue curve maps are established by analysis of the governing equation (Eq. 5.6). These properties are:

Property 1. The residue curve through any given liquid composition point is tangent to the vapor-liquid equilibrium tie-line through the same point.

Property 2. Residue curves do not cross each other, nor do they intersect themselves.

[7]This is perfectly consistent with our intuition because a binary mixture of methanol and ethanol will distill off the methanol, leaving the mixture ever richer in the heavier component, ethanol.

Property 3. The boiling temperature always increases along a residue curve (the only exception is at steady states where the boiling temperature remains constant because the composition remains constant).

Property 4. Steady-state solutions[8] of the equations occur at all pure components and azeotropes.

Property 5. Steady-state solutions are limited to one of the following types: (1) stable node, (2) unstable node, or (3) saddle point.

Property 6. Residue curves at nodes are tangent to a common direction. At pure component nodes this common direction must be one of the binary edges of the composition diagram, which is determined by the following rule. Vertex v ($v = 1, 2, \ldots, c$) is a stable node if $\alpha_{i,v} > 1$ ($i = 1, 2, \ldots, c; i \neq v$),[9] it is an unstable node if $\alpha_{i,v} < 1$ ($i = 1, 2, \ldots, c;\ i \neq v$); otherwise it is a saddle. At a pure component node, the residue curves are tangent to the edge with the relative volatility closest to unity, where the appropriate volatilities to check are $\alpha_{i,v}$ ($i = 1, 2, \ldots, c;\ i \neq v$).

These and other, more subtle, properties are proved in Doherty and Perkins (1978) and Van Dongen and Doherty (1984); Property 6 follows from the results in Doherty and Perkins (1978, Appendixes I and II).[10] Property 6 gives useful information about the product distribution at stable and unstable nodes. In ideal mixtures, for example, the lightest pure component (designated L) is always the one and only unstable node, so L will be the most plentiful component in the distillate. The next most plentiful component is determined by the shape of the residue curves near vertex $v = L$. For ideal mixtures $\alpha_{HL} < \alpha_{IL} < 1$;[11] therefore, the residue curves are always tangent to the I-L edge of the triangle at the unstable node, from which we conclude that the next most plentiful component is the intermediate-boiling component, i.e., the second lightest component. This is exactly what we would expect from intuition. The heaviest-boiling component is always the stable node (and therefore the most plentiful component is H); the next most plentiful component is determined by the shape of the residue curves near vertex $v = H$. For ideal mixtures $\alpha_{LH} > \alpha_{IH} > 1$, and the residue curves are always tangent to the I-H edge. Therefore, the next most plentiful component at the stable node is I (the second heaviest component), again as expected from intuition.

The volatilities at the three vertices in Fig. 5.5 are as follows:

Methanol Vertex: $\alpha_{n\text{-propanol,methanol}} = 0.25$, $\alpha_{\text{ethanol,methanol}} = 0.64$. Therefore, this vertex is an unstable node with methanol as the most plentiful component and ethanol as the next most plentiful component. The residue curves are tangent to the methanol-ethanol edge.

---

[8]These are also referred to as *singular points*.

[9]This means the local value of each relative volatility evaluated at the composition and temperature of the node, where $\alpha_{i,v} = P_i^{\text{sat}}\gamma_i / P_v^{\text{sat}}\gamma_v$.

[10]The basic mathematical fact that is needed to prove this property is that the trajectories (in our case residue curves) at a node are all tangent to the eigendirection with the smallest eigenvalue in absolute magnitude.

[11]Where I and H signify the intermediate-boiling and heavy components, respectively.

n-Propanol Vertex: $\alpha_{\text{methanol},n\text{-propanol}} = 4.0$, $\alpha_{\text{ethanol},n\text{-propanol}} = 2.1$. This vertex is a stable node with n-propanol as the most plentiful component and ethanol as the next most plentiful component. The residue curves are tangent to the n-propanol–ethanol edge.

Ethanol Vertex: $\alpha_{\text{methanol,ethanol}} = 1.56$, $\alpha_{n\text{-propanol,ethanol}} = 0.48$. This vertex is a saddle.

The calculated residue curves in Fig. 5.5 are in agreement with this analysis.

For ternary mixtures with no azeotropes there is only one kind of residue curve map (Fig. 5.5). For mixtures with a single minimum-boiling binary azeotrope, however, there are three kinds of map depending on which binary edge the azeotrope occurs. Figure 5.6 shows the three possibilities. In Fig. 5.6*a* and *c* the azeotrope is the unstable node (i.e., the lightest-boiling species in the mixture), the heavy pure component is the stable node, and the light (L) and intermediate (I) pure components are saddles (which means that they are *both* intermediate boilers for feeds that contain all three components). Therefore, in these mixtures the azeotrope has the effect of turning the lowest-boiling pure component (L) into an intermediate-boiling saddle. The residue curve map in Fig. 5.6*c* is the basis for extractive distillation, which is a widely practiced method for separating binary azeotropes. This map is usually called the *extractive map*, and we will have more to say about this map later in the chapter. Another new effect occurs in Fig. 5.6*b*. Here, the azeotrope turns the intermediate-boiling pure component (I) into a stable node. Mixtures of this type have two stable nodes, which means that some residue curves will go to one of them and the remaining residue curves will go to the other. This has the effect of dividing the composition triangle into two distinct regions divided by the single curve (shown in the figure) that connects the unstable node at L to the saddle at the azeotrope. This curve is called a *simple distillation boundary*.[12] All feed compositions that lie above the distillation boundary generate residue curves that go to I while all feed compositions below the boundary have residue curves that go to H. Therefore, the

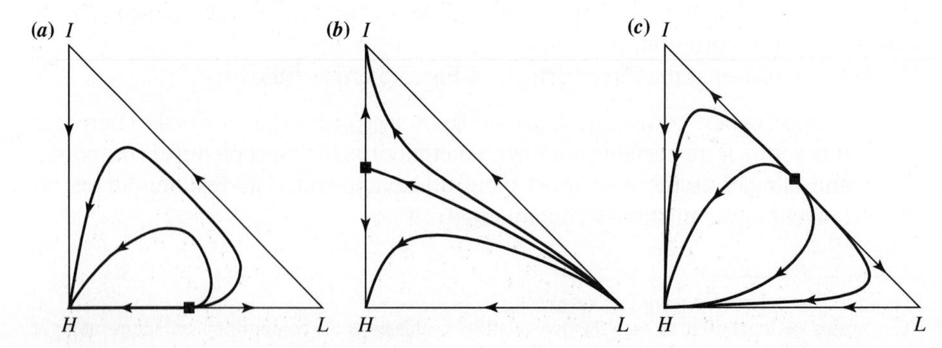

FIGURE 5.6
Residue curve maps (schematic) for a ternary mixture with one binary minimum-boiling azeotrope. The arrows point in the direction of increasing time. L, I, and H represent the light, intermediate, and heavy pure components, respectively.

[12] An equivalent technical term that you will read in the research literature is *separatrix*.

feed composition determines whether the final product (i.e., the last drop of liquid in the still) is pure component I or pure H. Distillation regions and distillation boundaries are defined as follows, and they generate two additional properties of residue curve maps.

Property 7. Each distillation region must contain one unstable node, one stable node, and at least one saddle. Distillation regions are divided by distillation boundaries.

Property 8. There must be a saddle on at least one end of every distillation boundary.

Examples of these residue curve maps (RCMs) for mixtures with a single binary minimum-boiling azeotrope are shown in Fig. 5.7.

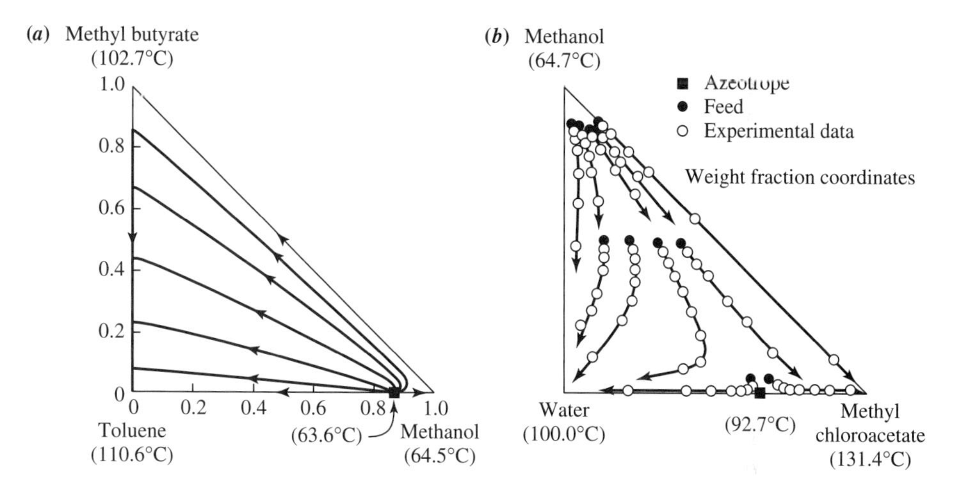

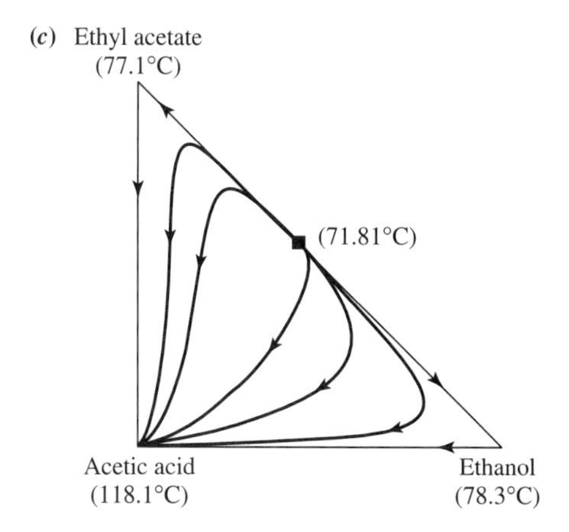

FIGURE 5.7
Residue curve maps for ternary mixtures with one binary minimum-boiling azeotrope. (*a*) Methanol-methyl butyrate-toluene calculated at 1 atm. (*b*) Methanol-water-methyl chloroacetate measured at 1 atm (Yamakita et al., 1983). Filled circles represent the composition of the liquid feeds; open circles are the measured liquid compositions along each residue curve. (*c*) Ethyl acetate-ethanol-acetic acid calculated at 1 atm.

It is common for ternary mixtures to exhibit more than one azeotrope, with the potential for multiple distillation boundaries and more complicated RCMs. A few calculated examples are shown in Fig. 5.8, and a beautiful set of experiments are shown in Fig. 5.9. Some points of interest about these RCMs are:

Fig. 5.8a. In this mixture there are two minimum-boiling binary azeotropes, one distillation boundary, and two distillation regions. All residue curves in the right-hand region are tangent to the n-propanol–isopropanol edge, which indicates $\alpha_{\text{benzene},n\text{-propanol}} > \alpha_{\text{isopropanol},n\text{-propanol}} > 1$ in the vicinity of pure n-propanol (the reader should check and see). Most feeds for this mixture produce n-propanol as the final product of the distillation, but some (lying

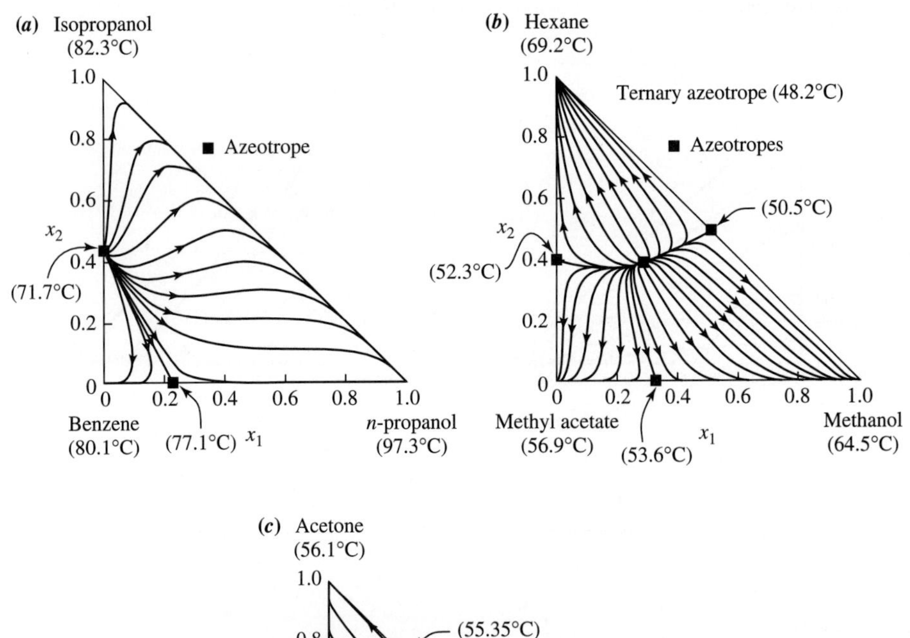

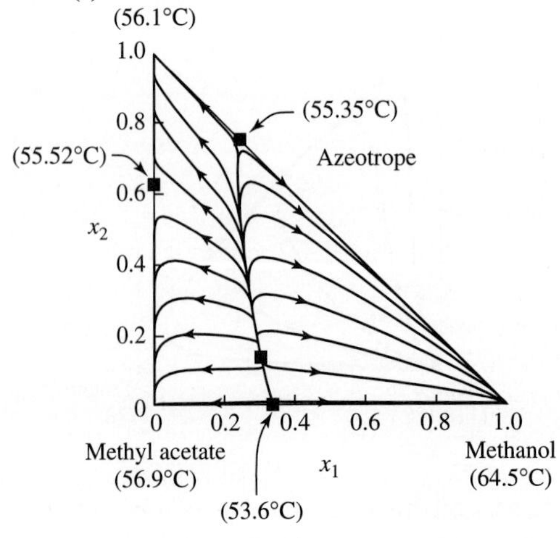

FIGURE 5.8
Selected residue curve maps for ternary mixtures with multiple azeotropes.

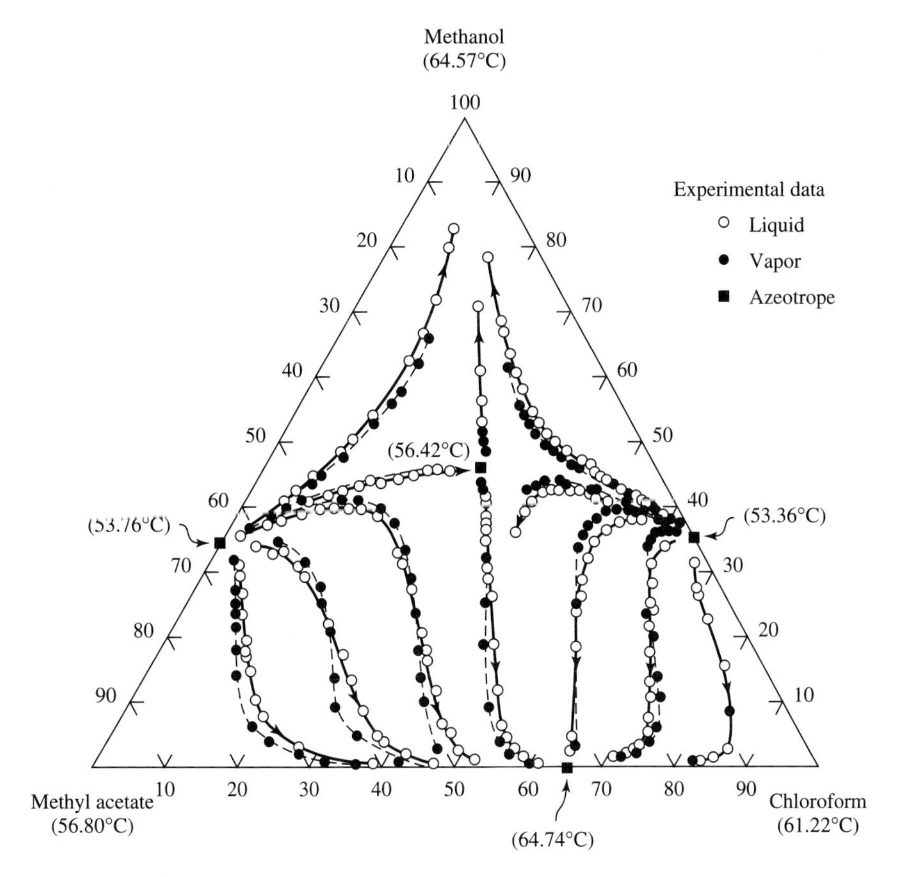

FIGURE 5.9
Measured residue curves (solid) and their vapor lines (dashed) for the ternary mixture methyl acetate-chloroform-methanol at 1 atm (Bushmakin and Kish, 1957). Open circles represent measured liquid compositions; the corresponding measured vapor compositions are filled circles.

in the smaller left-hand region) produce benzene. When benzene is the final product, the next most plentiful component in this product is *n*-propanol (the highest-boiling pure component!), not isopropanol, which might be the anticipated second most plentiful component because it is the intermediate-boiling pure component. This is another demonstration that azeotropes and nonidealities can cause unexpected behavior, which can be efficiently understood using these methods.

Fig. 5.8b. This mixture has three minimum-boiling binary azeotropes and an even lower-boiling ternary azeotrope. These produce three distillation boundaries and three distillation regions. All the residue curves start at the ternary azeotrope (there is only one unstable node in this mixture), but they end at either methyl acetate, methanol, or hexane depending on the feed composition (there are three stable nodes). This causes a very interesting and surprisingly common effect; the lightest pure component and the intermediate-boiling pure components can only be obtained as a residue of a distillation; i.e., they can only be obtained as "heavy" components out of the *bottom* of distillation columns! The same is true of the heaviest component, hexane, but this is not surprising.

Fig. 5.8c. This mixture has exactly the same characteristics as the previous one, but it is harder to see the three distillation regions because of the relative positions of the azeotropes. Study this map until you can see the three boundaries and the three regions.

Fig. 5.9. More than any other set of measurements, this confirms the validity of the modeling approach to simple distillation. The distillation boundaries are clearly visible, as are the unstable nodes at the two minimum-boiling binary azeotropes, the two stable nodes at pure methanol and at the methyl acetate-chloroform maximum-boiling binary azeotrope, and three saddles at pure methyl acetate, pure chloroform, and at the intermediate-boiling ternary azeotrope.[13] This ternary mixture has four distillation regions.

These RCMs are a small sample of the total number of possibilities for ternary mixtures. A more complete catalog of maps is given by Matsuyama and Nishimura (1977) and Doherty and Caldarola (1985). An even more extensive catalog has been compiled at Eastman Chemical Company along with the corresponding relative order of the boiling temperatures (Peterson and Partin, 1997). There are about 130 different maps for ternary mixtures that contain at least one minimum-boiling binary azeotrope,[14] and although this is a large number, it is much smaller than the number of possible ternary mixtures. Residue curve maps, therefore, provide an efficient way of seeing the similarities and differences between different ternary mixtures.

It is tempting to imagine that distillation boundaries in ternary mixtures are coincident with the projection of the locus of ridges and valleys in the boiling temperature surface onto the composition triangle. This seems reasonable for several reasons, including:

1. The temperature must always increase along a residue curve; therefore, we might expect that residue curves (and hence distillation boundaries) cannot cross over a ridge in the boiling-temperature surface.
2. For binary mixtures, the equivalent notions to ridges and valleys are maxima and minima in the boiling-temperature curve (i.e., maximum- and minimum-boiling azeotropes). These points certainly cannot be crossed in binary simple distillation. Again, this might lead to the conclusion that ridges and valleys in the boiling temperature are barriers to distillation.

Rather surprisingly, however, Van Dongen and Doherty (1984) showed that *ridges and valleys in the boiling-temperature surface do not coincide with simple distillation*

[13]Ternary azeotropes of this type are often called *saddle azeotropes*. The first azeotrope of this type was discovered by Ewell and Welch (1945).

[14]Maximum-boiling azeotropes are rare, and most known azeotropic mixtures contain at least one minimum-boiling binary azeotrope.

boundaries.[15] A later independent study (Rev, 1992) came to the same conclusion. Therefore, these ridges and valleys play no essential role in the behavior of azeotropic distillations operating at total reflux. This also holds at finite reflux, as shown by the experimental results reported by Li et al. (1999). They performed experiments on two separate azeotropic mixtures, chloroform-methanol-acetone, and chloroform-ethanol-acetone, in a 50-mm-diameter continuous distillation column consisting of 32 Oldershaw plates, a partial reboiler, and a total condenser operating at a reflux ratio of 6. The measured temperature and liquid composition profiles cross the ridges and valleys for both mixtures.

Residue curve maps can also be drawn for four-component mixtures, and two examples are shown in Fig. 5.10. Points of interest about these RCMs are as follows:

Fig. 5.10a. This mixture has two minimum-boiling binary azeotropes, one between methanol and methyl acetate (which is the lowest boiling point in the mixture and, therefore, this azeotrope is an unstable node), the other between water and methyl acetate. The methanol-methyl acetate azeotrope is the only unstable node, so all residue curves begin there. The acetic acid vertex is the only stable node, so all residue curves end there. This four-component mixture therefore has a single distillation region.

Fig. 5.10b. This mixture is more complicated because it has three minimum-boiling binary azeotropes (ethyl acetate-water, ethyl acetate-ethanol, ethanol-water) and a minimum-boiling ternary azeotrope between the components ethyl acetate-ethanol-water. In spite of this, however, the four-component mixture also has a single distillation region! All residue curves begin at the ternary azeotrope (the unstable node) and end at acetic acid (the stable node).

It is easy to find four-component mixtures with multiple distillation regions and distillation boundaries (which are *surfaces*). One example is shown in Fig. 7.21 in Chapter 7. Chapter 7 generalizes residue curve analysis and column synthesis and sequencing strategies to the case of multicomponent azeotropic mixtures.

Analogy Between Residue Curves and Continuous Distillation

Consider the rectifying section of a continuous distillation column with constant molar flows. For this analysis we count stages down from the top (stage 1 is the top stage of the rectifying section), or for a packed column we define the zero height of packing to be at the top of the packed section and measure positive distance down from that point. The steady-state material balances for these column sections are

Packed Column

$$\frac{d\mathbf{x}}{dh} = \mathbf{x} - \left(\frac{r+1}{r}\right)\mathbf{y} + \frac{\mathbf{x}_D}{r} \qquad \mathbf{x}(h=0) = \mathbf{x}_D \tag{5.14}$$

[15]Moreover, it has also been shown (Van Dongen, 1983) that simple distillation boundaries do not correspond to the locus of points where either $\alpha_{i,j} = 1$ or $K_i = 1$.

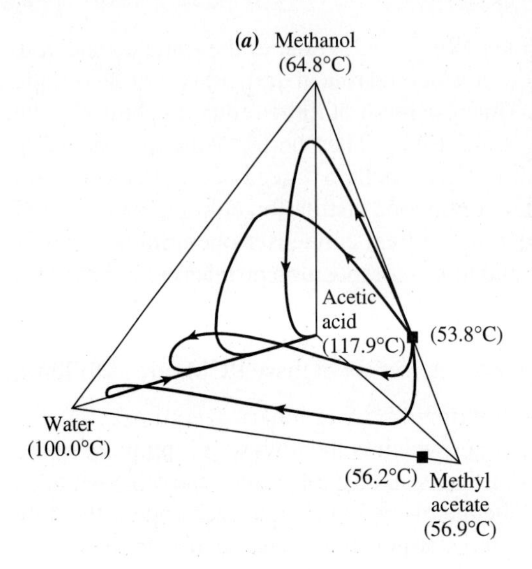

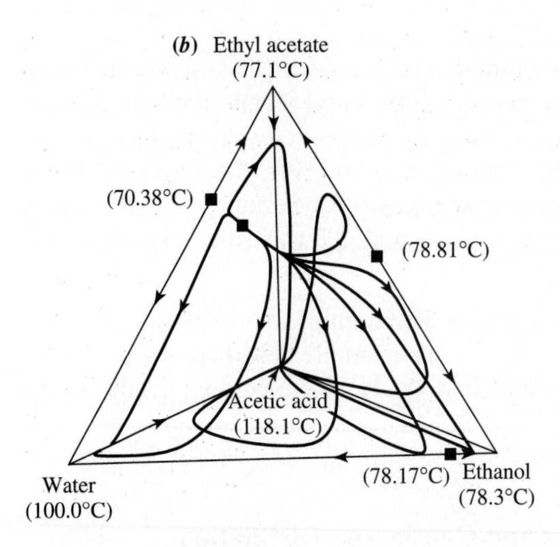

FIGURE 5.10
Calculated residue curve maps for four-component mixtures at 1 atm, (*a*) methanol-acetic acid-methyl acetate-water, (*b*) ethanol-acetic acid-ethyl acetate-water. The origin of the composition tetrahedron represents pure acetic acid in both maps.

Staged Column

$$\mathbf{y}_{n+1} = \left(\frac{r}{r+1}\right)\mathbf{x}_n + \left(\frac{1}{r+1}\right)\mathbf{x}_D \qquad \mathbf{x}_0 = \mathbf{x}_D \tag{5.15}$$

In the limiting case of infinite reflux these equations reduce to

Packed Column, $r \to \infty$

$$\frac{d\mathbf{x}}{dh} = \mathbf{x} - \mathbf{y} \qquad \mathbf{x}(h=0) = \mathbf{x}_D \tag{5.16}$$

Staged Column, $r \to \infty$

$$\mathbf{y}_{n+1} = \mathbf{x}_n \qquad \mathbf{x}_0 = \mathbf{x}_D \tag{5.17}$$

We see immediately that the steady-state composition profile in a packed column at total reflux is identical to a residue curve in simple distillation (where the height of packing is equivalent to warped time). There is also a close correspondence between the solutions of Eq. 5.17 (called *distillation lines*) and the simple distillation equation. Writing Eq. 5.6 or, equivalently, Eq. 5.16 as a difference equation using an implicit Euler representation of the derivative gives

$$\frac{\mathbf{x}_{n+1} - \mathbf{x}_n}{\Delta h} = \mathbf{x}_{n+1} - \mathbf{y}_{n+1} \tag{5.18}$$

If we take a unit step size (i.e., one stage), $\Delta h = 1$, Eq. 5.18 becomes

$$\mathbf{x}_n = \mathbf{y}_{n+1} \tag{5.19}$$

which is precisely the equation for the steady-state composition profile in a staged column at total reflux. Notice that the fixed points of Eq. 5.17 are identical to the singular points of Eq. 5.16 (i.e., all the pure components and azeotropes in the mixture). It can also be shown that the stability of these points is the same for each model, and the solutions are locally topologically equivalent (Julka and Doherty, 1990; Julka, 1995). In other words, the solutions of these equations are very similar, as the curves in Fig. 5.11 demonstrate. Therefore, we have the following important result: *The composition profiles in a single-feed packed continuous distillation column at total reflux are constrained to lie in the distillation regions defined by the simple distillation boundaries, or in the case of staged columns, the distillation line boundaries (i.e., the infinite reflux boundaries of Eq. 5.17).*

The relationship between the residue curves (or distillation lines) and the overall material balance is shown in Fig. 5.12. At total reflux, the feed and product flowrates are zero, but it is convenient to consider an asymptotic case where these flowrates are

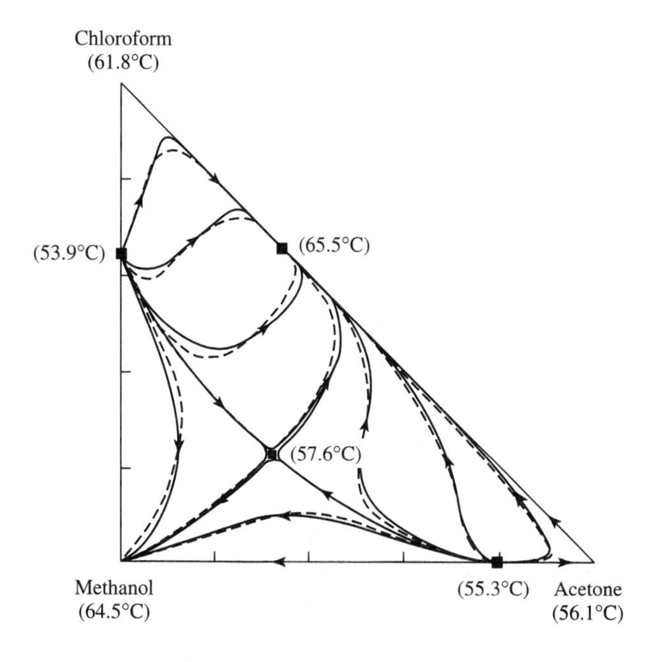

FIGURE 5.11
Comparison of residue curves (infinite reflux profiles in a packed column)—solid lines, and infinite reflux profiles in a staged column (distillation lines)—dashed lines, for the mixture acetone-chloroform-methanol at 1 atm.

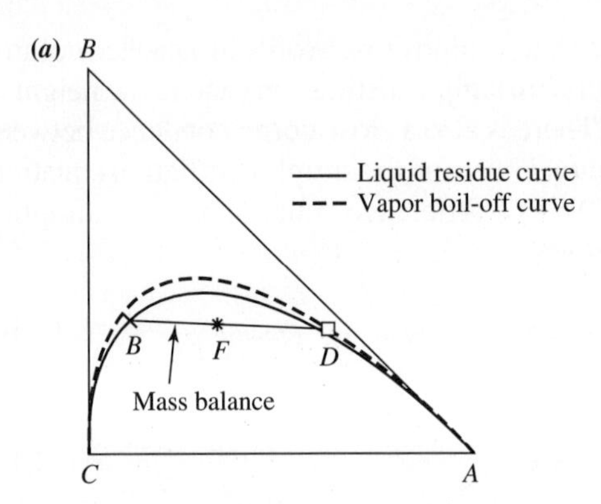

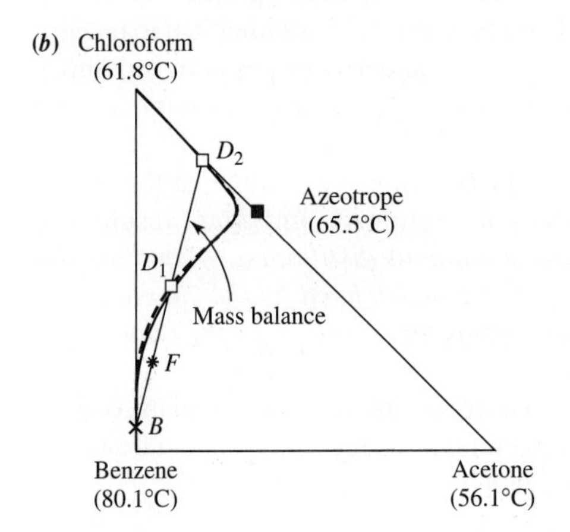

FIGURE 5.12
Relationship between the overall mass balance and residue curves for total reflux separations: (*a*) Ideal mixture. (*b*) Azeotropic mixture with an inflexion in the residue curves.

vanishingly small. Since the feed flowrate is infinitesimally small, the location of the feed stage is not relevant and the column can be treated as a single section. The liquid compositions throughout the column must be located on the same residue curve. The composition of the distillate corresponds to a vapor in phase equilibrium with a liquid on the top stage in the column and lies on the vapor boil-off curve corresponding to the residue curve. Since the feed and product compositions are collinear on account of the lever rule, the relationship between the feed and product compositions is as shown in Fig. 5.12. The design procedure at total reflux is quite simple. Knowing the feed composition $\mathbf{x}_F$ together with a specified composition at the bottom of the column $\mathbf{x}_B$, a trajectory of solutions of Eq. 5.17 can be computed. We make note of the point where the vapor trajectory intersects the material balance line. As a result, we obtain the distillate composition, and the minimum number of stages ($N_{\min}$) required for this separation. More than one intersection is possible when the distillation lines have inflections as shown in Fig. 5.12*b*. This situation corresponds to two different columns with different distillate compositions and different numbers of stages. When the feed composition, feed quality, and column pressure are known, there are four degrees of freedom in a design problem at finite reflux. At total reflux, both the reflux and reboil ratios are infinite (i.e., they are both specified), and there remain only two degrees of freedom. For the examples in Fig. 5.12 we chose to specify $x_{1,B}$, and $x_{2,B}$ in order to satisfy the two degrees of freedom. In the case of constant volatility mixtures, this is equivalent to Fenske's procedure.

Steady-state composition profiles in packed, staged, and wetted wall columns at total reflux have been measured and reported in the literature. Two sets of measurements, both for azeotropic mixtures, are described here. The first set of measurements, taken by Free and Hutchison (1960) on a seven-stage column at total reflux, is shown in Fig. 5.13*a*. The curves in this figure are the calculated distillation lines (total reflux composition profiles); the points are the measured compositions on each stage. The second data set is shown in Fig. 5.13*b* for the mixture benzene–1,2-dichloroethane–*n*-heptane (Lutugina et al., 1974). Here the points represent experimental measurements for the composition profile along the height of a packed column (∘) and a wetted wall column (×) at total reflux, the solid line represents the calculated distillation lines (plate column at total reflux), and the dotted lines represent the calculated simple distillation residue curves. These lines were calculated from the same initial conditions, which corresponded to the experimentally measured starting compositions. In both systems the models are remarkably close to the measurements (let's test this, see Exercise 8). Therefore, to experimentally measure the total reflux composition profiles we can perform either a column experiment or a simple distillation experiment—they both give essentially the same information.

Having established the constraints on column composition profiles at total reflux, the key question is, do these constraints deform much at finite reflux, i.e., do they give us any useful information about the feasible product compositions at finite reflux? It has been recognized for a long time that the finite reflux boundaries are not the same as the infinite reflux boundaries[16] (e.g., Nikolaev et al., 1979; Doherty and Perkins, 1979*a*; Van Dongen and Doherty, 1985), but the surprising thing is how close they often are. To demonstrate this, consider a mixture of hexane, methanol, and methyl acetate. The residue curve map for this mixture is shown in Fig. 5.8*b*. Van Dongen and Doherty (1985) computed the steady-state composition profiles for this mixture in a column with 20 theoretical stages operating with a reflux ratio of 3 (which is a long way from infinity!). The results are shown in Fig. 5.14. Figure 5.14*a* shows composition profiles for one feed in each of the three distillation regions. For a feed in the top region the bottom composition from the column is hexane (a stable node) and the distillate is near the ternary azeotrope (an unstable node). A feed in the lower left region, however, gives a bottom composition of methyl acetate (which is the stable node for this region) and a distillate near the ternary azeotrope, while a feed in the lower right region gives a bottom composition near methanol (which is the stable node for the lower right region) and a distillate toward the ternary azeotrope. With more stages and reflux these product compositions can be sharpened up (see Exercise 9). However, the main point is that the products of the distillation are different for feeds in different distillation regions. The profiles in Fig. 5.14*b* are for a series of seven different feed compositions in the top distillation region (the hexane feed

[16]The exact relationship is described in Sec. 5.4.

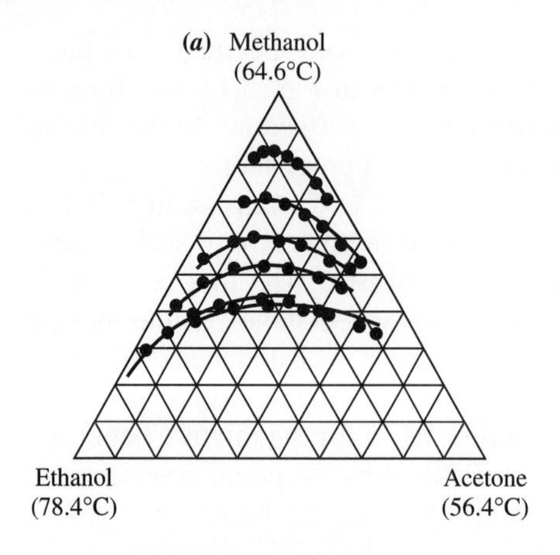

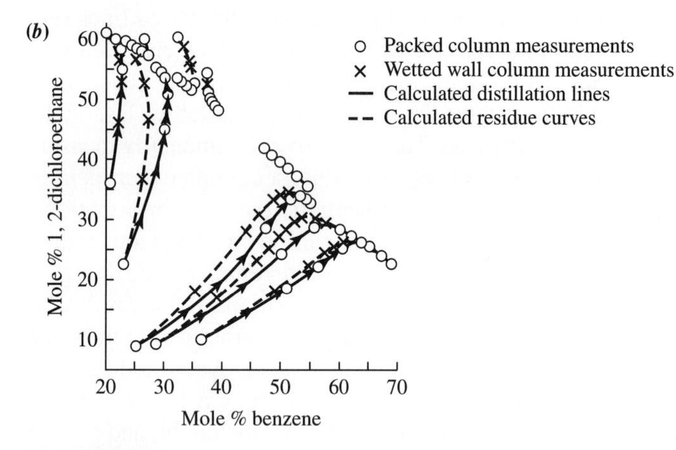

FIGURE 5.13
(a) Distillation lines in a seven-stage column measured at 1 atm for the mixture acetone-methanol-ethanol (Free and Hutchison, 1960). (b) Measured distillation lines for the mixture benzene-1,2-dichloroethane-n-heptane in a packed column and a wetted wall column (Lutugina et al., 1974). The points represent experimental measurements for the composition profile along the height of a packed column (o) and a wetted wall column (×) at total reflux, the solid line represents the calculated distillation lines (plate column at total reflux), and the dotted lines represent the calculated simple distillation residue curves.

composition is 60 mol % in each case). These profiles have an uncanny resemblance to the residue curves in this region. More extensive studies show that the product compositions can be made to extend beyond the simple distillation boundaries, but not by much, as discussed by Van Dongen and Doherty (1985) and Levy (1985, Chap. VI). We see, therefore, that *the structure of the residue curve map is the underlying thermodynamic principle that governs the shape of composition profiles and consequently the products that can be obtained from a distillation.*

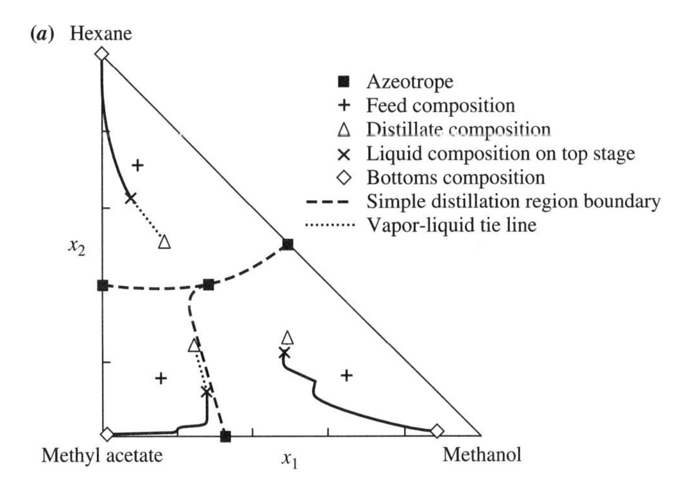

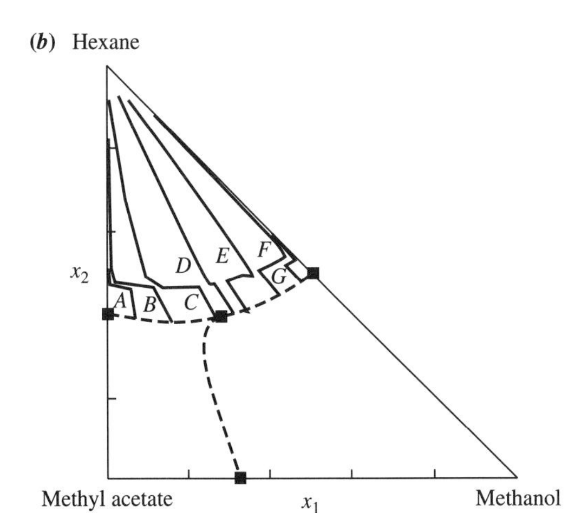

FIGURE 5.14
Finite reflux composition profiles calculated for a continuous stage column with 20 theoretical stages (one of them is a partial reboiler), a saturated liquid feed on stage 10, operating at a reflux ratio of 3 and a reboil ratio of 6, for the mixture hexane-methanol-methyl acetate at 1 atm pressure.

Mass Transfer Effects

All of the models developed in this book assume that mass transfer (within and across phases) is fast relative to other rate processes. This is not always true, of course, but it does allow for simple models with few parameters that enable us to develop conceptual designs quickly. The results of these calculations provide the incentive (or disincentive!) to develop more sophisticated models.

The significance of mass transfer in simple distillation and in columns operated at total reflux has been the subject of many investigations dating back at least 40 years (Free and Hutchison, 1960; Lutugina et al., 1974). Lutugina et al. (1974) derive the rate-based simple distillation equations, and show how the residue curves are related to the VLE tie-lines, for cases where: (1) resistance to mass transfer takes place only in the vapor phase, (2) resistance to mass transfer takes place only in the liquid phase, and (3) resistance to mass transfer takes place in both phases. Interest in this subject picked up again in the 1990s with additional experimental work and more sophisticated modeling (Agarwal and Taylor, 1994; Castillo and Towler, 1997; Castillo and Towler, 1998; Pelkonen et al., 1997; Baur et al., 1999).

Several general conclusions are clear from this work. As might be expected, more nonequilibrium stages are needed to reach the same compositions for a given mixture, e.g., Agarwal and Taylor (1994). In some cases, unequal mass transfer rates of the individual species between the phases can cause residue curves and distillation boundaries to exhibit sharper curvature, which may lead to novel designs in some cases. A less obvious and important conclusion is that the minimum reflux calculated from the nonequilibrium models is the *same* as that found from an equilibrium model (Agarwal and Taylor, 1994).

Figure 5.15 shows a set of experimental measurements together with calculations from two mass transfer models for the distillation of a quaternary azeotropic mixture at total reflux in a packed column (Pelkonen et al., 1997). The experiment is carried out close to a distillation boundary because this is where we expect the greatest sensitivity to modeling errors; this provides a good test to discriminate between models. One of the models (solid lines) predicts that the reboiler composition settles at the isopropanol stable node, while the other model (dashed lines) predicts it should be at the water stable node. Experiments indicate that the reboiler contains water.

There is a close and interesting relationship between simple distillation and the drying of a solid wetted with a liquid mixture by convective gas-phase-controlled evaporation. Luna and Martinez (1999) derive the governing equations and show the correspondences with simple distillation.

Flash Cascades

In Sec. 2.6 we showed that a distillation column can be approximated by two cocurrent cascades of flash separators, one for the rectifying section and one for the

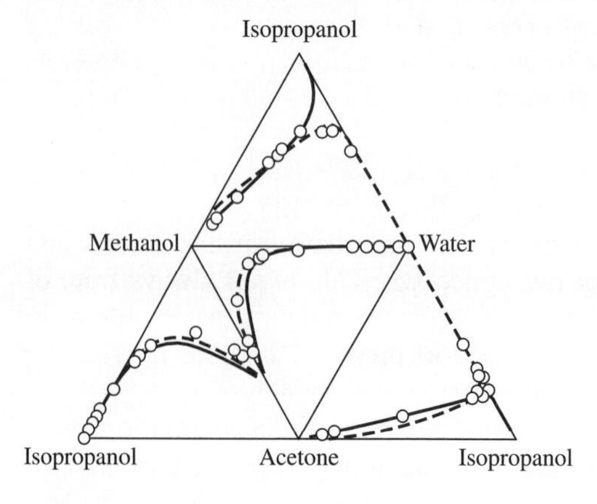

FIGURE 5.15
Experimental (circles) and calculated distillation lines for the mixture acetone-methanol-isopropanol-water at total reflux in a packed column. Two mass transfer models (dashed lines and solid lines) are represented. Each ternary diagram is represented by a side of an unfolded tetrahedron.

stripping section. With a sufficiently large number of flash stages these cascades can produce high-purity products, but with very low recovery. Nevertheless, the flash cascade model is easy to develop, and it is simple to find the product compositions when the cascades have a large number of stages. These results give useful information about the behavior of distillation columns, as you will see by solving Exercise 10.

Sketching Residue Curve Maps

We have seen that residue curve maps can be calculated from models or measured experimentally. But it is evident that the general features for many of the simpler mixtures can also be guessed just by knowing the boiling points of the pure components and azeotropes, and by using Properties 2 to 5, which are repeated below.

Property 2. Residue curves do not cross each other, nor do they intersect themselves.
Property 3. The boiling temperature always increases along a residue curve (the only exception is at steady states where the boiling temperature remains constant because the composition remains constant).
Property 4. Singular points occur at all pure components and azeotropes.
Property 5. Singular points are either: (1) stable nodes, (2) unstable nodes, or (3) saddle points.

Being able to sketch residue curve maps without the use of a detailed phase equilibrium model or experimental data is a tremendous advantage at the preliminary conceptual design stage. At this stage of a process design, models and data are normally incomplete and the key goal is to determine the incentive to develop them. Since a sketch of the residue curve map will provide knowledge of the stable and unstable nodes as well as the existence of distillation boundaries (but not their exact location), it is possible to determine the potential products that can be made by distillation separations. This is exactly the type of strategic information that is needed at the preliminary stages of a process design.

The general methodology is described in detail by Doherty (1985) and Foucher et al. (1991) and is summarized here briefly. The main idea is to classify each singular point in the map according to whether it is a node or a saddle. Then we sketch the map to be consistent with that classification. Define N_1, N_2, and N_3 to be the number of pure components that are nodes, the number of binary azeotropes that are nodes, and the number of ternary azeotropes that are nodes, respectively. Also define S_1, S_2, and S_3 to be the number of pure components that are saddles, etc. From just a knowledge of the boiling temperatures, we can always establish the values for N_1, N_3 and S_3. The value of N_1 is determined from Rule 1; N_3 and S_3 are determined from Rule 2.

Rule 1

If the arrows on the binary edges of the triangle both point toward a vertex or if they both point away from a vertex, then the vertex is a node; otherwise it is a saddle.

Each vertex is examined in turn and the total number of vertices that are nodes is established.

If the mixture has no ternary azeotrope, then $N_3 = S_3 = 0$. If a ternary azeotrope occurs, then its character is determined from Rule 2.

Notes

1. Let $\mathcal{B}$ = total number of binary azeotropes in the mixture.
2. A pure component saddle has no interior connections.
3. A ternary saddle *must* have four connections.
4. A necessary condition for a ternary azeotrope to be a saddle is the existence of two higher- and two lower-boiling species with which it *could* connect.

Rule 2

A ternary azeotrope is a *node* if:

1. $N_1 + \mathcal{B} < 4$, i.e., there are fewer than four potential connections.

or

2. Excluding the *pure component saddles*, the ternary azeotrope is the highest, second highest, lowest, or second lowest boiling species. In each of these cases there are fewer than four potential connections.

Otherwise, the ternary azeotrope is a *saddle*.

In addition, we know

$$N_2 + S_2 = \mathcal{B} \tag{5.20}$$

and

$$2N_3 - 2S_3 + N_2 - S_2 + N_1 = 2 \tag{5.21}$$

This last equation comes from a *topological conservation principle* for ternary azeotropic systems; see, for example, Doherty and Perkins (1979*b*). These equations can be rearranged to give expressions for the unknown number of binary nodes and saddles.

$$N_2 = (2 + \mathcal{B} - N_1 - 2N_3 + 2S_3)/2 \tag{5.22}$$

and

$$S_2 = \mathcal{B} - N_2 \tag{5.23}$$

EXAMPLE 5.1. Sketching A Residue Curve Map. Let's use this procedure to sketch the residue curve map for the mixture benzene–isopropanol–*n*-propanol at 1 atm pressure.

- First we collect the boiling point data. Good sources of azeotropic data are Horsley (1973), Gmehling et al. (1994*a*), and Gmehling et al. (1994*b*):

  Pure components:

  Benzene = 80.1°C

  Isopropanol = 82.3°C

  n-Propanol = 97.3°C

  Azeotropes:

  Benzene + isopropanol = 71.7°C

  Benzene + *n*-propanol = 77.1°C

- Draw a composition triangle and label the pure components and azeotropes with these boiling points. Then draw arrows on each of the binary edges in the direction of increasing temperature; see Fig. 5.16.
- Apply Rule 1, which tells us that benzene and *n*-propanol are both stable nodes[17] and isopropanol is a saddle.
- There are no ternary azeotropes, so $N_3 = S_3 = 0$. The above formulas give $N_2 = S_2$, $N_2 + S_2 = 2$. Therefore, $N_2 = 1$ and $S_2 = 1$. We can tell which azeotrope is the node and which is the saddle because the benzene-isopropanol azeotrope is the lowest boiler in the mixture. Therefore, this azeotrope must be an unstable node; the benzene–*n*-propanol azeotrope is a saddle. A sketch of the residue curve map gives a picture similar to Fig. 5.8*a*.

You can try your hand at sketching residue curve maps in Exercises 11–16. As you use the method, watch out for the possibility that the data are inconsistent. One advantage of using this procedure is that sometimes you can detect inconsistent data by applying the following test.

Inconsistency tests

The data are inconsistent whenever:

1. Either N_2 or S_2 are calculated to be negative or fractional.
2. The calculated value of N_2 must be greater than or equal to the total number of binary azeotropes that are either the highest- or the lowest-boiling species in the

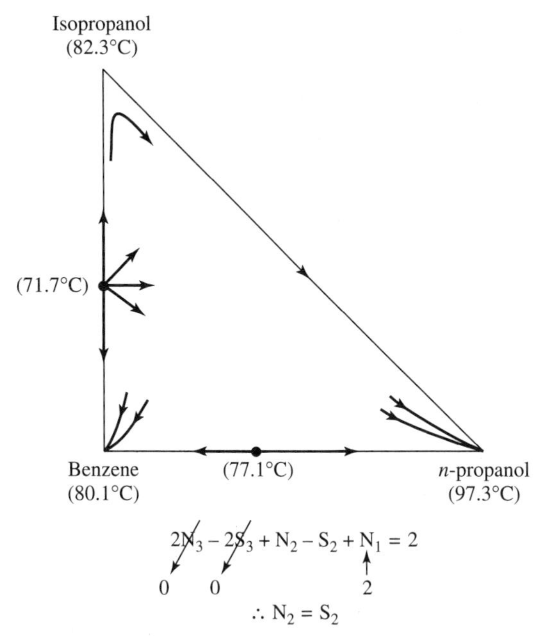

FIGURE 5.16
Sketching the residue curve map for the mixture benzene-isopropanol-*n*-propanol at 1 atm pressure.

[17]The lowest- and the highest-boiling pure components are both stable nodes in this mixture.

system (since thermodynamics requires that at *least* these binary azeotropes are nodes); otherwise the data are inconsistent.

5.4

FEASIBILITY, PRODUCT DISTRIBUTIONS, AND SEQUENCES

The method developed in Sec. 4.4 for calculating separation regions at finite reflux applies equally well to nonideal and azeotropic mixtures. The full treatment of this subject for ternary mixtures is given by Wahnschafft et al. (1992*a*), Fidkowski et al., (1993*a*), Poellmann and Blass (1994), and Davydyan et al. (1997). We will present a few examples and show some of the new, and unexpected, effects of liquid phase nonidealities.

Our first example is a mixture of acetaldehye (A)-methanol (B)-water (C) at 1 atm pressure, where A, B, and C are intended to represent the light, intermediate, and heavy components, respectively. The mixture is nonideal but has no azeotropes (the binary y-x diagrams and other information about this mixture can be found in Example 4.6). Its residue curve map is shown in Fig. 5.17. The separation regions for a saturated liquid feed of 30 mol % acetaldehyde, 30 mol % methanol, and 40 mol % water are shown in Fig. 5.18*a*. The transition line intersects the A-B and B-C edges of the triangle, and generates a large region of indirect splits and a small region of direct splits, as shown in the figure. As the feed gets richer in acetaldehye, a surprising thing happens; the transition line no longer intersects the A-B edge of the triangle but instead intersects the A-C edge[18], as shown in Fig. 5.18*b*. All the feed compositions that produce this behavior are shown as region II in Fig. 5.19. For these feeds, there are no direct split geometries; all splits must follow the indirect split geometry.[19] So, for example, the feed shown in Fig. 5.18*b* can be separated in a sequence that takes pure acetaldehyde as the distillate product from the first column,[20] but the shape of the column composition profiles follows the indirect geometry![21] This prediction can be tested quite easily, along with the consequences for minimum reflux calculations; see Fidkowski et al. (1991) and Exercise 17.

Mixtures with distillation boundaries present some additional challenges. A good example is the mixture acetone (A)-chloroform (B)-benzene (C) at 1 atm pressure. There is one azeotrope in the mixture, a maximum-boiling binary azeotrope between acetone and chloroform. The residue curve map is shown in Fig. 5.20*a*, where we can see the simple distillation boundary connecting the binary azeotrope to the benzene vertex. The classification of feeds for this mixture is shown in Fig. 5.20*b*. There are three different regions of feed compositions where the transition

---

[18]For saturated liquid feeds this can be anticipated from the shape of the residue curves because the transition line passes through a vapor-liquid tie-line at the feed composition. Since the tangent to a residue curve through the feed composition is a vapor-liquid tie-line, this tangent, when extended across the diagram, must be the transition line.

[19]Ideal mixtures cannot show this kind of behavior.

[20]This would conventionally be treated as a "direct" split or "lightest out first" split.

[21]This information is essential for minimum reflux calculations.

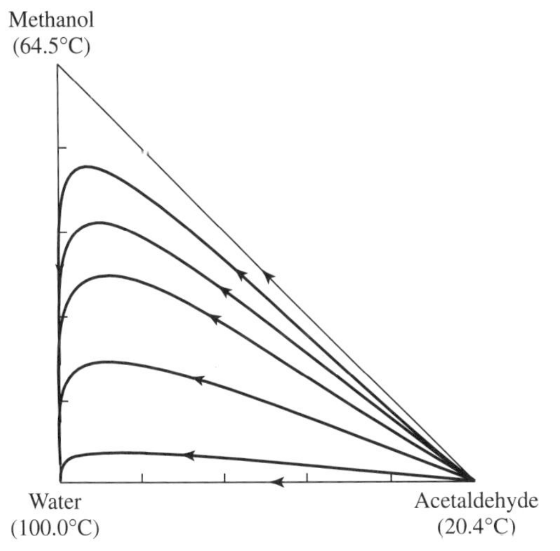

FIGURE 5.17
Residue curve map for acetaldehyde-methanol-water at 1 atm pressure.

line intersects the edges AB and AC (region I), AB and BC (region II), or BC and AC (region III). These regions can be anticipated by looking at the curvature of the residue curves.

We can find the feasible separation regions for any particular feed composition; these are shown in Fig. 5.21 for a saturated liquid feed containing 30 mol % acetone, 30 mol % chloroform, and 40 mol % benzene. However, the information in this picture is incomplete; if we attempt to separate acetone from benzene and chloroform (direct split) we know the region of possible distillate compositions (labeled D in the figure), but the bottom compositions (B in the figure) are harder to determine because of the boundary. The difficulty arises because we do not know exactly where the *continuous* distillation boundary is located. One way to *estimate* its location is with that of the simple distillation boundary (SDB), as shown in the figure. However, feeds on the concave side of the simple distillation boundary are able to produce products from a continuous distillation that lie on the opposite side of the boundary. Just how far these product compositions can extend beyond the SDB depends on the feed composition, the number of stages, and the reflux/reboil ratios. That is, the position of the continuous distillation boundary depends on the design parameters, so that there is a whole family of these boundaries (for a fixed feed composition) that are parameterized by the values of the design variables. It is not useful to calculate this family of curves, but it is useful to know the position of the envelope that contains them all. This has been solved by Davydyan et al. (1997), who have derived an equation for this envelope (see Eq. 8 in their paper). They call this envelope the *pitchfork distillation boundary* because it corresponds to the locus of pitchfork bifurcations in the distillation equations. Its position depends only on the feed quality q and the VLE model; it does not depend on feed composition or any of the design parameters for the column. The position of the pitchfork distillation boundary for the acetone-chloroform-benzene mixture is shown in Fig. 5.22 for $q = 1$. This

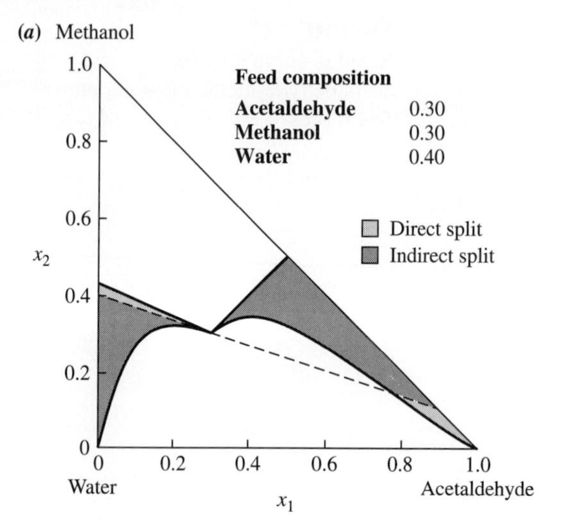

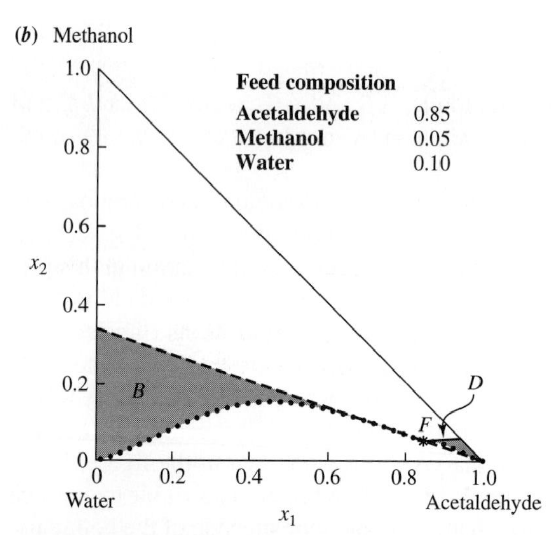

FIGURE 5.18
Feasible separation regions for saturated liquid feeds of acetaldehyde-methanol-water at 1 atm pressure. (*a*) Feed composition 30 mol % acetaldehyde, 30 mol % methanol, 40 mol % water. (*b*) 85 mol % acetaldehyde, 5 mol % methanol, 10 mol % water.

boundary actually intersects the chloroform-benzene edge of the triangular diagram at about 25 mol % chloroform.[22] We note the following important points:

1. Distillation regions at finite reflux are not disjoint; they *overlap*. The area of overlap is represented by the dotted region between the pitchfork and simple distillation boundaries in Fig. 5.22. The exact bounds (i.e., the pitchfork and simple distillation boundaries) are easy to calculate.
2. A given composition in the overlap region can be obtained as a bottom product from columns with a feed composition in the top distillation region (where chlo-

[22]This is in contrast to the simple distillation boundary, which ends at the benzene vertex where it is tangent to the chloroform-benzene edge.

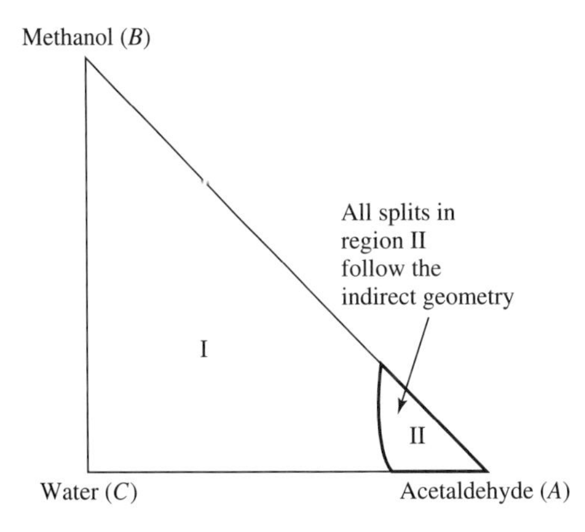

FIGURE 5.19
Regions of feed compositions for acetaldehyde-methanol-water, $q = 1$. In Region I the transition line intersects the AB and BC edges; in Region II it intersects the AC and BC edges.

roform is the distillate; see Fig. 5.23*a*), and from columns with a feed composition in the bottom distillation region (where acetone is the distillate; see Fig. 5.23*b*).

3. Because the pitchfork boundary intersects the side of the triangle, it is possible, for some feed compositions, to perform a separation where the distillate is acetone and the bottom product is a binary mixture of chloroform and benzene. This is shown in Fig. 5.24 along with a column sequence that will separate a ternary feed into its three constituent components. This sequence is discussed in more detail by Fidkowski et al. (1993*a*) together with other possibilities for exploiting the curvature of the boundaries.
4. The overlap region is vanishingly small unless the simple distillation boundary is strongly curved. A common case is represented by the acetone-isopropanol-water mixture in Fig. 5.25. The overlap region for this mixture is so small that no practical engineering advantage is possible. Therefore, realistic opportunities for devising novel sequences require very curved simple distillation boundaries. In such cases the model predictions *must be verified by experiment* because the risks are high and failure is assured if the boundary is not as curved as the model suggests. Also note that designs exploiting curved boundaries will not reach the target compositions at total reflux, so that their startup will be a complex task.

We use this information to devise the following heuristic:

Heuristic 5.1

For mixtures with distillation boundaries, choose product compositions from each column that lie within the feasible separation regions and on the same side of the simple distillation boundary (or on the same side of the infinite reflux boundary in the distillation line map).[23] Do not attempt to cross a simple distillation boundary unless all other alternatives have been explored, and then do so with caution. Many useful hints about how to do this are given in the papers by Wahnschafft et al.

[23]For mixtures with very curved distillation boundaries, this does not constrain the products from the entire sequence to lie within the same distillation region.

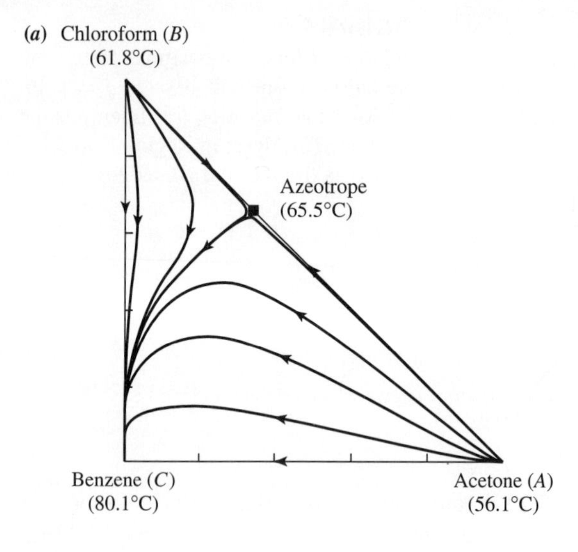

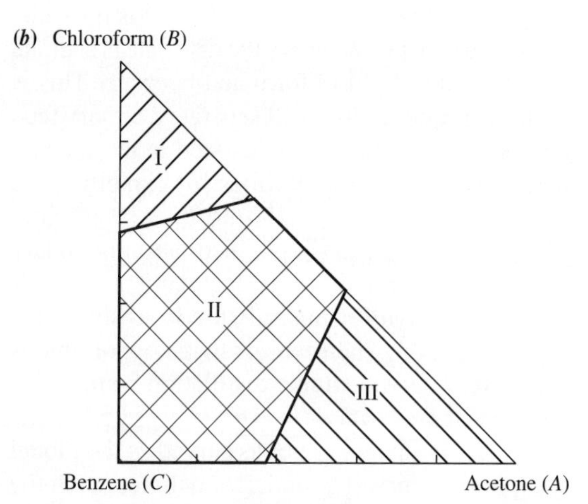

FIGURE 5.20
(*a*) Residue curve map for acetone-chloroform-benzene at 1 atm pressure. (*b*) Feed composition regions, $q = 1$. In Region I the transition line intersects the AB and AC edges; in Region II it intersects the AB and BC edges; in Region III it intersects the BC and AC edges.

(1992*a*), LaRoche et al. (1992*a*), LaRoche et al. (1992*b*), Fidkowski et al. (1993*a*), and Poellmann and Blass (1994).

Heuristic 5.1 tells us nothing about the region where the feed composition lies. There are two cases:

Straight Boundaries. The lever rule requires the feed composition to lie on the straight line connecting the two product compositions. When the boundaries are nearly straight (e.g., see Fig. 5.25), the lever rule together with Heuristic 5.1 require the feed to be in the same distillation region as the products. In this case, the products from the entire sequence must lie in the same distillation region (Doherty and Caldarola, 1985).

Curved Boundaries. In this case the feed may be in the same or in a different distillation region than the products. For example, the feed F_1 in Fig. 5.26

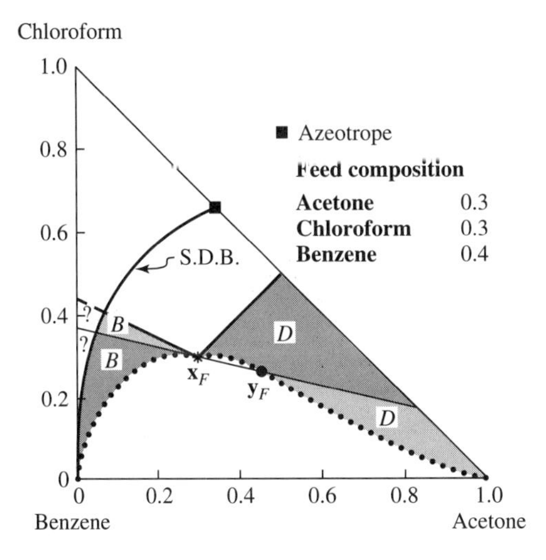

FIGURE 5.21
Feasible separation regions for a saturated liquid feed containing 30 mol % acetone, 30 mol % chloroform, and 40 mol % benzene 1 atm pressure. The simple distillation boundary is labeled SDB.

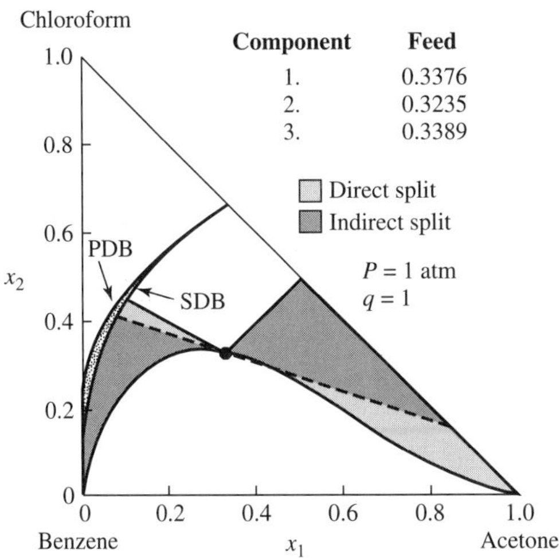

FIGURE 5.22
Pitchfork distillation boundary (PDB) and simple distillation boundary (SDB) for acetone-chloroform-benzene at 1 atm pressure, $q = 1$. The feasible separation regions are also shown for a saturated liquid feed containing 33.76 mol % acetone, 32.35 mol % chloroform, and 33.89 mol % benzene.

is in the same distillation region as the products D_1 and B_1, but the feed to the second column (stream B_1) is in a different distillation region than the products (streams B_2 and D_2). Stream B_1 is in the bottom distillation region, and streams B_2 and D_2 are in the top distillation region. The column composition profiles for feed stream B_1 are similar in shape to the composition profile shown in Fig. 5.12*b*. Also see Exercise 18. For mixtures with curved boundaries, the products from the entire sequence are not constrained to lie within the same distillation region. A classic example where a curved boundary is exploited in a commercial separation is the concentration of aqueous nitric acid mixtures[24] using sulfuric acid as the entrainer (Doherty and Knapp, 1993, Figs. 10, 11, and related text).

---

[24]Nitric acid forms a maximum-boiling azeotrope with water.

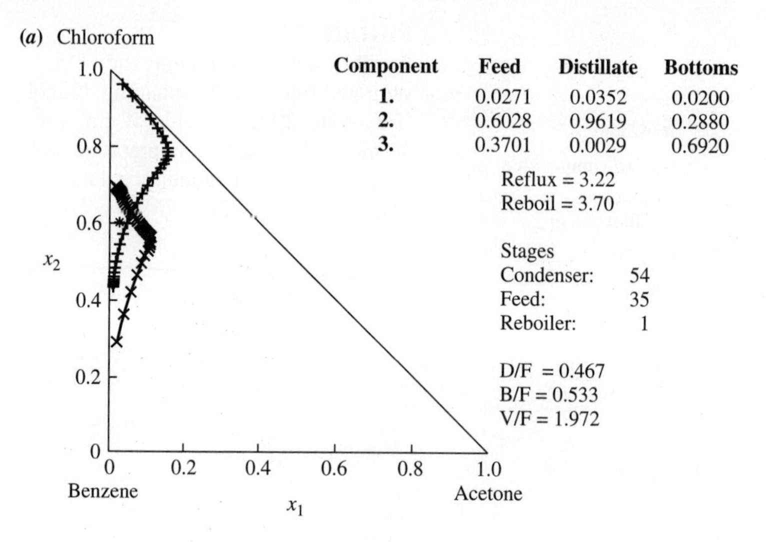

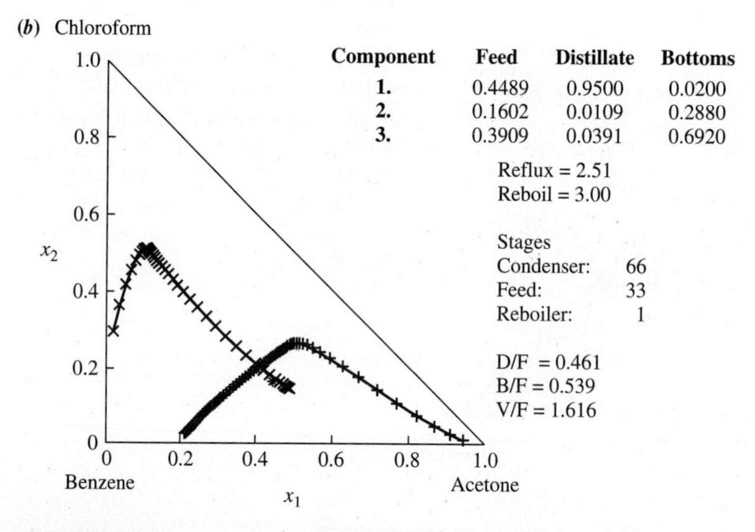

FIGURE 5.23
Column designs for two feed compositions of acetone-chloroform-benzene at 1 atm pressure, $q = 1$. (*a*) Chloroform as distillate. (*b*) Acetone as distillate. The bottom product composition is the same for each design.

Further implications of these results for synthesis of distillation systems to separate azeotropic mixtures are discussed in Sec. 5.6.

The first important idea for sequencing is to generalize the concept of "light," "middle," and "heavy" components. These are, respectively, unstable nodes, saddles, and stable nodes in the residue curve map. For mixtures with more than four components, these cannot be visualized, but all of the azeotropes as well as their

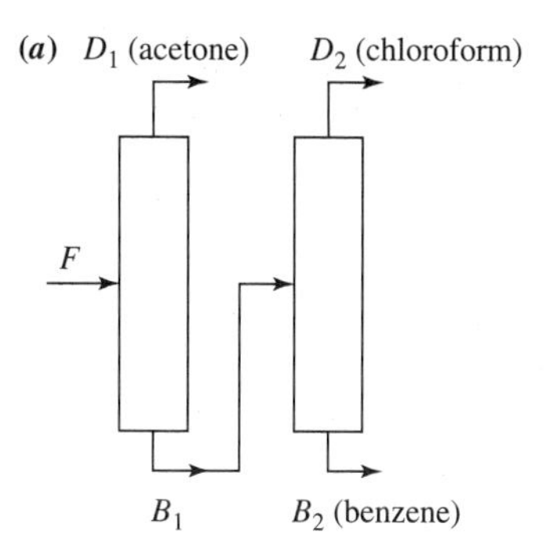

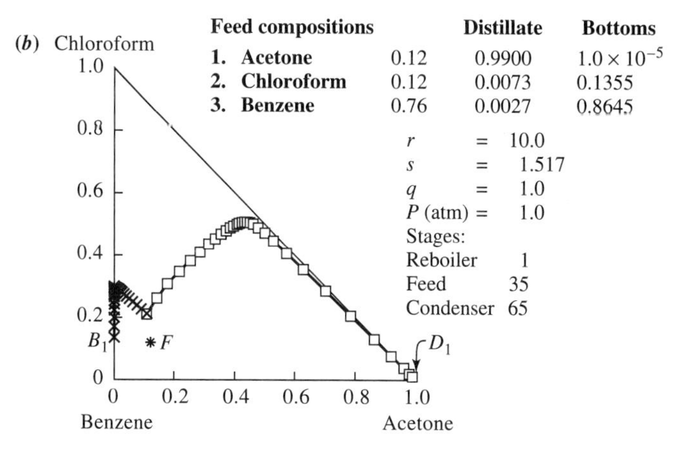

FIGURE 5.24
Column design that separates acetone from chloroform and benzene, exploiting the overlap region. The liquid composition profiles cross the simple distillation boundary into the overlap region.

stability classification and those of the pure components can be calculated from a vapor-liquid equilibrium model (Fidkowski et al., 1993*b*).

Once these classifications are known, any particular feed can be located in a distillation region and the nodes and saddles defining that region can be determined. When there are distillation boundaries, certain splits will be infeasible. Some of the feasible "sharp" splits will be

- Unstable node as the distillate with the bottoms determined by mass balance.
- Stable node as the bottoms with the distillate determined by mass balance.
- One or more of the saddles as a sidestream; only some saddles will be feasible depending on the feed composition (Rooks et al., 1996).

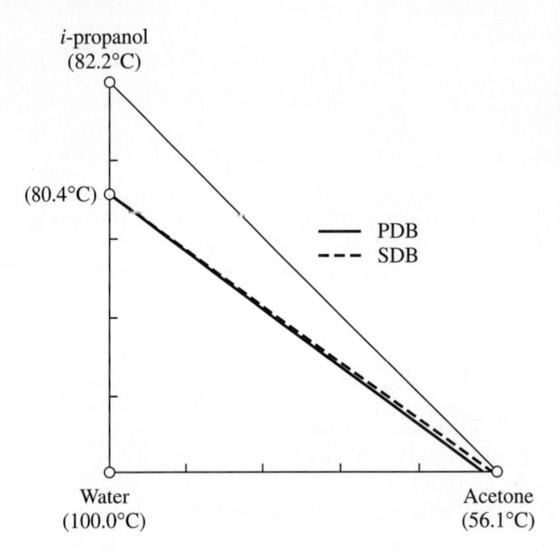

FIGURE 5.25
Pitchfork and simple distillation boundaries for acetone-isopropanol-water at 1 atm pressure, $q = 1$. The activity coefficients used in the VLE model are represented by the Wilson equation.

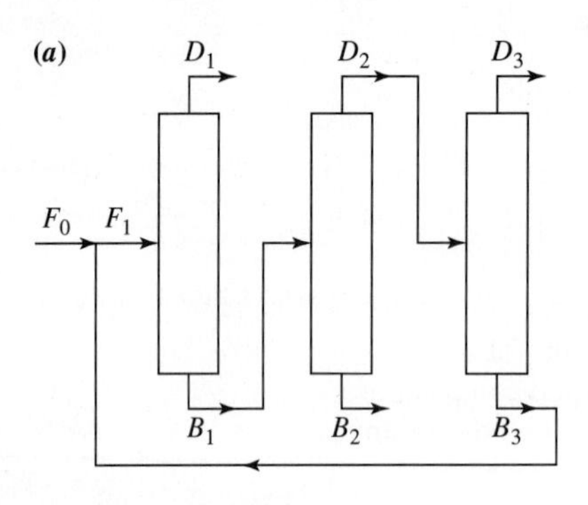

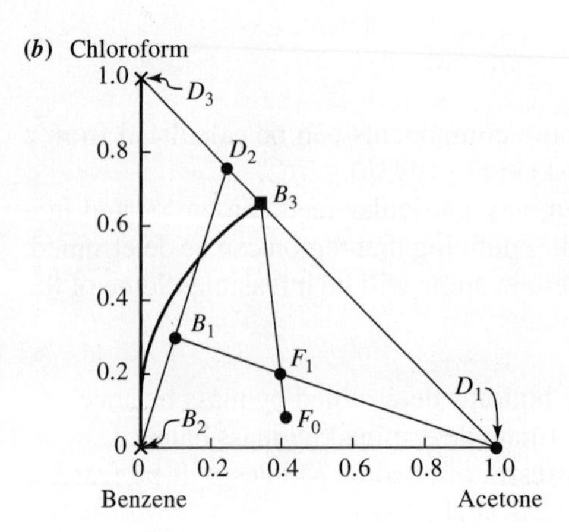

FIGURE 5.26
Sequence that exploits a curved distillation boundary without exploiting the overlap of distillation regions All three components in the feed F_0 are removed as pure products from the system.

5.5 CONCEPTUAL DESIGN METHOD

The pinch alignment technique described in Chap. 4 may be used to estimate minimum flows in mixtures containing tangent pinches and azeotropes (Fidkowski et al., 1991). Column designs at reflux ratios above the minimum are calculated using the ω-design method in Sec. 4.4.

Approach

We describe a hybrid procedure that combines heuristics and models for conceptual design. The models are nonlinear and make use of *geometric methods* to find solutions. The models are intended to preserve the essential nonlinearities and to address synthesis as well as analysis. On the other hand, for conceptual design, a complete first principles model is often too demanding of data, engineering time, and perhaps computational resources, so we will also use some heuristics to avoid expending resources to solve hard problems which are either not worth solving or which typically have good solutions that are not very different from one another.

The approach consists of the following steps.

1. *Specifications.* Choose a pressure, from the available utility levels, or from constraints such as the maximum and minimum temperatures of utilities or for thermal stability of the mixture. Determine the feed composition(s) available and the desired product purities.[25]
2. *Feasibility and Alternatives.* This requires VLE information (often forcing experiments) and consists of two parts.

   a. Construct *residue curve maps* by solving for the phase plane portrait of

   $$\dot{\mathbf{x}} = \mathbf{x} - \mathbf{y} \tag{5.24}$$

   where the liquid and vapor compositions $\mathbf{x}$ and $\mathbf{y}$ are related by a VLE model. If the VLE model is incomplete, sketch the residue curves based on boiling temperatures and compositions (Foucher et al., 1991). (If the data are not sufficient for a sketch, then do an experiment. Alternatively, use a guess or a group contribution method, and *then* do an experiment!) Use the following heuristics and rules to get a base-case sequence and, possibly, a few alternatives.

      i. The compositions of the desired products from each column should lie in the same distillation region, from Heuristic 5.1. If the distillation boundaries are linear, the products from the entire sequence must lie in the same

---

[25]The pressure level is certainly a candidate for optimization, provided that a reliable physical properties model is available for the pressure range of interest. Note that pressure changes can change the number of azeotropes and other important features of the vapor-liquid equilibrium behavior (Knapp and Doherty, 1992; Fidkowski et al., 1993*b*).

distillation region. If they are curved, the *products* from the entire sequence may lie in different regions.[26]

ii. Unstable nodes can be taken as distillate streams.

iii. Stable nodes can be taken as bottoms streams.

iv. Feeds and product flows and compositions must satisfy the overall material balance (lever rule).

b. For each feasible alternative, estimate the particular range of product compositions available for simple columns (Wahnschafft et al., 1992*a*; Fidkowski et al., 1993*a*; Sec. 7.5).

If no alternatives are feasible, adjust the pressure or add another component as an entrainer or "mass separating agent"and restart this step.[27]

3. *Flows, Energy, and Theoretical Stages*

a. Estimate the minimum reflux and, for extractive distillation, the minimum entrainer flow. This also gives the minimum energy.

b. Find the number of theoretical stages, using the heuristics.

i. The reflux should be 50% greater than the minimum. This also gives the heating and cooling loads.

ii. The entrainer flow should be two to four times greater than the minimum (Knapp and Doherty, 1994).

iii. In extractive distillation, the recycle purity of entrainer should be midway between the maximum and minimum values. The maximum value is either 100% purity or the composition of an azeotrope if one is used as the entrainer. The minimum value is determined from the amount of light impurity whose presence in the distillate from the azeotropic column would just meet the overhead purity specification (Knight and Doherty, 1989).

c. Unspecified compositions should be 99.5% of the maximum possible fractional recovery.

d. Beware of the possibility that the rectifying and stripping composition profiles may intersect twice! This does not happen in ideal mixtures but can happen in nonideal and azeotropic mixtures because of the variations in curvature introduced, especially by azeotropes. This means that there will be two designs with identical feed and product compositions, the same reflux and reboil ratios, but with a different number of stages and a different feed stage.[28] An example of this is reported for the mixture acetone-ethanol-water in Knapp and Doherty (1992, Figs. 9 and 10). Also see Exercise 19.

[26]Distillation boundaries can be routinely crossed by liquid-liquid phase separation (Chap. 8) or by reaction (Chap. 10).

[27]This procedure is not useful to *choose* candidate entrainers, but it is useful to *screen* candidates once they are selected. An initial short list of candidates can often be developed from components already present in the process or in a downstream or upstream process.

[28]This is a different phenomenon than varying the design by varying the value of ω in the ω-design method. In that case, the product distribution of the minor components changes as the design changes.

Often, alternatives can be eliminated at this point if the stage or energy requirements are excessive.

4. *Cost Estimates.* For comparison purposes among alternatives, first estimates are based on (see Chap. 6):

   a. A 50% stage efficiency and 0.5 m tray spacing for the column height.
   b. A vapor rate at 60% of flooding.
   c. Average overall heat transfer coefficients (see Table 6.2).
   d. A cost correlation for the equipment.
   e. Utility costs from the flows and site-specific information.

   For the least expensive alternative and any others with comparable costs, refine the design estimates and optimize pressure, reflux ratio, etc. (The meaning of "comparable" depends on the accuracy of the models, but within 25% is not atypical.)

5. *Complex Columns.* Sidestream columns are the most common, followed by side stream strippers or rectifiers. These configurations often lead to significant savings when volatility differences are large or when sidestream purity requirements are not extreme. Relatively ideal mixtures can be handled easily (Tedder and Rudd, 1978; Glinos and Malone, 1988) but less is known about mixtures containing azeotropes. It is feasible to reach a sidestream purity close to a saddle (middle-boiler) in the same distillation region as the other products (Rooks et al., 1996). In many cases this will not be economical, but if the sidestream purity requirements are not extreme, e.g., the sidestream is recycled, then the savings may be quite significant. There are certainly questions about the operability and control of complex columns, but our intent here is first to estimate the incentive to answer these questions concerning operability and control.
6. *Sensitivity.* It is particularly important to study sensitivities. These may be in the form of disturbances in process variables or parameters in physical property models.

Notes

1. For feasibility, we ask for a picture suited to design that also reflects the basic phase equilibrium information. For distillation, this is the *residue curve map.* The benefits of this visual representation should not be underestimated.
2. Residue curves for three- and four-component mixtures are straightforward to construct when a model is available. A more qualitative, but surprisingly useful, picture can often be deduced when only limited data on the boiling temperatures and approximate compositions of azeotropes is available (Foucher et al., 1991).
3. Even in mixtures with more components, it is quite useful to consider subsets of the components for feasibility. With five or more components, the full picture cannot be drawn, although it is straightforward to determine when particular compositions lie in the same distillation region. (This can be done by integrating the residue curve equations forward and backward, starting from the compositions in question). See Sec. 7.7 for additional information.

4. The minimum reflux can be estimated for nonideal and azeotropic mixtures quite closely from the "zero-volume" geometric construction (Julka and Doherty, 1990; Fidkowski et al., 1991).

EXAMPLE 5.2. Column Design. This example demonstrates the application of the ω-design method to an azeotropic mixture with a distillation boundary. We have a saturated liquid mixture of 20 mol % benzene, 40 mol % isopropanol, and 40 mol % n-propanol that we would like to separate into its pure components. Determine what splits are possible and judge whether the desired separation can be achieved.

We approach this task as follows:

- First we select a VLE model and check its fidelity against available VLE data. Once we are confident that it represents the binary pairs in a satisfactory way, we use it to draw the residue curve map in Fig. 5.27*a*.
- The mixture has two distillation regions with a boundary connecting the two binary azeotropes (marked A_1 and A_2 in the figure). The simple distillation boundary is quite straight, so we do not even attempt to calculate the pitchfork boundary. There is no opportunity here for crossing boundaries or exploiting curvature.
- The feed is in the left region, where n-propanol is the stable node and azeotrope A_1 is the unstable node. Two sharp splits are possible:

  Direct Split. Distillate: azeotrope A_1, bottom product: isopropanol + n-propanol.
  Indirect Split. Bottom product: n-propanol, distillate: benzene + isopropanol.

  The bottom product from the direct split can be further separated in a second column into isopropanol (distillate) and n-propanol (bottom). Also, the distillate from the indirect split can be separated in a second column into azeotrope A_1 (distillate) and isopropanol (bottom). Either way we have an azeotropic mixture of benzene and isopropanol to separate. This azeotrope can be broken using extractive distillation technology described in the next section.[29] The indirect split looks like a good place to start because it offers the possibility of fewer columns to separate the mixture.
- We use the ω-design method for the indirect split. First, assume that the feed and products are saturated liquids and that we operate the column at 1 atm pressure. We select the mole fraction of n-propanol in the two product streams as 0.9982 in the bottom (high purity) and 0.001 in the distillate (high recovery). We calculate the minimum reboil ratio by setting the mole fraction of benzene in the bottom product to a very small value (say 1×10^{-12}) and implementing the pinch alignment method using either the zero-volume method or the boundary value method.[30] A value of $r = 2.6$ is a good starting point for design. We now generate a spectrum of designs for different values of ω, shown in Fig. 5.28. Any value of ω in the range 0.2 to 0.8 generates a good design, and we choose $\omega = 0.5$. The corresponding design is shown in Fig. 5.27*b*. It has 14 theoretical stages in the stripping section and 23 stages in the rectifying section (rounding the fractional stages to the next higher integer value). The benzene composition in the bottom stream is in the ppb range. The number of stages in the column could be reduced by increasing the reflux ratio or by relaxing the purity and/or recovery of n-propanol. These alternatives represent opportunities for optimization and integration into the overall flowsheet.

---

[29]This suggests yet another alternative, which is to feed the distillate from the indirect split to the extractive distillation subsystem without first performing the binary separation.

[30]Do this yourself using the boundary-value method.

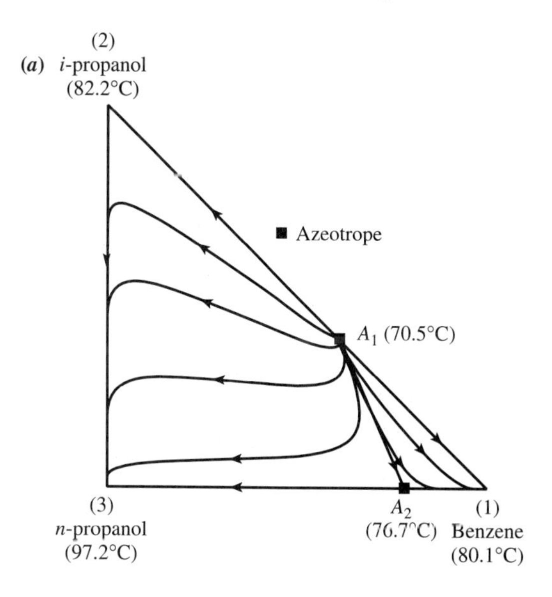

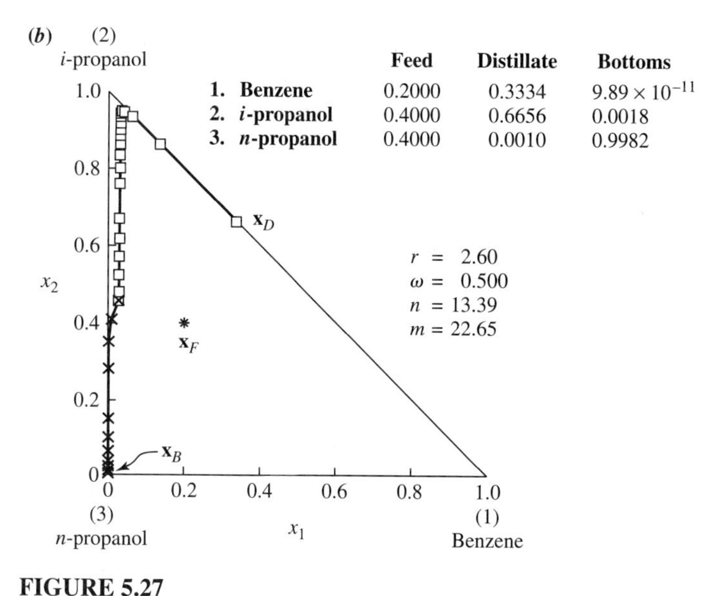

	Feed	Distillate	Bottoms
1. Benzene	0.2000	0.3334	9.89×10^{-11}
2. *i*-propanol	0.4000	0.6656	0.0018
3. *n*-propanol	0.4000	0.0010	0.9982

FIGURE 5.27
(*a*) Residue curve map. (*b*) Column composition profiles for the mixture benzene-isopropanol-*n*-propanol at 1 atm pressure.

EXAMPLE 5.3. The DeRosier Problem. The first part of this example is a mechanical application of the principles taught in class on the analysis and design of ternary azeotropic distillations. The second part challenges you to find a creative design to a typical engineering problem that, on face value, may seem impossible to achieve. This problem is named after Robert DeRosier, who was the first student to develop a successful design. We normally do this exercise in class using interactive design software, and although

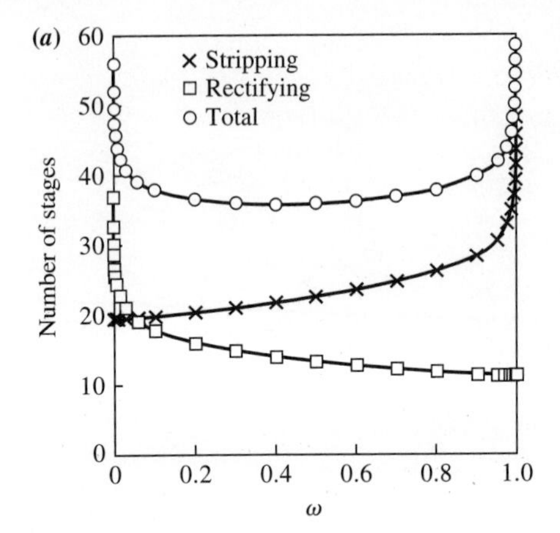

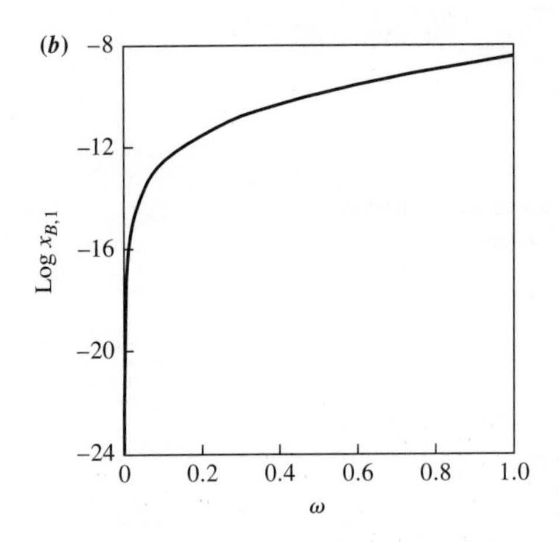

FIGURE 5.28
Spectrum of designs for benzene-isopropanol-*n*-propanol at 1 atm pressure, $r = 2.6$.

students often fail to solve Part 2, everybody benefits from the discussion that erupts in the attempt. The problem statement is as follows:

Part 1: Calculate the residue curve map for the mixture methanol-isopropanol-water at 1 atm pressure. Use the NRTL-ideal model to represent the VLE. Design a distillation column to separate a saturated liquid feed consisting of 40 mol % methanol, 40 mol % water, and 20 mol % isopropanol into saturated liquid products. The distillate is specified to contain 99 mol % methanol and 0.5 mol % water. The bottom product contains 0.5 mol % methanol. Use your engineering judgment to select a reflux ratio that trades off the vapor rate against the number of theoretical stages.

Part 2: After solving Part 1, you proudly show your design to the client who commissioned the job. She tells you that she wants as little water as possible in the methanol product because it kills the catalyst in her process. She will not accept methanol containing more than 50 ppm water. Since water is the heaviest component, this seems like an easy constraint to meet. Is it? What is the smallest composition of water that you can get in the methanol product? What is your design for the separation scheme?

The solution to this problem was prepared using the AEA software package DISTIL. Part 1 is straightforward; the calculated residue curve map is shown in Fig. 5.29. Figure 5.30 shows the column composition profiles at minimum reflux, which is determined by the presence of a node pinch in the stripping profile just below the feed stage. We find $r_{min} = 5.0$. As a first estimate, we let the operating reflux ratio be 50% larger than the minimum value, and obtain the design shown in Fig. 5.31. The column has 28 theoretical stages with a feed on stage 6 (numbering from the top of the column). Students typically begin solving Part 2 by successively lowering the mole fraction of water in the distillate and repeating the design strategy used for Part 1. You will find that when the distillate contains less than 3400 ppm water, the distillate composition lies in the lower distillation region and the rectifying profile does not intersect the stripping profile at any reflux ratio! This is shown in Fig. 5.32*a*. At this point many students feel smug that their superior engineering know-how has proved that the client cannot get what she wants. But in fact, she can, and the question is, how?

At this point students will suggest many alternative strategies that are confined within the exact constraints stated in the problem definition. Most alternatives involve

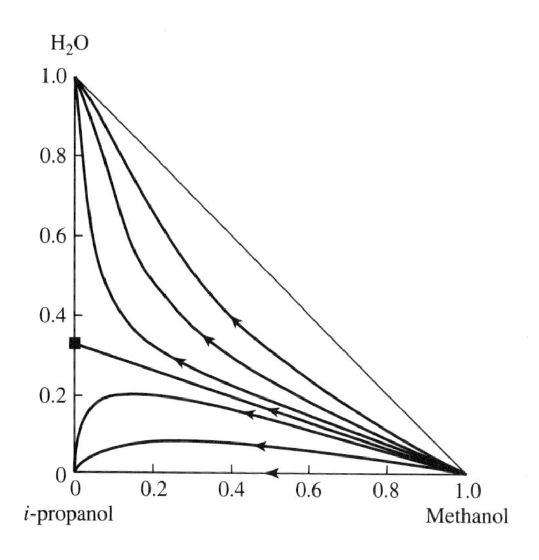

FIGURE 5.29
Residue curve map for the mixture methanol-water-isopropanol at 1 atm pressure.

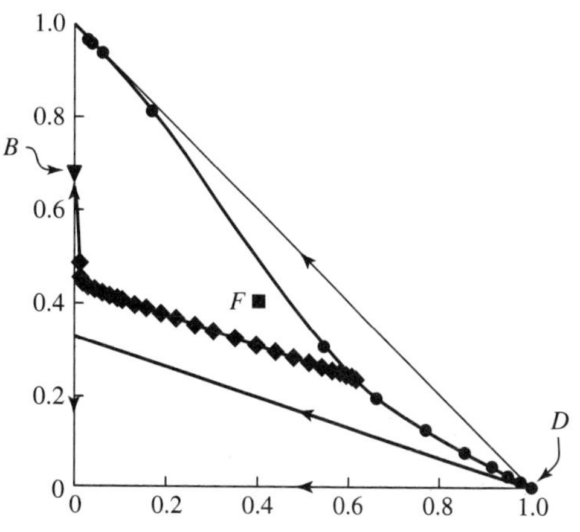

FIGURE 5.30
Minimum reflux composition profiles, $r_{min} = 5$.

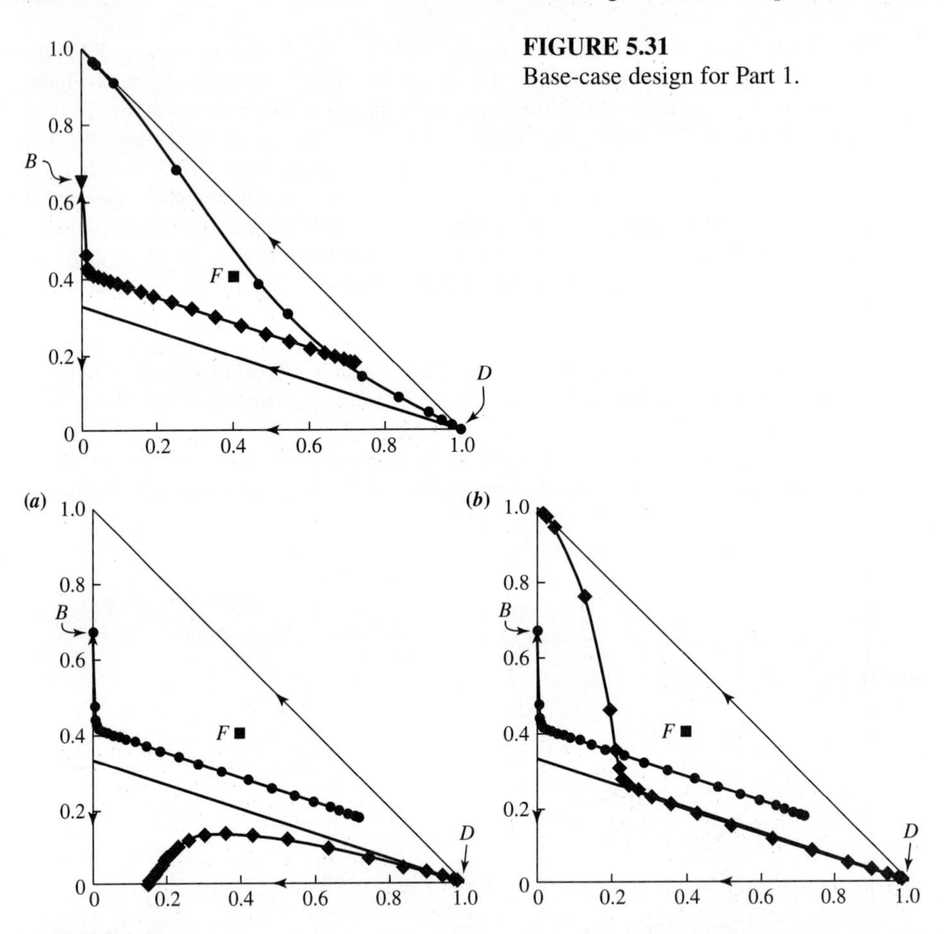

FIGURE 5.31
Base-case design for Part 1.

FIGURE 5.32
A water composition in the distillate of 3300 ppm (*a*) is below the minimum value of 3400 ppm (*b*).

adding more columns or other finishing separation steps. However, the key to achieving our goal is to get the distillate below the distillation boundary into the 50 ppm water range. The best way to do this is to recognize that as the distillation boundary approaches the pure methanol vertex it becomes tangent to the base of the triangle, i.e., free of water. Therefore, we cannot get below the distillation boundary at 99 mol % methanol, but we can drive the water content down by *increasing* the purity of the methanol product. The only remaining engineering issue is, how many extra stages are required to do this? Trying a methanol purity of 99.9 mol % does not do it, but 99.99 mol % does! The design is shown in Fig. 5.33, where a 33-stage column operating at the same reflux ratio as the previous (unsuccessful) design achieves a methanol purity of 99.99 mol % and a water content of 50 ppm. Perhaps we can even charge a premium price for our methanol because it is of such high quality! The important lesson to be learned here is that we have achieved our manufacturing goal by relaxing some of the (softer) constraints in order to meet a hard constraint. That is, we have translated the most important part of the customer's order into a redefinition of the problem statement. Breaking out of narrow problem definitions and thinking more broadly about the desired project goal is one of the most important aspects of engineering decision making.

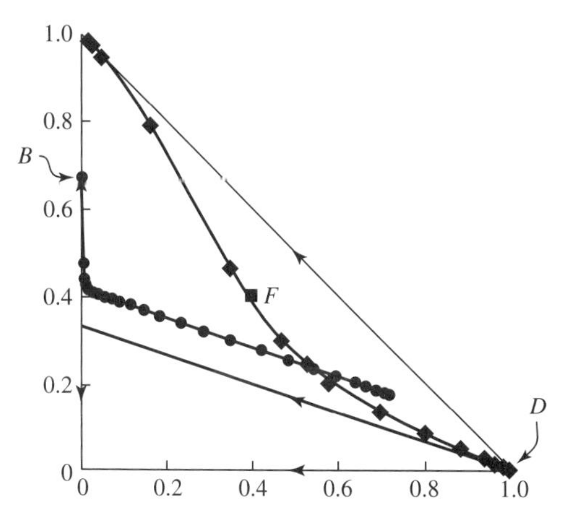

FIGURE 5.33
Column design for distillate stream containing 99.99 mol % methanol, 50 ppm water.

5.6 DISTILLATION SYSTEMS AND EXTRACTIVE DISTILLATION

Separation system synthesis for azeotropic mixtures is treated in Doherty and Caldarola (1985), Foucher et al. (1991), LaRoche et al. (1991, 1992*a*, 1992*b*), Barnicki and Siirola (1997), Wahnschafft et al. (1991, 1992*a*, 1992*b*, 1993), Fidkowski et al. (1993*a*), Safrit and Westerberg (1997), Rooks et al. (1998), Bauer and Stichlmair (1998), and Sargent (1998). The underlying principles governing these methods are described here.

Column Sequences

No distillation boundaries

Residue curve maps that have no distillation boundaries are usually the easiest to navigate around by distillation. For example, let's try to break the binary minimum-boiling azeotrope between methanol and toluene using methyl butyrate as an entrainer. The residue curve map is shown in Fig. 5.7*a*. The map has no distillation boundaries, and we can use the system shown in Fig. 5.34 to break the azeotrope. Toluene (stable node) is the bottom product from the first column, with methanol + methyl butyrate as distillate. The distillate is separated in a second column, from which the methyl butyrate is recycled to the process feed. The entrainer could also be recycled as a second feed stream to column 1 above the process feed. This class of entrainers is discussed in more detail by LaRoche et al. (1992*a*, 1992*b*), especially for the case of breaking the acetone-heptane minimum-boiling azeotrope using benzene as an intermediate-boiling entrainer.

Linear distillation boundaries

When the simple distillation boundaries are straight, the feed and product compositions from a continuous distillation column must lie in the same simple distillation region. It is not possible to devise internal recycles within the sequence to cross boundaries at recycle mixing points [see Doherty and Caldarola (1985) and

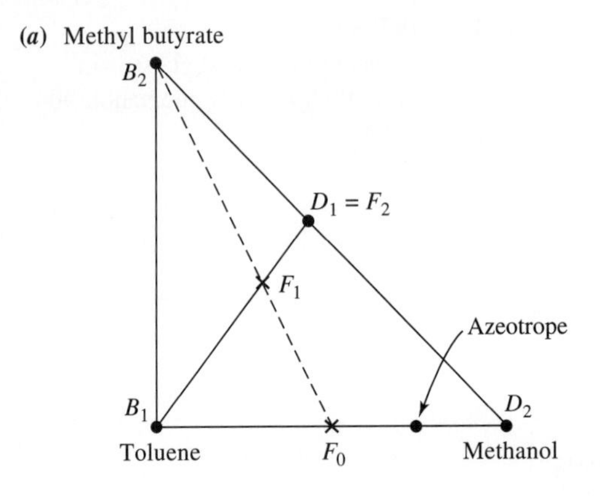

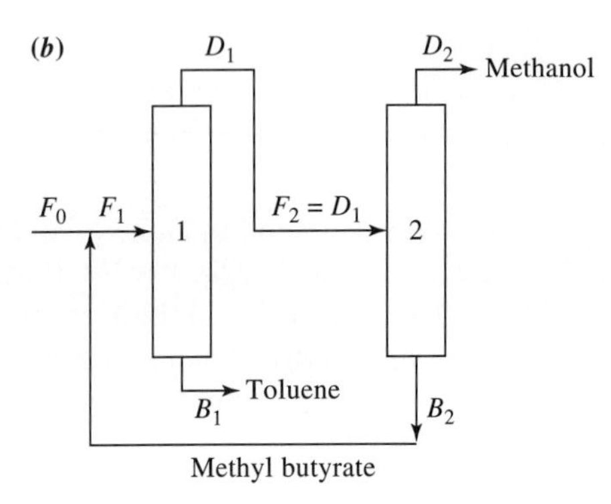

FIGURE 5.34
Sequence for an intermediate-boiling entrainer.

Exercise 20]. Therefore, we have the following rule for devising feasible sequences in mixtures with linear distillation boundaries.

Heuristic 5.2

For mixtures with linear simple distillation boundaries, the products from each column and the products from the entire sequence of columns must lie in the same distillation region as the process feed.

This heuristic is useful for selecting entrainers; one class of feasible entrainers are those which do not divide the components to be separated into different distillation regions.[31] A selection of residue curve maps that are favorable for homogeneous azeotropic distillation using this heuristic are shown in Fig. 5.35; see Exercises 23 and 24.

Curved distillation boundaries

In this case, distillation regions overlap between the pitchfork and simple distillation boundaries. Distillation systems can be invented that exploit this overlap (e.g., see Fig. 5.24). Even if we do not exploit the overlap, and follow Heuristic 5.1, the

[31] This rule also catches entrainers that introduce no distillation boundaries.

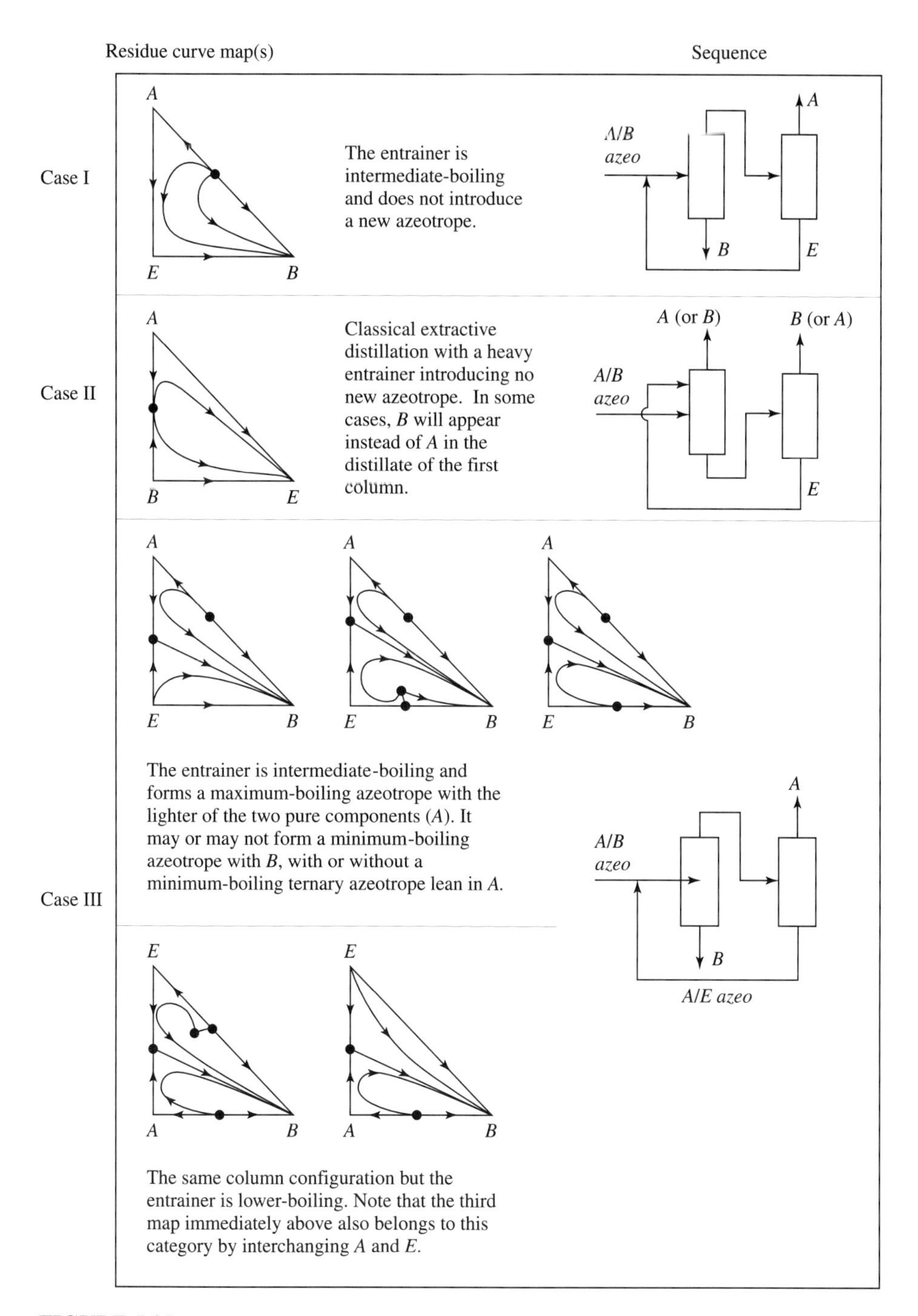

FIGURE 5.35
Favorable residue curve maps for breaking the A-B azeotrope using entrainer E by homogeneous azeotropic distillation. *[Taken from Foucher et al. (1991).]*

products from the entire sequence are not constrained to lie in the same distillation region; see Fig. 5.26.

As a word of warning, we should add that distillation systems that exploit curved boundaries are certainly possible, but this is a specialized technology best left to the expert. As noted above, the startup and operation of columns that exploit curved boundaries require careful study, since it is certain that they will not meet the desired product purity specifications at total reflux.

Extractive Distillation

Extractive distillation is a method of separating minimum-boiling binary azeotropes by use of an entrainer that is the heaviest species in the mixture, does not form any azeotropes with the original components, and is completely miscible with them in all proportions. The residue curve map for such mixtures will always look qualitatively like the schematic shown in Fig. 5.36. The minimum-boiling azeotrope is the unstable node, the entrainer vertex is the stable node, and the two components forming the azeotrope are both saddles. The extractive entrainer, being heavy, is continuously added near the top of the extractive distillation column so that an appreciable amount is present in the liquid phase on all of the stages below. The process feed, containing the azeotrope to be broken, is added through a second feed point lower down the column (see Case II, Fig. 5.35). In the extractive column the component having the greater volatility, not necessarily the component with the lowest boiling point, is taken overhead as a pure distillate (labeled "product" in Fig. 5.36). The other feed component (labeled "nonproduct") leaves with the entrainer via the column bottoms stream. The entrainer is separated from the remaining (nonproduct) component in a second distillation column (entrainer recovery column) and then recycled back to the first column.

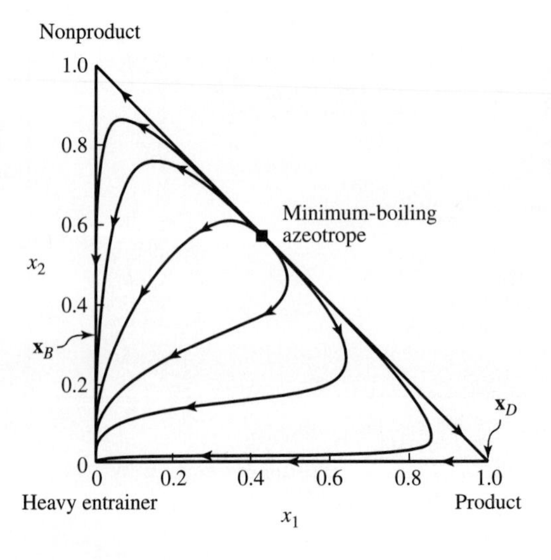

FIGURE 5.36
Residue curve map for extractive distillation. Typical distillate and bottoms compositions from the extractive column are also indicated.

Extractive distillation technology is the most widely used form of homogeneous azeotropic distillation in the chemical process industries. A selection of industrial applications includes (1) separation of the *n*-butane–butadiene azeotrope in mixed C_4 hydrocarbon streams using furfural as entrainer; (2) separation of toluene from an azeotrope formed with a nonaromatic hydrocarbon (e.g., octane); (3) the dehydration of ethanol using ethylene glycol; (4) breaking the acetone-methanol azeotrope using water; (5) separation of the pyridine-water azeotrope using bisphenol. Literature references to these and other applications can be found in Doherty and Knapp (1993). Many additional applications are described in the patent literature. Extractive distillation has been practiced on a commercial scale in the United States since at least the early 1940s. At the time, this technology was driven by the U.S. war effort to ramp up production of butadiene for the rubber industry, and toluene for military applications (Happel et al., 1946; Buell and Boatright, 1947; Lake, 1945; Benedict et al., 1945; and especially the classic paper by Benedict and Rubin, 1945). By 1950, it was already in textbooks (Robinson and Gilliland, 1950); also see the monograph by Hoffman (1964).

A typical extractive distillation is the dehydration of ethanol using ethylene glycol as entrainer. Ethanol forms a minimum-boiling azeotrope with water at about 90 mol % ethanol, so this azeotrope must be broken to achieve anhydrous ethanol. The residue curve map for the ternary mixture is shown in Fig. 5.37, and the separation regions for a single-feed column are shown in Fig. 5.38. The transition line always intersects the ethanol-water and water-ethylene glycol edges. Figure 5.38*a* is similar to the case of an ideal mixture, but with the azeotrope playing the role of the "lightest component." For this feed, the distillate can be made no richer in ethanol than the azeotrope, which is not very useful because we can achieve this composition without adding ethylene glycol. If we add more glycol, we can achieve distillate compositions closer to pure ethanol, but still they contain several percent water; see Fig. 5.38*b*.

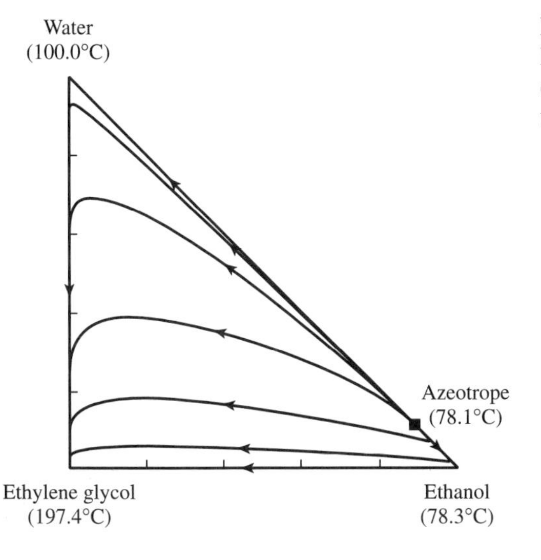

FIGURE 5.37
Residue curve map for extractive distillation of ethanol and water using ethylene glycol entrainer at 1 atm pressure.

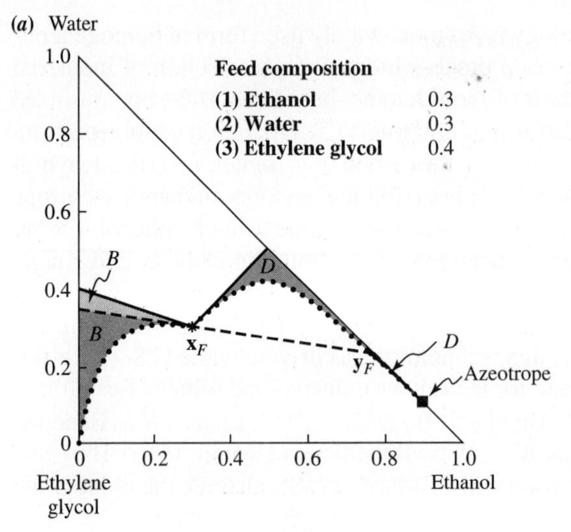

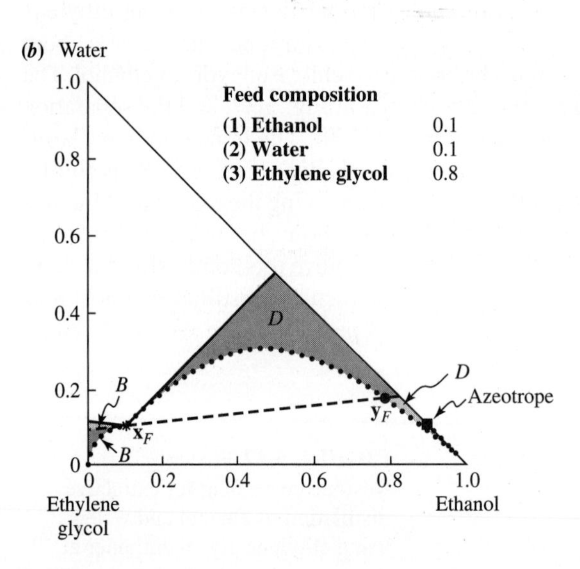

FIGURE 5.38
Separation regions for ethanol-water-ethylene glycol at 1 atm pressure.

The key to making the separation is to recognize that the ethylene glycol must be added as a separate feed near the top of the column, above the ethanol-water feed. This enables us to exploit the properties of four-sided distillation regions to produce products that are saddles in the residue curve map (the saddle is ethanol in this case). This works as follows. Choose a distillate composition near the product vertex on the product-entrainer side of the triangle in Fig. 5.36. The rectifying section profile starts at the distillate composition and moves along the base of the composition triangle toward the entrainer vertex (generally following the residue curves). The bottoms composition is located close to the vertical edge of the triangle with its position

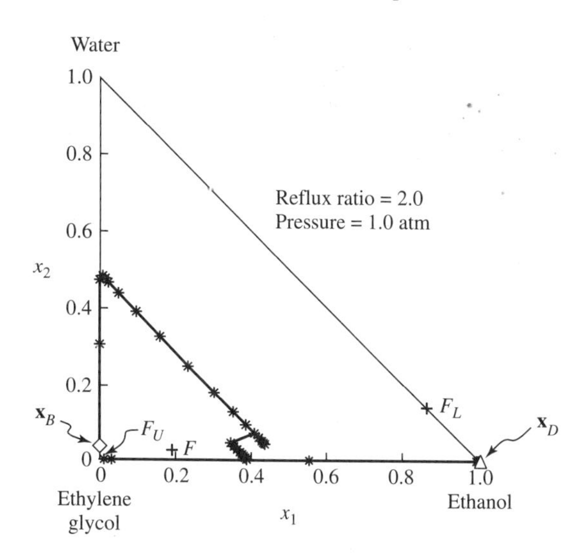

FIGURE 5.39
Composition profiles for ethanol-water-ethylene glycol extractive distillation at 1 atm pressure, $F_r = 3.5$. Note that values of F_r around 1 give better designs, which you can check with a computer-aided-design tool; also see Example 5.4.

determined by material balance;[32] a typical value for the bottom composition is shown in Fig. 5.36. The stripping section profile starts at $\mathbf{x}_B$ and moves up the edge of the triangle toward the nonproduct vertex. It turns around a saddle and then moves into the triangle, exactly as we have seen so many times for stripping profiles. The only way that the separation could be feasible in a single-feed column is if the stripping profile reaches the rectifying profile on the bottom edge. But this can't happen because the stripping profile reaches a node pinch first that originates from the azeotrope.[33] Computed profiles for this extractive distillation are shown in Fig. 5.39, and profiles for the acetone-methanol extractive distillation with water as entrainer are shown in Fig. 5.40; the profiles in this figure are labeled. Therefore, the purpose of the middle section is to provide a bridge connecting the stripping and rectifying profiles, as seen in Figs. 5.39 and 5.40.

The geometric behavior of the middle-section profile causes new features in extractive distillation that we have not seen so far.[34] At any value of the feed ratio F_r (defined as the ratio of entrainer flow to process feed flow), there is a minimum reflux ratio where the number of stages in the stripping section becomes infinite at a stable node that just touches the middle-section profile (Levy and Doherty, 1986), i.e., a feed pinch in the stripping profile. There is also a *maximum* reflux ratio where the number of stages in the the middle-section profile becomes infinite (Knapp and Doherty, 1994, especially Fig. 10). Figure 5.41 shows the number of stages versus reflux ratio for the acetone-methanol-water separation at two values of the feed ratio $F_r = 0.55$ and $F_r = 1.0$. In normal distillation, the number of stages decreases

[32]Lever rule for the distillate, bottoms, and overall feed composition. The latter depends on the flow and composition of the two feed streams.

[33]This tells us that extractive distillation will *never* work in a single-feed column at total reflux/reboil because the stripping section stable node is located at the azeotrope under these conditions.

[34]A very detailed study of this is given in Knapp and Doherty (1994).

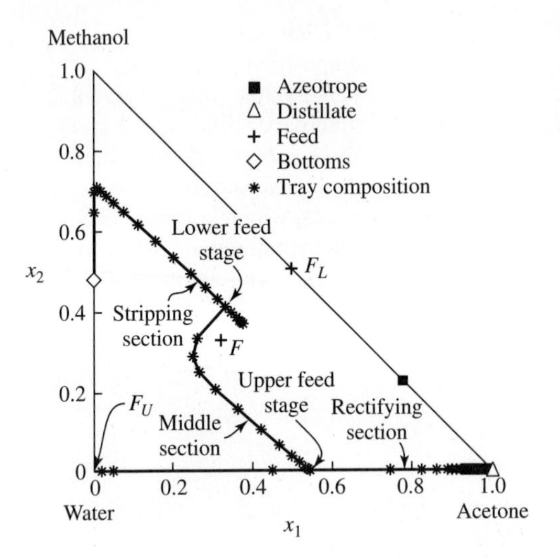

FIGURE 5.40
Composition profiles for acetone-methanol-water extractive distillation at 1 atm pressure, $F_r = 0.55$.

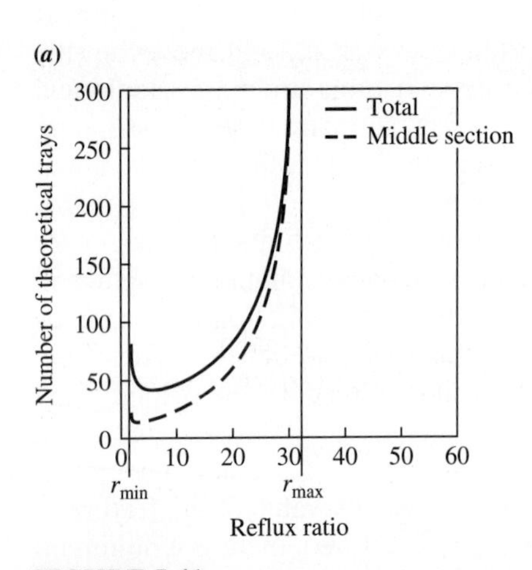

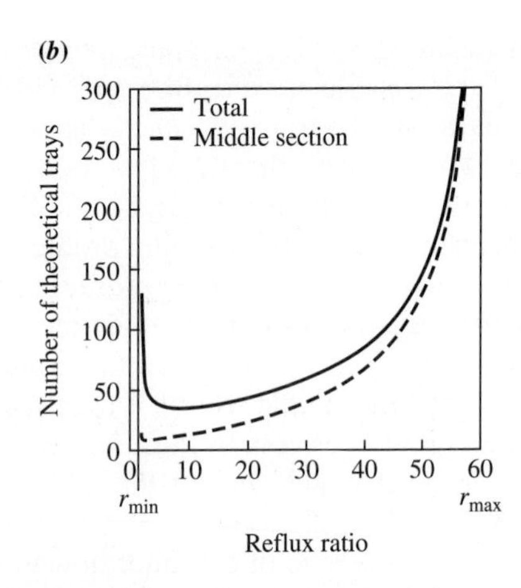

FIGURE 5.41
Number of theoretical stages versus reflux ratio for the acetone-methanol-water extractive distillation at 1 atm pressure. (*a*) $F_r = 0.55$. (*b*) $F_r = 1.0$.

monotonically as the reflux ratio increases from r_{min}, but in extractive distillation the number of stages starts to increase again because the middle-section profile approaches a pinch. Eventually there is an infinite number of stages at the pinch, corresponding to r_{max}. As seen from the figure, r_{min} and r_{max} get closer as the feed ratio gets smaller, i.e., as we add less entrainer. At small values of F_r there will be a narrow range of reflux ratios over which the desired separation is possible. In fact, there is a minimum value of F_r where the minimum and maximum reflux ratios occur simultaneously. Below this value, the separation is not possible. Figure 5.43 shows a similar plot for the ethanol-water-ethylene glycol separation.

A convenient way of representing the feasible range of design parameters is on a graph of reflux ratio versus feed ratio (Fig. 5.42). The maximum and minimum reflux ratios, as well as the minimum feed ratio, are all seen clearly.

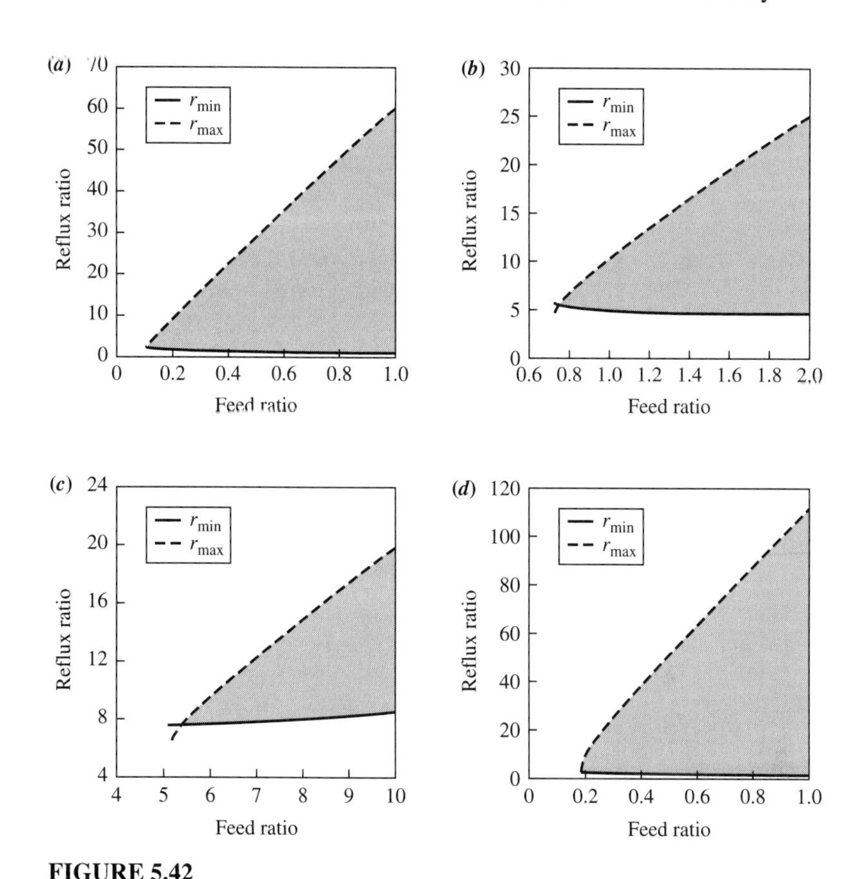

FIGURE 5.42
Reflux ratio versus feed ratio for extractive distillation. Shaded regions represent feasible design variables. (*a*) Acetone-methanol-water. (*b*) Methanol-acetone-methyl ethyl ketone (MEK). (*c*) Methanol-MEK-*sec*-butanol. (*d*) Ethanol-water-ethylene glycol.

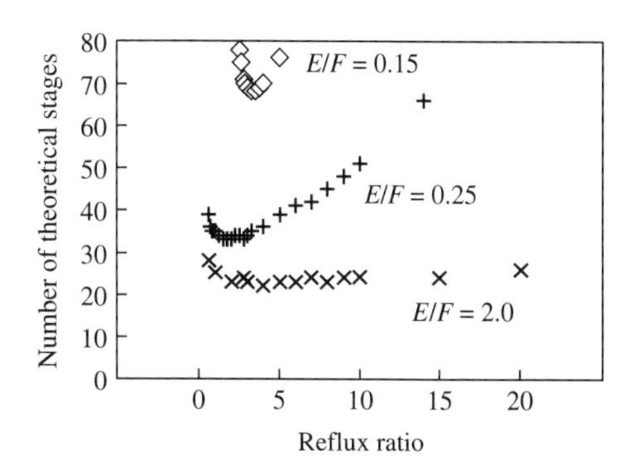

FIGURE 5.43
Number of stages as a function of reflux ratio and entrainer flow for extractive distillation of ethanol and water with ethylene glycol entrainer.

The designer is free to choose many variables in extractive distillation. Initial designs should be developed using the following heuristics:

Heuristics for extractive distillation

1. Set the feed ratio at three times its minimum value, $F_r = 3F_{r,\min}$ (Knapp and Doherty, 1994, especially Table 1 and Fig. 30). If the resulting value for F_r is bigger than 2 or 3, look for a better entrainer.
2. Set the reflux ratio in the extractive column to $r = 1.5r_{\min}$, where $r_{\min}$ is calculated at the value of F_r determined in Heuristic 1 (Knight and Doherty, 1989). In the entrainer recovery column also use a reflux ratio 50% bigger than the minimum.
3. Set the entrainer feed temperature to the extractive column to be 5 to 10°C below the boiling temperature of the distillate product from the extractive column (Knight and Doherty, 1989). This will make the entrainer feed a subcooled liquid. Additional results on feed qualities in extractive distillation are reported by Knapp and Doherty (1990).
4. Set the compositions in the bottom stream from the extractive column by taking 99.5 to 99.99% fractional recovery of the desired product (e.g., ethanol) in the the distillate stream from the extractive column.
5. Set the entrainer recycle purity (i.e., bottom composition from the entrainer recovery column) halfway between its minimum value and unity (or the composition of the azeotrope if the entrainer is a mixture at the azeotropic composition). The minimum value is calculated by assuming that all the impurity in the entrainer recycle stream leaves in the distillate from the extractive column. Knowing the flowrates and the desired product purity from the extractive column enables a simple calculation of the minimum allowable entrainer purity (Knight and Doherty, 1989). The allowable range is narrow, and typical recycle purities will be around 99.95 mol %. However, this is usually easy to achieve because the entrainer is very heavy relative to the other components present.

Once a base-case design is developed, it should be checked for sensitivity to variations in the choices made for the design variables and also to uncertainties in some of the key physical properties. The feed ratio is an important optimization variable, and if you only have time to vary one quantity it is always worth repeating the design at several values for the entrainer feed rate.

The designer is not free to pick which of the components in the azeotrope will be the distillate from the extractive column despite the apparent symmetry of the residue curve map (i.e., each of these components is a saddle in the map). For a given entrainer, one and only one of the feed components can be recovered in the distillate from the extractive column, and it is not always the component with the lowest boiling point. For example, the extractive distillation of ethanol and water using gasoline, some phenols, cyclic ketones, or cyclic alcohols causes *water* to be removed as the distillate from the extractive column and the lower-boiling ethanol to leave in the bottom stream with the entrainer. Other entrainers, such as ethylene glycol, cause ethanol to be removed as the distillate from the extractive column.[35]

---

[35]These and other instances are discussed in Knapp and Doherty (1994) and Doherty and Knapp (1993) together with more citations to the original literature.

It is important to know which of the feed components will appear in the distillate stream from the extractive column in order to design the equipment, and fortunately there are reliable ways to predict this. The easiest are:

1. Plot the pseudo-binary (entrainer free) equilibrium y_1-x_1 diagram for the binary azeotropic mixture with various constant amounts of entrainer present (e.g., 0%, 10%, 40%, etc.). If the curve moves above the 45° line, then component 1 has the lowest volatility and will be the distillate. If the curve moves below the 45° line, then component 2 has the lowest volatility, and it will be the distillate. The method is due to Scheibel (1948); see, e.g., Knapp and Doherty (1994) for further discussion.
2. Plot the isovolatility line $\alpha_{1,2} = 1$ on the composition triangle. This line begins at the binary azeotrope and intersects one of the other binary edges. If it intersects the 1-3 edge (3 represents entrainer), then component 1 has lower volatility than component 2, and 1 will be the distillate from the extractive column. If it intersects the 2-3 edge, then component 2 will be the distillate. The method was introduced by LaRoche et al. (1991) for homogeneous azeotropic distillations in general. For extractive distillation it gives identical results to Scheibel's method.

See Exercises 25, 26, and 27.

Most of these concepts that we have developed for extractive distillation also apply to mixtures with four-sided distillation regions where one or both of the saddles are azeotropes. For example, the mixture acetone-methanol-MEK has a four-sided distillation region where one of the saddles is the methanol-MEK binary azeotrope; see Fig. 5.44. If MEK is used as an entrainer to break the acetone-methanol azeotrope, then the methanol-MEK azeotrope is the distillate from the extractive column (not acetone, even though acetone boils at a lower temperature). Acetone and MEK leave in the bottom stream and are separated in a second column, from which the MEK is recycled (Knapp and Doherty, 1992). In some applications this type of separation

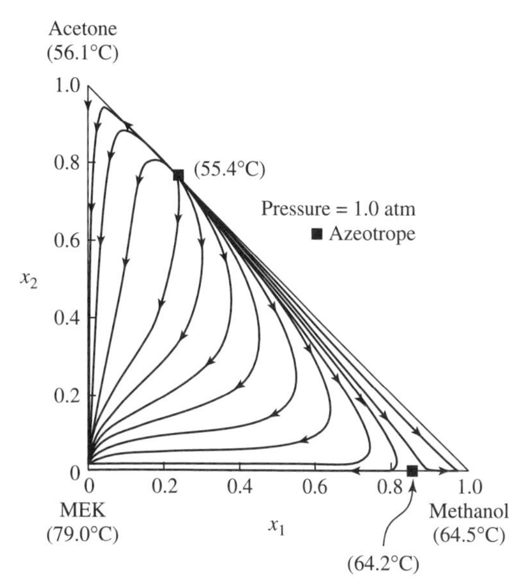

FIGURE 5.44
Residue curve map for acetone-methanol-MEK at 1 atm pressure.

is desired, for example, if methanol and MEK have the same destinations in the flowsheet, see Exercises 28 and 29.

Finally, it is sometimes possible to carry out extractive distillation in a single column, where the "product" leaves in the distillate, the "nonproduct" leaves in a sidestream, and the entrainer leaves in the bottom stream which is recycled. The concept was developed by Rooks et al. (1996); see Exercise 31.

EXAMPLE 5.4. Extractive Distillation. A two-column system for the separation of ethanol and water by extractive distillation is shown in Fig. 5.45; the residue curve map is shown in Fig. 5.37. It is straightforward to design the entrainer recovery column, since it treats what is basically a binary mixture of water and ethylene glycol. For the extractive column, we use the method described above to find the composition profiles shown in Fig. 5.46; the specifications and base-case design results are summarized in Table 5.1.

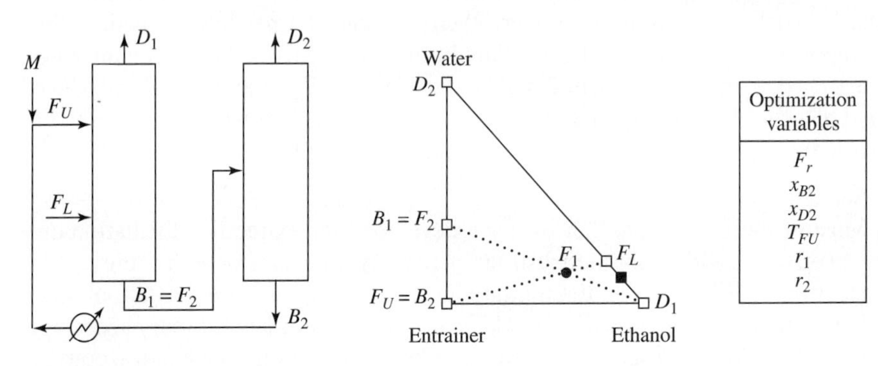

FIGURE 5.45
Extractive distillation system for ethanol and water. The extractive column (left) has two feeds; the feed ratio $F_r = F_U/F_L$ is an optimization variable. Several other variables can also be optimized, including the entrainer recycle purity, the temperature of the upper feed, the reflux ratios in each column, and the fractional recovery of entrainer in the entrainer recovery column (right). The entrainer makeup flow M is small compared to the total flow and must be balanced with the entrainer losses in streams D_1 and D_2.

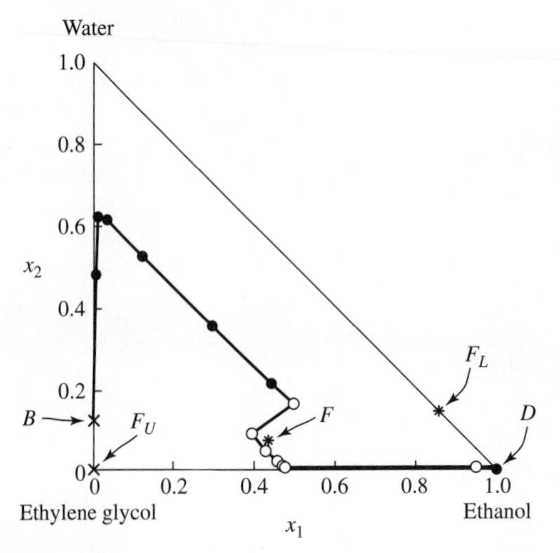

FIGURE 5.46
Extractive distillation profiles and design results for separation of ethanol and water using ethylene glycol as an entrainer.

TABLE 5.1
Summary of designs for Example 5.4. The ethanol-water feed is F_L

	Extractive column	Entrainer recovery column
Theoretical stages		
Condenser	22	7
Upper feed	14	
Lower feed	8	4
Reboiler	1	1
Entrainer/feed ratio	1.0	
Reflux ratio	1.0	1.0
Reboil ratio	1.4	0.284
D/F_L	0.858	0.124
B/F_L	1.142	0.876
D/B	0.751	0.142
V/F_L	1.60	0.249
Distillate mole fractions		
Ethanol	0.998	
Water	0.002	0.9994
Ethylene glycol	10^{-9}	0.0006
Upper feed mole fractions		
Ethanol	0	
Water	1.716×10^{-3}	
Ethylene glycol	0.9983	
Lower feed mole fractions		
Ethanol	0.8564	0
Water	0.1436	0.1258
Ethylene glycol	0	0.8742
Bottom mole fractions		
Ethanol	7.30×10^{-5}	0
Water	0.1257	1.716×10^{-3}
Ethylene glycol	0.8742	0.9983
Temperatures (°C)		
Distillate	78.29	100.02
Upper feed	78.29	
Lower feed	78.12	160.6
Bottom	160.6	196.6
Upper feed quality	1.41	
Lower feed quality	1.0	1.0
ω	0.744	0.15

Repeated application of the design method, while varying the reflux ratio and holding the other specifications constant at their base-case values, gives the number of stages shown in Fig. 5.47. Note the minimum number of stages at a reflux ratio of approximately 3. A one-variable study of the feed ratio gives the results shown in Fig. 5.48. Large feed ratios reduce the number of stages in the extractive column, but eventually lead to higher costs due to increased flows and more dilute feed to the entrainer recovery column.

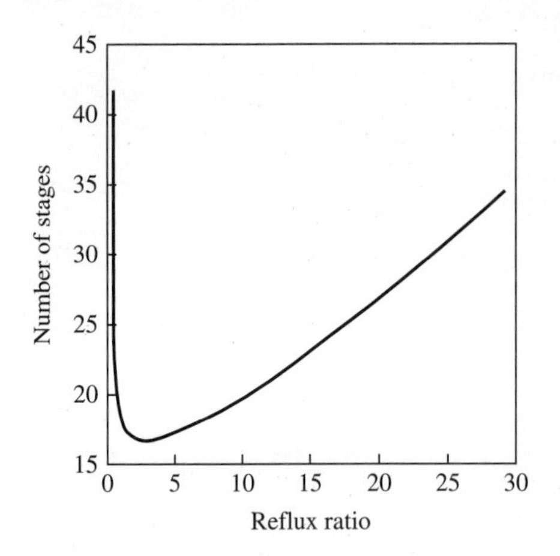

FIGURE 5.47
Number of theoretical stages in the extractive column as a function of reflux ratio for Example 5.4.

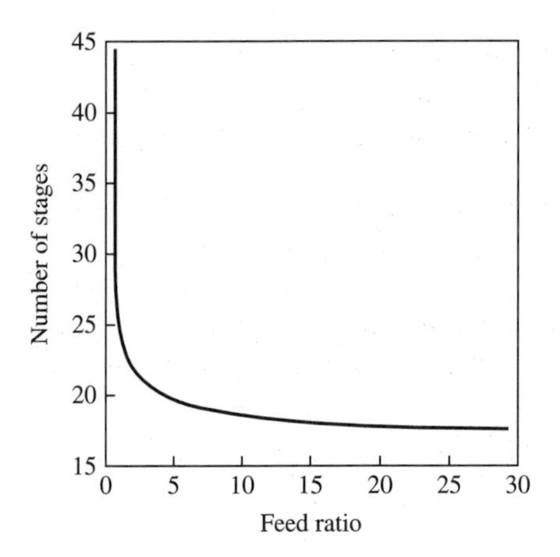

FIGURE 5.48
Number of theoretical stages in the extractive column as a function of entrainer to feed ratio for Example 5.4.

The existence of a maximum reflux in extractive distillation means that the product purity will change in unexpected ways with variations in the reflux ratio. This is most easily seen via performance simulations. In this case, we fix the number of stages in each column section, the entrainer to feed ratio and the quality of both feeds at their base-case values. The reflux ratio is varied, and the reboil ratio is adjusted to keep the ratio D/B constant at the base case value.[36] The resulting distillate mole fraction of ethanol is shown in Fig. 5.49. At low reflux ratios, near the minimum, the distillate purity is very sensitive to the reflux ratio, as might be expected. Beyond a reflux ratio of about 10, increasing the reflux ratio actually reduces product purity. This behavior is commonly found (sometimes unintentionally!) in many extractive distillations. The exact values of reflux ratio where the behavior changes character depend on the entrainer to feed ratio and other parameters in the particular problem. A sensitivity study of this type is valuable for existing columns and relatively simple to do with a simulator.

[36]If the reflux ratio is increased, the distillate flow will drop unless the reboil ratio is also increased to preserve the energy balance in the column.

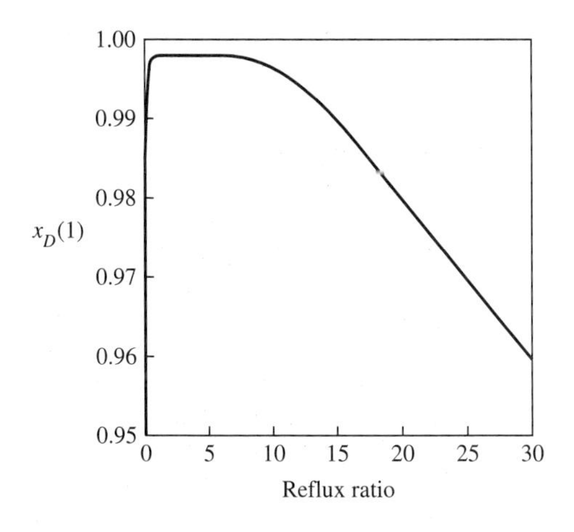

FIGURE 5.49
Distillate mole fraction of ethanol as a function of reflux ratio.

5.7 EXERCISES

1. Sketch the residue curve map for a ternary mixture of hexane, heptane, and octane at atmospheric pressure.
2. What is a steady-state solution of a differential equation? In simple distillation why do the steady states occur at all the pure components and azeotropes in the mixture?
3. For simple distillation, how many steady states are there in a mixture of hexane, heptane, and octane? How many in the mixture hexane-methanol-methyl acetate?
4. Explain why a residue curve through any given liquid composition point is tangent to the vapor-liquid equilibrium tie-line through the same point. How can you use this fact to show that residue curves cannot intersect themselves or each other?
5. Calculate the residue curve map for a mixture of methanol, ethanol, and *n*-propanol at 1 atm pressure. Integrate the simple distillation equation using the formula given in Eq. 5.9. Represent the vapor-liquid equilibrium with a constant relative volatility model.
6. Calculate residue curve maps for the four ternary mixtures on the faces of the tetrahedron in Fig. 5.10*a*.
7. Calculate residue curve maps for the four ternary mixtures on the faces of the tetrahedron in Fig. 5.10*b*.
8. Calculate the residue curve maps for a mixtures acetone-methanol-ethanol, and benzene–1,2-dichloroethane–*n*-heptane at 1 atm pressure and compare them with the measured column data shown in Fig. 5.13.
9. Consider a mixture of hexane-methanol-methyl acetate which was used as the basis for comparing residue curves and column composition profiles at 1 atm in Fig. 5.14. Use a simulation program to repeat the column simulations for feeds in each of the regions. Begin using the same reflux and reboil ratios used

to generate Fig. 5.14, i.e., $r = 3$ and $s = 6$. Try adjusting the total number of stages in the column, and the feed tray location in order to obtain bottom products that are close to the pure components. Use the Wilson equation to represent the liquid phase nonidealities, assume the vapor phase is ideal.

10. Consider a ternary mixture fed as a saturated liquid to the cocurrent flash cascades shown in Fig. 2.17. Assume that the fraction of feed that is vaporized in each flash is the same, and equal to ϕ. Derive the material balance equations for each cascade, and compare this model to the models for (*a*) distillation lines (i.e., a continuous staged column operating at infinite reflux and reboil, and (*b*) simple distillation. If each cascade has a large number of stages, what species will appear as distillate and bottom products from the cascades?
11. Sketch the residue curve map for the mixture methyl acetate-methanol-hexane at 1 atm pressure. Data: methyl acetate (56.9°C), methanol (64.5°C), hexane (69.2°C), methanol + hexane azeotrope (50.5°C), methyl acetate + hexane azeotrope (52.3°C), methyl acetate + methanol azeotrope (53.6°C), ternary azeotrope (48.2°C). The binary azeotropes are approximately midway along each of the binary edges, and the ternary azeotrope is roughly in the center of the diagram.
12. Sketch the residue curve map for the mixture benzene (B)-hexafluorobenzene (H)-methylcyclohexane (M) at 1 atm pressure. Data on boiling points and compositions: B (80.1°C), H (80.2°C), M (101°C), H-M azeotrope (79.9°C, 30 mol % M), B-H azeotrope (79.3°C, 80 mol % B), B-H azeotrope (80.3°C, 20 mol % B), B-H-M azeotrope (80.25°C, 30 mol % B, 20 mol % M). Note that the benzene-hexafluorobenzene mixture is known to have two binary azeotropes. Compare the sketch to the data in Fig. 5.50 from Wade and Taylor (1973, Fig. 1).
13. Sketch the residue curve map for the data shown in Fig. 5.51 (Hiaki et al., 1999). Compare the sketch and the data with the calculated residue curve maps if you have the necessary data and tools.
14. Sketch the residue curve map for the data shown in Fig. 5.52 (Carta et al., 1984). Compare the sketch and the data with the calculated residue curve maps if you have the necessary data and tools to make the calculations.
15. Sketch the residue curve map for the mixture methanol-water-methylchloroacetate at 1 atm pressure. Data: methanol (64.7°C), water (100°C), methylchloroacetate (131.4°C), water-methylchloroacetate azeotrope (92.7°C, 20 mol % water), methanol-water-methylchloroacetate azeotrope (67.85°C, 10 mol % water, 70 mol % methanol).
16. Write a response to the following e-mail.

Dear Mike,

Since you are a consultant for Bistro Chemical Company I presume we have secrecy agreements with you so I wanted to ask your opinion about a mixture that I am trying to separate.

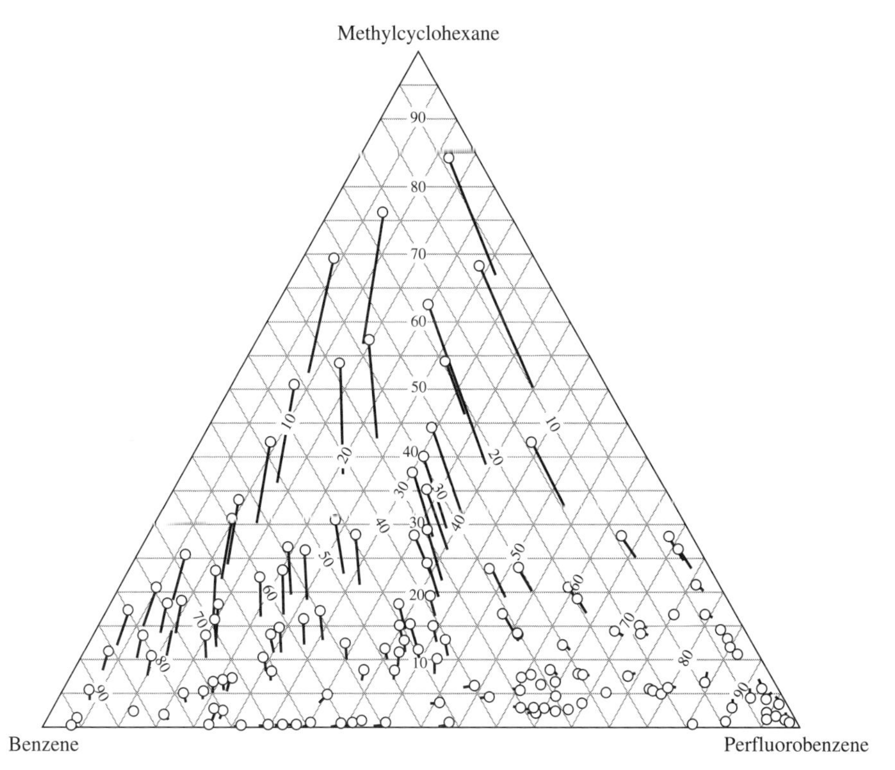

FIGURE 5.50
Data for Exercise 12 at 1 atm pressure.

I am looking at the acetic acid, water, TBM[37] system at 1 atm pressure. I believe there is one binary azeotrope which is the high-boiling H_2O/TBM azeotrope at 130°C. This is the highest-boiling mixture in the system. The lowest-boiling pure component in the system is TBM, and the highest-boiling pure component is acetic acid. Therefore, there is a distillation boundary that connects acetic acid to the H_2O/TBM azeotrope. The question is, what happens in the middle of the diagram? To the best of my knowledge, there are no other binaries. Is it possible to have a saddle azeotrope in the middle of the diagram? We have experimental data that seems to suggest this. Do you know of any systems that display this type of behavior?

With best regards,

Bill

Senior Engineer, Bistro Chemical Company

17. You want to separate a saturated liquid mixture containing 85 mol % acetaldehyde, 5 mol % methanol, and 10 mol % water at 1 atm pressure into saturated liquid products. The desired purity of acetaldehyde in the distillate is 99 mol %, and its desired composition in the bottom product is 0.1 mol %. The

[37] In the original e-mail, component TBM was identified but you are not allowed to know what it is!

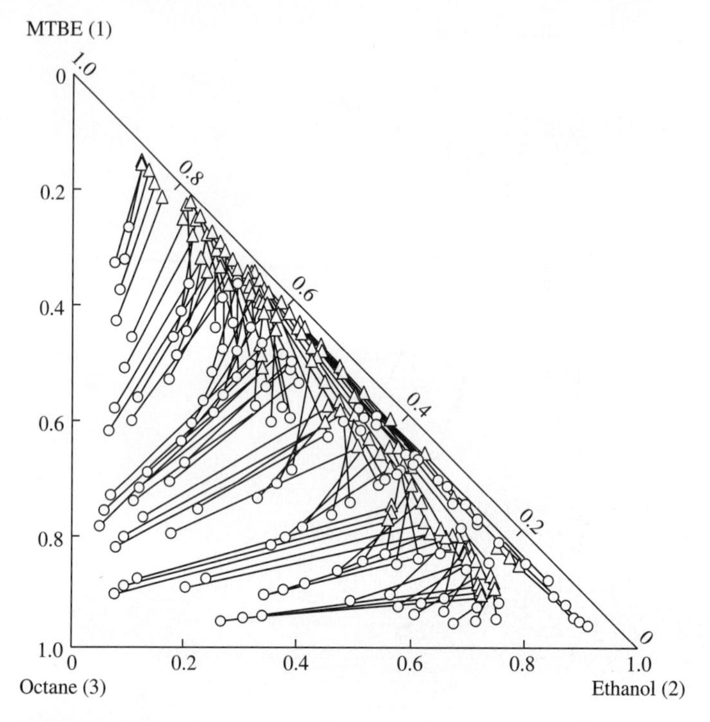

FIGURE 5.51
Data for Exercise 13 at 1 atm pressure. The components are (1) MTBE (methyl tertiary butyl ether), (2) ethanol, (3) octane. *[From Hiaki et al. (1999, Fig. 5).]* The open triangles indicate vapor compositions and the open circles show the corresponding liquid.

distillate should contain only trace amounts of water, so we will set its mole fraction in this stream to 1×10^{-6}. Use the boundary value design method to determine the minimum reflux and reboil ratios for this separation. You should notice that when the profiles just touch each other, the alignment of pinches is characteristic of an *indirect geometry* in spite of the fact that the split is a "lightest out first split." If you need physical property data for this exercise, you can find it in Fidkowski et al. (1991).

18. Explain how the sequence of distillation columns in Fig. 5.26 works. Sketch the composition profiles for each column. Check your answer by designing each of the columns using the boundary value design procedure. The flows and compositions of all the streams are given in Table 3 in Fidkowski et al. (1993*a*), and the column designs are given in Table 2 of the same paper. If you use any of these numbers, justify their values (or find better values). You can use the Wilson equation to represent the liquid phase activity coefficients (see Table 1 in the paper).
19. A mixture containing 39 mol % acetone, 55 mol % ethanol, and 6 mol % water is fed as a saturated liquid to a distillation column operating at 10 atm pressure. The products are withdrawn as saturated liquids having the following

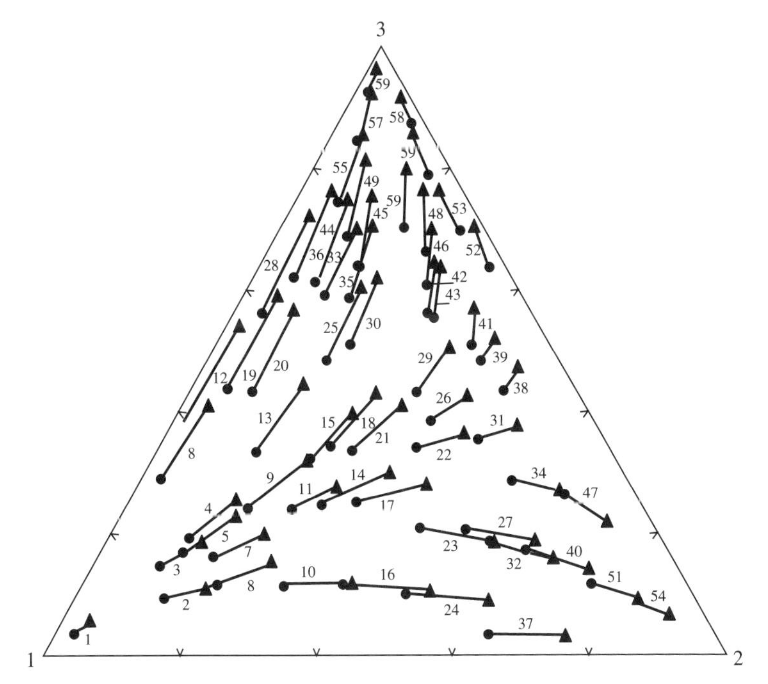

FIGURE 5.52
Data for Exercise 14 at 1 atm pressure. The components are (1) acetone, (2) ethyl acetate, and (3) ethanol. *[From Carta et al. (1984, Fig. 3).]* The solid triangles indicate liquid compositions and the filled circles show the corresponding vapor.

compositions: distillate—58.75 mol % acetone, 32.31 mol % ethanol, and 0.0894 mol % water; bottom—acetone mole fraction 1×10^{-9}, ethanol mole fraction 0.998. Use the boundary value design method to design the column. What is the minimum reflux ratio? Calculate designs at values of r above $r_{\min}$, and you should notice that for some values of r the composition profiles intersect twice. On one graph plot the total number of stages versus the reflux ratio for each of the two solutions. How large can you make r? Note, in order to see these effects you may have to use the thermodynamic models reported in Knapp and Doherty (1992).

20. This exercise concerns azeotropic mixtures with linear distillation boundaries. The purpose of the exercise is to show that for such mixtures it is impossible to devise distillation systems that produce pure products that lie in different distillation regions; i.e., it is not possible to isolate each pure component by distillation alone. Consider that we wish to separate a binary azeotropic mixture of ethanol and water (with composition F in Fig. 5.54*a*) by using an entrainer that azeotropes with water, giving the residue curve map in Fig. 5.53. Assume the distillation boundary is linear. Our plan is to try to develop a distillation system that feeds one of the columns with a composition F_1 in Region I, and another column with a composition F_2 in Region II. Feed F_1 can

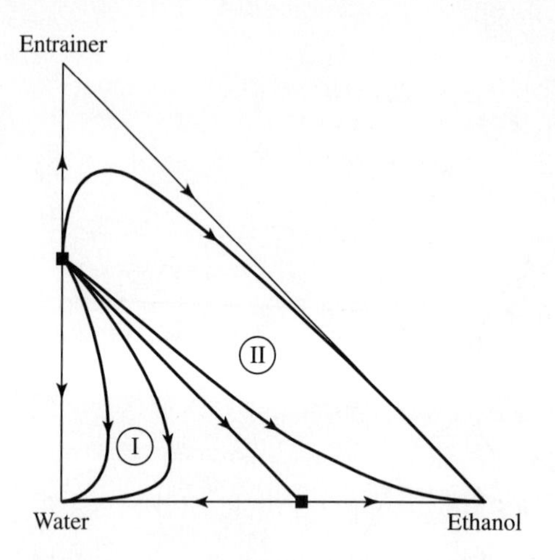

FIGURE 5.53
Residue curve map (schematic) for Exercise 20.

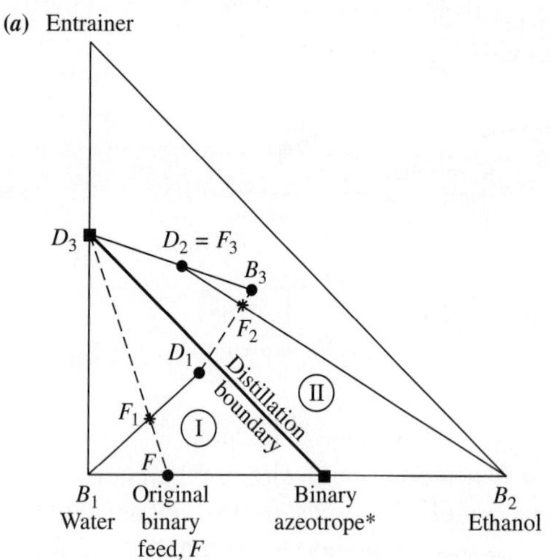

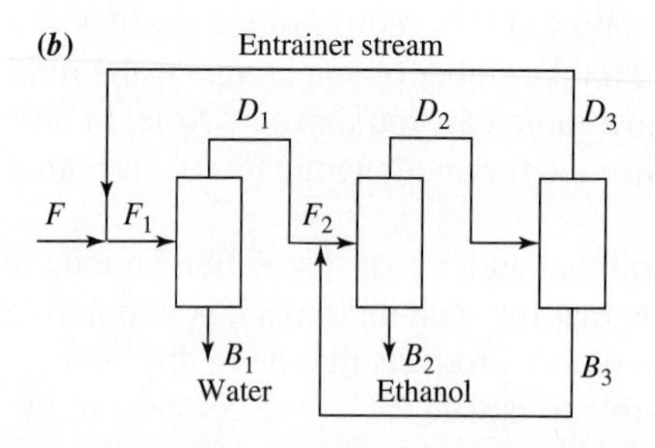

FIGURE 5.54
Mass balance and column sequence for Exercise 20.

* Note: Binary azeotropic composition not drawn to scale.

be split into pure water as bottom product, and feed F_2 can be split into pure ethanol as bottom product. If we can now find a way of creating these two feeds and simultaneously recycling the two distillate streams, then we will have invented the desired distillation system. Imagine that we have a stream of composition D_3 at the entrainer-water azeotrope (see Fig. 5.54). We mix this with stream F to produce F_1 in Region I. The first column in the sequence distills F_1 into streams B_1 and D_1. Stream D_1 is then mixed with a stream of composition B_3 to produce stream F_2 in Region II, which is split into streams B_2 and D_2. Stream D_2 is fed to a third column where it is split into distillate D_3 (the unstable node) and bottoms B_3. These are the two streams we need to generate feeds F_1 and F_2. The overall system is shown in Fig. 5.54*b*, and at first glance it looks good. Unfortunately, it won't work in spite of the fact that the material balance for each column is satisfied. Explain why it won't work. *Hint:* Check the overall material balance for a pair of columns.

21. In a chemical process, components A and B react to produce component C by the reaction

$$A + B \rightleftharpoons C \tag{5.25}$$

All three components are present in the reactor effluent stream, which is sent to a distillation system to separate the product from the reactants. The reactants are then recycled to the reactor. Given the pure component and azeotropic data provided below, determine which splits are feasible for each of the three candidate reactor effluents F_1, F_2, and F_3 to the separation system marked on Fig. 5.55. Draw the material balance line for each split on a triangular diagram as well as a schematic representation of each column sequence. Which one of these feeds do you recommend using in the process, and why? Data: Pure component boiling points: A (50°C), B (80°C), C (120°C). There is a binary azeotrope between components A and B containing 60 mol % A that boils at 30°C, and an azeotrope between A and C containing 30 mol % A that boils at 40°C. There are no other azeotropes in the mixture.

22. Acetone is made by dehydrogenation of isopropanol (IPA), see the flowsheet in Fig. 5.56. The isopropanol fresh feed to the process usually contains some water close to the azeotropic composition. Hydrogen is flashed off the reactor effluent, and the remaining components are sent to a distillation system. The feed to the distillation system contains 63 kmol/h acetone, 7 kmol/h isopropanol, and 30 kmol/h water. Devise a distillation system that has three product streams, one containing acetone, another containing water, and a third containing the isopropanol-water azeotrope for recycle to the reactor. Design

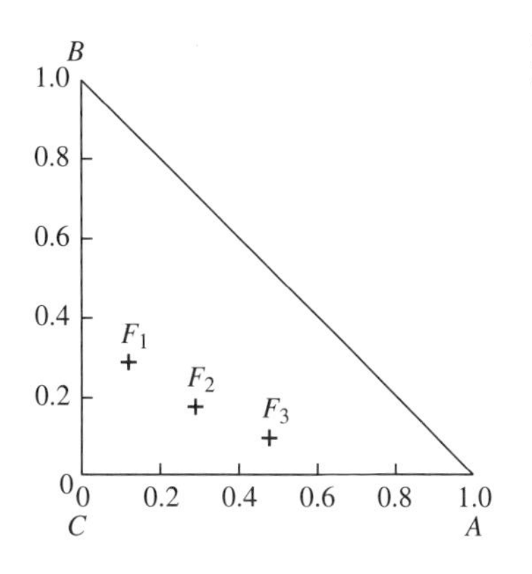

FIGURE 5.55
Triangular diagram for Exercise 21.

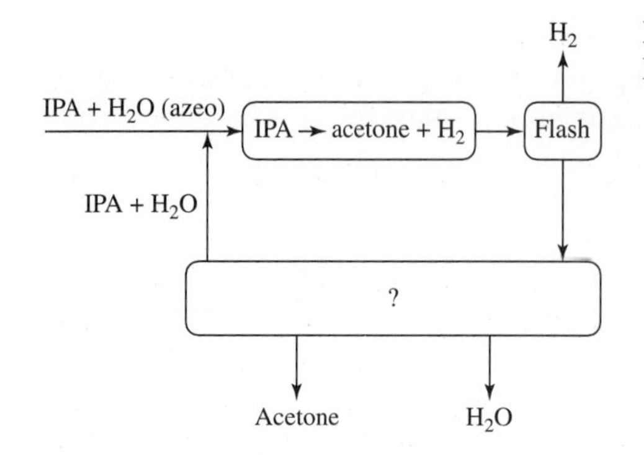

FIGURE 5.56
Flowsheet for acetone process.

each column and determine the number of stages and the vapor-to-feed rate in each column.

23. Explain how the Case III sequences work in Fig. 5.35. Draw material balance lines for each column on top of the residue curve maps to help with the explanation.
24. Ten potential entrainers have been identified to break the binary minimum-boiling azeotrope between butanol and butyl acetate. Azeotropic data have been collected for these systems and the residue curve maps have been prepared by sketching, shown in Fig. 5.57. When distillation boundaries occur, assume that they are linear. Decide which of the candidate entrainers are likely to break the azeotrope, and for each one draw a distillation system to achieve the separation. *Caution:* These data are almost certainly incomplete, which means that some of the residue curve maps are probably more complicated than shown. This means that some of the entrainers that you select will probably fail when more detailed studies are made—watch out for this possibility.
25. Check whether phenol will act as an extractive entrainer to break the ethanol-water azeotrope. If it does work, will ethanol or water leave as the distillate from the extractive column?
26. Cyclohexane and benzene have similar boiling points and form a minimum-boiling binary azeotrope which you would like to break using extractive distillation with aniline. Show that aniline behaves as an extractive entrainer, and determine which component in the azeotrope leaves the extractive column in the distillate.
27. Find an extractive entrainer to break the isopropanol-water azeotrope and determine which of the components in the azeotrope leaves the extractive column in the distillate.
28. Isopropanol and benzene form a minimum-boiling binary azeotrope which we want to break using extractive distillation with *n*-propanol; i.e., we want to recover all of the isopropanol in the feed. However, *n*-propanol also forms a minimum-boiling azeotrope with benzene, but we have a use for this azeotrope provided it contains no isopropanol. Can the desired separation be achieved?
29. In a process for making methanol, we need to separate an azeotropic mixture of methyl acetate and methanol. We would like to use water as the entrainer to

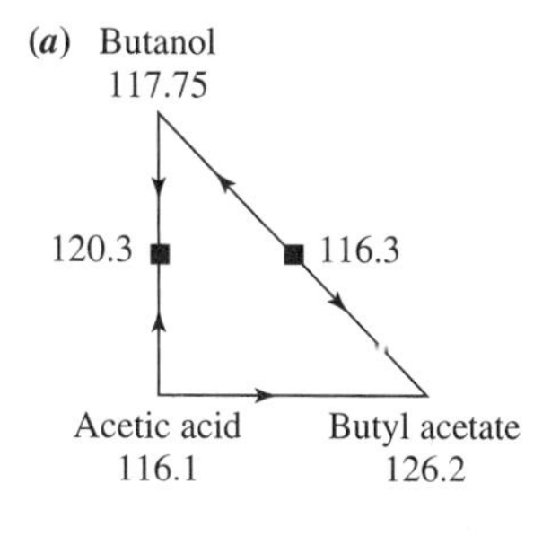

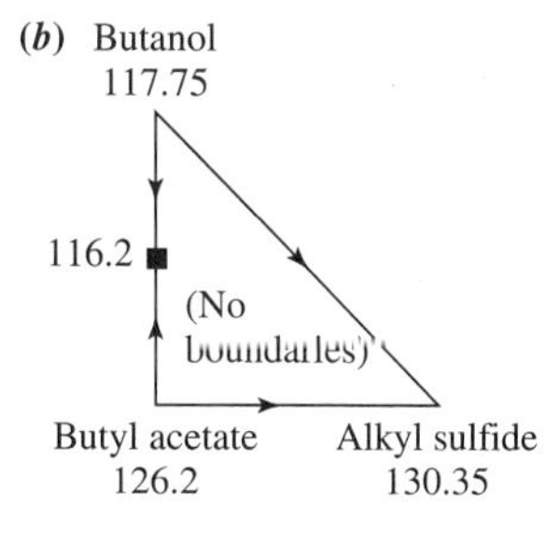

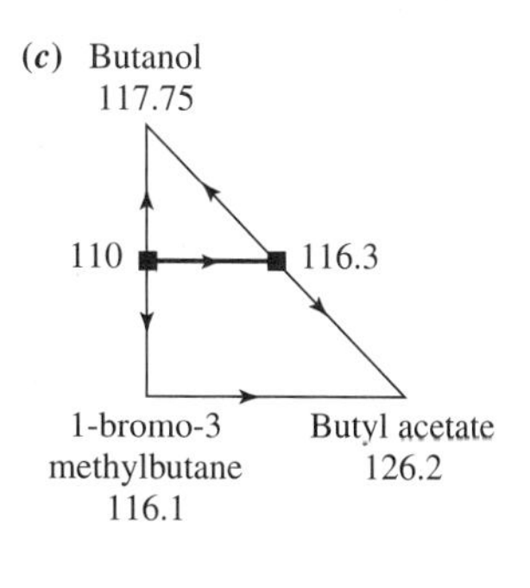

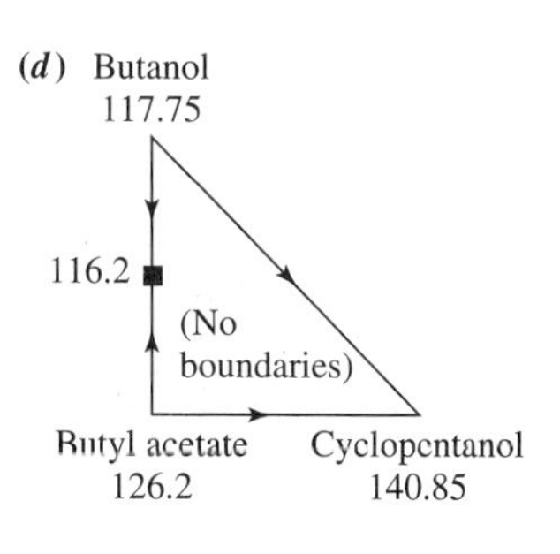

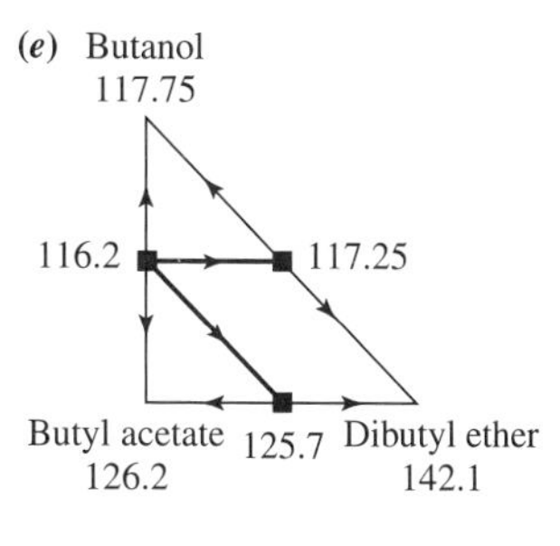

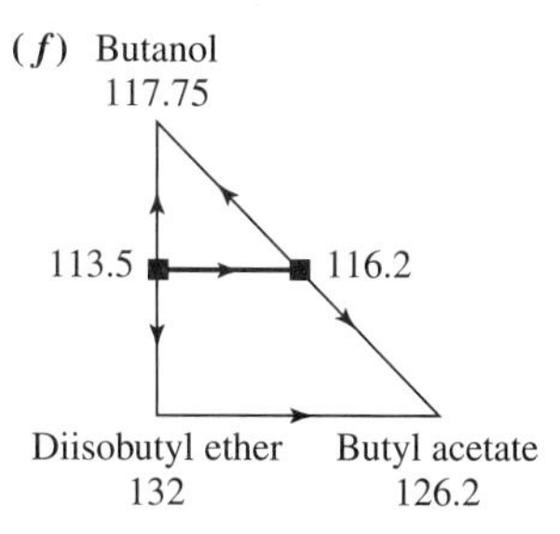

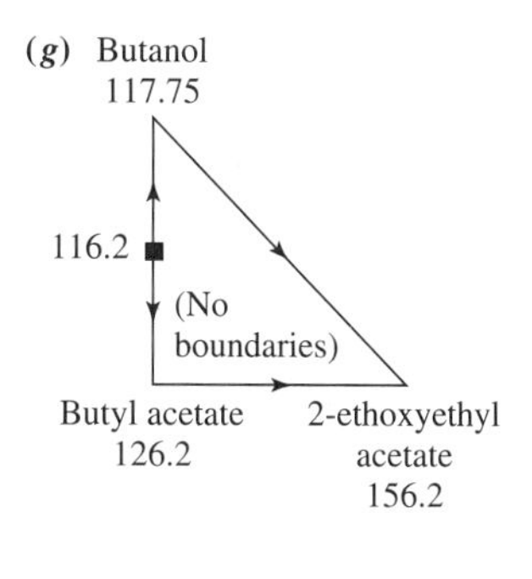

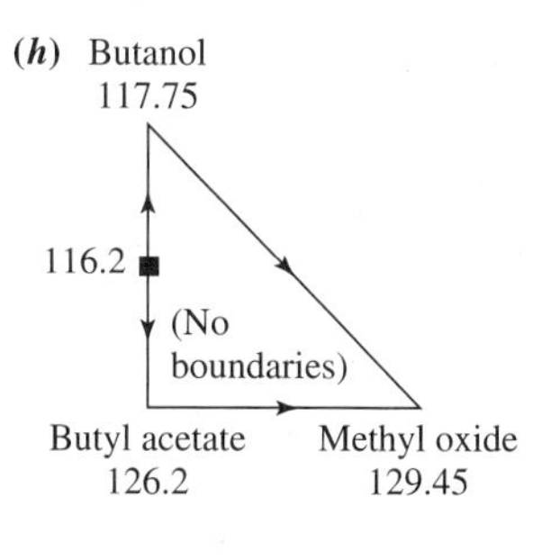

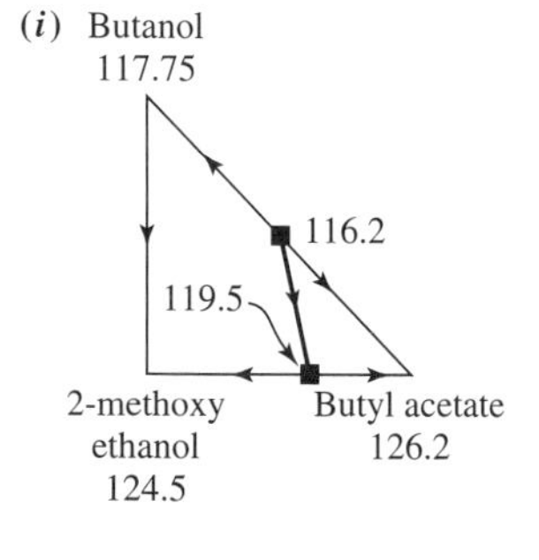

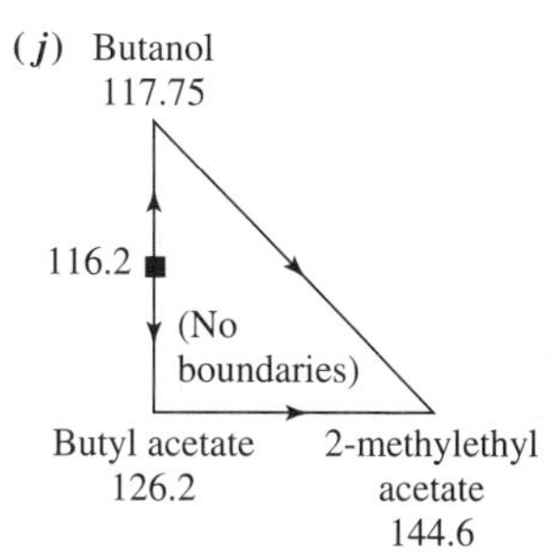

FIGURE 5.57
Candidate entrainers to break the binary azeotrope between butanol and butyl acetate.

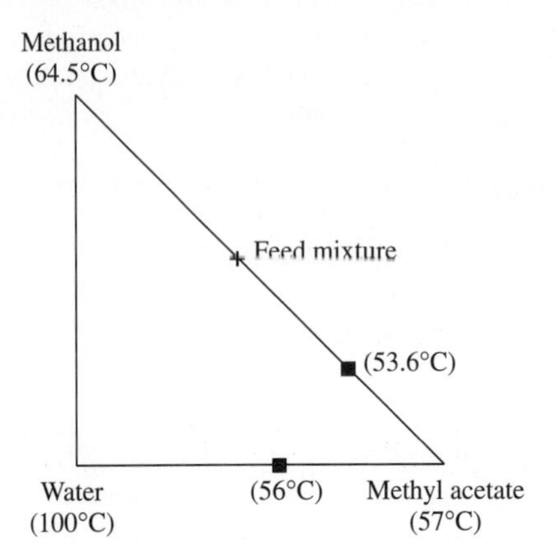

FIGURE 5.58
Data for Exercise 29.

break the azeotrope because water can be recycled to the reactor. Methyl acetate is also a reactant, and can also be recycled to the reactor.

To help invent this distillation system, you are given the data shown in Fig. 5.58. The numbers shown represent the boiling points of the pure components and azeotropes. Azeotropes are represented by filled square symbols, and there are two of them. Sketch the residue curve map for this mixture and use it to invent a distillation sequence to produce pure methanol product and a recycle stream of methyl acetate and water. Draw the sequence and also draw the material balance lines on a triangular diagram. Draw the flowsheet and identify the stream compositions on the triangular diagram.

30. Water is a feasible entrainer for the separation of a binary azeotropic mixture of methanol and acetone. Although a design is feasible, it is expensive because of the tangent pinch between acetone and water. It occurs to you that some other component might be a better entrainer, and you decide to consider components already in the process. These are ethylene glycol (EG), methyl ethyl ketone (MEK), and vinyl methyl ether (VME, C_3H_6O). Your supervisor is most enthusiastic about MEK because it is believed that trace amounts of this will be acceptable in the product.

    a. Your first task is to choose one of the entrainers for a more detailed design. Using only the following data, choose an entrainer and explain the reasons for your choice. Sketch the corresponding sequence of columns, and indicate the approximate compositions of the process streams on a triangular diagram.

Component or azeotrope	Boiling point at 1 atm (°C)	Composition
Methanol	64.48	
Acetone	56.06	
VME	11.86	
MEK	79.75	
EG	197	
Methanol-MEK azeotrope	63.09	29% MEK
Acetone-methanol azeotrope	55.35	79% methanol

   b. Using Distil or another computer-aided-design tool, confirm any information on residue curves needed in Part 30*a* and find a design (number of theoretical stages, reflux ratio, and reboil ratio) for at least the first column in your sequence.

31. Explain how to carry out extractive distillation in a single column; see Rooks et al (1996).

References

Agarwal, S., and R. Taylor, "Distillation Column Design Calculations Using a Nonequilibrium Model," *Ind. Eng. Chem. Research*, **33,** 2631–2636 (1994).

Barnicki, S. D., and J. J. Siirola, Separation Process Synthesis, in Kirk, R. E., and D. F. Othmer, editors, *Kirk-Othmer Encyclopedia of Chemical Technology,* vol. 21, pp. 923–962. Wiley, New York, 4th ed. (1997).

Bauer, M. H., and J. Stichlmair, "Design and Economic Optimization of Azeotropic Distillation Processes Using Mixed Integer Nonlinear Programming," *Computers Chem. Engng.,* **22,** 1271–1286 (1998).

Baur, R., R. Taylor, R. Krishna, and J. A. Copati, "Influence of Mass Transfer in Distillation of Mixtures with a Distillation Boundary," *TransIChemE,* **77A,** 561–565 (1999).

Benedict, M., C. A. Johnson, E. Solomon, and L. C. Rubin, "Extractive and Azeotropic Distillation II. Separation of Toluene from Paraffins by Azeotropic Distillation with Methanol," *Trans. AIChE,* **47,** 371–392 (1945).

Benedict, M., and L. C. Rubin, "Extractive and Azeotropic Distillation I. Theoretical Aspects," *Trans AIChE,* **41,** 353–370 (1945).

Buell, C. K., and R. G. Boatright, "Furfural Extractive Distillation for Separation and Purification of C_4 Hydrocarbons," *Ind. Eng. Chem.,* **39,** 695–705 (1947).

Bushmakin, I. N., and I. N. Kish, "Isobaric Liquid-Vapor Equilibirum in a Ternary System with an Azeotrope of the Saddlepoint Type," *J. Appl. Chem. USSR (Engl. Trans.),* **30,** 205–215 (1957).

Carta, R., S. Dernini, and P. Sanna, "Isobaric Vapor-Liquid Equilibria for the Ternary System Acetone-Ethyl Acetate-Ethanol," *J. Chem. Eng. Data,* **29**(4), 463–466 (1984).

Castillo, F. J. L., and G. P. Towler, "Azeotropic Distillation Design Considering Mass Transfer Rates," *Inst. Chem. Eng. Symp. Ser.,* (142), 625–634, part 2 (1997).

Castillo, F. J. L., and G. P. Towler, "Influence of Multicomponent Mass Transfer on Homogeneous Azeotropic Distillation," *Chem. Eng. Sci.,* **53,** 963–976 (1998).

Dalton, J., "Experimental Essays on the Constitution of Mixed Gases; On the Force of Steam or Vapour from Water and Other Liquids in Different Temperatures, Both in a Torricellian Vacuum and in Air; On Evaporation; and on the Expansion of Gases by Heat," *Mem. Proc. Manchr. Lit. Phil. Soc.,* **5,** 535 (1802).

Davydyan, A. G., M. F. Malone, and M. F. Doherty, "Boundary Modes in a Single Feed Distillation Column for the Separation of Azeotropic Mixtures," *Theor. Found. Chem. Engr.,* **31,** 327–338 (1997).

Doherty, M. F., "The Presynthesis Problem for Homogeneous Azeotropic Distillation Has a Unique Explicit Solution," *Chem. Eng. Sci.,* **40,** 1885–1889 (1985).

Doherty, M. F., and G. A. Caldarola, "Design and Synthesis of Homogeneous Azeotropic Distillation. 3. The Sequencing of Columns for Azeotropic and Extractive Distillations," *Ind. Eng. Chem. Fundam.,* **24,** 474–485 (1985).

Doherty, M. F., and J. P. Knapp, Distillation, Azeotropic and Extractive Distillation, in Kirk, R. E., and D. F. Othmer, editors, *Kirk-Othmer Encyclopedia of Chemical Technology,* pp. 358–398. Wiley, New York, 4th ed. (1993).

Doherty, M. F., and J. D. Perkins, "On the Dynamics of Distillation Processes—II. The Simple Distillation of Model Solutions," *Chem. Eng. Sci.,* **33,** 569–578 (1978).

Doherty, M. F., and J. D. Perkins, "The Behaviour of Multicomponent Azeotropic Distillation Processes," *Inst. Chem. Eng. Symp. Ser.,* **56,** 4.2/21–4.2/48 (1979*a*).

Doherty, M. F., and J. D. Perkins, "On the Dynamics of Distillation Processes—III. Topological Classification of Ternary Residue Curve Maps," *Chem. Eng. Sci.,* **34,** 1401–1414 (1979*b*).

Ewell, R. H., and L. M. Welch, "Rectification in Ternary Systems Containing Binary Azeotropes," *Ind. Eng. Chem.,* **37,** 1224–1231 (1945).

Fidkowski, Z. T., M. F. Doherty, and M. F. Malone, "Feasibility of Separations for Distillation of Nonideal Ternary Mixtures," *AIChE J.,* **39,** 1303–1321 (1993*a*).

Fidkowski, Z. T., M. F. Malone, and M. F. Doherty, "Nonideal Multicomponent Distillation: Use of Bifurcation Theory for Design," *AIChE J.,* **37,** 1761–1779 (1991).

Fidkowski, Z. T., M. F. Malone, and M. F. Doherty, "Computing Azeotropes in Multicomponent Mixutres," *Computers Chem. Engng.,* **17,** 1141–1155 (1993*b*).

Foucher, E. R., M. F. Doherty, and M. F. Malone, "Automatic Screening of Entrainers for Homogeneous Azeotropic Distillation," *Ind. Eng. Chem. Research,* **30,** 760–772 (1991).

Free, K. W., and H. P. Hutchison, Three Component Distillation at Total Reflux, in *Proceedings of the International Symposium on Distillation,* pp. 231–237, Brighton, England. Instn. Chem. Engrs. (1960).

Glinos, K. N., and M. F. Malone, "Optimality Regions for Complex Column Alternatives in Distillation Systems," *Chem. Engng. Res. Design,* **66,** 229–240 (1988).

Gmehling, J., J. Menke, K. Fischer, and J. Krafczyk, *Azeotropic Data. Part I.* VCH, New York (1994*a*).

Gmehling, J., J. Menke, K. Fischer, and J. Krafczyk, *Azeotropic Data. Part II.* VCH, New York (1994*b*).

Gmehling, J., U. Onken, and W. Arlt, *Vapour-Liquid Equilibrium Data Collection,* vol. 1/6a, Aliphatic Hydrocarbons C_4-C_6, of *Chemistry Data Series.* DECHEMA, Frankfurt/Main (1980).

Greenberg, M. D., *Foundations of Applied Mathematics.* Prentice-Hall, Englewood Cliffs, N.J. (1978).

Guckenheimer, J., and P. Holmes, *Nonlinear Oscillations, Dynamical Systems and Bifurcations of Vector Fields.* Springer-Verlag, New York (1983).

Hale J., and H. Koçak, *Dynamics and Bifurcations.* Springer-Verlag, New York (1991).

Happel, J., P. W. Cornell, D. Eastman, M. J. Fowle, C. A. Porter, and A. H. Schutte, "Extractive Distillation—Separation of C_4 Hydrocarbons Using Furfural," *Trans AIChE,* **42,** 189–214 (1946).

Hiaki, T., K. Tatsuhana, T. Tsuji, and M. Hongo, "Isobaric Vapor-Liquid Equilibria for 2-Methozy–2-methylpropane + Ethanol + Octane and Constituent Binary Systems at 101.3 kPa," *J. Chem. Eng. Data,* **44,** 323–327 (1999).

Hoffman, E. J., *Azeotropic and Extractive Distillation.* Interscience, New York (1964).

Horsley, L. H., *Azeotropic Data III,* number 116 in Advances in Chemistry Series. American Chemical Society, Washington, D.C. (1973).

Julka, V., *A Geometric Theory of Multicomponent Distillation,* Ph.D. thesis, University of Massachusetts, Amherst MA (1995).

Julka, V., and M. F. Doherty, "Geometric Behavior and Minimum Flows for Nonideal Multicomponent Distillation," *Chem. Eng. Sci.,* **45,** 1801–1822 (1990).

Knapp, J. P., *Exploiting Pressure Effects in the Distillation of Homogeneous Azeotropic Mixtures,* Ph.D. thesis, University of Massachusetts, Amherst MA (1991).

Knapp, J. P., and M. F. Doherty, "Thermal Integration of Homogeneous Azeotropic Distillation Sequences," *AIChE J.,* **36,** 969–984 (1990).

Knapp, J. P., and M. F. Doherty, "A New Pressure-Swing Distillation Process for Separating Homogeneous Azeotropic Mixtures," *Ind. Eng. Chem. Research,* **31,** 346–357 (1992).

Knapp, J. P., and M. F. Doherty, "Minimum Entrainer Flows for Extractive Distillation. A Bifurcation Theoretic Approach," *AIChE J.,* **40,** 243–268 (1994).

Knight, J. R., and M. F. Doherty, "Optimal Design and Synthesis of Homogeneous Azeotropic Distillation Sequences," *Ind. Eng. Chem. Research,* **28,** 564–572 (1989).

Kubierschky, K., Verfahren zur Gewinnung von hochprozentigem, bezw. absolutem Alkohol aus Alkohol-Wassergemischen in unterbrochenem Betriebe (1915). German Patent 287,897.

Lake, G. R., "Recovery of Toluene from Petroleum by Azeotropic Distillation," *Trans AIChE,* **41,** 327–352 (1945).

LaRoche, L., N. Bekiaris, H. W. Andersen, and M. Morari, "Homogeneous Azeotropic Distillation: Comparing Entrainers," *Canadian J. Chem. Engng.,* **69,** 1302–1319 (1991).

LaRoche, L., N. Bekaris, H. W. Andersen, and M. Morari, "Homogeneous Azeotropic Distillation: Separability and Flowsheet Synthesis," *Ind. Eng. Chem. Research,* **31,** 2190–2209 (1992*a*).

LaRoche, L., N. Bekiaris, H. W. Andersen, and M. Morari, "The Curious Behavior of Homogeneous Azeotropic Distillation—Implications for Entrainer Selection," *AIChE J.,* **38,** 1309–1328 (1992*b*).

Levy, S. G., *Design of Homogeneous Azeotropic Distillations,* Ph.D. thesis, University of Massachusetts, Amherst MA (1985).

Levy, S. G., and M. F. Doherty, "Design and Synthesis of Homogenous Azeotropic Distillations. 4. Minimum Reflux Calculations for Multiple Feed Columns," *Ind. Eng. Chem. Fundam.,* **25,** 269–279 (1986).

Li, Y., H. Chen, and J. Liu, "Composition Profile of an Azeotropic Continuous Distillation with Feed Composition on a Ridge or in a Valley," *Ind. Eng. Chem. Research,* **38,** 2482–2484 (1999).

Luna, F., and J. Martinez, "Stability Analysis in Multicomponent Drying of Homogeneous Liquid Mixtures," *Chem. Eng. Sci.,* **54,** 5823–5837 (1999).

Lutugina, N. V., O. F. Kovalichev, L. P. Shandalova, and I. V. Antipina, "Distribution of Liquid Composition along the Height of Columns of Various Types," *Theor. Found. Chem. Eng.,* **7,** 234–237 (1974).

Malesiński, W., *Azeotropy and Other Theoretical Problems of Vapour-Liquid Equilibrium.* Interscience, London (1965).

Matsuyama, H., and H. Nishimura, "Topological and Thermodynamic Classification of Ternary Vapor-Liquid Equilibria," *J. Chem. Eng. Japan,* **10,** 181–187 (1977).

Nikolaev, N. S., V. N. Kiva, A. S. Mozzhukin, L. A. Serafimov, and S. I. Goldoborodkin, "Rectification in Ternary Systems Containing Binary Azeotropes," *Theor. Found. Chem. Eng.,* **13,** 418 (1979).

Pelkonen, S., R. Kaesemann, and A. Gorak, "Distillation Lines for Multicomponent Separation in Packed Columns: Theory and Comparison with Experiment," *Ind. Eng. Chem. Research,* **36,** 5392–5398 (1997).

Peterson, E. J., and L. R. Partin, "Temperature Sequences for Categorizing All Ternary Distillation Boundary Maps," *Ind. Eng. Chem. Research,* **36,** 1799–1811 (1997).

Poellmann, P., and E. Blass, "Best Products of Homogeneous Azeotropic Distillation," *Gas Sep. & Purification,* **8,** 194–228 (1994).

Press, W. H., B. P. Flannery, S. A. Teukolsky, and W. T. Vetterling, *Numerical Recipes.* Cambridge University Press, New York (1986).

Prigogine, I., and R. Defay, *Chemical Thermodynamics.* Longmans, Green and Co., London, 4th ed. (1967).

Rayleigh, L., "On the Distillation of Binary Mixtures," *Phil. Mag.,* **S.6, 4**(23), 521–537 (1902).

Rev, E., "Crossing of Valleys, Ridges, and Simple Boundaries by Distillation in Homogeneous Ternary Mixtures," *Ind. Eng. Chem. Research,* **31,** 893–901 (1992).

Robinson, C. S., and E. R. Gilliland, *Elements of Fractional Distillation.* McGraw-Hill, New York, 4th ed. (1950).

Rooks, R. E., V. Julka, M. F. Doherty, and M. F. Malone, "Structure of Distillation Regions for Multicomponent Azeotropic Mixtures," *AIChE J.,* **44,** 1382–1391 (1998).

Rooks, R. E., M. F. Malone, and M. F. Doherty, "Geometric Design Method for Side-Stream Distillation Columns," *Ind. Eng. Chem. Research,* **35,** 3653–3664 (1996).

Rowlinson, J. S., *Liquids and Liquid Mixtures.* Butterworths, London, 2d ed. (1969).

Safrit, B. T., and A. W. Westerberg, "Algorithm for Generating the Distillation Regions for Multicomponent Mixtures," *Ind. Eng. Chem. Research,* **36,** 1827–1840 (1997).

Sargent, R. W. H., "A Functional Approach to Process Synthesis and Its Application to Distillation Systems," *Computers Chem. Engng,* **22,** 31–45 (1998).

Scheibel, E. G., "Principles of Extractive Distillation," *Chem. Engng. Prog.,* **44,** 927 (1948).

Schreinemakers, F. A. H., "Dampfdrucke Ternärer Gemische," *Zeitschrift f. Physik. Chemie.,* **XXXVI,** 257–289 (1901*a*).

Schreinemakers, F. A. H., "Dampfdrucke Terärer Gemische," *Zeitschrift f. Physik. Chemie.,* **XXXVI,** 413–449 (1901*b*).

Schreinemakers, F. A. H., "Einige Bemerkungen ber Dampfdrucke Ternärer Gemische," *Zeitschrift f. Physik. Chemie.,* **XLIII,** 671 (1903).

Smith, J. M., and H. C. VanNess, *Introduction to Chemical Engineering Thermodynamics.* McGraw-Hill, New York, 5th ed. (1996).

Swietoslawski, W., *Azeotropy and Polyazeotropy.* Pergamon Press, New York (1963).

Tedder, D. W., and D. F. Rudd, "Parametric Studies in Industrial Distillation, Parts I. and II.," *AIChE J.,* **24,** 303–334 (1978).

Van Dongen, D. B., *Distillation of Azeotropic Mixtures: The Application of Simple Distillation Theory to the Design of Continuous Processes,* Ph.D. thesis, University of Massachusetts, Amherst MA (1983).

Van Dongen, D. B., and M. F. Doherty, "On the Dynamics of Distillation Processes—V. The Topology of the Boiling Temperature Surface and Its Relation to Azeotropic Distillation," *Chem. Engng. Sci.,* **39,** 883 (1984).

Van Dongen, D. B., and M. F. Doherty, "Design and Synthesis of Homogeneous Azeotropic Distillations. I. Problem Formulation for a Single Column," *Ind. Eng. Chem. Fundam.,* **24,** 454–462 (1985).

Wade, J., and R. W. Merriman, "Influence of Water on the Boiling Point of Ethyl Alcohol at Pressures Above and Below the Atmospheric Pressure," *J. Chem. Soc. Trans.,* **99,** 997 (1911).

Wade, J. C., and Z. L. Taylor, Jr., "Vapor-Liquid Equilibrium in Perfluorobenzene-Benzene-Methylcyclohexane System," *J. Chem. Eng. Data,* **18**(4), 424 (1973).

Wahnschaff, O. M., T. P. Jurian, and A. W. Westerberg, "SPLIT: A Separation Process Designer," *Computers Chem. Engng.,* **15,** 565–581 (1991).

Wahnschafft, O. M., J. W. Koehler, E. Blass, and A. W. Westerberg, "The Product Composition Regions for Single-Feed Azeotropic Distillation Columns," *Ind. Eng. Chem. Research,* **31,** 2345–2362 (1992*a*).

Wahnschafft, O. M., J. P. L. Rudulier, P. Blania, and A. W. Westerberg, "SPLIT: II. Automated Synthesis of Hybrid Liquid Separation Systems," *Computers Chem. Engng.,* **16,** 305–312 (1992*b*).

Wahnschafft, O. M., J. P. L. Rudulierand, and A. W. Westerberg, "A Problem Decompostion Approach for the Synthesis of Complex Separation Processes with Recycles," *Ind. Eng. Chem. Research,* **32,** 1121–1141 (1993).

Yamakita, Y., J. Shiozaki, and H. Matsuyama, "Consistency Test of Ternary Azeotropic Data by Use of Simple Distillation," *J. Chem. Eng. Japan,* **16,** 145–146 (1983).

Young, S., Verfahren zur Gewinnung wasser freien Alkohols aus Spiritus mittels fraktionierter Destillation und ohne wasserentziehende Chemikalien (1903). German Patent 142,502.

6

Column Design and Economics

When the internal flows and the number of theoretical stages are known, the equipment size and the utility flows for a column and the auxiliary heat exchange equipment can be estimated. These form a basis for cost and optimization studies which are the main topics in this chapter. First, the equipment design is briefly discussed. The economics and cost estimates are described next with an emphasis on sensitivity; *these are not intended for use in final designs but for first estimates only*. The results of the first two sections are then used to formulate and solve some of the important optimizations that arise for single columns.

6.1
EQUIPMENT DESIGN

The operating temperature and pressure in a column are determined by the available utility temperatures, the boiling temperatures of the mixture, the desired purity of the separations, and any constraints on the stability of the mixture.[1] The material of construction and the thickness of the vessel walls are determined by the operating temperature and pressure and the corrosion properties of the mixture. The height of a column depends primarily on the number of trays, while the diameter depends mainly on the internal flows. The size of the heat exchangers and the utility flows depend primarily on the vapor rate. Other factors, such as the use of packed vs. tray columns, the type of packing or trays, the weir height, the tray hydraulics, etc., are important practical considerations in a final design or column operation, and more detailed information can be found in Fair et al. (1984, pp. 18–8 to 18–23) or van Winkle (1967, Chap. 13). Also see Fair (1987) or Humphrey and Keller (1997, Chap. 2) for a more detailed discussion of tray designs, mass transfer, and other factors. However, those detailed discussions are not necessary for understanding the conceptual design.

[1]For example, see King (1980, p. 248 and pp. 798*ff*.).

Pressure and Column Internals

The pressure in distillation can often be adjusted so that the least expensive utilities can be used for heating or cooling. For example, it is usually preferable to condense the overhead vapor stream with cooling water rather than to use chilled water or refrigeration, even though this may require increasing the column pressure in order to raise the boiling temperature of the distillate. This will also increase the temperature in the reboiler, but it is often more economical to supply a hotter utility to the reboiler, such as high-pressure steam, than to provide refrigeration in the condenser, although a small adjustment in the pressure may avoid both. This situation is common in petroleum refineries, where there may be a high-temperature source available from process streams outside the distillation system. The heat integration problem is discussed in Chap. 7. There are cases where refrigeration is necessary, e.g., when the critical temperature of the distillate is lower than the cooling water temperature. We may also operate a column under vacuum conditions to lower the temperature in cases where degradation or polymerization is encountered in the reboiler.

A pressure drop from the bottom to the top of the column is unavoidable.[2] The magnitude of the pressure drop depends on the type and the design of the column internals.

In many smaller columns or in the retrofit of existing columns, *packings* are used. There are many types of packings, with special geometric features. These may be small ceramic, metal, or plastic shapes filling the column shell at *random,* or they may be metal sheets, gauzes, or other *structured* packings. Some of these, especially the structured packings, have much higher efficiencies and/or lower pressure drops than others. These are most useful for feeds that are not too contaminated with solids. An excellent discussion of various packings, applications, and factors favoring packing or trays can be found in Humphrey and Keller (1997, especially pp. 45–72).

For conceptual design purposes, we will consider columns with *trays* or *plates* such as the one shown schematically in Fig. 6.1.[3] Estimates for the magnitude of the pressure drop can be made by several methods, each with different degrees of accuracy and complexity as discussed by Fair et al. (1984, pp. 18–8 to 18–12) or van Winkle (1967, pp. 506*ff*). A more thorough treatment is available in Lockett (1986). The pressure drop across a tray depends primarily on the resistance to flow from the vapor passage (bubble cap, hole, slot, etc.) and on the amount of the liquid-vapor mixture on the tray. The height of the liquid on the tray is dictated by the weir height, the tray hydraulics, and the type and design of the vapor passage (which, for instance, determines the size of bubbles). For preliminary designs, we will use a constant pressure drop on each tray, equivalent to approximately 4 cm of vapor-free

[2]On account of the pressure drop, the use of isobaric VLE is not strictly correct but usually introduces only a small error. However, when there are a very large number of stages, an upper limit on the temperature in the reboiler, or small boiling point differences, the pressure effects on the VLE from stage to stage may be significant.

[3]A very rough estimate of the height of packing that is equivalent to a theoretical plate or HETP for certain structured packings is $100/a$, where a is the specific surface area of the packing in m^2/m^3; this and a much better estimation method along with a comparison to more detailed methods is described by Lockett (1998).

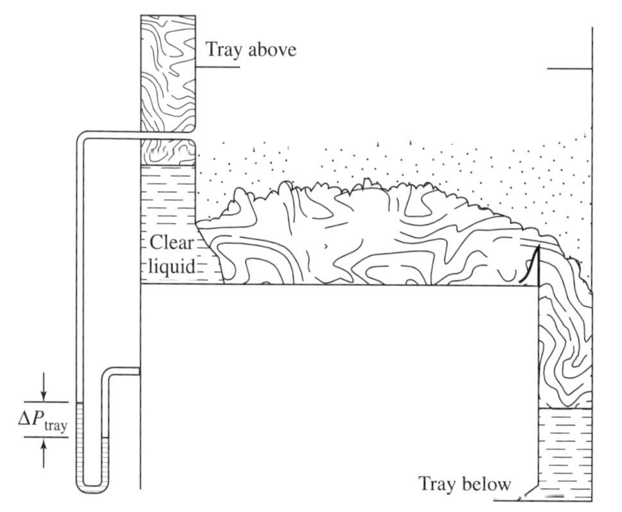

FIGURE 6.1
Schematic of liquid and vapor flows on a sieve tray. *[Adapted from Fair (1987, Fig. 5.7–3).]*

liquid. This corresponds approximately to 4×10^{-3} atm (0.06 psia) for typical liquid densities. It may be important to improve this estimate, e.g., in vacuum columns containing a large number of trays, and data may eventually be needed to complete the design.[4]

Once the column pressure is selected, the temperature range for operation can be estimated from the properties of the mixture and the desired purities. For example, the separation of a binary mixture into nearly pure distillate and bottoms streams results in condenser and reboiler temperatures that are close to the boiling temperature of the light and heavy components at the column pressure.

Mass Transfer and Efficiency

The rate of approach to equilibrium in a real column is finite, so one actual column stage does not provide one equilibrium stage or *theoretical plate*. Because the fluid mixtures that are separated by distillation typically have viscosities and flowrates in the column that result in a turbulent two-phase flow, exact solutions of the mass transfer problem are quite complicated. Imagine, for instance, the level of detailed modeling that would be required to describe a complete solution of the fluid flow and the heat and mass transfer on a single stage like the one shown in Fig. 6.1.

One alternative approach has been to define and develop correlations for an *efficiency*. This can have a separate value for each tray or it may be presented as an "overall" column efficiency E_o that relates the number of equilibrium or "theoretical" stages N_T to the total number of stages N by

$$N = \frac{N_T}{E_o} \tag{6.1}$$

[4]For more precise estimation, the method of Bennett, Agrawal, and Cook (1983) described by Fair et al. (1984) can be used.

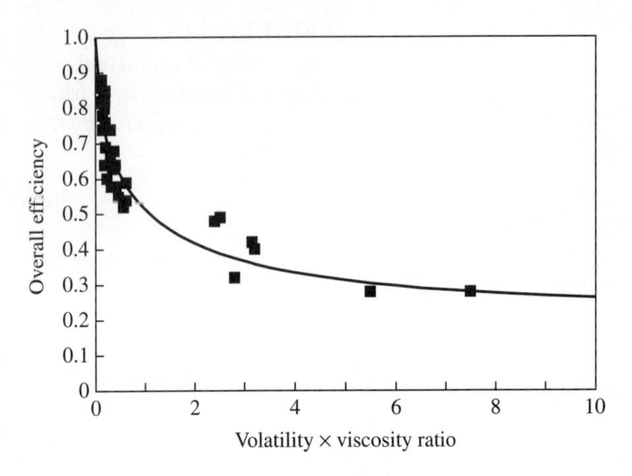

FIGURE 6.2
O'Connell's correlation for the overall column efficiency in distillation. Data from O'Connell (1946); the solid line is the correlation in Eq. 6.2.

For a conceptual design, it is typical to use a correlation like the one developed by O'Connell shown in Fig. 6.2.

An empirical description of these data is given by the expression

$$\frac{E_o - a}{1 - a} = \exp\left(-\sqrt{\alpha \frac{\mu}{\mu_0}}\right) \tag{6.2}$$

where $a = 0.24$ and $\mu_0 = 10^{-3}$ Pa·s (1 centipoise).[5] The viscosity of the liquid mixture at the feed composition is μ and α is the volatility between the key components; both are evaluated at the average temperature and pressure in the column.[6] Given the empirical nature of the correlation, it is typical to simply make a reasonable estimate of μ. For example, the viscosity of many organic liquids near the normal boiling point is 1×10^{-4} to 3×10^{-4} Pa·s (0.1 to 0.3 centipoise); see Douglas (1988) and Reid et al. (1977, Table 9–12).

O'Connell's correlation predicts that the overall efficiency decreases as the volatility or the viscosity increases. For example, for $\mu = 2 \times 10^{-4}$ Pa·s, Eq. 6.2 gives efficiencies that vary from 70 to 40% as the volatility changes from 1.03 to 12. In most cases, the conditions in the column will be substantially below the critical values for the mixture; the main effect of pressure should be on account of the shift in boiling temperature. As the pressure in the column is increased, the boiling temperature will generally increase and the viscosity will be lowered; therefore, the efficiency in distillation is expected to increase at higher pressures. For example, when the viscosity is 0.5×10^{-5} Pa·s (0.05 centipoise), the efficiency ranges from 90 to 60% for the same range of volatilities.

---

[5]There is no particular physical significance to parameters a or μ_0. They simply establish a numerical scale for the efficiency and the viscosity, and are convenient to make the expression dimensionless.

[6]It is important to remember that these efficiencies are based on a limited set of data and that even within this set there is substantial scatter around the values given by Eq. 6.2; the uncertainty in a is approximately ± 0.05. The correlation is better than the assumption of 100% efficiency, yet simpler, and requires less data than the more detailed procedures. Inaccuracies at the conceptual design stage are often overshadowed by uncertainties in other factors, and a study of the sensitivity of the final results to the assumed values can be used to rank these uncertainties in order of their importance, e.g., to plan and justify more detailed modeling and experiments.

We will generally use an overall efficiency for preliminary designs. However, this approach should be refined for final designs and for performance calculations where high accuracy is needed. For example, Fair et al. (1984) discuss correlations like this along with the procedures for more detailed designs that are recommended by the *American Institute of Chemical Engineers.* Most computer-aided design software has one or more of these procedures available. More accurate *nonequilibrium stage* models have also been developed. The best are based on the work of Taylor and Krishna (Krishnamurthy and Taylor, 1985*a*, 1985*b*; Taylor and Krishna, 1986, especially pp. 368–383). This approach is comprehensively described by Taylor and Krishna (1993). For a comparison of the equilibrium and nonequilibrium models for reactive distillation (Chap. 10) also see Baur et al. (2000). We recommend these approaches for performance studies where high accuracy is warranted, or for more detailed designs, but they require a lot more time and data than are generally practical or necessary for conceptual designs.

Recent progress in detailed simulation methods for fluid flow may significantly change the level of understanding and the extent of modeling that is practical; e.g., computational fluid dynamics (CFD) is now a useful research tool for hydrodynamic studies on trays, e.g., Krishna et al. (1999).

Internal Flows

There is a limited range of vapor and liquid flows over which a typical distillation column or gas absorber can operate. In fact, efficiency correlations like O'Connell's are developed for cases where the column flows are stable. Figure 6.3 shows a qualitative picture of the operability region for plate-type columns. The major design limitation is generally on the vapor velocity, on account of the flooding and weeping limits. To some extent, the flow characteristics depend upon the particular tray design selected, on the tendency of the liquid to bubble or foam, and on various other factors; see van Winkle (1967). However, our goals in a conceptual design can be met with a correlation of the behavior as follows.

The vapor velocity depends on the properties of the mixture, the column operating conditions, and the area available for flow, which dictates the column diameter. The simplest correlations are based on the hypothesis that the flooding limit is

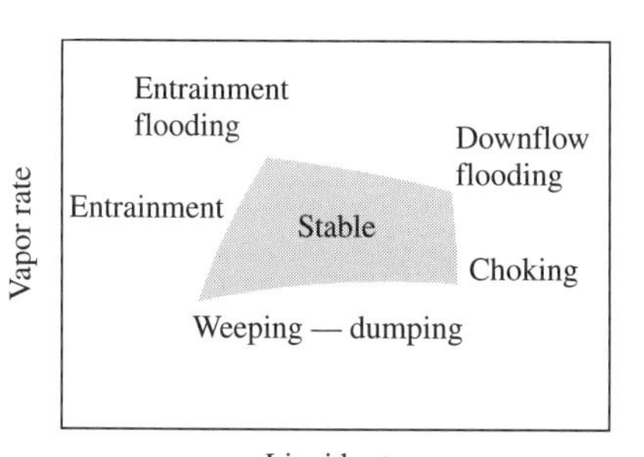

FIGURE 6.3
Schematic of the operating region for a plate column. *[After Fair et al. (1984, Sec. 18).]*

determined primarily by the average kinetic energy of the gas stream, $\frac{1}{2}\rho_v u^2$. Thus, the product of density and the square of the vapor velocity arises naturally. This quantity is often represented as an "F factor" given by

$$F = u\sqrt{\rho_v} \tag{6.3}$$

where ρ_v is the mass density of the vapor. The *superficial vapor velocity* u is the volumetric flowrate of vapor divided by the cross-sectional area of the column.

$$u \equiv \frac{VM_v/\rho_v}{A} \tag{6.4}$$

where M_v is the molecular weight of the vapor, V is the molar flowrate, and A is the cross-sectional area of the column. The velocity, and therefore the F factor, will have upper and lower bounds given by the flooding and weeping constraints, respectively. These limits are reflected experimentally in rapid decreases in the apparent column efficiency, as shown by the examples in Fair et al. (1984, Figs. 18–9, 18–30, 18–31, 18–32). For these experiments, F lies in the range 2000 to 9000 (m/h)(kg/m^3)$^{0.5}$ (which is 1800 to 7200 (ft/h)(lb/ft^3)$^{0.5}$ or 0.5 to 2.0 (ft/s)(lb/ft^3)$^{0.5}$).

If the F factor is known, the cross-sectional area of the column can be found from

$$A = \frac{M_v}{F\sqrt{\rho_v}}\,V \tag{6.5}$$

For existing columns, where performance calculations are required, a fairly precise knowledge of the limits in Fig. 6.3 may be required; sometimes these must be obtained from experiment. Alternatively, a conservative estimate of the vapor velocity at flooding can be made from the correlation developed by Fair (1961). The correlation improves somewhat on the hypothesis that the kinetic energy of the gas phase alone is important and includes the additional effects of the liquid flowrate and the surface tension, as well as the densities of both phases and the vapor flowrate. These are incorporated in a "capacity parameter" defined by

$$c \equiv u_{n,\text{flood}} \left(\frac{\sigma_0}{\sigma}\right)^{0.2} \left(\frac{\rho_v}{\rho_l - \rho_v}\right)^{0.5} \tag{6.6}$$

and a "flow parameter" given by

$$f \equiv \frac{L}{V} \left(\frac{\rho_v}{\rho_l}\right)^{0.5} \left(\frac{M_v}{M_l}\right)^{1.5} \tag{6.7}$$

In the correlation, the superficial vapor velocity is based on the net area for flow A_n rather than the total cross-sectional area, in an attempt to correct for the presence of downcomers on the trays. The velocities and areas are related by

$$A_n u_n = Au \tag{6.8}$$

and A_n/A is usually near 0.8. Besides the molar flowrates L and V, the mass densities and molecular weights of each phase are required. Finally, the surface tension of the liquid σ is normalized with the value $\sigma_0 = 20$ mN/m (20 dyn/cm). Estimates of the surface tension can sometimes be made using the methods described in Reid et al.

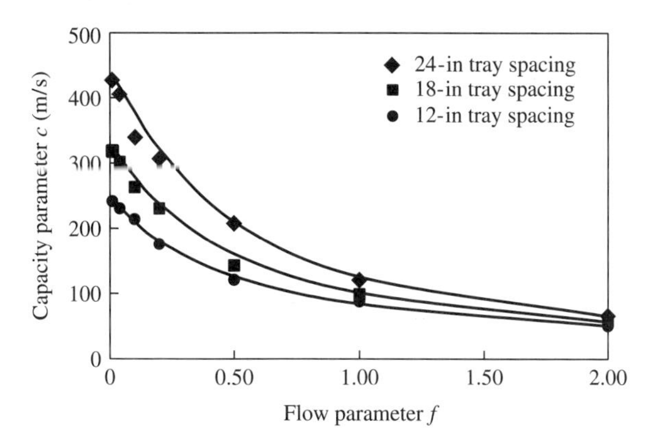

FIGURE 6.4
Fair's correlation for flooding velocity. Adapted from Fair et al. (1984). The solid lines are the predictions of Eq. 6.9.

TABLE 6.1
Constants for Fair's correlation

	Tray spacing		
	0.31 m (12 in)	**0.46 m (18 in)**	**0.61 m (24 in)**
c_0			
(m/h)	252	329	439
(ft/h)	828	1,080	1,440
c_1	2.0	2.3	2.5
c_2	1.0	1.1	1.2

(1977, Chap. 12), although the uncertainty is often quite large; fortunately the results are relatively insensitive to the value of σ. The correlation of Fair is reproduced in Fig. 6.4 and can also be represented by the expression

$$c = \frac{c_0}{1 + c_1 f^{c_2}} \tag{6.9}$$

where the constants are given in Table 6.1.

When the liquid and vapor flowrates are known (from the reflux and reboil ratios and material balances) and the physical properties such as molecular weight, densities, and surface tension are known, the flooding velocity can be estimated. For design purposes, a 24-in tray spacing can usually be assumed, and an appropriate F factor can be estimated for use in Eq. 6.5 by taking the superficial vapor velocity as 60 to 80% of the value at flooding. This is so that the overall efficiency will be near the maximum and that excessive entrainment will not occur.[7] The procedure for finding the column area can be summarized as follows.

1. Find the liquid and vapor molar flowrates in each column section corresponding to the reflux and reboil ratios and the product flows.
2. Estimate the mass density of both phases. Constant liquid densities can usually be assumed, while the ideal gas law result is often sufficient for the vapor.

[7] See the data for specific mixtures presented by Fair et al. (1984).

3. Compute the value of f using Eq. 6.7.
4. Use Eq. 6.9 or Fig. 6.4 and an estimate of the tray spacing (usually 24 in) to find the capacity parameter c.
5. Find the F factor at flooding from the expression

$$F_{\text{flood}} = c\left(\frac{A_n}{A}\right)\left(\frac{\sigma}{\sigma_0}\right)^{0.2}\sqrt{\rho_l - \rho_v} \tag{6.10}$$

Use $A_n/A = 0.8$ unless other information is available.
6. Use Eq. 6.5 to find the cross-sectional area. Design at 60% of the flooding velocity unless other data are available; i.e., use $F = 0.6F_{\text{flood}}$ in Eq. 6.5.

When there is a significant variation of pressure, temperature, or the internal flows inside the column, the estimates in steps 1 through 5 may be needed in each column section or even in subsections. Step 6 should then be based on 60% of the maximum value, and the proximity to flooding should be checked throughout the column.

A simplification of this procedure can be based on estimates for the magnitude of some of the terms in the development above; this was proposed by Douglas (1988). Typically, the density of the liquid is much greater than that of the vapor, and this simplifies the density term in Eq. 6.10. Furthermore, the "flow parameter" in Eq. 6.7 will be much less than unity if $L/V(M_v/M_l)^{1.5} << \sqrt{\rho_l/\rho_v}$ throughout the column.[8] When this is the case, $c \approx c_0$, i.e., the limit on the left in Fig. 6.4 is approached.

When $\sigma/\sigma_0 \approx 1$, we find

$$F_{\text{flood}} \approx \sqrt{\rho_l}\left(\frac{A_n}{A}\right)c_0 \tag{6.11}$$

A typical liquid density is 960 kg/m^3 (60 lb/ft^3). For a tray spacing of 0.6 m (2 ft) and $A_n/A = 0.8$, $F_{\text{flood}} \approx 10{,}900$ (m/h)(kg/m^3)$^{0.5}$ (which is 8,920 (ft/h)(lb/ft^3)$^{0.5}$ or 2.5 (ft/s)(lb/ft^3)$^{0.5}$).

With these approximations, the area of the column is

$$A = \frac{M_v}{\sqrt{\rho_l \rho_v}}\frac{1}{\phi_{\text{flood}} c_0}\left(\frac{A}{A_n}\right)V \tag{6.12}$$

where ϕ_{flood} is the fraction of flooding velocity desired in the design, approximately 0.6, and the fraction of the total area available for flow $A_n/A \approx 0.8$. When the tray spacing and operating conditions are fixed, the cross-sectional area is proportional to the vapor rate, which is determined by the reflux and reboil ratios and the material balances.

These approximations will overestimate F and underestimate the required area if L/V is large, and the applicability of the approximations should be checked in the individual cases.

EXAMPLE 6.1. We consider the separation of n-hexane and p-xylene from Example 3.1. For the given feed rate of 200 kmol/h, the liquid and vapor rates are $L_T = 113$, $V_T = 226$, $L_B = 213$, and $V_B = 126$ kmol/h.

[8] The range of values for the flowrate ratio is simple to estimate. In the rectifying section, $r_{\min}/(r_{\min}+1) < L/V < 1$ and in the stripping section, $1 < L/V < (s_{\min}+1)/s_{\min}$.

- The temperatures can be estimated as the normal boiling points of the pure components at the top and bottom of the column, $T_T = 68.7°C$ and $T_B = 138.3°C$. At the top of the column, where both phases are mostly hexane, we take $\rho_l = 659$ kg/m^3 and $\rho_v = 3.03$ kg/m^3. At the bottom, $\rho_l = 861$ kg/m^3 and $\rho_v = 3.10$ kg/m^3.
- The flooding parameter at the top of the column, where the phases are mostly hexane and the molecular weight ratio is close to 1, is approximately $f_T = \frac{113}{226}\left(\frac{3.03}{659}\right)^{0.5} = 0.034$. At the bottom, $f_B = \frac{213}{126}\left(\frac{3.10}{861}\right)^{0.5} = 0.101$.
- With a tray spacing of 0.6 m (24 in), Eq. 6.9 gives capacity parameters $c_T = 421$ m/h and $c_B = 379$ m/h.
- The F factor at flooding is given by Eq. 6.10; we use a net area for flow equal to 80% of the cross-sectional area, and assume a surface tension ratio of unity. $F_{T,\text{flood}} = 0.8\sqrt{659 - 3.03}(421) = 8{,}630$ and $F_{B,\text{flood}} = 0.8\sqrt{861 - 3.10}(379) = 8{,}880$, both with units of m/h (kg/m^3)$^{0.5}$.
- The areas corresponding to 60% of flooding are $A_T = \frac{86}{0.6(8{,}630)\sqrt{3.03}} 226 = 2.16$ m^2 and $A_B = \frac{106}{0.6(8{,}880)\sqrt{3.10}} 126 = 1.42$ m^2.

The difference in areas is primarily due to the partially vaporized feed and means that the column will operate at a different percentage of the flooding velocity in the top and bottom sections.[9] In cases where the difference is very large, a column with a constant diameter may be difficult to operate, e.g., Fig. 6.3. Occasionally, the top and bottom column sections may be constructed with different diameters. In this example, if the larger diameter is chosen so that the velocity in the top of the column is 60% of flooding, then the bottom section would operate at 40%, and this may be too low. Alternatively, a column with a 1.42 m^2 cross-sectional area would give 60% of flooding in the bottom section, but 91% for the top, which may be too large. An intermediate area of 1.6 m^2 would correspond to 80% and 53% of flooding from the top to the bottom, which is probably too extreme a variation. Most of the mismatch between the column sections could be eliminated and the potential for operability problems reduced if the feed were a saturated liquid.

The same example can be treated approximately with the simplified model in Eq. 6.11 and average values of the physical properties $A = \frac{96}{\sqrt{(756)(3.2)}} \frac{1}{0.6(439)}(1.25)(176) = 1.63$ m^2, which is close to the average of the results from the two more detailed calculations. In this case, the bottom of the column would operate at roughly 53% of flooding, and the top at 77%.

Column Height

The height can be estimated from the number of trays N *inside* the column as

$$H = H_{\min} + H_t N \tag{6.13}$$

The spacing between trays H_t is approximately 30 to 60 cm (12 to 24 in) but can be smaller or larger for some columns; e.g., columns with diameters larger than 4 or 5 m often require special supporting structures for the trays that result in larger

---

[9] This places some limitations on the difference in vapor and liquid rates that can be tolerated between column sections. For example, feeds that are very superheated or subcooled will be difficult to accommodate.

spacings. Other factors that are important in deciding the value are the flow and disengaging characteristics along with the cleaning and maintenance requirements. The additional height $H_{\min}$ allows for a liquid sump at the bottom of the column and for surge capacity and a demisting or vapor disengaging space that may be required at the top. We will take $H_{\min} = 3H_t$ for most columns.

Energy Requirements and Heat Exchanger Design

An energy balance for the condenser or reboiler gives the heat duty as a function of the specific enthalpy of the mixture and the vapor flows. In general,

$$Q_C = \left[(h_N^V - h_D) + (h_{N+1}^L - h_D)\frac{r}{r+1} \right] V_N \tag{6.14}$$

and

$$Q_R = [(h_0^V - h_1^L) + (h_B - h_1^L)s]\ V_0 \tag{6.15}$$

These energy balances take on important special forms in the case of a total condenser and a partial reboiler. When the CMO approximation is used and saturated liquid product and reflux are desired

$$Q_C = \lambda V_T = \lambda(r+1)\ D \tag{6.16}$$

For a partial reboiler with a saturated liquid bottoms and a saturated vapor reboil,

$$Q_R = \lambda V_B = \lambda s\ B \tag{6.17}$$

To account approximately for small heat effects the λ values in Eqs. 6.16 and 6.17 can be replaced by the heats of vaporization evaluated at the condenser and reboiler conditions, respectively.

The utility demands for the condenser and reboiler are major operating costs that can be determined from the heat loads. For example, when cooling water is used in the condenser, the inlet and outlet temperatures are approximately 30 to 50°C (90 to 120°F). When condensing steam is used to provide heating in the reboiler, there are often several pressure levels available, corresponding to different temperatures and heats of vaporization. In this case, we can find the utility flows from an energy balance for each utility stream. For the CMO case the flows of the heating and cooling utilities are

$$F_{\text{cool}} = \frac{Q_C}{c_p \Delta T_C} = \frac{\lambda}{c_p \Delta T_C} V_T \tag{6.18}$$

and

$$F_{\text{heat}} = \frac{Q_R}{\Delta H_{stm}} = \frac{\lambda}{\Delta H_{stm}} V_B \tag{6.19}$$

The heat duties and the utility flows increase linearly with the vapor rates and therefore with the reflux and reboil ratios. For a given feed purity and separation, the distillate and bottoms flows increase in proportion to the feed rate and the heat loads and utility flows do likewise. When the feed is a saturated liquid, and the heat of vaporization is constant, Eq. 2.30 shows that $D/B = s/(r+1)$; in this case, the heat duties for the condenser and reboiler are identical. Otherwise, the utility loads

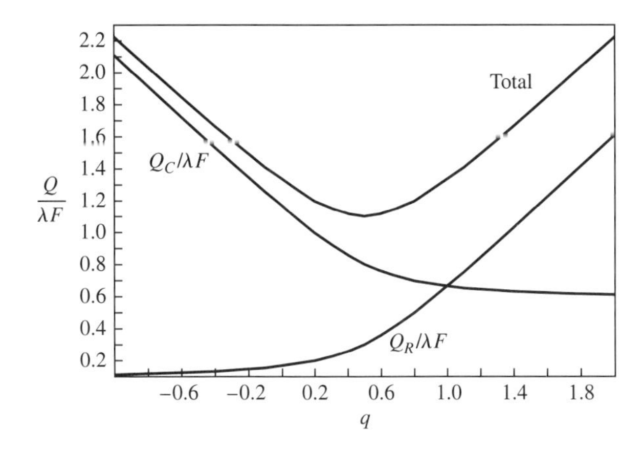

FIGURE 6.5
Heat duties and the effect of feed quality. Curves are for a binary mixture with a constant relative volatility of 3.0. The duties are evaluated for minimum reflux conditions and normalized to the latent heat, which is taken as a constant. The flows are based on 100 moles of an equimolar feed, and 99.5% recoveries of the components. Operability limitations such as flooding and weeping may restrict the design to values of q near 1.

depend on the feed quality, e.g., as shown in Fig. 6.5. There is generally some value of the feed quality q that results in a minimum total heat load for the column and there may therefore be an optimal value. However, extreme variations in the flows can result for q much outside the range $0 \leq q \leq 1$, and it may not be possible to design an operable column if the feed is too subcooled or superheated. Nevertheless, there may be some incentive to use feeds with $q > 1$ or < 0, especially when they are available in such a condition elsewhere in the process.

The sizes of the condenser and reboiler depend on the heat loads and also on the heat transfer coefficients. The simplest reasonable description relies on an estimate of the overall heat transfer coefficient U in the design equation

$$Q = U A_h \Delta T_{\text{avg}} \tag{6.20}$$

where A_h is the area for heat exchange and ΔT_{avg} is the appropriate average temperature driving force in the exchanger.[10] We use constant values for the overall transfer coefficients that depend only on the nature of the heat transfer taking place, e.g., between a liquid and a condensing vapor, or a boiling liquid and a condensing vapor. Table 6.2 shows typical values. More precise values can be estimated for any particular mixture; however, this is often of secondary importance at the conceptual design stage, and a sensitivity analysis of the results should be made.

The appropriate temperature driving force and heat transfer coefficient depend on the type of the heat exchange equipment; see Perry and Green (1984, Chap. 10) and references there for more details. For the countercurrent flow of well-mixed, single-phase fluids, the appropriate average driving force is the *log mean.* That is

$$\Delta T_{\text{avg}} = \Delta T_{lm} = \frac{(T_{\text{in}}^{\text{hot}} - T_{\text{out}}^{\text{cold}}) - (T_{\text{out}}^{\text{hot}} - T_{\text{in}}^{\text{cold}})}{\ln \frac{(T_{\text{in}}^{\text{hot}} - T_{\text{out}}^{\text{cold}})}{(T_{\text{out}}^{\text{hot}} - T_{\text{in}}^{\text{cold}})}} \tag{6.21}$$

[10] A detailed design should take into account the fact that multicomponent mixtures will condense or boil over a range of temperatures. Many studies and methods are available for the detailed calculations which are generally unnecessary for our conceptual designs.

TABLE 6.2
Typical heat transfer coefficients for shell and tube exchangers in W/m²K or (Btu/h ft² °F).

		Cold side fluid				
		1	2	3	4	5
	Hot side fluid	**Cooling water**	**Low-viscosity organic liquid**	**High-viscosity organic liquid**	**Boiling water**	**Boiling organic**
1	Low-viscosity organic liquid	800 (141)	550 (97)	150 (26)	700 (123)	550 (97)
2	High-viscosity organic liquid	140 (25)	130 (23)	80 (14)	140 (25)	130 (23)
3	Condensing steam (no dissolved air)	1600 (282)	820 (144)	170 (30)	1430 (252)	820 (144)
4	Condensing steam (with dissolved air)	1440 (254)	775 (136)	167 (29)	1300 (229)	775 (136)
5	Condensing hydrocarbon (no inert gas)	770 (136)	530 (93)	160 (28)	720 (127)	530 (93)
6	Condensing hydrocarbon (~10% inert gas)	350 (62)	280 (49)	220 (39)	330 (58)	280 (49)

Source: Adapted from the values given by Linnhoff et al. (1982) and Knudsen (1984). *These values should be used for estimation only*; final designs should be based on more detailed methods.

For a condenser using cooling water to produce a saturated liquid at a temperature T_D,

$$\Delta T_{\text{avg}} = \Delta T_{\text{cond}} = \frac{(T_D - T_{\text{out}}^{cw}) - (T_D - T_{\text{in}}^{cw})}{\ln \frac{(T_D - T_{\text{out}}^{cw})}{(T_D - T_{\text{in}}^{cw})}} \tag{6.22}$$

For a reboiler heated with steam

$$\Delta T_{\text{avg}} = \Delta T_{reb} = T_{stm} - T_B \tag{6.23}$$

where T_B is the average boiling temperature of the liquid in the reboiler.

An additional constraint on the reboiler arises with large driving forces and the associated thermal stresses incurred. The maximum heat flux $(Q/A)_{\text{max}}$ should be limited to approximately 32 kW/m² (~10,000 Btu/h ft²) to prevent film boiling. This means that a reboiler design based on the overall heat transfer coefficient should be limited to cases where $Q/A = U\Delta T_{\text{avg}}$ is less than $(Q/A)_{\text{max}}$. Thus,

$$A_R = \frac{Q_R}{\min\left[U_R \Delta T_R, (Q/A)_{\text{max}}\right]} \tag{6.24}$$

An upper limit on the temperature difference, e.g., 50°C, is sometimes used as an alternative design guideline.

For other practical considerations in the design of reboiler systems and a discussion of the various types and applications, see, e.g., McCarthy and Smith (1995).

6.2
COST MODELS

Using the methods summarized above, the size of the column and the associated heat exchangers, along with the utility loads, can be estimated. If heat integration is used, the loads for a collection of columns can be reduced by using the condenser in one column to reboil the bottoms of another, etc.; see Chap. 7. To estimate the cost of a single column and its associated equipment is straightforward. However, it is often impossible to make these estimates with high precision at the conceptual design stage. Fortunately, factors that are difficult to estimate precisely in the costs are often not important for relative comparisons and for design decisions such as the optimal operating conditions for a column.

The cost of a distillation is made up of the utility charges and other *operating costs* along with the installed costs of the column, internals, heat exchange equipment, and auxiliaries. The latter investment costs or *capital costs* have to be compared to the operating costs and must be put on a yearly basis (or the operating costs must be converted to an equivalent dollar figure in capital). We will use a *total annual cost* (TAC) described by Douglas (1988, Chap. 2).

$$\text{TAC} = [2.43K_{cap} + 0.19]C_{cap} + C_{op} \tag{6.25}$$

where C_{cap} is the total installed capital cost of the equipment and C_{op} is the sum of the heating and cooling costs and the difference in value between the products and the raw materials, unless otherwise noted. The *capital charge factor* K_{cap} accounts for the time value of money; typical values are 1/3 to 1/4 year^{-1} for an expansion of a continuous plant; i.e., no large offsite costs are included. For higher-risk projects, a larger capital charge factor corresponding to a higher rate of return should be used and vice versa for ventures with lower risk. For final designs, a more detailed study of the cash flows, rate of return, depreciation, taxes, insurance, etc., will eventually be done. The TAC objective is useful for first estimates where a simple model that accounts for both capital and operating costs is essential.

Capital Costs

The installed cost of the column and trays, as well as the heat exchangers, can be estimated from the correlations such as those developed by Guthrie (1974). There are many other sources for cost correlations and there can be a significant difference in the estimates. The capital cost correlations should be updated as frequently as possible, based on data for actual costs. When current data are unavailable, the capital costs determined in past years can be adjusted using "escalation factors" such as the *Marshall and Swift index* shown in Fig. 6.6. The ratio of the index in any particular year to the index for the year in which the capital costs were correlated is used as a multiplier to account for inflation. This means that the capital cost correlation cannot be used without knowledge of the base year and also that an estimate of the "*M&S*" index is needed for the time period when the equipment purchase is anticipated.

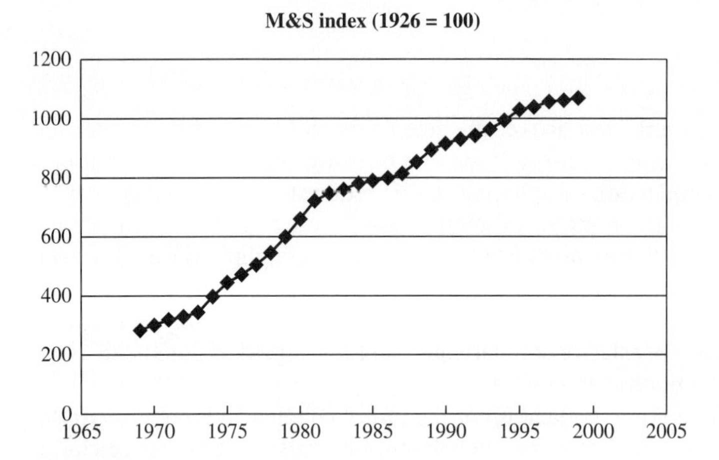

FIGURE 6.6
Marshall and Swift index for cost escalation in chemical process equipment. The current values are published in *Chemical Engineering* magazine.

The general form for capital cost correlations accounts for the fact that there is generally some "economy of scale" in the relationship between cost and equipment size. That is, the cost is proportional to the size raised to a power less than unity, e.g., $Cost \propto Size^a$ with $a < 1$. Therefore, larger equipment is *less expensive per unit size* up to some maximum size that is dictated by the materials of construction or the technology for fabrication or installation.

Heat exchangers

A common class of heat exchanger for vapor-liquid systems is the shell and tube heat exchanger, which can be manufactured in several different design types. Guthrie provides a correlation for the cost of a floating head design as a function of the total heat transfer area.[11]

$$C_h = C_{0,h}\left(\frac{A_h}{A_0}\right)^{a_h} \tag{6.26}$$

The various design types, material(s) of construction, larger pressures, and installation costs are taken into account with separate factors in the expression

$$C_{0,h} = \left(\frac{M\&S}{M\&S_{\text{base}}}\right)\left[(F_d + F_p)F_m - 1 + F_I F_D\right] c_{0,h} \tag{6.27}$$

The value of $M\&S$ in the first term must be estimated for the year of interest; some previous yearly averages are shown in Fig. 6.6. Table 6.3 gives estimates for the values of the rest of the parameters for various conditions.

[11]The basis for the correlation is an exchanger with a carbon steel shell, 1.9 cm by 4.9 m (0.75 in by 16 ft) containing tubes on a 2.5 cm (1 in) square pitch, for service up to 150 psig.

TABLE 6.3
Parameters in the heat exchanger cost correlation. Also see Guthrie (1969, 1974) for more detailed estimates. ***These are for preliminary estimates only and should be updated with current cost data***

Parameter	Value	Unit/type	Comment
$c_{0,h}$	8,700	\$ (1970)	±30%
A_0	93 1,000	m^2 ft^2	Data correlated for 5 times larger or smaller areas
a_h	0.65	—	
$M\&S_{base}$	301	—	1970
F_d	1.00 1.35 0.85 0.80	— — — —	Floating head Kettle reboiler U-tube Fixed tube sheet
F_p	0.00 $0.10\frac{P-P_0}{P_0}$	— —	$P \leq P_0 = 11$ bar (150 psig) $P_0 < P \leq 70$ bar (1000 psig)
F_m^a	1.0 1.2 1.8 1.6 3.0 2.2 3.9 2.7 12.0 7.0	CS/CS CS/Adm Mol/Mol CS/Mol SS/SS CS/SS Mon/Mon CS/Mon Ti/Ti CS/Ti	Shell/tube materials: Adm, admiralty CS, carbon steel Mol, molybdenum Mon, monel SS, stainless steel Ti, titanium
F_D^b	2.30	—	Direct cost factor
F_I^c	1.38	—	Indirect cost factor

[a]These are nominal values at A_0; Guthrie gives values that increase with surface area in the range of ±20%.
[b]This factor accounts for additional direct costs including labor and materials for installation. A breakdown is provided in Guthrie (1974, p. 144). Note that this factor is a multiple of the current purchased cost of the *base* equipment without allowance for material, pressure, or design factors.
[c]This factor accounts for the indirect costs of installation as a multiple of the total direct costs of the *base* equipment. This can vary strongly according to local conditions but is generally the same for all of the capital equipment in the design.

Column and internals

The column is a "pressure vessel," and the purchased cost is sensitive to factors such as the thickness of the vessel walls, the height, and the diameter. A correlation similar to the one given above for heat exchangers can be used to estimate the cost of the vessel "shell," and a separate correlation is available for the column internals. The installed cost for a vertical, carbon steel pressure vessel including support hardware and access openings can be correlated as

$$C_s = C_{0,s}\left(\frac{d}{d_0}\right)\left(\frac{H}{H_0}\right)^{a_s} \tag{6.28}$$

where H is the height and d is the diameter; the units and subscripts are defined in Table 6.4. The prefactor $C_{0,s}$ is given by

$$C_{0,s} = \left(\frac{M\&S}{M\&S_{\text{base}}}\right)\left[F_m F_p - 1 + F_I F_D\right] c_{0,s} \tag{6.29}$$

Similarly, the trays and internals that are added to the shell cost can be estimated from

$$C_t = C_{0,t}\left(\frac{d}{d_0}\right)^{a_t}\left(\frac{H}{H_0}\right) \tag{6.30}$$

with

$$C_{0,t} = \left(\frac{M\&S}{M\&S_{\text{base}}}\right)[F_s + F_t + F_m]\, c_{0,t} \tag{6.31}$$

The total installed capital cost of the column is then

$$C_{col} = C_s + C_t \tag{6.32}$$

The nominal values of the coefficients and factors are summarized in Table 6.4. In most cases the cost of the trays and internals will be much less than the cost of the shell; for the base case in the correlation there is an order of magnitude difference.

Operating Costs

Significant operating costs for most distillation systems are associated with the heating and cooling utilities. Although these utilities will vary somewhat according to local conditions, we present some guidelines here that are adequate in the absence of more detailed, site-specific information.

It is often economical to generate process steam at a high pressure in a central facility and then to generate electricity by expanding the steam through turbines in order to reduce the pressure to the levels required. The cost of steam at any given pressure level depends on the cost of the fuel used to generate the high-pressure steam and on the value of the electricity that is generated by the turbines. We can develop a correlation for the steam costs as follows.

We assume that superheated steam is generated at 1,000 psia and 1100°F (69 bar, 594°C), and that the turbines can be treated as isentropic. On a Mollier diagram for water, the isentrope through this temperature and pressure is roughly linear and intersects the saturated vapor envelope at a pressure of 30 psia (e.g., Perry and Green, 1984, pp. 3–242). This isentrope is approximated quite well by the expression

$$\ln P = 0.003656\, h - 6.3967 \tag{6.33}$$

where P is the pressure in psia and h is the enthalpy in kJ/kg.[12] The enthalpy difference between any two pressure levels is thus simple to estimate, and it is a straightforward exercise in thermodynamics to deduce the functional form for the

[12] In the United States, steam pressures are still reported in psia or psig, and we have chosen to use this, despite the peculiar mixture of units that results.

TABLE 6.4

Parameters in the column cost correlations. Also see Guthrie (1969, 1974) for more detailed estimates. ***These are for preliminary estimates only and should be updated with current cost data***

Parameter	Value	Unit/type	Comment
d_0	1 3.28	m ft	Data correlated for 3 times larger or smaller diameters
H_0	6.1 20	m ft	Data correlated for 5 times larger or smaller heights
$M\&S_{base}$	301	—	1970
F_D	3.00	—	Direct cost factor
F_I	1.38	—	Indirect cost factor
		Shell cost	
$c_{0,s}$	5,000	$ (1970)	
a_s	0.82	—	
F_p	1 $1 + t[1 + e^{-t/2}]$	— —	$P \leq P_0 = 4.5$ bar (50 psig) $P > P_0$
t	$0.13 \frac{P - P_0}{P_0}$	—	$P < 346$ bar (5,000 psig)
F_m	Clad	Solid	Materials:
	1.00 2.30 2.60 4.50 4.89	1.00 3.50 4.25 9.75 10.6	Carbon steel Stainless 304 Stainless 316 Monel Titanium
		Tray cost	
$c_{0,t}$	500	$ (1970)	
a_t	1.8	—	
F_s	1.0 1.3 2.0	60 cm (24 in) 45 cm (18 in) 30 cm (12 in)	Tray spacing
F_t	0.0 0.0 0.0 0.3 1.6 3.2	Grid Plate Sieve Valve Bubble cap Koch	Tray type
F_m	0.0 1.5 8.5	Carbon steel Stainless Monel	Tray material

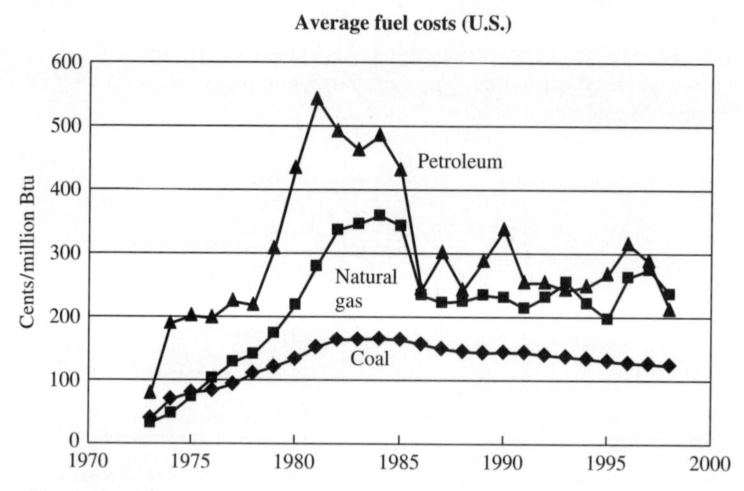

FIGURE 6.7
Historical average cost of fuels in the United States. *(Monthly Energy Review, 2000).*

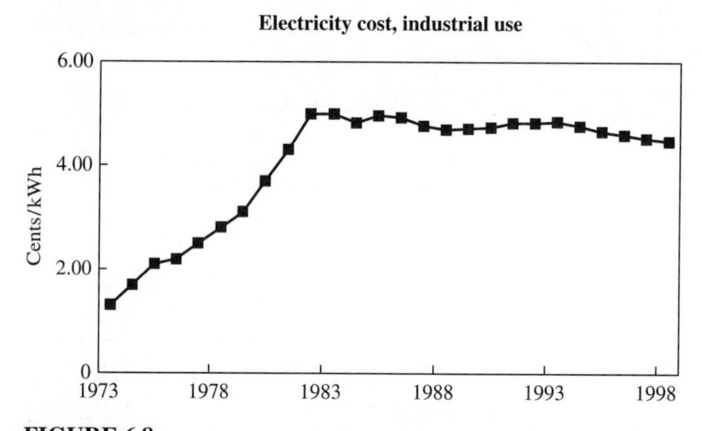

FIGURE 6.8
Historical average electricity costs for industrial use in the United States. *(Monthly Energy Review, 2000).*

cost of steam. For example, correlating a particular set of 1985 steam costs and assuming that the steam is generated in a coal-fired boiler, we find

$$C_{stm} = (31 \ln P - 214) P_E + 4.82 P_C \tag{6.34}$$

where C_{stm} is the cost of steam in \$/1,000 kg, P is the steam pressure in psia, P_E is the price of electricity in \$/kWh, and P_C is the cost of coal in \$/MM Btu.[13] Historical data for the cost of coal delivered to steam-electric utility plants and for the price of electricity supplied to industrial facilities are given in Figs. 6.7 and 6.8.

[13]This is how the cost of fuel is reported in the primary U.S. source (*Monthly Energy Review,* 2000) published by the Energy Information Administration of the United States government.

Equation 6.34 gives reasonable results for the historical costs of steam at different pressure levels. If a fuel other than coal is used as the primary energy source, it is expected that the numerical coefficients in Eq. 6.34 may change, although the functional form should remain the same. The historical cost of other fuels is also shown in Fig. 6.7.

The steam produced at each pressure level along the isentrope will be superheated, which is convenient for avoiding condensation in the steam lines throughout the plant. It is normal practice to desuperheat the steam isobarically at the point of use, and therefore the steam temperature at each pressure level should be taken as the saturation value. The following simple expression fits the saturation data from the steam tables quite accurately, as shown in Fig. 6.9*a*.

$$\ln P = 15.5 - 4{,}770/T \tag{6.35}$$

This equation has the same functional form as the Clausius-Clapeyron equation, with constants determined by a least-squares regression of the experimental data; the pressure is expressed in psia and the temperature in Kelvin.

The steam cost increases linearly with the logarithm of pressure, and comparison of Eqs. 6.34 and 6.35 reveals that it is therefore also linear in $1/T$. Table 6.5 shows some typical values for the steam costs at various temperature and pressure levels.

Although the cost of steam increases with pressure or temperature, its heating value diminishes since the latent heat decreases as shown in Table 6.5, eventually

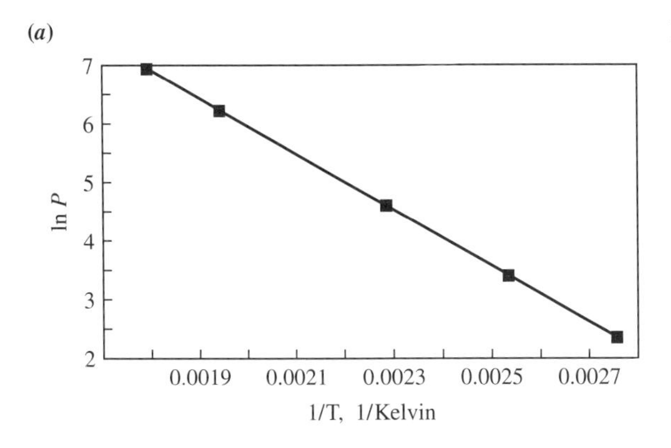

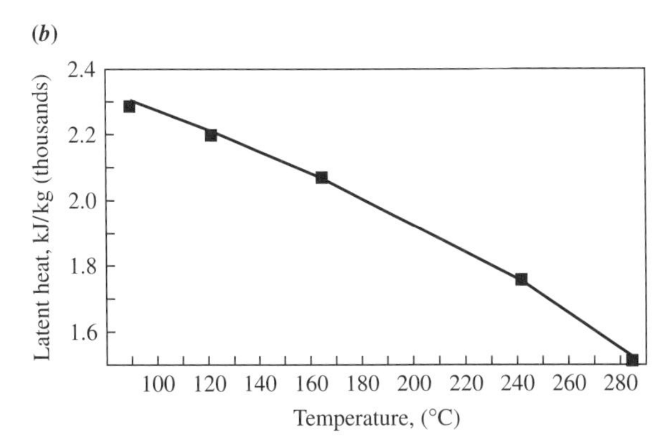

FIGURE 6.9
(*a*) Saturation vapor pressure curve for steam. Comparison of Eq. 6.35 with data taken from the steam tables.
(*b*) Latent heat as a function of temperature. Comparison of Eq. 6.36 with data taken from the steam tables.

TABLE 6.5
Typical steam costs, based on Eq. 6.34 using a price of 0.05 $/kWh for electricity and 1.62 $/MM Btu for coal

Pressure (psia)	Temperature (°C)	Cost ($/1,000 kg)	ΔH_{stm} (kJ/kg)
30	121	2.38	2,213
50	138	3.17	2,159
100	165	4.25	2,067
200	194	5.32	1,960
500	242	6.74	1,755
750	266	7.37	1,634

becoming zero at the critical pressure. Of course, at each pressure or temperature level, the appropriate latent heat must be used in the energy balances. With another regression of data from the steam tables, we find the following equation gives an accurate representation of the change in the latent heat of steam as it varies with temperature

$$\Delta H_{stm} = 2{,}069 \left(\frac{647.1 - T}{209.4} \right)^{0.356} \tag{6.36}$$

where ΔH_{stm} is the latent heat in kJ/kg and T is the saturation temperature in Kelvin. This equation has the same functional form as the Watson equation (see Smith et al., 1996, sec. 4.3) with a modified exponent. A comparison of Eq. 6.36 with data from the steam tables is shown in Fig. 6.9*b*.

Cooling water is often used to provide a relatively low temperature heat sink, and it may be treated to prevent scaling so that it can be recirculated through cooling towers; in other cases it may be available from an offsite source. In any event, the inlet and outlet temperatures, or at least their nominal values, are generally known. For example, the largest inlet temperature is often taken to be near 30°C and the maximum permissible outlet temperature near 50°C. Conservative values should be chosen for the initial design to allow for seasonal fluctuations. The cost of cooling water can be more sensitive to local fluctuations than the steam cost and may thus be very site-specific. We use an estimate of 20% of the cost of 50 psia steam (per kJ) for the cooling water. For example, 50 psia steam at a cost of 3.17 $/1,000 kg corresponds to a price of $(3.17/2{,}159) \times 10^{-3} = 1.47 \times 10^{-6}$ $/kJ of energy. Cooling water available at 30°C and rejected at 50°C requires $1(50 - 30) = 20$ kcal/kg or 83.7 kJ/kg. Thus, the cooling water cost would be 20% of $1.47 \times 10^{-6} \times (83.7)$ $/kg or 0.0246 $/1,000 kg.

Accuracy and Sensitivity

In view of the uncertainty in the cost correlations, efficiencies, and other parameters in the models, it is important that an estimate be made of the sensitivity of the design results to variations in these quantities. Fortunately, this is a relatively simple matter in most cases, if the calculations can be done automatically.

The cost typically depends on several physical properties and design variables. It is useful to estimate the sensitivity of the cost to variation in each of these quantities as well as the total uncertainty in the estimate. Uncertainties in the physical properties often arise from a limited database, and a sensitivity analysis can often be used to plan further experiments. Design variables may fluctuate unavoidably during the operation of a process (e.g., utility temperatures) or may be optimization variables that we are free to select over a range of values at the design stage. It is interesting to estimate the sensitivity of these quantities to identify important control variables or important optimizations. Cost parameters generally have large uncertainties, but the cost is sometimes not very sensitive to some of these; see Table 6.6.

We incorporate the dependence of the cost C on the M variables[14] p_i in a Taylor series and retain the linear terms as follows:

$$C = C_0 + \sum_{i=1}^{M} \delta C_i = C_0 + \sum_{i=1}^{M} \left(\frac{\partial C}{\partial p_i}\right)_0 \delta p_i \tag{6.37}$$

where the subscript 0 indicates the average or "base-case" conditions. The uncertainty in each of the variables δp_i leads to a corresponding uncertainty in the cost δC_i. It is convenient to rewrite this relationship in a dimensionless and scaled form

$$\delta\left(\frac{C}{C_0}\right) = \sum_{i=1}^{M} \left(\frac{\partial \ln C}{\partial \ln p_i}\right)_0 \delta\left(\frac{p_i}{p_{i,0}}\right) \tag{6.38}$$

The design and performance models described in this book can be used to estimate the sensitivity coefficient for each variable, $\partial \ln C/\partial \ln p_i$, but the uncertainty $\delta(p_i/p_{i,0})$ must be known or estimated from other sources. (Here, $\delta p = p_{\max} - p_0$.)

The maximum error in the total cost can be estimated from the sum of the absolute values of the terms in the summation in Eq. 6.38. However, it is unlikely that all of the variables will attain the maximum or minimum values simultaneously, so this estimate is not informative when it is large. For a known distribution of the errors in each variable, an estimate of the uncertainty in the cost can be made. In practice, these distributions are rarely known accurately, and the root mean squared (rms) "error" is often used as a practical estimate of the uncertainty in the cost (as if the errors followed a normal distribution). The rms value is simply the square root of the average of the squared deviations from the mean cost. That is, rms $= \sqrt{\sum_{i=1}^{M} \delta(C_i/C_0)^2/M}$.

Table 6.6 and Fig. 6.10 show the results of a sensitivity study for the binary separation of n-hexane and p-xylene discussed above. The sensitivity coefficients were estimated numerically from a central difference formula while the uncertainty in each variable was estimated from the available data or by experience on a few designs. The prefactors in the capital cost correlations for the heat exchange equipment and

[14] Here, we consider any of the important parameters that enter the cost calculation directly, e.g., as cost coefficients, or indirectly, e.g., in the calculation of number of stages, heat exchanger size, etc. It is the impact of the variation in these quantities that must be estimated.

TABLE 6.6
Sensitivity analysis of distillation costs for the distillation of 200 kmol/h of a mixture of *n*-hexane and *p*-xylene. *C* is the Total Annual Cost; for the average values of the parameters $C = C_0 = 516{,}785$ \$/year. An upper bound on the error in this cost estimate is 68% but the rms error is 7%

	Variable p	Average p_0	Range min	Range max	Unit	$\frac{\partial \ln C}{\partial \ln p}$	$\frac{\delta p}{p_0} \times 100$	$\frac{\delta C}{C_0} \times 100$
1	$c_{0,h}$	8,600	3,500	13,920	\$(1970)	0.4833	60	28.9
2	$c_{0,shell}$	5,000	2,000	8,000	\$(1970)	0.3609	60	21.7
3	T_{stm}	149	142	156	°C	−2.7239	5	−12.8
4	$M\&S$	1,068	961	1,175	(1999)	0.8570	10	8.6
5	K_{cap}	0.33	0.30	0.36	year^{-1}	0.6928	9	6.3
6	λ	32,420	29,178	35,662	kJ/kmol	0.4746	10	4.7
7	μ	0.33	0.17	0.50	cp	0.08576	48	4.2
8	q	1.0	0.9	1.1		0.4014	10	4.0
9	F	200	190	210	kmol/h	0.6491	5	3.2
10	U_{reb}	800	680	920	W/m^{2}K	−0.1977	15	−3.0
11	ϕ_{flood}	0.70	0.60	0.80		−0.1919	14	−2.7
12	U_{cond}	600	480	720	W/m^{2}K	−0.1162	20	−2.3
13	a_{shell}	0.82	0.74	0.90		0.2361	10	2.3
14	x_D	0.995	0.993	0.997		10.624	0.2	2.1
15	$T_{c,in}$	30	23	38	°C	0.08638	23	2.0
16	z_F	0.55	0.52	0.58		0.3579	5	2.0
17	c_0	439	395	483	m/h	−0.1919	10	−1.9
18	A_n/A	0.80	0.72	0.88		−0.1919	10	−1.9
19	a_{he}	0.65	0.55	0.75		0.1054	15	1.6
20	a	0.24	0.216	0.264		−0.1272	10	−1.3
21	σ	20	14	26	dyn/cm	−0.03837	30	−1.2
22	c_{heat}	3.17	2.85	3.49	\$/1,000 kg	0.1192	10	1.2
23	α	7.00	6.79	7.21		−0.3885	3	−1.2
24	f_l	0.995	0.993	0.997		5.683	0.2	1.1
25	$T_{c,\text{out}}$	50	43	58	°C	0.07752	14	1.1
26	$r/r_{\min}$	1.5	1.05	1.95		−0.03113	30	−0.9
27	$H_{\min}$	2.4	1.8	3.0		0.03283	25	0.8
28	$c_{0,\text{tray}}$	500	200	800	\$(1970)	0.01273	60	0.8
29	H	8,200	7,790	8,610	h/year	0.1430	5	0.7
30	ΔH_{stm}	2,159	2,051	2,267	kJ/kg	−0.1191	5	−0.6
31	c_{cool}	0.0246	0.0209	0.0283	\$/1,000 kg	0.02387	15	0.4
32	c_2	1.20	1.14	1.26		−0.0473	5	−0.2
33	c_1	2.50	2.38	2.63		0.01540	5	0.1
34	a_{tray}	1.82	1.64	2.00		0.006606	10	0.1

the column shell are important primarily because the uncertainty is large. On the other hand, even though the cost is nearly as sensitive to the relative volatility, the value of α is known with enough precision so that there is not much uncertainty in the cost on account of this variable. In contrast, even a 60% uncertainty in the prefactor for the tray cost (no. 28) results in only a 0.8% uncertainty in the total cost because of a lack of sensitivity. It is useful to examine the entire list in order to focus the limited time and resources available to refine a design.

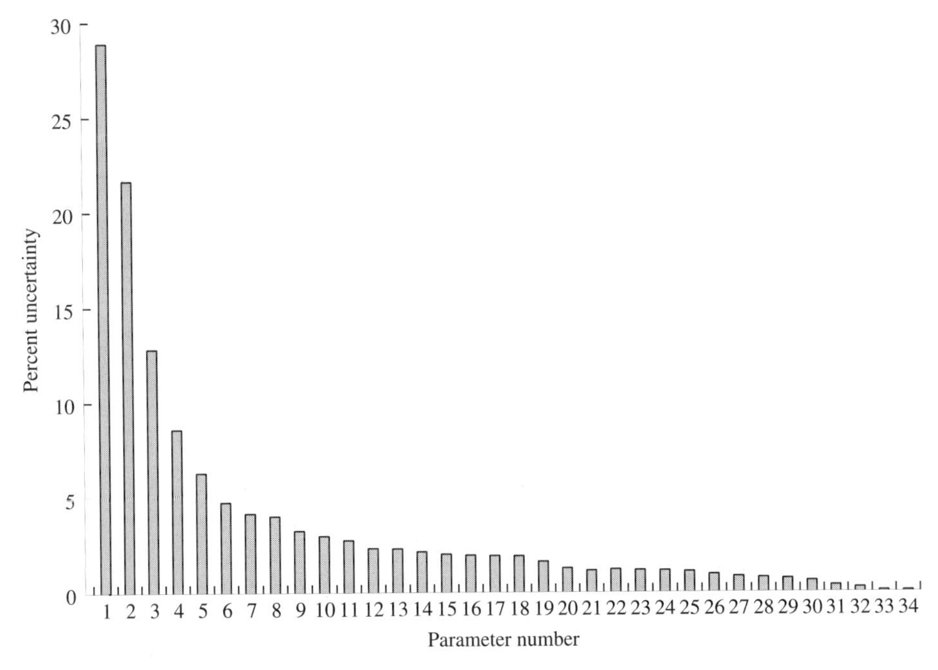

FIGURE 6.10
Parametric sensitivity of distillation costs. The numbers correspond to the variables listed in Table 6.6 and the absolute value of $\delta C/C_0$.

6.3 OPTIMAL DESIGN OF SINGLE COLUMNS

Design problems are always underspecified, and engineering decisions aided by optimization studies or heuristics are recommended.[15]

Optimum Reflux Ratio

A classic example in distillation design is the choice of the vapor rate, usually represented in terms of the reflux ratio. Many combinations of stages and reflux can be found to achieve a given separation, e.g., Fig. 6.11. The best value of reflux and

[15] *Heuristic,* from the Greek *heuriskein, to discover,* refers here to problem solving based on experience rather than a detailed analysis. In optimization studies, *heuristic methods* use simple rules to get good but suboptimal solutions quickly. These are often effective when no exact method is known or when exact methods are impractical to implement. Often, heuristic methods are much faster and need less data than exact methods. However, all heuristics have exceptions, and they should not be relied upon exclusively. They are more useful in conceptual designs than in final designs. Some heuristics are based on laboratory experience and some on computer simulations. When a heuristic can be *derived,* this has the great advantage of revealing its limitations (Douglas, 1988).

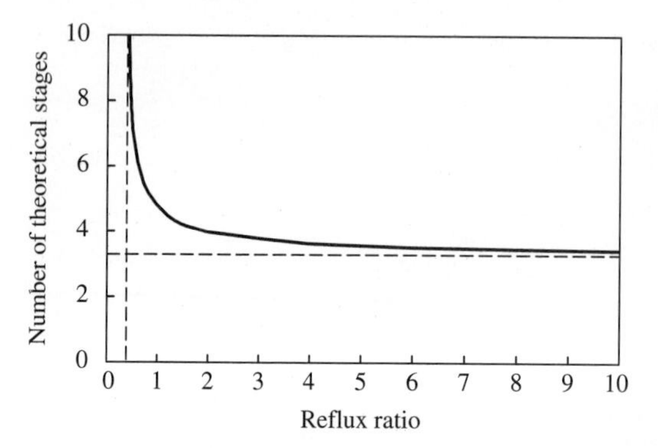

FIGURE 6.11
Number of theoretical stages as a function of reflux ratio for the hexane-xylene example. The minimum reflux ratio is 0.398 and the minimum number of stages is 3.3.

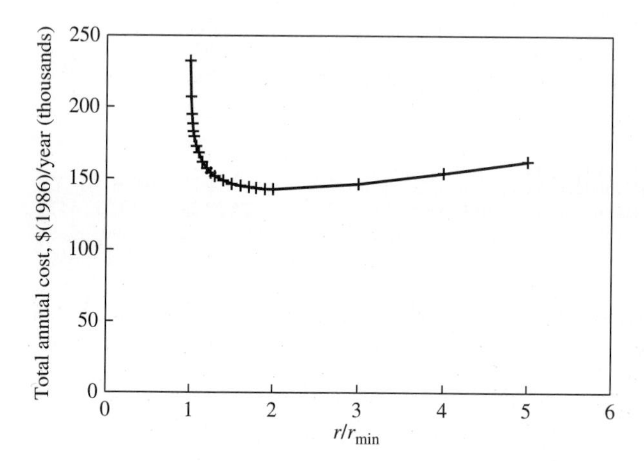

FIGURE 6.12
Total annual cost as a function of reflux ratio for the hexane-xylene example. The feed rate is 200 kmol/h (approximately 150×10^6 kg/year).

the corresponding number of stages is a *trade-off.* This will balance the investment cost associated with a large number of stages as the reflux ratio approaches its minimum value against large operating (energy) and investment costs (column diameter and heat exchange equipment) when the reflux ratio is large. As the reflux ratio approaches the minimum value, the number of stages grows unbounded, while at large values of the reflux ratio, the number of stages approaches a minimum.[16]

Based on this relationship between the reflux ratio and the number of stages, and using the design and cost models developed earlier in this chapter, the optimum design can be determined. Figure 6.12 shows the total annual cost for a single feed rate. The capital cost related to the column height decreases rapidly at values of the reflux ratio only slightly above the minimum due to the extreme sensitivity of the number of stages as a pinch point is approached. As the reflux ratio increases, so does the vapor rate. Therefore, the operating costs for heating and cooling increase. The capital cost first decreases due to the reduction in the number of stages, but then

[16] An important exception to this picture can occur for some distillations of nonideal mixtures, where there can also be a *maximum* reflux. For example, see the discussion of extractive distillation in Chap. 5 or reactive distillation in Chap. 10.

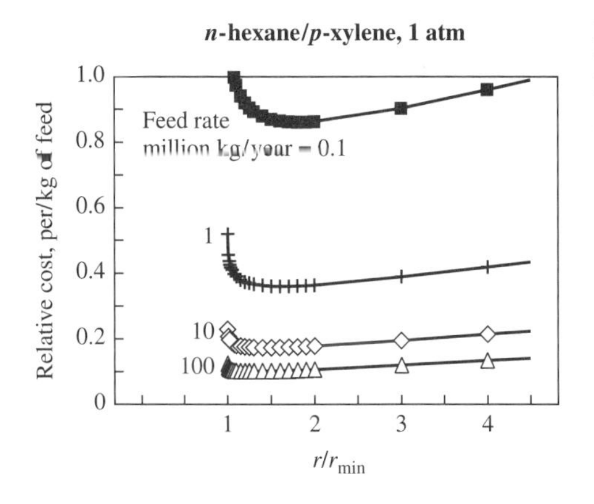

FIGURE 6.13
Relative costs per kg of feed as a function of reflux ratio and feed rate for the hexane-xylene example.

eventually grows due to the increased column diameter and the increased areas for the condenser and reboiler. The total annual cost increases relatively slowly at larger values of the reflux ratio. The one-sided nature of the optimum means that there is a relatively large risk in using a heuristic that places the optimum close to $r_{\min}$, and a relatively small risk for larger values.[17]

When the relative costs of capital and energy change, or when some of the costs for the heat exchangers and utilities are reduced by heat integration or increased by very expensive utilities, the optimum will increase or decrease, respectively. If the materials of construction are very expensive, relative to energy, then much larger reflux ratios may be preferred.

Figure 6.13 shows the effect of feed rate for the hexane-xylene example. The optima are at reflux ratios slightly above the minimum for large production rates up to nearly twice the minimum at low production rates. This is on account of the economy of scale in the capital investment; also note the large *decrease in unit cost* for larger production rates. The larger absolute costs for larger flows and the position of the optimum closer to the minimum justifies more detailed data, modeling, and optimization for these larger flows.

Gathering all of the information required to prepare Fig. 6.13 is time-consuming, and it is common to use a heuristic to choose the reflux ratio; a typical one is $r = 1.2r_{\min}$.[18] However, from the shapes of these curves, and in view of the uncertainties involved in the design and cost models, you can probably see why it is prudent to overestimate rather than to underestimate the reflux ratio. For instance, $r = 1.5r_{\min}$ is also a sensible and more conservative choice. For larger production rates, rather then relying on heuristics, this may be an optimization worth doing. However, we recommend preparing a first design or two using a reflux range of 20 to 50% larger than the minimum, rather than beginning with an optimization. Also see Exercise 9.

---

[17]For instance, the VLE models are imperfect, so that the exact value of the minimum reflux ratio is uncertain. If the true minimum reflux ratio is greater than, say, $1.2r_{\min}$, the column may not meet the anticipated purities.

[18]You can think of this heuristic as the approximate solution to an optimization problem that is usually insensitive to the details of the problem.

Since there are exceptions to any heuristic, it is *not* possible to give a universal factor for the best value of the reflux ratio or other design variables, e.g., the fractional recovery (see below). The economic trade-offs in deciding the optimum reflux ratio have been understood for a long time, (Lewis, 1922, p. 497; McCabe and Thiele, 1925, p. 608). To the best of our knowledge, the insensitivity of the optimum and the heuristic for $r/r_{\min}$ appears quite a bit later (Robinson and Gilliland, 1950, pp. 129–131).

Optimum Fractional Recovery

When product purity specifications or constraints do not fix the product composition, there is another opportunity for optimization. A heuristic approach is to choose the fractional recovery for valuable components to be approximately 99.5% (Douglas, 1988, p. 77). The economic trade-offs typically involve the costs of raw materials and waste treatment and the annual costs for one or more columns in the separation system. Normally, the processing costs in the remainder of the process such as reaction, recycle heating and cooling, etc., are also important. Rapid estimates are helpful in the latter case, because it is usually necessary to make repeated designs of the separation system as part of an optimization study.

Suppose, for instance, that methanol with a purity of 99 mol % is desired and that a binary mixture of methanol and water containing 70 mol % methanol is available. Imagine also that the water recovered is not recycled. The purity requirement for methanol sets one degree of freedom for the flows. However, the best choice for the fraction of methanol to recover from the feed is an optimization variable. As a larger fraction of the methanol in the feed is recovered in the distillate, more and more stages are required in the stripping section of the column; a larger reboil ratio and more energy may also be needed. With more stages, though, the purity of water in the bottoms product is increased and the costs for treatment of the water stream are decreased. Thus the optimum fractional recovery trades off the incremental cost for stages and energy in the bottom of the distillation column for the incremental value of the additional methanol recovered and the reduction in cost of water treatment. See Exercises 11 and 12.

You can also imagine an important exception to this heuristic. Imagine that the column in question is large and that the capacity for methanol treatment in waste streams at the plant site is operating at or near its maximum rate. For a new column, the incremental cost for methanol treatment will be relatively small initially but will soon be prohibitive because an expansion of the treatment capacity would be necessary. This sort of *system* interaction is often a very important factor in design. We will study collections of columns, sometimes thought of as *subsystems,* in the next chapter.

6.4 EXERCISES

1. How do the results in Example 6.1 change if the feed is a saturated liquid? What if the pressure changes to 5 atm? Compare the results with those from the shortcut model in Eq. 6.12 discussed in the text.
2. Find the size of the column and the heat exchangers for Example 6.1. Also calculate the utility flows and costs. How do these change if the feed is a saturated liquid; a saturated vapor?
3. (*Strongly Recommended.*) Prepare a spreadsheet to do the following. Be sure that you separate all of the inputs from the quantities that are calculated. Also, be sure to label the variables carefully and indicate units. The idea is that you will want to use this spreadsheet in other problems to get equipment sizes and costs. *Note:* We suggest that this be done in groups.
   a. Calculate the diameter and height of a column, given the vapor rate in each column section, the number of stages, and compositions. Note that other quantities such as flows, as well as numerical values of data and parameters in correlations, will also be needed. You should enter these in your spreadsheet so that they can be easily found and changed, e.g., in order to make a sensitivity calculation.
   b. Calculate the area required for the condenser and reboiler of a column and the flowrates of steam and cooling water given the same information used in Question 3*a*.
   c. Estimate the investment cost, the operating costs, and the total annual cost, based on the solutions of Questions *a* and *b* and any additional information needed (for instance, the material of construction, which will be an additional input).
4. Estimate the current capital costs for the column and heat exchange equipment corresponding to Examples 3.1 and 6.1. If the feed rate doubles, how do these capital costs change? Estimate the total annual cost for the separation. *Note:* Use a pressure of 1 atm, and carbon steel construction. What is the total annual cost for each feed rate? *Notes:* If you solved Exercise 2, you already have the utility costs and flows for the lower feed rate. If you solved Exercise 3, simply use the spreadsheet. In that case, also make some estimates of the sensitivity similar to the calculations done for Table 6.6.
5. Suppose that a distillation column with a 4-ft diameter and 2-ft tray spacing is available. You are trying to determine the maximum amount of feed that can be processed without flooding the column. The feed is a mixture of water with alcohols, mostly methanol. Assume that the feed is a saturated liquid at 1 atm containing 75 mol % methanol. Be sure to indicate any assumptions or additional information you use. *Hint:* Assume that there are enough stages to make a high-purity separation.
6. This question concerns the distillation of a mixture containing chloroform and acetic acid, available as a saturated liquid mixture at 1 atm pressure and containing 40 mol % chloroform. See Table 2.5 for the boiling points and relative volatility; note the uncertainty in the relative volatility.

a. For product purities of 95 mol % and using the Fenske and Underwood equation-based method as described in Sec. 4.3 find:
   i. The minimum number of equilibrium stages
   ii. The minimum reflux and reboil ratios
   iii. The minimum vapor flows in each section of the column, and the corresponding energy requirements
   iv. The number of equilibrium stages in each section of the column and the energy requirements at a vapor rate 30% larger than the minimum

b. Suppose that the chloroform is a solvent that is recycled within a process, but the acetic acid is a product. The acetic acid purity must be larger than 95%, but the actual value has not been decided. To aid in this decision, prepare a graph of the number of stages and the energy requirements as a function of the mole fraction of chloroform in the acetic acid. Discuss how you would use this result to choose the optimum purity. What other information do you need? *Hint:* If chloroform is recycled, the cost of makeup chloroform elsewhere in the process is saved. Use a value for the cost of chloroform from a recent issue of the *Chemical Marketing Reporter* or a similar source.

7. Consider the hexane-xylene separation discussed in Example 6.1. Suppose that the feed rate doubles, from 200 to 400 kmol/h, but that the feed composition and enthalpy do not change and the reflux ratio and reboil ratio are also constant.
   a. Estimate the height and diameter of the new column.
   b. Estimate the size of the condenser and reboiler.
   c. Estimate the installed capital cost of the column in current dollars. Use stainless steel construction materials for the estimates.
   d. Calculate the total annual cost and do some sensitivity calculations like those shown in Table 6.6.
8. Do Exercise 7, but keep the feed flowrate unchanged from the example. Instead, increase the pressure from 1 to 5 atm.
9. For one of the problems you solved in Chap. 3, plot the capital cost of the column and the capital cost of the condenser and reboiler vs. the reflux ratio. Also plot the heating and cooling costs. Using these results, make a graph of total annual cost vs. reflux ratio. What is the optimum value? What is the uncertainty in this result?
10. If you solved Exercise 3.9, calculate the total annual cost of a single column to process all of the feed. What is the value of the energy savings in the thermally integrated system, compared to the single column? Compare the total annual costs for the single column and the thermally integrated system. For a small group miniproject, find the optimum reflux and reboil ratios and the optimum pressure in the high-pressure column. (Assume that the low-pressure column operates at 1 atm.)
11. Consider the case of alcohol recovery from an alcohol-water mixture as discussed for methanol in the section above on Optimal Design of Single Columns. Suppose that the alcohol is isopropanol instead of methanol and that the feed contains 30% alcohol. The distillate should have an alcohol composition of 99% of that at the azeotrope. The feed rate is 10 million kg/year.

a. Estimate the flows and design a column to separate this mixture; assume 99.5% of the maximum possible recovery of the isopropanol and a reflux ratio 20% above the minimum. The isopropanol-rich distillate is recycled and the bottoms stream, containing mostly water, is processed as waste. Use a water purity of 0.998 mol%.

b. Estimate the size and cost of the column and utilities.

c. For a fixed value of the fractional recovery at 99.5% of the maximum possible, find the total annual cost at 1.01, 1.2, 1.5, 2, and 3 times the minimum reflux ratio. Use these values to make a graph and estimate the optimum reflux. *Note:* Include the value of the isopropanol recovered as a credit (a negative cost). Use a value for the cost of isopropanol from a recent issue of the *Chemical Marketing Reporter* or a similar source.

d. Keeping the reflux ratio constant at the value estimated in the last part of this problem, plot the total annual cost for the separation vs. the fractional recovery of isopropanol. Estimate the optimum value.

Hints: Work in groups, and automate the calculations as far as possible. Use the results of Exercise 3 for the costs.

12. Repeat Exercise 11, but instead of varying the fractional recovery of isopropanol, use a constant value of 99% of the maximum possible. Also, use a reflux ratio of $1.5r_{\min}$. Instead of estimating the optimum values for those variables, examine the purity of the water-rich stream produced as a bottoms product. It is too costly to make the stream completely free of isopropanol. The optimum value of purity trades off waste treatment costs related to the amount of isopropanol left in the water against the value of the isopropanol and the incremental cost of stages to increase its recovery. Estimate the optimum purity of the water stream if the treatment charges are \$0.6/lb of isopropanol. How does your answer change if the treatment costs increase by a factor of 10?

References

Baur, R., A. P. Higler, R. Taylor, and R. Krishna, "Comparison of Equilibrium Stage and Nonequilibrium Stage Models for Reactive Distillation," *Chem. Eng. J.,* **76,** 33–47 (2000).

Bennett, D. L., R. Agrawal, and P. J. Cook, "New Pressure Drop Correlation for Sieve Tray Distillation Columns," *AIChE J.,* **29,** 434–442 (1983).

Douglas, J. M., *The Conceptual Design of Chemical Processes.* McGraw-Hill, New York (1988).

Fair, J. R., "How to Predict Sieve Tray Entrainment and Flooding," *Petro/Chem. Engng.,* **33**(10), 45–52 (1961).

Fair, J. R., D. E. Steinmeyer, W. R. Penney, and B. B. Crocker, Liquid-Gas Systems, in Perry, R. H., and D. W. Green, editors, *Perry's Chemical Engineers' Handbook.* McGraw-Hill, New York, 6th ed. (1984).

Fair, J. R., Distillation, in Rousseau, R. W. editor, *Handbook of Separation Process Technology.* Wiley-Interscience, New York (1987).

Guthrie, K. M., "Captial Cost Extimating," *Chem. Eng.,* **76,** 114 (1969).

Guthrie, K. M., *Process Plant Estimating, Evaluation and Control.* Craftsman Book Company of America, 542 Stevens Avenue, Solana Beach, CA 92075 (1974).

Humphrey, J. L., and G. E. Keller, III, *Separation Process Technology.* McGraw-Hill, New York (1997).

King, C. J., *Separation Processes.* McGraw-Hill, New York, 2d ed. (1980).

Knudsen, J. G., Heat Transmission, in Perry, R. H., and D. W. Green, editors, *Perry's Chemical Engineers' Handbook.* McGraw-Hill, New York, 6th ed. (1984).

Krishna, R., J. M. van Baten, J. Ellenberger, A. P. Higler, and R. Taylor, "CFD Simulations of Sieve Tray Hydrodynamics," *Chem. Engng. Res. Design,* **77**(A7), 639–646 (1999).

Krishnamurthy, R., and R. Taylor, "Nonequilibrium State Model of Multicomponent Separation Processes. Part I: Model Description and Method of Solution," *AIChE J.,* **31,** 449–456 (1985*a*).

Krishnamurthy, R., and R. Taylor, "Nonequilibrium Stage Model of Multicomponent Separation Processes. Part II: Comparison with Experiment," *AIChE J.,* **31,** 456–465 (1985*b*).

Lewis, W. K., "The Efficiency and Design of Rectifying Columns for Binary Mixtgures," *J. Ind. En. Chem.,* **14**(6), 492–497 (1922).

Linnhoff, B., D. W. Townsend, D. Boland, and G. F. Hewitt, *A User Guide on Process Integration for the Efficient Use of Energy.* Institution of Chemical Engineers, Rugby (Warwickshire) (1982).

Lockett, M. J., *Distillation Tray Fundamentals.* Cambridge University Press, Cambridge, UK (1986).

Lockett, M. J., "Easily Predict Structured-Packing HETP," *Chem. Engng. Prog.,* **94**(1), 60–66 (1998).

McCabe, W. L., and E. W. Thiele, "Graphical Design of Fractionating Columns," *Ind. Eng. Chem.*, **17,** 605–611 (1925).

McCarthy, A. J., and B. R. Smith, "Reboiler System Design; The Tricks of the Trade," *Chem. Engng. Prog.*, **91**(5), 34–47 (1995).

Monthly Energy Review. U.S. Department of Energy (2000). http://www.eia.doe.gov/mer.

O'Connell, H. E., "Plate Efficiency of Fractionating Columns and Absorbers," *Trans. AIChE,* **42,** 741–755 (1946).

Perry, R. H., and D. W. Green, *Perry's Chemcial Engineers' Handbook.* McGraw-Hill, New York, 6th ed. (1984).

Reid, R. C., J. M. Prausnitz, and T. K. Sherwood, *The Properties of Gases and Liquids.* McGraw-Hill, New York, 3d ed. (1977).

Robinson, C. S., and E. R. Gilliland, *Elements of Fractional Distillation.* McGraw-Hill, 4th ed. (1950).

Smith, J. M., H. C. Van Ness, and M. M. Abbott, *Introduction to Chemical Engineering Thermodynamics.* McGraw-Hill, New York, 5th edition (1996).

Taylor, R., and R. Krishna, Multicomponent Mass Transfer: Theory and Application, in *Handbook of Heat and Mass Transfer,* Chap. 7. Gulf Publishing, Houston (1986).

Taylor, R., and R. Krishna, *Multicomponent Mass Transfer.* Wiley, New York (1993).

van Winkle, M., *Distillation.* McGraw-Hill, New York (1967).

7

Column Sequencing and System Synthesis

7.1 INTRODUCTION

Many successful separation systems have been invented by experience. This leaves open the questions of what potential there may be for improving them as well as what methodology may be useful to devise new and improved systems. Furthermore, new technologies, along with environmental and economic conditions, present problems for which there is a relatively small base of experience. The recent development of systematic design methods and tools, especially for nonideal mixtures, can be used to address these questions.

One school of thought is that the purpose of process synthesis is to provide the optimal specification of equipment based on known physical properties and specifications. We believe differently—that larger benefits from computing tools come when they are used in conjunction with experimental studies. For separations, this is partly because there remains a significant uncertainty in thermodynamic predictions made for very nonideal systems, despite the availability of large databases for vapor-liquid and liquid-liquid equilibrium. The data available for reacting mixtures is even more sparse. Furthermore, synthesis results for separation systems are often coupled to effects on a higher level, connected to the basic chemistry and reactor systems that give rise to the separations. Therefore, it is important to consider a *systems* approach to chemical processing, building on the ideas developed for single columns in earlier chapters. A major advantage of the systems approach lies in the fact that the typical goals for a separation often cannot be achieved in a single unit, while there may be many different systems of interconnected columns that can satisfy the requirements.

A second issue in any synthesis exercise is the number of alternative solutions. There may be many alternatives and there is no general rule to choose among them. Many may have comparable costs, or all but a few may be economically unattractive so some analysis for the ranking of alternatives is useful. Fortunately, this analysis need not be a rigorous design. When the differences among alternatives are small, high accuracy is required to identify the true optimum, but the choice among neighboring alternatives is not so important because the differences are small. Conversely, larger differences among the alternatives mean that less accurate models will not lead to bad decisions. In other words, models need to be sufficiently accurate to discard poor alternatives and yield a smaller number of candidate designs worthy of further attention. A well-known example is the large number of alternative distillation sequences for ideal mixtures. For very nonideal mixtures many of these alternative sequences are infeasible. However, this does not necessarily mean that there are no alternatives, but only that they are different structures and more work is required to find them.

In this chapter, we describe an approach for finding and comparing systems based on distillation that will attain certain design goals. For systems without azeotropic behavior, many separations are feasible for multicomponent mixtures. The task in conceptual design is typically to select from among these alternatives the sequence or sequences that deserve more detailed study. There is a large incentive for speed in this sequencing exercise because the separation system is generally part of a larger processing system that includes a reactor system and possibly additional upstream and downstream plants, e.g., Douglas (1988).

The first part of the chapter considers mixtures without azeotropes, which can be separated in sequences of simple columns. Following that, we discuss sequences containing "complex" columns, which involve multiple feeds, sidestream product withdrawal, or combinations of interconnected rectifying and stripping sections. This is followed by a discussion of sequences for separating azeotropic mixtures, and heat integration.

7.2 SEQUENCES OF SIMPLE COLUMNS

Table 7.1 shows some sequences of splits that can be used to separate nonazeotropic mixtures. These can all be accomplished in "simple" columns, which have a single feed and two products. Each of the splits listed corresponds to one simple column; e.g., each of the 14 simple distillation sequences for 5-component mixtures has 4 columns. The general result is that $n - 1$ simple columns are sufficient to separate an n-component mixture into nearly pure streams. Actually, more columns can be used, and this is sometimes more economical, as discussed later in the context of heat integration. It is not difficult to see that the number of simple column sequences increases dramatically with the number of components (Exercise 3).

Heuristic Approaches

Several early studies on distillation sequencing suggested heuristics for the choice of a sequence. Some of these heuristics are given in Table 7.2.

Some of these heuristics are simply intuitive; others have been suggested by patterns in simulation case studies [e.g., Tedder and Rudd (1978), Nishida et al. (1981)] or can be derived from known assumptions (Wahnschafft et al., 1993). Heuristics have the advantage of speed and minimal data requirements. However, heuristics can

TABLE 7.1
Simple distillation sequences for mixtures without azeotropes. The components are *A*, *B*, *C*, . . . in order of increasing boiling point

Sequences for three components

	Column 1	Column 2
1	A/BC	B/C
2	AB/C	A/B

Sequences for four components

	Column 1	Column 2	Column 3
1	A/BCD	B/CD	C/D
2	A/BCD	BC/D	B/C
3	AB/CD	A/B	C/D
4	ABC/D	A/BC	B/C
5	ABC/D	AB/C	A/B

Sequences for five components

	Column 1	Column 2	Column 3	Column 4
1	$A/BCDE$	B/CDE	C/DE	D/E
2	$A/BCDE$	B/CDE	CD/E	C/D
3	$A/BCDE$	BC/DE	B/C	D/E
4	$A/BCDE$	BCD/E	B/CD	C/D
5	$A/BCDE$	BCD/E	BC/D	B/C
6	AB/CDE	A/B	C/DE	D/E
7	AB/CDE	A/B	CD/E	C/D
8	ABC/DE	A/BC	D/E	B/C
9	ABC/DE	AB/C	D/E	A/B
10	$ABCD/E$	A/BCD	B/CD	C/D
11	$ABCD/E$	A/BCD	BC/D	B/C
12	$ABCD/E$	AB/CD	A/B	C/D
13	$ABCD/E$	ABC/D	A/BC	B/C
14	$ABCD/E$	ABC/D	AB/C	A/B

TABLE 7.2
Selected heuristics for distillation sequencing

1. **General heuristics**
   (*a*) Remove corrosive or reactive components as early as possible
   (*b*) Remove final products as distillates or as vapor streams from total reboilers
   (*c*) Prefer the direct sequence
   (*d*) Prefer to reduce the number of columns in a recycle loop
   (*e*) Lump pairs of components with relative volatilities less than 1.1 and remove these as a single product to be separated using another technology
2. **Heuristics for simple columns**
   (*a*) Remove the most plentiful component first
   (*b*) Remove the lightest component first
   (*c*) Make splits with the highest recoveries last
   (*d*) Make the most difficult splits last
   (*e*) Favor splits which give molar flows of distillate and bottoms with the smallest difference
   (*f*) Make the cheapest split next in selecting a sequence of columns

conflict with one another; e.g., it is not clear how to treat a difficult split involving the most plentiful component (Heuristics 2*a* vs. 2*d*). This is simply because neither intuition nor patterns in simulation are complete. We recommend a more quantitative approach if the time and the data are available.

Design and Cost Estimates

One approach for simple sequences is to design and optimize all of the sequences, then choose the lowest-cost alternative. However, this is quite a lot of work and is often unjustified for a conceptual design. A more rapid, approximate method for nonazeotropic mixtures involves the partial use of heuristics in combination with some design estimates to find the sequence(s) with a low total vapor rate. The logic is that the number of stages is not very sensitive to the arrangement of the columns (see Exercise 1) but that the vapor rate requirements do depend more strongly on the feed rate and compositions. This is a reasonable approach in situations where the following conditions hold:

- All columns use the same materials of construction.
- All columns can he heated or cooled with utilities of comparable unit cost.
- The number of stages in a column depends primarily on the key components; i.e., the mixtures are not very nonideal.
- The number of products is not too large, e.g., no more than four or five.

This approach gives roughly the same results as sequencing for minimum steam consumption, e.g., Morari and Faith (1980).

The simplicity of this approach can be seen from a comparison of the direct and indirect sequences for a ternary ideal mixture. This approach has been described in more detail by Malone et al (1985).

EXAMPLE 7.1. A Ternary Mixture. Suppose that we have a ternary mixture of pentane (A), hexane (B), and heptane (C). We take the relative volatilities as constant $\alpha_{AB} = 2.59$ and $\alpha_{BC} = 2.45$ (Table 2.5) so $\alpha_{AC} = 6.35$. The feed mole fractions are $z_{A,F}$, $z_{B,F}$, and $z_{C,F}$. For relatively high-purity splits, the product flows for each column and the feed flow to the second column can be estimated easily by assuming complete recoveries. Then, using the minimum reflux expressions in Table 4.1 for the first column in each sequence and Eq. 3.58 for the second column, it is straightforward to estimate the vapor rate for each column in each sequence. For a feed containing equal molar flows of all three components, these estimates are shown below.

	Direct sequence			Indirect sequence		
	Column 1	**Column 2**	**Total**	**Column 1**	**Column 2**	**Total**
Split	A/BC	B/C		AB/C	A/B	
F	1	0.667		1	0.667	
D	0.333	0.334	0.667	0.667	0.333	1.000
r_{min}	1.442	1.377		0.705	1.260	
V	0.909	0.886	1.795	1.231	0.836	2.068

The indirect sequence has a total vapor boilup requirement more than 25% greater than the direct sequence for this feed. On the other hand, if the feed contains a large amount of heptane, say 90%, and equal amounts of the lighter components, then the direct sequence has a vapor boilup approximately 17% larger than the indirect. If pentane is in large excess, say 90% of the feed with equal amounts of hexane and heptane, then the total vapor boilup for the indirect sequence is more than 100% larger than the direct. A simple spreadsheet can be prepared to make these estimates (see Exercise 2).

Using heuristics like those in Table 7.2 generates a sequence rapidly, but design estimates are normally more reliable. For instance, suppose we have a ternary mixture with volatilities $\alpha_{AB} = 2.59$ and $\alpha_{BC} = 1.2$ and a feed containing 90% component C. One heuristic from Table 7.2 suggests that the difficult B/C split should be last, i.e., the direct sequence, while another heuristic suggests that component C should be removed first by the AB/C split in the indirect sequence. It turns out that the vapor rate for the indirect sequence is more than 60% lower than for the direct sequence. Note that in this case, the difficult split should be done first!

A similar analysis for mixtures containing up to five components can also be done, though it is worth preparing a special-purpose program for this exercise. Glinos (1984) developed shortcut design methods and cost models for constant volatility mixtures, starting with the Fenske-Underwood-Gilliland design method (Glinos and Malone, 1984). This approach has been used for sequencing of simple columns (Malone et al., 1985) and also for more complicated "complex column" configurations (Glinos and Malone, 1988) discussed later. We illustrate the idea and the results for simple columns in two examples.

EXAMPLE 7.2. Alcohol Separation. A commonly studied example in distillation sequencing is a five-component mixture of alcohols: ethanol, *i*-propanol, *n*-propanol, *i*-butanol and *n*-butanol. [This example was originally suggested by King (1971) and was studied by many others subsequently, e.g., Andrecovich and Westerberg (1985*b*).] The feed composition typically examined is 25:25:35:10:15 on a molar basis. The relative costs and vapor rates for the simple sequences according to the method suggested by Glinos (1984) are given in Fig. 7.1. (The sequence numbers correspond to the entries in the third section of Table 7.1.)

There is a reasonable correlation between the relative cost and the vapor rate; the lowest-cost sequences have the lowest vapor rates and the highest-cost sequences have somewhat higher vapor rates. Thus, seeking one or two sequences with low total vapor rates, which is a lot less time-consuming than estimating all of the costs, would be a good strategy for selecting a sequence.

A remarkable feature of the last example is the fact that there are no striking differences in cost or vapor rates among the sequences. In that sense, this is an unfortunate test problem for studies of sequencing heuristics! One might incorrectly conclude that the sequencing problem is not so important, but it turns out that this depends on the feed composition and the nature of the particular mixture. For example, mixtures with relatively large amounts of some components and small amounts of others are often encountered in chemical processing. This is typically because of the process chemistry and economics; e.g., it is never economical to produce large amounts of by-products, or large excess quantities of some component or solvent are

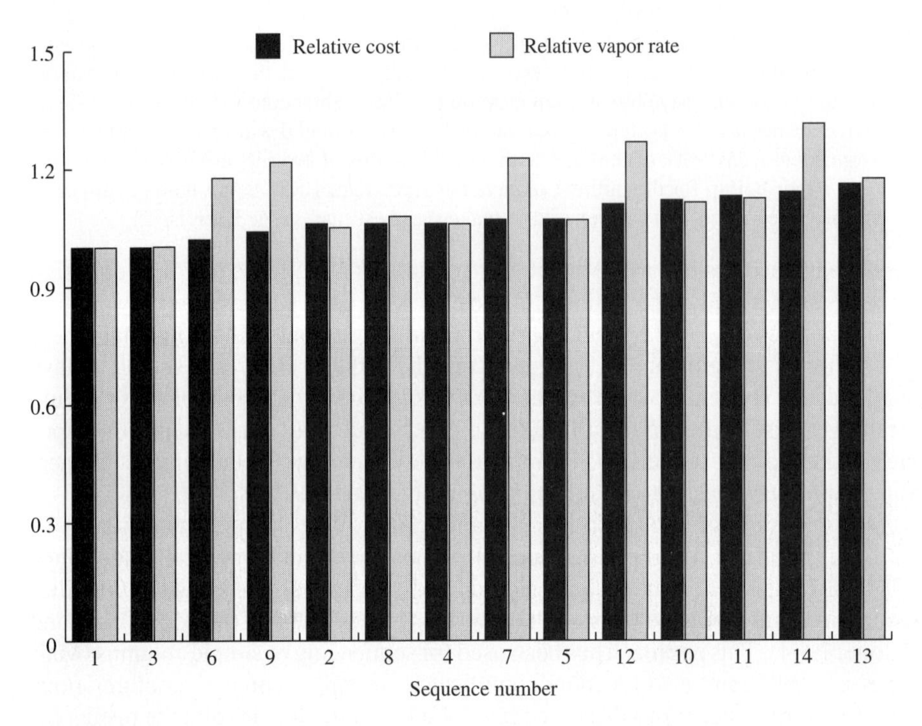

FIGURE 7.1
Comparison of simple sequences for a five-component mixture of alcohols. Costs and total vapor rates are reported relative to the values for sequence number 1.

required to improve selectivity or temperature control, etc. We examine this aspect in another example.

EXAMPLE 7.3. Butane Alkylation. We consider a process for the alkylation of butane to make *i*-octane (Douglas, 1988, pp. 213*ff*). The process chemistry includes many reactions, but the essential conceptual design must include at least the main step for product formation and a second reaction to capture selectivity losses to the formation of heavy components, e.g.,

$$\underset{\text{(1-butene)}}{C_4H_8} + \underset{(i\text{-butane})}{C_4H_{10}} \longrightarrow \underset{(i\text{-octane})}{C_8H_{18}}$$

$$\underset{(i\text{-octane})}{C_8H_{18}} + \underset{\text{(1-butene)}}{C_4H_8} \longrightarrow \underset{\text{(dodecane)}}{C_{12}H_{26}}$$

There will also be a small amount of propane and some *n*-butane to be processed. The normal boiling points and a typical feed composition to the separation system are as follows:

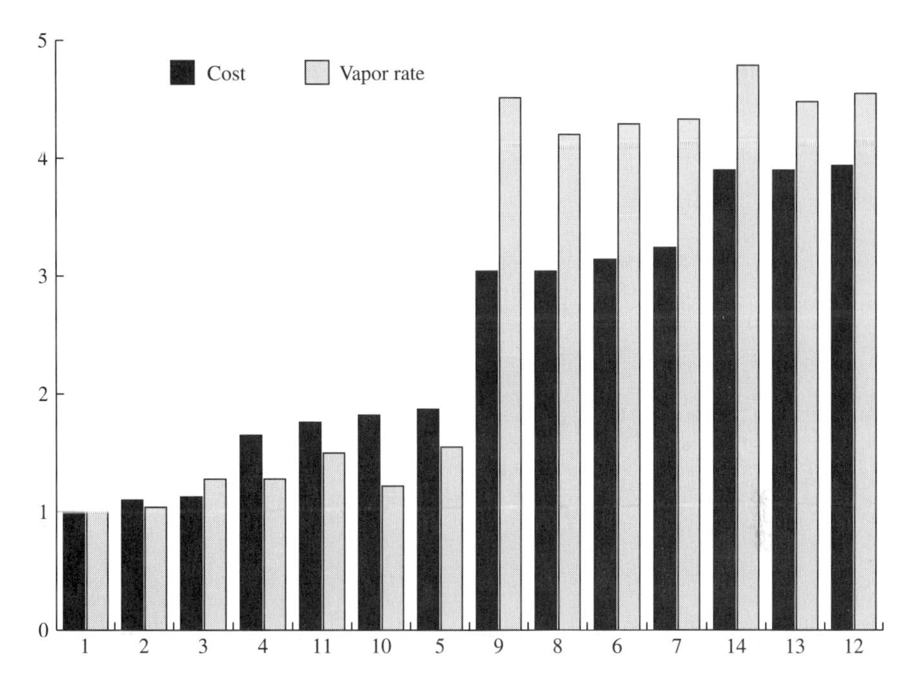

FIGURE 7.2
Comparison of simple sequences for butane alkylation. Costs and vapor rates are reported relative to the values for sequence number 1.

	Species	n.b.p. (°C)	Feed (mol%)	Destination
A	Propane	−42.1	5.0	Purge
B	*i*-Butane	−11.7	69.0	Recycle
B	1-Butene	−6.25	1.0	Recycle
C	*n*-Butane	−0.50	10.0	Purge
D	*i*-Octane	99.4	10.0	Product
E	Dodecane	216.3	5.0	Fuel by-product

Since 1-butene and isobutane have adjacent boiling points and are both recycled to the reactor system, we will lump these into a single hypothetical component B. In view of the large excess of isobutane,[1] we take the properties of B as those of isobutane for the sequencing. Note that even though propane and n-butane are both purged, they must first be separated in order to isolate B.

The relative costs and vapor rates are shown in Fig. 7.2 for each sequence (the sequence numbers correspond to those in Table 7.1).

Unlike the alcohol example, there is a dramatic difference in the sequences, with nearly a fourfold difference in costs and an even larger range in the total vapor boilup. Once again, the total vapor boilup correlates reasonably well with the total annual cost of the sequences. Part of this large range is on account of the imbalance in feed compositions.

[1] The molar ratio is approximately 10:1 at the reactor inlet to promote selectivity.

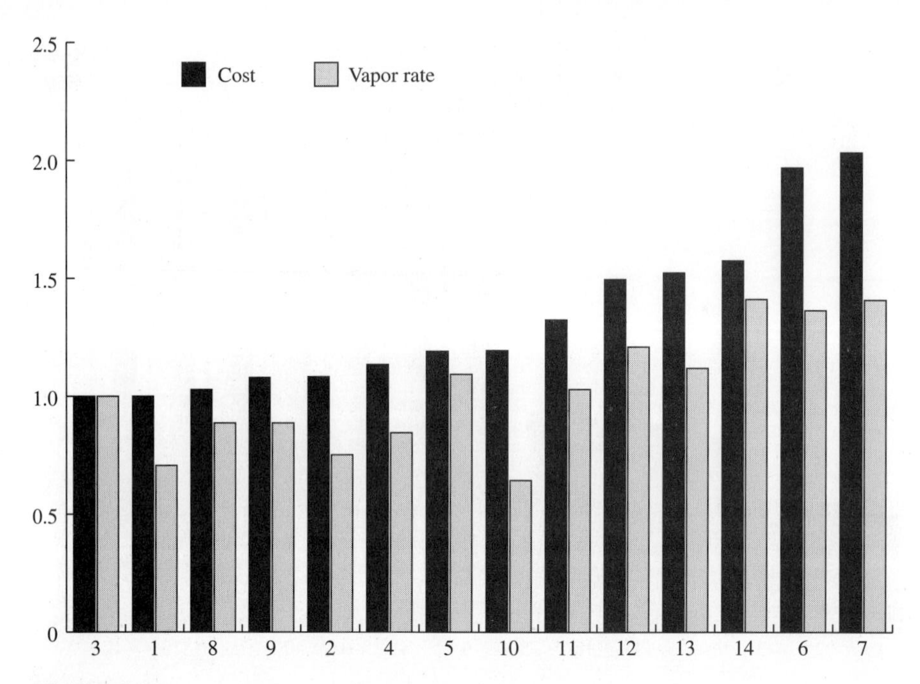

FIGURE 7.3
Comparison of simple sequences for butane alkylation with an equimolar feed. Costs and total vapor rates are reported relative to the values for sequence number 3.

Another factor is the large difference in boiling points among the components; for an equimolar feed the range is smaller, though still significant, as shown in Fig. 7.3.

The difference in boiling points accounts for roughly a factor of 2 in the total annual cost and a somewhat smaller difference in total vapor boilup. Sequences with small total vapor boilup are decent candidates, though there is not such a good correlation with cost.

7.3 COMPLEX COLUMN CONFIGURATIONS

There are many alternatives for the configuration of distillation units and systems besides the simple columns described above. Almost all of these seem more difficult to operate, but some provide significant savings in cost, e.g., substituting a single "complex" column for two simple columns.

A common "complex" column is the sidestream column shown in Fig. 7.4. The sidestream can be located above or below the feed; the basic idea is to have the sidestream enriched in the middle-boiling component. If the sidestream is above the feed, the major impurity is the lighter component (or components). The sidestream below the feed should be mostly the intermediate-boiling (saddle) along with heavier components. The sidestream can actually be made quite pure in the middle-boiler, but sometimes not without large increases in the energy input and the number of stages compared to a simple 2-column sequence. The design of the sidestream column is

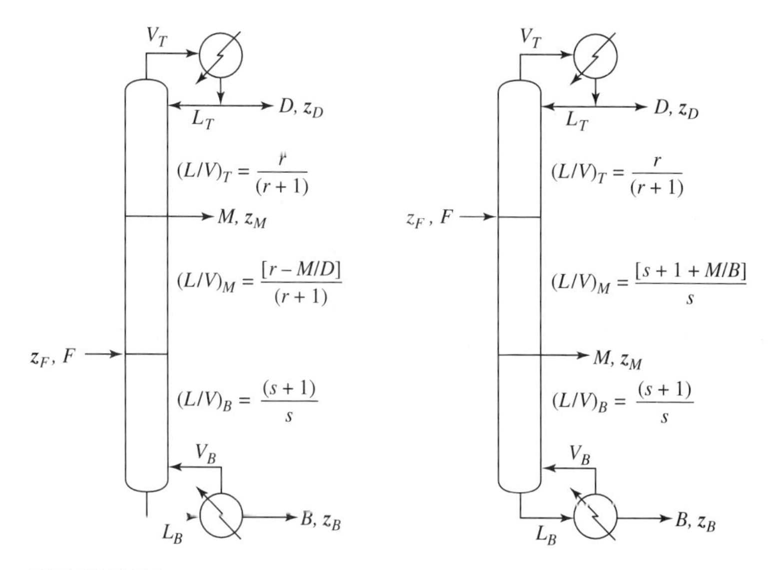

FIGURE 7.4
Sidestream columns with a side draw above the feed (left) or below the feed (right). The internal flows for the CMO case are also given.

typically sensitive to the sidestream composition. Thus, the sidestream column may or may not be a good alternative and the desired purity of the sidestream is important. For instance, sidestream products that are recycled may be more attractive than final products.

A few of the other alternative complex column configurations are shown schematically in Fig. 7.5.

Because there is usually some concern about the sensitivity of complex column configurations to disturbances, *it is a good strategy to search for the best simple sequence first and then determine the incentive to add complexity.* This gives an estimate for the incentive to study the sensitivity, operability, and control in more detail. There are several strategies for developing complex column alternatives such as the heuristics in Table 7.3. Another approach is to begin with a simple sequence and combine pairs of adjacent columns using alternatives from Fig. 7.5. Design methods similar to the one developed for sidestream columns can also be developed for these other complex columns, e.g., Glinos and Malone (1988). Using those methods, we revisit the butane alkylation example.

EXAMPLE 7.4. **Complex Columns for Butane Alkylation.** If we take each of the simple sequences already generated for the butane alkylation example and consider combining each of the 3 pairs of adjacent columns, using the 11 complex columns shown in Fig. 7.5, there are $11 \times 3 \times 14 = 462$ alternatives. Of course, there are many more, since further combinations and other complex columns could also be examined.[2]

[2]Computer-aided conceptual design tools are very useful in this sort of exercise, though no commercial tools yet have this functionality as far as we know. We discuss a more systematic approach to generate complex column alternatives in the next section.

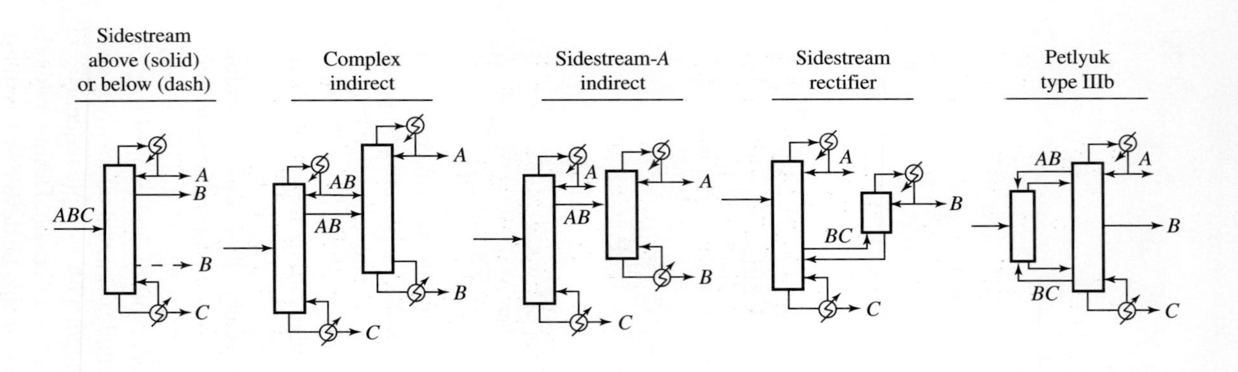

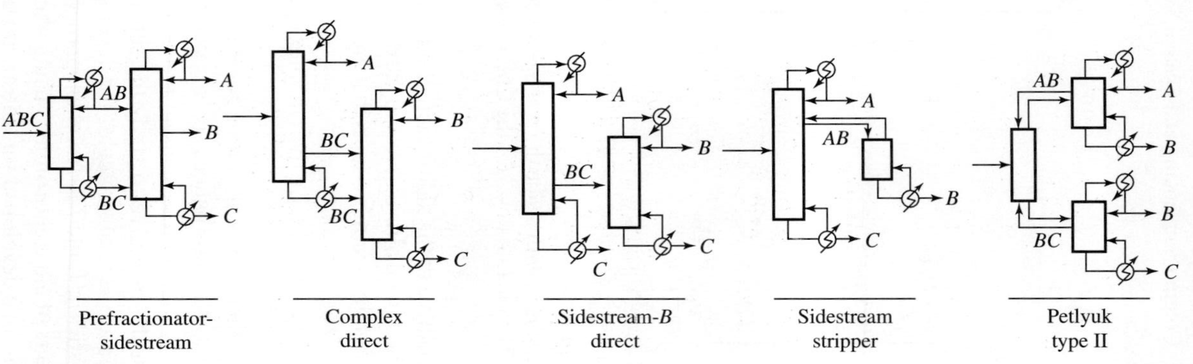

FIGURE 7.5
Selected complex column configurations for the distillation of ternary mixtures. The feed components are A, B, and C in order of increasing boiling point (decreasing volatility).

TABLE 7.3
Selected heuristics for complex columns in distillation sequencing. The first three are from Tedder and Rudd (1978)

1. Use a sidestream column if the mole fraction of the middle-boiler is large
2. Use a sidestream stripper or sidestream rectifier when the mole fraction of the middle-boiler is small and the relative volatility of the lightest to middle-boiling component is more than 60% larger than the relative volatility of the lightest component to the heaviest component
3. Use a Petlyuk configuration when the amount of the middle-boiler is large but at least 20% of the feed is light plus heavy component
4. Consider a sidestream column when the sidestream can be recycled

After searching through each of the simple sequences and combining adjacent columns one pair at a time, the most interesting result is found beginning from simple sequence number 5. When columns 2 and 3 in that sequence are combined using complex column alternatives, we find the following results, depending on the arrangement chosen from Fig. 7.5.

Configuration	Sidestream purity of D	Cost relative to simple sequence 1
Sidestream above feed	0.36	1.366
Sidestream below feed	0.94	1.441
Prefractionator-sidestream	0.99	1.560
Complex indirect		1.813
Complex direct		0.702
Sidestream A indirect		1.416
Sidestream B direct		1.502
Sidestream stripper		1.516
Sidestream rectifier		1.542
Petlyuk type IIIb	0.92	0.946
Petlyuk type II		0.934

Most of the sequences with complex columns are significantly *more* expensive in this case. The two Petlyuk columns are comparable in cost, though this does not include any increased cost for control, so these do not appear promising. The single alternative that appears most promising uses the "complex direct" sequence and offers a potential cost reduction of about 30% below the cost of simple sequence 1 (and nearly 60% lower than simple sequence 5). The configuration is shown in Fig. 7.6.

Although this configuration does not reduce the number of columns, it does accomplish some of the D/E split in the lower part of the second column. The heating for this column can be done with a less expensive (lower-temperature) utility than the multiple-feed column where the final D/E split is done.

It is interesting that the costs of the complex column alternatives listed above are *all* lower than simple sequence number 5, which was used to generate them. The alternative shown in Fig. 7.6 is better than complex alternatives generated from the best simple sequence, number 1. It is tempting, but not advisable, to first find that solution and discard all the other simple sequences.

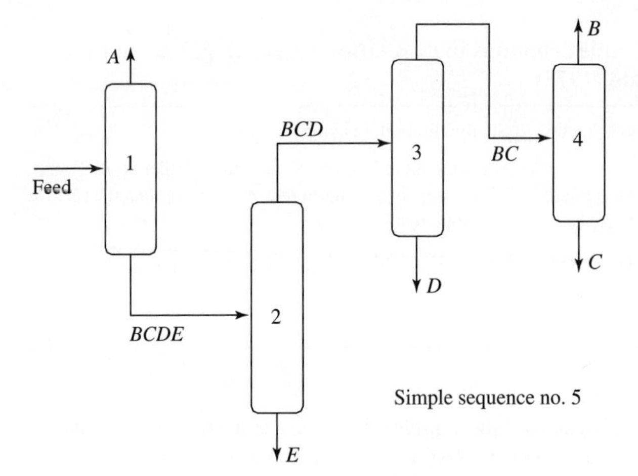

FIGURE 7.6
Column sequences for butane alkylation. Simple sequence number 5 is not the most economical. However, a complex sequence generated from it by replacing simple columns 2 and 3 with the complex direct arrangement is significantly less expensive than the best simple sequence.

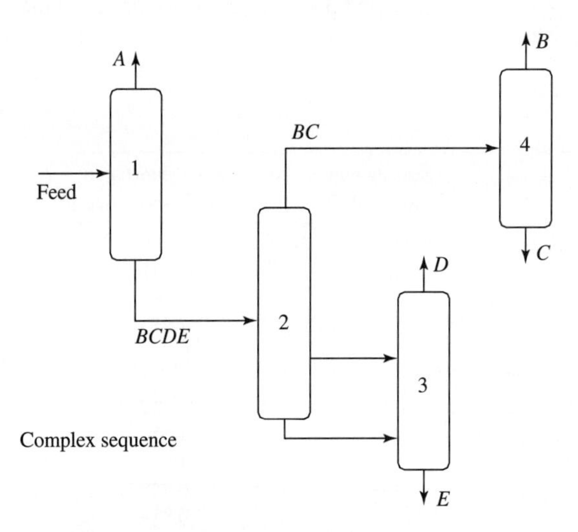

7.4 STATE -TASK NETWORK REPRESENTATION

One of the chief difficulties with the design of complex columns is identifying all of the possible sequences and keeping track of them in a systematic manner. The simplest known way of doing this is to use the "state-task network" representation (STN), as introduced by Sargent and co-workers; e.g., Sargent (1998) gives a general description of this idea with specific application to distillation problems. The STN representation gives explicit recognition to the fact that the feed(s) to a process typically undergo a series of transformations, creating a number of intermediate *states*, and that a *task* may be defined as a device for transforming material from one set of states to another. The state-task network is represented as a set of nodes (which represent the states, including the feeds, products, and intermediate states) which

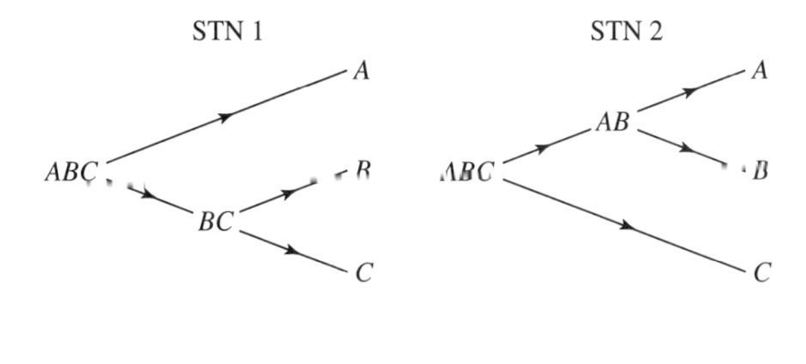

FIGURE 7.7
State-task-network representations for the separation of a three-component mixture. Arrows inclined upward represent a net upward flow in the column section while arrows inclined downward represent a net downward flow in the section.

are interconnected by lines that represent the tasks needed to transform one state into another. In distillation the basic module chosen for the task is a column *section,* which is defined as a countercurrent sequence of stages (or packing) with liquid feed at the top and vapor feed at the bottom which is not interrupted by entering or exiting streams, and with well-defined rules for stream mixing (i.e., mixing of states) and the inclusion of heat transfer devices. The net flow through a column section may be up or down, depending on the relative flowrates of the passing liquid and vapor streams[3] (and by mass balance is constant at every level in the section). We define the function of a column section to be the effective elimination of the least volatile component from the net flow if this flow is upward, or the effective elimination of the most volatile component from the net flow if this flow is downward. We restrict attention to mixtures that obey ideal or near ideal VLE and base the discussion largely on Sargent (1998) and Agrawal (1996).

Using this module alone, without allowing mixing of states, yields mixtures with consecutive volatilities as possible intermediate states. For a three-component mixture, e.g., we can generate a stream containing only $A+B$ using a column section with a net upward flow, or $B+C$ using a column section with a net downward flow, etc. STN representations to separate a ternary mixture into its pure components are shown in Fig. 7.7. In any given network, the feed mixture is represented by a root node on the far left-hand side of the diagram, and the desired products are represented by terminal nodes on the far right-hand side of the diagram. The line connecting a node with a successive node represents a section of a distillation column, and the arrows indicate the direction of net flows through the network. Both networks in Fig. 7.7 have four lines connecting different nodes, and, therefore, in each of the sequences that can be generated from these networks four column sections are needed to achieve the overall separation. *In general,* $2(c-1)$ *column sections are needed to separate a c component ideal mixture into its constituent pure components.*

Now that we have identified the number of sections needed, the next steps are:

Step 1. Decide the number of heat transfer devices (reboilers and condensers), and assign them to particular column sections, and

Step 2. Group the sections together into columns so that each section has a source of vapor boilup (either from the reboiler in that section or from a vapor stream

[3]Clearly, the net flow is always upward in a rectifying section and downward in a stripping section.

leaving the top of another section) and a source of liquid reflux (either from the condenser in that section or from the liquid stream leaving the bottom of another section). Note that there is often more than one way of grouping the sections together, so that the same STN represents multiple distillation sequences.

This determines the number of distillation columns needed as well as their material interconnections; i.e., this determines the sequence. Alternative sequences are generated systematically by applying the rules assigned to step 1. It is possible to limit the set of alternative sequences by assigning simple rules to this step; by assigning more complex rules we can systematically generate more complex sequences. The simplest approach is:

Rule 1 (Simple Sequences). Assign one reboiler to each column section with a net downward flow and one condenser to each column section with a net upward flow.

In this case there is only one way of grouping sections together, which is to group the two column sections leading out of the same node to form a single distillation column. Just to reinforce this we have drawn in Fig. 7.8 (*a*) STN 1 with the heat transfer devices shown on the sections, (*b*) each individual section for STN 1 with each stream labeled with its components, and (*c*) the resulting simple sequence of distillation columns. This leads to a total of $2(c - 1)$ heat transfer devices and $(c - 1)$ simple distillation columns (i.e., each column has one feed, two products, one reboiler, and one condenser). With these rules the two state-task networks shown in Fig. 7.7 correspond to the two simple sequences given in the top block of Table 7.1:

Network representation	Sequence in Table 7.1
STN 1	1 (direct sequence)
STN 2	2 (indirect sequence)

The simplest complex column configurations (e.g., a side rectifier and/or a side stripper) are generated by recognizing that the vapor and liquid flows from one column section can be shared with another section in a different column. For example, the vapor generated by a reboiler can be shared by a stripping section and a side rectifier placed above the stripping section. Reflux in the side rectifier is provided by a condenser, and the liquid leaving the bottom of the side rectifier is returned to the stripping section of the original column (see Fig. 7.9). This eliminates one reboiler from the sequence.[4] In a similar way, condensers can be eliminated by using side strippers. This implies that for any two sections associated with a component

[4]The capital and operating cost savings for eliminating this reboiler must be traded off against the increased capital and operating cost of the reboiler in the main column. Moreover, the incremental vapor generated in this reboiler for the side rectifier is generated at a higher temperature than would otherwise be necessary, which may require a more expensive utility.

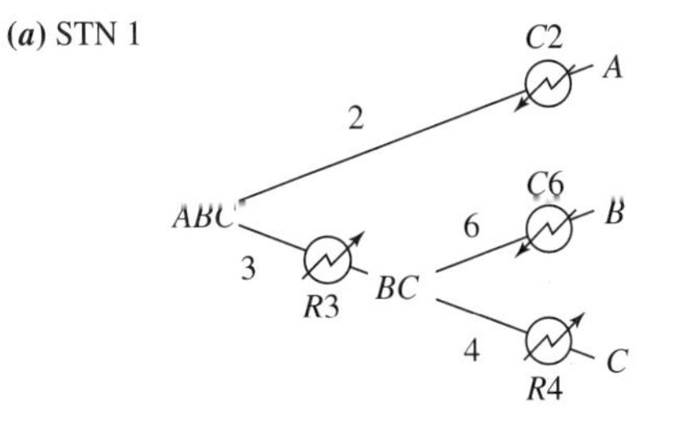

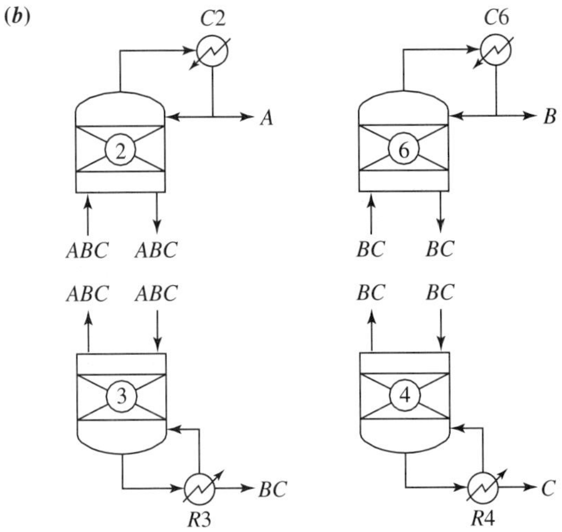

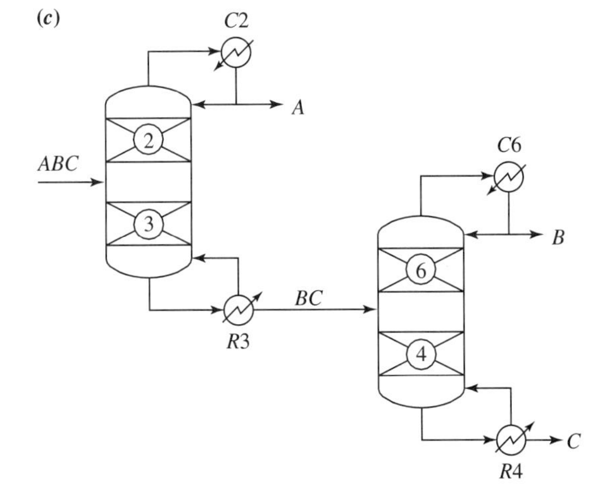

FIGURE 7.8
Separation of a three-component mixture using the minimum number of column sections (4) according to STN 1: (*a*) STN 1 with the heat transfer devices shown on the sections, (*b*) each individual section with each stream labeled with its components, (*c*) simple sequence of distillation columns.

of intermediate volatility only one reboiler or one condenser is needed. A condenser is always needed for the lightest component, and a reboiler is always needed for the heaviest. Therefore, *the minimum number of heat exchangers needed for a distillation sequence which uses the minimum number of column sections is equal to the number of components in the feed mixture.* This leads to:

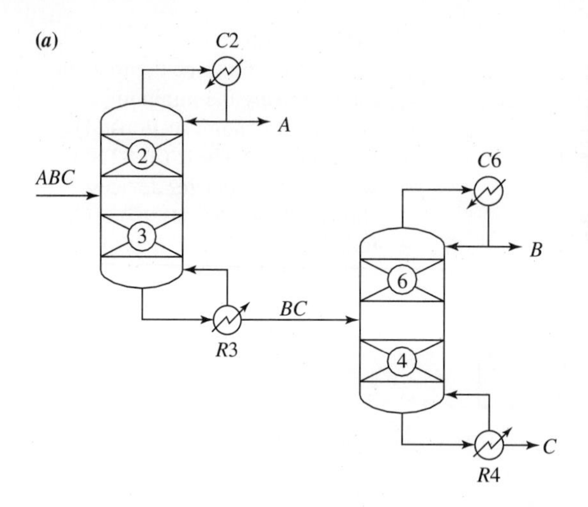

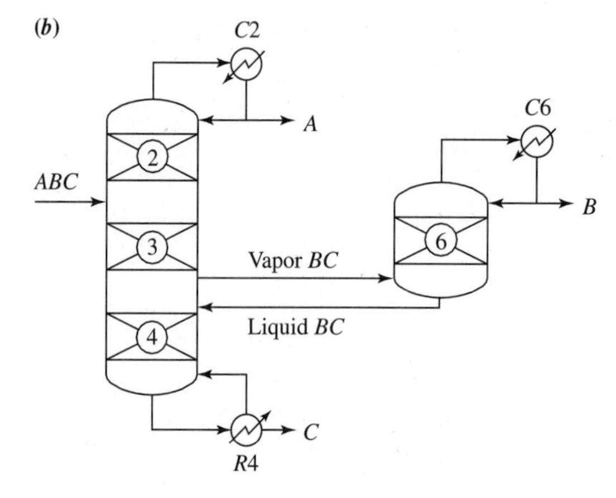

FIGURE 7.9
Separation of a three-component mixture using the minimum number of column sections (4) according to STN 1: (*a*) simple sequence generated by Rule 1, (*b*) complex sequence generated by Rule 2 using the minimum number of exchangers allowed for a network with four sections.

Rule 2 (Partially Coupled Sequences). Form the parent STN by assigning one reboiler to each column section with a net downward flow and one condenser to each column section with a net upward flow. Define an *intermediate section* as one that connects two nodes that each contain a mixture of components with a component of intermediate volatility in common (e.g., STN 1 in Fig. 7.7 has one intermediate section connecting node ABC to node BC). Systematically eliminate the reboilers from intermediate column sections with net downward flows and the condensers from intermediate column sections with net upward flows. Do this one at a time, then two at a time, three at a time, etc., in all possible combinations. Repeat this procedure for each STN. Each intermediate column section with a missing reboiler must have vapor provided by a lower section (in the STN), and each intermediate column section with a missing condenser must have liquid provided by an upper section.

Applying Rule 2 to the STNs in Fig. 7.7 gives the STNs and their corresponding sequences shown in Fig. 7.10. A column sequence for STN 1a is derived as follows.

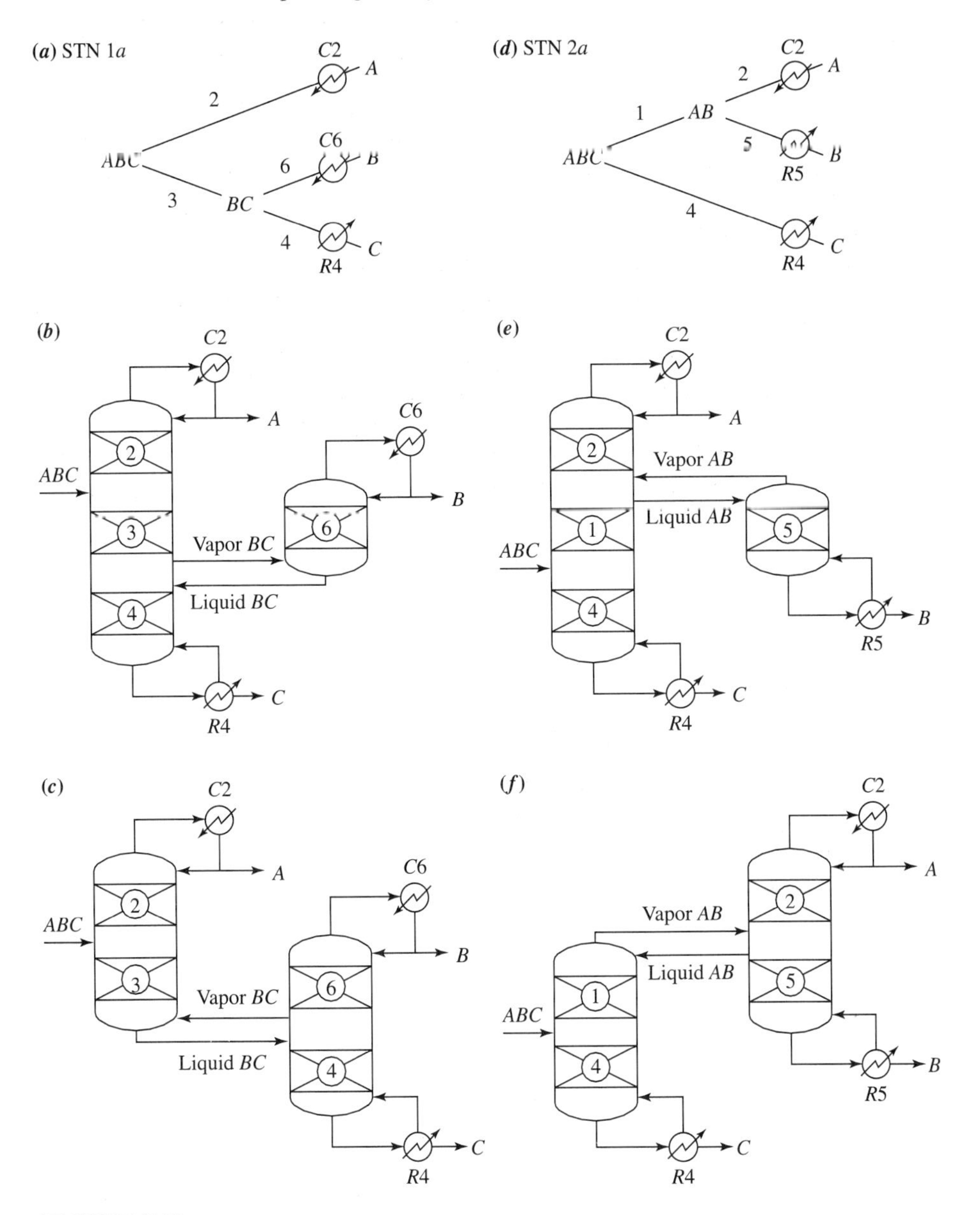

FIGURE 7.10
Partially coupled complex sequences generated by Rule 2 for the separation of a ternary mixture using the minimum number of column sections (4).

Section 3 has a net downward flow but has no reboiler to generate vapor. Vapor must be generated for this section by reboiler $R4$; therefore, sections 3 and 4 are grouped together in the same column. Section 2 generates liquid reflux via condenser $C2$, and this provides liquid flow for sections 2, 3, and 4, which are all grouped together in the same column. This arrangement also provides a vapor stream for section 2.

Section 6 has a net upward flow and generates liquid reflux via condenser $C6$. Vapor for this section must be provided by reboiler $R4$, and therefore, section 6 is a side rectifier. The resulting sequence of columns is shown in Fig. 7.10*b*. Closer inspection of STN 1*a*, however, reveals an alternative sequence. There is no necessity to group sections 3 and 4 together in the same column, the only requirement is that section 3 must get vapor from section 4. Therefore, we can group sections 2 and 3 together in the same column, and sections 6 and 4 together in the same simple column. In this embodiment section 3 is connected to section 4 via liquid and vapor sidestreams, and the sequence is shown in Fig. 7.10*c*. The material interconnections of the sections in these two alternative sequences are identical (because they share the same STN), but the sequences are different. We can expect the sequences to have different cost and operability characteristics.

The column sequences for STN 2*a* are derived along similar lines. Therefore, Rule 1 provides one simple sequence for STN 1, and Rule 2 provides two complex alternatives. The same is true for STN 2. Therefore, there are two simple sequences for separating a ternary mixture and an additional four complex sequences at the Rule 2 level.[5]

Now that we have determined the minimum number of column sections and the associated minimum number of reboilers and condensers, let us attempt to reduce the number of heat transfer devices even further by eliminating reboilers and condensers associated with components of intermediate volatility, e.g., condenser $C6$ from STN 1*a* and reboiler $R5$ from STN 2*a*. This is the point where the state-task network representation becomes an indispensable tool for deriving what must happen in order to successfully eliminate these heat transfer devices. Figure 7.11*a* shows STN 1*a* with condenser $C6$ eliminated. This is an infeasible arrangement because there is no means of providing liquid reflux to section 6. In order to provide this reflux, we must add a section to the network (directly above section 6) with a net downward flow. The modified network is shown in Fig. 7.11*b*. However, this network is also infeasible because the added section which terminates at B (section 5) has no origin to start from. The only way that we can provide this origin is to *add a node*; i.e., we *must add an additional split to the network* that was not previously required in order to accomplish our goal of eliminating condenser $C6$. We add the AB node to the network, which introduces yet another new section (section 1), and the resulting feasible fully coupled STN is shown in Fig. 7.11*c*. Therefore, the rule for generating fully coupled sequences is:

> *Rule 3 (Fully Coupled Sequences).* Start with a partially coupled network generated by Rule 2. Systematically eliminate the reboilers and condensers from column sections associated with pure component nodes of intermediate volatility. For each eliminated heat transfer device add one new node and its corresponding two new sections to the network. Eliminate the heat transfer devices one at a time, then two at a time, three at a time, etc., in all possible combinations until the fully coupled network is obtained. Repeat this procedure for each STN.

---

[5]Sidestream columns are special cases of the sidestream stripper or rectifier, when the desired purity for the intermediate component can be met without any stages and without reflux or reboil in the sidestream rectifier or stripper, respectively.

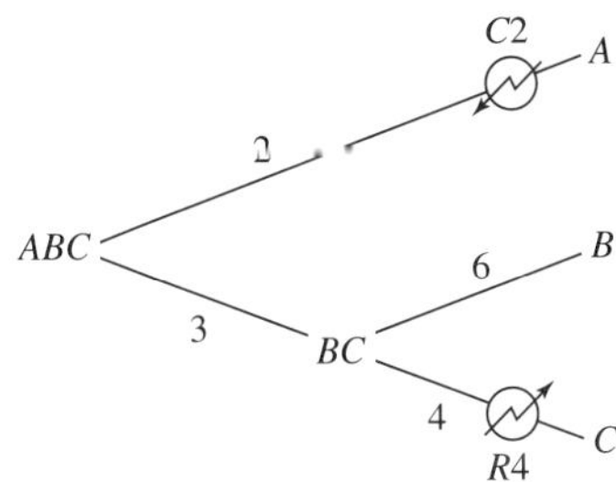

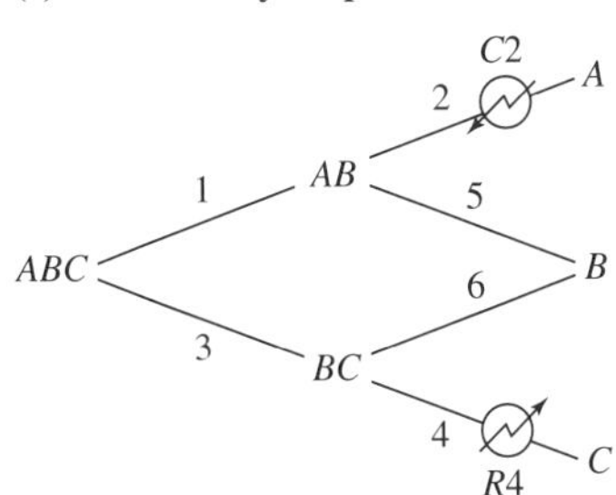

(c) Feasible fully coupled STN

C2
A
2
1
AB
5
ABC
B
6
3
BC
4
C
R4

FIGURE 7.11
Fully coupled STN for separating a ternary mixture.

Applying this rule to the two partially coupled STNs shown in Fig. 7.10 leads to the same fully coupled STN, shown in Fig. 7.11*c*. This STN has six column sections, one reboiler, and one condenser. Inspection of this network leads to four alternative fully coupled sequences:

1. Group sections 1 and 3 in the first column, and sections 2, 5, 6, and 4 in the second; shown in Fig. 7.12*a*
2. Group sections 1, 3, and 4 in the first column, and sections 2, 5, and 6 in the second; shown in Fig. 7.12*b*
3. Group sections 2, 1, and 3 in the first column, and sections 5, 6, and 4 in the second; shown in Fig. 7.12*c*
4. Group sections 2, 1, 3, and 4 in the first column, and sections 5 and 6 in the second; shown in Fig. 7.12*d*

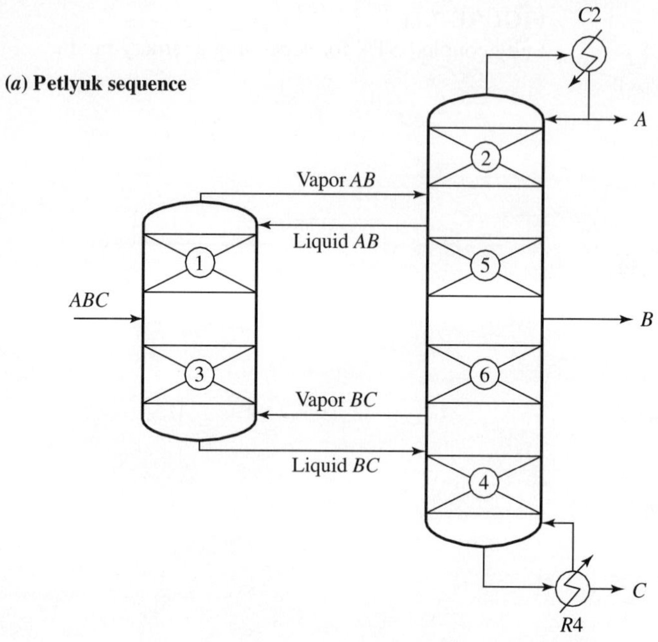

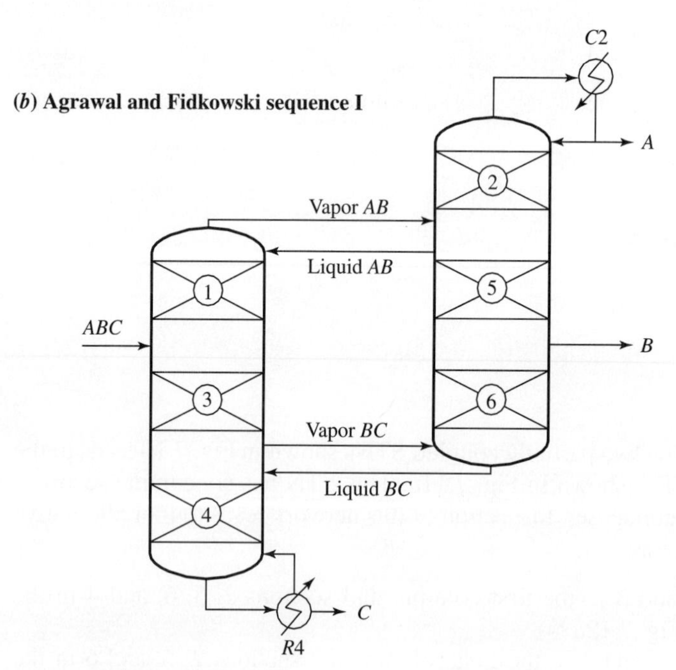

FIGURE 7.12
The four alternative fully coupled column sequences corresponding to the STN in Fig. 7.11*c* for the separation of a ternary mixture.

(*c*) Agrawal and Fidkowski sequence II

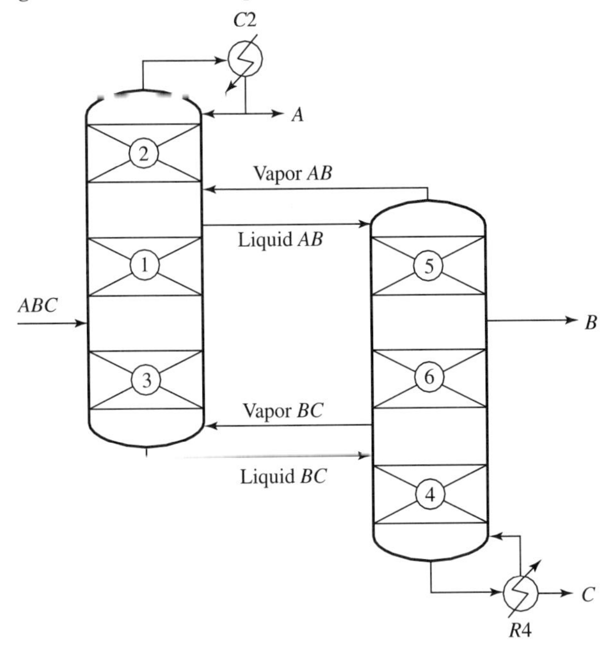

(*d*)

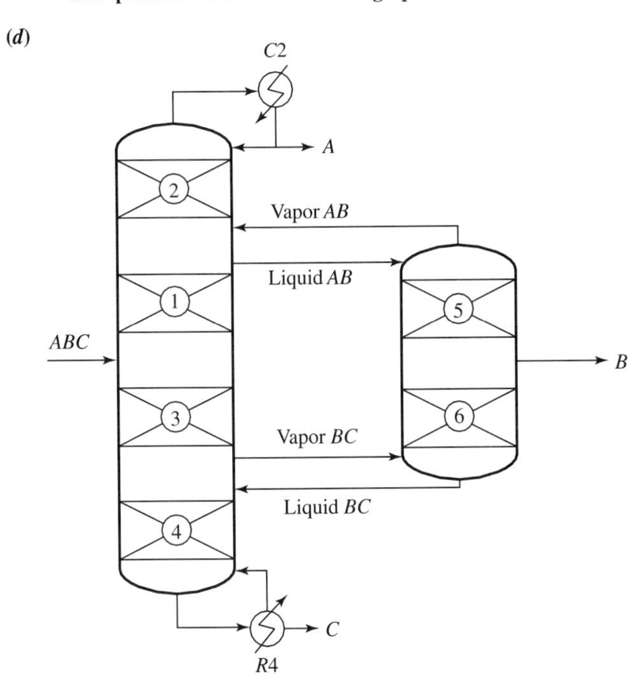

FIGURE 7.12 (*Continued*)

The column arrangement shown in Fig. 7.12*a* is a classical Petlyuk sequence (Petlyuk et al., 1965) that has been studied extensively in the literature.[6] The main advantage of this sequence is that it has the lowest vapor flow for a given separation of all the possible column sequences that distill an ideal ternary mixture into pure product streams,[7] as proved by Fidkowski and Krolikowski (1987). Its disadvantage is that vapor must be transported back and forth between the columns, which normally cannot be done without the use of a compressor,[8] and this destroys the advantage of the sequence. The column arrangements in Fig. 7.12*b* and *c* were recently discovered by Agrawal and Fidkowski (1998*b*) and have the major advantage of a *unidirectional flow of vapor*. This means that the column providing the source of vapor can be pressurized slightly so that the lowest pressure in this column is slightly higher than the highest pressure in the low-pressure column. This removes the operability disadvantage of the Petlyuk sequence. The sequence shown in Fig. 7.12*d* has not been reported before in the literature but suffers from the same disadvantage as the Petlyuk sequence.

The focus so far has been on devising sequences from the simple to the complex using the minimum number of column sections and the associated minimum number of heat exchangers. It is possible, however, to devise sequences with more sections and/or more heat exchangers; see, for example, the Petlyuk type II sequence in Fig. 7.5 which has six column sections and four heat exchangers. We will not find these sequences from the STNs developed so far. *We can find them, however, together with all the sequences so far developed by drawing the maximally interconnected STN containing all possible nodes (and therefore all possible splits and all possible column sections) and by putting a heat transfer device on every column section.* The resulting STN is shown in Fig. 7.13*a*. It has six column sections and six heat exchangers. We can devise all possible sequences[9] by applying:

---

[6]One embodiment of this sequence in a single column shell, the so-called *dividing wall column*, has been known since at least 1949 (Wright, 1949). Further analysis of the dividing wall arrangement is given by Kabel (1987), Christiansen et al. (1997), Mutalib and Smith (1998), and Mutalib et al. (1998).

[7]Of course, the other sequences shown in Fig. 7.12 have an identical vapor flow because they all share the same STN. A systematic study of ternary distillations (Agrawal and Fidkowski, 1998*a*) showed that the total vapor flow in the fully coupled sequence is lower than that of conventional direct and indirect sequences by 10 to 50%. Contrary to conventional wisdom, however, the thermodynamic efficiency of the fully coupled sequence is often worse than in other configurations, since all the heat has to be supplied at the highest temperature (boiling point of the least volatile product C) and rejected at the lowest temperature (boiling point of component A). A very thorough treatment of this topic is given by (Agrawal and Fidkowski, 1998*a*) together with a worked example from the cryogenic air separation industry that shows that in the production of argon from air, the thermodynamic efficiency of the fully coupled configuration is 14.9% while that of a conventional configuration is 44.5%.

[8]The compressor is not needed if the pressure at the top of section 4 is greater than the pressure at the bottom of section 3, which in turn is greater than the pressure at the top of section 1, and this must be greater than the pressure at the bottom of section 2, i.e.,

$$P_4^{\text{top}} > P_3^{\text{bottom}} > P_1^{\text{top}} > P_2^{\text{bottom}} \tag{7.1}$$

Therefore, the pressure in the prefractionator is neither uniformly higher nor uniformly lower than the pressure in the product column. In order for these inequalities to be satisfied, the pressure drop in sections 5+6 must be *higher* than the pressure drop in sections 1+3 for all operating conditions.

[9]Within the design space allowed by the constraint imposed earlier on the function of a column section.

(*a*)

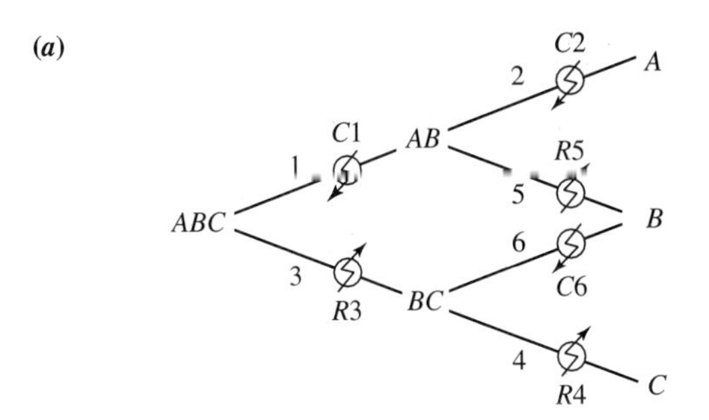

(*b*)

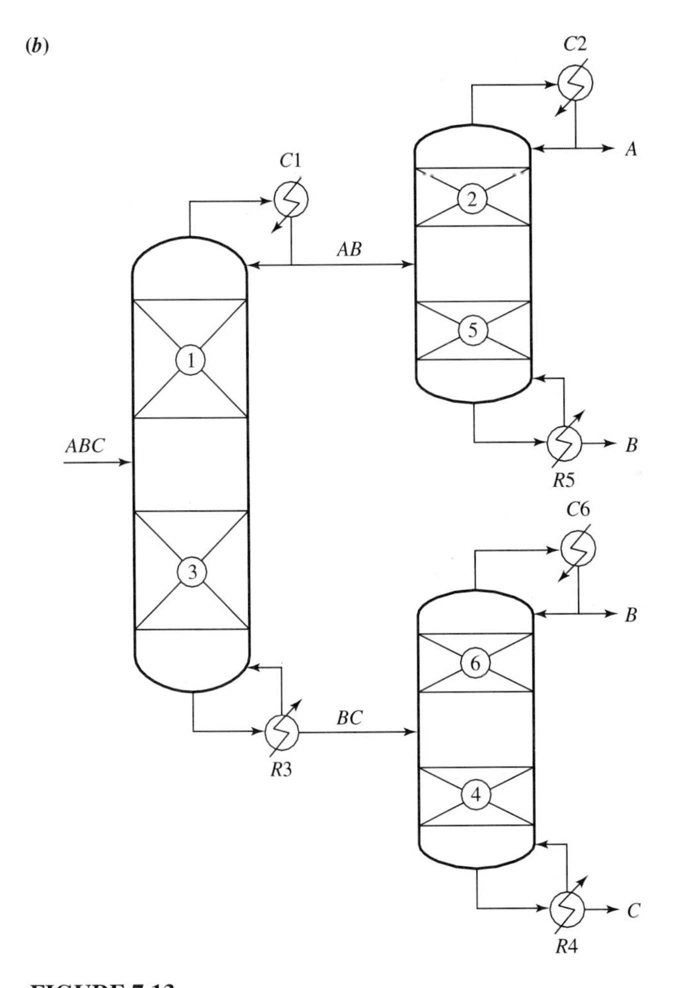

FIGURE 7.13
Maximally interconnected STN for the separation of a ternary mixture.

Rule 4

Step 1. Start with the maximally interconnected STN, including a heat exchanger for each task, and draw the column sequence.

Step 2. Successively eliminate heat exchangers (except $C2$ and $R4$) to get all possible coupled systems at this level of nodes.

Step 3. Successively eliminate nodes and repeat step 2.

The column sequence that corresponds to step 1 for a ternary mixture is given in Fig. 7.13*b*. We have not seen this sequence before in this chapter, but we can anticipate that it will be an attractive alternative when the AB/BC transition split has a very low value for the minimum reflux ratio (remember that the transition split always has the lowest possible minimum reflux ratio for an A from C split in the presence of B). The other possible sequences are explored in Exercise 5.

As the number of components in the mixture increases, the number of alternative column sequences expands rapidly. New structural features start to appear in the STNs, and many different classes of fully coupled sequences are possible. The new features start to appear for four-component mixtures, which we briefly survey.

The state-task network (STN) representations to separate a four-component mixture into its pure components using simple columns are shown in Fig. 7.14. All of the networks in Fig. 7.14 have six lines connecting different nodes, and therefore, in each of the sequences that can be generated from these networks six column sections and six heat exchangers are needed to achieve the overall separation. This agrees with the general rule that for simple column sequences, $2(c-1)$ column sections, $2(c-1)$ heat exchangers, and $c-1$ simple columns are needed to separate a c component ideal mixture into its constituent pure components. The five state-task networks shown in Fig. 7.14 correspond to the following simple sequences given in the middle block of Table 7.1:

Network representation	Sequence in Table 7.1
STN 1	1 (direct sequence)
STN 2	2
STN 3	3
STN 4	4
STN 5	5 (indirect sequence)

Applying Rule 2 to STN 2 in Fig. 7.14 gives the STNs and their corresponding sequences shown in Fig. 7.15. This figure shows the parent STN with all its reboilers and condensers marked on the column sections, together with the three complex alternatives derived by eliminating one intermediate reboiler (STN 2*a*), one intermediate condenser (STN 2*b*), and both at once (STN 2*c*). A column sequence for STN 2*a* is derived as follows. Section 2 has a net downward flow but has no reboiler to generate vapor. Vapor must be generated for this section by reboiler $R6$; therefore, sections 4 and 6 are grouped together in the same column. Section 3 generates liquid reflux via condenser $C3$, and this provides liquid flow for sections 3, 4, and 6, which are all grouped together in the same column. This arrangement also provides a vapor stream for section 3. Section 8 has a net upward flow and generates liquid

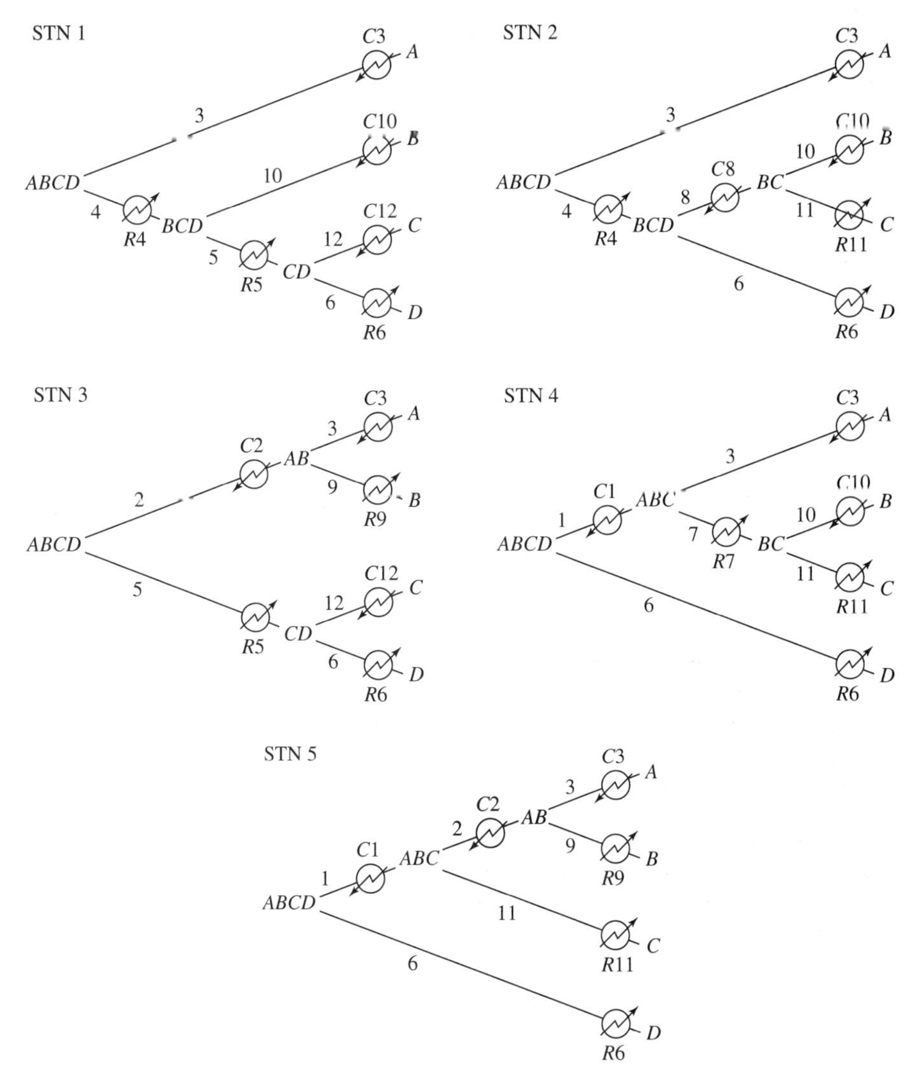

FIGURE 7.14
State-task-network representations for the separation of a four-component mixture.

reflux via condenser $C8$. Vapor for this section must be provided by reboiler $R6$, and therefore, section 8 is a side rectifier. The BC product from section 8 is split in a simple column consisting of sections 10 and 11 together with their associated heat transfer devices $C10$ and $R11$. The resulting sequence of columns is shown in Fig. 7.15iv. There is, however, an alternative sequence. There is no necessity to group sections 4 and 6 together in the same column; the only requirement is that section 4 must get vapor from section 6. Therefore, we can group sections 3 and 4 together in the same column, sections 8 and 6 together in the same simple column, and sections 10 and 11 together in the same simple column. Section 4 is connected to section

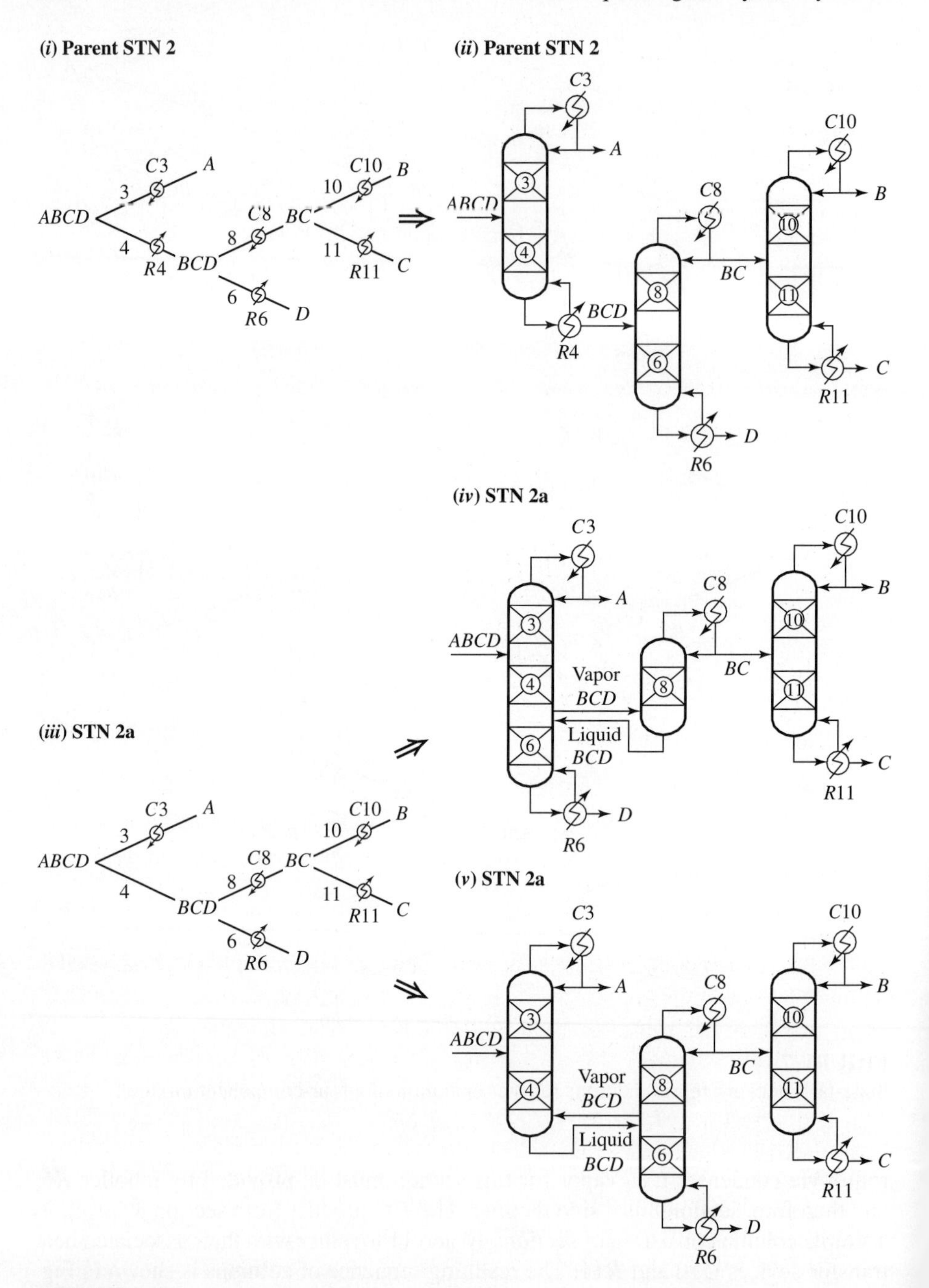

FIGURE 7.15
Partially coupled sequences generated by Rule 2 for the separation of a four-component mixture using the minimum number of column sections (6) according to STN 2.

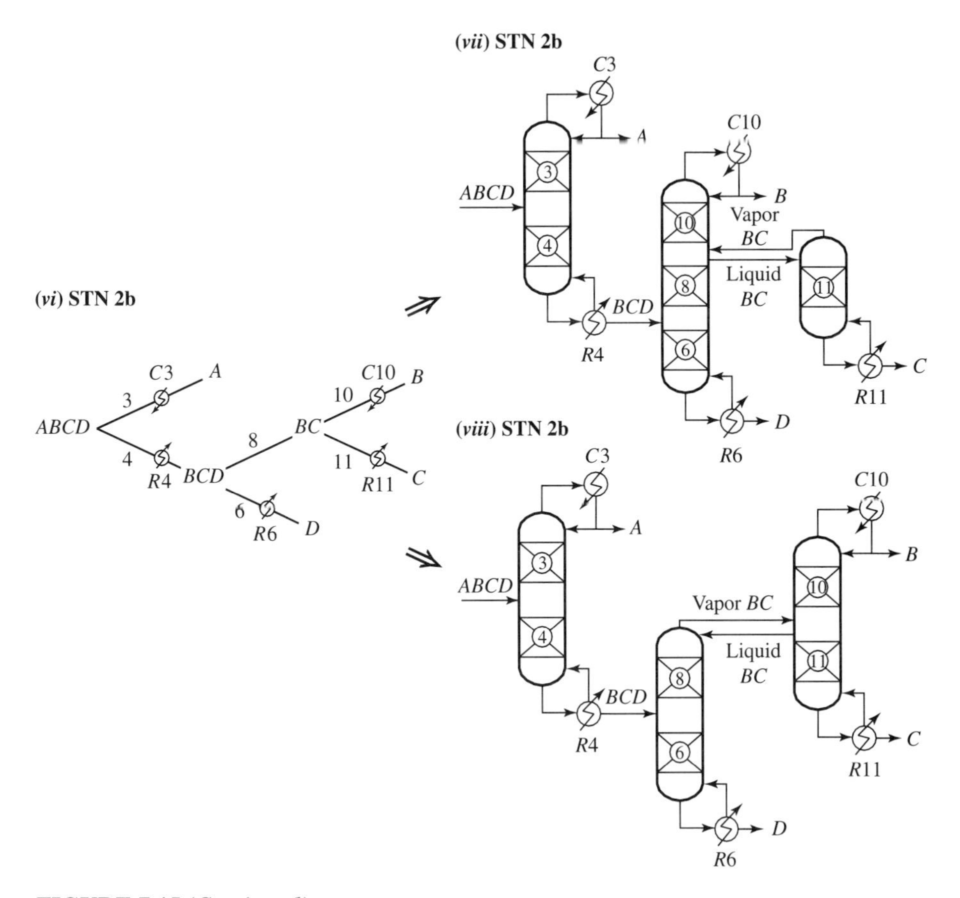

FIGURE 7.15 (*Continued*)

6 via liquid and vapor sidestreams, and the sequence is shown in Fig. 7.15v. The interconnections of the sections in these two alternative sequences are identical, but the sequences will have different cost and/or operability characteristics.

The column sequences for STN 2*b* and 2*c* are derived along similar lines. There are two sequences for STN 2*b* and three for STN 2*c*, as shown in Fig. 7.15. Therefore, Rule 1 provides one simple sequence for STN 2, and Rule 2 provides seven partially coupled complex alternatives. Rules 3 and 4 generate the corresponding fully coupled sequences.

Applying Rule 3 to the partially coupled STNs leads to the three different fully coupled networks shown in Fig. 7.16. STN 1 and STN 2 both lead to the fully coupled network STN-*FC*1 in Fig. 7.16*a*; STN 3 leads to STN-*FC*2 in Fig. 7.16*b*; and STN 4 and STN 5 both lead to STN-*FC*3 in Fig. 7.16*c*. Each of these networks has the minimum number of column sections, which is 10, as well as 1 condenser and 1 reboiler. There are many alternative column sequences that can be derived from these networks; one of the alternatives for STN-*FC*2 is shown in Fig. 7.17. This was discovered by Agrawal (1996), who calls it the "satellite column" arrangement. It is

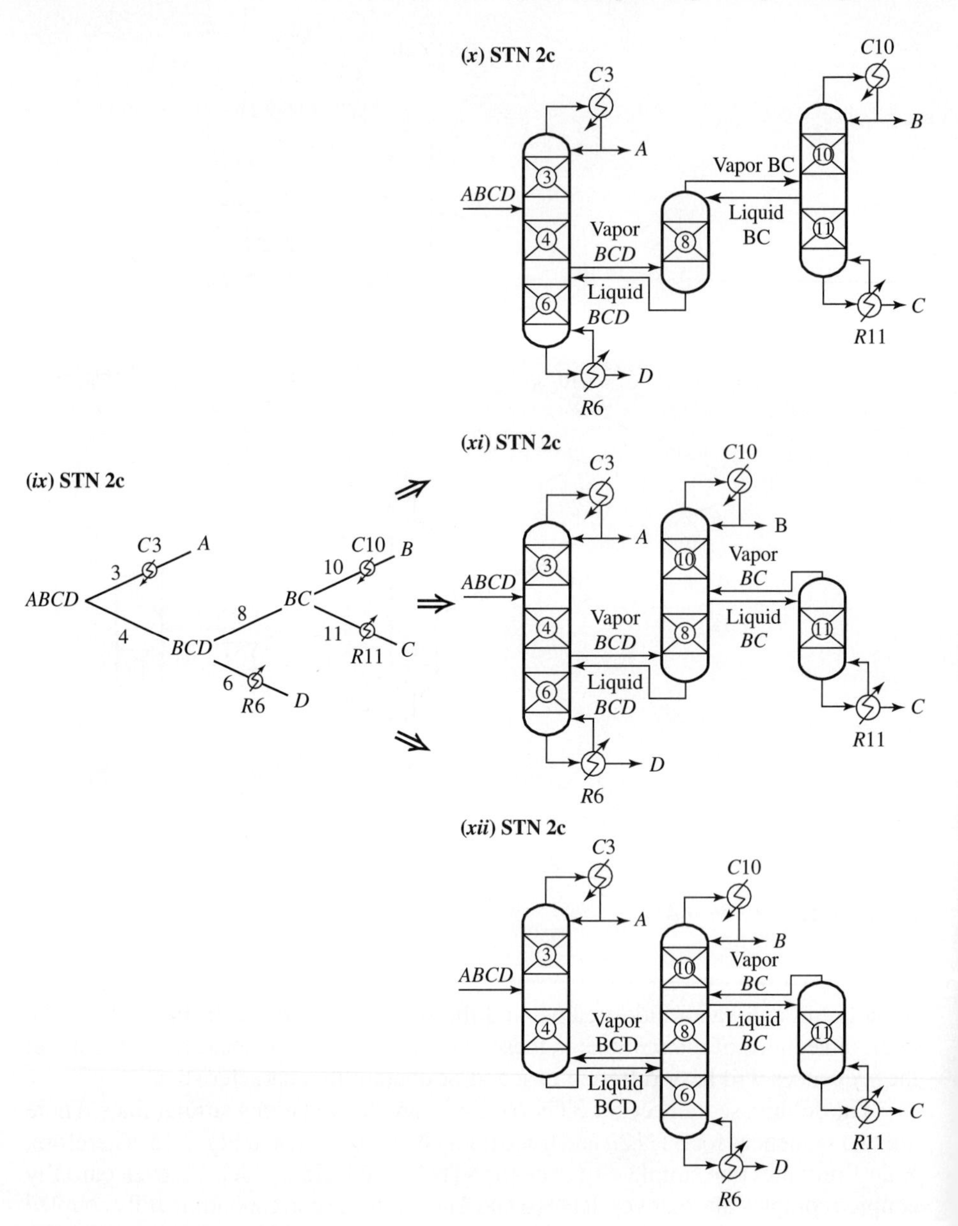

FIGURE 7.15 (*Continued*)

interesting to note that the fully coupled STNs shown in Fig. 7.16 do not use all the nodes. If we draw the maximally interconnected STN, as shown in Fig. 7.18, and apply Rule 4, we can develop many additional sequences that contain more than the minimum number of column sections. One of these is the fully coupled sequence that uses all the nodes. It has 12 column sections instead of the minimum number of 10, and one particular column arrangement that can be derived from this STN is the

(*a*) **STN-FC1**

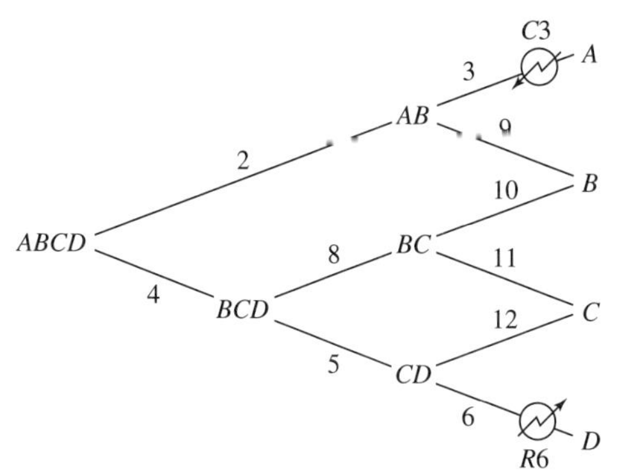

(*b*) **STN-FC2**

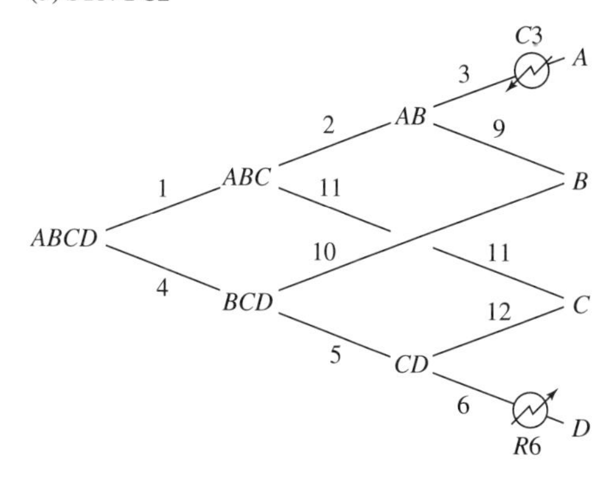

(*c*) **STN-FC3**

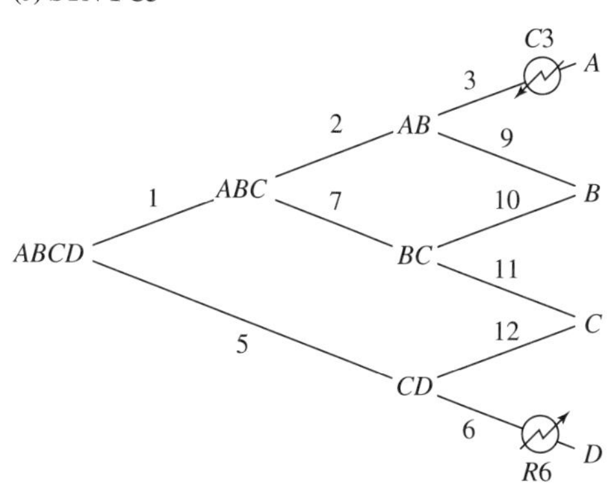

FIGURE 7.16
Three alternative fully coupled STNs for separating a four-component mixture.

Sargent and Gaminibandara sequence (Gaminibandara and Sargent, 1976) shown in Fig. 7.19. A detailed discussion on the relationship among these sequences, together with a case study on cryogenic air separation, is given in Agrawal (1996).

It is clear that an enormous number of column sequences can be devised to separate a multicomponent mixture. Agrawal (1999) gives a detailed account of the

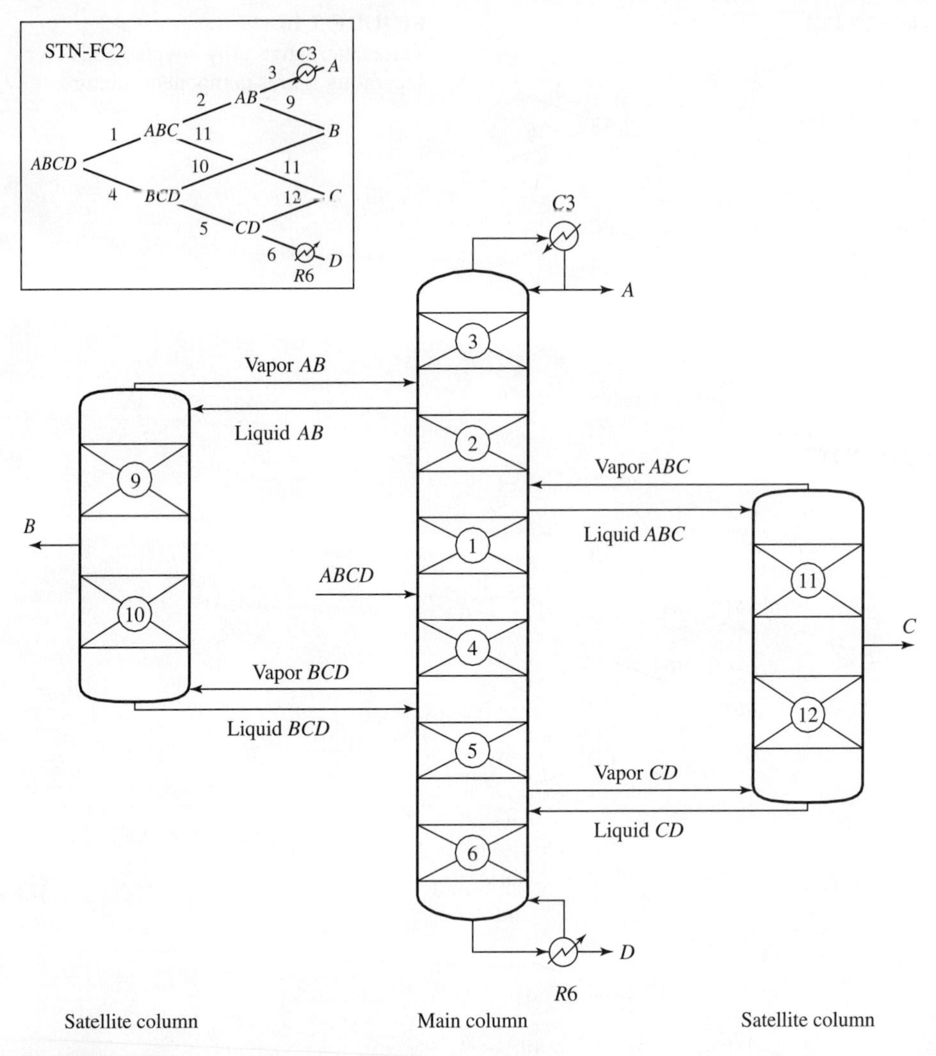

FIGURE 7.17
Agrawal's satellite column configuration for STN-FC2.

fully coupled sequences that are possible[10] and calculates that there are over 10,000 of these alone to separate a six-component mixture! The best sequence to use in any particular application depends on the feed composition, the boiling points of the pure components, and the values for the relative volatilities. Agrawal and Fidkowski (1998*a*, 1999, 2000) studied this in detail for ternary mixtures using thermodynamic efficiency as an objective function to discriminate among the alternatives. Figure 7.20*a* and *b* shows some of their results. The comparison was carried out for various

[10]The numbering scheme for column sections that we have used is identical to Agrawal's for ease of comparison.

ABCD
1
ABC
4
BCD
2
AB
7
8
BC
5
CD
3
A
9
10
B
11
12
C
6
D

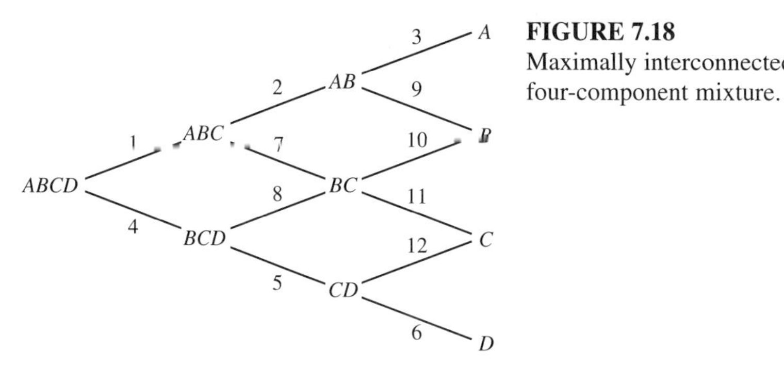

FIGURE 7.18
Maximally interconnected STN for a four-component mixture.

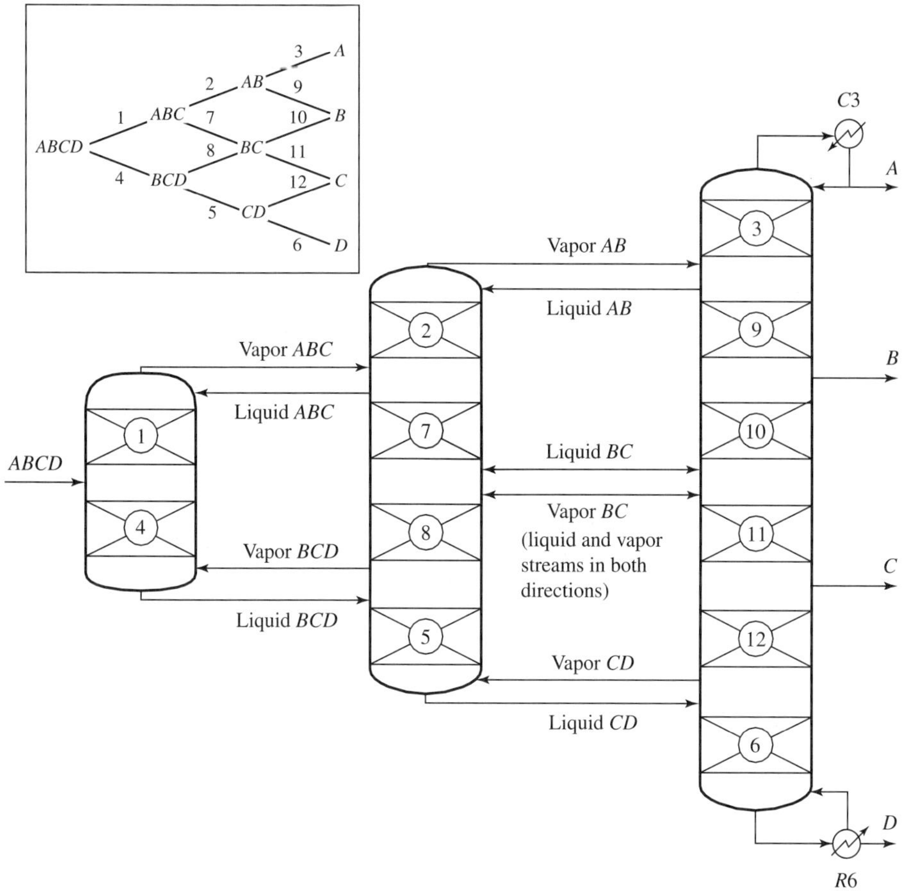

ABCD
Vapor ABC
Liquid ABC
Vapor BCD
Liquid BCD
Vapor AB
Liquid AB
Liquid BC
Vapor BC
(liquid and vapor streams in both directions)
Vapor CD
Liquid CD
C3
A
B
C
D
R6

FIGURE 7.19
Sargent and Gaminibandara sequence.

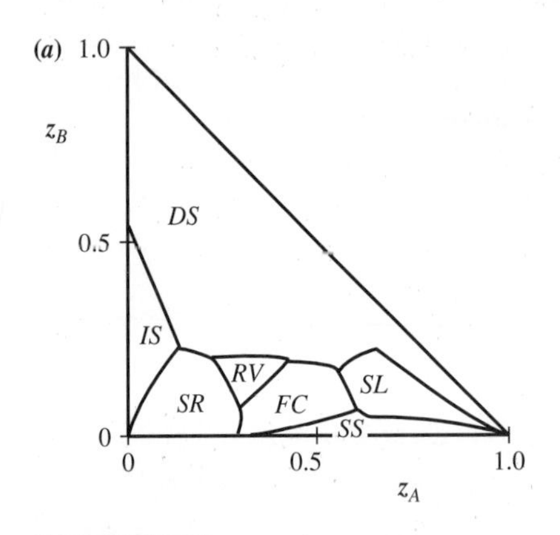

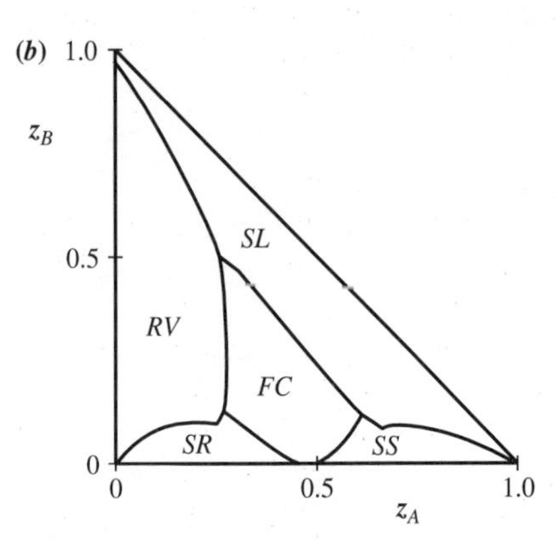

FIGURE 7.20
Optimum efficiency regions for separating ternary mixtures shown on a feed composition triangle. (*a*) Easy-easy splits, $\alpha_{AC} = 6.25$, $\alpha_{BC} = 2.5$. (*b*) Difficult-difficult splits, $\alpha_{AC} = 1.21$, $\alpha_{BC} = 1.1$. Legend: DS = direct sequence, IS = indirect sequence, FC = fully coupled, SR = side-rectifier column, SS = side-stripper column, SL = side-stripper prefractionator (performing the AB/BC split) with liquid connection, RV = side-rectifier prefractionator (performing the AB/BC split) with vapor connection.

types of splits, assuming that both the A/B split and, independently, the B/C split may be easy or difficult. Relative volatilities of 2.5 and 1.1 were chosen to represent the easy split and the difficult split, respectively. Figure 7.20*a* shows the optimal regions for the easy-easy splits. The optimal regions for the partially coupled sequences are adjacent to the fully coupled region. They more than double the region of feed compositions where complex columns are the most efficient. The improvement in efficiency is not very great (about 2 or 3%), but the reduction in the minimum vapor rate is as large as 20% in comparison to the direct or indirect sequences.

For difficult-difficult splits, however, the complex sequences eliminate entirely the conventional direct and indirect configurations (see Fig. 7.20*b*). The relative efficiency increase due to the use of complex alternatives is about 10 to 20%, with a reduction of the minimum vapor flow by as much as 40%. Similar results have also been obtained by others using different approaches (Tedder and Rudd, 1978; Glinos and Malone, 1988).

7.5 SYSTEM SYNTHESIS FOR AZEOTROPIC MIXTURES

The principal focus for nonideal and azeotropic mixtures so far has been on three- and four-component mixtures because it is possible to visualize the results. For example, in ternary mixtures, distillation boundaries or their limits can be calculated for both infinite and finite reflux (Krolikowski et al., 1996; Davydyan et al., 1997). The feasible products for distillation of ternary homogeneous mixtures have been studied in detail (Wahnschafft et al., 1992; Fidkowski et al., 1993*a*). Methods are also available for assessing the structure of distillation regions in systems with four or five components based on a topological analysis of the separating manifolds (Baburina and Platonov, 1987; Baburina et al., 1988; Baburina and Platonov, 1990*a*, 1990*b*). Progress continues to be made on treating mixtures containing any number of components (Safrit and Westerberg, 1997; Ahmad and Barton, 1996), and it remains an important area of research.

In this section we describe a method for evaluating distillation feasibility based on Rooks et al. (1998). The method identifies the most important splits and applies to all homogeneous mixtures, regardless of the number of azeotropes or components. First we describe the data structures and a method for calculating their properties from a vapor-liquid equilibrium model. Next, we relate these results to feasibility of various splits. This is followed by several examples. A systematic procedure for generating alternative sequences of distillation columns can be developed by combining the feasibility methods from this section with the state-task network formalism described earlier (also see Sargent, 1998).

Representation of the Residue Curve Structure

Pure components and azeotropes are the singular points for the residue curve map equation.

$$d\mathbf{x}/d\xi = \mathbf{x} - \mathbf{y}(\mathbf{x}) \tag{7.2}$$

Model equations for continuous column sections have fixed points at total reflux which are the same as the singular points of Eq. 7.2. This allows the representation of certain limiting cases based on the structure of the residue curves. In mixtures with more than four components, where graphical methods fail, the *directed adjacency* matrix **A** and a related *reachability* matrix **R** (Knight and Doherty, 1990) can be used to represent this structure.

We list these singular points in order of their boiling temperatures, since the residue curves connect these points in the direction of increasing temperature (Doherty and Perkins, 1978). The adjacency matrix is defined by $a_{i,j} = 1$ if a residue curve joins i to j; otherwise $a_{i,j} = 0$. The reachability matrix **R** is defined by $r_{i,j} = 1$ if there is *any* path from i to j; i.e., j is "reachable" from i; otherwise $r_{i,j} = 0$. Note that a path may include intermediate singular points and there may be more than one path when j is reachable from i.

Fidkowski et al. (1993*b*) describe a homotopy continuation method for computing *all* the azeotropes (and their stability) from a vapor-liquid equilibrium model. In addition to the temperatures and compositions of the azeotropes, their stability as well as that of each pure component is required here. This is based on a linearization of Eq. 7.2 at each singular point and the properties of the Jacobian matrix

$$J_{i,j} = \delta_{i,j} - \partial y_i/\partial x_j \tag{7.3}$$

The eigenvectors of the Jacobian show the characteristic directions of the residue curves near a singular point. The eigenvalues are always real and fixed points are either saddles or nodes (Doherty and Perkins, 1978). Node-to-node connections do not form distillation boundaries (Foucher et al., 1991), which leaves saddle-saddle and saddle-node connections to form distillation boundaries.

Connections between singular points on the binary edges of the composition space can be determined from a knowledge of the boiling temperatures for binary pairs in the mixture. If there is no azeotrope, then the lower-boiling pure component is adjacent to the higher-boiling component. For binary pairs with a single azeotrope, a minimum-boiling azeotrope is adjacent to the two pure species; a maximum-boiling azeotrope has both pure components connected to it.[11]

Once the binary connections have been made, residue curves are computed starting at all binary and higher-dimensional saddles to complete the construction of the adjacency matrix. These curves initially follow the eigenvectors, either forward in time for a positive eigenvalue or backward in time for a negative eigenvalue. The endpoints of these residue curves are "adjacent" to the saddle or the saddle is adjacent to them. The elements of the adjacency matrix are determined simply by repeating this procedure until there are no saddles remaining.

The general procedure for computing the adjacency matrix can be summarized as follows:

Algorithm 1: Adjacency

1. Given: a mixture, a column pressure, and a VLE model.
2. Compute all azeotropes and determine the stability of all the singular points (pure components and azeotropes) using the method of Fidkowski et al. (1993*a*). Number the singular points in order of increasing boiling temperature.
3. For pairs of components without binary azeotropes, indicate a connection from the lower-boiling component; set $a_{i,j} = 1$, where i is lower-boiling.
4. For pairs of components that do exhibit binary azeotropes, set $a_{az,i} = 1$ and $a_{az,j} = 1$, where az refers to the index of the minimum-boiling azeotrope. For a maximum-boiling azeotrope, set $a_{i,az} = 1$ and $a_{j,az} = 1$.
5. For each saddle:

   a. Compute the eigenvectors.

   b. For each eigenvector corresponding to a positive eigenvalue, integrate Eqs. 7.2 forward in time from the saddle:

      i. In a direction along the eigenvector and

      ii. In a direction opposite to the eigenvector,

   omitting directions in the integration that point outside the composition space.

   c. For each eigenvector corresponding to a negative eigenvalue, repeat the previous step, but integrate backward in time.

   d. Integration is stopped when the residue curve approaches any singular point (other than the one it started from) within a Euclidean distance of 0.001. Indicate a connection from the lower-boiling singular point i, to the higher-boiling singular point j by setting $a_{i,j} = 1$.

[11]We do not consider the possibility of double azeotropes in binary mixtures because it is rare, though not unknown (Gmehling et al., 1994*a*, 1994*b*).

The reachability matrix can be computed from the adjacency matrix via

$$\mathbf{R} = \text{boolean}[\mathbf{A} + \mathbf{I}]^{\nu-1} \tag{7.4}$$

where ν is the number of singular points and $\mathbf{I}$ is the identity matrix. (The boolean operation replaces nonzero elements by unity while zero elements remain zero.) This provides a simple and effective method to represent the structure (Knight and Doherty, 1990).

The adjacency matrix has a property that makes it easy to determine which points are stable nodes and unstable nodes. Because the points are connected in the direction of increasing temperature, *an unstable node is indicated by a column of zeros* since no other point can connect to it. Likewise, *a stable node is identified by a row of zeros* since a stable node cannot connect to any other point.

Distillation Regions

A distillation region is a family of curves connecting one unstable node to one stable node. A region must also have the same dimension in composition space as the original mixture. Therefore, a distillation region can be identified from a collection of points with one unstable node, one stable node, and a number of intermediate-boiling saddles which satisfy the following properties:

Property 1. The stable node is reachable from the unstable node; $r_{UN,SN} = 1$.
Property 2. The nodes and saddles define a $c - 1$ dimensional composition space.

The number of distillation regions is no greater than the number of pairs of unstable and stable nodes for which $r_{UN,SN} = 1$. For example, consider a mixture of methyl acetate, ethyl acetate, methanol, and ethanol which has the residue curve map shown in Fig. 7.21. There are three binary azeotropes for a total of seven singular points numbered as in Fig. 7.21.

The adjacency and the reachability matrices are:

$$\mathbf{A} = \begin{bmatrix} 0 & 1 & 1 & 1 & 0 & 0 & 0 \\ 0 & 0 & 0 & 0 & 1 & 1 & 1 \\ 0 & 0 & 0 & 1 & 1 & 1 & 0 \\ 0 & 0 & 0 & 0 & 0 & 0 & 1 \\ 0 & 0 & 0 & 0 & 0 & 1 & 1 \\ 0 & 0 & 0 & 0 & 0 & 0 & 0 \\ 0 & 0 & 0 & 0 & 0 & 0 & 0 \end{bmatrix}$$

$$\mathbf{R} = \begin{bmatrix} 1 & 1 & 1 & 1 & 1 & 1 & 1 \\ 0 & 1 & 0 & 0 & 1 & 1 & 1 \\ 0 & 0 & 1 & 1 & 1 & 1 & 1 \\ 0 & 0 & 0 & 1 & 0 & 0 & 1 \\ 0 & 0 & 0 & 0 & 1 & 1 & 1 \\ 0 & 0 & 0 & 0 & 0 & 1 & 0 \\ 0 & 0 & 0 & 0 & 0 & 0 & 1 \end{bmatrix}$$

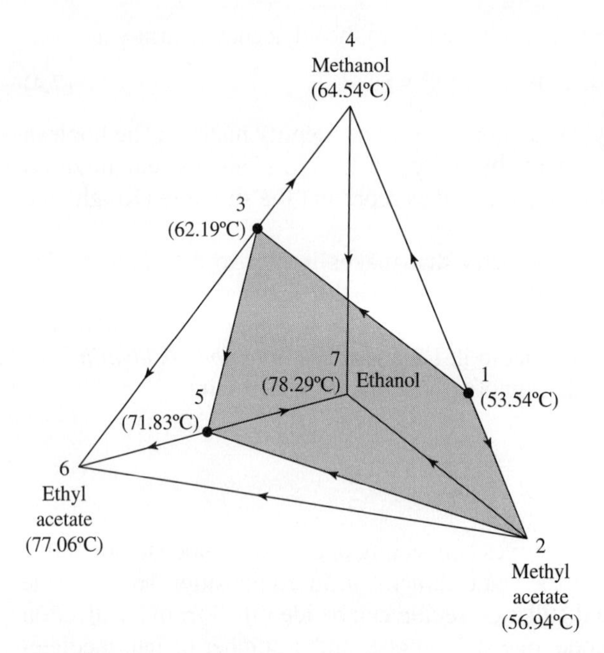

FIGURE 7.21
Distillation boundary and residue curve map structure for the mixture of methanol, methyl acetate, ethyl acetate, and ethanol. The distillation boundary is the surface (shaded) that splits the mixture into two distillation regions.

Matrix **A** shows two stable nodes (rows 6 and 7, corresponding to pure ethyl acetate and pure ethanol, respectively) and one unstable node (column 1, corresponding to the methanol-methyl acetate azeotrope). There are two combinations of stable and unstable nodes (1 to 6 and 1 to 7) and no more than two distillation regions, since $r_{1,6} = 1$ and $r_{1,7} = 1$.

It is also necessary to determine all of the saddle points which bound a distillation region. This is the set of saddle points for which $r_{i,SN} = 1$ and $r_{UN,i} = 1$. Each region is enclosed by either composition or distillation boundaries connecting these points.[12] Distillation boundaries are created by binary or higher-dimensional saddles, which cause residue curves to break into different families. The vertices that define a distillation boundary can be found by grouping those singular points from a given region which are shared by another region. For example, the mixture of methyl acetate, methanol, and ethyl acetate (Fig. 7.23) has two saddles (points 2 and 3). A distillation boundary joins the methyl acetate-methanol unstable node to the methanol-ethyl acetate saddle and divides the mixture into two distillation regions. The lower region has three composition boundaries defined by the lines joining points (1)-(2), (2)-(5), and (3)-(5). The (1)-(3) connection forms a distillation boundary, since those two points are shared by both the lower and upper regions. Distillation boundaries are of dimension $(c - 2)$, and the distillation boundaries of lower-dimensional mixtures form edges of the higher-dimensional boundaries. The structure of the regions can be determined with the following algorithm.

[12] A composition boundary is simply the limit of composition space defined by $0 \leq x_i \leq 1$ and $\sum x_i = 1$.

Algorithm 2: Distillation Regions

1. *Candidate Regions and Points.*

   a. Find all pairs of an unstable node (UN) and a stable node (SN) for which $r_{UN,SN} = 1$. The number of pairs is the number of candidate regions.
   b. For each candidate region identified in (*a*), generate a list of saddle points i for which $r_{i,SN} = 1$ and $r_{UN,i} = 1$, that is, saddles which can reach the stable node and which can be reached from the unstable node.

2. *Rooks et al. (1998) conjectured the following test for Property 2.*

   a. Form a separate adjacency matrix for each candidate region using only the vertices identified in step 1. For each candidate region find the number of *distinct paths* joining the UN to the SN. Paths are distinct if they differ by at least one vertex. The number of distinct paths can be found by recursive application of either breadth first search or depth first search algorithms on the adjacency matrix for the candidate region (Corman et al., 1990; McHugh, 1990).
   b. If the number of distinct paths is at least $(c - 1)$, accept the candidate region. Otherwise reject it.

3. *Identify the boundaries of each region.*

   a. *Composition boundaries.* Group the vertices that are missing one of the pure components. Each group defines one composition boundary. For example, the lower region of the mixture in Fig. 7.23 has three composition boundaries. The first, (1)–(2), is found by grouping those vertices in the lower region that do not contain ethyl acetate (5). The other two composition boundaries (3)–(5) and (2)–(5) are found by grouping vertices in the region that are missing methyl acetate (2) and methanol (4), respectively.
   b. *Distillation boundaries.* Group those vertices that are shared by another region. Each group defines one distillation boundary.

Frequently, Property 2 is satisfied by all candidate regions, and step 2 in Algorithm 2 may seem unnecessary. However, Fig. 7.22 gives an example (hypothetical) where Property 2 is *not* satisfied by some of the candidate regions. This example has two unstable nodes (vertices 1 and 2) and three stable nodes (vertices 5, 6, and 7), giving six combinations. All six satisfy Property 1, i.e., $r_{1,6} = r_{1,7} = r_{2,5} = r_{2,7} = r_{1,5} = r_{2,6} = 1$, but only the first four satisfy Property 2 (i.e., by inspection there are only four distillation regions). Step 2 of Algorithm 2 detects the faulty candidates and correctly rejects them.

Moreover, when Algorithm 2 is applied to the examples in Figs. 7.21 and 7.23, it correctly predicts two distillation regions in each case.

Feasible Splits

The feasible products for mixtures with three components has been worked out in detail (Wahnschafft et al., 1992; Fidkowski et al., 1993*a*). The most important splits

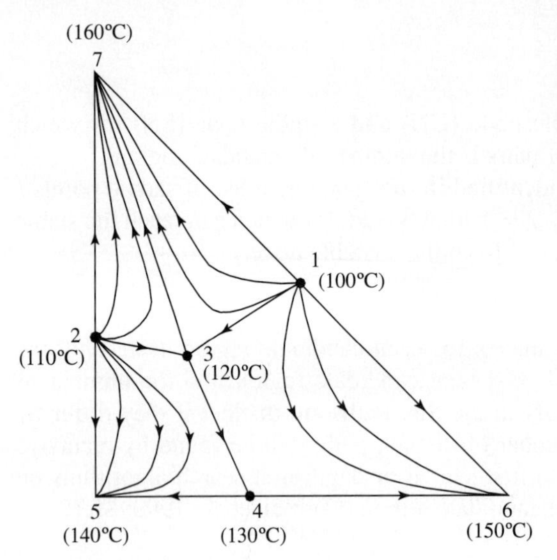

FIGURE 7.22
Hypothetical residue curve map with six candidate regions. Two of the candidate regions are rejected, 1-3-4-5 and 2-3-4-6, leaving four, 1-3-4-6, 1-3-7, 2-3-7 and 2-3-4-5.

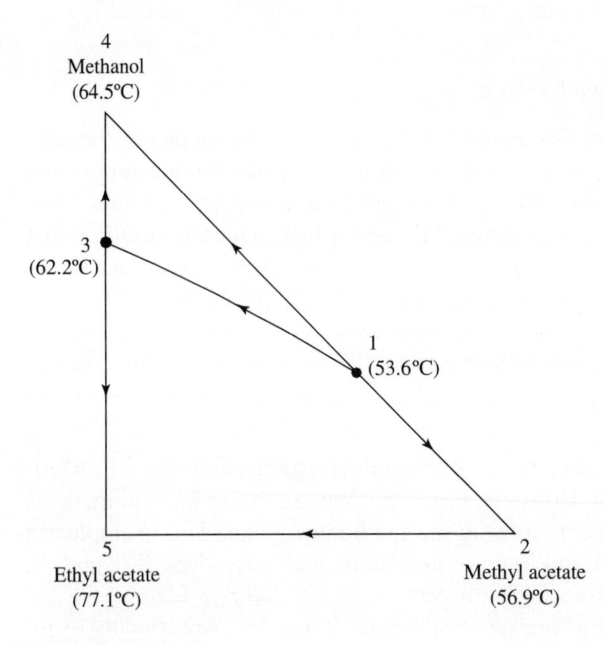

FIGURE 7.23
Distillation boundary and residue curve map structure for the mixture of methyl acetate, methanol, and ethyl acetate.

occur very close to the edges of the composition triangle where one of the components has been nearly eliminated from a product stream. These often correspond to "sharp" splits between components with adjacent boiling points. Components with nonadjacent boiling points can also be split; these are sometimes called "nonsharp" splits. We also use these terms to refer to splits with compositions on distillation boundaries in azeotropic mixtures. To determine these, we need a feasibility test which can be used in higher dimensions, and this test should also be practical to compute.

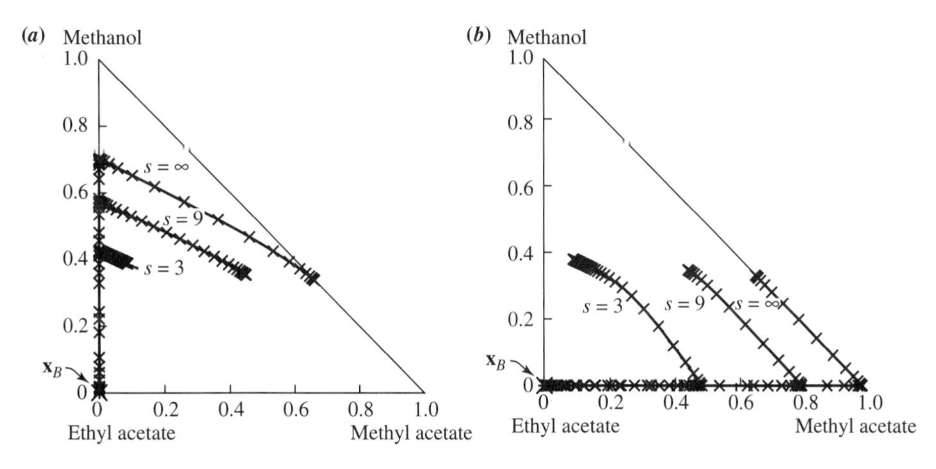

FIGURE 7.24
Series of composition profiles for the stripping section of a column for a mixture of methyl acetate, methanol, and ethyl acetate (see Fig. 7.23 for the residue curve map and the boiling temperatures). Part (*a*) represents those profiles which have a bottoms product on the methanol-ethyl acetate binary edge. Part (*b*) represents those profiles which have a bottoms product on the methyl acetate-ethyl acetate binary edge.

We consider a somewhat simpler test than the full analysis described in the references mentioned above, which is much easier to implement. A ternary mixture of methyl acetate, methanol, and ethyl acetate has the residue curve map structure shown in Fig. 7.23 at atmospheric pressure. For feed compositions in the lower distillation region, the bottoms composition can lie either below the azeotrope on the methanol-ethyl acetate binary edge or along the methyl acetate-ethyl acetate binary edge essentially free of methanol. Stripping profiles for the first case are shown in Fig. 7.24*a*. As the reboil ratio is increased, the stripping profile approaches the distillation boundary. For the second case (Fig. 7.24*b*) the stripping profile approaches the binary edge between pure methyl acetate and the methyl acetate-methanol azeotrope as the reboil ratio is increased.

For a feed composition as shown in Fig. 7.25, suppose we seek a distillate composition very close to the distillation boundary. The rectifying profile will first approach the saddle (3), and then the stable node (5). The feasible bottoms composition (mass balance indicated by the solid line) lies on the (3-5) edge. A bottoms composition (mass balance indicated by the dashed line) on the (2-5) binary edge is infeasible because the stripping profile approaches the saddle at methyl acetate (2) and then the unstable node (1). *For feasible sharp splits, it is necessary for the rectifying and stripping profiles to approach the same saddle at large reflux.* We call this the *common saddle* test.

These ideas extend to four-component mixtures using the concept of a rectifying manifold (Julka and Doherty, 1993). For the general case of n components, the feasible products can be determined in the following manner starting from the unstable node (other splits can be found starting from the stable node):

Algorithm 3: Feasibility

1. Specify a feed composition, a pressure, and a VLE model.
2. Compute the adjacency matrix using Algorithm 1 and the reachability matrix using Eq. 7.4.
3. Determine the singular points that bound each of the regions using Algorithm 2.
4. Determine which region contains the feed by integrating a residue curve through the feed and matching the stable and unstable nodes of the residue curve to the nodes of each region determined in step 3.
5. Select a distillate composition between the unstable node and an adjacent saddle.
6. Find a mass balance through the distillate and feed compositions and extend this until a composition boundary is reached at point $\mathbf{x}$. If $\mathbf{x}$ is in the same distillation region as the distillate (integrate a residue curve from the bottoms composition to check this), then $\mathbf{x}$ is a potential bottoms composition and it lies on a composition boundary of the distillation region. Otherwise search over the material balance line from $\mathbf{x}$ to the feed composition until a point is found (very close to a distillation boundary) that first produces a residue curve in the same region as the distillate. This point is a potential bottoms composition.
7. The last step does not tell us *which* composition or distillation boundary contains the bottoms composition. We determine this as follows:

   a. If the bottoms composition is free of one of the pure components, then the singular points that compose the composition boundary are the set of singular points enclosing the region that do not contain that pure component *and* which form a distillation region in the lower dimensional mixture which also does not contain that pure component.

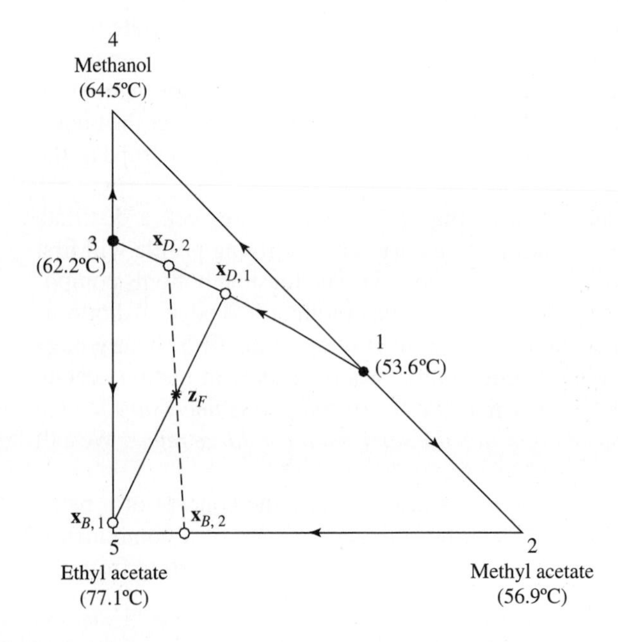

FIGURE 7.25
Two candidate splits for the same feed. The dashed line is infeasible and the solid line is feasible. Point 3 is the common saddle.

b. Otherwise compute two residue curves, starting at points along the mass balance line on either side of the bottoms composition (which is on a distillation boundary). Each of these residue curves will lie in a different distillation region. The distillation boundary is enclosed by the set of vertices shared by both regions.

8. If the lowest-boiling saddle on the bottoms boundary is reachable from the distillate composition, then this saddle is common to both the distillate and bottoms, and the split is feasible.
9. Repeat step 5 for each saddle adjacent to the unstable node.

This method provides a sufficient condition for product feasibility. It is not a necessary condition, and there are other alternative splits that are feasible and which do not fulfill these criteria. Fortunately, many common splits are covered by this method.

Implementing Algorithm 3 for feed *A* in Fig. 7.26, we find the range of sharp splits shown in the figure. Adding methyl acetate to the feed alters the range of feasible splits. For instance, with feed *B* it is possible to obtain complete recovery of methanol in the distillate, and ethyl acetate in the bottoms simultaneously. This is not possible with feed *A*.

EXAMPLE 7.5. Four-Component Mixture. A mixture of methyl acetate, ethyl acetate, methanol, and ethanol has three binary azeotropes at atmospheric pressure as shown in Fig. 7.21. These are the methanol-methyl acetate unstable node, the methanol-ethyl

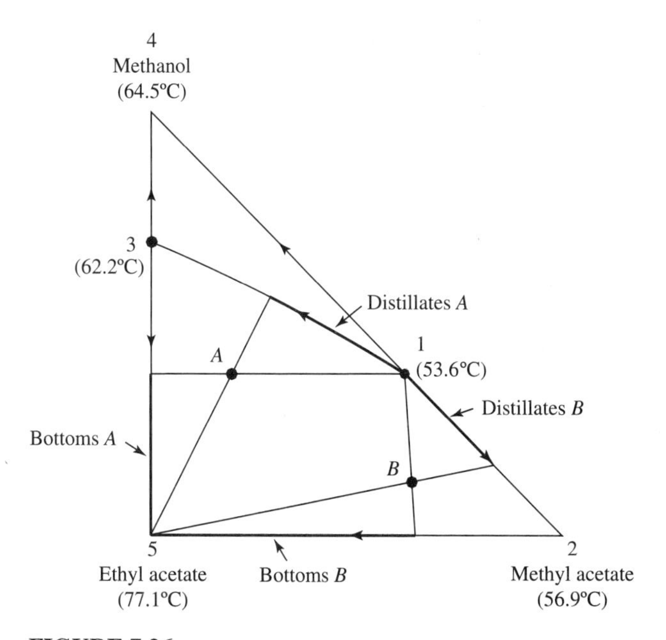

FIGURE 7.26
Feasible splits for two different feeds *A* and *B* in a mixture of methyl acetate, methanol, and ethyl acetate.

acetate saddle, and the ethyl acetate-ethanol saddle. This results in two distillation regions, one with a stable node at ethyl acetate and another with a stable node at ethanol. The distillation boundary (shaded) has vertices at each of the binary azeotropes, plus pure methyl acetate.

Consider a feed at the approximate center of the region behind the distillation boundary. One potential split removes the methanol-methyl acetate azeotrope as a distillate and leaves methanol, ethyl acetate, and ethanol in the bottoms (split 1). This is the analog of a "direct" split in simpler mixtures. For a high recovery of methyl acetate, the bottoms composition can be chosen near the composition boundary, defined by points (7-4-3-5). The lowest-boiling saddle in this set is (3), which we must reach at step 8 of Algorithm 3. There are three saddles adjacent to the unstable node, i.e., (2), (3), and (4). When saddle (2) or (4) is selected in step 5, the reachability matrix shows that saddle (3) is not reachable from (2) or (4) since $r_{2,3} = 0$ and $r_{4,3} = 0$. Therefore, the distillate cannot be along the (1-2) or (1-4) edges. The remaining saddle (3) is a common saddle since $r_{1,3} = 1$. Therefore, this is a feasible split and the distillate composition lies on the (1-3) edge very close to vertex 1; possible stream compositions are given in Table 7.4. Note that other feasible splits can be found for distillate compositions along the (1-3) edge of the distillation boundary, but not along the (1-2) or (1-4) edges.

A second feasible split is the analog of the "indirect" split, where a composition near pure ethanol (the stable node) is taken out as a bottoms stream (split 2). The mass balance and the geometry of the mixture confine the distillate composition to lie near the distillation boundary, because we do not attempt to cross these (see Chap. 5 for a

TABLE 7.4
Feed, distillate, and bottoms compositions for the mixture of methyl acetate, methanol, ethyl acetate, and ethanol. The numbering of the components in the table is methyl acetate (1), methanol (2), ethyl acetate (3), and ethanol (4)

	Split 1	Split 2
Feed		
$z_{F,1}$	0.25	0.25
$z_{F,2}$	0.25	0.25
$z_{F,3}$	0.25	0.25
$z_{F,4}$	0.25	0.25
Distillate		
$x_{D,1}$	0.6622	0.2767
$x_{D,2}$	0.3378	0.2767
$x_{D,3}$	6.519×10^{-6}	0.2767
$x_{D,4}$	4.412×10^{-8}	0.17
Bottoms		
$x_{B,1}$	1.0×10^{-6}	4.659×10^{-10}
$x_{B,2}$	0.1967	5.551×10^{-5}
$x_{B,3}$	0.4016	1.255×10^{-7}
$x_{B,4}$	0.4016	0.9999

TABLE 7.5
Singular points for Example 7.6

Number	Component or azeotrope	Boiling point, °C	Stability
1	Methyl acetate-methanol	53.5	UN
2	Methyl acetate-water	56.4	SA
3	Methyl acetate	56.9	SA
4	Methanol-ethyl acetate	62.2	SA
5	Methanol	64.5	SA
6	Ethyl acetate-ethanol-water	70.7	SA
7	Ethyl acetate-ethanol	71.8	SA
8	Ethyl acetate-water	71.9	SA
9	Ethyl acetate	77.1	SA
10	Ethanol-water	78.27	SA
11	Ethanol	78.29	SA
12	Water	100.0	SA
13	Acetic acid	117.9	SN

discussion of crossing boundaries). Therefore, the distillate contains all four components, and high recovery of ethanol in the bottoms product is not possible. Saddle (5) is a common saddle, indicating that the split is feasible.

More feasible splits can be found. For instance, a distillate can be taken along the (1-3) boundary. For certain feed compositions, the bottoms can be close to the (5-7) binary edge and nearly free of methyl acetate. Saddle (3) is adjacent to the distillate. Saddle (5) is adjacent to the bottoms composition and is also reachable from (3) so (5) is the common saddle. Note that these splits, though feasible, may not be very practical because variations in the feed composition may be difficult to accommodate.

EXAMPLE 7.6. Six-Component Example. We consider a six-component mixture of methanol, methyl acetate, ethanol, ethyl acetate, water, and acetic acid at 1 atm pressure modeled with the NRTL equation. Table 7.5 shows the stability characteristics and boiling temperatures for each singular point. For details of the vapor-liquid equilibrium model and parameters see Rooks (1997).

The adjacency and reachability matrices, numbered according to the singular points listed in Table 7.5, are

$$
\mathbf{A} = \begin{bmatrix}
0 & 1 & 1 & 1 & 1 & 0 & 0 & 0 & 0 & 0 & 0 & 0 & 0 \\
0 & 0 & 1 & 0 & 0 & 1 & 0 & 1 & 0 & 1 & 0 & 1 & 0 \\
0 & 0 & 0 & 0 & 0 & 0 & 1 & 0 & 1 & 0 & 1 & 0 & 1 \\
0 & 0 & 0 & 0 & 1 & 1 & 1 & 1 & 1 & 0 & 0 & 0 & 0 \\
0 & 0 & 0 & 0 & 0 & 0 & 0 & 0 & 0 & 1 & 1 & 1 & 1 \\
0 & 0 & 0 & 0 & 0 & 0 & 1 & 1 & 0 & 1 & 0 & 0 & 0 \\
0 & 0 & 0 & 0 & 0 & 0 & 0 & 0 & 1 & 0 & 1 & 0 & 0 \\
0 & 0 & 0 & 0 & 0 & 0 & 0 & 0 & 1 & 0 & 0 & 1 & 0 \\
0 & 0 & 0 & 0 & 0 & 0 & 0 & 0 & 0 & 0 & 0 & 0 & 1 \\
0 & 0 & 0 & 0 & 0 & 0 & 0 & 0 & 0 & 0 & 1 & 1 & 0 \\
0 & 0 & 0 & 0 & 0 & 0 & 0 & 0 & 0 & 0 & 0 & 0 & 1 \\
0 & 0 & 0 & 0 & 0 & 0 & 0 & 0 & 0 & 0 & 0 & 0 & 1 \\
0 & 0 & 0 & 0 & 0 & 0 & 0 & 0 & 0 & 0 & 0 & 0 & 0
\end{bmatrix}
$$

and

$$\mathbf{R} = \begin{bmatrix} 1 & 1 & 1 & 1 & 1 & 1 & 1 & 1 & 1 & 1 & 1 & 1 & 1 \\ 0 & 1 & 1 & 0 & 0 & 1 & 1 & 1 & 1 & 1 & 1 & 1 & 1 \\ 0 & 0 & 1 & 0 & 0 & 0 & 1 & 0 & 1 & 0 & 1 & 0 & 1 \\ 0 & 0 & 0 & 1 & 1 & 1 & 1 & 1 & 1 & 1 & 1 & 1 & 1 \\ 0 & 0 & 0 & 0 & 1 & 0 & 0 & 0 & 0 & 1 & 1 & 1 & 1 \\ 0 & 0 & 0 & 0 & 0 & 1 & 1 & 1 & 1 & 1 & 1 & 1 & 1 \\ 0 & 0 & 0 & 0 & 0 & 0 & 1 & 0 & 1 & 0 & 1 & 0 & 1 \\ 0 & 0 & 0 & 0 & 0 & 0 & 0 & 1 & 1 & 0 & 0 & 1 & 1 \\ 0 & 0 & 0 & 0 & 0 & 0 & 0 & 0 & 1 & 0 & 0 & 0 & 1 \\ 0 & 0 & 0 & 0 & 0 & 0 & 0 & 0 & 0 & 1 & 1 & 1 & 1 \\ 0 & 0 & 0 & 0 & 0 & 0 & 0 & 0 & 0 & 0 & 1 & 0 & 1 \\ 0 & 0 & 0 & 0 & 0 & 0 & 0 & 0 & 0 & 0 & 0 & 1 & 1 \\ 0 & 0 & 0 & 0 & 0 & 0 & 0 & 0 & 0 & 0 & 0 & 0 & 1 \end{bmatrix}$$

There is one six-component distillation region for this mixture. For an equimolar feed, there are four splits shown in Table 7.6. It is possible to remove methyl acetate with high recoveries as a distillate (split 1) but not in high purity because of the methanol-methyl acetate azeotrope. It is also possible to completely recover pure acetic acid in an indirect split (split 3). Starting with these four splits, and applying the algorithms recursively, many alternative sequences to separate the original mixture into its pure components can be generated.

For example, if all of the methyl acetate is removed in the distillate as an azeotrope with methanol, the resulting bottoms will contain the remaining components, which

TABLE 7.6
Four splits for an equimolar feed of methanol, ethanol, methyl acetate, ethyl acetate, acetic acid, and water

Split	Distillate	Bottoms
1	Methyl acetate Methanol	Methanol Ethyl acetate Ethanol Water Acetic acid
2	Methyl acetate Methanol Ethyl acetate	Ethyl acetate Ethanol Water Acetic acid
3	Methyl acetate Methanol Ethyl acetate Ethanol Water	Acetic acid
4	Methyl acetate Methanol Ethyl acetate Ethanol Water	Acetic acid Water

form a single distillation region. Instead, if the acetic acid is removed as a bottoms product, the remaining components will appear in the distillate. These components form three distillation regions which can be determined as follows. Delete row 13 and column 13 in the adjacency matrix above. Three singular points (9, 11, and 12) become stable nodes, since those rows in the reduced adjacency matrix contain only zeros. Point 1 remains the only unstable node and all three stable nodes can be reached from this unstable node, giving three candidate regions. (Generally, this requires recalculation of the reachability matrix.) All three of these satisfy Property 2, and therefore there are three distillation regions. This split may seem undesirable because of the complicated structure of the distillate mixture and because at least 83% of the feed mixture must be vaporized. However, this split avoids the difficult separation of water from acetic acid, which will be expensive. A design and cost estimate for some or all of the columns in the sequences is needed to rank them, e.g., using a branch-and-bound approach.

To limit the search space, we apply the following heuristics:

- Avoid splits that lead to mixtures containing distillation boundaries that must be crossed in subsequent separations.
- Avoid splits that lead to mixtures containing severe tangent pinches that must be broken in subsequent separations (e.g., avoid the acetic acid-water mixture in split 4). Therefore, delete split 4 from the search tree.
- Avoid splits that lead to ternary or higher order azeotropes because they cannot be broken using known distillation technology (e.g., avoid splits leading to the ethyl acetate-ethanol-water azeotrope). The bottom stream from split 2 has one distillation region with acetic acid as the stable node and the ternary azeotrope of ethyl acetate-ethanol-water as the unstable node (see the residue curve map in Fig. 5.10). Since this azeotrope must eventually be a distillate product from one of the columns in the sequence, we delete split 2 from the search tree.

After applying the above heuristics, only two splits remain (splits 1 and 3). In addition, we should consider the possibility of moving the original feed composition by selectively adding one or more of the feed components (the excess feed components are recovered and recycled) so as to generate additional splits that may lead to attractive sequences. One possibility is to add methyl acetate to the feed mixture in order to recover all the methanol as a methyl acetate-methanol azeotrope in the distillate stream from the first column (see Branch 2, Fig. 7.27). For this example we limit our search tree to these three branches (see Fig. 7.27).

Sequence 1

Starting with split 1, which appears to be a good initial candidate, we develop the subtree of alternative splits that follow from this starting point (labeled Branch 1 in Fig. 7.27). A selection of the resulting splits are shown in Fig. 7.28. For simplicity we show only those splits where either the distillate or bottoms product is a node (i.e., the nonideal analogs of direct or indirect splits, respectively). Moreover, we show only a selection of these splits![13]

One sequence that can be generated from this branch is shown schematically in Fig. 7.29 (we label this Sequence 1). The first column recovers methyl acetate as an

---

[13]For example, the methanol, ethanol, water mixture has two distillation regions. Feeds in the water-rich region can be split into a bottom stream of water and a distillate containing all three components (indirect split), or a distillate stream of methanol and a bottom stream containing ethanol and water (direct split). Other splits are possible for feeds in the ethanol-rich region. For simplicity, we show only one of these splits in Fig. 7.28.

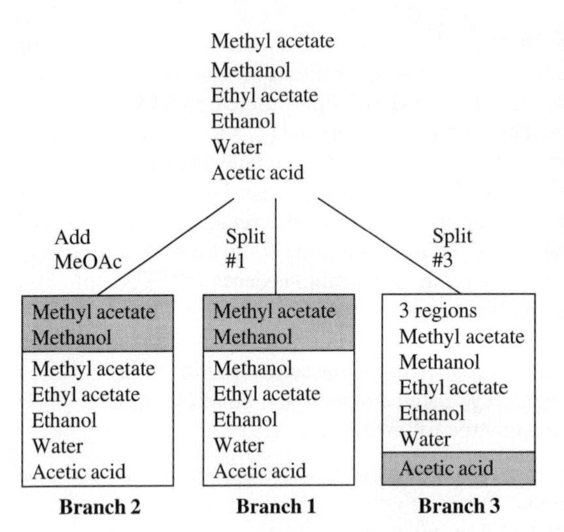

FIGURE 7.27
Three branches in the tree of alternative splits for the six-component mixture of methyl acetate, methanol, ethyl acetate, ethanol, water, and acetic acid at 1 atm pressure. The shaded boxes are either final products or binary mixtures to be split using known technology such as pressure shifting or extractive distillation.

azeotrope with methanol in the distillate. This mixture can be separated using known technology such as extractive, azeotropic, or perhaps pressure-swing distillation (we do not follow these binary separations any further). Methanol is added in sufficient quantity to the resulting bottoms product so that all of the ethyl acetate can be recovered in the distillate as an azeotrope with methanol (Rooks et al., 1998, Appendix). This methanol addition avoids a split involving the ethyl acetate-ethanol-water azeotrope.

The bottoms stream from the second split is sent to another column, where pure methanol can be recovered in the distillate. A fourth split removes acetic acid as the bottoms product; this is liable to be expensive if high purities are desired. The ethanol/water mixture can be split in a conventional extractive distillation subsystem, e.g., using ethylene glycol as entrainer.

Sequence 2

A second class of sequences are generated by starting with Branch 2 in Fig. 7.27. A selection of the resulting splits are shown in Fig. 7.30 (we apply similar restrictions as in Fig. 7.28).

One sequence that can be generated from this branch is shown schematically in Fig. 7.31 (we label this Sequence 2). Methyl acetate is added to the process feed in order to recover all the methanol as an azeotrope with methyl acetate in the distillate from the first column. The remaining five components leave in the bottoms stream and are fed to a second column along with more methyl acetate in order to recover all the water as an azeotrope with methyl acetate in the distillate from this column. The remaining four components leave in the bottoms stream from the second column and are fed to a third column which performs an indirect split to recover all the acetic acid in the bottoms stream, leaving methyl acetate, ethyl acetate, and ethanol in the distillate. This stream is fed to a fourth column where all the methyl acetate is recovered in the distillate, leaving a binary azeotropic mixture of ethyl acetate and ethanol in the bottoms stream. The various binary azeotropes are broken using known technology and are not considered further here.

A third class of sequences can be generated by following Branch 3 in Fig. 7.27. A selection of the resulting splits are shown in (Rooks et al., 1998, Fig. 4.6). Clearly,

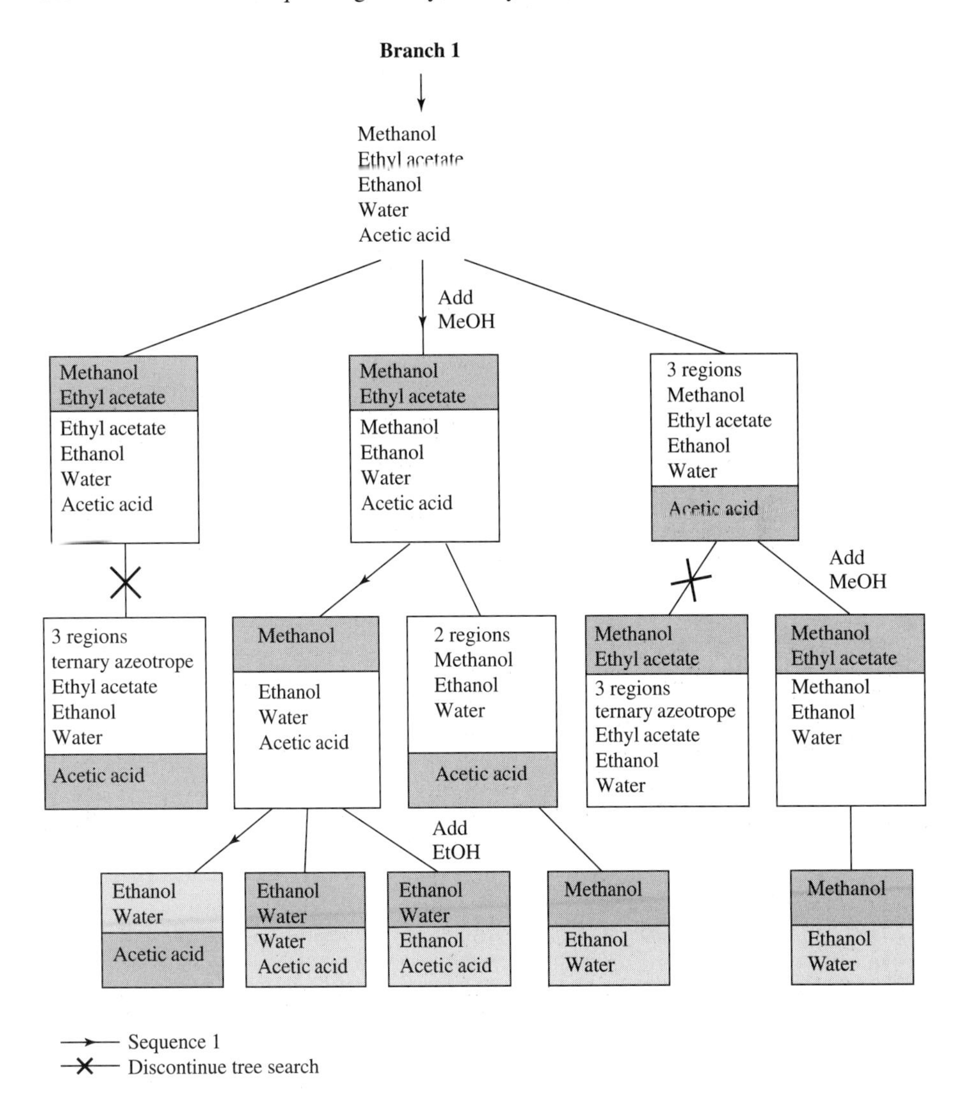

FIGURE 7.28
Branch 1. Subtree of splits for the five-component mixture of methanol, ethyl acetate, ethanol, water, and acetic acid 1 atm pressure. The shaded boxes are either final products or binary mixtures to be split using known technology.

there are many feasible ways to separate this six component mixture and Rooks et al. (1998) determines that at least 26 alternative sequences can be generated, not taking into account sidestream columns, entrainer selection, and nonsharp splits!

Evaluation

To evaluate the sequences, we set target product compositions on each column (these targets are easy to set once we know the splits from the methods just described).

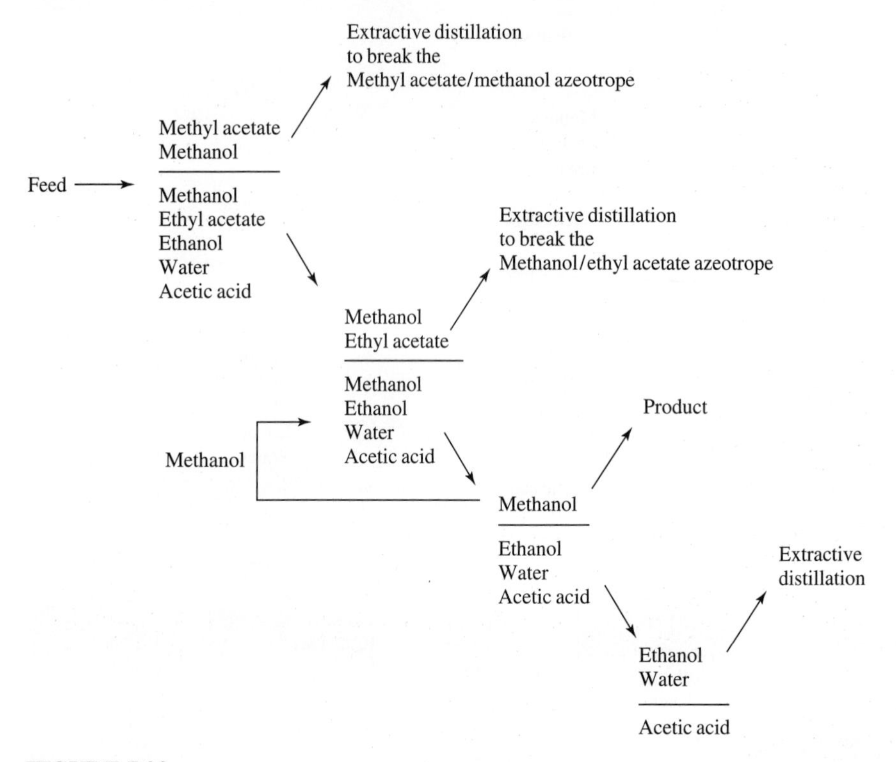

FIGURE 7.29
Sequence 1 to separate an equimolar feed of methyl acetate, methanol, ethyl acetate, ethanol, water, and acetic acid at 1 atm pressure.

Each column in each sequence is designed using the methods in Chap. 4 and a vapor rate of 1.5 times the minimum value ($s = 1.5s_{\text{min}}$). We can then compare sequences on the basis of cost (using the methods in Chap. 6), or total vapor rate, etc. For this example we compare on the basis of total vapor rate.

Designs for each column in Sequence 1 for an equimolar process feed with flowrate F_0 are given in Table 7.7. The flowrate of additional methanol to the second column is $0.5F_0$ in order to recover all the ethyl acetate overhead in an azeotrope with methanol.

Designs for each column in Sequence 2 are given in Table 7.8. The flowrate of additional methyl acetate to the first column is $0.19F_0$ to allow for the complete recovery of methanol overhead as an azeotrope with methyl acetate. Unfortunately, a large amount of additional methyl acetate ($1.533F_0$) must be added to the feed to the second column in order to recover all the water overhead. The corresponding vapor rate in this column is enormous ($V/F_0 = 17.5$).

Each sequence requires three extractive distillations (other sequences may require a different number). We prefer to discard the unattractive sequences as early as possible in order to avoid designing all the extractive distillations. At this point we have a good basis for abandoning Sequence 2 and concentrating our effort on other alternatives. Sequence 1 has a total of 230 stages and $\sum V/F_0 = 11.97$. The second sequence also has 230 stages (a coincidence!) and $\sum V/F_0 = 26.51$. The second sequence has the

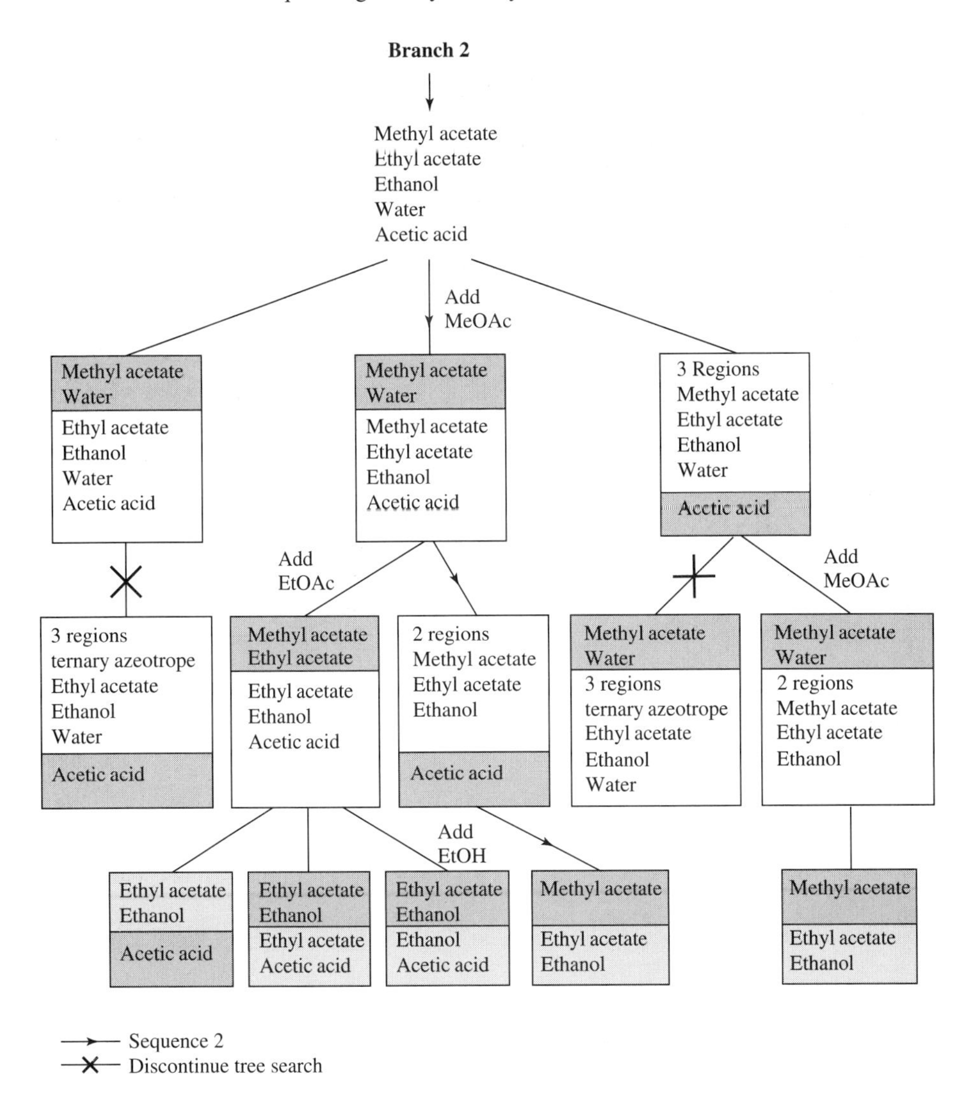

FIGURE 7.30
Branch 2. Subtree of splits for the five-component mixture of methyl acetate, ethyl acetate, ethanol, water, and acetic acid 1 atm pressure. The shaded boxes are either final products or binary mixtures to be split using known technology.

same number of stages but consumes over twice as much energy (it will also require larger-diameter columns and larger heating and cooling equipment). The columns that contribute most to this high vapor flowrate are the first two, where large amounts of methyl acetate are added so that the desired splits can be performed. Although these are not optimized sequences, it is unlikely that optimization of Sequence 2 will produce a design that is competitive with Sequence 1; therefore, we would abandon Sequence 2 at this point.

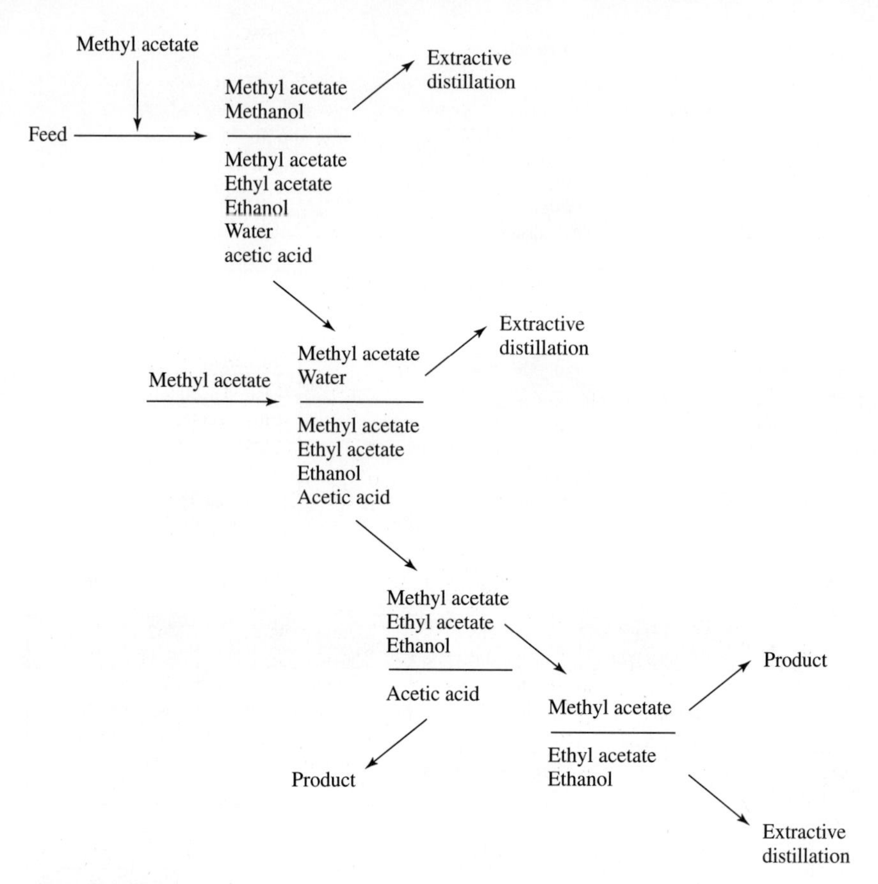

FIGURE 7.31
Sequence 2 to separate a mixture of methyl acetate, ethyl acetate, ethanol, water, and acetic acid 1 atm pressure.

The methods described here provide feasibility tests and sequencing of distillation columns to separate nonideal mixtures containing many components. With these tools, the residue curve map structure of complex multicomponent mixtures can be calculated rapidly. The development avoids crossing distillation boundaries, and avoids exploiting liquid-liquid phase behavior.[14] Either of these two effects can give rise to additional alternatives, with the latter typically being of greatest practical importance. The methods described here apply to homogeneous mixtures but also provide a basis for studying heterogeneous mixtures, chemically react-

[14] It should be noted that the six-component mixture studied in the above example exhibits liquid-liquid phase behavior that was not exploited in the synthesis of the separation system.

TABLE 7.7
Column designs for Sequence 1 to separate a six-component mixture (1-methyl acetate, 2-methanol, 3-ethyl acetate, 4-ethanol, 5-water, 6-acetic acid). The specifications appear in italics. The stages are numbered up the column starting with the reboiler as stage 1

	Column 1	Column 2	Column 3	Column 4
Feed composition				
$z_{F,1}$	*0.1666*	*0.0*	*0.0*	*0.0*
$z_{F,2}$	*0.1666*	*0.4660*	*0.280*	*0.0*
$z_{F,3}$	*0.1666*	*0.1335*	*0.0*	*0.0*
$z_{F,4}$	*0.1666*	*0.1335*	*0.240*	*0.3333*
$z_{F,5}$	*0.1666*	*0.1335*	*0.240*	*0.3333*
$z_{F,6}$	0.1666	0.1335	0.240	0.3333
Feed flow rate	$1.0F_0$	$1.248F_0$	$0.694F_0$	$0.4997F_0$
Distillate composition				
$x_{D,1}$	*0.6624*	0.0	0.0	0.0
$x_{D,2}$	*0.3376*	*0.6992*	*0.9999*	0.0
$x_{D,3}$	1.858×10^{-6}	*0.3008*	0.0	0.0
$x_{D,4}$	2.045×10^{-8}	4.545×10^{-6}	1.066×10^{-4}	0.5003
$x_{D,5}$	1.918×10^{-5}	7.513×10^{-6}	4.056×10^{-7}	0.4997
$x_{D,6}$	1.128×10^{-33}	2.656×10^{-28}	2.670×10^{-23}	1.0×10^{-7}
Bottoms composition				
$x_{B,1}$	1.0×10^{-7}	0.0	0.0	0.0
$x_{B,2}$	0.1092	0.280	1.0×10^{-7}	0.0
$x_{B,3}$	0.2227	1.0×10^{-7}	0.0	0.0
$x_{B,4}$	0.2227	0.240	0.3333	9.374×10^{-6}
$x_{B,5}$	0.2227	0.240	0.3333	0.001
$x_{B,6}$	0.2227	0.240	0.3333	*0.999*
Reflux ratio	12.83	9.53	9.41	*0.92*
Reboil ratio	*4.65*	*8.4*	*4.05*	3.82
Number of stages	50	37	71	72
Feed tray location	12	11	17	19
V/F_0	3.48	5.83	2.025	0.637

ing mixtures (and even to solid-liquid equilibrium systems for multicomponent crystallization).

Because of the difficulty of constructing feasible sequences of distillation columns for azeotropic mixtures, it is easy to conclude that there may not be many alternatives, and that as soon as one feasible alternative is found it is less important to develop others. The examples above demonstrate that there can be a surprisingly large number of alternatives for the distillation of azeotropic mixtures, and we have seen that some of them are much more attractive than others.

TABLE 7.8
Column designs for Sequence 2 to separate a six-component mixture (1-methyl acetate, 2-methanol, 3-ethyl acetate, 4-ethanol, 5-water, 6-acetic acid). The specifications appear in italics. The stages are numbered up the column starting with the reboiler as stage 1

	Column 1	Column 2	Column 3	Column 4
Feed composition				
$z_{F,1}$	*0.3*	*0.701*	*0.187*	*0.2564*
$z_{F,2}$	*0.14*	*0.0*	*0.0*	*0.0*
$z_{F,3}$	*0.14*	*0.07475*	*0.271*	*0.3718*
$z_{F,4}$	*0.14*	*0.07475*	*0.271*	*0.3718*
$z_{F,5}$	*0.14*	*0.07475*	*0.0*	*0.0*
$z_{F,6}$	0.14	0.07475	0.271	0.0
Feed flow rate	$1.19F_0$	$2.23F_0$	$0.614F_0$	$0.447F_0$
Distillate composition				
$x_{D,1}$	*0.6624*	*0.8968*	0.2564	*0.999*
$x_{D,2}$	*0.3376*	0.0	0.0	0.0
$x_{D,3}$	3.386×10^{-7}	1.24×10^{-7}	0.3718	0.0
$x_{D,4}$	5.166×10^{-9}	4.363×10^{-6}	0.3718	5.49×10^{-6}
$x_{D,5}$	1.375×10^{-5}	*0.1032*	0.0	0.001
$x_{D,6}$	1.036×10^{-35}	1.596×10^{-27}	1.0×10^{-7}	0.0
Bottoms composition				
$x_{B,1}$	0.0432	0.187	7.24×10^{-9}	1.0×10^{-7}
$x_{B,2}$	1.0×10^{-7}	0.0	0.0	0.0
$x_{B,3}$	0.2392	0.271	2.64×10^{-6}	0.0
$x_{B,4}$	0.2392	0.271	2.214×10^{-4}	0.5002
$x_{B,5}$	0.2392	1.0×10^{-7}	0.0	0.4998
$x_{B,6}$	0.2392	0.271	*0.9999*	0.0
Reflux ratio	11.91	9.85	0.784	15
Reboil ratio	*9.15*	*28.5*	*4.8*	*5.52*
Number of stages	72	68	21	69
Feed tray location	20	16	8	50
V/F_0	6.37	17.5	0.8	1.84

7.6
HEAT INTEGRATION

Energy use is often an issue in distillation. This is particularly the case in locations where utility generation is near capacity, where energy is expensive relative to capital, and for products that are fuels. Fortunately, it is possible to set targets for minimum energy use and to find designs which approach these using a combination of pressure-shifting and multieffect distillation. For distillation, the basic idea was explained in King (1980). The main conceptual design approach is the "column stacking" method (Andrecovich and Westerberg, 1985*b*). It is important to keep in mind that heat integration with the rest of the process and utility systems can also be very important

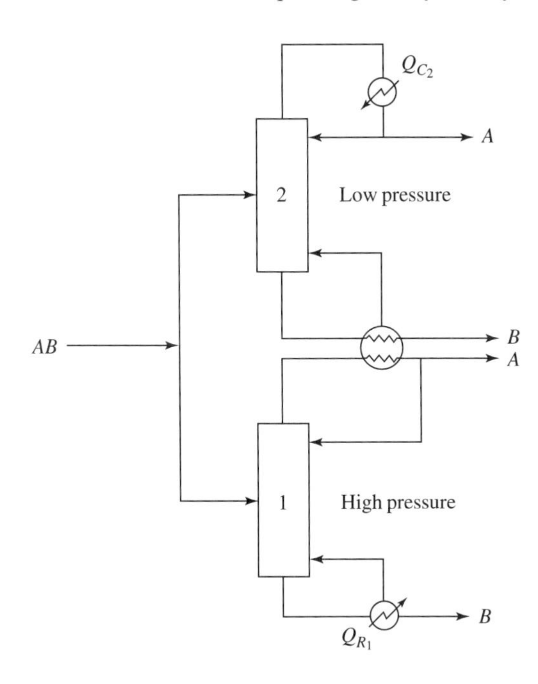

FIGURE 7.32
Multieffect distillation. A binary mixture of A and B is split into the correct proportions so that the overhead from the high-pressure column provides energy in an amount and at a temperature sufficient to drive the reboiler of the low-pressure column. The low-pressure column is "stacked" on the high-pressure column. *(Andrecovich and Westerberg, 1985b, Fig. 1.)*

in reducing energy use, e.g., Douglas (1988, Chap. 8), Smith (1995, Chaps. 6, 7, and 14), Linnhoff et al. (1982). The main ideas in the method are as follows.

By raising the pressure in a *system* of columns, each performing a *different* separation, the condenser in one column can be operated at a temperature higher than the reboiler of another column. Thus, that condenser can also serve as the reboiler of the second column. At the expense of raising the pressure and the temperatures for the high-pressure column, a single heat exchanger can replace one condenser and one reboiler as well as some of the associated external utilities for heating and cooling. It is also sometimes economical to split a given feed into two streams in order to distill one at high pressure and the other at lower pressure, so that the total energy use can be reduced, by coupling together two columns which perform the *same* separation. This is also sometimes called "multieffect distillation," and the column configuration for a two-stage system is shown in Fig. 7.32.

One general approach to heat integration is "pinch technology," e.g., Linnhoff et al. (1982).[15] In that approach, the temperature level and enthalpies of streams are represented on a "T-H" diagram. Each distillation task is represented on such a diagram by a box with width Q and height ΔT. Here Q is the reboiler or condenser duty; these duties are the same for CMO, adiabatic columns, and saturated liquid feeds. ΔT is the temperature difference between the condenser and the reboiler. Single-effect, double-effect, and a general multieffect distillation are shown in Fig. 7.33.[16]

[15]Don't confuse the use of the word "pinch" here with the composition pinch used throughout this book.

[16]A single column appears as a rectangle in this figure. For cases where the reboiler and condenser duties differ, a single column is represented by a quadrilateral with parallel top and bottom edges.

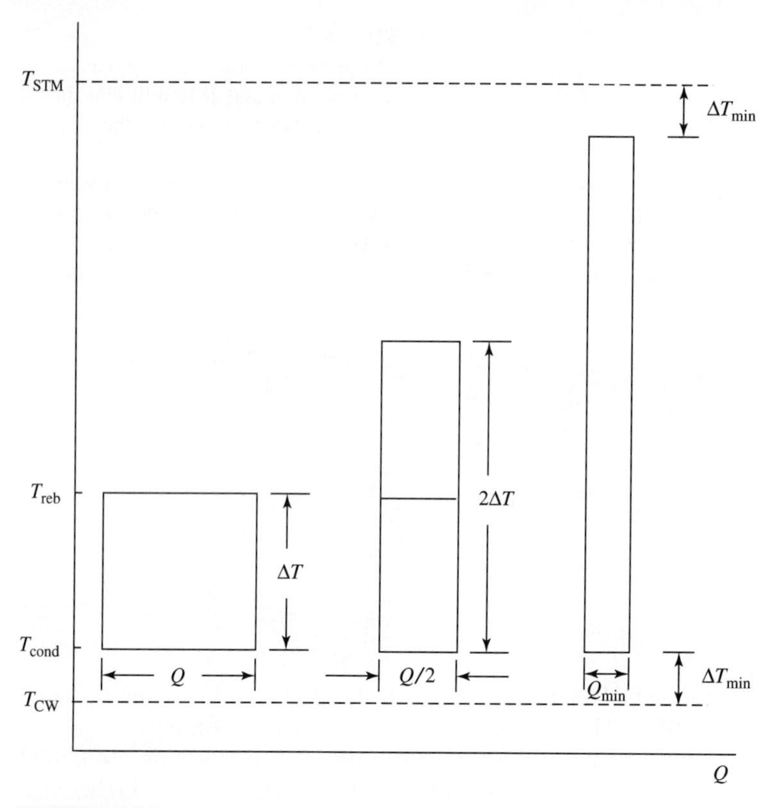

FIGURE 7.33
Temperature-enthalpy diagram for multieffect distillation. The minimum temperature driving force $\Delta T_{\min}$ along with the highest and lowest temperatures available for heating and cooling, determine the maximum column stacking possible. *(Andrecovich and Westerberg, 1985b, Fig. 2.)*

The heating and cooling requirements for columns making different separations will generally be different. Therefore, the heat available from condensing the overhead of one column will generally not match exactly with the heat required for another column in the system. In fact, without adjusting the pressures in the columns, it may be that no stacking is possible, e.g., Fig. 7.34*a*, where the heating and cooling requirement is $Q_1 + Q_2 + Q_3$.

By adjusting the column pressures, at least part of the energy load can generally be matched between columns making different separations. Sometimes, all of the columns in a system can be "stacked," within the available utility temperatures and without violating any temperature constraints on the mixture. In this case, the column with the largest heat load will determine the net energy requirement for the system. This is shown in Fig. 7.34*b*, where the heating and cooling are reduced to Q_3.

The order of column stacking is not unique, but stacking of the original columns in any order will eventually reach a limit where the utility loads are controlled by the column with the largest heating and cooling. Figure 7.35*a* shows an alternative arrangement for stacking the three columns discussed above. In addition to adjusting

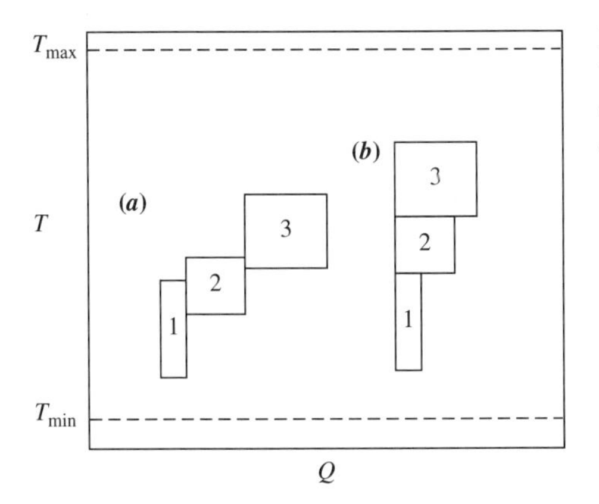

FIGURE 7.34
Temperature-enthalpy diagrams for a distillation system containing three columns. See the text for a discussion.

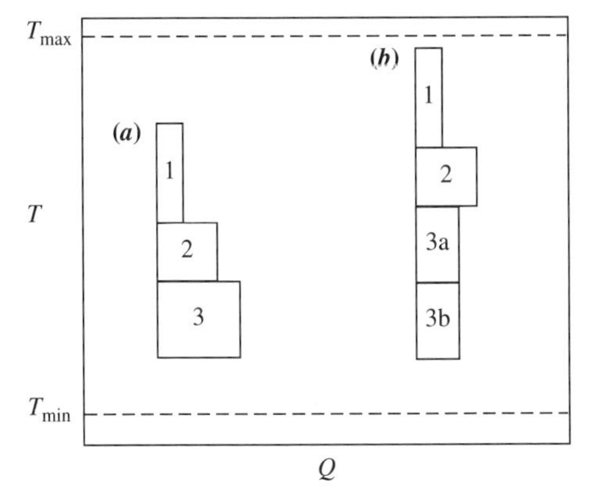

FIGURE 7.35
Alternatives for column stacking in distillation system containing three columns. See the text for a discussion.

pressures, the separation task in a single column can be done in a multieffect system by adding one or more *additional* columns to the system. Note that this will usually mean an increased capital cost, to be balanced against a reduced operating cost. Figure 7.35*b* shows the result when column 3 from part *a* of the figure is replaced with a double-effect distillation.

It is particularly simple to apply this method for heat integration if the quantity $Q\Delta T$ is taken as constant, which was suggested as a first approximation by Andrecovich and Westerberg (1985*b*). For instance, the double-effect distillation alternative in Fig. 7.35*b* shows the third column split into two columns, each with the same ΔT and processing half the original feed with half the utility loads. Therefore, to a first approximation, N effects in multieffect distillation are expected to reduce energy use by $1/N$.

The constancy of $Q\Delta T$ is not always an accurate approximation (Andrecovich and Westerberg, 1985*c*; Glinos et al., 1985). In those cases, utility bounds can still be found with linear programming methods (Andrecovich and Westerberg, 1985*a*).

One alternative configuration found by application of column stacking without multiple effect distillation to the mixture discussed in Example 7.2, is shown in Fig. 7.36. Note that the energy use is reduced substantially, but the total annual cost

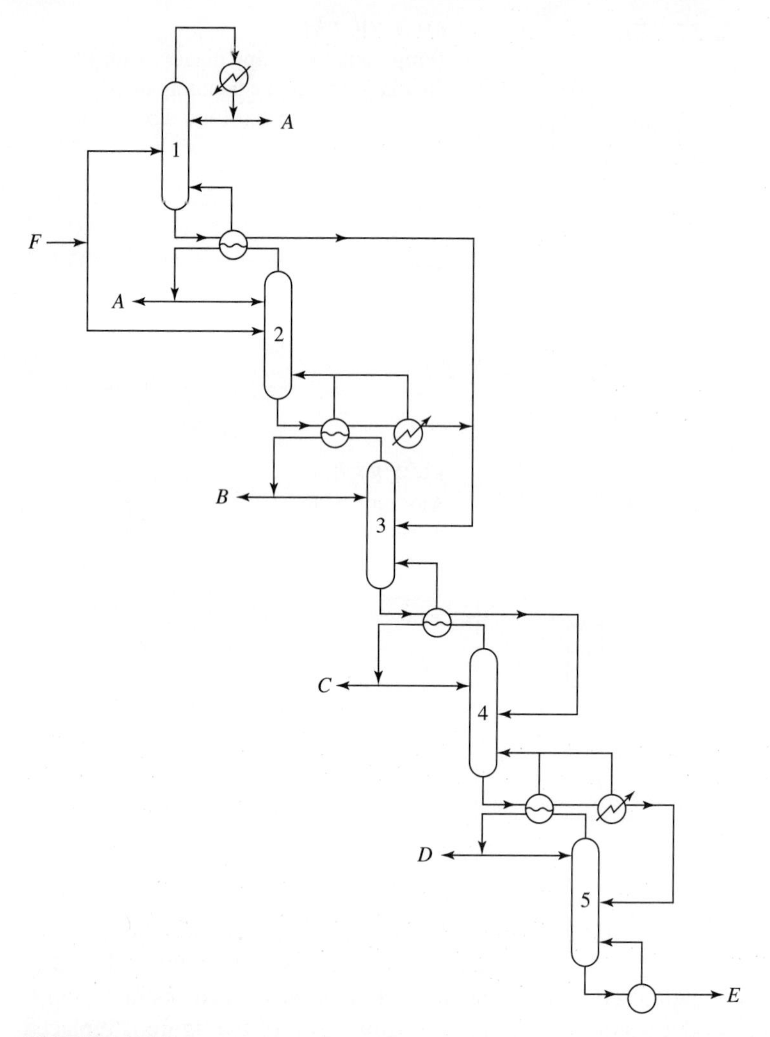

FIGURE 7.36
One heat-integrated system for Example 7.2. The energy use is reduced by 70% over the simple sequence, but the total annual cost is reduced by only 10%.

is not lowered in proportion, because the columns operate at higher pressures and because the heating for the highest-pressure column must be done with a much more expensive utility. For more details on this example and other alternatives, including multiple effects, see Andrecovich and Westerberg (1985*b*).

Multieffect distillation uses a number of columns that is larger than the minimum so that higher capital costs should be expected. The interconnection of columns is also more complex, which may lead to higher costs for control and operations. Despite these factors, multieffect distillation can yield substantial savings in energy use and costs. Certain combinations of volatilities, feeds, compositions, and flows have been shown to give energy reductions of 50% and cost reductions up to 35% (Siirola, 1996; Blakely, 1984) through the use of two coupled columns in place of a single column for binary distillation.

For nonideal systems, it is very important to recognize that changing the pressure can sometimes lead to significant changes in the phase behavior. For instance, azeotropes may appear (or disappear) as the pressure is changed. Knapp (1991, Chap. II) developed a systematic method to study the effect of pressure on the composition and boiling temperatures of azeotropes. Some azeotropes show substantial changes as the pressure varies, but a general rule for anticipating the changes is not known.

This means that a heat-integrated system or multiple effects derived via column stacking may not even be feasible. When heat integration using substantial pressure changes is contemplated, a study of the effects of pressure on the phase behavior is required. A good starting point is to compute all of the azeotropes in the mixture for the pressures contemplated, e.g., using the method described by Fidkowski et al. (1993*b*). Note that this may also require experimental study of the phase behavior at elevated pressures.

7.7 EXERCISES

1. Estimate the vapor rates and the number of stages corresponding to the first three sequences in Fig. 7.1. What is the most expensive split? Why does sequence 6 have a much higher total vapor rate than the others?
2. Prepare a spreadsheet or computer program to estimate the designs and costs for Example 7.1. Compare the best of these sequences to a solution found using heuristics.
3. How many different *splits* are involved in the 14 simple sequences for a five-component mixture listed in Table 7.1? Beginning with sequence 6, draw some sequences that use complex columns.
4. This question concerns a distillation sequence for a ternary mixture of ethanol, isopropanol, and *n*-propanol.
   a. Consider an equimolar feed at 1 atm pressure. Draw the two simple sequences and estimate the flows and temperatures of each product stream from each column.
   b. For each of the heuristics in Table 7.2, which sequence is favored? Which sequence would you choose, based only on the heuristics?
   c. Estimate the vapor rate and number of stages for each column in each sequence. Which sequence would you choose based on these results?
   d. If you developed a spreadsheet model for costs as an exercise in Chap. 6, use this to estimate the cost of each sequence. (If you do not have the spreadsheet or another automated way to make cost estimates, skip this part of the problem!)
   e. How do your answers to these questions change if there is only 5 mol% ethanol in the feed, with the balance equal parts of isopropanol and *n*-propanol?

5. Start with the maximally interconnected STN for a ternary mixture given in Fig. 7.13*a* and devise all possible STNs and the corresponding column sequences. If there are multiple column sequences for a given STN, just draw one of them and make a statement that others are possible. Identify which STNs correspond to the column sequences shown in Fig. 7.5.
6. Develop the alternative STNs and the associated sequences for STN 3 in Fig. 7.14 by eliminating condensers and reboilers for intermediate sections; i.e., develop the partially coupled systems at the Rule 2 level.
7. Suppose that a binary mixture of methanol and water is separated by distillation. The feed contains 20 mol % methanol and is available at 1 atm pressure as a liquid near its boiling point. For a high-purity product, say 99.5% methanol and 99.5% fractional recovery, when is it economical to use a double-effect distillation?
8. Draw complex column arrangements that you think might be attractive alternatives to the simple sequences for Exercise 4. First, consider the equimolar feed and then the feed containing only a small amount of ethanol. Explain why you think the complex column arrangements might be attractive. Now estimate the difference in number of stages and vapor rates using a simulation.
9. If you solved Exercise 4, take the best simple sequence and estimate how much the energy use and costs could be reduced by "column stacking" without using multiple effects. If only double-effect distillation is used for the column with the largest utility loads, how much can the energy use be reduced?

References

Agrawal, R., "Synthesis of Distillation Column Configurations for a Multicomponent Separation," *Ind. Eng. Chem. Research,* **35,** 1059–1071 (1996).

Agrawal, R., "More Operable Fully Thermally Coupled Distillation Column Configurations for Multicomponent Distillation," *TransIChemE,* **77**(6), 543–553 (1999).

Agrawal, R., and Z. T. Fidkowski, "Are Thermally Coupled Distillation Columns Always Thermodynamically More Efficient for Ternary Distillation?" *Ind. Eng. Chem. Research,* **37,** 3444–3454 (1998*a*).

Agrawal, R., and Z. T. Fidkowski, "More Operable Arrangements of Fully Thermally Coupled Distillation Columns," *AIChE J.,* **44,** 2565–2568 (1998*b*).

Agrawal, R., and Z. T. Fidkowski, "New Thermally Coupled Schemes for Ternary Distillation," *AIChE J.,* **45,** 485–496 (1999).

Agrawal, R., and Z. T. Fidkowski, Improving Efficiency of Distillation with New Thermally Coupled Configurations of Columns, in Malone, M. F., and J. A. Trainham, editors, *Foundations of Computer-Aided Process Design,* number 323 in AIChE Symp. Ser., pp. 381–384, New York. AIChE (2000).

Ahmad, B. S., and P. I. Barton, "Homogeneous Multicomponent Azeotropic Batch Distillation," *AIChE J.,* **42,** 3419–3433 (1996).

Andrecovich, M. J., and A. W. Westerberg, "An MINLP Formulation for Heat-Integrated Distillation Sequence Synthesis," *AIChE J.,* **31,** 1461–1474 (1985*a*).

Andrecovich, M. J., and A. W. Westerberg, "A Simple Synthesis Method Based on Utility Bounding for Heat-Integrated Distillation Sequences," *AIChE J.,* **31,** 363–375 (1985*b*).

Andrecovich, M. J., and A. W. Westerberg, "Utility Bounds for Nonconstant $Q\Delta T$ for Heat-Ingegrated Distillation Sequence Synthesis," *AIChE J.,* **31,** 1475–1479 (1985*c*).

Baburina, L. V., and V. M. Platonov, "Analysis of Separating Manifolds in Multidimensional Simplices of Polyazeotropic Mixutres," *Khimicheskya Promyshlennost,* **19,** 35–37 (1987).

Baburina, L. V., and V. M. Platonov, "Application of Theory of Conjugate Tie Lines to Structural Analsyis of Three- and Four-Component Mixtures," *Theor. Found. Chem. Engng.,* **24,** 287–291 (1990*a*).

Baburina, L. V., and V. M. Platonov, "Application of Theory of Conjugate Tie Lines to Structural Analysis of Five-Component Polyazeotropic Mixtures," *Theor. Found. Chem. Engng.,* **24,** 382–387 (1990*b*).

Baburina, L. V., V. M. Platonov, and M. G. Slin'ko, "Classification of Vapor-Liquid Phase Diagrams for Homoazeotropic Systems," *Theor. Found. Chem. Engng.,* **22,** 390–396 (1988).

Blakely, D. M., Cost Savings in Binary Distillation through Two-Column Designs, Master's thesis, Clemson Univeristy, Clemson SC (1984).

Christiansen, A. C., S. Skogestad, and K. Lein, “Complex Distillation Arragnements: Extending the Petlyuk Ideas,” *Computers Chem. Engng.,* **21** (Suppl. S), S237–S242 (1997).

Corman, T., C. Leiserson, and R. Rivest, *Introduction to Algorithms.* McGraw-Hill, New York (1990).

Davydyan, A. G., M. F. Malone, and M. F. Doherty, “Boundary Modes in a Single Feed Distillation Column for the Separation of Azeotropic Mixtures,” *Theor. Found. Chem. Engr.,* **31,** 327–338 (1997).

Doherty, M. F., and J. D. Perkins, “On the Dynamics of Distillation Processes—I. The Simple Distillation of Multicomponent Non-Reacting Homogeneous Liquid Mixtures,” *Chem. Engng. Sci.,* **38,** 281–301 (1978).

Douglas, J. M., *The Conceptual Design of Chemical Processes.* McGraw-Hill, New York (1988).

Fidkowski, Z. T., M. F. Doherty, and M. F. Malone, “Feasibility of Separations for Distillation of Nonideal Ternary Mixutres,” *AIChE J.,* **39,** 1303–1321 (1993*a*).

Fidkowski, Z. T., and L. J. Krolikowski, “Minimum Energy Requirements of Thermally Coupled Distillation Systems,” *AIChE J.,* **33,** 643–653 (1987).

Fidkowski, Z. T., M. F. Malone, and M. F. Doherty, “Computing Azeotropes in Multicomponent Mixtures,” *Computers Chem. Engng.,* **17,** 1141–1155 (1993*b*).

Foucher, E. R., M. F. Doherty, and M. F. Malone, “Automatic Screening of Entrainers for Homogeneous Azeotropic Distillation,” *Ind. Eng. Chem. Research,* **29,** 760–772 (1991).

Gaminibandara, K., and R. W. H. Sargent, Optimal Design of Plate Distillation Columns, in Dixon, L. W. C., editor, *Optimization in Action,* p. 267, London. Academic Press (1976).

Glinos, K., *A Global Approach to the Preliminary Design and Synthesis of Distillation Trains,* Ph.D. thesis, University of Massachusetts, Amherst, MA (1984).

Glinos, K., and M. F. Malone, “Minimum Reflux, Product Distribution and Lumping Rules for Multicomponent Distillation,” *Ind. Eng. Chem. Process Design Dev.,* **23,** 764 (1984).

Glinos, K., M. F. Malone, and J. M. Douglas, “Shortcut Evaluation of ΔT and $Q\Delta T$ for the Synthesis of Heat Integrated Distillation Sequences,” *AIChE J.,* **31,** 1039 (1985).

Glinos, K. N., and M. F. Malone, “Optimality Regions for Complex Column Alternatives in Distillation Systems,” *Chem. Engng. Res. Design,* **66,** 229–240 (1988).

Gmehling, J., J. Menke, K. Fischer, and J. Krafczyk, *Azeotropic Data. Part I.* VCH, New York (1994*a*).

Gmehling, J., J. Menke, K. Fischer, and J. Krafczyk, *Azeotropic Data. Part II.* VCH, New York (1994*b*).

Julka, V., and M. F. Doherty, “Geometric Nonlinear Analysis of Multicomponent Nonideal Distillation: A Simple Computer-Aided Design Procedure,” *Chem. Engng. Sci.,* **48,** 1367–1391 (1993).

Kabel, G., “Distillation Columns with Vertical Partitions,” *Chem. Eng. Technol.,* **10,** 92 (1987).

King, C. J., *Separation Processes.* McGraw-Hill, New York (1971).

King, C. J., *Separation Processes.* McGraw-Hill, New York, 2d ed. (1980).

Knapp, J. P., *Exploiting Pressure Effects in the Distillation of Homogeneous Azeotropic Mixtures,* Ph.D. thesis, University of Massachusetts, Amherst, MA (1991).

Knight, J. R., and M. F. Doherty, Systematic Approaches to the Synthesis of Separation Schemes for Azeotropic Distillation, in Siirola, J. J., I. E. Grossmann, and G. Stephanopoulos, editors, *Foundations of Computer-Aided Process Design.* Elsevier, New York (1990).

Krolikowski, L. J., A. G. Davidyan, M. F. Doherty, and M. F. Malone, Exact Bounds on the Feasible Products for Distillation of Ternary Azeotropic Mixtures (1996). Paper 90f, 1996 AIChE Annual Meeting, Chicago, IL.

Linnhoff, B., D. W. Townsend, D. Boland, and G. F. Hewitt, *A User Guide on Process Integration for the Efficient Use of Energy.* Institution of Chemical Engineers, Rugby (Warwichshire) (1982).

Malone, M. F., K. Glinos, F. E. Marquez, and J. M. Douglas, "Simple, Analytical Criteria for the Sequencing of Distillation Columns," *AIChE J.,* **31,** 683 (1985).

McHugh, J. A., *Algorithmic Graph Theory.* Prentice-Hall, Englewood Cliffs, NJ (1990).

Morari, M., and D. C. Faith, III, "Synthesis of Distillation Trains with Heat Integration," *AIChE J.,* **26,** 916–928 (1980).

Mutalib, M. I. A., and R. Smith, "Operation and Control of Dividing Wall Distillation Columns—Part 1: Degrees of Freedom and Dynamic Simulation," *Chem. Engng. Res. Design,* **76**(A3), 308–318 (1998).

Mutalib, M. I. A., A. O. Zeglam, and R. Smith, "Operation and Control of Dividing Wall Distillation Columns—Part 2: Simulation and Pilot Plant Studies Using Temperature Control," *Chem. Engng. Res. Design,* **76**(A3), 319–334 (1998).

Nishida, N., G. Stephanopoulos, and A. W. Westerberg, "A Review of Process Synthesis." *AIChE J.,* **27,** 321 (1981).

Petlyuk, F. B., V. M. Platonov, and D. M. Slavinskii, "Thermodynamically Optimal Method for Separating Multicomponent Mixtures," *Intl. Chem. Eng.,* **5,** 555–561 (1965).

Rooks, R. E., *Feasibility and Column Sequencing for the Distillation of Homogeneous Multicomponent Azeotropic Mixtures,* Ph.D. thesis, University of Massachusetts, Amherst, MA (1997).

Rooks, R. E., V. Julka, M. F. Doherty, and M. F. Malone, "Structure of Distillation Regions for Multicomponent Azeotropic Mixtures," *AIChE J.,* **44,** 1382–1391 (1998).

Safrit, B. T., and A. W. Westerberg, "Algorithm for Generating the Distillation Regions for Multicomponent Mixtures," *Ind. Eng. Chem. Research,* **36,** 1827–1840 (1997).

Sargent, R. W. H., "A Functional Approach to Process Synthesis and Its Application to Distillation Systems," *Computers Chem. Engng.,* **22,** 31–45 (1998).

Siirola, J. J., "Industrial Applications of Chemical Process Synthesis," *Adv. Chem. Engng.,* **23,** 1–62 (1996).

Smith, R., *Chemical Process Design.* McGraw-Hill, New York (1995).

Tedder, D. W., and D. F. Rudd, "Parametric Studies in Industrial Distillation, Parts I. and II.," *AIChE J.,* **24,** 303–334 (1978).

Wahnschafft, O. M., J. W. Koehler, E. Blass, and A. W. Westerberg, "The Product Composition Regions for Single-Feed Azeotropic Distillation Columns," *Ind. Eng. Chem. Research,* **31,** 2345–2362 (1992).

Wahnschafft, O. M., J. P. L. Rudulierand, and A. W. Westerberg, "A Problem Decomposition Approach for the Synthesis of Complex Separation Process with Recycles," *Ind. Eng. Chem. Research*, **32,** 1121–1141 (1993).

8

Heterogeneous Azeotropic Distillation

8.1 INTRODUCTION

Heterogeneous azeotropic distillation, or simply *azeotropic distillation*, is used widely for separating nonideal mixtures. The technique uses minimum-boiling azeotropes and liquid-liquid immiscibilities in combination to defeat the presence of other azeotropes or tangent pinches in the mixture that would otherwise prevent the desired separation. The azeotropes and liquid heterogeneities that are used to make the desired separation feasible may either be induced by the addition of a separating agent, usually called an *entrainer*, or they may be already present, in which case the mixture is sometimes called *self-entrained*. The most common case is the former; it includes such classic separations as ethanol dehydration using either benzene, heptane, ethyl ether, etc., as the entrainer, and acetic acid recovery from water using either ethyl acetate, 1-propyl acetate, or 1-butyl acetate as the entrainer. In ethanol dehydration the entrainer is used to break the homogeneous minimum-boiling azeotrope between ethanol and water, but in the acetic acid recovery process the entrainer is used to overcome the tangent pinch between acetic acid and water.

Heterogeneous entrainers have been used to separate binary azeotropic mixtures into their constituent pure components since the turn of the century. The first successful application was discovered by Young (1902), who prepared absolute alcohol from a binary mixture of ethanol and water using benzene as the entrainer; he received a patent for this batch distillation process (Young, 1903). This was later converted to a continuous process and patented by Kubierschky in 1915 (Kubierschky, 1915). Keyes (1929) describes several modifications of this continuous process, and Guinot and Clark (1938) give a good review of the early development and widespread application of continuous azeotropic distillation in the prewar chemical industry. One of the main contributors to the development and application of this technique was Donald Othmer, who wrote many papers and patents in the 1930s and 1940s (Othmer, 1982). Many of his processes are still used today.

Until recently azeotropic distillation processes were developed on an individual basis using experimentation to guide the design. While there is no substitute for experimental verification of a process, it is not an efficient method for generating initial designs. Following the treatment given previously for extractive distillation, we will use residue curve maps as a vehicle to explain the behavior of entire systems of heterogeneous azeotropic distillation columns as well as the individual columns that make up the system. This approach provides a unifying framework for the subject, it can be applied rapidly, and it produces an excellent starting point for detailed simulations and experiments. Several leading chemical companies now follow this procedure for designing azeotropic distillations.

8.2
PHASE DIAGRAMS

For binary mixtures, it is well known that when a liquid-liquid envelope merges with a minimum-boiling vapor-liquid phase envelope, the resulting heterogeneous azeotropic phase diagram has the form shown in Fig. 8.1. When the liquid composition $x_1 = x_1^{Az}$ in Fig. 8.1*a* then the vapor composition y_1 is also equal to x_1^{Az} and the mixture boils at constant temperature and at constant (and equal) composition in each phase. When the overall liquid composition $x_1^0 = (x_1^0)^{Az}$ in Fig. 8.1*b*, then y_1 is also equal to $(x_1^0)^{Az}$ and again the mixture boils at constant temperature and at constant composition in each phase. However, the liquid of composition $(x_1^0)^{Az}$ splits into two liquid phases so that the *three* coexisting equilibrium phases have different compositions. Therefore, homogeneous and heterogeneous azeotropes share the common property that the overall liquid composition is equal to the vapor composition. This characteristic property of azeotropes provides a means for finding them experimentally and computationally. At homogeneous azeotropes, the equilibrium surfaces are tangent and stationary. This provides a second method for locating the azeotropic point. No such condition exists at heterogeneous azeotropes.

The properties of ternary heterogeneous vapor-liquid-liquid equilibrium (VLLE) phase diagrams are not well documented, but they are all-important for

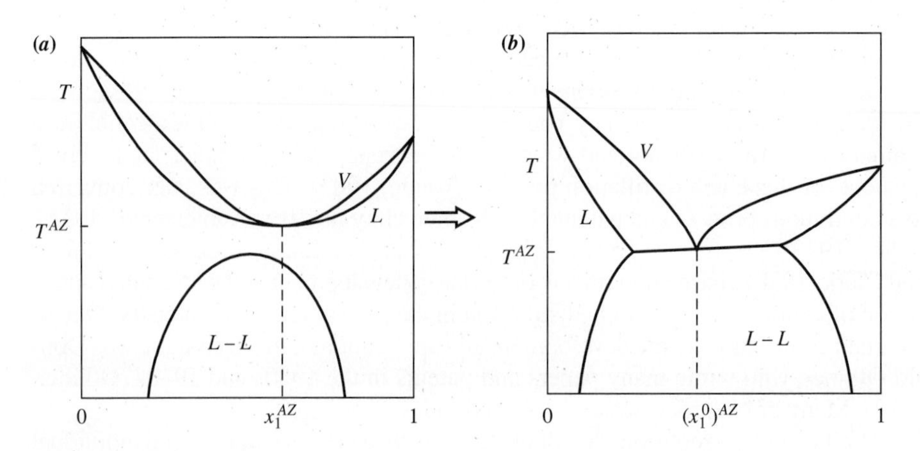

FIGURE 8.1
Schematic isobaric phase diagrams for binary azeotropic mixtures. (*a*) Homogeneous azeotrope. (*b*) Heterogeneous azeotrope.

understanding azeotropic distillation. We will describe the simplest phase diagram, shown in Fig. 8.2. This figure shows what happens when a liquid-liquid envelope merges with a vapor-liquid equilibrium (VLE) surface containing a single minimum-boiling binary azeotrope. The characteristic feature of Fig. 8.2*b* is the existence of a *heterogeneous liquid boiling surface*, as shown on the figure. This surface is defined as follows. When the overall liquid composition lies inside the heterogeneous boiling envelope, and the temperature is raised to the point where it lies on the heterogeneous boiling surface, then the liquid will boil and split into two equilibrium liquid phases and one coexisting vapor phase. An important point to note is that when the overall liquid composition lies on the boiling surface, the Gibbs phase rule requires that the locus of all the corresponding equilibrium vapor compositions forms a *curve* in T-$\mathbf{y}$ space and *not* a surface, as they do in the homogeneous region. The *critical line* is the locus of liquid-liquid critical points.

A convenient way of representing the T-$\mathbf{x}$-$\mathbf{y}$ phase diagram in Fig. 8.2*b* is by projection onto the composition triangle at the base of the figure. It is understood that the temperature varies from point to point on the projected vapor line and on the projected boiling envelope. The latter looks like an isothermal liquid-liquid binodal envelope, but we should emphasize that it is not. Each tie-line across the boiling envelope is associated with a different boiling temperature (see Fig. 8.3).

Experimental vapor-liquid-liquid phase equilibrium data for ternary mixtures are scarce. A list of all the data sources that we are aware of is given in Table 8.1. Little has been published on the experimental determination of low-pressure VLLE phase behavior in the last twenty years. This is unfortunate because during the same period our ability to model these systems has improved greatly, but without experimental

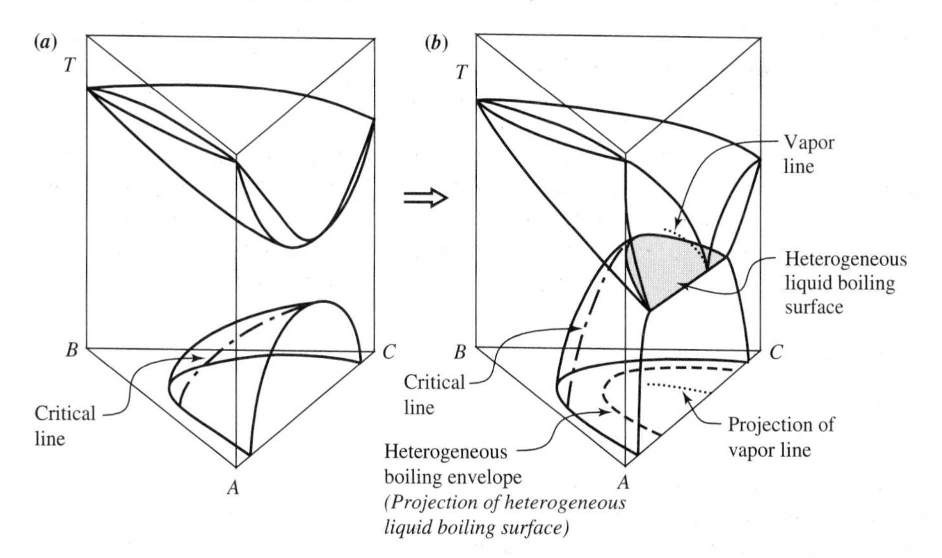

FIGURE 8.2
Schematic isobaric phase diagrams for ternary azeotropic mixtures. (*a*) Homogeneous liquid. (*b*) Partially miscible liquids. Note that the bubble-point line for the BC binary mixture is not visible because it is hidden behind the dew point surface.

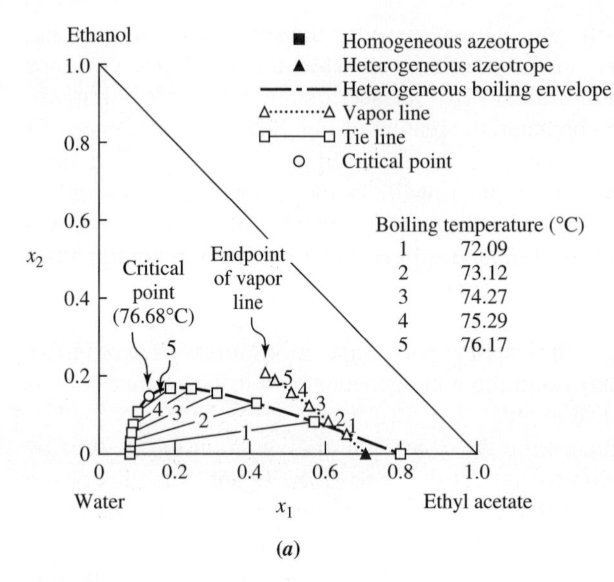

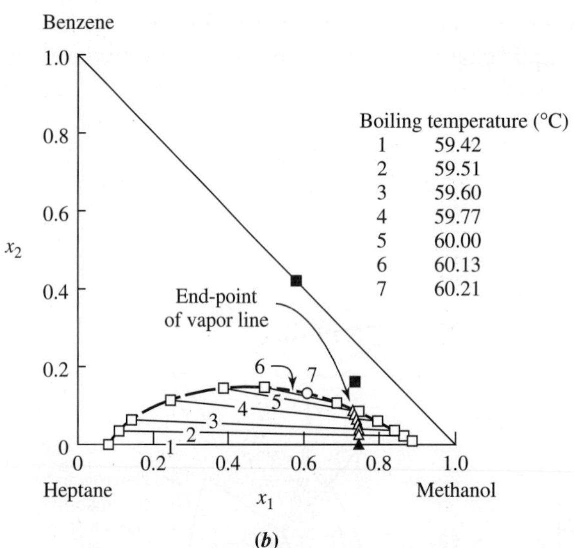

FIGURE 8.3
Isobaric vapor-liquid-liquid (VLLE) phase diagrams calculated at 1 atm for the systems. (*a*) Ethanol-water-ethyl acetate. (*b*) Benzene-heptane-methanol. Figures (*a*) and (*b*) were calculated using the two-parameter Margules model to represent the liquid phase activity coefficients. *(Taken from Pham and Doherty, 1990a.)*

data it is not possible to acquire good model parameters nor is it possible to test the reliability of the predictions after the modeling effort has taken place. Figure 8.3 gives a selection of computed phase diagrams for progressively more complex mixtures. Figure 8.3*a* and *b* shows the phase diagrams for ethanol-water-ethyl acetate and benzene-heptane-methanol at 1 atm pressure, respectively. In both these figures the vapor line begins at the binary heterogeneous azeotrope (as it must) and moves inside the triangle. It may be entirely contained within the boiling envelope, as in Fig. 8.3*b*, or it may move outside the envelope, as in Fig. 8.3*a*. In either case the

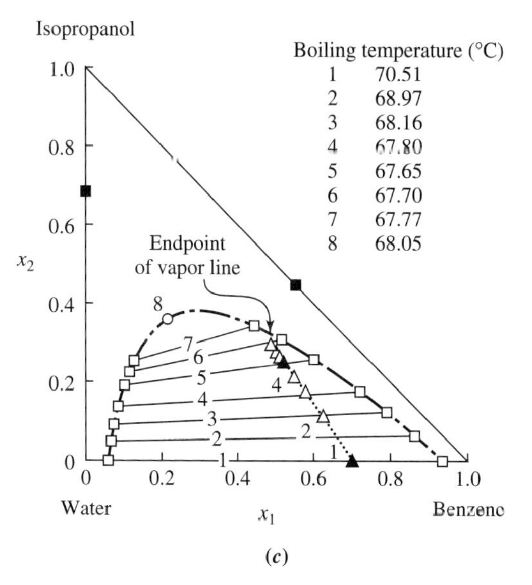

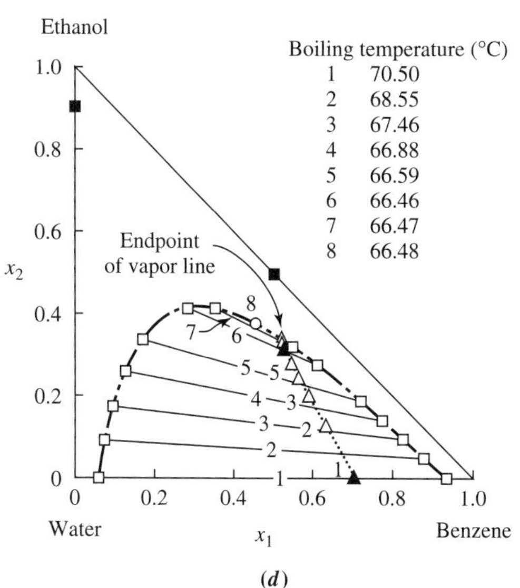

FIGURE 8.3 (*Continued*) Isobaric vapor-liquid-liquid (VLLE) phase diagrams calculated at 1 atm for the systems. (*c*) Isopropanol-water-benzene. (*d*) Ethanol-water-benzene. Figures (*c*) and (*d*) were calculated using the regular solution model. *(Taken from Pham and Doherty, 1990a.)*

endpoint of the vapor line is in equilibrium with the liquid at the critical point on the boiling envelope. On these figures, liquids and vapors in equilibrium with each other are signified by a common number. For example, in Fig. 8.3*a* the coexisting liquids on tie-line number 1 are in equilibrium with vapor number 1 at a boiling temperature of 72.09°C.

In the three-dimensional temperature-composition diagram, the vapor line must lie above the liquid boiling surface. It may, however, touch the boiling surface at a point. At the point of contact, the overall liquid composition is equal to the

TABLE 8.1
Sources of data for ternary vapor-liquid-liquid equilibrium systems

System	Reference
Methanol-water-ethyl acetate	van Zandijcke and Verhoeye (1974)
Ethanol-water-benzene	Barbaudy (1927), Norman (1945)
Ethanol-water-trichloroethylene	Reinders and DeMinjer (1947*a*)
Ethanol-water-ethyl acetate	van Zandijcke and Verhoeye (1974)
1-Propanol-water-1-butanol	Newsham and Vahdat (1977)
Isopropanol-water-diisopropylamine	Komarov and Krichevtsov (1970)
Allyl alcohol-water-trichloroethylene	Hands and Norman (1945)
Allyl alcohol-water-carbon tetrachloride	Hands and Norman (1945)
Formic acid-water-1,2 dichloroethane	Bushmakin and Molodenko (1964*b*)
Formic acid-water-*m*-xylene	Reinders and DeMinjer (1947*b*)
Acetic acid-water-*p*-xylene	Murogova et al. (1973)
Acetone-water-chloroform	Reinders and DeMinjer (1947*c*)
Acetone-water-phenol	Schreinemakers (1901)
Acrylonitrile-acetonitrile-water	Blackford and York (1965)
Isopropylamine-water-diisopropylamine	Komarov and Krichevtsov (1970)
Nitrocyclohexane-cyclohexanone oxime-water	Lutugina and Soboleva (1970)
Hexane-benzene-sulpholane	Rawat et al. (1980)
Ethanol-benzene-carbon tetrachloride	Schreinemakers (1903*a*)

equilibrium vapor composition, and consequently the liquid must boil at constant composition in each of the three equilibrium phases. This is precisely what we mean by a heterogeneous azeotrope. Figure 8.3*c* and *d* shows phase diagrams for mixtures that display ternary heterogeneous azeotropes; experimental VLLE data for the ethanol-water-benzene mixture is shown in Fig. 8.4. For each of these mixtures the ternary azeotrope is minimum-boiling, as can be seen by following the temperature of successive tie-lines. Heterogeneous saddle azeotropes are also possible, e.g., formic acid-water-*m*-xylene (Reinders and DeMinjer, 1947*b*), acetone-water-chloroform (Reinders and DeMinjer, 1947*c*) (see Figs. 8.7*d* and 8.9); however, maximum-boiling heterogeneous azeotropes cannot exist (Matsuyama, 1978). This is not the case for homogeneous azeotropes, where all three types are found in nature.

It is worth pointing out an interesting property of VLLE systems which at first sight seems quite unusual. Prokopakis and Seider (1983*b*, Fig. 2) noted that a large number of homogeneous liquid compositions around the perimeter of the liquid boiling envelope give rise to equilibrium vapor compositions that lie along a narrow path inside the heterogeneous region. In fact, this is a generic property of VLLE systems. All liquid compositions lying *on* the liquid boiling envelope will produce equilibrium vapor compositions *on* the vapor line. As the liquid composition moves away from the liquid boiling envelope into the homogeneous region, the corresponding equilibrium vapor composition moves off the vapor line. However, the cusp in the vapor surface on either side of the vapor line ensures that a band of homogeneous liquid compositions around the entire perimeter of the liquid boiling envelope will produce a corresponding set of equilibrium vapor compositions that lie inside a strip containing the vapor line. The width of the liquid band and the narrowness of the vapor strip are determined by the steepness of the cusp in the vapor surface. This

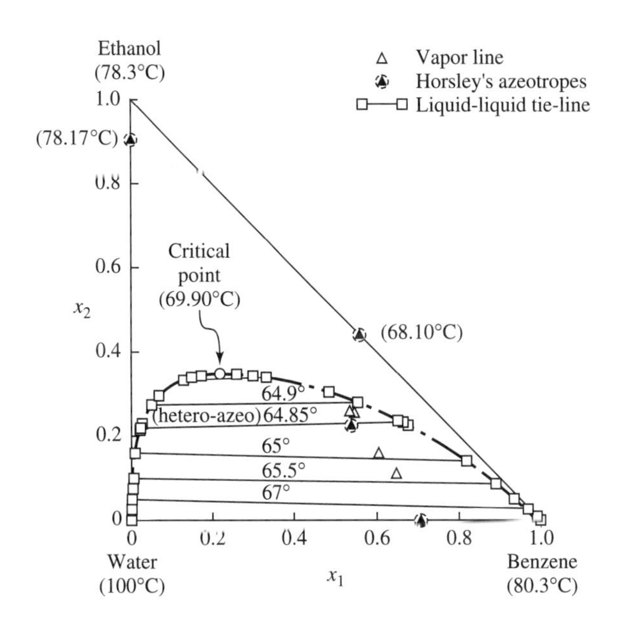

FIGURE 8.4
Experimental vapor-liquid-liquid equilibrium data for the system ethanol-water-benzene at 1 atm. Compositions are in mole fractions. The VLLE data are taken from Barbaudy (1927) and Norman (1945); see Pham and Doherty (1990*a*) for additional commentary on these data. The azeotropic data are taken from Horsley (1973).

is a useful observation that we will return to when setting composition targets for azeotropic distillation columns.

The typical phase equilibrium problem encountered in distillation is to calculate the boiling temperature and the vapor composition in equilibrium with a liquid phase of specified composition at a given pressure. If the liquid phase-separates, then the problem is to calculate the boiling temperature and the compositions of the two equilibrium liquid phases plus the coexisting vapor phase at the specified *overall* liquid composition. Therefore, given P and $\mathbf{x}^0$, the first task is to decide whether the liquid is homogeneous or heterogeneous. The question cannot be answered until the boiling temperature is known. The typical method of attack is to guess the boiling temperature, perform a liquid-phase stability check, and on the basis of the outcome to perform either a VLE calculation step or a VLLE calculation step. At the end of either of these steps, a new estimate of the boiling temperature will be available, and the process is repeated until the equilibrium equations are satisfied. During the course of these calculations, liquid phases may come and go, but ultimately the correct number of equilibrium phases will be found (with respect to the model) if a reliable stability check is used.

If a poor stability check or if no stability check is used, this may lead to the conclusion that there is one liquid phase present at equilibrium when in fact the model predicts there are two or more. Even worse, when this happens, it is normally accompanied by the existence of multiple spurious solutions to the phase equilibrium equations (Baker et al., 1982; Van Dongen et al., 1983). The difficulty is caused by a classical problem in thermodynamics, how to distinguish between unstable, metastable, and absolutely stable phases. This problem was actually solved over a hundred years ago by Gibbs (1873, 1875), who formulated the *Gibbs tangent plane test*. However, it is only in recent years that a robust and practical numerical implementation of this test has been developed (Baker et al., 1982; Michelsen, 1982*a*, 1982*b*). It is now generally agreed that Michelsen's implementation (or variations on it) of the Gibbs tangent plane test is the best method for solving multiphase equilibrium problems (Swank and Mullins, 1986; Cairns and Furzer, 1990*b*; Widagdo et al., 1992; Wasylkiewicz et al., 1996; McDonald and Floudas, 1995; McDonald and Floudas, 1997).

Armed with these techniques and a nonideal solution model that is capable of predicting multiple liquid phases, it is a routine exercise to produce phase diagrams like those shown in Fig. 8.3. It is an entirely different matter, however, to make the predictions quantitatively accurate. Pham and Doherty (1990*a*) compare predictions made by the regular solution and NRTL models to VLLE data for ternary mixtures of ethanol, water, and benzene at 1 atm pressure. They conclude that when the model parameters are based on either binary VLE data alone or on ternary LLE data alone neither of these models gives satisfactory performance over the full range of temperatures and compositions encountered in azeotropic distillation. The UNIQUAC model does not seem to perform any better. Prausnitz et al. (1980, p. 66) show that the UNIQUAC equation with parameters fitted to binary VLE data gives unsatisfactory ternary liquid-liquid predictions. In addition, Prokopakis and Seider (1983*b*) and Rovaglio and Doherty (1990) show that the UNIQUAC model, with parameters based on ternary VLE data alone (parameters taken from Gmehling and Onken, 1977*a*, p. 642), also gives unsatisfactory predictions for mixtures of ethanol, water, and benzene. Cairns and Furzer (1990*c*) examined the mixture 1-propanol, water, and 1-butanol and also concluded that none of the models studied (including the NRTL, UNIFAC, and UNIQUAC models with interaction parameters based on either VLE or LLE data alone) gave satisfactory predictions of the liquid boiling envelope. These results are in general agreement with the advice given by Prausnitz et al. (1980, pp. 63-71), who recommend fitting ternary LLE data simultaneously with binary VLE data in order to obtain the best overall model parameters. To date, the best conventional model for predicting ternary VLLE phase behavior in organic-water mixtures is the modified UNIQUAC equation with parameters based on both binary VLE and ternary LLE data (Prausnitz et al., 1980, Chap. 4).[1] Parameters for such a model are reported in Prausnitz et al. (1980, p. 65) for 10 ternary mixtures, including the system ethanol-water-benzene. This model is now used in most of the recent literature on azeotropic distillation (Prokopakis and Seider, 1983*a*; Prokopakis and Seider, 1983*b*; Venkataraman and Lucia, 1988; Ryan and Doherty, 1989; Rovaglio and Doherty, 1990). See Fig. 8.5 for a comparison with experimental data. Unfortunately, the most extensive source of VLE and LLE data, the DECHEMA Chemistry Data Series, reports parameters for the original UNIQUAC model by fitting binary VLE data alone, ternary VLE data alone, or ternary LLE data alone. However, the models reported in the DECHEMA Chemistry Data Series were not necessarily intended for predicting ternary VLLE phase behavior and should be used with caution for such calculations.

---

[1]More recent studies offer improved models, such as Talley et al. (1993), at the expense of extensive experimental data.

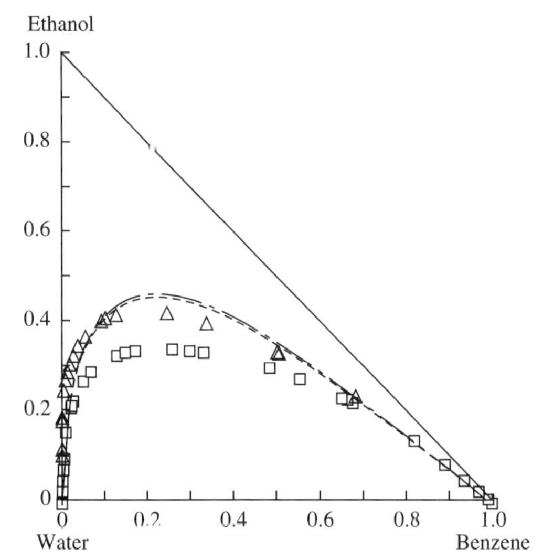

FIGURE 8.5
Comparison of the heterogeneous boiling envelope and binodal curve calculated using the modified UNIQUAC model (Prausnitz et al., 1980, p. 65) with experimental data at 1 atm. Model parameters taken from Prokopakis and Seider (1983*b*, Table 1).

8.3
RESIDUE CURVE MAPS

In a heterogeneous simple distillation process, a multicomponent partially miscible liquid mixture is vaporized in a still, and the vapor that is boiled off is treated as being in phase equilibrium with all the coexisting liquid phases. The vapor is then withdrawn from the still as distillate (see Fig. 8.6). Because there is no feed and no reflux, the composition of each individual liquid phase changes continuously with time. The changing liquid composition is most conveniently described by following the trajectory (or residue curve) of the overall composition of all the coexisting liquid phases. Pioneering experimental and theoretical studies of such systems were carried out by Schreinemakers (1901, 1903*b*), Reinders and DeMinjer (1940, 1947*c*), and Matsuyama (1978). Reinders and De Minjer (1947*c*) report an extensive amount of valuable experimental data for the mixture acetone-water-chloroform, including binary and ternary LLE, VLE, and VLLE data, and both simple distillation and batch distillation residue curves. Experimentally determined simple distillation residue curves are also reported by Bushmakin and Molodenko (1964*a*)[2] for the heterogeneous

[2]Note, in the Russian literature simple distillation residue curves are normally referred to as "distillation lines," and batch distillation residue curves are referred to as "rectification lines."

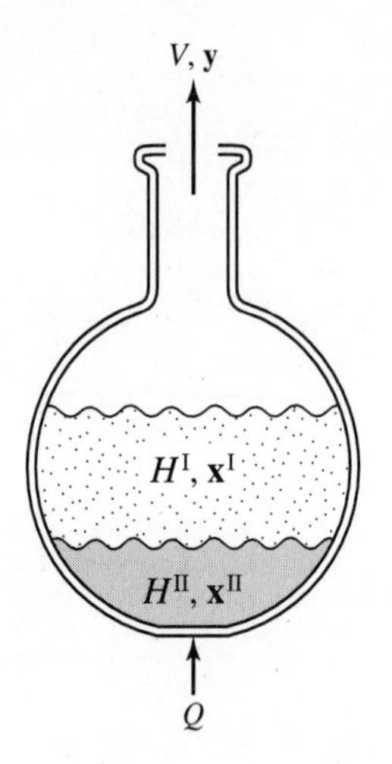

FIGURE 8.6
Schematic representation of the open evaporation of a partially miscible liquid.

system formic acid-water-1,2,-dichloroethane, and infinite reflux distillation paths have been measured by Furzer (1985) and Cairns and Furzer (1990*a*).

It is also possible to calculate residue curves by integrating the differential equation describing their trajectories. This equation is derived by performing an overall material balance about the still shown in Fig. 8.6 to give

$$\frac{dH}{dt} = -V \tag{8.1}$$

where the total molar holdup H of the two liquid phases is

$$H = H^I + H^{II} \tag{8.2}$$

and V is the molar flowrate of escaping vapor. Defining x_i^0 to be the overall liquid phase mole fraction of component i, then

$$Hx_i^0 = H^I x_i^I + H^{II} x_i^{II} \quad i = 1, 2, \ldots, c-1 \tag{8.3}$$

A material balance on component i yields

$$\frac{d(Hx_i^0)}{dt} = -Vy_i \quad i = 1, 2, \ldots, c-1 \tag{8.4}$$

which can be written in the form (using Eq. 8.1)

$$\frac{dx_i^0}{dt} = \frac{V}{H}(x_i^0 - y_i) \quad i = 1, 2, \ldots, c-1 \tag{8.5}$$

We now apply the same time transformation as used earlier for homogeneous liquids (Eq. 5.5); thus Eq. 8.5 becomes

$$\frac{dx_i^0}{d\xi} = x_i^0 - y_i \quad i = 1, 2, \ldots, c-1 \tag{8.6}$$

where ξ is a dimensionless time ranging from 0 to $+\infty$. At the initial time ($t = 0$) the value of ξ is 0, and at the final time when the still runs dry the value of ξ is $+\infty$.

When one of the liquid phases in the still disappears, the overall liquid composition is simply the liquid composition of the remaining homogeneous liquid and Eq. 8.6 becomes the equation describing homogeneous simple distillation. Therefore, the form of Eq. 8.6 does not change as the number of coexisting liquid phases increases or decreases.

For isobaric simple distillation, Eq. 8.6 can be integrated if we can solve the phase equilibrium problem for $\mathbf{y}$, given values for P and $\mathbf{x}^0$. As discussed in a number of sources (Michelsen, 1982*a*, 1982*b*; Pham and Doherty, 1990*a*; Cairns and Furzer, 1990*b*), this is a well-posed phase equilibrium calculation, although it necessarily involves the added complication of finding the individual liquid phase compositions $\mathbf{x}^I$ and $\mathbf{x}^{II}$ even though they do not appear explicitly in Eq. 8.6.

The singular points of Eq. 8.6 consist of all the pure components, homogeneous azeotropes, and heterogeneous azeotropes present in the mixture. The patterns of the residue curves around pure components and homogeneous azeotropes have been studied quite extensively and are restricted to be either nodes or saddles. These restrictions also apply to heterogeneous azeotropes as shown by Matsuyama (1978). For homogeneous azeotropes the nodes may be either stable (maximum-boiling) or unstable (minimum-boiling). However, heterogeneous azeotropes can be either unstable nodes or saddles but not stable nodes; i.e., they cannot be maximum-boiling.

Using the new generation of VLLE techniques described earlier, it is now possible to calculate residue curve maps for heterogeneous liquid systems. A selection of examples is given in Fig. 8.7. In each map the liquid boiling envelope has been superimposed on the residue curves in order to distinguish the homogeneous and heterogeneous regions of the triangular diagram.

Systems with a ternary homogeneous azeotrope. This type of system has a ternary azeotrope that lies outside the heterogeneous region, as shown in Fig. 8.7*a* for the mixture methanol-benzene-heptane. The phase diagram for this mixture is shown in Fig. 8.3*b*. The mixture has two binary homogeneous azeotropes, one binary heterogeneous azeotrope, and one ternary homogeneous azeotrope. The homogeneous azeotrope between benzene and heptane is undetectable by the model. However, this azeotrope lies so close to the pure benzene vertex (x^{Az} = 99.45 mol % benzene) that the residue curves are essentially the same as in Fig. 8.7*a*. Residue curves move toward either pure heptane or pure methanol according to the initial overall liquid composition in the pot. The composition triangle is divided into distinct distillation regions, and one of the boundaries runs through the heterogeneous region. We can exploit the fact that the heterogeneous region straddles the boundary to devise distillation systems that separate the mixture into its pure components.

Given an initial overall composition near the binary heterogeneous azeotrope, the residue curve moves through the heterogeneous region until it hits the liquid boiling envelope. Beyond this envelope, the mixture is homogeneous and the residue curve continues its travels toward either pure heptane or pure methanol. Inside the heterogeneous region, each overall liquid composition is in equilibrium with a vapor whose composition is restricted to lie on the vapor line. The position of the vapor line relative to the liquid boiling envelope is an important factor in the design of azeotropic distillation columns.

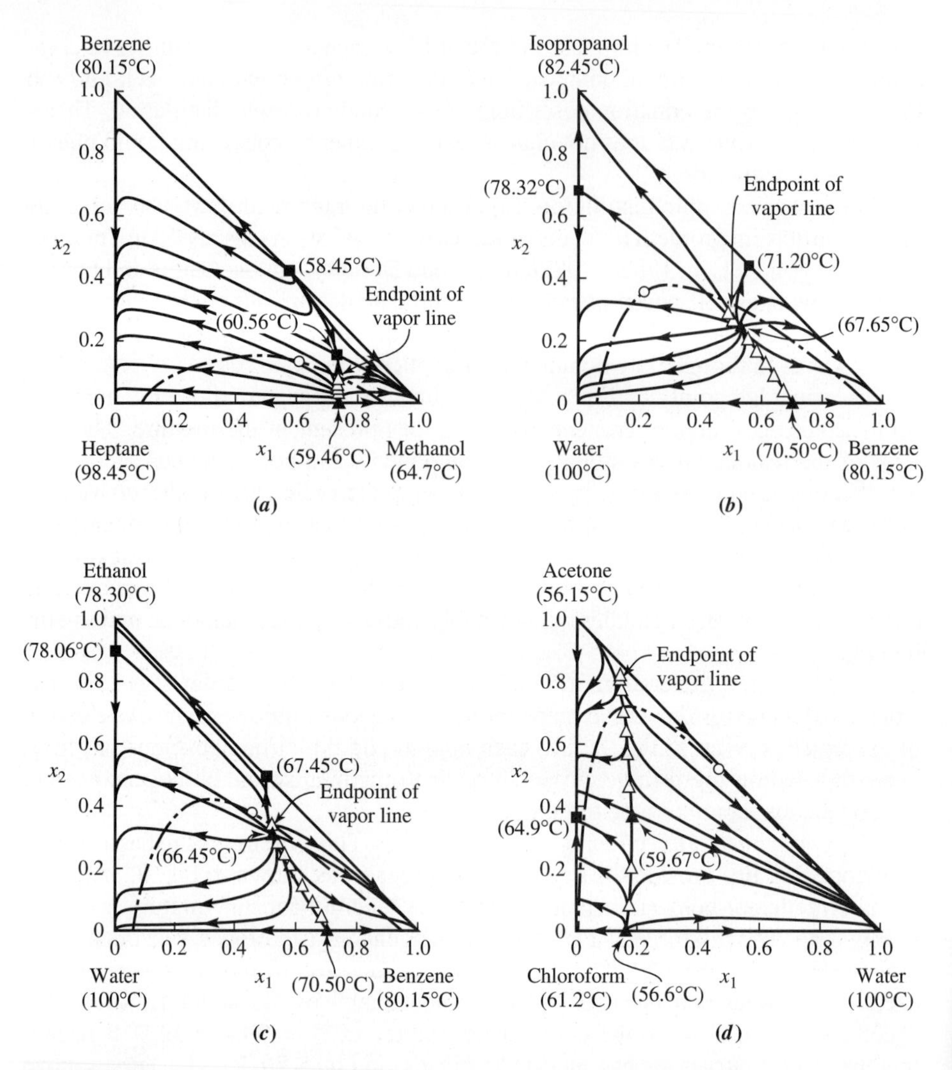

FIGURE 8.7
Selection of residue curve maps for ternary mixtures with partially miscible liquids at 1 atm. (*a*) Methanol-benzene-heptane. (*b*) Isopropanol-water-benzene. (*c*) Ethanol-water-benzene. (*d*) Acetone-water-chloroform. Figure (*a*) was calculated using the two-parameter Margules model to represent the liquid phase activity coefficients; figures (*b*), (*c*), and (*d*) were calculated using the regular solution model. Parameter values used in these calculations are reported in Pham and Doherty (1990*a*, 1990*b*).

Systems with a ternary heterogeneous azeotrope. These systems contain a ternary azeotrope that lies inside the heterogeneous region. Binary or ternary heterogeneous azeotropes are restricted to be either unstable nodes or saddles in the residue curve map (Matsuyama, 1978). Representative examples are shown in Fig. 8.7*b*, *c*, and *d*. Figure 8.7*b* and *c* shows the computed residue curve maps for two mixtures which exhibit ternary minimum-boiling heteroazeotropes. The computed phase diagrams are shown in Fig. 8.3*c* and *d*. Each of these maps exhibits three distillation regions. The three distillation boundaries all begin at the minimum-boiling ternary heteroazeotrope and end at each of the binary azeotropes. If either isopropanol or ethanol is the desired product of the separation, the initial condition for the simple distillation *must* lie in the upper region of Fig. 8.7*b* or *c*, respectively. A more accurate representation of the residue curve map for the ethanol-water-benzene mixture calculated using the modified UNIQUAC model is shown in Fig. 8.8.

The acetone-water-chloroform system exhibits a ternary heterogeneous saddle azeotrope. This was detected experimentally by Reinders and DeMinjer (1947*c*), who report extensive measurements. Ternary VLLE data are shown in Fig. 8.9. The computed residue curve map for this system is shown in Fig. 8.7*d* and is qualitatively similar to the experimentally measured residue curve map (Reinders and DeMinjer, 1947*c*, Fig. 16).

In all the experimentally measured and computed heterogeneous residue curve maps that have been reported in the literature, it is observed that the boiling temperature uniformly increases along residue curves. For homogeneous liquids this property always holds (Doherty and Perkins, 1978; Van Dongen and Doherty, 1984); however, nobody has succeeded in extending the proof to heterogeneous liquids. Until shown otherwise, we conjecture that the liquid temperature always increases along residue curves in the heterogeneous region also. This conjecture, coupled with the fact that all singular points on heterogeneous residue curve maps are restricted to be either unstable nodes or saddles, means that the entire set of machinery for constructing homogeneous residue curve maps from boiling temperature data alone (Foucher et al., 1991) extends verbatim to heterogeneous mixtures.

Significance of the vapor line. The existence of a vapor line instead of a vapor surface for heterogeneous liquid mixtures means that all residue curves inside the heterogeneous region have vapor boil-off curves that are confined to lie on some portion of this single line. If the phase equilibrium calculations are simplified by

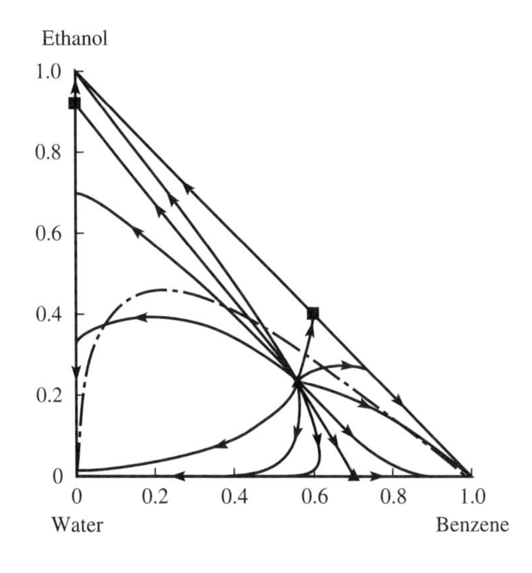

FIGURE 8.8
Residue curve map for the ternary mixture ethanol-water-benzene at 1 atm pressure calculated using the modified UNIQUAC model of Prausnitz et al. (1980, Chap. 4). Model parameters taken from Prokopakis and Seider (1983*b*, Table 1).

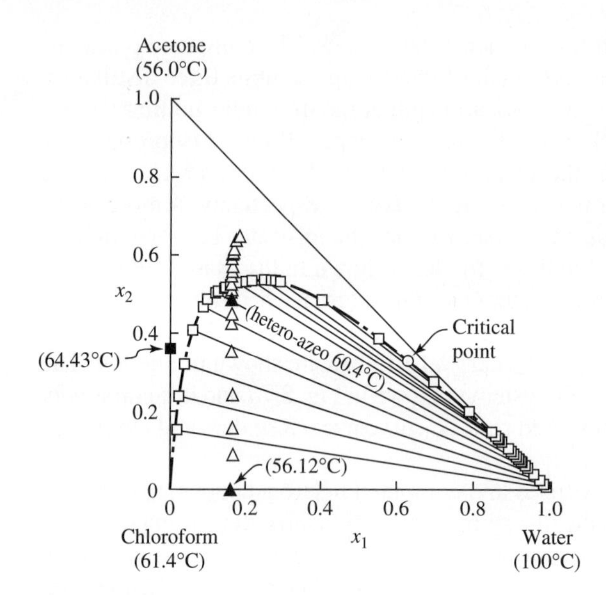

FIGURE 8.9
Experimental vapor-liquid-liquid equilibrium data for the mixture acetone-water-chloroform at 1 atm. Data from Reinders and DeMinjer (1947*c*, Table XI and related text) have been converted from wt % to mole fractions.

ignoring the liquid-liquid phase separation, the vapor compositions are no longer restricted to lie on a single line and may exhibit spurious behavior, including effects such as self-intersecting vapor boil-off curves. These pathologies cannot occur in nature and may lead to incorrect results for feasibility and design of heterogeneous distillation systems. Pham and Doherty (1990*b*) warn against the practice of ignoring the existence of multiple liquid phases and discuss this and other aspects of the vapor line in more detail.

8.4
DISTILLATION SYSTEM SYNTHESIS

The analysis of residue curve maps for homogeneous azeotropic mixtures provides a simple and useful technique for distinguishing between feasible and infeasible systems of distillation columns (Secs. 5.4 and 5.6). The method is not exact, since it recommends that simple distillation boundaries should not be crossed during continuous distillation. The advantage of this approach is that it requires only a knowledge of the presence or absence of distillation boundaries in the composition simplex, and this information is readily available *without* the aid of a VLE model using the methods developed in Sec. 5.3. More exact methods for checking feasibility of distillation systems (such as the distillation limits described in Sec. 5.4) have not yet been extended to heterogeneous systems.

For homogeneous mixtures, simple distillation boundaries cannot be crossed by residue curves, and for all practical purposes they can rarely be crossed to any advantage by the steady-state liquid composition profile in a continuous distillation column. If in addition the distillation boundaries are close to linear, then in order to isolate two pure components that lie in two different simple distillation regions it is necessary to have two different feed compositions (one in each of the two regions) and two distillation columns. With these assumptions such an arrangement is impossible to construct by internal recycles alone. Therefore, under the above restrictions, it is impossible to isolate by distillation two components that are divided by a (linear) simple distillation boundary.

Heterogeneous mixtures also exhibit distillation boundaries. Residue curves cross continuously through the liquid boiling envelope from one side to the other. Thus, a simple distillation boundary inside the heterogeneous region will not stop abruptly or exhibit a discontinuity at the liquid boiling envelope but will pass continuously through it, becoming a homogeneous distillation boundary thereafter (see Fig. 8.7 for examples). As with homogeneous systems, residue curves cannot cross heterogeneous distillation boundaries. However, if the two individual equilibrium liquid phases resulting from a point $\mathbf{x}^0$ on a residue curve inside the heterogeneous region lie in two different distillation regions, then we can exploit a liquid-liquid phase separation to "jump" across heterogeneous distillation boundaries in a way that is not possible for homogeneous systems. This is the key fact that is used to devise feasible systems of columns for separating heterogeneous mixtures.

Binary Mixtures

During the distillation of a binary mixture containing a homogeneous azeotrope, it is possible to separate up to, but not beyond, the azeotropic composition. Because of this, special techniques are deployed to separate azeotropic mixtures into their pure components. If the binary azeotrope is heterogeneous, however, the situation is more favorable and a system of two columns, shown in Fig. 8.10, is capable of isolating each pure component. The process feed is fed as a saturated liquid to a decanter where it phase-separates into an A-rich phase and a B-rich phase. The B-rich phase is fed to column 1 and the A-rich phase is fed to column 2. Because of the liquid-liquid phase split, the compositions of these two feed streams lie on either side of the azeotrope. Therefore, column 1 produces pure B as a bottoms product and the azeotrope as distillate, whereas column 2 produces pure A as a bottoms product and the azeotrope as distillate. The two distillate streams are fed to the decanter along with the process feed to give an overall decanter composition partway between the azeotropic composition and the process feed composition according to the lever rule. This arrangement is well suited to purifying water-hydrocarbon mixtures (e.g., water with any one of the following components: C_4-C_{10}, benzene, toluene, xylene, etc.), water-alcohol mixtures (e.g., butanol, pentanol, etc.), as well as other immiscible systems. There are several process alternatives for this distillation system, including:

1. A better arrangement is to operate each column in Fig. 8.10b as a *stripping column*. This is accomplished by removing the reflux streams (which are two-phase) and replacing them with the feed streams to each column (see

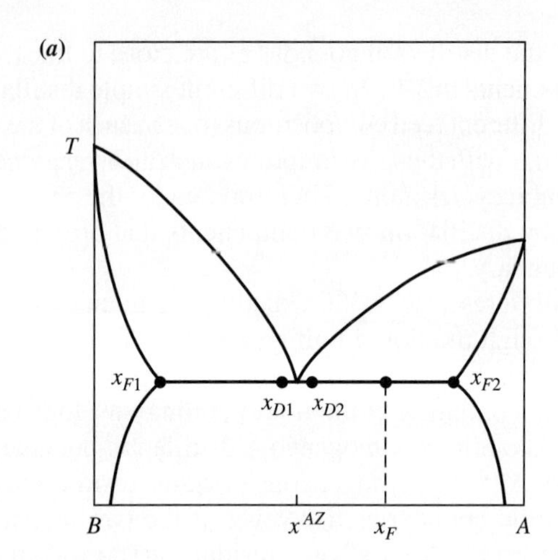

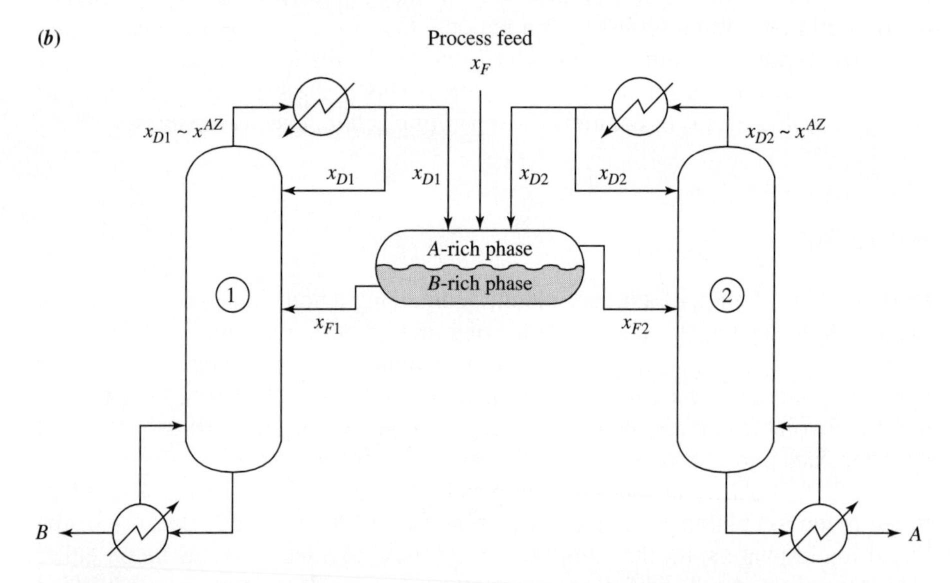

FIGURE 8.10
Separation of a binary heterogeneous azeotropic mixture. (*a*) Phase diagram. (*b*) Distillation system.

Exercise 2). Othmer (1978) describes a process very similar to this for separating water-butanol mixtures.

2. If water is one of the components in the feed mixture, then open steam could be used to provide reboil in the water-removal column.
3. The liquid in the decanter need not necessarily be saturated; it could be subcooled.
4. If the process feed does not lie in the liquid-liquid region, it can be made to do so by deliberately feeding either pure A or pure B to the decanter, as required.

This may only be necessary during startup or for control purposes, since the recycled azeotrope has the beneficial effect of dragging the decanter composition toward or further into the liquid-liquid region. Another alternative is to feed the (homogeneous) process feed directly to the first column (see Exercise 3).

Ternary Mixtures

When the binary mixture containing the minimum boiling azeotrope is completely homogeneous, i.e., the liquid is homogeneous for all compositions, the methods described above will not work without some modification. In this case a third component, called the *entrainer,* is added, which induces a liquid-liquid phase separation over a limited portion of the ternary composition diagram. The literature describes many options for sequencing ternary heterogeneous azeotropic distillation systems. These systems generally consist of two, three, or four columns with various techniques for handling the entrainer recycle stream. The feasibility of such systems rests on the use of a liquid-liquid phase split to provide each column with a feed composition in a different distillation region. In this regard the systems for ternary mixtures resemble the systems for binary mixtures. In all cases the heart of the process is the azeotropic column and its decanter, which we will describe first.

In order to fix ideas, we begin by treating the classical separation of ethanol from water using benzene as the entrainer. Many other separations are similar, e.g., ethanol-water-carbon tetrachloride, isopropanol-water-benzene, and several others mentioned later in Table 8.2. The task is to separate a homogeneous binary mixture of ethanol and water into its pure components given that the mixture has a minimum-boiling binary azeotrope. We use an entrainer, benzene, which has limited miscibility with one of the components (water in this case). In addition, entrainers in this class cause two more minimum-boiling binary azeotropes to form (one homogeneous, the other heterogeneous) together with a ternary minimum-boiling heterogeneous azeotrope. The resulting residue curve map will be similar to those shown in Fig. 8.7*b* and *c*, in which there are three distillation regions, as noted in Fig. 8.11. The first thing we observe about such a map is that the components to be separated lie in two different distillation regions, I and II.

We now specify target compositions that the azeo-column must be designed to meet. The bottoms composition is specified to be almost pure ethanol (i.e., $\mathbf{x}_B$ is specified to lie close to the vertex of distillation region II). The overhead vapor composition leaving the last vapor-liquid equilibrium stage of the column $\mathbf{y}_N$ is specified to lie in the wedge-shaped portion of region II *inside* the heterogeneous region near the ternary azeotrope. Typical target values for $\mathbf{x}_B$ and $\mathbf{y}_N$ are shown in Fig. 8.11. The final values of $\mathbf{x}_B$ and $\mathbf{y}_N$ will be subject to optimization and may differ slightly from those shown in the figure; however, we are not concerned with that level of detail at this stage. All values of $\mathbf{y}_N$ inside the wedge-shaped region, except those special values which lie on the vapor line, will be in equilibrium with a *homogeneous liquid.* Therefore, the liquid composition leaving the top tray $\mathbf{x}_N$ lies in the homogeneous portion of region II as shown in Fig. 8.11. The azeo-column

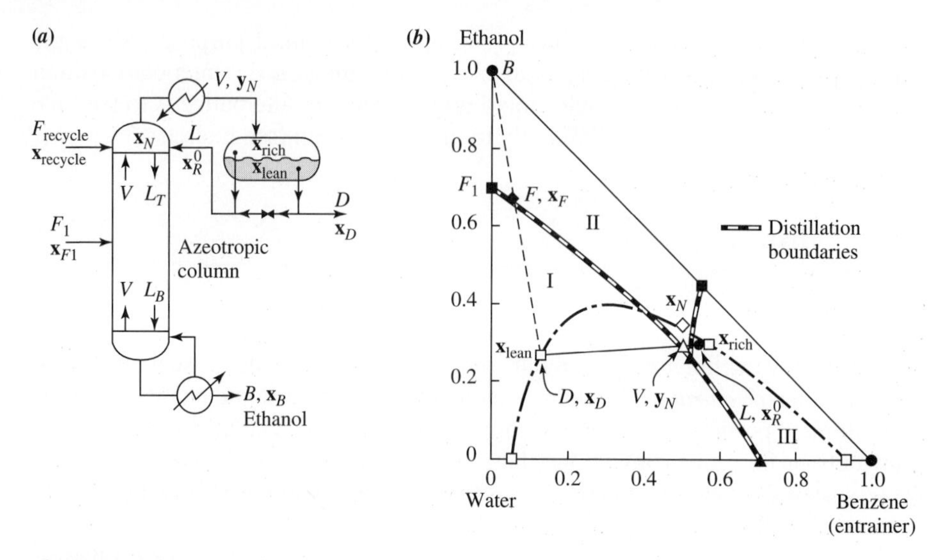

FIGURE 8.11
(a) Schematic representation of a distillation column, and (b) Material balance lines for separating ethanol from a mixture of ethanol, water, and benzene.

must therefore be designed so that the steady-state liquid composition profile runs from $\mathbf{x}_B$ to $\mathbf{x}_N$, and this can normally be done in such a way that every stage inside the column is homogeneous, i.e., has only one liquid phase.[3] A method for doing this is given by Pham et al. (1989).

Continuing the analysis, the overhead vapor, of composition $\mathbf{y}_N$, is totally condensed into two equilibrium liquid phases, an entrainer-rich phase of composition $\mathbf{x}_{\text{rich}}$ and an entrainer-lean phase of composition $\mathbf{x}_{\text{lean}}$. The relative proportion of these two liquid phases in the condenser ϕ is given by the lever rule

$$\phi = \frac{y_{i,N} - x_{i,\text{lean}}}{y_{i,N} - x_{i,\text{rich}}} \tag{8.7}$$

where ϕ represents the molar ratio of the entrainer-rich phase to the entrainer-lean phase in the condensate.

The two condensate liquids must now be used to provide reflux and distillate streams. There are two reflux options: $r = L/D \geq \phi$ and $r \leq \phi$. The first option requires that the reflux rate be *greater* than the condensation rate of entrainer-rich phase and that the distillate rate be correspondingly less than the condensation rate of entrainer-lean phase. This means that the distillate stream consists of pure entrainer-lean phase, i.e., $\mathbf{x}_D = \mathbf{x}_{\text{lean}}$, and the reflux stream consists of *all* the entrainer-rich phase plus the balance of the entrainer-lean phase. Thus, the overall composition of

[3]This does not mean that all such columns have actually been designed this way nor does it imply that columns that have been will remain homogeneous under the action of disturbances; see Prokopakis and Seider (1983*a*, 1983*b*), Kovach and Seider (1987*a*, 1987*b*), Widagdo et al. (1989), Cairns and Furzer (1990*a*, 1990*b*, 1990*c*), Rovaglio and Doherty (1990), Wong et al. (1991), and Widagdo et al. (1992).

the reflux stream $\mathbf{x}_R^0$ lies on the tie-line between the points $\mathbf{x}_{\text{rich}}$ and $\mathbf{y}_N$, as shown in Fig. 8.11. The reflux ratio is given by

$$r = \frac{L}{D} = \frac{y_{i,N} - x_{i,D}}{y_{i,N} - x_{i,R}^0} \geq \phi \tag{8.8}$$

where $x_{i,D} = x_{i,\text{lean}}$. The equality in Eq. 8.8 is satisfied as $\mathbf{x}_R^0 \to \mathbf{x}_{\text{rich}}$, i.e., when there is a clean split of the two liquid phases between reflux and distillate streams. The fact that $\mathbf{x}_R^0$ lies on the tie-line, i.e.,

$$\frac{x_{2,\text{lean}} - x_{2,R}^0}{x_{1,\text{lean}} - x_{1,R}^0} = \frac{x_{2,\text{rich}} - x_{2,R}^0}{x_{1,\text{rich}} - x_{1,R}^0} \tag{8.9}$$

together with Eq. 8.8 indicates that once we pick a value for r, the composition of the reflux stream becomes fixed. However, we do have the flexibility to manipulate the reflux ratio at the design stage (with corresponding changes in the reflux composition, as dictated by Eqs. 8.8 and 8.9) at fixed values of $\mathbf{x}_B$, $\mathbf{x}_F$, $\mathbf{y}_N$ and $\mathbf{x}_D$. This physically corresponds to varying the amount of entrainer-lean phase in the reflux stream at fixed material balance points for the other key input and output streams. If the reflux flow is written as the sum of the entrainer-rich reflux (L_{rich}) and the entrainer-lean reflux (L_{lean}), then the reflux ratio can be written in the following useful way:

$$r = \phi + r_{\text{lean}}(1 + \phi) \tag{8.10}$$

where $r_{\text{lean}} = L_{\text{lean}}/D$. It is evident that for such a policy the operating reflux ratio is bounded as follows:

$$\infty \geq r > r_{\min} \geq \phi \tag{8.11}$$

Finally, we note that overall material balance requires that the points $\mathbf{x}_B$, $\mathbf{x}_F$, and $\mathbf{x}_D$ be collinear, as shown in Fig. 8.11.

The second reflux option, $r \leq \phi$, requires that the reflux rate be *less* than the condensation rate of the entrainer-rich phase, i.e., $\mathbf{x}_R^0 = \mathbf{x}_{\text{rich}}$, and that the distillate rate be correspondingly greater than the condensation rate of entrainer-lean phase. In this option, some of the entrainer-rich phase leaves with the distillate. Thus, the overall distillate composition lies in the two-phase region on the tie-line between the points $\mathbf{x}_{\text{lean}}$ and $\mathbf{y}_N$. For such a reflux policy with specified values for $\mathbf{x}_B$, $\mathbf{x}_F$ and $\mathbf{y}_N$ there is a *unique* value of $\mathbf{x}_D^0$ and $\mathbf{x}_R (= \mathbf{x}_{\text{rich}})$, thus a *unique* value of r. Therefore, with such a reflux policy we have lost the ability to manipulate r at fixed distillate and decanter compositions. Little is known about this policy (or variations of it, such as manipulating $\mathbf{y}_N$ in order to vary r), and it will not be considered here further.

Completing the Separation System

The purpose of the remainder of the separation system is to separate the distillate stream leaving the azeotropic column (column 2) into a product stream and a recycle stream so that the entire system is closed with respect to the entrainer.

Kubierschky Three-Column System. If only simple columns are used, i.e., no sidestreams, siderectifiers/strippers, etc., then the system can be completed by adding an entrainer recovery column (column 3) to recycle the entrainer and a preconcentrator (column 1) to bring the feed to the azeotropic column up to the composition of the binary azeotrope.

The entrainer recovery column takes the distillate stream D_2 from the azeo-column and separates it into a bottoms stream of pure water and a ternary distillate stream for recycle to column 2. The overall material balance line for column 3 is shown in Fig. 8.12. This system was one of the two original continuous processes disclosed by Kubierschky in 1915 (see Fig. 3 in Keyes, 1929). In recent years it has been applied to azeotropic separations by Black (1980), Townsend (1982, ICI), and Ryan and Doherty (1989).

Ryan and Doherty (1989) carried out extensive design and optimization studies for this system. The principal optimization variables (i.e., the design variables that have the largest impact on the economics of the process) are the reflux ratio in the azeo-column, the position of the tie-line for the mixture in the decanter (this is determined by the temperature and overall composition of the mixture in the decanter), the position of the decanter composition on the decanter tie-line [see Pham et al. (1989) for a discussion of the importance of these variables], and the distillate composition from the entrainer recovery column.

The reflux ratio is important because it determines the vapor rate, and therefore the capital and operating costs for the azeo-column. The position of the decanter tie-line influences two quantities:

1. The minimum reflux ratio in the azeo-column.
2. The composition and flowrate of distillate from the azeo-column to the entrainer recovery column. This in turn is a major factor in determining the entrainer recycle composition and flowrate from the entrainer recovery column back to the azeo-column. The entrainer flowrate is normally the most important optimization variable in extractive distillation and plays an important role in heterogeneous systems also.

The position of the decanter composition along the decanter tie-line often has a strong influence on the value of r_{min} for the azeo-column (Pham et al., 1989). Finally, the distillate composition from the entrainer recovery column is an important variable because it determines:

1. The overall feed composition and flowrate to the azeo-column.
2. The distillate and bottoms flowrate from the azeo-column.
3. The distillate and bottoms flowrate from the entrainer recovery column.

Figure 8.13 shows material balance lines for three different decanter tie-lines. The process feed to each system is a binary mixture of 4.2 mol % ethanol and 95.8 mol % water. The product purity from the azeo-column is set at 99.9 mol % ethanol, and the water purity leaving the entrainer recovery column is set at 99.5 mol % water. These specifications are essentially identical to those used by Knight and Doherty (1989) and Knapp and Doherty (1990) in their study of optimal extractive distillation systems. For design 1, the decanter tie-line is set at the bubble point of the mixture

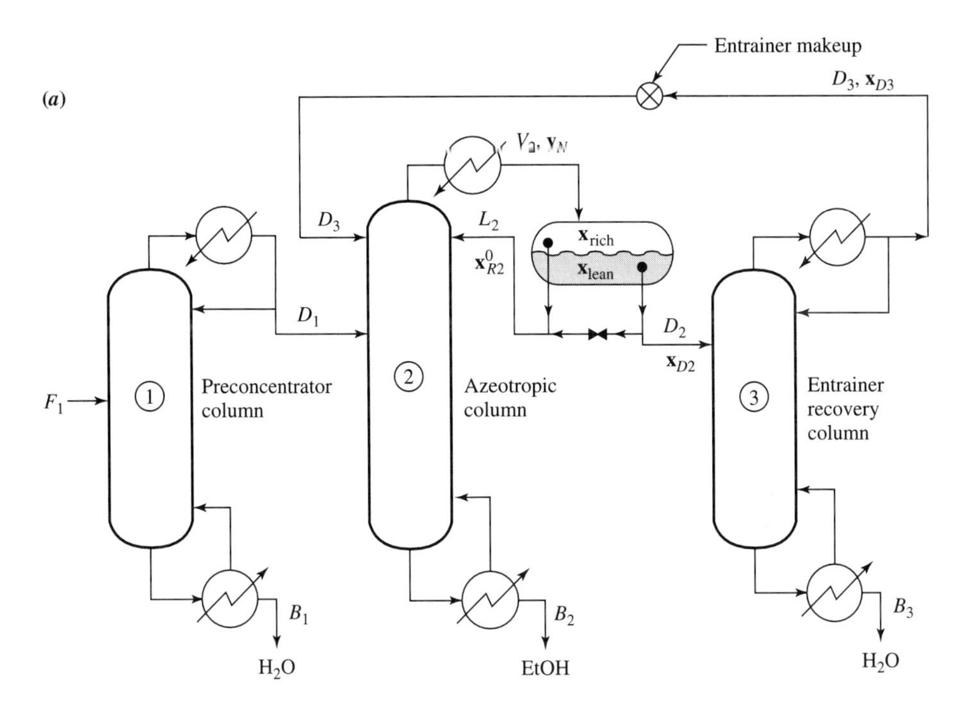

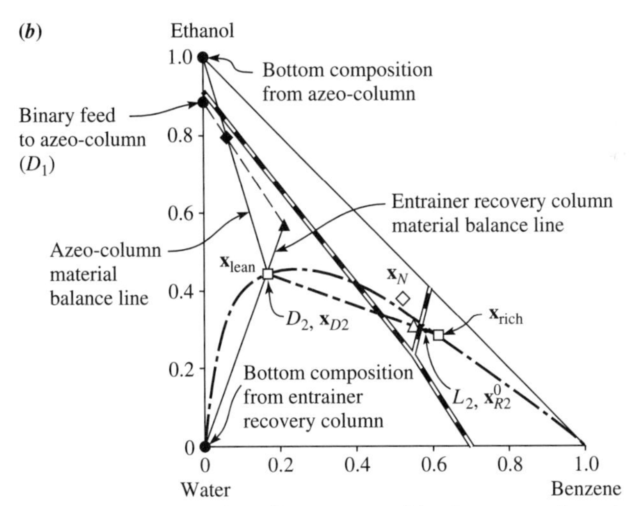

△ Overall vapor composition from azeo-column ($\mathbf{y}_N$)
◇ Liquid in equilibrium with overhead vapor from azeo-column
▲ Distillate composition from entrainer recovery column ($\mathbf{x}_{D3}$)
◆ Overall feed composition to azeo-column ($D_1 + D_3$)

FIGURE 8.12
(*a*) Kubierschky three-column system. (*b*) Material balance lines for separating ethanol and water from a mixture of ethanol, water, and benzene.

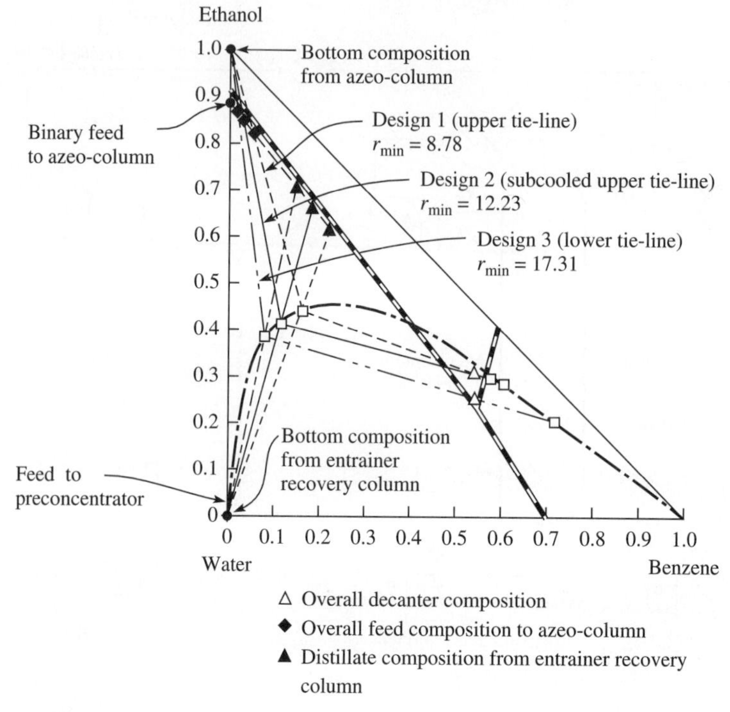

FIGURE 8.13
Three sets of material balance lines for the Kubierschky three-column system.

leaving the top of the azeo-column (of composition $\mathbf{y}_N$). This temperature is 337.57 K. For design 2, the decanter composition is the same as for design 1 but subcooled to 298.0 K, and for design 3 the decanter composition is placed nearer to the ternary azeotrope than for designs 1 and 2, and the decanter temperature is set at the bubble point of the mixture. In each design, the distillate composition from the entrainer recovery column is placed close to the distillation boundary. Note what a significant influence the position of the tie-line has on the distillate composition from the azeo-column and how this influences the position of the material balance line for the entrainer recovery column. Note also how the position of the distillate composition from the entrainer recovery column influences the overall feed composition to the azeo-column.

Optimization studies indicate that the distillate composition from the entrainer recovery column does indeed have a strong influence on the process economics and that its optimum position is always close to the distillation boundary. This decreases the amount of water being recycled, or equivalently, makes the overall feed to the azeo-column richer in ethanol. For each tie-line, the optimal position of the decanter liquid composition is found by the method proposed by Pham et al. (1989). The minimum reflux ratio (for a given tie-line) can be reduced by as much as 50% by making small changes in the decanter composition. Calculations indicate that the optimal reflux ratio for the azeo-column is normally in the range 1.1 to 1.5 $r_{\min}$ and that the cost of the system is insensitive to this factor. This leaves the position of the decanter tie-line as the sole remaining optimization variable.

The intrasystem flows, compositions, and reflux ratios are quite sensitive to relatively small changes in the position of the decanter tie-line. Thus, for design 1, the minimum reflux ratio for the azeo-column is $r_{\min} = 8.78$; for design 2 it is 12.23, and for design 3, 17.31. Intuition suggests that the vapor rate and total annualized cost for system 1 are the lowest, both rising for system 2, and higher still for system 3. This intuition is based on the fact that for homogeneous distillations at the design stage the feed and product flowrates, as well as their compositions, can be held constant as the reflux ratio is changed from one design to another. Thus, there is a direct relationship between increased minimum reflux ratio and increased costs. Such a relationship does *not* occur for heterogeneous distillations, and our intuition can be faulty.

For columns with a saturated liquid feed, an approximate expression for the vapor rate leaving the reboiler V is

$$V = (r + 1)D \tag{8.12}$$

where r is the reflux ratio and D is the distillate flowrate. For homogeneous distillations D is constant so that V increases as r increases. For azeotropic distillation, however, both r *and* D change from one tie-line to another. These effects may tend to reinforce each other or cancel each other, depending on the mixture. There is no general rule, and each mixture must be treated separately. In the ethanol-benzene-water system the reflux ratio increases from one design to another, but the distillate flowrate decreases, as can be seen from the material balance lines for the azeo-column in Fig. 8.13. The net effect is that the vapor rate in the azeo-column hardly changes from one design to another.[4] All the systems shown in Fig. 8.13 have approximately the same cost. This fortuitous cancellation of effects cannot be expected to occur in general, and it is always worthwhile exploring the economic impact of variations in the position of the decanter tie-line.

The Kubierschky system is not the only way to perform the separation. Alternatives include:

1. If the process feed already has a composition at or near the composition of the binary azeotrope, then the preconcentrator will not be needed.
2. A common alternative is to recycle the distillate stream from the entrainer recovery column directly to the decanter (presumably by analogy with the binary process shown in Fig. 8.10). Since the composition of this stream is not normally close to the composition of the overhead vapor from the azeotropic column[5] (unlike the binary case shown in Fig. 8.10), such a policy adversely affects the operation of the decanter. Pham and Doherty (1990*c*) show that this recycle alternative causes

---

[4]An alternative way to see this is to recognize that $V = s \times B$. The flowrate B does not change from one design to another, since it is fixed by the amount of ethanol entering with the process feed. Values for the operating reboil ratio s are 4.26, 4.26, and 4.94, respectively, for designs 1, 2, and 3. Therefore, the vapor rate hardly changes from one design to another.

[5]In fact, the composition of the entrainer recycle stream often lies in the homogeneous liquid region, as in Fig. 8.12, and will not separate into two liquid phases on cooling.

the reflux ratio in the azeotropic column to be much larger than necessary, and it should be avoided even though it has been studied extensively in the literature, e.g., Norman (1945), Robinson and Gilliland (1950, pp. 312–334), Bril' et al. (1975, 1977), Prokopakis et al. (1981), and Prokopakis and Seider (1983*a*), (1983*b*).

3. *Kubierschky Two-Column System.* A novel two-column system was also disclosed by Kubierschky (1915) in his original patent (see Keyes, 1929, Fig. 2). In this alternative the preconcentrator is eliminated and the process feed is sent directly to the entrainer recovery column. The entrainer recycle stream is returned as the only feed to the azeo-column. Ryan and Doherty (1989) studied this alternative for the ethanol-water-benzene system and concluded that it has lower capital costs but higher operating costs than the Kubierschky three-column system so that the total annualized cost is about the same for both systems. No generalizations should be made from this one example other than that this system is worth exploring as a sensible process alternative for other azeotropic distillations.
4. *Steffen Three-Column System.* The basic layout of this alternative is the same as the Kubierschky three-column system. The essential new feature is to replace the single decanter in the Kubierschky system by multiple decanters, adding fresh or recycled water to each (Keyes, 1929, Fig. 4). The entrainer-rich phase from each decanter is returned to the azeo-column as reflux while the aqueous-rich phase from each decanter is sent as feed to the next decanter in the train. This effectively replaces the decanter by a liquid-liquid extraction step which pushes the "effective decanter tie-line" deeper into the two-phase region than it could possibly go by distillation alone. This has the beneficial effect of reducing the amount of entrainer in the feed to the entrainer recovery column, but this must be balanced against the extra costs associated with the additional water and the liquid-liquid extraction equipment.

   Note, the actual Steffen process is slightly different; reflux to the azeo-column is provided by condensing the overhead vapor and returning part of the ternary liquid mixture before it goes to the decanters. The entrainer-rich phases from the decanters then get returned to the azeo-column as a second feed stream.
5. *Ricard-Allenet Four-Column System.* Another common alternative described in the literature is a four-column system which is normally credited to Guinot and Clark, 1938) [see various editions of Perry's Handbook or Holland et al. (1981), etc.]. However, it was first proposed by Ricard-Allenet in 1923 and is reported in the more accessible article by Keyes (1929). In this alternative the entrainer-lean phase from the decanter is fed to an entrainer recovery column which is designed to produce a bottoms stream that is entrainer-free, i.e., containing only ethanol and water. The distillate from this column has a composition close to the overhead vapor from the azeo-column; therefore, it is recycled back to the decanter. The bottoms stream is sent to a final (fourth) column where it is separated into pure water and a distillate close to the ethanol-water binary azeotrope which is recycled to the feed of the azeo-column. This alternative is described by Keyes (1929, especially Fig. 9), King (1980, pp. 345–349), Ryan and Doherty (1989), and Pham and Doherty (1990*c*).

A variation on this system was also proposed by Ricard-Allenet (Keyes, 1929, Fig. 10) in which the overhead vapors from the entrainer recovery column are provided with a separate condenser-decanter system, similar to the one provided for the overhead vapors from the azeo column.

6. *Ricard-Allenet Three-Column System.* The original Ricard-Allenet four-column system is now mostly of historical interest because it requires unnecessary duplication of columns (the first and last) that perform essentially the same task. However, an attractive process alternative is to remove the last column and recycle the ethanol-water mixture from the bottom of the third column (the entrainer recovery column) to the preconcentrator, making it a double-feed column. The resulting three-column system (which was not actually proposed by Ricard-Allenet but is such an obvious variation on their idea that we will give it their name) has good economic possibilities. Unfortunately, there are no studies of this alternative in the literature, although it has been noted in passing by Prokopakis et al. (1981), Prokopakis and Seider (1983*b*) and Pham and Doherty (1990*c*). This system is worth exploring as a sensible process alternative for azeotropic distillation.

In summary, for systems of the ethanol-water-benzene type, there are three attractive systems for carrying out azeotropic distillation: the Kubierschky three-column system, the Kubierschky two-column system, and the Ricard-Allenet three-column system. For each of these there is the added possibility of putting a liquid-liquid extraction step after the azeo-column.

8.5
OTHER CLASSES OF ENTRAINERS

Not all azeotropic mixtures are of the ethanol-water-benzene type, see Fig. 8.14 and Table 8.2. The number of azeotropes in the mixture as well as their character (i.e., maximum-boiling or minimum-boiling, heterogeneous or homogeneous) varies from system to system. In addition, the size and shape of the liquid-liquid region varies greatly from system to system. The feasibility and sequencing strategy for each new system is most conveniently established with the use of residue curve maps.

To begin, any entrainer that induces a liquid-phase heterogeneity over a portion of the composition triangle *and* which does not divide the components to be separated into different distillation regions is automatically a feasible entrainer. This follows from the properties of heterogeneous residue curve maps described earlier. Examples are shown in Fig. 8.14*a* and *b*, and it is possible to construct many more like them by strategically placing heterogeneous regions on the feasible residue curve maps given in Chap. 5. Systems based on these maps will normally have multiple liquid phases

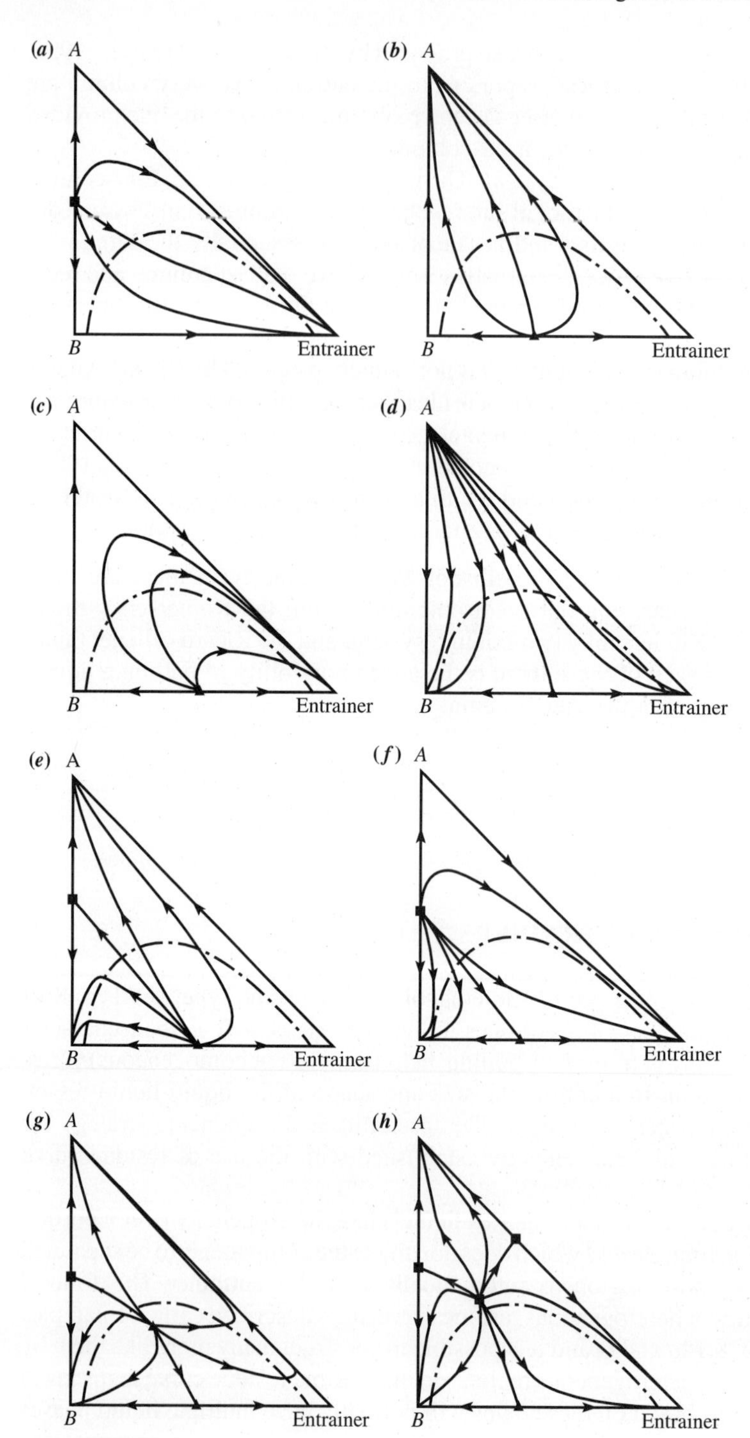

FIGURE 8.14
Selection of residue curve maps for ternary heterogeneous mixtures.

TABLE 8.2
Examples of heterogeneous azeotropic separations (entrainer italicized)

Map type[a]	System	References
Fig. 8.8	Ethanol-water-*benzene*	Robinson and Gilliland (1950), Prokopakis and Seider (1983*a*, 1983*b*)
Fig. 8.8	Ethanol-water-*cyclohexane*	Furzer (1984)
Fig. 8.8	Ethanol-water-*iso-octane*	Furzer (1985), Cairns and Furzer (1990*a*, 1990*c*)
Fig. 8.8	Ethanol-water-*pentane*	Black (1980)
Fig. 8.8	Ethanol-water-*carbon tetrachloride*	Keyes (1929)
Fig. 8.8	Ethanol-water-*trichloroethylene*	Keyes (1929), Colburn and Phillips (1944), Reinders and DeMinjer (1947*a*)
Fig. 8.8	Isopropanol-water-*benzene*	Bril' et al. (1975, 1977)
Fig. 8.8	Isopropanol-water-*cyclohexane*	Prokopakis and Seider (1983*a*)
Fig. 8.8	Allyl alcohol-water-*carbon tetrachloride*	Hands and Norman (1945)
Fig. 8.8	Allyl alcohol-water-*trichloroethylene*	Hands and Norman (1945)
Fig. 8.14*b*	Acetic acid (A)-water (B)-*ethylene dichloride*	Clarke and Othmer (1931*b*), Othmer (1941, 1978)
Fig. 8.14*b*	Acetic acid (A)-water (B)-*ethyl acetate*	Siirola (1996)
Fig. 8.14*b*	Acetic acid (A)-water (B)-*n-propyl acetate*	Othmer (1941, 1978)
Fig. 8.14*c*	Acetic acid (A)-water (B)-*n-butyl acetate*	Othmer (1941, 1978)
Fig. 8.14*c*	Acetic acid (A)-water (B)-*furfural*	Othmer (1941)
Not shown	Acetone-water-*ethyl bromide*[b]	Iino et al. (1971)
Fig. 8.14*d*	Acetone (A)-water (B)-*methyl isobutyl ketone*	Harrison (1990)
Fig. 8.14*d*	Acetone (A)-butanol (B)-water[c]	Pucci et al. (1986)
Fig. 8.14*e*	Ethanol (A)-water (B)-*ethyl ether*	Wentworth and Othmer (1940), Othmer and Wentworth (1940), Wentworth et al. (1943)
Fig. 8.14*e*	Acetonitrile (A)-water (B)-*acrylonitrile*	Blackford and York (1965)
Fig. 8.14*f*	Ethanol (A)-water (B)-*1-butanol*	Ross and Seider (1980), Schuil and Bool (1985)
Fig. 8.14*f*	n-propanol (A)-water (B)-*1-butanol*[d]	Block and Hegner (1976), Ross and Seider (1980), Buzzi Ferraris and Morbidelli (1981), Schuil and Bool (1985), Cairns and Furzer (1990*c*)

[a] Separation systems associated with the maps shown in Fig. 8.14*c, d,* and *f* have heterogeneous bottoms streams; all others have heterogeneous overhead streams. This knowledge is vital for the correct placing of the decanter.
[b] This map is similar to Fig. 8.14*c* but with component B (water) as the highest-boiling component.
[c] This mixture is self-entrained; i.e., all components are present in the process feed to the separation system.
[d] This mixture may be separated using 1-butanol as entrainer by employing a novel separation system (invent it!) not mentioned in any of the references cited. These references are mainly concerned with simulation algorithms.

TABLE 8.2 (*Continued*)
Examples of heterogeneous azeotropic separations (entrainer italicized)

Map type[a]	System	References
Fig. 8.14*h*	Ethanol (*A*)-diethoxymethane (*B*)-water[e]	Martin and Raynolds (1988)
Fig. 8.14*h*	Ethanol (*A*)-water (*B*)-*ethyl acetate*	Keyes (1929), Bril' et al. (1975)
Not shown	Formic acid-water-*n-propyl formate*[f]	Clarke and Othmer (1931*a*)
See Exercise 8	*sec*-Butyl alcohol-water-*di-sec-butyl ether*	Kovach and Seider (1987*a*, 1987*b*)
Unknown	Butylene-ammonia-*water*	Poffenberger et al. (1946)
Unknown	Butadiene-butylene-*ammonia*	Poffenberger et al. (1946)

[a] Separation systems associated with the maps shown in Fig. 8.14*c, d,* and *f* have heterogeneous bottoms streams; all others have heterogeneous overhead streams. This knowledge is vital for the correct placing of the decanter.
[e] This mixture is self-entrained. However, a special "trick" is required in order to achieve the separation; see Exercise 7.
[f] Formic acid and water form a maximum boiling azeotrope.

on some of the stages in the column, which may lower the mass transfer efficiency on those stages.[6] However, this is of secondary importance to the feasibility issue.

The map shown in Fig. 8.14*b* is relevant to the separation of acetic acid and water, which is of great commercial significance. Although this binary mixture does not form an azeotrope, it does have a severe tangent pinch at high compositions of water, which prevents the distillate from being acid-free. It is not economical to separate this mixture into pure product streams without the aid of an entrainer. All known commercial entrainers for this separation are heterogeneous and produce residue curve maps similar to the one shown in Fig. 8.14*b*.

Clarke-Othmer Process for Acetic Acid-Water Separation. Large amounts of dilute acetic acid need to be purified in the manufacture of volatile acetates for auto lacquers, cellulose acetate for rayon and filter materials, and in the manufacture of other important materials. The entrainer (e.g., ethylene dichloride, ethyl acetate, 1-propyl acetate, 1-butyl acetate, etc.) is charged to the azeo-column, which has a process feed consisting of acetic acid and water. The bottoms stream from the column is pure acetic acid, and the overhead vapor is close to the composition of the

[6] Practitioners of the art seem to agree that mass transfer efficiency and hydrodynamic performance of three-phase columns does not present a problem in practice. As early as 1938, Guinot and Clark (1938, p. 197) stated that, "The efficiency of the plates was apparently undiminished by the heterogeneity of the boiling liquid and that was undoubtedly due to the violent agitation produced on the plates by the rapid bubbling of the vapours through the liquid. . . ." Experiments indicate that stages with compositions well inside the two liquid phase regions do not exhibit foaming or loss of capacity. However, foaming and loss of capacity does occur on stages with compositions near the liquid-liquid boiling envelope, e.g., Davies et al. (1987), Herron et al. (1988). An assessment of the literature from a practitioner's viewpoint is given by Harrison (1990).

minimum-boiling heterogeneous binary azeotrope formed by the entrainer and water (the residue curve map is similar to Fig. 8.14*b*). The azeotropic vapors are condensed and decanted into a water stream which leaves as distillate, and an entrainer-rich stream which is returned to the column as reflux. Additional reflux, if needed, is achieved by returning some of the water stream. It is typical for these systems to have multiple liquid phases present on many stages in the rectifying section. This process was invented by Clarke and Othmer (1931*b*); more detailed operating information and entrainer comparison for this separation is given by Othmer (1941) and Othmer (1978, pp. 365–368).

Wentworth Process for Ethanol-Water Separation. In this process ethyl ether is used as the entrainer, which produces a residue curve map similar to the one shown in Fig. 8.14*e*. Ethyl ether and water form a minimum-boiling heterogeneous azeotrope at 34.15°C containing 98.75 wt % ether and 1.25 wt % water at atmospheric pressure. There is no azeotrope between ethyl ether and ethanol, and no ternary azeotrope. The dilute ethanol process feed is first preconcentrated up to the composition of the ethanol-water azeotrope. This stream is fed to the azeo-column, which produces pure ethanol as bottoms product and an overhead vapor close to the composition of the ethyl ether-water azeotrope. The overhead vapors are condensed and decanted into an ether-rich layer which is returned to the column as reflux, and a water-rich layer which leaves as distillate. The distillate contains no alcohol and very little ether. Pure water may be obtained from this stream by sending it to a stripping column where water is the bottoms product and the overhead vapor has a composition near the ethyl ether-water azeotrope. These vapors are condensed and recycled to the decanter. The azeo-column is normally operated at about 7 atm pressure, since this increases the amount of water in the ether-water azeotrope, thereby reducing the amount of ether needed in the system.

This process was invented by Wentworth and is described in more detail by Othmer and Wentworth (1940), Wentworth and Othmer (1940), and Wentworth et al. (1943). When it was invented, the main advantage of this process was its low energy consumption (about 20,000 Btu/gal ethanol product)[7] relative to the benzene process (43,000 Btu/gal ethanol product) (Wentworth and Othmer, 1940). Since then the gap has been narrowed by better designs for the benzene process, which is capable of producing 99.8 mol % ethanol with an energy consumption of 30,000 Btu/gal ethanol product (Ryan and Doherty, 1989); also see Knapp and Doherty (1990, Table 3) or with thermally integrated columns for 18,000 Btu/gal of ethanol product, Katzen et al. (1980) and Knapp and Doherty (1990, Table 3). In recent years homogeneous entrainers have shown great promise for this separation, and extractive distillation processes (using ethylene glycol as the entrainer) have been designed with energy consumptions of 22,000 Btu/gal ethanol product (Knapp and Doherty, 1990) or with thermally integrated columns for 8,000 to 12,000 Btu/gal ethanol product (Lynn and Hanson, 1986; Knapp and Doherty, 1990).

[7]This number, as well as other specific energy consumptions reported here for comparison, are based on a process feed (beer feed) of 4.2 mol % ethanol and 95.8 mol % water (10 wt % ethanol), an ethanol product containing 99.8 mol % ethanol, and a water product containing 99.5 mol % water. The process feed is assumed to be at 38°C and the feed preheater is taken into account when reporting the specific energy consumed by the process. The specific energies reported can be translated into SI units by using the factor 1 kJ/kg ethanol = 2.833 Btu/gal ethanol.

Rodebush System for Ethanol-Water Separation. When ethyl acetate is used as the entrainer to break the ethanol-water azeotrope, the residue curve map is similar to the one shown in Fig. 8.14*h*, i.e., the ternary azeotrope is homogeneous; otherwise the map is the same as for ethanol-water-benzene. In such cases the liquid leaving the condenser from the azeo-column will not separate into two liquid phases, and the system is infeasible unless special "tricks" are employed. In the Rodebush system, water is continuously added to the decanter in order to shift the overall composition into the two-liquid phase region. Each of the liquid phases from the decanter is fed to a separate distillation column which produce pure water and pure ethyl acetate, respectively. Some of the water is recycled to the decanter and all of the ethyl acetate is recycled to the azeo-column. Keyes (1929, Fig. 5) describes the system in a little more detail. A clever variation on this system was patented by Martin and Raynolds (1988) for separating a ternary feed consisting of ethanol, water, and diethoxymethane (which also has a residue curve map similar to the one shown in Fig. 8.14*h*) into its constituent pure components.

More Complex Mixtures. All the systems described above applied to type I liquid systems, i.e., mixtures in which only one of the binary pairs shows liquid-liquid behavior. Many mixtures of commercial interest display liquid-liquid behavior in two of the binary pairs (type II systems), e.g., *sec*-butyl alcohol-water-di-*sec*-butyl ether (SBA-water-DSBE), and water-formic acid-metaxylene (Reinders and DeMinjer, 1947*b*). Systems for these separations can be devised on the basis of residue curve maps along the lines proposed above. The SBA-water-DSBE separation is practiced by ARCO and is considered in detail by Kovach and Seider (1987*a*, 1987*b*, 1988), and by Widagdo et al. (1989); see Exercise 8.

In each case discussed above, the decanter is fed with an overhead stream from the column. However, for mixtures with maps like the one shown in Fig. 8.14*d*, novel systems can be devised where the decanter is fed with a sidestream from the middle of the column (Ciric et al., 2000).

8.6
EXERCISES

1. It is proposed to separate a binary mixture containing 40 mol % water and 60 mol % butyl acetate using the system shown in Fig. 8.10. Using the binary y-x diagram given in Fig. 8.15, draw a McCabe-Thiele construction to design the butyl acetate column assuming constant molar overflow. Take the butyl acetate as a 99 mol % pure saturated liquid bottoms product (the distillate should contain 70 mol % water), and assume the reboil ratio is 50% above the minimum value. Assume the phases in the decanter are in liquid-liquid equilibrium at their bubble point. How many equilibrium stages are required in each column section and what is the vapor-to-feed ratio?
2. Figure 8.10 shows a system of distillation columns for separating a partially miscible binary liquid mixture with a heterogeneous azeotrope into its constituent pure components. A better alternative is to eliminate the reflux streams and replace them with the feed streams so that each column operates as

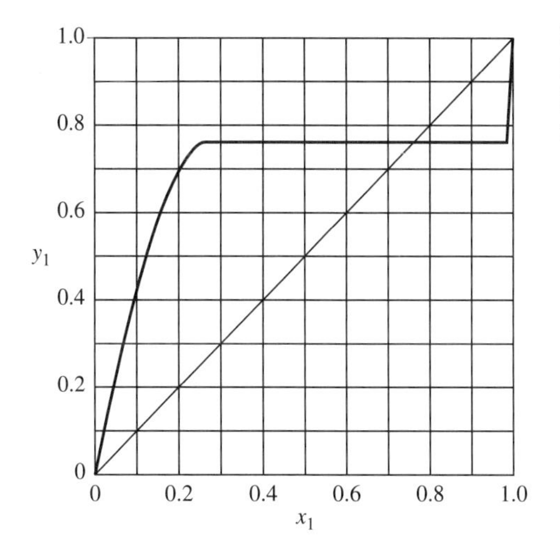

FIGURE 8.15
Binary *y-x* diagram for water (1) and butyl acetate (2) at 1 atm pressure. Taken from Gmehling et al. (1981, p. 417).

a stripping column, as shown in Fig. 8.16. Assume that the process feed is a saturated liquid with known flowrate F and composition of component A, x_F, and that each of the bottoms product streams is specified to be high-purity. Assume that the two liquid phases in the decanter are in liquid-liquid equilibrium at the boiling temperature of the azeotrope and that they are perfectly decanted. Therefore, the feed compositions to each column are fixed at the equilibrium values shown on Fig. 8.16. For simplicity assume that constant molar overflow prevails throughout the system. Draw a schematic binary y-x diagram for the mixture and show the operating lines and distillate compositions for each stripping column at selected reboil ratios between the minimum and maximum values. By making material balances show that the minimum feed flowrate to column 1 $F_{1,\min}$, the minimum vapor flowrate in column 1 $V_{1,\min}$, and the minimum reboil ratio in column 1 $s_{1,\min}$ are given by the expressions:

$$F_{1,\min} = \frac{F x^{Az}(1 - x_F)}{x^{Az} - x_{F1}} \tag{8.13}$$

$$V_{1,\min} = \frac{F x_{F1}(1 - x_F)}{x^{Az} - x_{F1}} \tag{8.14}$$

$$s_{1,\min} = \frac{V_1}{B_1} = \frac{x_{F1}}{x^{Az} - x_{F1}} \tag{8.15}$$

3. Consider a binary liquid mixture with a heterogeneous azeotrope in which there is a relatively narrow region of immiscibility at the boiling point, as shown in

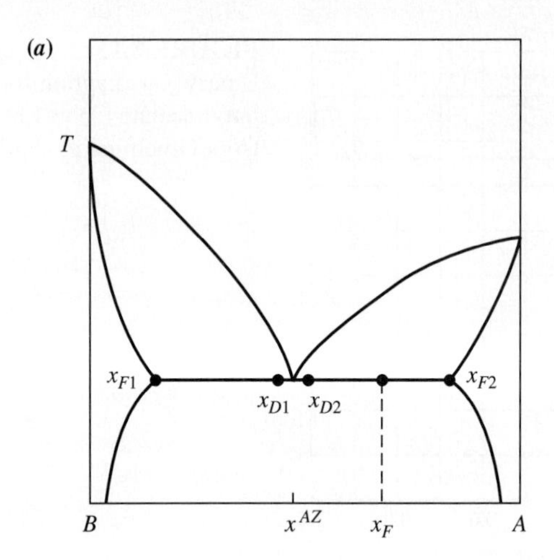

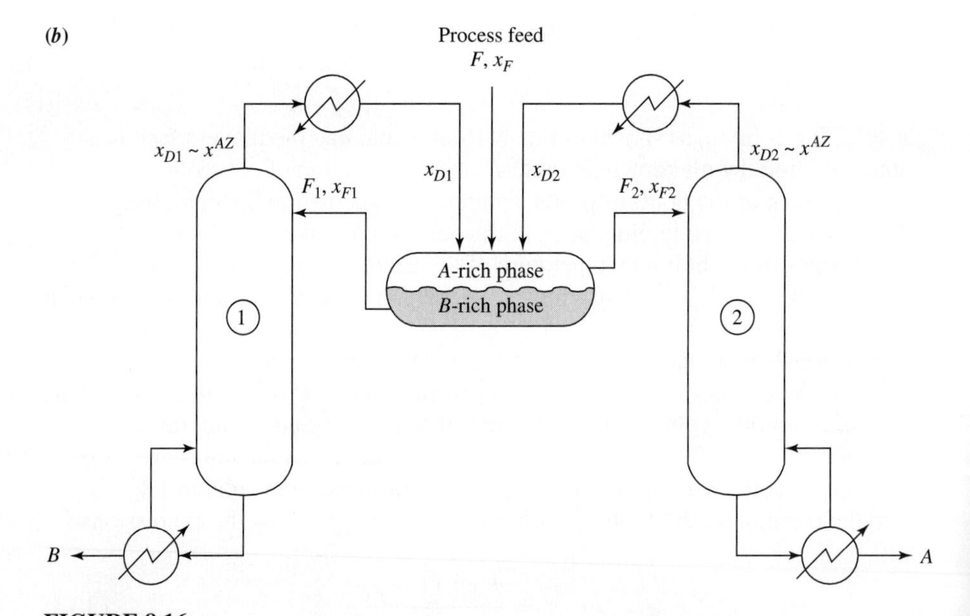

FIGURE 8.16
A system of stripping columns for separating binary heterogeneous mixtures.

Fig. 8.17 for water and 1-butanol at 760 mmHg. If the process feed is a homogeneous liquid mixture at its boiling point containing 20 mol % water, then the distillation system shown in Fig. 8.10 should be replaced with the one shown in Fig. 8.18 in which the homogeneous liquid feed is fed directly to column 1 and the reflux to each column is provided by streams from the decanter. Using the binary *y-x* diagram given in Fig. 8.17*b*, draw a

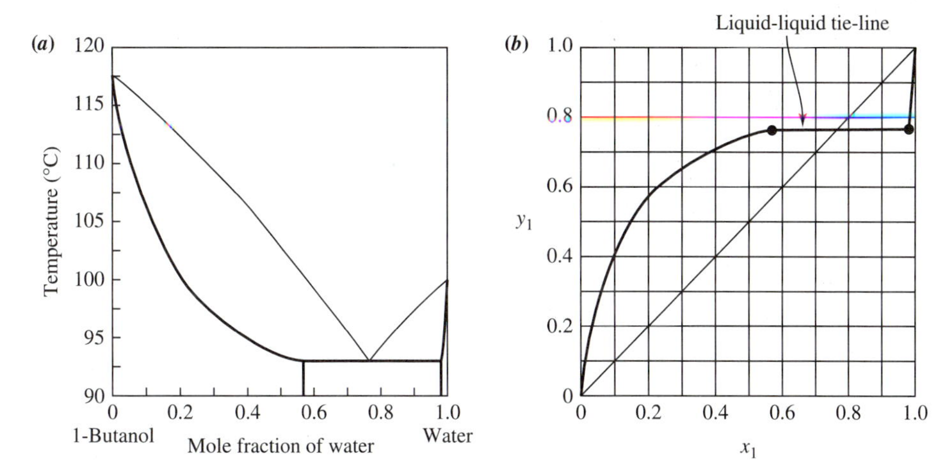

FIGURE 8.17
Binary VLE diagrams for water (1) and 1-butanol (2) at 1 atm pressure. (*a*) T-x_1-y_1 diagram. (*b*) y_1-x_1 diagram. The diagrams were calculated using the NRTL equation with model parameters taken from Gmehling and Onken (1977*a*, p. 406).

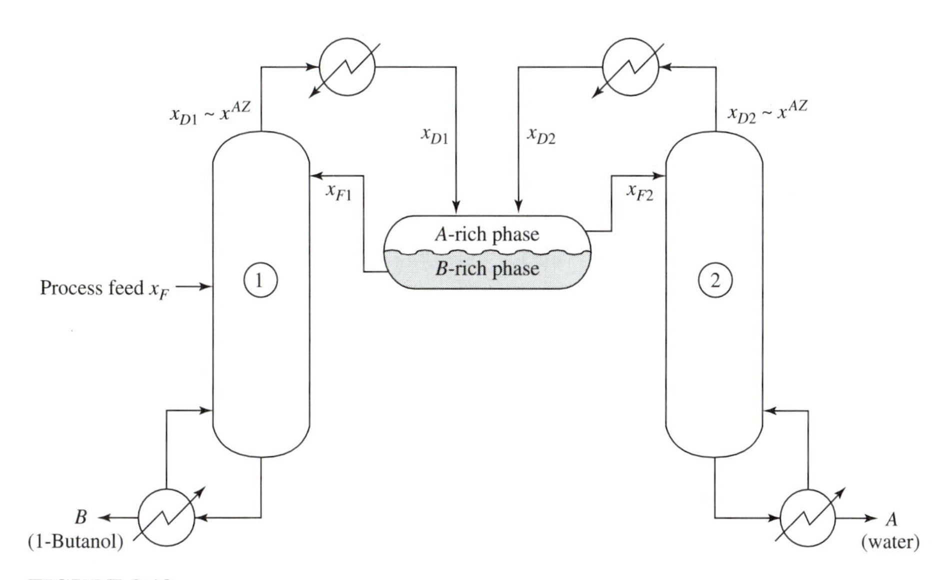

FIGURE 8.18
Distillation system for the separation of water (A) and 1-butanol (B).

McCabe-Thiele construction to design the 1-butanol column assuming constant molar overflow. Take the 1-butanol as a 99 mol % pure saturated liquid bottoms product, assume the distillate contains 70 mol % water and that the reflux ratio is 50% above the minimum value. Assume the phases in the decanter are in

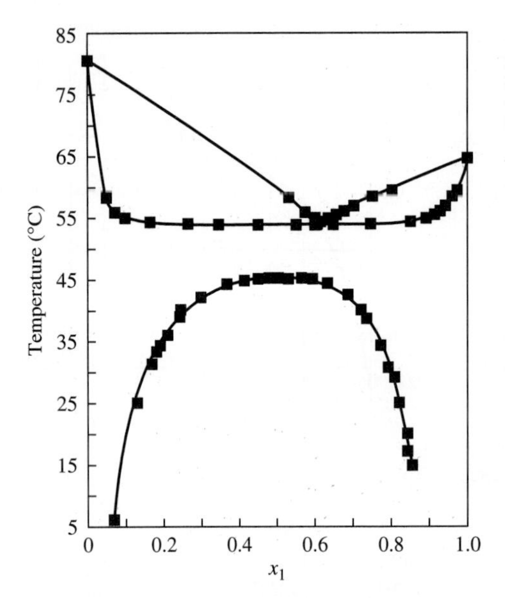

FIGURE 8.19
Binary phase diagram for methanol (1) and cyclohexane (2) at 1 atm pressure. Selected experimental data taken from Gmehling and Onken (1977*b*, p. 239) and Sorenson and Arlt (1979, pp. 89–91).

liquid-liquid equilibrium at their bubble point. How many equilibrium stages are required in each column section and what is the vapor-to-feed ratio?

4. Methanol and cyclohexane form a minimum-boiling binary azeotrope at 1 atm pressure. The liquid mixture is miscible in all proportions at the boiling point, but at lower temperatures it is only partially miscible. A phase diagram for the system is shown in Fig. 8.19. Devise a distillation system to separate this mixture into pure component product streams *without* employing a third component as a separating agent.
5. The Kubierschky two-column system for separating ethanol-water mixtures with benzene as entrainer is described in the text and shown in Fig. 8.20. On a triangular diagram draw the material balance lines for this system and mark the most important compositions similar to those shown on Fig. 8.12*b*.
6. Invent a distillation system to produce three pure products given the residue curve map and feed composition shown in Fig. 8.21. On a triangular diagram draw the material balance lines and mark the key compositions for your system.
7. Invent a distillation system to separate a ternary mixture of ethanol, water, and diethoxymethane (DEM) into its constituent pure components. Take the feed composition to be in region II on Fig. 8.22. On a triangular diagram draw the material balance lines and mark the key compositions for your system.
8. Methyl ethyl ketone is produced commercially by the catalytic dehydrogenation of *sec*-butyl alcohol (2-butanol). An important processing step is the separation of unreacted *sec*-butyl alcohol (SBA) from water. Invent a distillation system to dehydrate SBA using di-*sec*-butyl ether (DSBE) as entrainer at 1 atm pressure. The process feed consists of 38 mol % SBA and 62 mol % water; the product

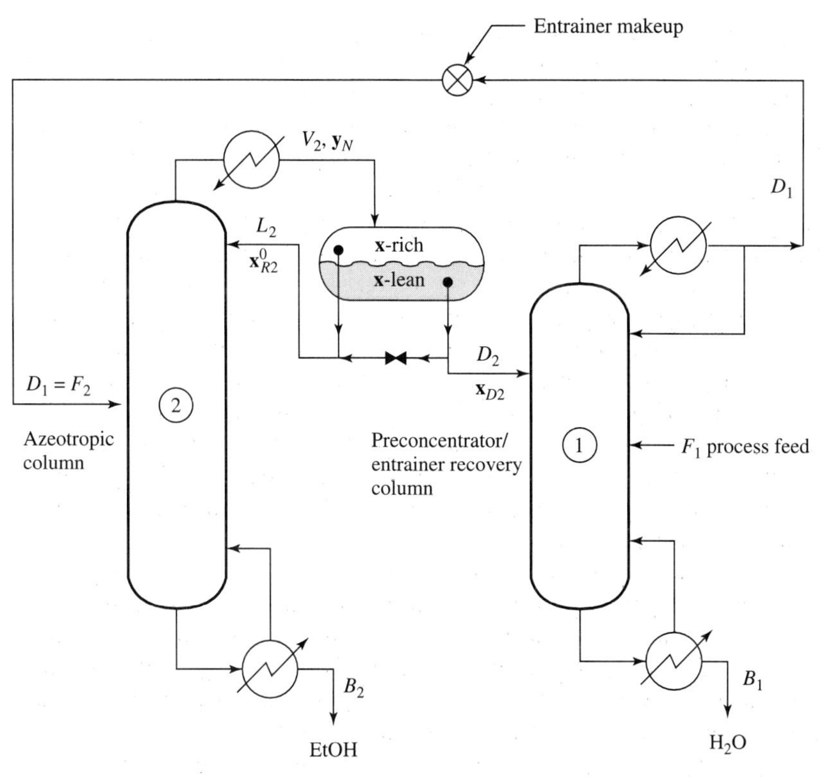

FIGURE 8.20
Kubierschky two-column system for separating ethanol and water using benzene as entrainer.

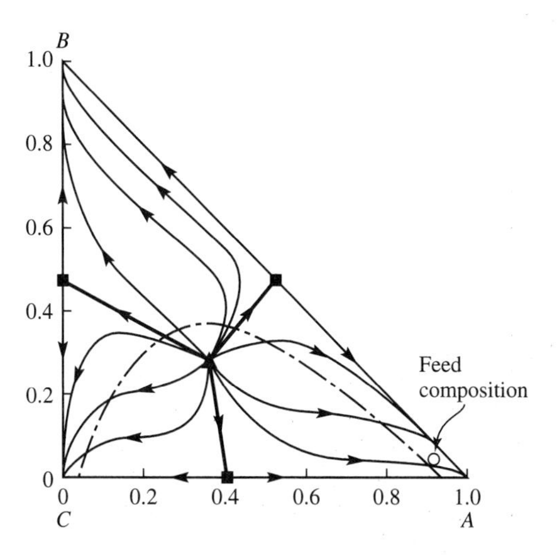

FIGURE 8.21
Residue curve map and feed composition for Exercise 6.

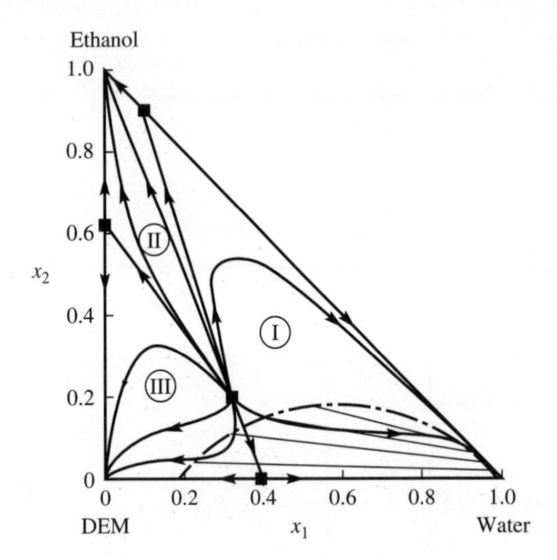

FIGURE 8.22
Schematic residue curve map for the mixture ethanol-water-DEM based on the patent of Martin and Raynolds (1988).

should consist of 99.15 mol % SBA, negligible amounts of DSBE, and the balance water. You are provided with the following data.

Pure component data:
SBA b.p. 98°C, water b.p. 100°C, DSBE b.p. 122°C

Azeotropic data:

SBA-water azeotrope	b.p. 87.37°C	Composition 38 mol % SBA[a]
Water-DSBE azeotrope	Minimum-boiling	Composition ≈9.5 mol % DSBE[b]
SBA-DSBE azeotrope	Minimum-boiling	Composition 85 mol % SBA[c]
SBA-DSBE-water azeotrope	b.p. 83°C	Composition ≈33 mol % SBA, ≈7 mol % DSBE[d]

Liquid-liquid equilibrium data at 38°C are given in Fig. 8.23. This exercise is based on papers by Kovach and Seider.

---

[a] Data taken from Gmehling and Onken (1977*a*, p. 419). The binary azeotrope is homogeneous but lies close to a region of liquid-liquid immiscibility which occurs over the composition range of approximately 5 to 30 mol % SBA.
[b] Horsley (1973, system number 736). Data estimated.
[c] Data taken from Kovach and Seider (1988). The y-x diagram reported by Kovach and Seider shows clearly that the binary azeotrope is minimum-boiling. Its temperature is difficult to establish accurately, but a value that is 1 or 2°C below the boiling point of SBA seems reasonable.
[d] Data estimated from Horsley (1973, system numbers 16168 and 16169).

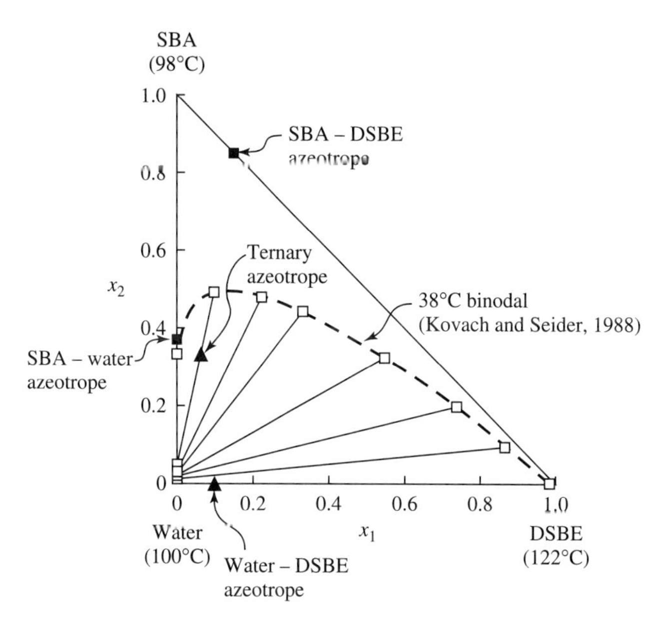

FIGURE 8.23
Liquid-liquid equilibrium data for the mixture SBA-water-DSBE at 38°C. Taken from Kovach and Seider (1988). Azeotropic data also shown.

References

Baker, L. E., A. C. Pierce, and K. D. Luks, "Gibbs Energy Analysis of Phase Equilibria," *Soc. Petrol. Engrs. J.,* **22,** 731–742 (1982).

Bancroft, W. D., and S. S. Hubbard, "A New Method for Determining Dineric Distribution," *J. Am. Chem. Soc.,* **64,** 347–353 (1942).

Barbaudy, J., "Contribution a L'Étude de la Distillation des Mélanges Ternaires Hétérogènes. II. Systèm Alcool Ethylique-Benzène-Eau," *J. Chim. Phys.,* **24,** 1–23 (1927).

Black, C., "Distillation Modeling of Ethanol Recovery and Dehydration Processes for Ethanol and Gasohol," *Chem. Engng. Prog.,* **76**(9), 78–85 (1980).

Blackford, D. S., and R. York, "Vapor-Liquid Equilibria of the System Acrylonitrile-Acetonitrile-Water," *J. Chem. Engng. Data,* **10,** 313–318 (1965).

Block, U., and B. Hegner, "Development and Application of a Simulation Model for Three-Phase Distillation," *AIChE J.,* **22,** 582–589 (1976).

Bril', Z. A., A. S. Mozzhukhin, F. B. Petlyuk, and L. A. Serafimov, "Mathematical Simulation and Investigation of the Heteroazeotropic Fractionation Process," *Theo. Found. Chem. Eng. (Engl. Trans.),* **9,** 761–770 (1975).

Bril', Z. A., A. S. Mozzhukhin, F. B. Petlyuk, and L. A. Serafimov, "Investigation of Optimal Conditions of Heteroazeotropic Rectification," *Theo. Found. Chem. Eng. (Engl. Trans.),* **11,** 675–681 (1977).

Bushmakin, I. N., and P. Y. Molodenko, "Distillation and Rectification in the System Water-Formic Acid-1, 2-Dichloroethane," *Russ. J. Phys. Chem. (Engl. Trans.),* **37,** 2618–2624 (1964*a*).

Bushmakin, I. N., and P. Y. Molodenko, "Method for Selection of the Separating Agent for Azeotropic Rectification of Binary Mixutres," *J. Appl. Chem. USSR,* **37,** 2609–2617 (1964*b*).

Buzzi Ferraris, G., and M. Morbidelli, "Distillation Models for Two Partially Immiscible Liquids," *AIChE J.,* **27,** 881–888 (1981).

Cairns, B. P., and I. A. Furzer, "Multicompoent Three-Phase Azeotropic Distillation. 1. Extensive Experimental Data and Simulation Results," *Ind. Eng. Chem. Research,* **29,** 1349–1363 (1990*a*).

Cairns, B. P., and I. A. Furzer, "Multicomponent Three-Phase Azeotropic Distillation. 2. Phase-Stability and Phase-Splitting Algorithms," *Ind. Eng. Chem. Research,* **29,** 1364–1382 (1990*b*).

Cairns, B. P., and I. A. Furzer, "Multicomponent Three-Phase Azeotropic Distillation. 3. Modern Thermodynamic Models and Multiple Solutions," *Ind. Eng. Chem. Research,* **29,** 1383–1395 (1990*c*).

Ciric, A. R., H. S. Mumtaz, G. Corbett, M. Reagan, W. D. Seider, L. A. Fabiano, D. M. Kolesar, and S. Widagdo, "Azeotropic Distillation with an Internal Decanter," *Computers Chem. Engng.*, in press (2000).

Clarke, H., and D. F. Othmer, Method for the Dehydration of Formic Acid (1931*a*). U.S. Patent 1,816,302.

Clarke, H. T., and D. F. Othmer, Process of Removing Water from Aqueous Acetic Acid (1931*b*). U.S. Patent 1,804,745.

Colburn, A. P., and J. C. Phillips, "Experimental Study of Azeotropic Distillation—Use of Trichloroethylene in Dehydration of Ethanol," *Trans. Am. Inst. Chem. Engrs.*, **40,** 333–359 (1944).

Davies, B., Z. Ali, and K. E. Porter, "Distillation of Systems Containing Two Liquid Phases," *AIChE J.*, **33,** 161–163 (1987).

Doherty, M. F., and G. A. Caldarola, "Design and Synthesis of Homogeneous Azeotropic Distillation. 3. The Sequencing of Columns for Azeotropic and Extractive Distillations," *Ind. Eng. Chem. Fundam.*, **24,** 474–485 (1985).

Doherty, M. F., and J. D. Perkins, "On the Dynamics of Distillation Processes—I. The Simple Distillation of Multicompoent Non-Reacting Homogeneous Liquid Mixtures," *Chem. Engng. Sci.*, **33,** 281–301 (1978).

Foucher, E. R., M. F. Doherty, and M. F. Malone, "Automatic Screening of Entrainers for Homogeneous Azeotropic Distillation," *Ind. Eng. Chem. Research,* **29,** 760–772 (1991).

Furzer, I. A., "Vapor-Liquid-Liquid Equilibria Using UNIFAC in Gasohol Dehydration Systems," *AIChE J.*, **30,** 826–829 (1984).

Furzer, I. A., "Ethanol Dehydration Column Efficiences Using UNIFAC," *AIChE J.*, **31,** 1389–1392 (1985).

Gibbs, J. W., "A Method of Geometrical Representation of the Thermodynamic Properties of Substances by Means of Surfaces," *Trans. Conn. Acad.*, **II,** 382–404 (1873). Reprinted in *The Scientific Papers of J. Willard Gibbs,* vol. 1, Dover, New York (1961).

Gibbs, J. W., "On the Equilibrium of Heterogeneous Substances," *Trans. Conn. Acad.*, **III,** 108–248 (1875). Reprinted in *The Scientific Papers of J. Willard Gibbs,* vol. 1, Dover, New York (1961); see pp. 55–129, especially the section "Geometrical Illustrations," pp. 115–129.

Gmehling, J., and U. Onken, *Vapour-Liquid Equilibrium Data Collection,* vol. 1/1, Aqueous-Organic Systems, of *Chemistry Data Series.* DECHEMA, Frankfurt/Main (1977*a*).

Gmehling, J., and U. Onken, *Vapour-Liquid Equilibrium Data Collection,* vol. 1/2a, Organic Hydroxy Compounds: Alcohols, of *Chemistry Data Series.* DECHEMA, Frankfurt/Main (1977*b*).

Gmehling, J., U. Onken, and W. Arlt, *Vapor-Liquid Equilibrium Data Collection,* vol. 1/1a (Supplement 1) Aqueous-Organic Systems, of *Chemistry Data Series.* DECHEMA, Frankfurt/Main (1981).

Guinot, H., and F. W. Clark, "Azeotropic Distillation in Industry," *Trans. Inst. Chem. Engrs.*, **16,** 189–199 (1938).

Hands, C. H. G., and W. S. Norman, "The Dehydration of Allyl Alcohol by Azeotropic Distillation," *Trans. Inst. Chem. Engrs.*, **23,** 76–88 (1945).

Harrison, M. E., "Consider Three-Phase Distillation in Packed Columns," *Chem. Eng. Progr.*, **86**(11), 80–85 (1990).

Herron, C. C., Jr., B. K. Kruelskie, and J. R. Fair, "Hydrodynamics and Mass Transer on Three-Phase Distillation Trays," *AIChE J.*, **34,** 1267–1273 (1988).

Holland, C. D., S. E. Gallun, and M. J. Lockett, "Modeling Azeotropic and Extractive Distillations," *Chem. Engng.*, **88,** 185–200 (1981).

Horsley, L. H., *Azeotropic Data III,* number 116 in Advances in Chemistry Series. American Chemical Society, Washington, DC (1973).

Iino, M. A., A. Nakae, J. Sudoh, Y. Hirose, and M. Hirata, "Removal of a Small Amount of Water in Acetone by Azeotropic Distillation Employing Ethylbromide as Separating Agent," *J. Chem. Engng. Japan,* **4,** 33–36 (1971).

Katzen, R., G. D. Moon, Jr., and J. D. Kumans, Distillation System for Motor Fuel Grade Anhydrous Alcohol (1980). U.S. Patent 4,217,178.

Keyes, D. B., "The Manufacture of Anhydrous Ethyl Alcohol," *Ind. Eng. Chem.,* **21,** 998–1001 (1929).

King, C. J., *Separation Processes.* McGraw-Hill, New York, 2d ed. (1980).

Knapp, J. P., and M. F. Doherty, "Thermal Integration of Homogeneous Azeotropic Distillation Sequences," *AIChE J.,* **36,** 969–984 (1990).

Knight, J. R., and M. F. Doherty, "Optimal Design and Synthesis of Homogeneous Azeotropic Distillation Sequences," *Ind. Eng. Chem. Research,* **28,** 564–572 (1989).

Komarov, V. M., and B. K. Krichevtsov, "Liquid-Vapor Equilibrium in the Ternary Systems Isopropylamine-Diisopropylamine-Water and Isopropyl Alcohol-Diisopropylamine-Water," *J. Appl. Chem. USSR (Engl. Trans.),* **43,** 301–306 (1970).

Kovach, J. W. III, and W. D. Seider, "Heterogeneous Azeotropic Distillation: Experimental and Simulation Results," *AIChE J.,* **33,** 1300–1314 (1987*a*).

Kovach, J. W. III, and W. D. Seider, "Heterogeneous Azeotropic Distillation-Homotopy-Continuation Methods," *Comput. Chem. Engng.,* **11,** 593–605 (1987*b*).

Kovach, J. W. III, and W. D. Seider, "Vapor-Liquid and Liquid-Liquid Equilibria for the System *sec*-Butyl Alochol-Di-*sec*-Butyl Ether-Water," *J. Chem. Engng. Data,* **32,** 16–20 (1988).

Kubierschky, K., Verfahren zur Gewinnung von hochprozentigem, bezw. absolutem Alkohol aus Alkohol-Wassergemischen in unterbrochenem Betriebe (1915). German Patent 287,897.

Lutugina, N. V., and I. N. Soboleva, "Study of Liquid-Liquid-Vapor Equilibrium in the System Nitrocyclohexane-Cyclohexanone Oxime-Water," *J. Appl. Chem. USSR (Engl. Trans.),* **43,** 112–17 (1970).

Lynn, S., and D. N. Hanson, "Multieffect Extractive Distillation for Separating Aqueous Azeotropes," *Ind. Eng. Chem. Process Design Dev.,* **25,** 936–941 (1986).

Martin, D. L., and P. W. Raynolds, Process for the Purificaiton of Diethoxymethane from a Mixture with Ethanol and Water (1988). U.S. Patent 4,740,273 assigned to Eastman Kodak Company.

Matsuyama, H., "Restrictions on Patterns of Residue Curves around Heterogeneous Azeotropes," *J. Chem. Eng. Japan,* **11,** 427–431 (1978).

McDonald, C. M., and C. A. Floudas, "Global Optimization for the Phase Stability Problem," *AIChE J.,* **41,** 1798 (1995).

McDonald, C. M., and C. A. Floudas, "GLOPEQ: A New Computational Tool for the Phase and Chemical Equilibrium Problem," *Computers Chem. Engng.,* **21,** 1–23 (1997).

Michelsen, M. L., "The Isothermal Flash Problem. Part I. Stability," *Fluid Phase Equilibria,* **9,** 1–19 (1982*a*).

Michelsen, M. L., "The Isothermal Flash Problem. Part II. Phase-Split Calculation," *Fluid Phase Equilibria,* **9,** 21–40 (1982*b*).

Murogova, R. A., G. L. Tudorovskaya, N. I. Pleskach, N. A. Safonova, I. D. Gridin, and L. A. Serafimov, "Liquid-Vapor Equilibrium in the System Water-Acetic Acid-*p*-Xylene at 760 mm," *J. Appl. Chem. USSR (Engl. Trans.),* **46,** 2615–2617 (1973).

Newsham, D. M. T., and N. Vahdat, "Prediction of Vapour-Liquid-Liquid Equilibria from Liquid-Liquid Equilibria. Part I. Experimental Results for the Systems Methanol-Water-*n*-Butanol, Ethanol-Water-*n*-Butanol and *n*-Propanol-Water-*n*-Butanol," *Chem. Eng., J.,* **13**(1), 27–31 (1977).

Norman, W. S., "The Dehydration of Ethanol by Azeotropic Distillation," *Trans. Instn. Chem. Engrs,* **23,** 66–75 (1945).

Othmer, D. F., "Azeotropic Distillation for Dehydrating Acetic Acid," *Chem. Metall. Engng.,* **40,** 91–95 (1941).

Othmer, D. F., Azeotropic and Extractive Distillation, in Kirk, R. E., and D. F. Othmer, editors, *Kirk-Othmer Encyclopedia of Chemical Technology,* pp. 352–377. Wiley, New York, 3d ed. (1978).

Othmer, D. F., Distillation—Some Steps in Its Development, in Furter, W. F., editor, *A Century of Chemical Engineering.* Plenum Press, New York (1982).

Othmer, D. F., and T. O. Wentworth, "Absolute Alcohol. An Economical Method for Its Manufacture," *Ind. Eng. Cem.,* **32,** 1588–1593 (1940).

Pham, H. N., and M. F. Doherty, "Design and Synthesis of Heterogeneous Azeotropic Distillations—I. Heterogeneous Phase Diagrams," *Chem. Engng. Sci.,* **45,** 1823–1836 (1990*a*).

Pham, H. N., and M. F. Doherty, "Design and Synthesis of Heterogeneous Azeotropic Distillations—II. Residue Curve Maps," *Chem. Engng. Sci.,* **45,** 1837–1843 (1990*b*).

Pham, H. N., and M. F. Doherty, "Design and Synthesis of Heterogeneous Azeotropic Distillations—III. Column Sequences," *Chem. Engng. Sci.,* **45,** 1845–1854 (1990*c*).

Pham, H. N., P. J. Ryan, and M. F. Doherty, "Design and Minimum Reflux for Heterogeneous Azeotropic Distillation Columns," *AIChE J.,* **35,** 1585–1591 (1989).

Poffenberger, N., L. H. Horsley, H. S. Nutting, and E. C. Britton, "Separation of Butadiene by Azeotropic Distillation with Ammonia," *Trans. Am. Inst. Chem. Engrs.,* **42,** 815–826 (1946).

Prausnitz, J. M., T. F. Anderson, E. A. Grens, C. A. Eckert, and R. Hsieh, *Computer Calculations for Multicomponent Vapor-Liquid and Liquid-Liquid Equilibria.* Prentice-Hall, Englewood Cliffs, NJ (1980).

Prokopakis, G. J., and W. D. Seider, "Dynamic Simulation of Azeotropic Distillation," *AIChE J.,* **29,** 1017–1029 (1983*a*).

Prokopakis, G. J., and W. D. Seider, "Feasible Specifications in Azeotropic Distillation," *AIChE J.,* **29,** 49–60 (1983*b*).

Prokopakis, G. J., W. D. Seider, and B. A. Ross, Azeotropic Distillation Towers with Two Liquid Phases, in Mah, R. S., and W. D. Seider, editors, *Foundations of Computer-Aided Process Design,* pp. 239–272, New York (1981).

Pucci, A., P. Mikitenko, and L. Asselineau, "Three-Phase Distillation. Simulation and Application to the Separation of Fermentation Products," *Chem. Engng. Sci.,* **41,** 485–494 (1986).

Rawat, B. S., A. N. Goswami, and S. Krishna, "Isobaric Vapour-Liquid Equilibria of the Ternary System Hexane-Benzene-Sulpholane," *J. Chem. Tech. Biotechnol.,* **30,** 557–562 (1980).

Reinders, W., and C. H. DeMinjer, "Vapour-Liquid Equilibria in Ternary Systems. I. The Course of the Distillation Lines," *Rev. Trav. Chim.,* **59,** 207–230 (1940).

Reinders, W., and C. H. DeMinjer, "Vapor-Liqud Equilibria in Ternary Systems IV. The System Water-Ethanol-Trichloroethene," *Rev. Trav. Chim.,* **66,** 522–563 (1947*a*).

Reinders, W., and C. H. DeMinjer, "Vapour-Liquid Equilibria in Ternary Systems V. The System Water-Formic Acid-meta-Xylene," *Rev. Trav. Chem.,* **66,** 564–572 (1947*b*).

Reinders, W., and C. H. DeMinjer, "Vapour-Liquid Equilibria in Ternary Systems VI. The System Water-Acetone-Chloroform," *Rev. Trav. Chim.,* **66,** 573–604 (1947*c*).

Robinson, C. S., and E. R. Gilliland, *Elements of Fractional Distillation.* McGraw-Hill, 4th ed. (1950).

Ross, B. A., and W. D. Seider, "Simulation of Three-Phase Distillation Towers," *Computers Chem. Engng.,* **5,** 7–20 (1980).

Rovaglio, M., and M. F. Doherty, "Dynamics of Heterogeneous Azeotropic Distillation Columns," *AIChE J.,* **36,** 39–52 (1990).

Ryan, P. J., and M. F. Doherty, "Design/Optimization of Ternary Heterogeneous Azeotropic Distillation Sequences," *AIChE J.,* **35,** 1592–1601 (1989).

Schreinemakers, F. A. H., "Dampfdrucke im System: Wasser, Aceton und Phenol. II.," *Z. Phys. Chem.*, **39,** 440 (1901).

Schreinemakers, F. A. H., "Dampfdrucke im System: Benzol, Tetrachlorkohlenstoff und Äthylaklkohol. II.," *Z. Phys. Chem.*, **47,** 257 (1903*a*).

Schreinemakers, F. A. H., "Einige Bemerkungen ber Dampfdrucke Ternärer Gemische," *Zeitschrift f. Physik. Chemie.*, **XLIII,** 671 (1903*b*).

Schuil, J. A., and K. K. Bool, "Three-Phase Flash and Distillation," *Computers Chem. Engng.*, **9,** 295–300 (1985).

Siirola, J. J., "Industrial Applications of Chemical Process Synthesis," *Adv. Chem. Engng.*, **23,** 1–62 (1996).

Sorenson, J. M., and W. Arlt, *Liquid-Liquid Equilibrium Data Collection: Binary Systems,* vol. V/1 of *Chemistry Data Series.* DECHEMA, Frankfurt/Main (1979).

Swank, D. J., and J. C. Mullins, "Evaulation of Methods for Calculating Liquid-Liquid Phase-Splitting," *Fluid Phase Equilibria,* **30,** 101–110 (1986).

Talley, P. K., J. Sangster, C. W. Bale, and A. D. Pelton, "Prediction of Vapor-Liquid and Liquid-Liquid Equilibria and Thermodynamic Properties of Multicomponent Organic Systems from Optimized Binary Data Using the Kohler Method," *Fluid Phase Equilibria,* **85,** 101–128 (1993).

Townsend, D. W., Private Comunication (1982). ICI.

Van Dongen, D. B., M. F. Doherty, and J. R. Haight, "Material Stability of Multicomponent Mixtures and the Multiplicity of Solutions to Phase-Equilibrium Equations I. Nonreacting Mixtures," *Ind. Eng. Chem. Fundam.*, **22,** 472–485 (1983).

Van Dongen, D. B., and M. F. Doherty, "On the Dynamics of Distillation Processes—V. The Topology of the Boiling Temperature Surface and Its Relation to Azeotropic Distillation," *Chem. Engng. Sci.*, **39,** 883–892 (1984).

van Zandijcke, F., and L. Verhoeye, "The Vapour-Liquid Equilibrium of Ternary Systems with Limited Miscibility at Atmospheric Pressure," *J. Appl. Chem. Biotechnol.*, **24,** 709–729 (1974).

Venkataraman, S., and A. Lucia, "Solving Distillation Problems by Newton-Like Methods," *Computers Chem. Engng.*, **12,** 55–69 (1988).

Wasylkiewicz, S. K., L. N. Sridhar, M. F. Doherty, and M. F. Malone, "Global Stability Analysis and Calculation of Liquid-Liquid Equilibrium in Multicomponent Mixtures," *Ind. Eng. Chem. Research,* **35,** 1395–1408 (1996).

Wentworth, T. O., and D. F. Othmer, "Absolute Alcohol. An Economical Method for Its Manufacture," *Trans. Am. Inst. Chem. Engrs.*, **36,** 785–799 (1940).

Wentworth, T. O., D. F. Othmer, and G. M. Pohler, "Absolute Alcohol, an Economical Method for Its Manufacture. II. Plant Data," *Trans. Am. Inst. Chem. Engrs.*, **39,** 565–578 (1943).

Widagdo, S., W. D. Seider, and D. H. Sebastian, "Bifurcation Analysis in Heterogeneous Azeotropic Distillation," *AIChE J.*, **35,** 1457–1464 (1989).

Widagdo, S., W. D. Seider, and D. H. Sebastian, "Dynamic Analysis of Heterogeneous Azeotropic Distillation," *AIChE J.*, **38,** 1229–1242 (1992).

Wong, D. S., S. S. Jang, and C. F. Chang, "Simulation of Dynamics and Phase Pattern Changes for an Azeotropic Distillation Column," *Computers Chem. Engng.*, **15,** 325–335 (1991).

Young, S., "The Preparation of Absolute Alochol from Strong Spirit," *J. Chem. Soc.*, **81,** 707–717 (1902).

Young, S., Verfahren zur Gewinnung wasser freien Alkohols aus Spiritus mittels fraktionierter Destillation und ohne wasserentziehende Chemikalien (1903). German Patent 142,502.

9

Batch Distillation

Batch processes are sometimes preferable to the continuous processes discussed in earlier chapters. For example, batch processes can be developed faster, which is important for new products to establish a market share rapidly. Batch processes allow flexibility in scheduling demands, which is also important for new products or any product with uncertain demands, and also for the production of multiple products using the same equipment. Even when a continuous process offers reduced labor and investment costs per unit of production, batch processing is more flexible and can be more effective. Furthermore, batch experiments are often used to collect data, and we should understand how to interpret these results and how to plan further experiments, even if the goal is the design and development of continuous processes.

Batch distillation is a common operation in speciality and fine chemicals as well as in the production of flavors, fragrances, pharmaceuticals, dyes, and other relatively small volume products. This chapter is focused on feasibility and process synthesis. A comprehensive treatment of batch distillation is beyond the scope of this chapter; see Diwekar (1995) for a thorough treatment. Although they are not discussed here, simulation tools for batch distillation are useful, both commercial, *BATCHFRAC* (Aspen Technolgoy, 2000) and Chemcad-BATCH (Chemstations, Inc., 2000) and academic *MultiBatchDS* (Diwekar, 1996).

In the first two sections of this chapter we develop a basic model for batch distillation and examine its behavior for simple mixtures. The approach is developed further for azeotropic mixtures in the third section, followed by a section using the results for system synthesis. Targets for operating policies are discussed in Sec. 5, and developing work on novel configurations completes the chapter.

9.1
SIMPLE MODEL FORMULATION

The isobaric open evaporation discussed in Sec. 5.2 usually gives products of low purity.[1] Fortunately, substantial improvements in purity can be made by adding a

[1]Another option, which is useful for the treatment of temperature-sensitive materials, is to adjust the pressure to keep the temperature constant or to limit the increase. The ideas discussed here for the isobaric case are also useful in these cases, with a few modifications.

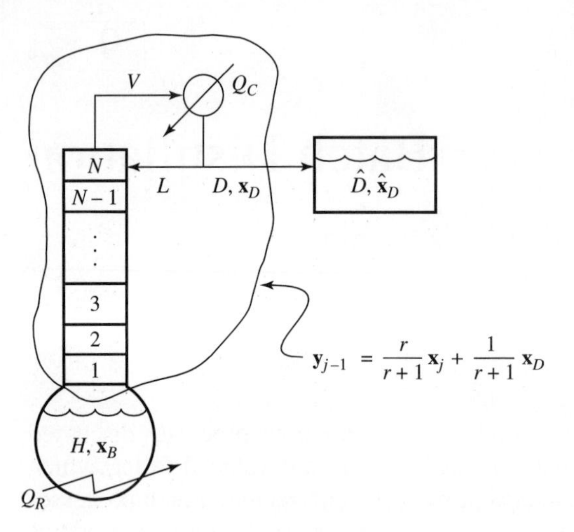

FIGURE 9.1
A batch rectifier.

column section to further purify the vapor as shown in Fig. 9.1. The *still* is charged with a liquid feed which is then boiled; vapor from the still enters the bottom of a rectification column with N theoretical stages and a total condenser which yields a saturated liquid product along with the reflux.[2]

A useful limiting case is for a holdup in the column that is small compared to that in the still. This is a model favored since the first studies of batch distillation (Robinson and Gilliland, 1950, Chap. 14) and captures many important features. The simplified model that results is based on unsteady-state mass balances for the still combined with a steady-state description of the column, or quasi-steady-state (QSS) model.[3]

The overall QSS material balance is

$$\frac{dH}{dt} = -D \tag{9.1}$$

The *holdup* H is the total number of moles in the still at time t and D is the instantaneous molar flowrate of the distillate. We may also be interested in the cumulative amount of distillate collected $\hat{D}$, which can be found by integrating $D(t)$ from the start of the batch until the time of interest. Balances for the individual components

[2] A partial condenser can also be used.

[3] It is important to note that this model is not accurate for startup, for rapid transitions between cuts, or near the end of a batch cycle if very little material remains in the still (the distillation will often be stopped for operability or safety reasons before this point). Generally, when the volume in the column and condenser is a few percent of each cut anticipated, the QSS model is accurate. An estimate for the size of the cuts from the QSS model and a retrospective check of the relative volumes is an efficient approach. For a detailed model and results for rapid transitions, see, e.g., Kruel et al. (1999). Detailed dynamic models compare well with the simplified model for the conceptual design studies we consider here (Cheong and Barton, 1999*c*).

can be written

$$\frac{d(H\mathbf{x}_B)}{dt} = -D\mathbf{x}_D \tag{9.2}$$

where $\mathbf{x}_B$ and $\mathbf{x}_D$ are the $c - 1$ independent mole fractions of still and distillate compositions, respectively. The "warped time" introduced in Chap. 5 to describe the simple open evaporation is also useful here to find

$$\frac{d\mathbf{x}_B}{d\xi} = \mathbf{x}_B - \mathbf{x}_D \tag{9.3}$$

where the warped time ξ is

$$\xi = \ln\left(\frac{H_0}{H}\right) \tag{9.4}$$

This is the same idea used for *simple* distillation in Chap. 5, as in that case the transformation uncouples the composition variables from the flowrates, holdups, and utility requirements.[4] Physically, $e^{(-\xi)}$ is the fraction of the original mixture remaining in the still so $1 - e^{(-\xi)}$ is the fraction of the original feed that has been collected overhead (neglecting the holdup in the column section). A complete solution to the problem eventually requires a relationship of these quantities to the real time of operation and to the heating policy. However, we do not need this complete solution to decide the sequence of product compositions that can be achieved, regardless of the heating policy and the actual batch times. This is an enormous advantage in the analysis of complicated mixtures because it allows for partial solutions that reveal limits on the compositions. To find those, we focus on solutions of Eqs. 9.3.

Starting from an initial composition in the still $\mathbf{x}_{B,0}$, Eqs. 9.3 can be integrated once a relationship between the product composition and the still composition is known. For the *simple* distillation considered in Chap. 5 this relationship was just the vapor-liquid equilibrium. When a column is added, the relationship is more complicated, involving the mass balances, reflux ratio, and number of stages for the column. Since we are considering the dynamics of the rectifier to be much faster than those of the reboiler, we use steady-state relationships for the column. For example, with the constant molar overflow assumption, these are

$$\mathbf{y}_{j-1} = \frac{r}{r+1}\mathbf{x}_j + \frac{1}{r+1}\mathbf{x}_D \tag{9.5}$$

where $j = 1, \ldots, N$ is the stage number. There are two degrees of freedom and, for instance, the number of stages and a constant reflux ratio can be specified; then all of the compositions can be found as a function of the warped time by integrating Eqs. 9.3, using Eqs. 9.5 to relate the still to the product compositions. The latter is an iterative calculation except at total reflux when Eqs. 9.5 are simply $\mathbf{y}_{j-1} = \mathbf{x}_j$; these are independent of the distillate composition.

[4]This also means that the response of the column is instantaneous for this model, which is a set of differential equations (for the still) coupled to algebraic equations (for the column and the vapor-liquid equilibrium). These "DAE" systems can be difficult to integrate accurately for the full solution. For example, see Logsdon and Biegler (1993).

9.2 SOLUTIONS FOR SIMPLE MIXTURES

Two Components

A simple but instructive example is the limit of very large reflux for binary mixtures with constant volatility. The Fenske equation 3.48, or the earlier equation 3.45, applied to the N stages in the rectifying section (recognizing that the distillate is the same composition as y_N for a total condenser) can be used in Eq. 9.3 for binary mixtures to give the following single equation.

$$\frac{dx_B}{d\xi} = x_B - \frac{\alpha^{N+1} x_B}{1 + (\alpha^{N+1} - 1)x_B} \tag{9.6}$$

Within the limitations of this quasi-steady-state model, the effect of adding stages at total reflux is identical to simply increasing the relative volatility. This model for batch distillation is the same as for simple distillation, with the relative volatility replaced by α^{N+1}.

Results from the integration of Eq. 9.6 are shown in Fig. 9.2. With few stages, the distillate purity is poor and gradually diminishes as the composition of the light component in the batch is depleted. The addition of stages leads to a much more sharply defined variation in the distillate composition. The distillate mole fraction of the light component is quite pure for a time and then drops precipitously after the light component in the still is depleted. This behavior defines a "cut," or product fraction. This sort of sharp definition in the change of product composition is often desirable. Because of this rapid change, batch distillation models, especially for cases with a large number of stages, can be difficult to integrate accurately. Exercise 1 gives an example.

Note also that the quasi-steady-state model is not accurate near a very sharp transition in the distillate composition. This is because the column and condenser (and the reflux accumulator, if used) do have some finite holdup and the associated dynamics that will slow the rate of change. An accurate model of such sharp changes must be more detailed and is beyond the scope of this chapter. See Diwekar (1995) for a more detailed treatment of batch distillation, including column dynamics.

When the reflux ratio is not so large, Eq. 9.5 must be used to relate the product and still compositions at each step in the integration of Eqs. 9.3. This generally requires an iterative solution of the VLE and the operating relationship.[5] More importantly, the reflux ratio *policy* must be specified for the integration.

One possibility is to keep the reflux ratio constant, which leads to a decreasing distillate purity. This can be seen from the McCabe-Thiele diagram as shown in Fig. 9.3*a*. Note that the number of stages is constant and is shown for only the first operating line in the figure. Alternatively, the reflux ratio can be manipulated

---

[5]For constant volatility mixtures, Smoker's equation can be used to relate the product and still compositions. See Exercise 8 in Chap. 3, and especially Eq. 3.70.

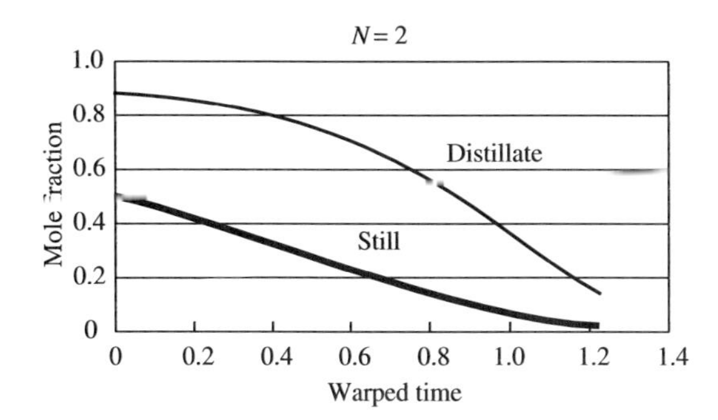

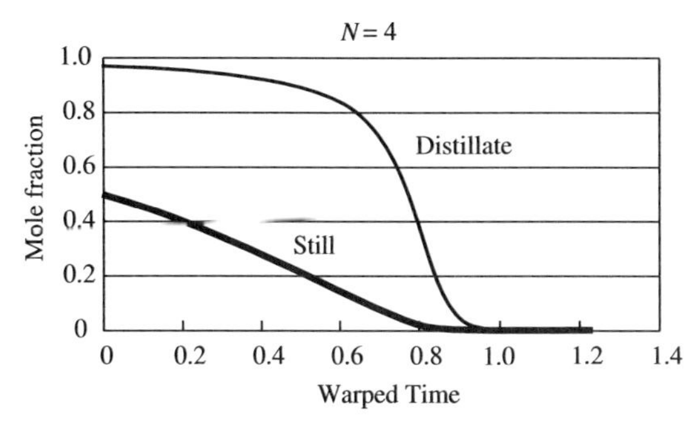

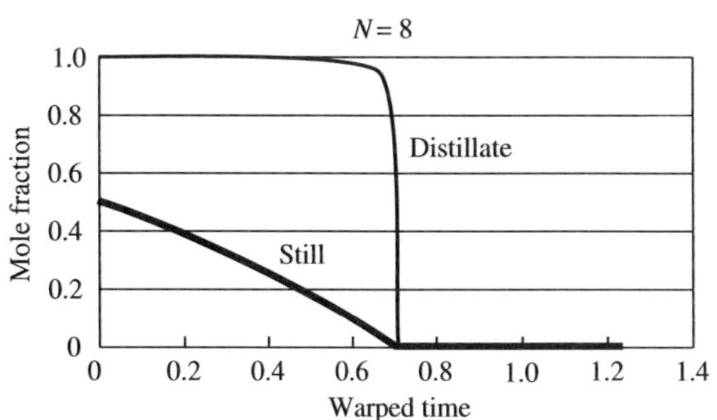

FIGURE 9.2
Distillate and still compositions for batch rectification of a binary mixture. The relative volatility is constant at a value of 2.0; the initial composition is 50 mol % light. N is the number of theoretical stages in the rectifying section.

during the process to compensate for the change in still composition. A common and intuitive approach is to increase the reflux ratio to keep the distillate composition constant; the graphical interpretation is shown in Fig. 9.3*b*. Note that the slope of the operating line is $r/(r+1)$, so small changes in the slope correspond to large changes in r when the slope approaches unity. This means that the increase in reflux ratio will be quite sharp as the still composition of the light component is depleted to low levels. As the reflux ratio is increased, the energy and time required for the distillation grows. In practice, it may be less costly to accommodate a gradual change in the distillate purity, as long as the average purity over time $\hat{x}_D$ meets the desired purity specification. In practice, the reflux ratio may be increased in discrete steps as the batch proceeds. Sometimes there are also intermediate periods of total reflux as well, in order to allow for column dynamics, purity measurements, etc.

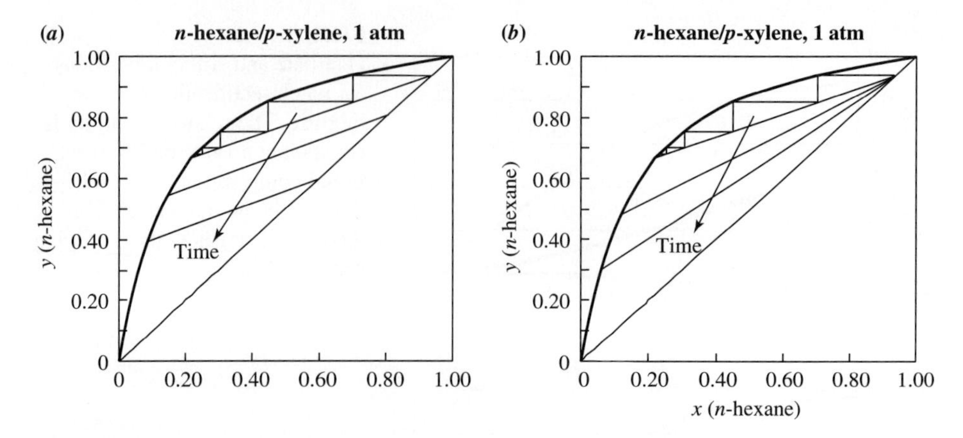

FIGURE 9.3
Constant reflux (*a*) and constant distillate policies (*b*) for a batch rectifier on a McCabe-Thiele diagram.

Three or More Components

The results from the QSS model are shown as a function of the warped time for constant volatility ternary and quaternary mixtures in Fig. 9.4. The basic results agree with intuition. Most of the lowest-boiling component is removed in the first cut with a purity that is determined by the number of stages and the reflux ratio. As the lightest component is removed, the mole fractions of the others in the still increase. Toward the end of the first cut, the mole fraction of the lightest component in the product drops as the still becomes more dilute in this component.[6] The lightest component is eventually depleted and the scenario is repeated for the next lowest boiler. An estimate of the warped time corresponding to the different cuts can be made using Eq. 9.4 and a mass balance based on complete recoveries; the warped time at the end of cut i would be $\xi_i = -\ln(1 - \sum_{k=1}^{i} x_{k,0})$.

The cuts shown in Fig. 9.4 are not very high in purity and they do not change composition very rapidly between cuts; i.e., they are not very "sharp." The design or operation can be changed to sharpen the cuts at the expense of stages and reflux. This requires increased investment due to increased stages, increased energy due to utility use, and also investment due to increased time for collection of the cuts (causing decreased productivity). It may not be best to seek sharp cuts; lower costs may be found when "intermediate" or "sloppy" cuts with lower purity are taken and then recycled to subsequent batches; see cuts 2 and 4 in Fig. 9.5. This depends on the desired product purity and on the cost of reflux and stages that would be required to eliminate the intermediate cut. The intermediate cuts can be recycled to subsequent batches if more than one batch is processed. In this case, subsequent batches will have a different composition and amount than the first batch, so the design and/or the cuts and operating times may change. See also Luyben (1988).

---

[6]It is useful and common to increase the reflux ratio to postpone such a decrease of the distillate purity.

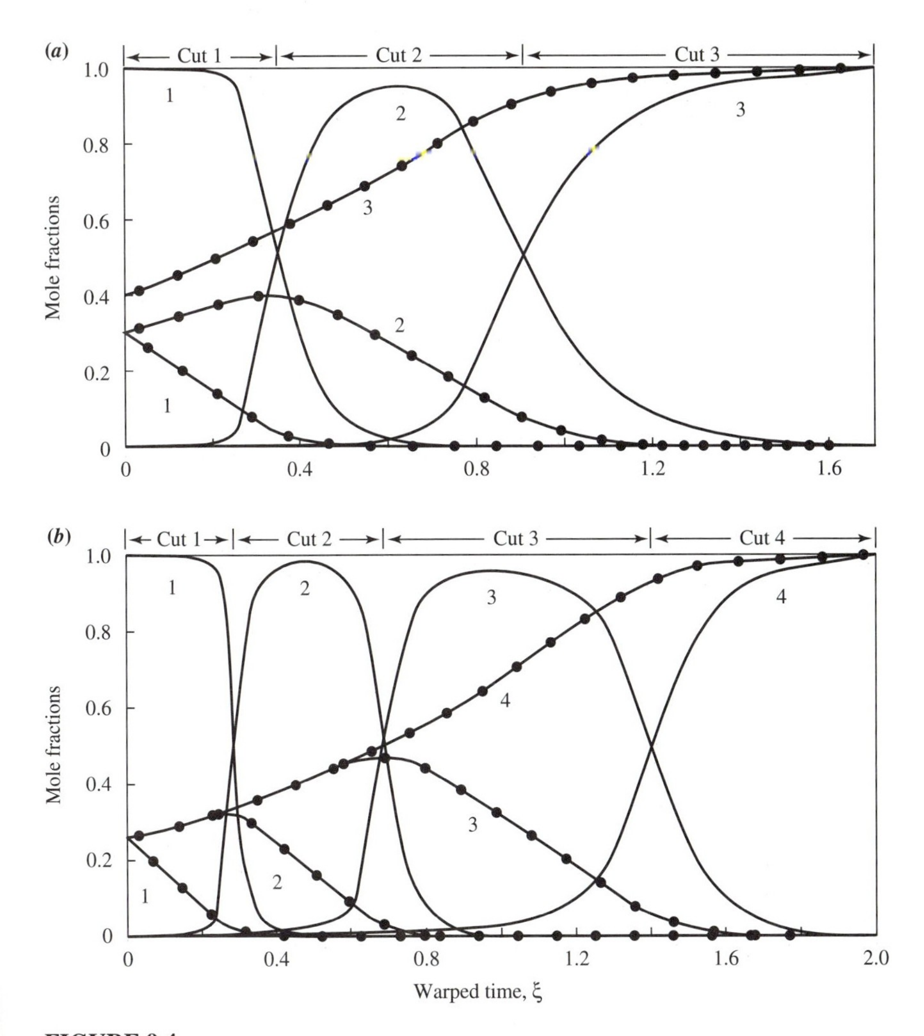

FIGURE 9.4
Distillate (——) and still (–•–•–•–) compositions for batch rectification of ideal mixtures. Hexane, heptane, and octane (*a*) with 7 theoretical stages and $r = 5$; the warped time at the end of the cuts can be approximated by mass balance $\xi_1 = \ln(1/0.7) = 0.36$ and $\xi_2 = \ln(1/0.4) = 0.92$. For pentane, hexane, heptane, and octane (*b*) with 5 theoretical stages and a constant reflux ratio of 8, the warped times at the end of the cuts are approximately $\xi_1 = \ln(1/0.75) = 0.29$, $\xi_2 = \ln(1/0.50) = 0.69$, $\xi_3 = \ln(1/0.25) = 1.39$. The pressure is 1 atm.

9.3 AZEOTROPIC MIXTURES

It is more difficult to understand the behavior of multicomponent azeotropic mixtures. For instance, there may be a sensitive dependence of the product purity on the feed composition when distillation boundaries are present and it may actually be

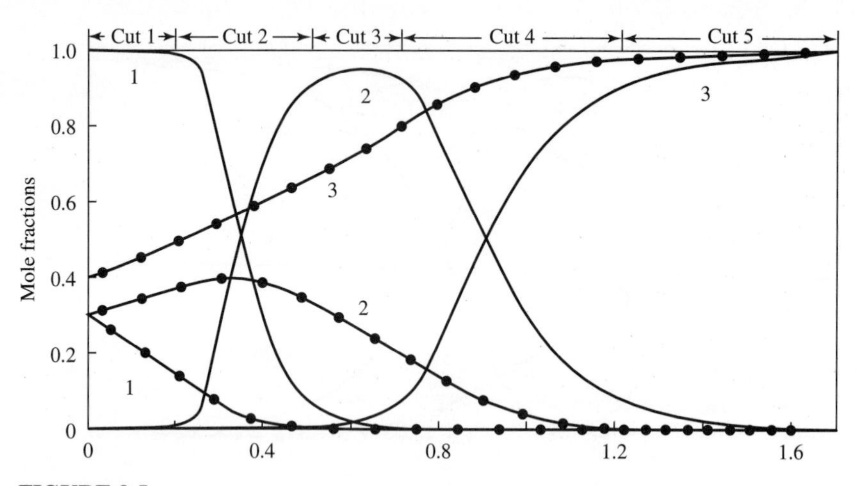

FIGURE 9.5
Distillate (——) and still (–•–•–•–) compositions for batch rectification of ideal mixtures. The example is identical to Fig. 9.4*a* except that sloppy cuts 2 and 4 are included.

impossible to produce certain high-purity cuts. Furthermore, while the still temperature always increases monotonically, the distillate temperature may actually *decrease* during part of the batch. The first experiments showing this "distillation anomaly" were described by Reinders and DeMinjer (1940*a*) and Ewell and Welch (1945). We explain below that such behavior is actually *not* anomalous, and why it is, in fact, unavoidable in certain cases. The best way to understand these phenomena is to consider residue curves and their consequences for the batch distillation.

Residue Curves and Product Compositions

The trajectories of the still composition in the phase plane are called *batch distillation residue curves,* to distinguish them from the *simple distillation residue curves.* Figures 9.6*a*, *b* and *c* show batch distillation residue curves for a mixture of methanol, acetone, and chloroform at large reflux and for various numbers of stages. These should be compared to the simple distillation residue curves shown in Fig. 9.6*d*.

With a large number of stages, the batch distillation residue curves approach straight lines in the phase plane until an edge of the triangle or the stable separatrix from the simple distillation residue curves is encountered. This behavior is generic and can most easily be explained by considering the limiting case of total reflux. In this argument, it is important to distinguish the *simple* from the *batch* distillation residue curves. The essential idea of feasibility is this: *At total reflux, the simple distillation residue curves correspond to profiles of the liquid composition inside a column.* That is, at every instant of time

- The distillate composition must lie on the *simple* distillation residue curve that passes through the corresponding still composition.

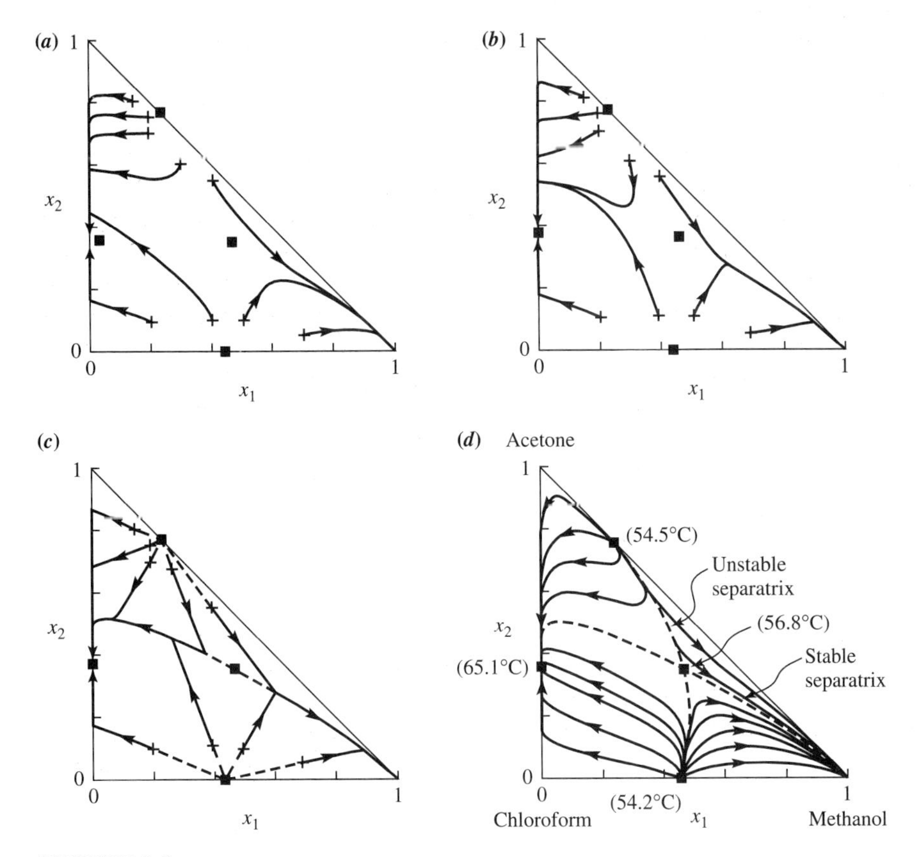

FIGURE 9.6
Predicted still compositions in a batch rectifier for a mixture of methanol, acetone, and chloroform at 1 atm pressure. The trajectories originating from various initial still compositions (+) are shown for total reflux and (*a*) $N = 3$, (*b*) $N = 8$, and (*c*) $N = 20$. The *simple* distillation residue curves are shown in (*d*).

- The distillate composition must approach the unstable node for the *simple* distillation residue curve or another saddle singular point for that curve if we consider the limit of a sufficiently large number of stages.

The first statement follows from the analogy between simple distillation residue curves and the composition profiles in continuous columns at total reflux. This is useful here because we employ a quasi-steady-state model for the column. It can also be shown that

- The tangent to the *batch* distillation residue curves must pass through the distillate composition. This can be seen for ternary mixtures from Eqs. 9.3, which can be rearranged to

$$\frac{dx_{2,B}}{dx_{1,B}} = \frac{x_{2,B} - x_{2,D}}{x_{1,B} - x_{1,D}} \tag{9.7}$$

There are two possibilities for the variation of distillate composition. First, it may be a constant, e.g., at or very near the composition of the unstable node. In that case, a mass balance demands that the batch composition move away from the unstable node in a straight line. Consequently, the *batch* distillation residue curve can exhibit a linear or nearly linear portion when the number of stages and the reflux are large. These straight-line segments must pass through the unstable node in the simple distillation region that contains the feed composition. This explains the first part of the batch distillation residue curves shown in Fig. 9.6*c*.

A second possibility is that the distillate composition is not constant. In this case, the linear behavior ends when the batch distillation residue curves reach a stable separatrix or the edge of the composition space, which they follow in the direction of increasing temperature. Equation 9.7 demands that the tangent to the batch distillation residue curve pass through the product composition, but not necessarily that the unstable node be on that tangent. This means that there can be a pinch point *inside* the column section, rather than at the top, when we consider the limiting case of a large number of stages.

Using these properties, once the simple distillation residue curves are known, the behavior of the batch distillation residue curves can be anticipated without much additional work for the limiting case of total reflux and a large number of stages.

EXAMPLE 9.1. A ternary mixture of acetone, chloroform, and benzene contains a single maximum-boiling binary azeotrope between acetone and chloroform at a pressure of 1 atm. The temperatures and compositions are as shown in Fig. 9.7. Without a detailed VLE model, the simple distillation residue curves can be sketched using the methods from Chap. 5. The azeotrope turns out to be a saddle, and there is a distillation boundary (a stable separatrix) joining the azeotrope to pure benzene as shown in the figure.

- For initial compositions on the chloroform side of the separatrix, the still path is initially on a straight line through, and moving away from, pure chloroform. The still path then turns toward the benzene vertex along the stable separatrix. This is because crossing into the adjacent region would mean that a new distillate of nearly pure acetone would appear, acting to move the still trajectory back toward the separatrix.

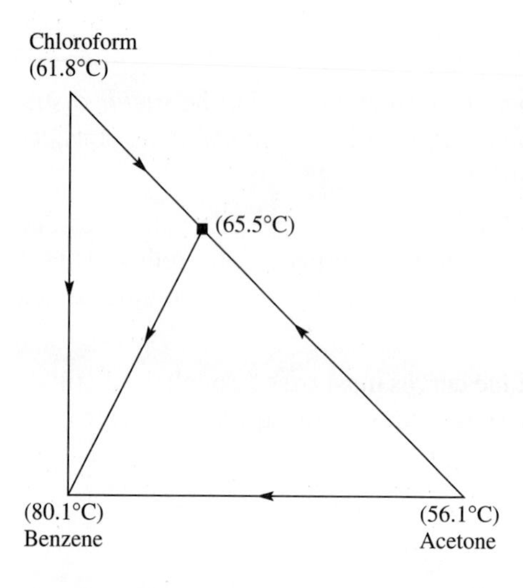

FIGURE 9.7
Sketch of the simple distillation boundaries for a mixture of acetone, chloroform, and benzene at a pressure of 1 atm. Acetone and chloroform are unstable nodes, benzene is a stable node, and the azeotrope is a saddle.

- The distillate would initially be nearly pure chloroform, followed by the acetone-chloroform azeotrope until only benzene was left in the still. This is easily determined from the various tangents to the batch distillation residue curves.
- Similarly, the behavior of a batch with an initial composition on the acetone side of the separatrix can also be sketched. The resulting distillate sequence would be *acetone* and then the *acetone-chloroform azeotrope,* leaving nearly pure *benzene* in the still.

Effects of Curvature

The separatrix in the simple distillation residue curve map may have a significant curvature leading to somewhat different results.[7] For example, Fig. 9.8 shows the simple distillation residue curves computed using Eq. 2.20 with the Wilson model for the activity coefficients and the parameters in Chap. 2. Although this is not an important mixture in technology, these results are in good agreement with the experiments of Reinders and DeMinjer (1940*b*), and the curvature is large enough to illustrate the major effects.

The batch distillation residue curves still display an initial linear portion before following the separatrix to the benzene vertex. However, because of the curvature, the tangent and thus the distillate compositions change gradually as the still composition moves along the separatrix. For feeds on the upper left side of the boundary, the distillate will be nearly pure chloroform, followed by a binary mixture of acetone and chloroform of variable composition, until pure benzene is left in the still. On the other hand, initial compositions in the other region yield a distillate of acetone, followed by a binary mixture of acetone and chloroform of variable composition

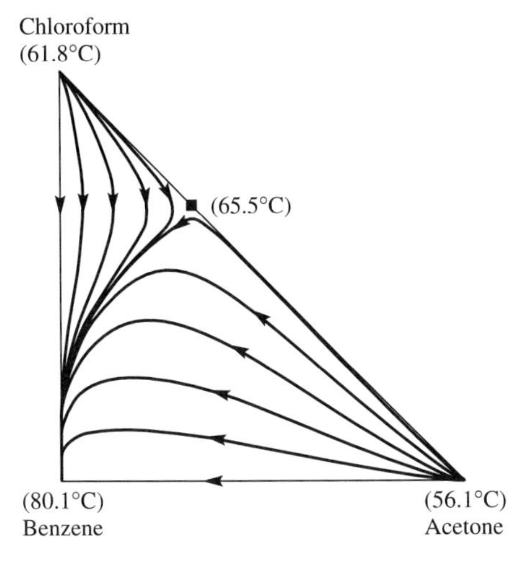

FIGURE 9.8
Simple distillation residue curves for acetone, chloroform, and benzene at 1 atm.

[7] It may be expensive in practice to develop sufficient VLE data to accurately describe such curvature. Nevertheless, when the mixture exhibits a curved separatrix, this may have important consequences for the design and operation of the batch distillation.

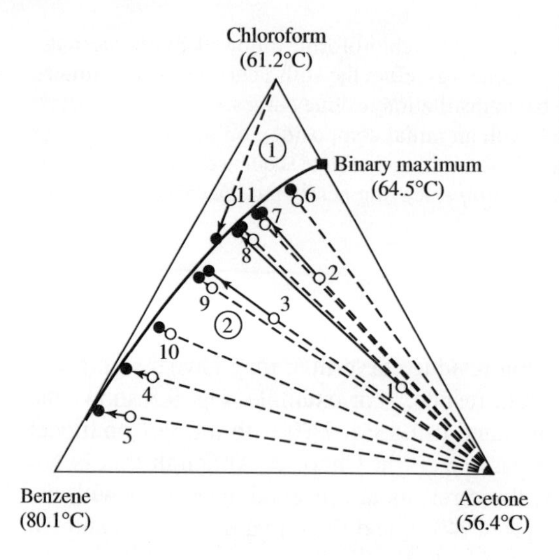

FIGURE 9.9
Batch rectification paths for acetone, chloroform, and benzene (Ewell and Welch, 1945). The measured temperatures reported in this figure are slightly different from the values calculated from VLE models and reported in Fig. 9.8.

initially close to the azeotrope and gradually increasing *toward pure chloroform,* until pure benzene is left in the still. (Contrast this with the results for linear boundaries and the corresponding cuts described in Example 9.1.) Thus, the separatrix presents a fundamental boundary if pure acetone is desired as a distillate cut when the initial feeds are in the upper left region. Feeds in the other region give distillates of pure acetone, then a sloppy acetone-chloroform cut with a gradual transition to high-purity chloroform, leaving benzene in the still. The sloppy cut *cannot* be made sharp or eliminated by adjusting the reflux or the number of stages. Also, note that the *distillate temperature drops* after the sloppy cut in agreement with the experiments of Reinders and DeMinjer (1940*a*), Reinders and DeMinjer (1940*b*), Reinders and DeMinjer (1940*c*), and Ewell and Welch (1945), who refer to this as a "distillation anomaly." The measured batch distillation residue curves are shown in Fig. 9.9; only the linear portions of the curves were studied experimentally.

For more information on batch distillation regions, see Safrit and Westerberg (1997*a*, 1997*c*), Ahmad and Barton (1996), Bernot et al. (1990, 1991).

Sketching Batch Distillation Regions

From a knowledge of the simple distillation residue curves, the residue curves for batch distillation can be estimated for the limit of total reflux and a large number of stages. For any given feed composition, a straight line is drawn through the "local" unstable node.[8] The still trajectory proceeds along this straight line away from the unstable node until it reaches either an edge or a stable separatrix, whereupon it turns abruptly in the direction of increasing temperature. The distillate cuts can be determined as follows. For every point on the still trajectory, the distillate lies along its tangent line. The distance along the tangent is set by the number of stages and can

[8]This is the unstable node in the simple distillation region that contains the feed. The point can be found by integrating the *simple* distillation residue curve backward in time from the feed composition.

be found by stepping back along the simple distillation residue curve. If the simple distillation residue curve approaches a singular point, the number of stages grows very large. This can happen at either a saddle point along the simple distillation residue curve or eventually at the unstable node. Those cases correspond to a pinch inside or at the top of the rectifying column, respectively.

The feed compositions can be divided into different *batch distillation regions* using the procedure described by Bernot et al. (1990). For a ternary mixture, this is as follows.

1. Determine the simple distillation residue curve map; see Chap. 5.
2. Partition the composition triangle using the stable separatrix in the simple distillation residue curve map (if any); this isolates the unstable nodes.
3. Subdivide each partition into regions by joining the unstable node to any other saddles and stable nodes using a straight line.
   a. If the stable separatrix is linear, proceed to step 4.
   b. If the stable separatrix is curved, then one or more of the straight lines forming the subdivisions will cross this separatrix before ending at a stable node or saddle. Replace these with straight lines from the unstable node to the point of tangency with the stable separatrix that is closest to the crossing.
4. Find the sequence of distillate compositions for a region as follows.
   a. Sketch a linear still path from inside the region up to and then along the boundary in the direction of increasing temperature.
   b. The distillate compositions lie on the tangents to the still path at the edge of the composition triangle or at the unstable separatrix.

In quaternary mixtures, the triangle is replaced with the tetrahedron, the stable separating *manifold(s)* partition the composition space, and the distillate lies on the face of the tetrahedron or on the unstable separating manifold.

EXAMPLE 9.2. There are more distillation regions and more complicated effects of curvature when there are higher-order azeotropes. For example, there is a ternary saddle azeotrope as well as three binary azeotropes in a mixture of acetone, chloroform, and methanol as shown in Fig. 9.6*d*. Using the procedure above, we find the following.

- There are six distillation regions for a batch rectifier, as shown in Fig. 9.10.
- The distillate cuts for each region are summarized in Table 9.1.
- Feeds in Region 5 result in distillate temperatures that can show an "anomalous" decrease in temperature during part of the batch.
- The *unstable separatrix* is a boundary for the distillate composition, and distillate cuts sometimes contain substantial amounts of all three components.
- The batch distillation residue curves may cross the stable separatrix slightly *from the concave side*.

Ewell and Welch (1945) studied the batch distillation of acetone, methanol, and chloroform experimentally. The measured still compositions are shown in Fig. 9.11. See Van Dongen and Doherty (1985) or Bernot et al. (1990) for a more detailed discussion of this example.

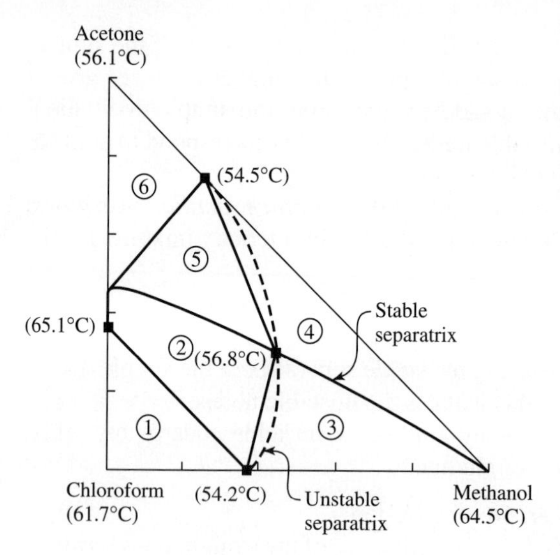

FIGURE 9.10
Distillation regions for batch rectification of methanol, acetone, and chloroform at a pressure of 1 atm.

TABLE 9.1
The distillate cuts for batch rectification of methanol (*M*), acetone (*A*), and chloroform (*C*) depend on which of the six batch distillation regions shown in Fig. 9.10 contains the feed

Region	Cuts	Distillate compositions
1	3	*MC* azeo, *C*, *AC* azeo
2	5	*MC* azeo, *MAC* azeo (sloppy), *MA* azeo (trace) *A*, *AC* azeo
3	3	*MC* azeo, *MAC* azeo (sloppy), *M*
4	3	*MA* azeo, *MAC* azeo (sloppy), *M*
5	5	*MA* azeo, *MAC* azeo (sloppy), *MA* azeo (trace), *A*, *AC* azeo
6	3	*MA* azeo, *A*, *AC* azeo

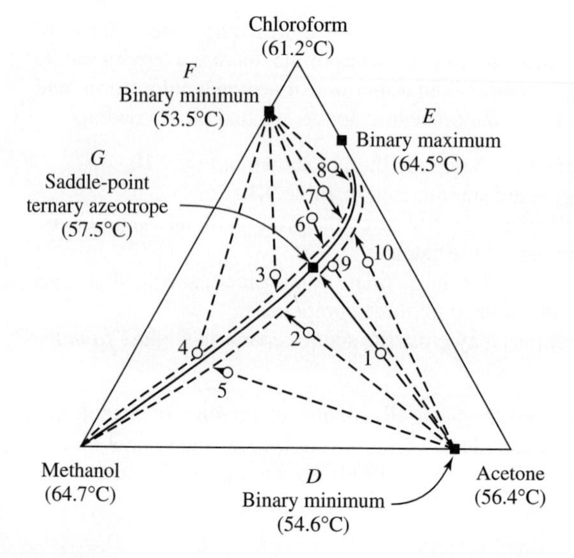

FIGURE 9.11
Measured still compositions in batch distillation for acetone, methanol, and chloroform (Ewell and Welch, 1945). Note the different orientation of labels from Fig. 9.10.

The Batch Stripper

In the next section, we illustrate the use of these ideas to synthesize a batch distillation system. For that, we also need to summarize the behavior of a *batch stripper* or *inverted* batch distillation where the product cuts are withdrawn as liquids after stripping; Fig. 9.12 shows a schematic. The analysis for this case is similar to the approach for the conventional batch column, but we follow the dynamics of the composition in the reflux accumulator and seek to determine the sequence of compositions for the bottoms product. The approach is described in detail by Bernot et al. (1991) and is summarized here.

We call the batch composition variations in the composition triangle the "condenser paths." For a large reboil ratio and a large number of stages, these have significant linear portions for reasons similar to those described for the batch rectifier. That is, the liquid composition profiles in the column follow the simple distillation residue curves according to the quasi-steady description, and the bottoms composition is on a tangent to the condenser path. These two facts are sufficient to ensure that the condenser path follows a straight line from the feed away from the *stable* node. This is in contrast to the conventional column where the still composition moves away from the *unstable* node.

Figure 9.13 shows the limiting batch trajectories for both configurations in the case of a ternary mixture with no azeotropes. The batch rectifier, with large reflux and stages, would produce A and B in two high-purity distillate cuts, leaving C in the bottoms; the inverted configuration produces C and then B as bottoms products with A concentrated in the reflux accumulator. When there are boundaries in the simple distillation residue curve map, the condenser path follows either the edge of the triangle or the unstable separatrices in ternary mixtures.

When the separatrices are curved, the batch distillation regions for the conventional and inverted configurations differ. For example, Fig. 9.14 shows the distillation regions for batch stripping of a ternary mixture of methanol, acetone, and

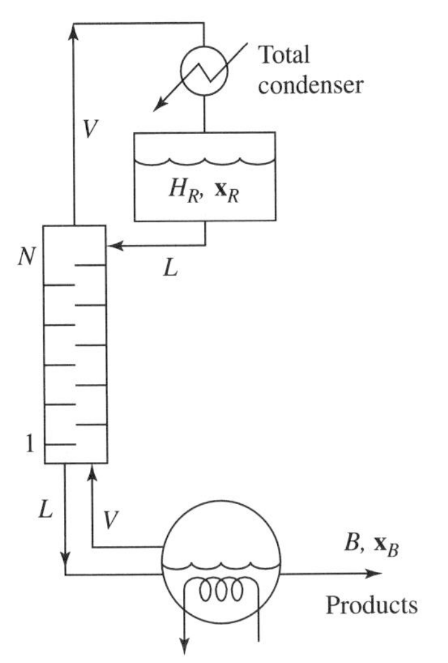

FIGURE 9.12
Schematic of the batch stripper or *inverted* batch column.

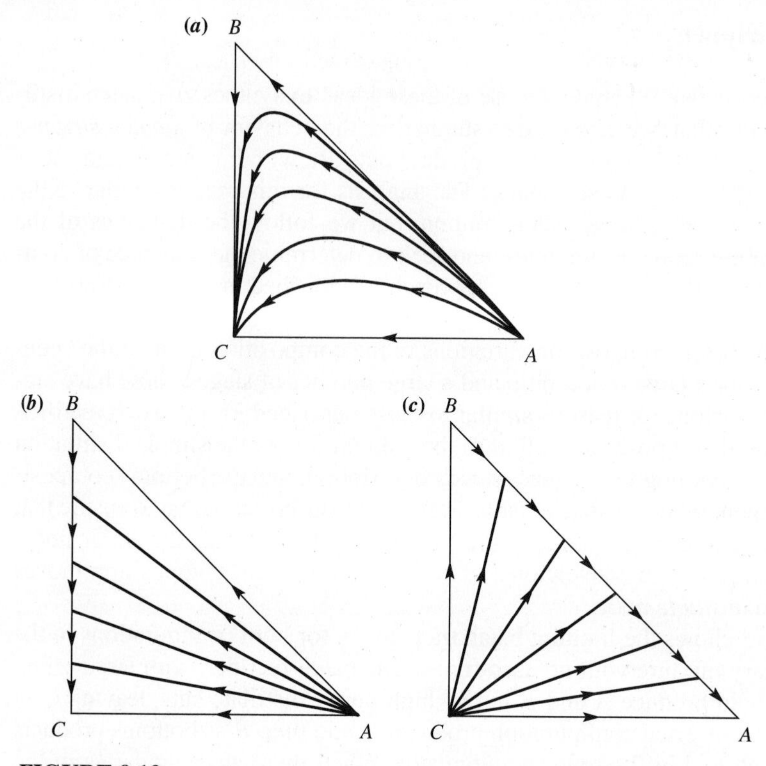

FIGURE 9.13
The simple distillation residue curve map (*a*) for a ternary mixture containing no azeotropes. The batch distillation residue curves are shown in (*b*) for a rectifier where the distillate cuts are $A - B - C$. For a stripper (*c*) the batch distillation residue curves originate at C and end at A so the bottoms cuts are $C - B - A$.

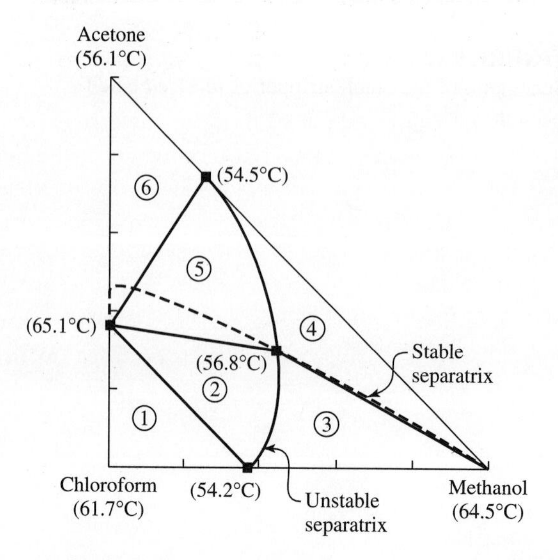

FIGURE 9.14
Batch distillation regions in the inverted configuration for the mixture methanol, acetone, and chloroform.

TABLE 9.2
Bottoms cuts for batch stripping of methanol (M), acetone (A) and chloroform (C) depend on which of the six distillation regions shown in Fig. 9.14 contains the feed

Region	Cuts	Bottoms compositions
1	3	AC azeo, C, MC azeo
2	3	AC azeo, MAC azeo (sloppy), MC azeo
3	3	M, MAC azeo (sloppy), MC azeo
4	3	M, MAC azeo (sloppy), MA azeo
5	3	AC azeo, MAC azeo (sloppy), MA azeo
6	3	AC azeo, A, MA azeo

chloroform. The sequence of cuts in Table 9.2 for the batch stripper should be compared to those for the batch rectifier in Table 9.1. Note that *the inverted configuration does not simply produce the same cuts as the conventional configuration in the reverse order.*

9.4 SYSTEM SYNTHESIS

Feasibility

A batch rectifier always produces distillate products in order of increasing boiling point when there are no distillation boundaries, while the batch stripper produces bottoms products in the reverse order. However, the product compositions in azeotropic mixtures generally depend on the feed composition, and it is necessary to check the feasibility of the desired separation.

For example, imagine that we desire to separate a binary azeotropic mixture into nearly pure products by the addition of an entrainer to the batch. Several schemes for batch distillation may be feasible, depending on the nature of the azeotrope and the entrainer. For instance, the addition of an intermediate-boiling entrainer E that forms no new azeotropes to a binary mixture containing a maximum-boiling azeotrope results in the simple distillation residue curves shown schematically in Fig. 9.15. In this case, there is no batch distillation boundary and the conventional batch rectifier can produce distillate cuts with a high purity of A, then E, for recycle, leaving primarily B in the still.[9] If the entrainer introduces another azeotrope, this configuration may still be feasible with recycle of an azeotropic mixture as outlined in Exercise 2.

[9] You might expect that a third distillate cut would contain mostly B, and that the still would eventually attain the composition of the $A - B$ azeotrope. However, the amount of A would be negligible in this limiting case of total reflux and an indefinitely large number of stages. The cost of achieving such a small residual amount of A in a real column may be large.

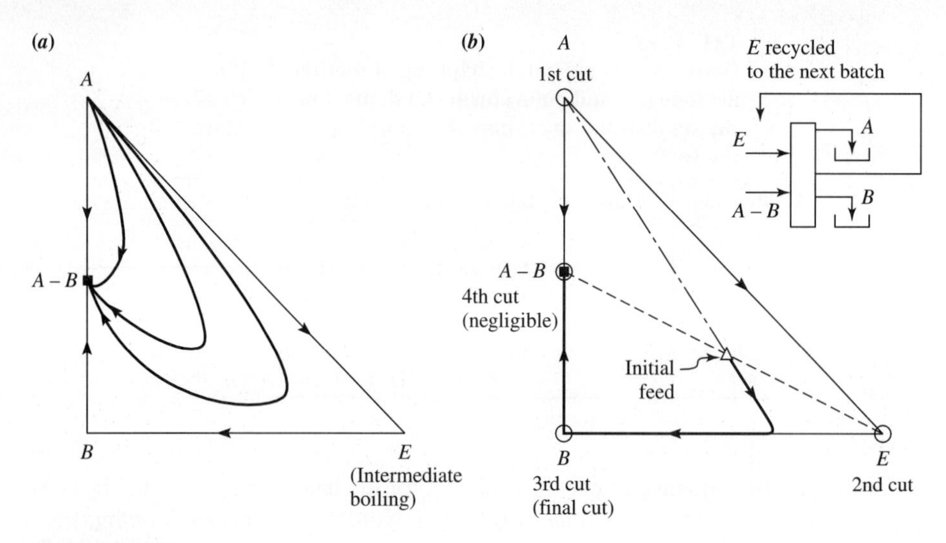

FIGURE 9.15
An intermediate-boiling entrainer E can be used to break a maximum-boiling binary azeotrope between A and B in the batch rectifier. The distillate cuts are A, then E, leaving a batch rich in B.

There are many other simple distillation residue curve maps that can be analyzed by this approach. Most of the azeotropes encountered will be minimum-boiling; a few examples (e.g., Exercise 3) show that the conventional batch rectifier cannot be used to separate these azeotropes for most entrainers. It turns out that *the inverted configuration described above is effective for separating minimum-boiling binary azeotropes by adding a suitable entrainer to the batch.* For example, consider a binary mixture with a minimum-boiling azeotrope and an intermediate-boiling entrainer that gives the simple distillation residue curve map shown in Fig. 9.16. The condenser paths at total reboil, and a large number of stages are as shown. The bottoms cuts will have a high purity in B, followed by E, leaving primarily A in the batch tank.[10]

Sometimes the curvature of separatrices in the simple distillation residue curve map can influence the batch distillation regions and the feasibility for the reasons pointed out earlier in the methanol, acetone, chloroform mixture. This is discussed in the process synthesis example below and in Exercise 4.

In many cases, mixtures with ternary azeotropes can be separated into pure components and binary azeotropes with either the conventional or inverted batch distillation. It may be necessary to add one of the pure components in the mixture to place the feed in some particular batch distillation region, and in fact there may be several alternatives. The resulting binary azeotropes can then be broken by new entrainers or pressure shifting, etc. In 4 of the 67 maps for ternary mixtures containing a ternary azeotrope, the addition of a fourth component is required to achieve a feasible distillation configuration.[11]

---

[10]The trace amount of the $A - B$ azeotrope remaining is ignored, as it was for the conventional column.
[11]See Bernot et al. (1991); the four maps are 222-m, 444-m, 010-s, and 300-s using the classification of Doherty and Caldarola (1985).

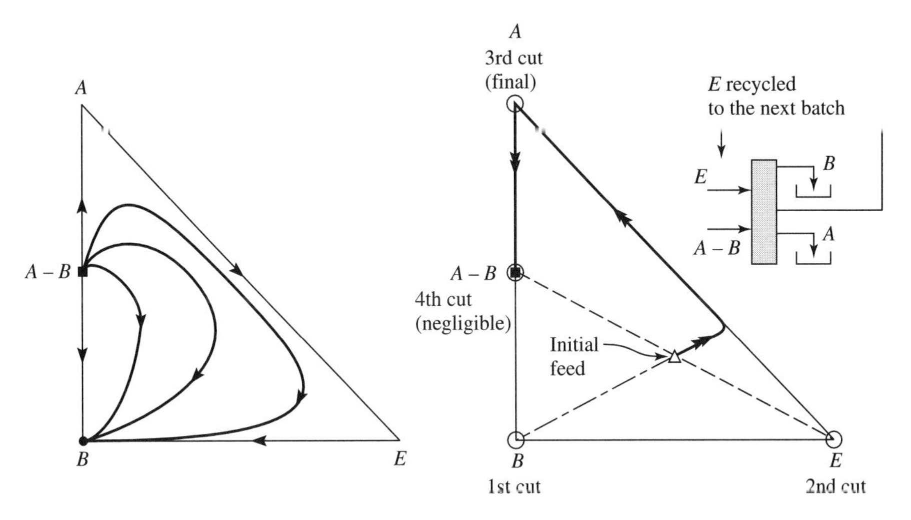

FIGURE 9.16
An intermediate-boiling entrainer E can be used to break a minimum boiling binary azeotrope between A and B in the batch stripper. The bottoms cuts are B, then E, leaving a batch rich in A.

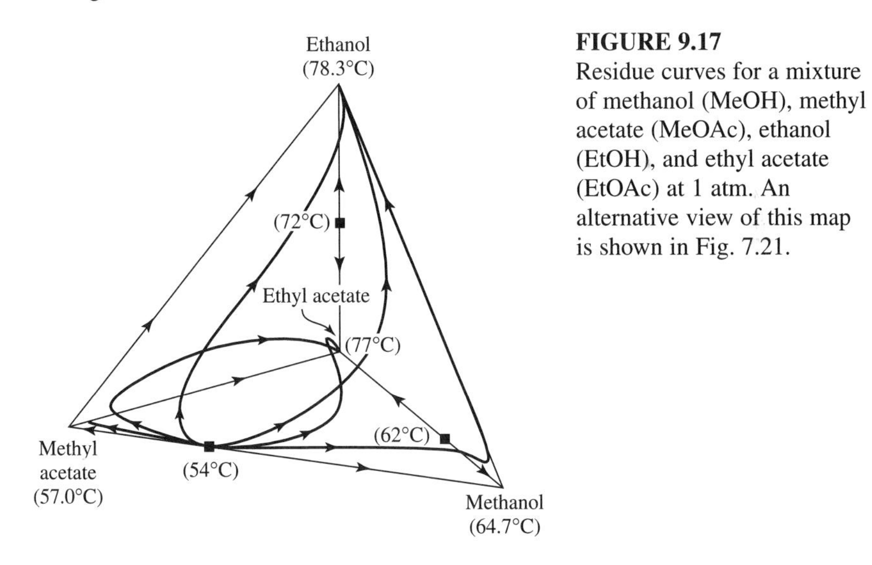

FIGURE 9.17
Residue curves for a mixture of methanol (MeOH), methyl acetate (MeOAc), ethanol (EtOH), and ethyl acetate (EtOAc) at 1 atm. An alternative view of this map is shown in Fig. 7.21.

EXAMPLE 9.3. Transesterification. The simple distillation residue curves for a mixture of methanol (MeOH), methyl acetate (MeOAc), ethanol (EtOH), and ethyl acetate (EtOAc) are shown in Fig. 9.17.[12] We imagine that this mixture is the product of the liquid phase, catalytic transesterification reaction

$$\text{MeOH} + \text{EtOAc} \rightleftharpoons \text{EtOH} + \text{MeOAc}$$

which is carried out in a separate reactor.[13]

[12] Based on Bernot et al. (1991).

[13] We examine the option of combining reaction and separation in Chap. 10.

There are five batch distillation regions shown in Fig. 9.18; Table 9.3 lists the corresponding product cuts from a batch rectifier.

If the reactor is fed with methanol and ethyl acetate, all of the possible feed mixtures for subsequent distillation have a composition that is constrained by the reaction stoichiometry to satisfy $x_{\mathrm{MeOAc}} = x_{\mathrm{EtOH}}$. This is a linear constraint shown by the "stoi-

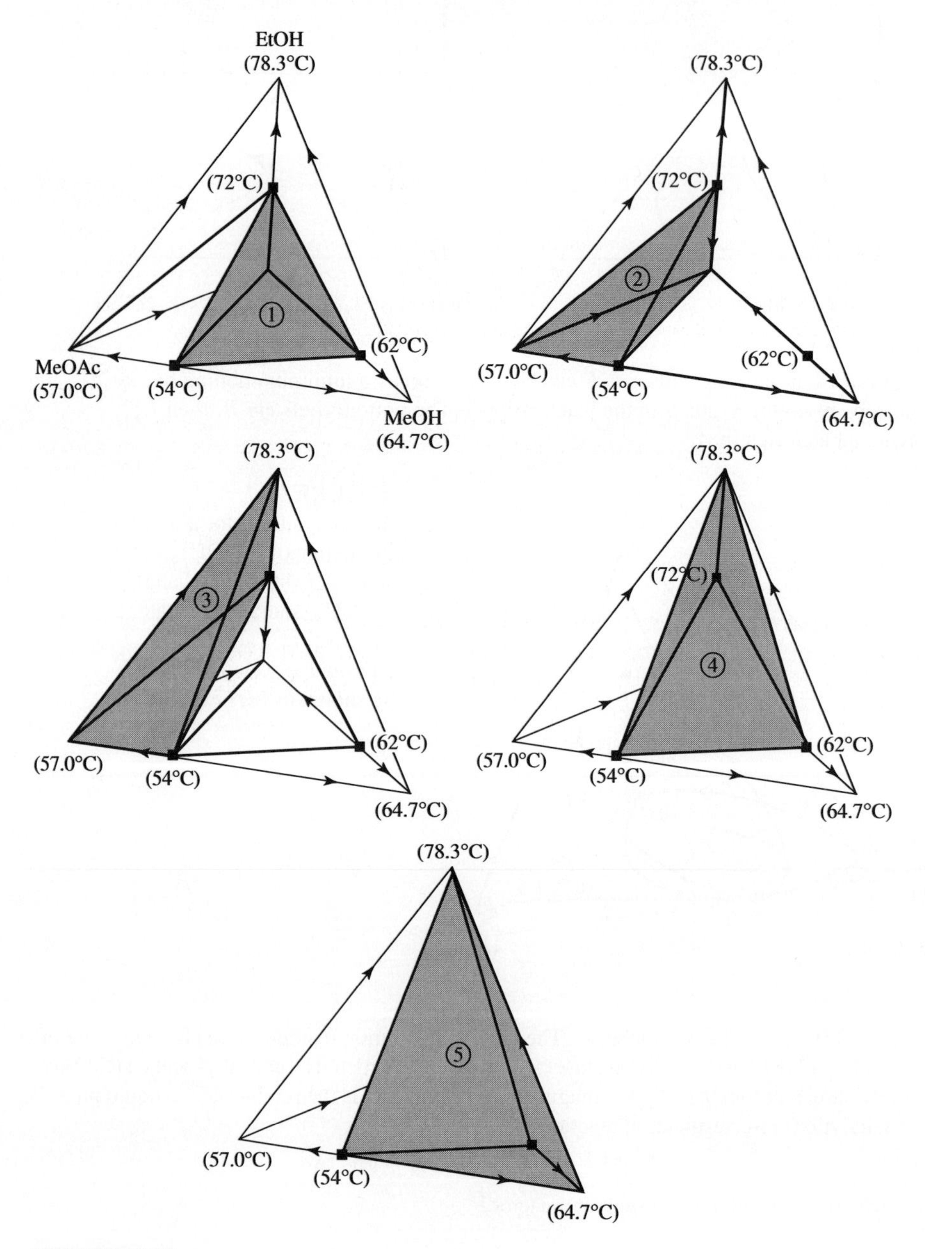

FIGURE 9.18
Batch distillation regions (schematic) for a mixture of methanol, methyl acetate, ethanol, and ethyl acetate at 1 atm.

TABLE 9.3
Distillate cuts for methanol (MeOH), ethanol (EtOH), methyl aceate (MeOAc), and ethyl acetate (EtOAc) at 1 atm

Region	Cuts	Distillate compositions
1	4	MeOAc-MeOH azeo, MeOH-EtOAc azeo, EtOAc-EtOH azeo, EtOAc
2	4	MeOAc-MeOH azeo, MeOAc, EtOAc-EtOH azeo, EtOAc
3	4	MeOAc-MeOH azeo, MeOAc, EtOAc-EtOH azeo, EtOH
4	4	MeOAc-MeOH azeo, MeOH-EtOAc azeo, EtOAc-EtOH azeo, EtOH
5	4	MeOAc-MeOH azeo, MeOH-EtOAc azeo, MeOH, EtOH

chiometric plane" in Fig. 9.19. The plane crosses all five batch distillation regions, and the initial feed to the still can be in any region, depending on the conversion and the amount of excess reactant chosen. Some of the distillate cuts are more favorable than others, e.g., different azeotropes are obtained, some cuts can be recycled, etc. The result is a strong coupling between the reactor and the separation system.

Three batch distillation regions (2, 3, and 5) contain only two azeotropic cuts. Both of the azeotropic cuts in regions 2 and 3 must be broken to produce product and recycle streams. However, in Region 5 only one cut (MeOAc-MeOH azeotrope) needs to be broken; the MeOH-EtOAc azeotrope can be recycled. So an interesting alternative is to operate the reactor to produce a feed in this region. This requires excess methanol, with a molar ratio of methanol to ethyl acetate in the reactor feed greater than approximately 2.5 to 1, or perhaps a bit smaller depending on the conversion in the reactor (see Fig. 9.19).

Figure 9.20 shows two alternative processing schemes. The feed is in Region 5, and the cuts are listed in the last row of Table 9.3. In alternative (*a*), a batch rectifier ($C1$) removes the first cut, the MeOAc-MeOH azeotrope. This is subsequently separated in a batch stripper using methyl formate as an entrainer (see Exercise 6) to produce the methyl acetate product and methanol for recycle. The second and third cuts from the batch rectifier contain MeOH and EtOAc, which are recycled to the reactor. The last cut in the batch rectifier is the ethanol product.

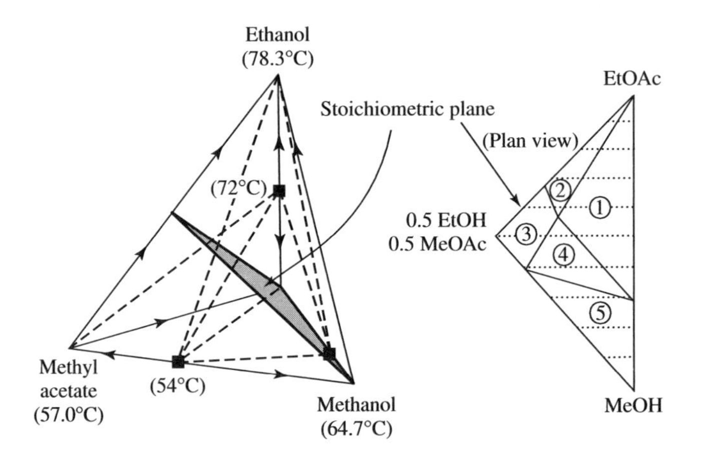

FIGURE 9.19
Stoichiometric plane and batch distillation regions for transesterification. Dotted lines represent the liquid composition paths for the reactor.

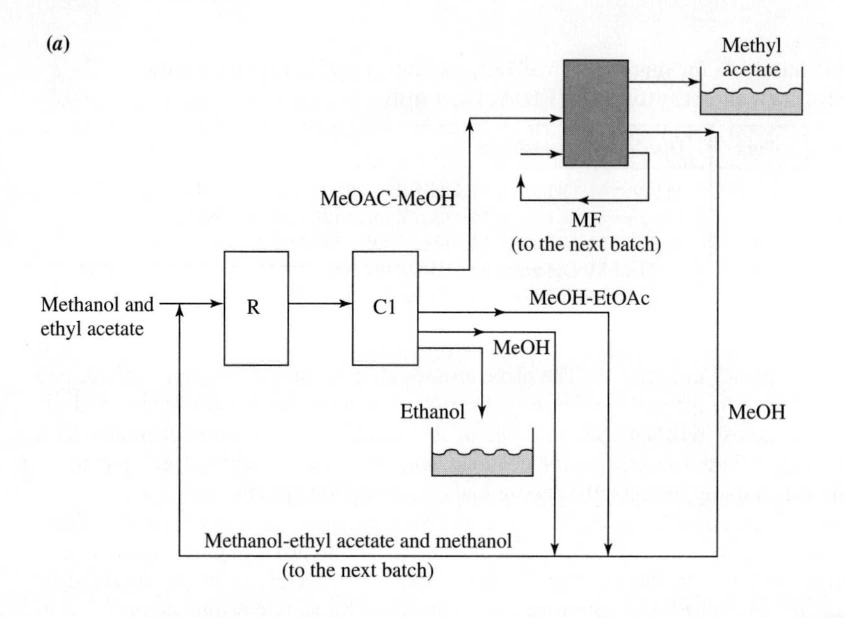

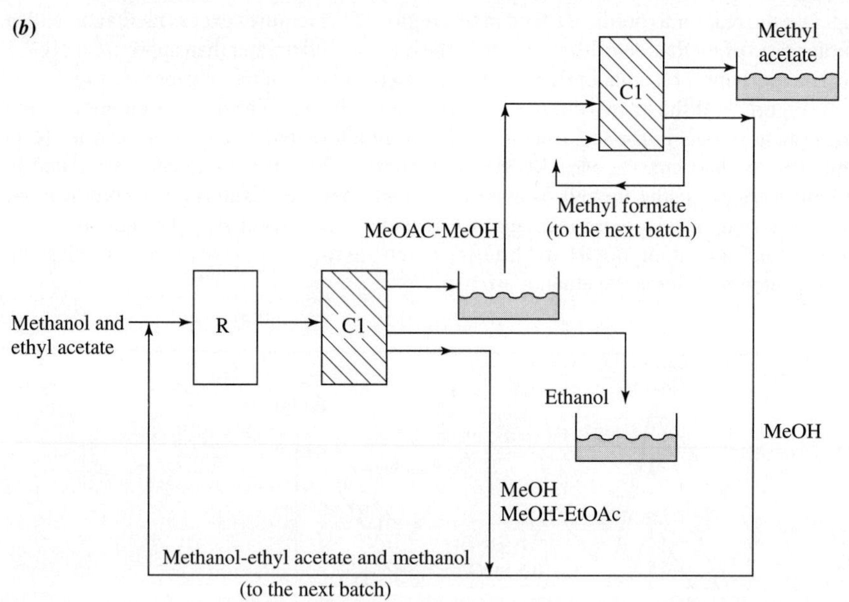

Legend: R Reactor C Batch rectifier Batch stripper C Batch rectifier/stripper

FIGURE 9.20
Alternative flowsheets for batch transesterification.

Alternative (*b*) is based on a hybrid operation that alternates between rectifying and stripping in a single multipurpose device. This takes advantage of product cut sequence MeOAC-MeOH from the rectifier, followed by EtOH from the stripper (the last cut listed for the rectifier in region 5, Table 9.3). The remaining MeOH and EtOAc are recycled without further separation. This multipurpose device can do both batch rectification and stripping (and potentially also the MeOAc-MeOH separation using methyl formate as an entrainer). Although they are not in wide use, such multipurpose devices have been studied for some time; these are discussed later in this chapter.

9.5
TARGETS FOR OPERATING POLICIES

The dynamic nature of batch distillation means that the product compositions change with time, as discussed in Sec. 9.2. For example, the batch rectifier with a constant reflux ratio has a decreasing product purity. There are several ways to respond to these changes.[14]

1. One is to provide sufficient stages and a *constant reflux* so that the *average* composition of the product collected meets specifications and the amount collected is reasonable. The best choice of reflux is a trade-off. The distillate purity will decrease with time, beginning at a value above the specification and ending at a value below. When the *average* purity of the cut collected reaches the specification, the balance of the major product in the first cut may be collected in an intermediate cut until the next high-purity cut begins. The intermediate cut may be re-processed with the next batch, collected and processed separately, recycled, or treated, depending on the size of the cut and the costs of these options.
2. The reflux ratio can be increased during the cut to increase the amount of the primary component recovered in satisfactory purity, at the expense of energy, time, and labor. This reduces or perhaps eliminates the intermediate cut and the costs of handling that fraction. A commonly studied limiting case is the *constant distillate composition* policy, where the reflux ratio vs. time policy is found to keep the composition of the major component in the cut constant for as long as possible. The duration obviously depends on the number of stages, the feed composition, and other parameters but eventually drops essentially to zero as the component in question is depleted.
3. Between these two limiting cases, there is an optimal policy that trades off the additional time, energy, and labor needed in the constant distillate case against the benefits of reducing or eliminating slop cuts. A precise calculation of the optimal policy is possible with sufficient data. Even when this is not available, an increase of the reflux ratio during a batch distillation is common. This is sometimes described as "squeezing" the batch. This can be done gradually but is more often done in practice using a series of constant reflux periods with increasing values of the reflux ratio, perhaps also including periods of total reflux. This also has the advantage of allowing time for column dynamics or composition measurements.

It is useful to have a rapid estimate of the incentive to manipulate the reflux as a function of time. Provided there are a sufficient number of stages in the column

[14]The case of a batch rectifier is discussed here. Similar manipulation of the reboil ratio is also advantageous for the batch stripper.

to obtain the desired separation (Robinson, 1969), a good estimate of the incentive can be made from the constant product composition policy (Bernot et al., 1993).

For example, Eq. 9.3 for a binary mixture with a constant distillate composition is

$$\frac{dx_B}{d\xi} = x_B - x_D \tag{9.8}$$

which has the solution

$$x_B(\xi) = x_D + (x_{B,0} - x_D)e^{\xi} \tag{9.9}$$

This gives the still composition as a function of warped time and the desired constant distillate composition. For each ξ, there is a minimum reflux $r_{\min}(\xi)$ and, using the heuristic developed for continuous columns, a target for the batch rectifier is $r(\xi) = 1.5r_{\min}(\xi)$.

EXAMPLE 9.4. Reflux Policy for a Simple Binary Mixture. We consider an example from Kim (1985), who found the optimum design and reflux policy to minimize the total annual cost using a detailed and time-consuming integration of the full column model. The feed is a binary mixture for which a constant volatility of 2.0 adequately describes the vapor-liquid equilibrium. The feed contains 30 mol % of the light component. The distillate is to contain 95% of the light component originally present in the feed at a purity of 99 mol %. By material balance, this fractional recovery will be reached at a warped time of 0.34. (If all of the light component could be removed, the warped time would be 0.36, but this would be indefinitely expensive.) From Eq. 9.9, the constant distillate composition would correspond to a batch composition that obeys $x = 0.99 - 0.69e^{\xi}$.

For this simple mixture, the reflux is approximately

$$r = 1.5\left(\frac{0.99 - y}{y - x_B}\right)$$

with $y = 2x_B/(1 + x_B)$ from the VLE. This gives the reflux policy as a function of ξ. This approximation is compared in Fig. 9.21 with the detailed optimization results of

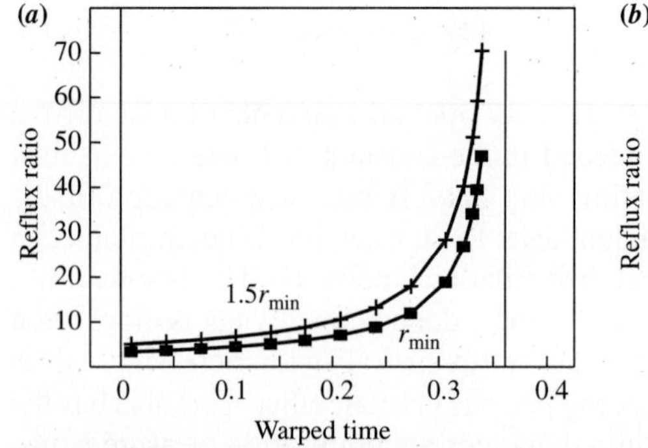

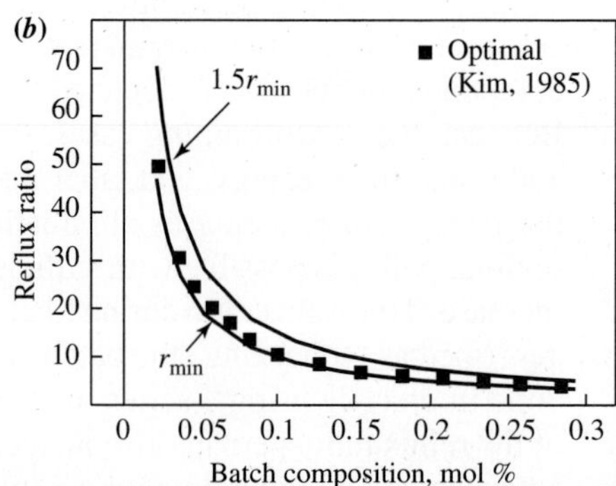

FIGURE 9.21
Reflux ratios for Example 9.4. The instantaneous minimum reflux and a target of 50% above the minimum (*a*) are compared with detailed optimization results (*b*) (Kim, 1985). In part (*b*), note that the still composition has been used as the independent variable instead of the warped time, via Eq. 9.9.

Kim (1985). The number of stages can also be found easily for this example, using the Fenske and Smoker equations, or from McCabe-Thiele constructions. The procedure suggested here gives 14 theoretical stages; Kim reports an optimal value of 18 but with a reflux ratio that is substantially closer to the minimum. Even for this very simple mixture, there is a substantial savings in engineering time and the results are in good agreement with the detailed modeling.

Extension of this targeting idea to multicomponent, nonideal mixtures is described in detail by Bernot et al. (1993).

9.6 NOVEL CONFIGURATIONS

There are several more complicated configurations than the traditional batch rectifier. One of these, the batch stripper, is discussed above in Sec. 9.3, and both arrangements are shown in Fig. 9.22. Other arrangements of equipment configuration and operation are active areas of study. However, they do not seem to have wide use, presumably on account of their complexity.

Rectification to remove light compounds and stripping to remove heavy compounds can be combined in a more complex operation sometimes called a "middle vessel" column; see Fig. 9.23. This was suggested at least as early as 1950 by Robinson and Gilliland, 1950, p. 388, who did no analysis but did point out that this arrangement "can save heat but requires additional equipment." The first

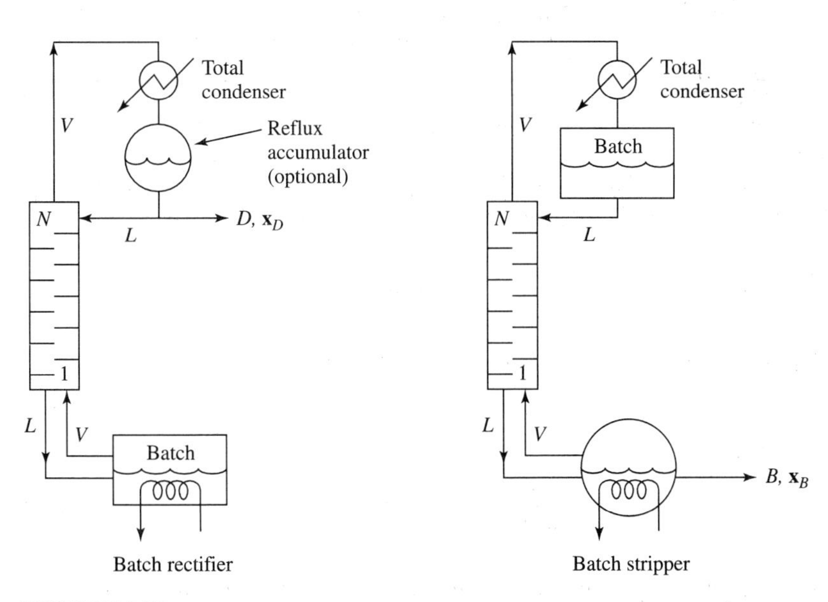

FIGURE 9.22
Batch distillation configurations: traditional or batch rectifier (left) and inverted configuration or batch stripper (right).

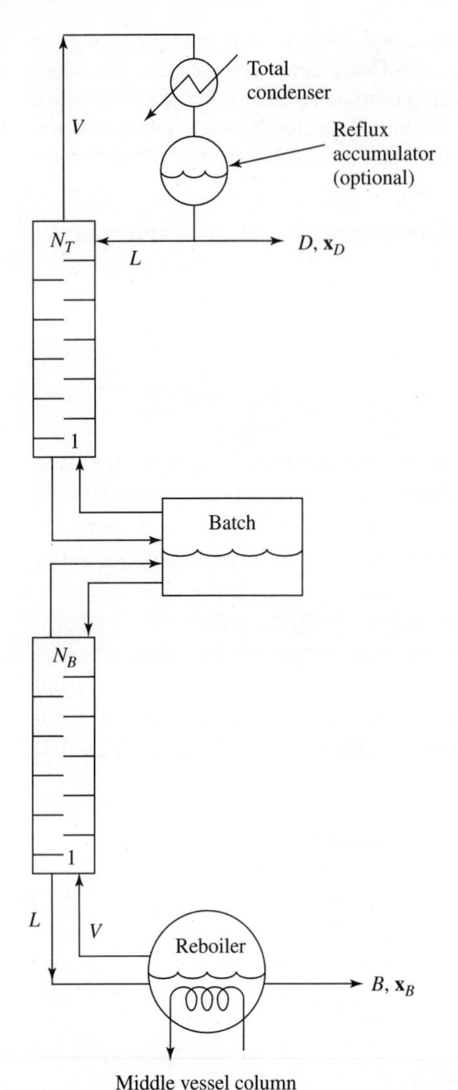

FIGURE 9.23
Middle vessel batch distillation.

detailed study was not published until 1970 (Bortolini and Guirase, 1970). More recently, this configuration has been the subject of many simulation studies, including interesting work for azeotropic mixtures. See Cheong and Barton (1999*a*, 1999*b*, 1999*c*) and references there as well as Barolo et al. (1996) and Davidyan et al. (1994). It is this sort of middle vessel column that might be used in Fig. 9.20*b*.

The operation of the middle vessel column as a completely closed batch system, i.e., at total reflux and total reboil, will concentrate a ternary mixture into low-, middle-, and high-boiling fractions (Barolo and Botteon, 1997).

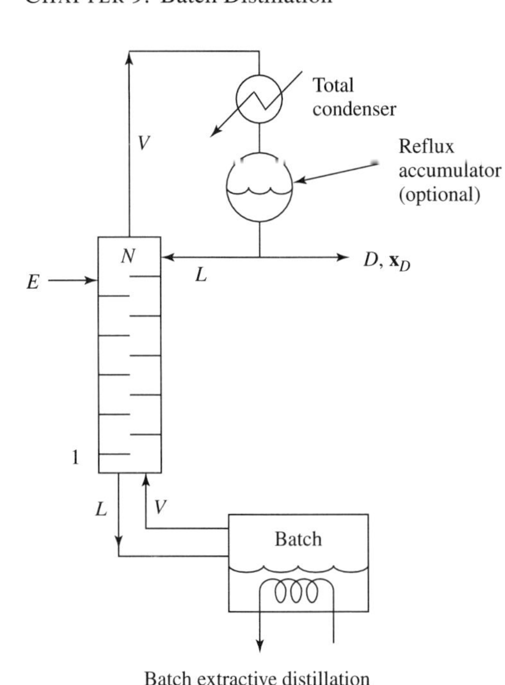

FIGURE 9.24
Batch extractive distillation.

The generalization of the middle vessel idea to include more intermediate vessels in a "multieffect batch distillation" has been studied by Hasebe et al. (1995, 1996, 1999). In this arrangement, intermediate boilers can be concentrated and removed simultaneously. The dynamics and control of such systems, including experimental studies, have been examined by Wittgens et al. (1996), Skogestad et al. (1997), and Wittgens and Skogestad (1997).

Batch Extractive Distillation

The QSS model is also useful to give a framework for thinking about other configurations, such as batch extractive distillation, where entrainer is added toward the top of the rectifying column to break an azeotrope (Fig. 9.24). With this configuration, however, the extractive agent accumulates in the still, diluting the composition and raising the boiling point. Therefore, there is an incentive for the simultaneous use of a batch stripper to remove and recycle the heavy entrainer, i.e., batch extractive distillation in a middle vessel column (Fig. 9.25).

These configurations have been the subject of several recent studies, e.g., Safrit et al. (1995), Safrit and Westerberg (1997*b*), and Lelkes et al. (1998*a*, 1998*b*).

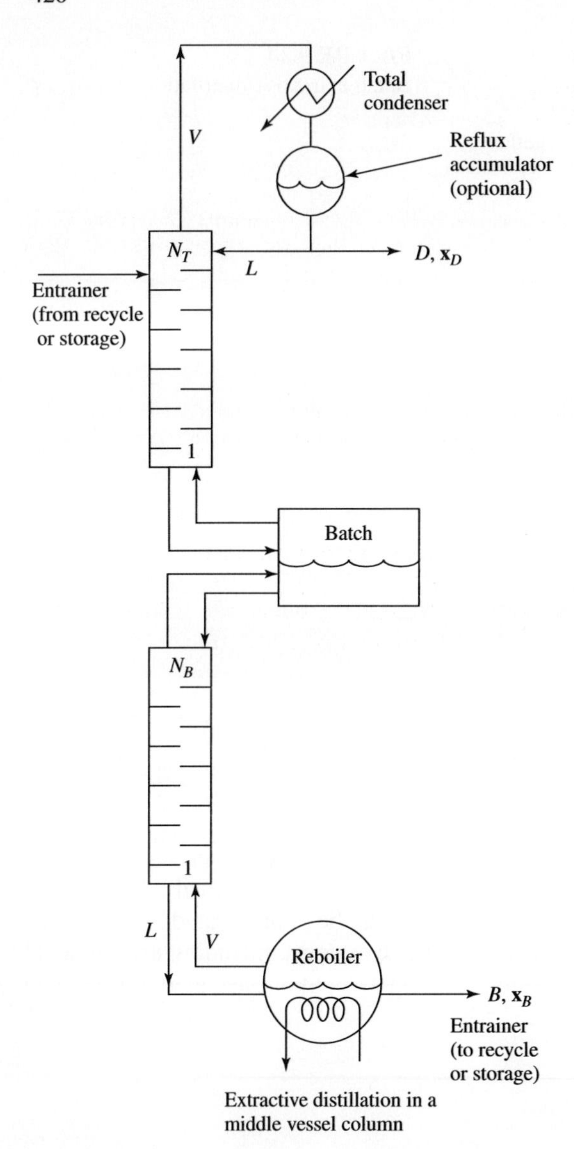

FIGURE 9.25
Batch extractive distillation in a middle vessel column.

9.7 EXERCISES

1. Integrate the quasi-steady-state model in Eq. 9.6 for $\alpha = 2$ and $x_{B,0} = 0.5$ to reproduce curves like those in Fig. 9.2 as follows.
   a. Take a simple Euler step as suggested for simple distillation in Eq. 5.9. For a fixed step size, say $h = 0.001$, the Euler integration will fail for a large number of stages. Why? Can you simply make h smaller to treat a large number of stages? (You may need to examine more than eight stages).

   b. Use a more sophisticated method, preferably one designed specifically for "stiff" systems, such as Gear's method (see the discussion of Eq. 5.9) to solve the cases for which Euler's method is not accurate.
2. Figure 9.26 shows the simple distillation residue curve map for a ternary mixture of an entrainer E with two compounds A and B which are desired as high-purity products. Sketch the still paths for a batch rectifier (Fig. 9.1) starting with a batch composition of 20 % A and 20% E. What are the product cuts? How do your answers change when a batch stripper (Fig. 9.12) is used?
3. Find the distillation regions for a batch rectifier (Fig. 9.1) for the two mixtures with the simple distillation residue curves shown in Fig. 9.27. Assume that the reflux ratio and the number of stages are both large.
4. Suppose that a candidate entrainer gives the simple distillation residue curves shown in Fig. 9.28*a*. A batch stripper gives a sequence of products rich in A,

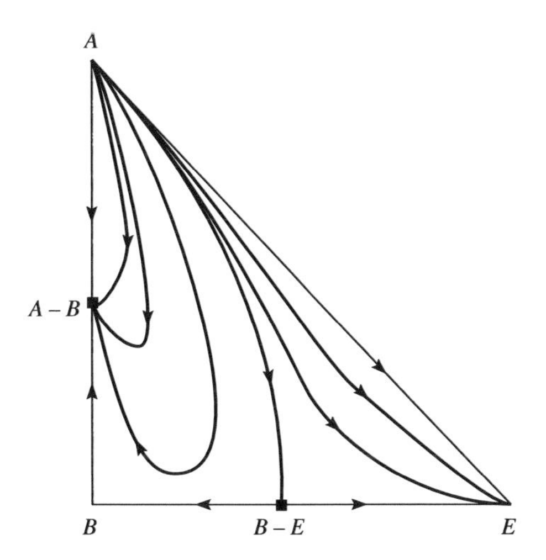

FIGURE 9.26
Simple distillation residue curves for entrainer E plus a binary mixture of A and B containing a maximum-boiling azeotrope.

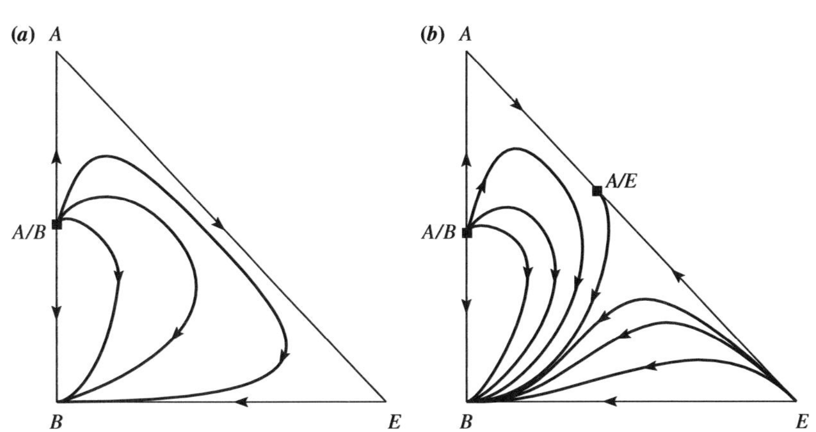

FIGURE 9.27
Simple distillation residue curves for Exercise 3.

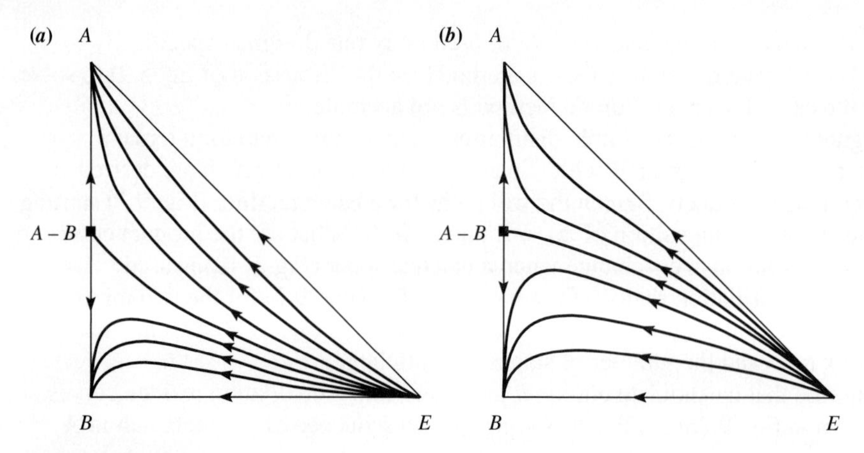

FIGURE 9.28
Simple distillation residue curves for Exercise 4.

followed by a fraction containing both A and B having a variable composition, leaving the batch with a high purity of E. Explain why this is so, and whether or not the purity of the second cut can be made constant. What if the simple distillation residue curves were as shown in part (*b*) of the figure? What if the separatrix in the simple distillation residue curves were straight?

5. When two components have close boiling points, an azeotrope is often found (and even without an azeotrope it may be prohibitively expensive to separate the mixture by distillation). It is sometimes possible add a "facilitator" F to the mixture, which may also form an azeotrope. Sketch the residue curves for batch distillation in the conventional and inverted configurations, if the simple distillation residue curve map is the one in Fig. 9.29. How would you choose the facilitator to feed ratio for a batch rectifier? Sketch a complete sequence for the separation of A and B using two conventional columns.[15]
6. A light entrainer like methyl formate can be used to separate methanol and methyl acetate. What are the product cuts for a batch rectifier and for a batch stripper? Describe the process and make a *rough estimate* of the amount of methyl formate needed. See the residue curve map in Fig. 9.30. Would methyl formate work if the distillation boundary were not so curved? Can you find another entrainer, e.g., one that is heavy and does not rely on curvature of the boundary?
7. The measured batch distillation residue curves for a mixture of acetone, chloroform, and isopropyl ether are shown in Fig. 9.31. There are four batch distillation regions I to IV. Sketch (qualitative) pictures of the distillate temperature for feeds at points 1, 2, 4, and 7. Also sketch the simple distillation residue curves. How many simple distillation regions are there?

[15]This problem is discussed in more detail by Bernot, Doherty, and Malone (1991); Bushmakin and Molodenko (1964) discussed similar ideas based on the experiments of Lutugina et al. (1960) for heterogeneous mixtures.

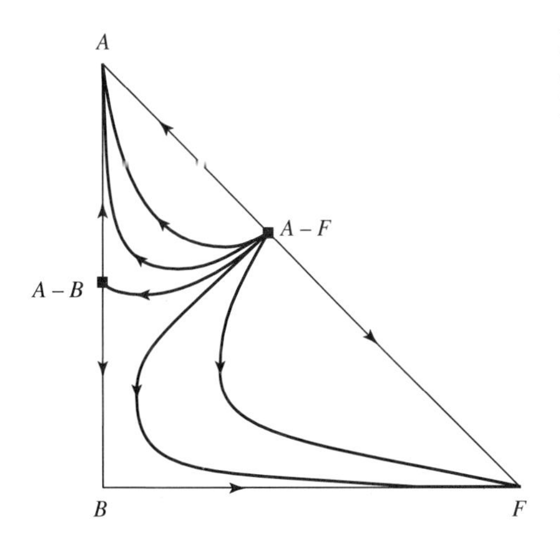

FIGURE 9.29
Simple distillation residue curves for Exercise 5.

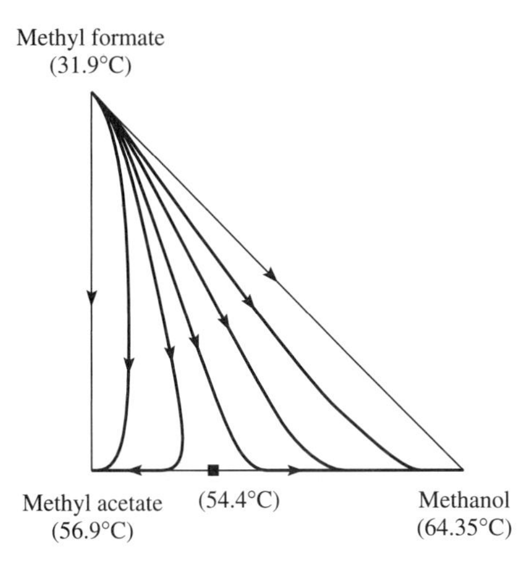

FIGURE 9.30
Simple distillation residue curve map for methyl acetate, methanol, and methyl formate at 1 atm.

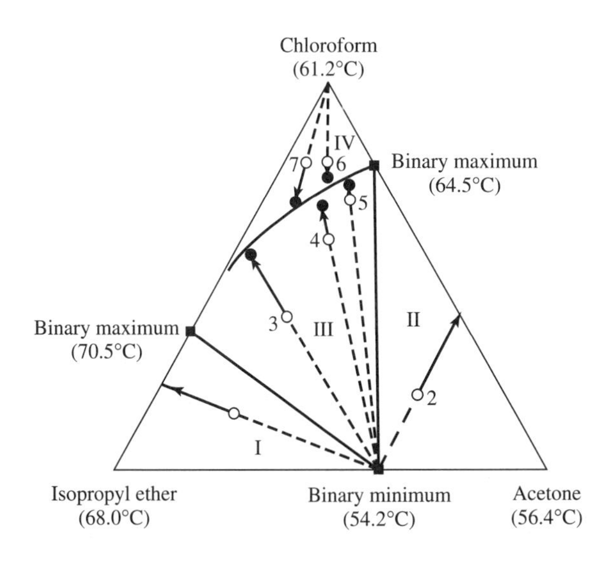

FIGURE 9.31
Measured batch distillation residue curves for acetone, chloroform, and isopropyl ether at 1 atm pressure; from Ewell and Welch (1945).

References

Ahmad, B. S., and P. I. Barton, "Homogeneous Multicomponent Azeotropic Batch Distillation," *AIChE J.,* **42,** 3419–3433 (1996).

Aspen Technology, BATCHFRAC (2000). www.aspentech.com.

Barolo, M., and F. Botteon, "Simple Method of Obtaining Pure Products by Batch Distillation," *AIChE J.,* **43,** 2601–2604 (1997).

Barolo, M., G. B. Guarise, S. A. Rienzi, A. Trotta, and S. Machietto, "Running Batch Distillation in a Column with a Middle Vessel," *Ind. Eng. Chem. Research,* **35,** 4612–4618 (1996).

Bernot, C., M. F. Doherty, and M. F. Malone, "Patterns of Composition Change in Multicomponent Batch Distillation," *Chem. Engng. Sci.,* **45,** 1207–1221 (1990).

Bernot, C., M. F. Doherty, and M. F. Malone, "Feasibility and Separation Sequencing in Multicomponent Batch Distillation," *Chem. Engng. Sci.,* **46,** 1311–1326 (1991).

Bernot, C., M. F. Doherty, and M. F. Malone, "Design and Operating Targets for Nonideal Multicomponent Batch Distillation," *Ind. Eng. Chem. Research,* **32,** 293–301 (1993).

Bortolini, P., and G. B. Guirase, "Un nuovo metodo di distillazione discontinua," *Quad. Ing. Chim. Ital.,* **6,** 150 (1970).

Bushmakin, I. N. and D. Ya. Molodenko, "Method for Selection of the Separating Agent for Azeotropic Rectification of Binary Systems," *J. Appl. Chem. USSR,* **37,** 2609–2617 (1964).

Chemstations, Inc., Chemcad-BATCH (2000). www.chemstations.net.

Cheong, W., and P. I. Barton, "Azeotropic Distillation in a Middle Vessel Batch Column. 1. Model Formulation and Linear Separation Boundaries," *Ind. Eng. Chem. Research,* **38,** 1504–1530 (1999*a*).

Cheong, W., and P. I. Barton, "Azeotropic Distillation in a Middle Vessel Batch Column. 2. Nonlinear Separation Boundaries," *Ind. Eng. Chem. Research,* **38,** 1531–1548 (1999*b*).

Cheong, W., and P. I. Barton, "Azeotropic Distillation in a Middle Vessel Batch Column. 3. Model Validation," *Ind. Eng. Chem. Research,* **38,** 1549–1564 (1999*c*).

Davidyan, A. G., V. N. Kiva, G. A. Meski, and M. Morari, "Batch Distillation in a Column with a Middle Vessel," *Chem. Engng. Sci.,* **49,** 3033–3051 (1994).

Diwekar, U. M., *Batch Distillation: Simulation, Optimal Design and Control.* Taylor and Francis, Washington DC (1995).

Diwekar, U. M., "Understanding Batch Distillation Process Principles with MultiBatchDS," *Comp. Appl. Engng. Educ.,* **4,** 275–284 (1996). See http://www.che.utexas.edu/cache/default.html for software.

Doherty, M. F., and G. A. Caldarola, "Design and Synthesis of Homogeneous Azeotropic Distillation. 3. The Sequencing of Columns for Azeotropic and Extractive Distillations," *Ind. Eng. Chem. Fundam.,* **24,** 474–485 (1985).

Ewell, R. H., and L. M. Welch, "Rectfication in Ternary Systems Containing Binary Azeotropes," *Ind. Eng. Chem.,* **37,** 1224–1231 (1945).

Hasebe, S., T. Kurooka, B. Balukrishnan, A. Aziz, I. Hashimoto, and T. Watanabe, "Simultaneous Separation of Light and Heavy Impurities by a Complex Batch Distillation Column," *J. Chem. Engng. Japan,* **29,** 1000 (1996).

Hasebe, S., T. Kurooka, and I. Hashimoto, Comparison of the Separation Performances of a Multi-Effect Batch Distillation System and a Continuous Distillation System, in *Proceedings of DYCORD+95,* pp. 249–254, Helsingor, Denmark (1995).

Hasebe, S., M. Noda, and I. Hashimoto, "Optimal Operation Policy for Total Reflux and Multi-Effect Batch Distillation Systems," *Computers Chem. Engng.,* **23,** 523–532 (1999).

Kim, Y. S., *Optimal Control of Time-Dependent Processes,* Ph.D. thesis, University of Massachusetts, Amherst, MA (1985).

Kruel, L. U., A Gorak, and P. I. Barton, "Dynamic Rate-Based Model for Multicomponent Batch Distillation," *AIChE J.,* **45,** 1953–1962 (1999).

Lelkes, Z., P. Lang, B. Benadda, and P. Moszkowicz, "Feasibility of Extractive Distillation in a Batch Rectifier," *AIChE J.,* **44,** 810–822 (1998*a*).

Lelkes, Z., P. Lang, P. Moszkowicz, B. Benadda, and M. Otterbein, "Batch Exractive Distillation: The Process and the Operational Policies," *Chem. Engng. Sci.,* **53,** 1331–1348 (1998*b*).

Logsdon, J. S., and L. T. Biegler, "Accurate Determination of Optimal Reflux Policies for the Maximum Distillate Problem in Batch Distillation," *Ind. Eng. Chem. Research,* **32,** 692–700 (1993).

Lutugina, N. V. and K. S. Tavatsherna and V. M. Kalyustinyi, "Investigation of the Ternary System Methyl Acetate-Chloroform-Water by a Rectification Method," *J. Appl. Chem. USSR,* **33,** 248–251 (1960).

Luyben, W. L., "Multicomponent Batch Distillation. I. Ternary System with Slop Recycle," *Ind. Eng. Chem. Research,* **27,** 642–647 (1988).

Reinders, W., and C. H. DeMinjer, "Vapour-Liquid Equilibria in Ternary Systems. I. The Course of the Distillation Lines," *Rev. Trav. Chim.,* **59,** 207–230 (1940*a*).

Reinders, W., and C. H. DeMinjer, "Vapour-Liquid Equilibria in Ternary Systems. II. The System Acetone-Chloroform-Benzene," *Rev. Trav. Chim.,* **59,** 369–391 (1940*b*).

Reinders, W., and C. H. DeMinjer, "Vapour-Liquid Equilibria in Ternary Systems. III. The Course of the Distillation Lines in the System Acetone-Chloroform-Benzene," *Rev. Trav. Chim.,* **59,** 392–406 (1940*c*).

Robinson, C. S., and E. R. Gilliland, *Elements of Fractional Distillation.* McGraw-Hill, 4th ed. (1950).

Robinson, E. R., "The Optimisation of Batch Distillation Operation," *Chem. Engng. Sci.,* **24,** 1661–1668 (1969).

Safrit, B. T., A. W. Westerberg, U. Diwekar, and O. M. Wahnschafft, "Extending Continuous Conventional and Extractive Distillation Feasibility Insights to Batch Distillation," *Ind. Eng. Chem. Research,* **34,** 3257–3264 (1995).

Safrit, B. T., and A. W. Westerberg, "Algorithm for Generating the Distillation Regions for Multicomponent Mixtures," *Ind. Eng. Chem. Research,* **36,** 1827–1840 (1997*a*).

Safrit, B. T., and A. W. Westerberg, "Improved Operational Policies for Batch Extractive Distillation Columns," *Ind. Eng. Chem. Research,* **36,** 463–443 (1997*b*).

Safrit, B. T., and A. W. Westerberg, "Synthesis of Azeotropic Batch Distillation Regions," *Ind. Eng. Chem. Research,* **36,** 1841–1854 (1997*c*).

Skogestad, S., B. Wittgens, and E. Sorensen, "Multivessel Batch Distillation," *AIChE J.,* **43,** 971–978 (1997).

Van Dongen, D. B., and M. F. Doherty, "On the Dynamics of Distillation Processes—VI. Batch Distillation," *Chem. Engng. Sci.,* **40,** 2087–2093 (1985).

Wittgens, B., R. Litto, E. Sorensen, and S. Skogestad, "Total Reflux Operation of Multivessel Batch Distillation," *Computers Chem. Engng.* (Suppl. pt. B), S1041–S1046 (1996).

Wittgens, B., and S. Skogestad, "Multivessel Batch Distillation—Experimental Verification," *IChemE Symp. Ser.,* No. 142, 239–248 (1997).

10

Reactive Distillation

10.1 INTRODUCTION

Combining reaction and distillation is an old idea that has received renewed attention. Figure 10.1 shows the rate of growth in the literature and in U.S. patents.[1] Reactive or catalytic distillation has captured the imagination of many because of the demonstrated potential for capital productivity improvements (from enhanced overall rates, by overcoming very low reaction equilibrium constants, and by avoiding or eliminating difficult separations), selectivity improvements (which reduce excess raw materials use and by-product formation), reduced energy use, and the reduction or elimination of solvents. Some of these advantages are realized by using reaction to improve separation, e.g., overcoming azeotropes or reacting away contaminants; others are realized by using separation to improve reactions, e.g., overcoming reaction equilibrium limitations, improving selectivity, or removing catalyst poisons. The potential for major improvements is greatest when several aspects are important.[2]

A large number of systems have been studied in the papers and patents counted in Fig. 10.1. Sample chemistries and components are listed in Table 10.5 at the end of this chapter.

The Eastman Chemical Company's methyl acetate reactive distillation process, discussed in Chap. 1, is a classic success story in reactive distillation. That process was a radical departure from traditional technology that had been a genuine economic success for over 15 years. One hybrid reactive distillation device replaced an entire flowsheet consisting of 11 major units plus all their heat exchangers, control systems,

[1]Ten years ago it was possible to give a short summary of this literature. This is no longer possible, since several hundred papers and patents have been published in recent years. These are not cited here in the interest of brevity but many recent papers have been reviewed (Taylor and Krishna, 2000).
[2]Similar progress has been made in other reaction-separation technologies, such as simulated moving bed reactors, reactive membranes, reactive crystallization, etc., and several interesting applications have been commercialized.

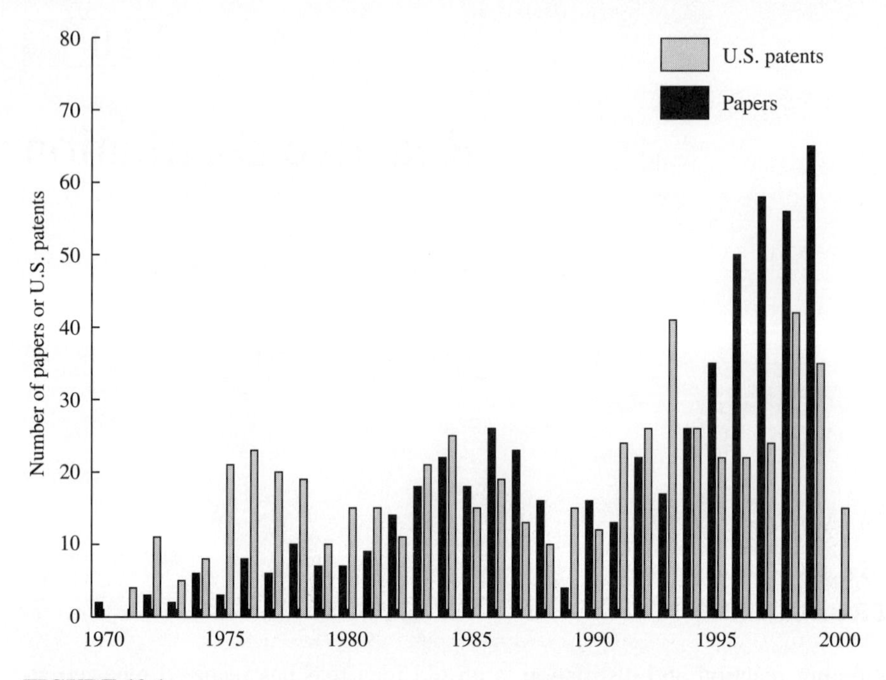

FIGURE 10.1
Publications or U.S. patents including reactive or catalytic distillation; 562 publications for the period 1970–1999, as listed in the Engineering Index; May 2000; 571 U.S. Patents for the period 1971–2000 (through June 26, 2000).

pumps, intermediate storage tanks, etc. Similar successes have been achieved by other companies making other products, but they have often chosen to keep the technology a trade secret. The main reason that the old technology to make methyl acetate was so complicated and expensive is the existence of azeotropes in the reactor effluent mixture, consisting of methanol + methyl acetate (one reactant and one product), and methyl acetate + water (both products). The azeotrope between methyl acetate and water is of vital concern because this must be broken in order to obtain pure products. However, in equilibrium reactive distillation this azeotrope disappears because it reacts into a four-component mixture, thereby "destroying" the azeotrope. This is an enormous benefit of carrying out reaction simultaneously with distillation that cannot be achieved by carrying out these steps sequentially. The end result is a process with five times lower investment and five times lower energy use as well as other benefits (Siirola, 1995).

Processes for the synthesis of fuel ethers such as MTBE, ETBE, and TAME have also met with commercial success, and the MTBE example is discussed in some detail below, starting with Example 10.2.

The improvements in some processes via reactive distillation are dramatic, and we might expect a much greater number of chemical processes to be candidates for simultaneous reaction and separation instead of more traditional separate steps.

However, combining reaction and distillation is not always advantageous, and in fact in some cases it may not even be feasible. For example, the temperatures for reaction and phase equilibrium may have a large mismatch, some systems may have large energy requirements and costs, or the interactions of phase behavior and reaction may make separation more difficult in some cases. A key question is "How can we decide quickly whether reactive distillation is a good process concept?" Despite some useful guidelines, e.g., Xu et al. (1985), no comprehensive set of general rules is known for deciding feasibility and incentives for reactive distillation. Generally, *methods* are necessary for decision making, as discussed in earlier chapters for systems without chemical reactions.

Conceptual design should focus on feasibility, process alternatives, and methods to estimate equipment sizes, energy use, and cost. There is much more to successful development of a reactive distillation process beyond the conceptual design; we offer a few comments on that near the end of the chapter.

The feasible product compositions from a reactive distillation are determined for a given feed, column pressure, and rate of reaction. It is often possible to vary these quantities parametrically to enlarge the design space. In this step, knowledge of the basic process chemistry and the phase equilibrium is vital, and feasibility studies often suggest experiments.

Feasibility analysis in reactive distillation must incorporate all of the features discussed in earlier chapters for ideal and azeotropic mixtures, as well as new phenomena caused by the introduction of chemical reactions. Surprisingly, reaction can *induce* the formation of azeotropes that were not there to begin with! This phenomenon has now been well documented from theory, models, and experiments (Song et al., 1997) and is a major consideration for feasibility analysis in reactive distillation.

For large rates of reaction, the chemical equilibrium limit is approached; this limiting case is discussed in Sec. 10.4. The effects of chemical reaction can be captured effectively using residue curve maps along with pinch tracking methods that are especially tailored to this case, from which it is also possible to devise sequences. Although a close approach to reaction equilibrium can be achieved in theory for large residence times it is not normally desirable to operate commercial devices under these conditions, and so the question of what happens to the feasible splits in the finite rate regime is critical. Consequently, recent research has focused on the important question: "What is the effect of chemical kinetics on feasibility?" (e.g., Bessling et al., 1998), and this is addressed in the first part of Sec. 10.5.

Although it is possible to construct hypothetical systems where nothing unexpected happens in the kinetic regime,[3] this is not the norm. Based on the results of well–established methods for assessing feasibility in the limit of no reaction, and in the limit of equilibrium reaction, we know that the feasible separation systems

[3]For example, an ideal mixture of A, B, C, and D reacts according to the chemistry $A + B \rightleftharpoons C + D$. If C is the lightest component and D is the heaviest, no new *reactive* azeotropes are introduced. The feasible splits at reaction equilibrium are exactly what we would expect; C is produced as distillate and D as bottoms.

are generally different in these cases.[4] Thus, there must be one or more transitions in the kinetic regime, and different separation system structures appear for different ranges of the reaction rate. The practical implication is that the feasibility and product purities for a reactive distillation may depend on the production rate, the liquid holdup, and the catalyst concentration. For example, in some applications the desired product purities are not feasible if there is too little catalyst in the column, or if there is *too much!* This means that there can be a finite range of feasibility in terms of the residence time, which can give fundamental *upper and lower* limits to the production rate. It is vital to have sufficient data and analysis at the conceptual design stage to provide a good estimate of the range of feasibility before a process alternative is selected for further study.

At the moment, no single approach is completely satisfactory for application to realistic column configurations and for any number of components and reactions. Side reactions and selectivity effects are especially important issues, and these have no counterpart in the sequencing of nonreactive columns. Results from feasibility studies are used to identify and organize alternative column sequences, which generally include fully reactive columns, nonreactive columns, and/or hybrid columns (both reactive and nonreactive sections in the same device). There should be at least as many alternatives as for nonideal and azeotropic distillation, and probably a lot more, so systematic methods are vital.

Design methods estimate equipment sizes (number of reactive stages, number of nonreactive stages, column diameter), feed flows and locations, heating and cooling loads, catalyst concentrations, and liquid holdups. This provides the basis for an economic evaluation and ranking of the process alternatives.

It is a common exercise to seek designs for reactive distillation using simulation tools and adjusting design variables. This sort of "design by simulation" approach is frequently frustrating without guidance from systematic methods. This chapter describes systematic methods to find improved designs like those shown in the examples in Sec. 10.2. We begin with residue curve maps for simple reactive distillation in Sec. 10.3, which makes a close connection to the methods used in earlier chapters. These residue curve maps motivate the treatment of the limit for very fast reactions as a special case of equilibrium reactive distillation in Sec. 10.4. In Sec. 10.5, we connect results from this case with results from earlier chapters for distillation without reaction, by a systematic treatment for the effects of chemical kinetics. This is followed by a complete example of conceptual design using a case study of methyl acetate synthesis. A short discussion of open questions and important issues for the further development of reactive distillation closes the chapter.

---

[4]This is caused by several effects in isolation or combination:

1. The reaction stoichiometry. For example, the reaction $A + B \rightleftharpoons C$ cannot have pure C as a feasible product because as C is enriched it will decompose to form A and B. This happens in the MTBE chemistry.
2. The disappearance of azeotropes between two reactants or two products because of chemical reaction. This happens in methyl acetate chemistry.
3. The appearance of reactive azeotropes. This happens in isopropyl acetate chemistry.

10.2 EXAMPLES

EXAMPLE 10.1. A Constant Volatility Mixture. Consider the equilibrium-limited chemical reaction $A + B \rightleftharpoons C$ with an equilibrium constant $K_{eq} = 2$ and a rate of reaction per mole of liquid given by $r = k(x_A x_B - x_C/K_{eq})$. The components have normal boiling points $T_{b,A} = 51.6°C$, $T_{b,B} = 65.5°C$, and $T_{b,C} = 100°C$ with constant relative volatilities $\alpha_A = 5$, $\alpha_B = 3$, and $\alpha_C = 1$. The molecular weights are 86.18, 100.21, and 186.39 for A, B, and C, respectively. A traditional process would include one or more reactors to make a mixture of all three components, probably followed by a distillation to recover the ingredients A and B. Those ingredients need not be further separated, since they have adjacent boiling points and both should be recycled to the reactor. We suppose that the molar ratio of reactants is unity and that there are no by-product reactions. In this process, reaction equilibrium limits the conversion and a significant amount of A and B must be separated and recycled. Instead of reaction followed by distillation, it might be attractive to combine the two steps and use the column needed to separate C from the reaction mixture to carry out the reaction simultaneously. For a mixture containing 40 mol % each of A and B and 20 mol % C, the methods described in Chap. 4 can be used to find the design summarized in Table 10.1. The corresponding composition profiles are shown in Fig. 10.2*a*.

If chemical reaction is carried out along with the distillation using the same design, the profiles in Fig. 10.2*b* and *c* result. In these figures, the reaction is characterized by the magnitude of a characteristic reaction time relative to a characteristic residence time, a "Damköhler" number Da. Small values of Da correspond to little or no reaction, and large values correspond to a close approach to reaction equilibrium. Da can be increased at the cost of larger holdup, greater catalyst concentrations, higher temperatures, or any combination of these. Figure 10.2*b* corresponds to an average residence time per stage equal to approximately 25% of the characteristic reaction time. Neither the conversion nor the product purities are very attractive. Figure 10.2*c* corresponds to forty times larger Da than that in part *b*, but the conversions and product purities remain low. Based on

TABLE 10.1
Design based on the CMO model for Example 10.1, to separate a saturated liquid feed with $\mathbf{x}_F = (0.4, 0.4, 0.2)$; 20% reactor conversion for an equimolar feed of A and B

Variable	Value	Note
Total stages	27	Total condenser = stage 0 Equilibrium reboiler = 27 Feed stage = 20
$\mathbf{x}_D$	$0.5055, 0.4994, 10^{-4}$	
D	$0.8F$	
$\mathbf{x}_B$	$10^{-4}, 4.9 \times 10^{-3}, 0.995$	
B	$0.2F$	
$r_{\min}$	0.59	Minimum reflux ratio
$r = 1.2r_{\min}$	0.71	Reflux ratio
s	6.80	Reboil ratio

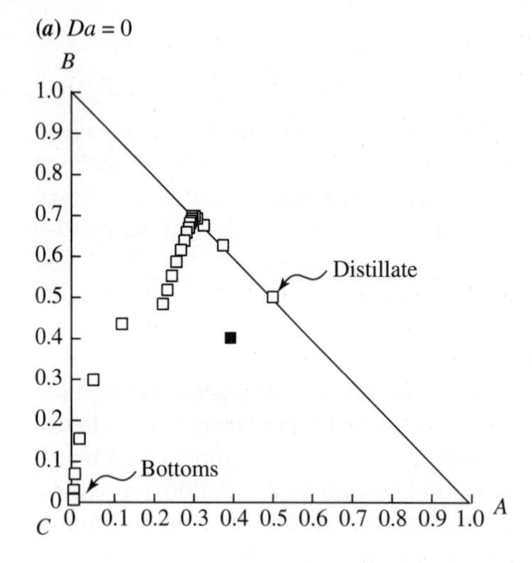

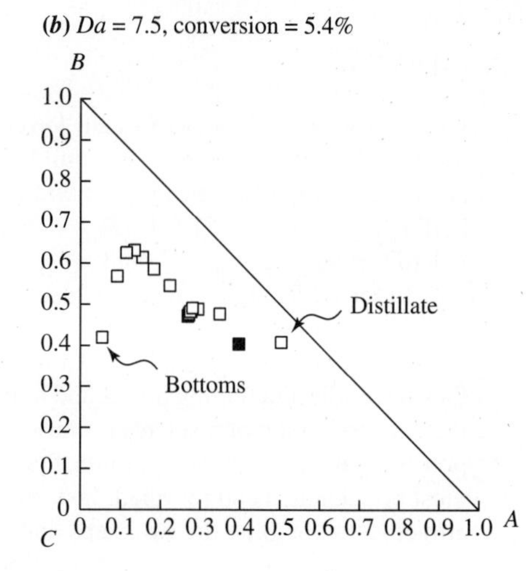

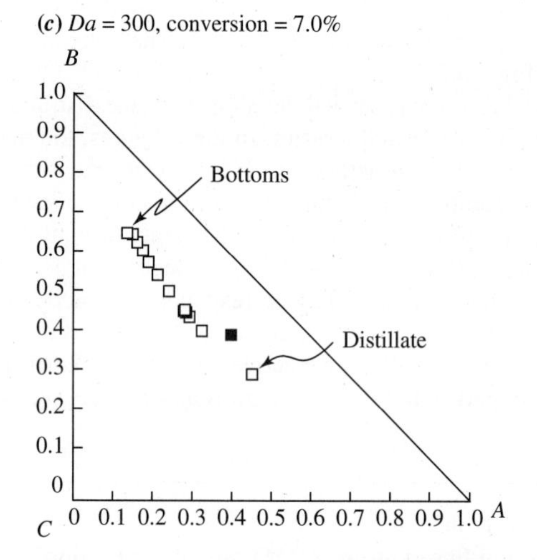

FIGURE 10.2
Composition profiles for Example 10.1. Without chemical reaction, the indirect split (*a*) produces high-purity C as a bottoms product, with a high fractional recovery. If reaction is present throughout the column, (*b*) shows composition profiles for $Da = 7.5$, and (*c*) for $Da = 300$. Neither case provides high purities, recoveries, or conversions. The liquid compositions in the column are shown by □ and the feed composition by ■.

these simulation results, one might (incorrectly) conclude that reactive distillation is not an alternative worth further study.

It is possible to find better designs like the one summarized in Fig. 10.3. In this case, the column has 14 stages rather than 27; the feed is on stage 5. Of these 14 stages, only stages 2 to 7 have reaction with $Da = 15$. The bottoms flow is $0.597F$ and the distillate $0.0072F$, with a conversion of 98.9%. The feed contains equal parts of A and B, the reflux ratio is 1000, and the reboil ratio is 12. The bottoms product contains 99.7 mol % C, and the small distillate flow containing 0.7 mol % C can be recycled.

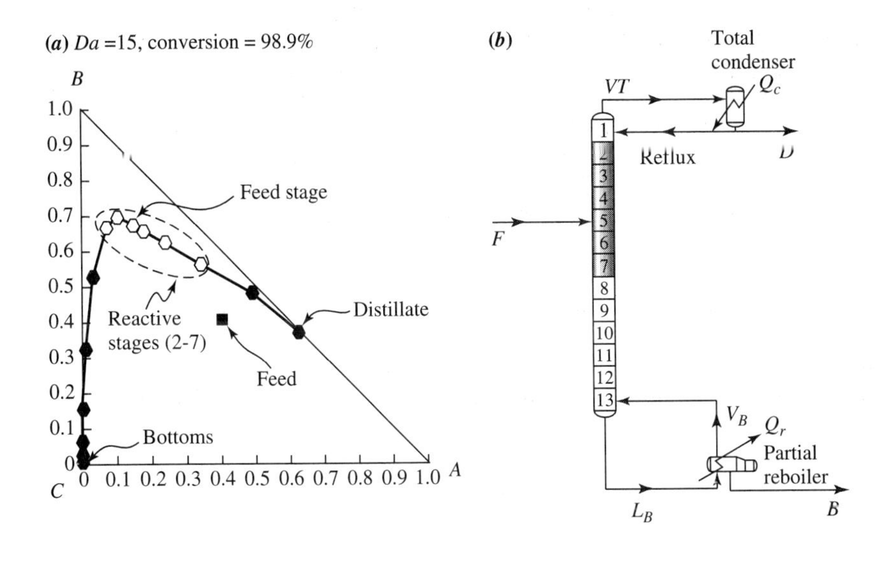

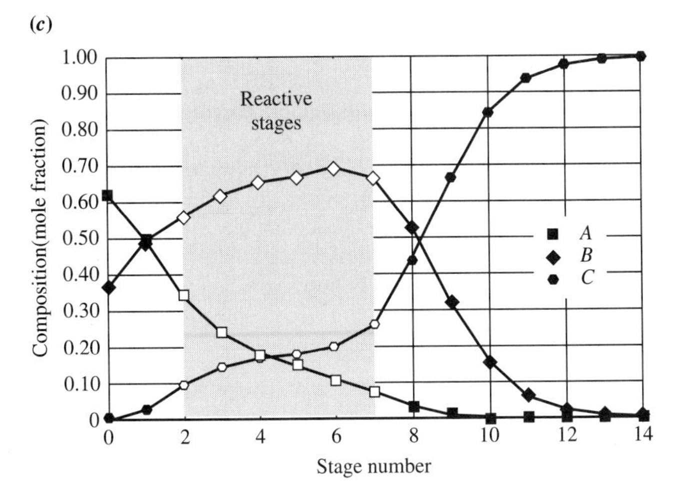

FIGURE 10.3
An improved design for Example 10.1. The column has 14 stages, with reaction on stages 2–7 and a feed enters on stage 5. The other parameters in the design are summarized in the text. The product purities, fractional recovery of C, and conversion are all high. The compositions on reactive stages are shown with open symbols and closed symbols show the nonreactive stages. The column was drawn as part of a HYSYS simulation.

In Example 10.1, despite the fact that the AB/C separation is relatively easy, simply adding simultaneous reaction everywhere within the distillation column does not lead to a useful design. The hybrid column is a better alternative, and a similar design also applies when the phase equilibrium is more complex, as shown in Example 10.2 for MTBE production.

EXAMPLE 10.2. MTBE Process. Methyl tertiary butyl ether (MTBE) is made by reacting together methanol and isobutene in the liquid phase. The reaction is equilibrium limited and has the following stoichiometry:

$$\underset{\text{(Methanol)}}{CH_3OH} \quad + \quad \underset{\text{(Isobutene)}}{i - C_4H_8} \rightleftharpoons \underset{\text{(MTBE)}}{t - C_4H_9OCH_3} \tag{10.1}$$

The reactor effluent stream is fed to a distillation column where it is desired to separate MTBE from the mixture. The other product from the column is recycled to the reactor.

The pure component and azeotropic boiling data for the mixture at 1 atm pressure are as follows:

- Pure components
  Isobutene = −6.9°C
  MTBE = 55.1°C
  Methanol = 64.7°C
- Azeotropes
  Methanol + MTBE = 51.2°C
  Isobutene + methanol = −7.1°C

Using the methods from Chap. 5, we sketch the residue curve map shown in Fig. 10.4. A major difficulty with this process is caused by the distillation boundary that connects the two azeotropes. In order to get pure MTBE from the distillation column, it is essential to arrange for the reactor effluent to be located in the lower distillation region, where MTBE is the stable node. For a particular isobutene-methanol feed containing excess isobutene, the reactor effluent follows a stoichiometric line as shown in Fig. 10.5. Higher conversions move the effluent composition further along the line until the equilibrium conversion is reached. For a particular conversion we obtain the effluent composition shown in the figure at point F. This mixture is fed to the distillation column, where it is separated into pure MTBE as a bottoms product, and a mixture of all three components as distillate D, with a composition close to the distillation boundary. Since

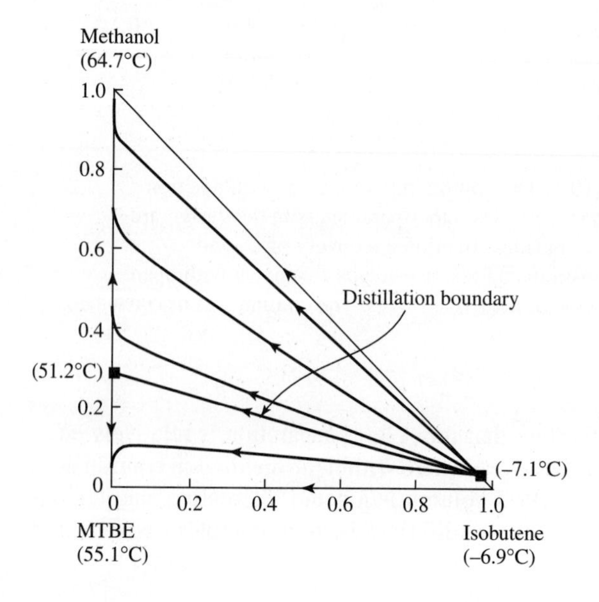

FIGURE 10.4
Residue curve map for the mixture isobutene-methanol-MTBE at 1 atm pressure.

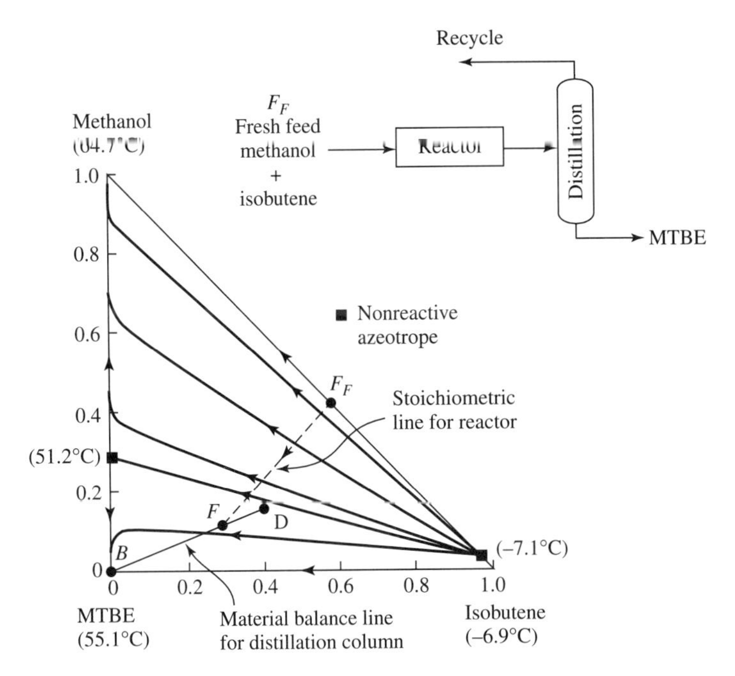

FIGURE 10.5
Flowsheet and material balance lines for the MTBE process, without closure of the recycle loop. It is interesting to close the material balances as an interactive exercise.

the distillate contains all the unreacted isobutene and methanol, we recycle this stream to the reactor, as shown in Fig. 10.5.

This arrangement may have a large inventory of MTBE in the recycle system, which increases investment and operating cost. Conventional alternatives seek to add more separation devices to reduce the amount of MTBE in the recycle stream (think of some yourself), but a better alternative is to achieve this goal with less equipment or recycle. Reactive distillation provides a way of accomplishing this by placing a reactive catalytic section *above* the stripping section of the column; see Fig. 10.6. In this arrangement, the methanol and isobutene that rise from the stripping section get reacted in the catalytic section to make more MTBE. Such an arrangement achieves close to 100% conversion.

10.3
SIMPLE REACTIVE DISTILLATION

The first questions in conceptual design are feasibility and process alternatives. Answers to these questions for reactive distillation are a bit more complicated than for nonreactive mixtures but build on the ideas used there.

A basic model for simple distillation with one reaction is useful; Fig. 10.7 shows a schematic. We consider the reaction only in the liquid phase, which is well mixed. It is necessary to describe the rate of reaction, and we need a description suited to nonideal liquids, for use over a wide range of conversions, e.g., to recover correctly

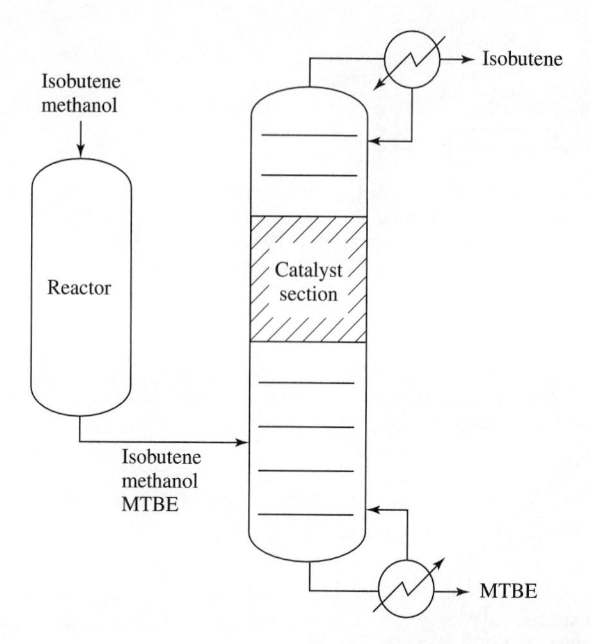

FIGURE 10.6
Reactive distillation column for making MTBE. The process feed contains a slight excess of isobutene. Methanol, the limiting reactant, has nearly complete conversion.

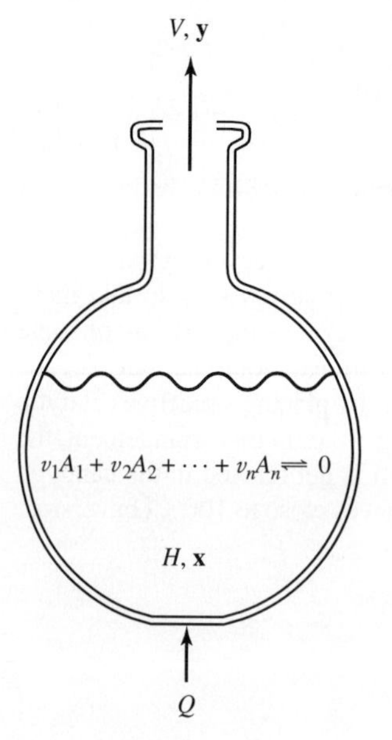

FIGURE 10.7
Schematic of simple reactive distillation. The chemical species are A_i with stoichiometry coefficients v_i, positive for products and negative for reactants. H is the molar holdup in the liquid and V is the molar flowrate of vapor.

the reaction equilibrium at very long times. A variety of rate expressions could be developed, but the most effective and economical in terms of data regression seem to be those based on activities rather than concentrations.[5] For a homogeneous liquid

[5]This idea dates to at least the 1940s (Hougen and Watson, 1947, Chap. XVII) and can be used to derive many other forms for constant volume, variable volume, ideal mixtures, etc.

phase reaction, we write the rate per mole of liquid as

$$r = k\left(\prod_{i=1}^{n_R} a_i^{-\nu_i} - \prod_{i=1}^{n_P} a_i^{\nu_i}/K_{eq}\right) \tag{10.2}$$

where the stoichiometry coefficients ν_i are negative for the n_R reactants and positive for the n_P products. K_{eq} is the thermodynamic equilibrium constant, discussed in more detail in Sec. 10.4, and k is the rate constant. The activities a_i for each component i in the liquid are $a_i = \gamma_i x_i$ as described in Chap. 2. This approach gives accurate results when sufficient data on the constituent binary pairs are available to obtain interaction parameters for the activity coefficient models. Results from such models will be described below for several cases. The rate constant k follows an Arrenhius temperature dependence

$$k = k_0 \exp(-E/RT) \tag{10.3}$$

where the pre-exponential factor k_0 and the activation energy E are generally found from regression of kinetic data. For some systems, $k_0 = A$, a constant is sufficient. For systems with a homogeneous catalyst, the dependence on catalyst concentration is often described with $k_0 = WA$, or $k_0 = (a + bW)A$, etc., where W is the concentration of catalyst, and the constants a, b, and A are determined by regression of kinetic data. This form of rate expression seems to fit data for a variety of interesting systems, but more complex forms are needed, e.g., for heterogeneous catalysts. Specific cases are discussed below in several examples.

The development of a model for simple reactive distillation is similar to the treatment of simple distillation without reaction in Chap. 5. However, the number of moles in the liquid can change both by vaporization and on account of reaction. The rate of change for the moles of i in the liquid is

$$\frac{d(Hx_i)}{dt} = -Vy_i + \nu_i r H \qquad (i = 1, \ldots, c) \tag{10.4}$$

where y_i is the vapor phase mole fraction of component i in phase equilibrium with the liquid. The overall balance, obtained by summing these equations, is

$$\frac{dH}{dt} = -V + \nu_T r H \tag{10.5}$$

were $\nu_T \equiv \sum_{k=1}^{c} \nu_i$ is the net change in moles on complete reaction, and c is the number of components. Combination and rearrangement of these equations gives the generalization of Eq. 5.6.

$$\frac{dx_i}{d\xi} = x_i - y_i + Da(\nu_i - \nu_T x_i)\mathcal{R} \qquad (i = 1, \ldots, c-1) \tag{10.6}$$

The warped time ξ and its physical interpretation are given in Sec. 5.2. $\mathcal{R} \equiv r/r_0$ is the dimensionless rate of reaction, scaled by a reference rate r_0. The dimensionless group Da is a Damköhler number (Damköhler, 1939), giving the ratio of the characteristic process time to the characteristic reaction time.

$$Da \equiv \frac{t_{\text{process}}}{t_{\text{reaction}}} = \frac{H/V}{1/r_0} \tag{10.7}$$

Rev (1994), Venimadhavan et al. (1994), and Thiel et al. (1997) discuss the use of Da and residue curves in more detail.

The choice of reference quantities for scaling is arbitrary, but some choices are more useful than others when approximations of limiting cases are desired. Sensible choices for r_0 can be the rate itself or the rate constant $r_0 = k_0$, evaluated at some standard temperature such as the lowest boiling point among the pure components and azeotropes. When the reaction is catalyzed, r_0 will typically also include a choice of a reference catalyst concentration. It is desirable, but not always possible, that a choice of a scaling factor for dimensionless groups be made so that the magnitude of the terms in a model is reflected by the numerical value of the group (Lin and Segel, 1974, Chap. 6). For instance, the magnitude of $x_i - y_i$ on the right-hand side of Eq. 10.6 is between 0 and 1, and it would be useful if the relative importance of the reaction term were given by Da. However, the reaction rate is difficult to scale uniformly. For example, the magnitude of $x_i - y_i$ is very small near a pure component or an azeotrope, but the reaction term can be either large or small there, depending on the chemistry. Also, when the rate varies strongly with temperature over the range of boiling temperatures, it may not be possible to scale the equation as desired. A relevant quantity in this case is an Arrhenius number.[6]

$$Ar = \frac{E}{R}\left(\frac{1}{T_{b,\min}} - \frac{1}{T_{b,\max}}\right) \tag{10.8}$$

For the special case of $Da >> 1$, there is a very useful theory that combines both chemical reaction equilibrium and vapor–liquid equilibrium; this is discussed in Sec. 10.4.

Solutions of Eq. 10.6 depend on the value of Da. Unlike the residue curve maps for simple distillation without reaction, consideration of the vapor rate and holdup relative to the rate of reaction is required.[7] The holdup and vapor rate each change with time, and the characteristic rate of reaction r_0 may also change due to changes in temperature or catalyst concentration. For different heating policies (which may include cooling for exothermic reactions), Da will vary differently with time. However, a relatively simple and physically appealing case is to take Da to be a constant. For example, in uncatalyzed or homogeneously catalyzed reactions, if V is reduced in proportion to H, and if $Ar << 1$, then $Da = \frac{H_0/V_0}{1/k_0}$ is a constant. When Da is not constant, Eqs. 10.6 are nonautonomous and the behavior is substantially more complex. (Venimadhavan et al., 1994 discuss a nonautonomous case in more detail.)

The singular points of Eqs. 10.6 satisfy

$$Da\mathcal{R} = -\frac{x_i - y_i}{\nu_i - \nu_T x_i} \tag{10.9}$$

---

[6]Sundmacher et al. (1994) discuss this and several dimensionless numbers for characterizing reactive distillation processes.

[7]For simple distillation without reaction, residue curve maps are independent of the heating policy, since V and H appear only in the warped time and the residue curve map depends only on the pressure and the phase equilibrium parameters. The values of V and H as a function of time do depend on the heating policy, as well as the pressure and the mixture characteristics, but these can be determined separately from the residue curve map.

Equation 10.9 is true for any component, and choosing component 1 arbitrarily, we can write

$$\frac{x_1 - y_1}{\nu_1 - \nu_T x_1} = \frac{x_i - y_i}{\nu_i - \nu_T x_i} \qquad i = 2, 3, \ldots, c - 1 \tag{10.10}$$

For the ternary system discussed in Example 10.1, the effects of Da on the residue curve maps are shown in Fig. 10.8. For small Da, the residue curves are identical to those for simple distillation without reaction, as might be expected; see Fig. 10.8*a*. Residue curves for intermediate rates are shown in Fig. 10.8*b* and *c*. For large Da, the residue curves are essentially straight lines for an initial portion and move rapidly to approach the chemical equilibrium curve. This is shown in Fig. 10.8*d*. All of the trajectories approach a stable node at a composition of approximately 11% A and 72% B. A mixture of that composition and temperature will boil and react at

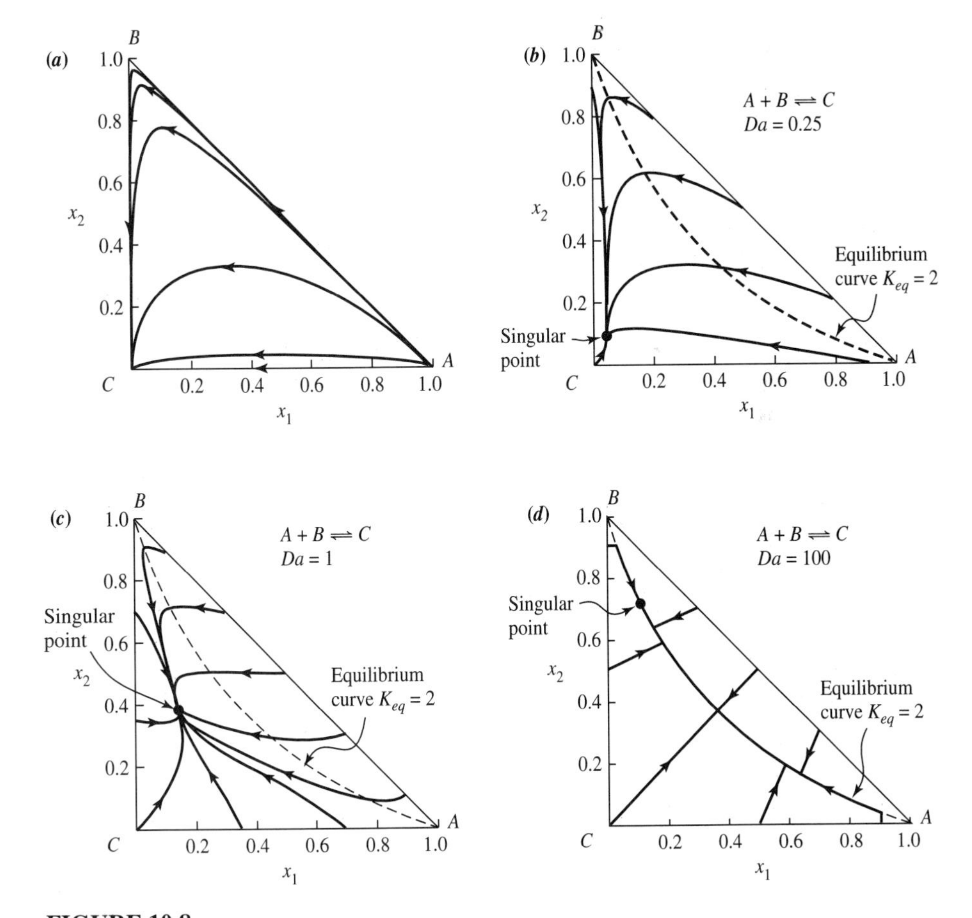

FIGURE 10.8
Simple distillation residue curve maps for (*a*) no reaction ($Da = 0$), (*b*) slow reaction rate ($Da = 0.25$), (*c*) intermediate reaction rate ($Da = 1$), and (*d*) fast reaction rate ($Da = 100$, which rapidly approaches chemical equilibrium).

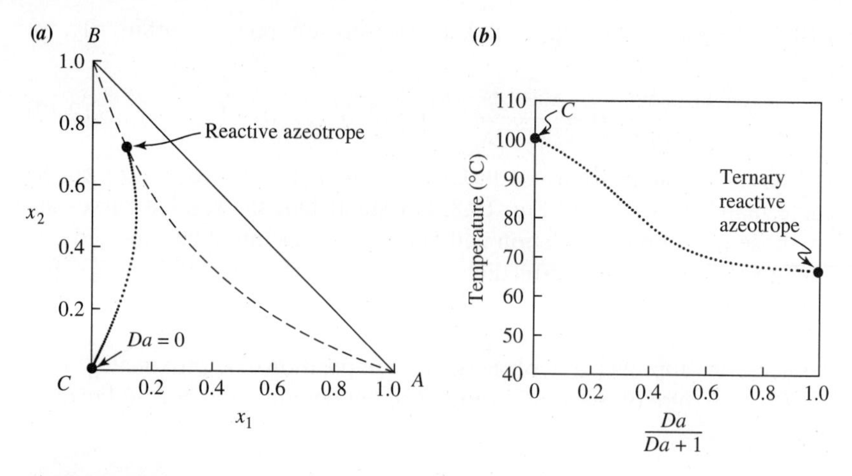

FIGURE 10.9
Composition (*a*) and boiling temperature (*b*) of the singular points in Example 10.1 as a function of *Da*.

constant composition and temperature. We refer to this point in the limit of large *Da* as a "reactive azeotrope." Knowledge of reactive azeotropes is very important for distillation, since they can limit the product purity in the same way that azeotropes in a nonreactive mixture limit distillation. At intermediate values of *Da*, the composition and temperature of these singular points depend on *Da* as shown in Fig. 10.9.

The behavior at large values of *Da* in Fig. 10.8*d* is an interesting and important limiting case. Equation 10.6 in this limit can be written for $\epsilon \equiv 1/Da << 1$ as

$$\frac{dx_i}{d\zeta} = \epsilon(x_i - y_i) + (\nu_i - \nu_T x_i)\mathcal{R} \qquad i = 1, \ldots, c-1 \tag{10.11}$$

The dimensionless time ζ is relative to the characteristic reaction time r_0 instead of the residence time H/V. The singular points are as follows.

- Any points where $\nu_i - \nu_T x_i = 0$. These are "pole points" outside the triangle like the one shown in Fig. 10.10, which is an unstable node.[8] If ϵ is small, the vaporization term can be ignored relative to the reaction term, unless the initial composition is close to the reaction equilibrium curve. In Fig. 10.10, the model becomes one for a well-mixed batch reactor at short times. The reaction stoichiometry alone determines the residue curves at short times, which move along straight lines away from the pole point through the initial compositions. This behavior holds for a time on the order of $1/r_0$, the characteristic reaction time.
- At longer times, the rate decreases as the liquid composition approaches the reaction equilibrium curve, which is defined by $\mathcal{R} = 0$. When $\mathcal{R}$ diminishes to the same

[8]In general, it can be shown that for a chemistry $\sum_{i=1}^{c} \nu_i A_i = 0$ the reaction pole point is given by $x_i = \nu_i/\nu_T$, where $\nu_T = \sum_{i=1}^{c} \nu_i$ (Hugo, 1965; Hauan and Lien, 1998; Frey and Stichlmair, 1999). This pole point is the place where all the stoichiometric lines intersect. Note that this point occurs at a physically impossible point in mole fraction coordinates. Nevertheless, singular points located outside the composition space, like this one, can have a profound effect on the behavior inside the composition space.

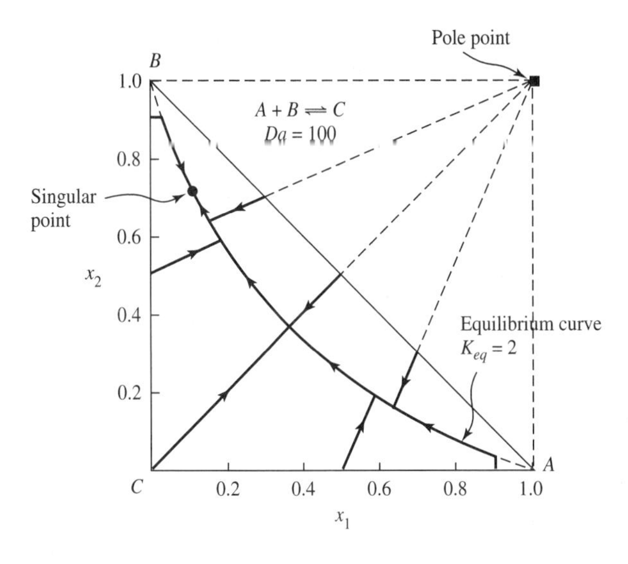

FIGURE 10.10
Simple distillation residue curve maps for very fast reaction ($Da = 100$, which rapidly approaches chemical equilibrium). Dashed lines are stoichiometric lines, which meet at the pole point.

order as ϵ, the vaporization term $\epsilon(x_i - y_i)$ can no longer be ignored. Both terms in Eq. 10.11 are necessarily of the same order in this asymptotic case. The larger the value of Da, the more closely the residue curves approach the reaction equilibrium curve. The composition change along the equilibrium curve is driven by both vaporization and reaction. Vaporization has a slow time scale $t_{\text{residence}} = H/V$ (measured by ξ), slower than the reaction time scale $t_{\text{reaction}} = 1/r_0$ (measured by ζ). In this case, the reaction adjusts quickly to changes in the liquid composition caused by vaporization. Equation 10.9 is true at all singular points regardless of the value for Da. At large Da, the left-hand side of this equation is finite and on the order of unity, since $\mathcal{R} = O(\epsilon)$ and $Da = 1/\epsilon$. As the residue curves approach singular points on the equilibrium curve, the compositions satisfy both reaction equilibrium as well as Eq. 10.10. Such points are pure components or reactive azeotropes.

The limit of reaction equilibrium is important in determining the ultimate behavior of the residue curves, and many other aspects of reactive distillation. The next section of the chapter treats this special case of simultaneous reaction and phase equilibrium.

10.4
EQUILIBRIUM REACTIVE DISTILLATION

Phase Diagrams and Degrees of Freedom

We consider the case of a single chemical reaction expressed in the form

$$\sum_{i=1}^{c} \nu_i A_i = 0 \tag{10.12}$$

where A_i represents component i and ν_i is the stoichiometric coefficient for component i. By convention, $\nu_i < 0$ for reactants, $\nu_i > 0$ for products, and $\nu_i = 0$ if component i does not take part in the reaction, i.e., inert components or solvents. We treat the reaction as occurring only in the liquid phase; this assumption has no effect on the equilibrium conditions but will influence the kinetic modeling. The general equations for phase and reaction equilibrium are:

Phase Equilibrium

$$\mu_i^L(T, P, \mathbf{x}) = \mu_i^V(T, P, \mathbf{y}) \qquad i = 1, \ldots, c \tag{10.13}$$

Reaction Equilibrium

$$\sum_{i=1}^{c} \nu_i \mu_i^L = 0 \tag{10.14}$$

In addition, the mole fractions must obey

$$\sum_{i=1}^{c} x_i = 1 \qquad \sum_{i=1}^{c} y_i = 1 \tag{10.15}$$

A degrees of freedom analysis shows that there are $2c + 2$ variables (T, P, x_i, y_i, $i = 1, \ldots, c$) related by $c+3$ equations (Eqs. 10.13, 10.14, 10.15). This leaves $c-1$ degrees of freedom, in agreement with the Gibbs phase rule for reacting systems

$$\mathcal{P} + \mathcal{F} = c + 2 - R \tag{10.16}$$

where R is the number of independent chemical reactions. For two phases $\mathcal{P} = 2$ and one reaction $R = 1$, the degrees of freedom are $\mathcal{F} = c - 1$. This expresses the fact that we must specify one less variable than for the corresponding calculation without chemical reaction because we have one additional equation, Eq. 10.14. For example, in an isobaric bubble–point calculation we specify the system pressure and the values for $c - 2$ liquid compositions, $x_1, x_2, \ldots, x_{c-2}$; the system temperature, the composition of the vapor phase, and the remaining liquid compositions may be calculated by solving the phase and reaction equilibrium equations. The resulting phase diagrams are quite different from those for nonreacting systems in several respects, including: (1) the equilibrium surfaces have one less dimension than the corresponding surfaces for nonreacting mixtures; e.g., a two-dimensional bubble-point surface for a nonreacting ternary mixture becomes a one-dimensional curve for a reacting mixture; (2) ordinary azeotropes in nonreacting mixtures can be eliminated by a chemical reaction; (3) new azeotropes, called *reactive azeotropes*, can be introduced by a chemical reaction.

In order to produce a selection of phase diagrams we must first introduce the models that are commonly used. As discussed in Chap. 2, for low-pressure applications it is common to represent each phase by a different model[9]. When the vapor is modeled as a mixture of perfect gases (Eq. 2.6) and the liquid is treated as a nonideal

[9]An alternative approach is to represent each phase by the same model, i.e., an equation of state.

mixture using the activity coefficient formalism,[10] the resulting equilibrium equations are:

Phase Equilibrium

$$Py_i = P_i^{\text{sat}}\gamma_i x_i \qquad i = 1, \ldots, c \tag{10.17}$$

Reaction Equilibrium

$$K_{eq} = \prod_{i=1}^{c} (x_i\gamma_i)^{\nu_i} = \frac{\prod_{i=1}^{n_P} (x_i\gamma_i)^{\nu_i}}{\prod_{i=1}^{n_R} (x_i\gamma_i)^{-\nu_i}} \tag{10.18}$$

where n_R is the number of reactants and n_P is the number of products. If we let n_I represent the number of inert components,[11] then $n_R + n_P + n_I = c$.

The symbol K_{eq} is called the *reaction equilibrium constant* and represents the quantity

$$K_{eq}(T) \equiv \exp\left(-\frac{\Delta G^0(T)}{RT}\right) \tag{10.19}$$

where $\Delta G^0(T) = \sum_{i=1}^{c} \nu_i\mu_i^{0L}(T)$. The standard Gibbs free energy of reaction $\Delta G^0(T)$ represents the difference between the free energy of the products and the free energy of reactants at temperature T. Since ΔG^0 and ΔH^0 typically have the same sign and the same order of magnitude, we find:

Exothermic Reactions

$$\Delta H^0 < 0;\ \Delta G^0 < 0 \Rightarrow K_{eq} > 1 \tag{10.20}$$

Endothermic Reactions

$$\Delta H^0 > 0;\ \Delta G^0 > 0 \Rightarrow K_{eq} < 1 \tag{10.21}$$

A typical plot of the temperature dependence of the reaction equilibrium constant is shown in Fig. 10.11 for the MTBE reaction in the temperature range 290 to 370 K. There is quite a bit of scatter in the experimental data, which is typical for these types of measurements. For example, at 347 K the measured values of K_{eq} vary from 30 to 50. This raises the important question of sensitivity of design calculations[12] to uncertainties in the value of K_{eq}.

The reaction equilibrium equation is more conveniently expressed in the following form by dividing Eq. 10.18 throughout by $(K_{eq} + 1)$ and rearranging

$$\left(\frac{K_{eq}}{K_{eq}+1}\right)\prod_{i=1}^{n_R} (x_i\gamma_i)^{-\nu_i} - \left(\frac{1}{K_{eq}+1}\right)\prod_{i=1}^{n_P} (x_i\gamma_i)^{\nu_i} = 0 \tag{10.22}$$

---

[10]That is, $\mu_i^L(T, \mathbf{x}) = \mu_i^{0L}(T) + RT \ln x_i\gamma_i$, where μ_i^{0L} is the chemical potential of pure liquid i at the temperature of the mixture. (The pressure dependence is neglected for liquids.)

[11]These components do not take part in the chemical transformation, but they do affect the phase equilibrium, and hence the reaction equilibrium.

[12]K_{eq} enters into the rate models as well as reaction equilibrium models. Therefore, uncertainties in K_{eq} affect (1) the regression and interpretation of reaction rate data, (2) the equilibrium conversion and shape of the reaction equilibrium diagram, (3) the structure and properties of combined phase and reaction equilibrium diagrams, and hence (4) the conceptual design of reactive distillation systems.

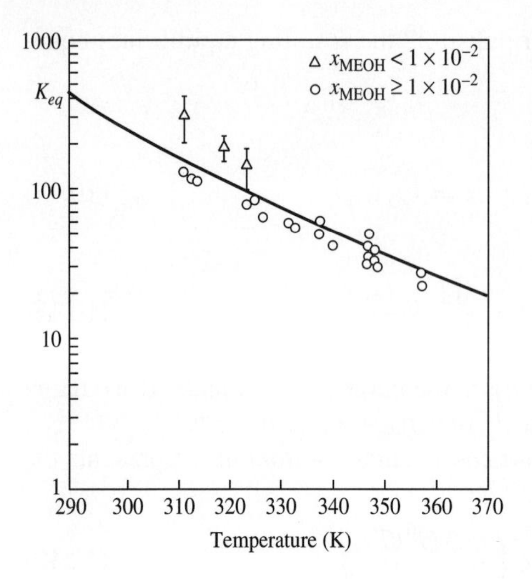

FIGURE 10.11
Plot of K_{eq} vs. T for the MTBE reaction. Solid line calculated from Eq. 10.19; symbols are experimentally measured values. *(Taken from Colombo et al., 1983.)*

This equation has limits that are easy to identify, it also has numerical advantages at large and small values of K_{eq} and at small equilibrium mole fractions; e.g., division by zero is avoided. The limits are

$K_{eq} \rightarrow 0$ *Complete Back Reaction*

This limit occurs when the standard Gibbs free energy of reaction is large and positive, $\Delta G^0 \rightarrow +\infty$, i.e., highly endothermic reactions. There is no forward reaction because the free energy of the products is enormous compared to the free energy of the reactants. For this case Eq. 10.22 becomes

$$\prod_{i=1}^{n_P} (x_i \gamma_i)^{-\nu_i} \rightarrow 0 \tag{10.23}$$

which is satisfied when some or all of the *product mole fractions are zero at equilibrium*, indicating that complete back reaction occurs. Irreversible backward reactions correspond to these conditions.

$K_{eq} \rightarrow \infty$ *Complete Forward Reaction*

This limit occurs when the standard Gibbs free energy of reaction is large and negative, $\Delta G^0 \rightarrow -\infty$, i.e., highly exothermic reactions. In this situation there is complete forward reaction because the free energy of the products is vanishingly small compared to the free energy of the reactants. For this case Eq. 10.22 becomes

$$\prod_{i=1}^{n_R} (x_i \gamma_i)^{-\nu_i} \rightarrow 0 \tag{10.24}$$

which is satisfied when some or all of the *reactant mole fractions are zero at equilibrium*, indicating that complete (irreversible) forward reaction occurs.

The shape of the reaction equilibrium curves in these limits and at various values of K_{eq} in between is shown in Fig. 10.12 for a ternary ideal mixture undergoing the

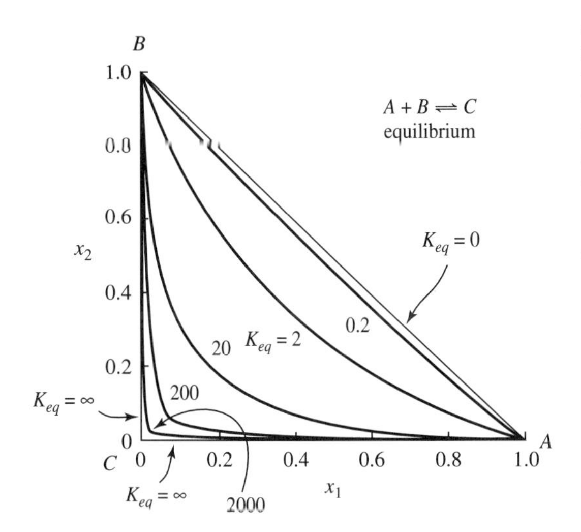

FIGURE 10.12
Reaction equilibrium curves for an ideal reacting liquid mixture at selected values of K_{eq}. The stoichiometry is $A + B \rightleftharpoons C$, and the equilibrium equation is $K_{eq} = \frac{x_C}{x_A x_B}$ with $x_A + x_B + x_C = 1$.

chemistry $A + B \rightleftharpoons C$. Note that the case $\Delta G^0 \rightarrow 0$ gives $K_{eq} \rightarrow 1$, and reaction equilibrium is roughly midway between reactants and products because the free energies are the same. Typical values for K_{eq} for a selection of reactions are given in Table 10.2. The normal range of values in reactive distillation applications involving ethers and esters is $1 < K_{eq} < 50$. However, reactive distillation can also be very effective at much larger or much smaller values.

When the equilibrium constant is large, the heat of reaction is normally large and reactive distillation is attractive because it provides a way of moderating the heat release. This is very attractive if the reactants are less volatile than the products. When one or more of the reactants is very volatile,[13] the advantage of reactive distillation is lost, because large internal flows are needed to keep a sufficient amount of the reactant in the liquid phase. When the equilibrium constant is very small, the reaction is endothermic, management of the heat release is not a factor, and reactive distillation is attractive when it provides a way of removing the product(s) as they are formed and overcoming the unfavorable reaction equilibrium (Block, 1978; Mayer and Wörz, 1980; Xu et al., 1985). For example, aqueous solutions of formaldehyde contain very little free formaldehyde because it hydrolyzes to form methylene glycol

$$\underset{\text{(Methylene glycol)}}{CH_2(OH)_2} \rightleftharpoons \underset{\text{(Formaldehyde)}}{HCHO} + H_2O \qquad (10.25)$$

The reaction has a very small equilibrium constant $K_{eq} \sim 5 \times 10^{-4}$. However, when this mixture is distilled, removal of formaldehyde, which has a large volatility relative to methylene glycol, causes the reaction to shift to the right and the yield of formaldehyde is significantly increased (e.g., Xu et al., 1985).

[13]For example, ethylene oxide in the reaction ethylene oxide + water $\rightarrow$ ethylene glycol ($\Delta H = -20$ kcal/mol) (Ciric and Gu, 1994; Okasinski and Doherty, 1998).

TABLE 10.2
Selected reactions and equilibrium constants

Reaction	T(°C)	K_{eq}	References
Methanol + acetic acid $\rightleftharpoons$ methyl acetate + water	$30 < T < 120$	$30 > K_{eq} > 17$	Song et al., 1998
Isopropanol + acetic acid $\rightleftharpoons$ isopropyl acetate + water	$50 < T < 110$	$K_{eq} \sim 8.7$	Lee and Kuo, 1996
Butanol + acetic acid $\rightleftharpoons$ butyl acetate + water	$100 < T < 120$	$15 > K_{eq} > 11$	Venimadhavan et al., 1999*b*
2 Methanol $\rightleftharpoons$ dimethyl ether + water	$62 < T < 100$	$87 > K_{eq} > 58$	Nisoli et al., 1997
Methanol + isobutene $\rightleftharpoons$ MTBE	$-10 < T < 67$	$1911 > K_{eq} > 53$	Colombo et al., 1983
Methanol + 2-methyl-1-butene $\rightleftharpoons$ TAME	$30 < T < 100$	$102 > K_{eq} > 9$	Syed et al., 2000
Methanol + 2-methyl-2-butene $\rightleftharpoons$ TAME	$30 < T < 100$	$8 > K_{eq} > 1.3$	Syed et al., 2000
Methylene glycol $\rightleftharpoons$ formaldehyde + water	$T \sim 20 - 25$	$K_{eq} \sim 5 \times 10^{-4}$	Xu et al., 1985
Ethylene oxide + water $\rightleftharpoons$ ethylene glycol	$T \sim 50$	$K_{eq} > 1000$	
Acetic anhydride + water $\rightleftharpoons$ 2 acetic acid	$T \sim 50$	$K_{eq} > 1000$	

A note on the temperature dependence of K_{eq} and sensitivities

From Eq. 10.19, $\ln K_{eq} = -\Delta G^0(T)/RT$ and it is conventional to plot $\ln K_{eq}$ vs. $1/T$. Since K_{eq} is very sensitive to small changes in the value of ΔG^0, and that quantity depends on temperature, the graph is not linear. For instance, at $T = 300K$, when $\Delta G^0 = -2$kcal/mol, $K_{eq} = 30$; but if $\Delta G^0 = -3$ kcal/mol, $K_{eq} = 150$. To find a simple correlation for K_{eq}, the plot can be linearized about a typical temperature of interest. This gives the form $\ln K_{eq} = a/T + b$, which can be interpreted as $\ln K_{eq} = -\Delta H^0/RT + \Delta S^0/R$, because $\Delta G^0 = \Delta H^0 - T\Delta S^0$. If ΔH^0 and ΔS^0 are close to constant over the temperature range of interest, this gives a useful and fundamentally correct representation for the temperature dependence of the equilibrium constant. The magnitude of K_{eq} is determined by ΔG^0, but the slope of the graph depends only on ΔH^0.

EXAMPLE 10.3. Reaction Equilibrium in the Limit $K_{eq} \to 0$. Consider the reactions $A + B \rightleftharpoons C$, and $C \rightleftharpoons A + B$ in the limit as $K_{eq} \to 0$. For each initial composition given in Fig. 10.13, show the path followed by the reaction and the final equilibrium composition.

The path followed by the reaction from each initial condition will be a straight line determined by the stoichiometry. Consider the reaction $A + B \rightleftharpoons C$. In this reaction there is a change in the total number of moles as the reaction takes place because two moles (one of A and one of B) are converted to one mole of C. Therefore, the basis for calculating mole fractions changes as the extent of reaction changes. We consider a feed

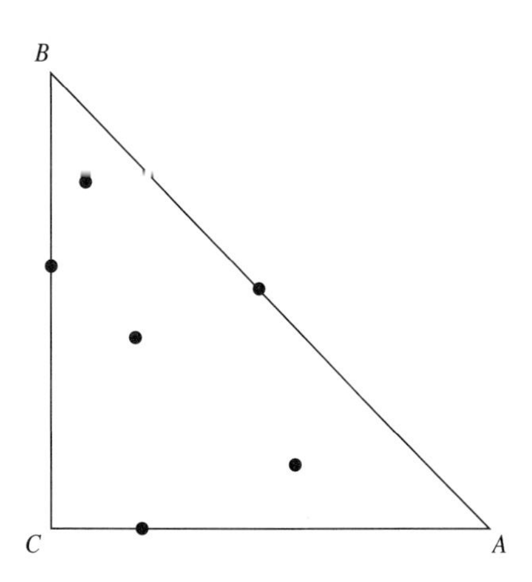

FIGURE 10.13
Initial compositions for the reactions in Example 10.3.

containing only A and B, and develop equations that relate the allowable mole fractions of A, B, and C for all extents of reaction as constrained by the reaction stoichiometry. If we know the total moles of A, B, and C after reaction n_A, n_B, and n_C, respectively, we can determine the amount of A in the feed as simply the moles of A remaining after reaction plus the moles of A used to make C. Determining the moles of B in the feed is carried out similarly.

Let N be the total number of moles of mixture after reaction $N = N_A + N_B + N_C$, and N_A^0 be the total number of moles of A in the feed; then

$$N_A^0 = Nx_A + Nx_C \tag{10.26}$$

where x_A and x_C are the mole fractions of A and C in the mixture after reaction. Nx_A is the total moles of A after reaction and Nx_C is the total moles of A used to make C. Likewise for B

$$N_B^0 = Nx_B + Nx_C \tag{10.27}$$

where Nx_B is the total moles of B after reaction and Nx_C is the total moles of B used to make C. The total number of moles in the feed mixture is $N^0 = N_A^0 + N_B^0$, or

$$N^0 = (Nx_A + Nx_C) + (Nx_B + Nx_C) \tag{10.28}$$

Let the mole fractions in the feed be $X_A = N_A^0/N^0$ and $X_B = N_B^0/N^0$. Then using Eqs. 10.26, 10.27, and 10.28 together with $\sum_{i=1}^{3} x_i = 1$, we find

$$X_A = \frac{x_A + x_C}{1 + x_C} \tag{10.29}$$

and

$$X_B = \frac{x_B + x_C}{1 + x_C} \tag{10.30}$$

Eqs. 10.29 and 10.30 define straight lines across the mole fraction triangle passing through the points $x_A = X_A$, $x_B = X_B$, $x_C = 0$ on the A-B edge, and the reaction pole point $x_A = 1$, $x_B = 1$ where the stoichiometric lines all intersect. These stoichiometric lines are the dashed lines in Figs. 10.10, 10.14*a*, and 10.14*b*.

The reaction paths follow the stoichiometric lines through the initial compositions until there is complete conversion of all the product(s) into reactant(s). For the reaction

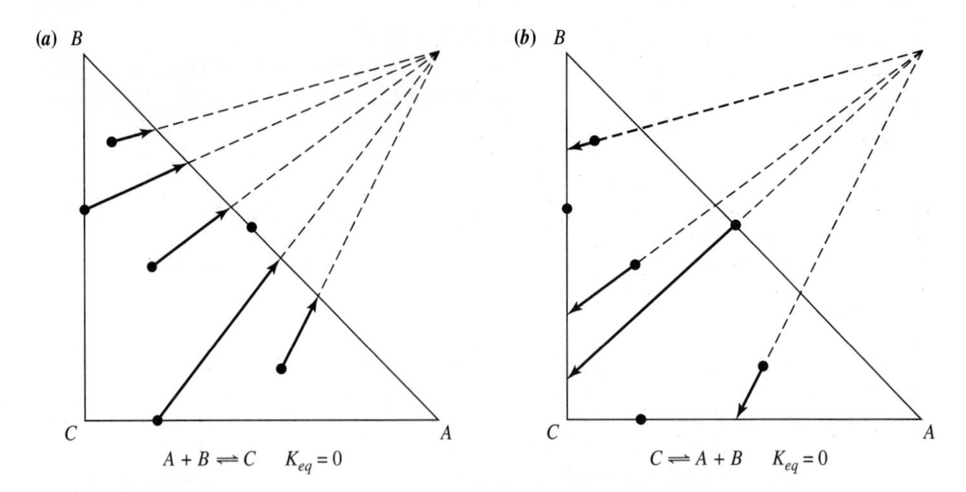

FIGURE 10.14
(a) Paths followed by the reaction $A + B \rightleftharpoons C$. ($b$) Paths followed by the reaction $C \rightleftharpoons A + B$. $K_{eq} \rightarrow 0$.

$A + B \rightleftharpoons C$, this corresponds to converting all the C to A and B; see Fig. 10.14a. Therefore, the equilibrium curve coincides with the line $x_C = 0$, in agreement with Eq. 10.23, which demands that some or all of the product mole fractions are zero at equilibrium. Since there is only one product, it follows that the only solution to the equation is $x_C = 0$. The reaction $C \rightleftharpoons A + B$ produces equilibrium compositions on the other two edges of the triangle, as shown in Fig. 10.14b. In this case there are three solutions to Eq. 10.23: (1) $x_A = 0$, $(x_B, x_C \neq 0)$. These solutions lie on the ordinate and correspond to all the A reacting with some of the B to form C together with leftover B. Initial conditions that lead to these solutions have excess B. (2) $x_B = 0$, $(x_A, x_C \neq 0)$. These solutions lie on the abcissa and correspond to excess A in the feed. (3) $x_A = x_B = 0, x_C = 1$. This solution corresponds to feeds with equal molar amounts of A and B. See Exercises 6 and 7.

There are many algorithms in the literature for solving the phase and reaction equilibrium equations, Eqs. 10.17, 10.22, and 10.15. One of the earliest and widely used is due to Sanderson and Chien (1973). Other approaches solve the problem using optimization-based methods, e.g., Gautam and Seider (1979), Castier et al. (1989), McDonald and Floudas (1997). Excellent accounts of reaction equilibria and computational methods are given by van Zeggeren and Storey (1970) and Smith and Missen (1982).

The following examples demonstrate the types of phase behavior that occur in multiphase reacting mixtures. They are taken from Barbosa and Doherty (1988c), who also report the models and data used in the calculations.

EXAMPLE 10.4. Phase and Reaction Equilibrium Diagrams for an Ideal Ternary Mixture. We first consider a ternary mixture in which the reaction $A + B \rightleftharpoons C$ occurs in the liquid phase. We treat the vapor as a mixture of perfect gases, and the liquid as an ideal mixture. Therefore, the activity coefficients are always unity and phase equilibrium is governed by Raoult's law. The pure component vapor pressures are represented by the Antoine equation and are chosen so that the volatilities of the reactants A and B relative to the product C are approximately constant at 4 and 2, respectively. Therefore, A is the lightest component, B is the intermediate-boiling component, and C is the heaviest. The standard Gibbs free energy of reaction is taken to be constant at a value $\Delta G^0 = -8.314$ kJ/mol. This makes the reaction equilibrium constant approximately constant at a value of $K_{eq} \sim 14$; over the boiling temperature range of the mixture $12 \leq K_{eq} \leq 15$.

The isobaric temperature-composition phase diagram for this system at 1 atm pressure is shown in Fig. 10.15. Figure 10.15*a* shows the liquid and vapor chemical equilibrium curves in the full temperature-composition space, and Fig. 10.15*b* shows their orthogonal projection onto the temperature-(A, B) composition face. The projection of these curves onto the composition triangle (i.e., the base triangle) is also shown in Fig. 10.15*a*. These curves show that while the mole fractions of A and B vary between zero and unity, the mole fraction of C first increases and then decreases, going through a maximum; see Fig. 10.16*c*.

Figure 10.16 shows the y-x diagrams for each component at three values of ΔG^0, corresponding to small, medium, and large values of K_{eq}. Note that the point where curve 2 in Fig. 10.16*b* crosses the 45° line does not correspond to an azeotrope. The

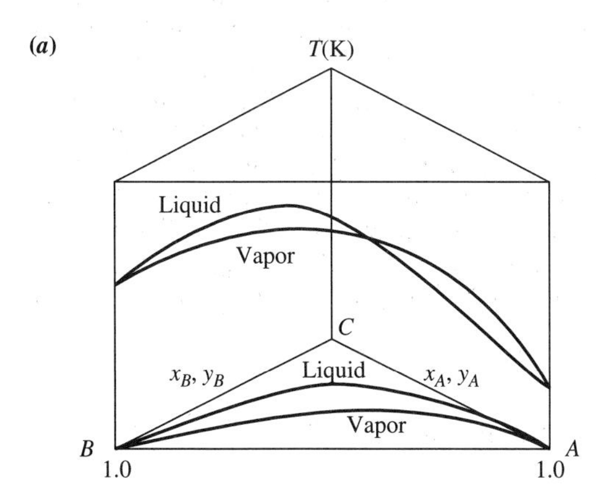

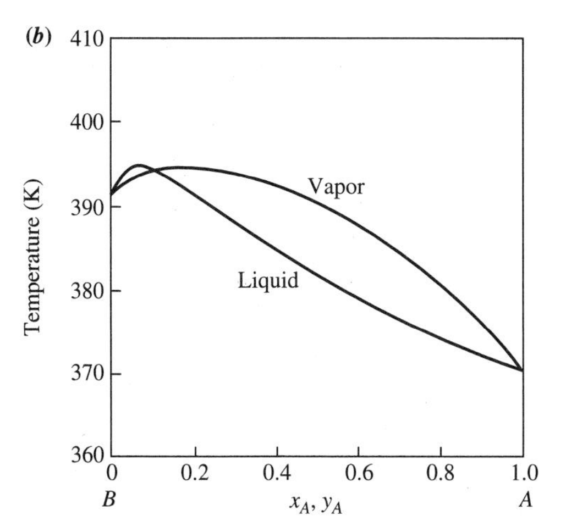

FIGURE 10.15
Reactive phase diagrams for an ideal ternary mixture at 1 atm pressure. $A + B \rightleftharpoons C$, with $\Delta G^0 = -8.314$ kJ/mol ($K_{eq} \sim 14$), $\alpha_{A,C} \sim 4$, $\alpha_{B,C} \sim 2$, and $\alpha_{C,C} = 1$. (*a*) Temperature (T)-composition (A, B, C). (*b*) Orthogonal projection onto the temperature (T)-composition (A, B) face.

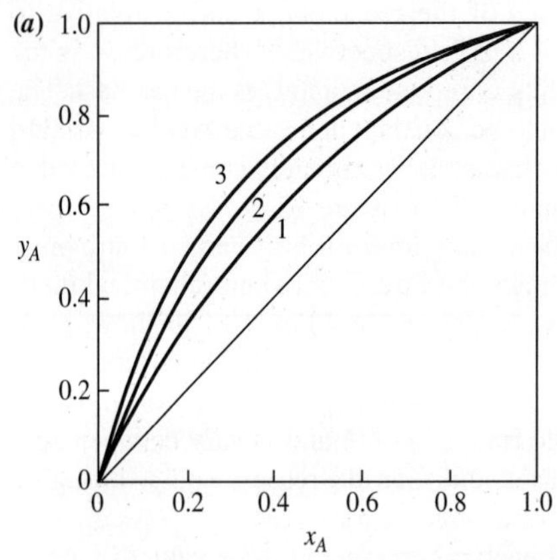

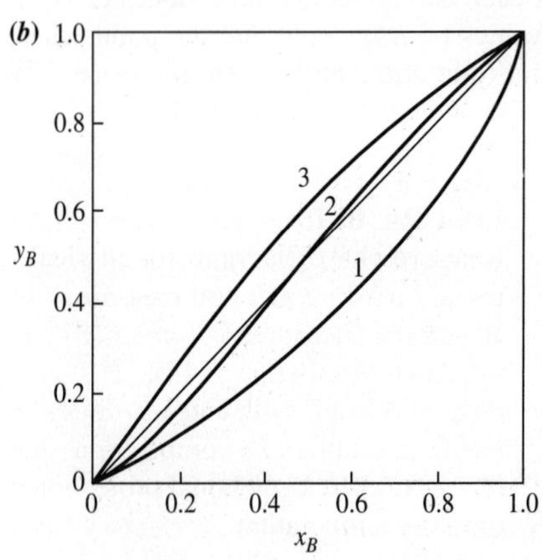

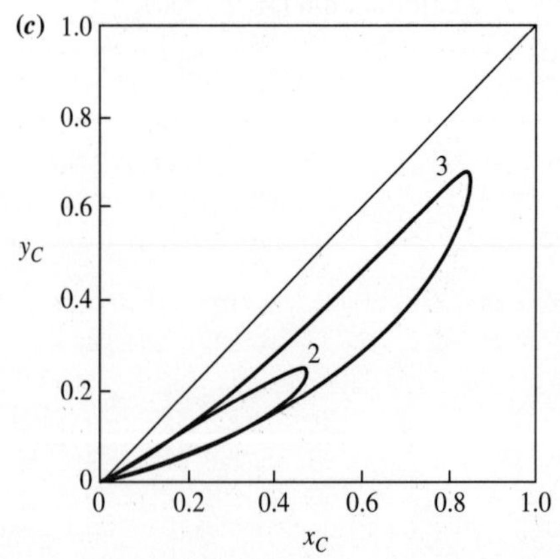

FIGURE 10.16
y-x diagrams for the ideal ternary mixture at three values of ΔG^0. Curve (1) $\Delta G^0 = +8.314$ kJ/mol ($K_{eq} \sim 0.07$), Curve (2) $\Delta G^0 = -6.236$ kJ/mol ($K_{eq} \sim 7$), Curve (3) $\Delta G^0 = -16.628$ kJ/mol ($130 < K_{eq} < 230$). Curve (1) is not visible in part (*c*) because $(x_C)_{\max} = 0.02$; see Eq. 10.31.

maximum allowable liquid and vapor compositions in Fig. 10.16*c* depend on the value of the equilibrium constant. For the special case where K_{eq} is constant, the maximum liquid composition is given by (see Exercise 8)

$$(x_C)_{\max} = 1 + \frac{2}{K_{eq}}[1 - (1 + K_{eq})^{1/2}] \tag{10.31}$$

TABLE 10.3
Compositions at the maximum temperature in Fig. 10.15

x_A	y_A	x_B	y_B	x_C	y_C
0.0697	0.1738	0.4955	0.5519	0.4348	0.2743

As expected, in the limits of $K_{eq} = 0$ and $K_{eq} = \infty$, this equation gives $(x_C)_{\max} = 0$ and $(x_C)_{\max} = 1$, respectively. Phase diagrams like the ones shown in Fig. 10.16 have also been reported and used to analyze reactive distillation columns by Holve (1977), Terrill et al. (1985), Cleary and Doherty (1985), and Grosser et al. (1987).

An unexpected feature of the phase diagrams in Fig. 10.15 is the appearance of a stationary maximum in the T-(x, y) curves. This normally indicates the presence of an azeotrope, which is not expected in an ideal mixture. Such stationary points also indicate the presence of azeotropes in reactive mixtures, and they occur when the following condition is satisfied:

$$\frac{y_1 - x_1}{\nu_1 - \nu_T x_1} = \frac{y_i - x_i}{\nu_i - \nu_T x_i} \qquad i = 2, \ldots, c-1 \tag{10.32}$$

This is the necessary and sufficient condition for azeotropes to occur in two-phase reactive systems (Barbosa and Doherty, 1987). Such azeotropes are normally called *reactive azeotropes*, and they impose the same kinds of limitations on equilibrium reactive distillation as regular azeotropes do on ordinary distillation. They also leave a fingerprint on reactive distillation in the kinetically-controlled regime, as discussed in Secs. 10.3 and 10.5.

The compositions at the maximum temperature in the liquid and vapor curves in Fig. 10.15 are given in Table 10.3. These values satisfy Eqs. 10.32, indicating that the maximum in temperature is indeed a reactive azeotrope. This example shows that chemical reaction can have a significant impact on vapor-liquid equilibrium and can even cause azeotropes that are not there when the reaction is turned off.

The physical explanation for the formation of reactive azeotropes (in both ideal and nonideal mixtures) is quite simple. A temperature is reached where the rate of vaporization (or condensation) and the rate of reaction for each species are such that phase change occurs without change of composition in either phase. This is precisely what is meant by an azeotropic state or transformation. A simple graphical representation of this is given in the section below on Composition Variable Transformations.

EXAMPLE 10.5. Phase and Reaction Equilibrium Diagrams for MTBE Chemistry. We now consider the nonideal vapor-liquid system of isobutene-methanol-MTBE in which the following liquid phase reaction occurs (typically with an acid catalyst such as sulfuric acid or over an ion exchange resin).

$$\underset{\text{(Methanol)}}{CH_3OH} \quad + \quad \underset{\text{(Isobutene)}}{i\text{-}C_4H_8} \rightleftharpoons \underset{\text{(MTBE)}}{t\text{-}C_4H_9OCH_3} \tag{10.33}$$

This chemistry differs from the previous one in several respects: (1) the liquid mixture is strongly nonideal and the activity coefficients are not close to unity, (2) the normal boiling point of the product is between the boiling temperature of the reactants. In the previous case A was light, B was intermediate, and C was heavy; in this case A (isobutene) is light, C (MTBE) is intermediate, and B (methanol) is heavy. (3) the standard Gibbs free energy of reaction is (~ -3 kcal/mol) and the heat of reaction is (~ -8.3 kcal/mol), which implies that the equilibrium constant is high and that it varies strongly with temperature [$-10 < T(°C) < 67 \Rightarrow 1911 > K_{eq} > 53$].

We model the vapor-liquid equilibrium by Eq. 10.17 with the activity coefficients represented by the Wilson equation. The pure component vapor pressures are represented by the Antoine equation. The standard Gibbs free energy of reaction is treated as a function of temperature, and the equilibrium constant varies significantly over the boiling temperature range at 1 atm pressure.

The isobaric temperature-composition phase diagram for this system at 1 atm pressure is shown in Fig. 10.17. Figure 10.17*a* shows the liquid and vapor chemical equilibrium curves in the full temperature-composition space, and Fig. 10.17*b* shows their orthogonal projection onto the temperature-(isobutene, methanol) composition face. The projection of these curves onto the composition triangle (i.e., the base triangle) is also shown in Fig. 10.17*a*. These phase diagrams seem to indicate that the two azeotropes present in the nonreactive mixture (shown in Fig. 10.4) have disappeared because of the chemistry, i.e., they have "reacted away." It is hard to tell whether any reactive azeotropes have been created because it is difficult to interpret the meaning of the shoulder in

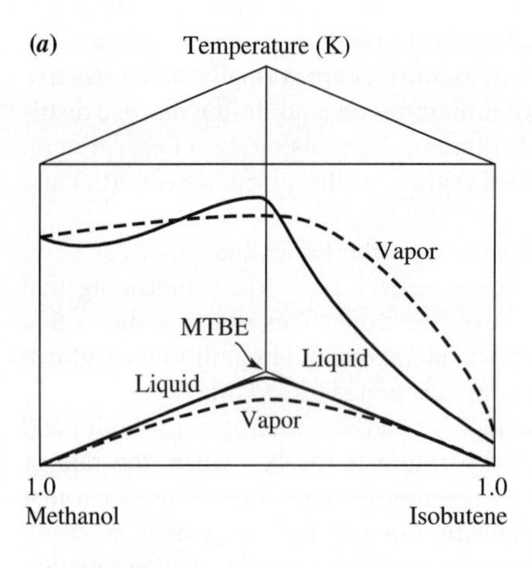

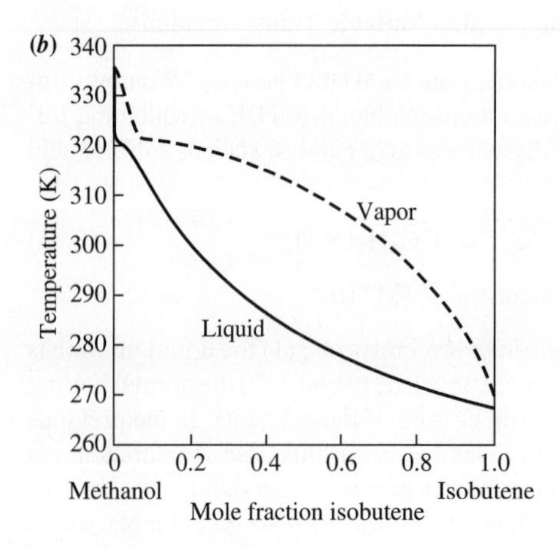

FIGURE 10.17
Reactive phase diagrams for the reaction methanol + isobutene $\rightleftharpoons$ MTBE at 1 atm pressure. (*a*) Temperature (T)-composition (isobutene-methanol-MTBE). (*b*) Orthogonal projection onto the temperature (T)-composition (isobutene, methanol) face.

Fig. 10.17*b*. It is easier to interpret these diagrams when they are plotted in a new composition coordinate system that is introduced in the next section.

Azeotropes in reactive mixtures correspond to solutions of Eq. 10.32. There are three possibilities: (1) nontrivial solutions where $x_i \neq y_i$. These correspond to reactive azeotropes created by the chemistry and not present in the mixture when reaction is extinguished. (2) Solutions where $x_i = y_i$. These correspond to ordinary azeotropes that survive the chemistry and are present whether reaction occurs or not. These types of azeotropes do not occur in reactions involving two or three components because they always react away, but they do occur in four-component reactive mixtures, see Exercise 9. (3) No solution, indicating that there are no azeotropes of any kind at phase and reaction equilibrium.

There are several ways of establishing whether Eq. 10.32 has solutions, including bifurcation/continuation methods (Venimadhavan et al., 1999*a*); (Okasinski and Doherty, 1997); global optimization methods (Harding and Floudas, 2000); and graphical methods (Frey and Stichlmair, 1999). Reactive azeotropes have been predicted to occur in a wide variety of ideal and nonideal mixtures. The first experimentally measured reactive azeotrope was reported to occur in the esterification of acetic acid with isopropanol at 1 atm pressure by Song et al. (1997).

For the special case of reactions occurring in ideal mixtures with constant relative volatilities and constant K_{eq}, reactive azeotropes can only occur when all reactants are either heavier or lighter than the products ("segregated α's") (Barbosa and Doherty, 1988*c*). This means, for example, that if the volatility of a product lies between the volatilities of the reactants ("mixed α's"), then reactive azeotropes will not occur (see Exercise 10). The formation of reactive azeotropes in systems with segregated α's depends on the value of the equilibrium constant and on the relative volatilities. The components in Example 10.4 have segregated α's (the reactants are both lighter than the product), and when $K_{eq} = 14$ a reactive azeotrope occurs, but when K_{eq} is smaller it disappears.[14] This is to be expected because as $K_{eq} \to 0$ the reaction equilibrium curve in the triangle approaches the *A*-*B* edge. Therefore, the ternary reactive phase diagram approaches the binary phase diagram for the nonreactive *A*-*B* mixture, which has no azeotropes.

Composition Variable Transformations

So far we have represented reactive phase diagrams using either the full space of mole fractions or the orthogonal projection onto an edge of the triangle. There is, however, a third way which is the best of all. The idea can be explained with reference to Fig. 10.18, which shows the vapor and liquid curves at phase and reaction equilibrium for the ideal mixture in Fig. 10.15. Tie-line 1 connects a liquid of composition $\mathbf{x}_1$ in phase and reaction equilibrium with a vapor of composition $\mathbf{y}_1$. The mole fractions of fresh feed *A* and *B* needed to produce the equilibrium liquid are found by following the stoichiometric line from $\mathbf{x}_1$ to the *A*-*B* edge. The resulting mole fractions X_A

[14]Barbosa and Doherty (1988*c*) computed the phase diagram at $K_{eq} = 1.3$, and there is no azeotrope.

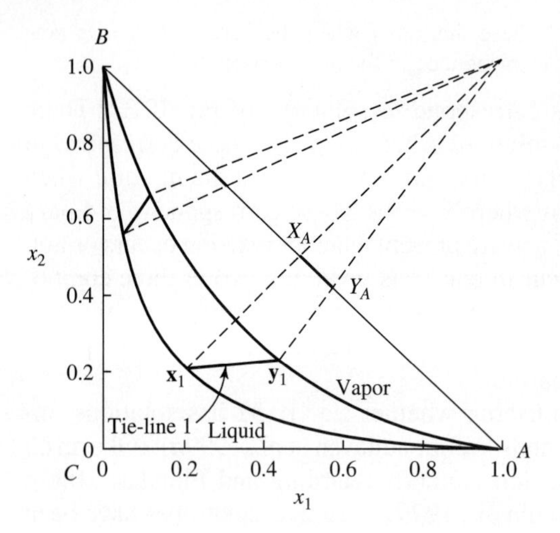

FIGURE 10.18
Vapor and liquid curves at phase and reaction equilibrium for the ideal mixture in Fig. 10.15. Two tie-lines are shown along with the stoichiometric lines and the reaction pole point.

and X_B are given by

$$X_A = \frac{x_A + x_C}{1 + x_C} \tag{10.34}$$

and

$$X_B = \frac{x_B + x_C}{1 + x_C} \tag{10.35}$$

These are not independent, since they sum to unity, so we need only one of them to define the mole fractions of A and B in the fresh feed. The mole fractions of A and B needed to produce $\mathbf{y}_1$ are found in a similar way, giving

$$Y_A = \frac{y_A + y_C}{1 + y_C} \tag{10.36}$$

and

$$Y_B = \frac{y_B + y_C}{1 + y_C} \tag{10.37}$$

These also sum to unity. Therefore, tie-line 1 in the mole fraction triangle can be accurately represented by the tie-line connecting X_A to Y_A, as shown on the figure. If we know the values of X_A and Y_A, we can get the values of $\mathbf{x}_1$ and $\mathbf{y}_1$ from Eqs. 10.34 and 10.36 together with the equations for phase and reaction equilibrium (Ung and Doherty, 1995*d*). Therefore, it is possible to go back and forth between tie-line 1 and its image under the composition transforms given by Eqs. 10.34 and 10.36. The entire phase diagram can be projected in this way to give the phase diagram shown in Fig. 10.19. This projection has two major advantages over using mole fraction coordinates: (1) The diagram has one independent composition, namely, X_A, which is in exact agreement with the phase rule for this chemistry. Therefore, the phase diagram has exactly the same number of dimensions as required by the phase rule. In mole fraction coordinates the number of dimensions of the space is larger than the dimension of the equilibrium surfaces. This mismatch in dimensions gets larger

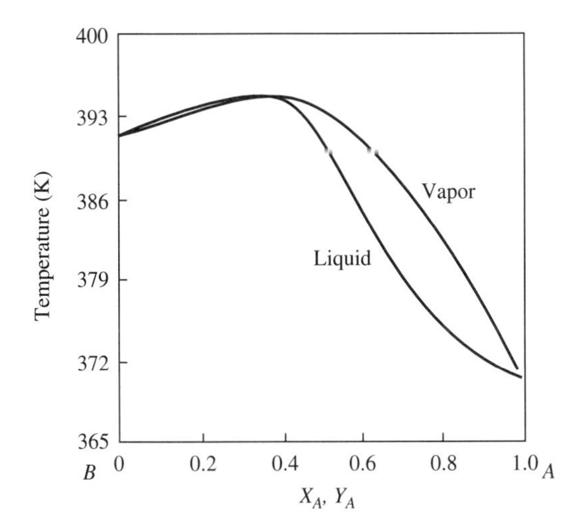

FIGURE 10.19
Phase diagram of temperature versus transformed compositions (X_A, Y_A) for the ideal mixture in Figs. 10.15 and 10.18.

as the number of reactions increases. There are many complex reaction systems that can be easily and faithfully represented with only one or two independent X_i's yet require four, five, or six mole fractions, thus preventing effective visualization of the phase behavior. (2) At the reactive azeotrope the liquid and vapor equilibrium curves touch each other at a stationary point, thus making the reactive phase diagram appear like any "normal" phase diagram. This implies that the reactive azeotrope corresponds to the condition

$$X_A = Y_A \tag{10.38}$$

Therefore, using the transformed compositions makes the shape of the phase diagram and the defining equation for an azeotrope have the same forms as we have seen already for nonreactive mixtures. In fact, the general theory of phase and reaction equilibrium in transformed compositions is essentially identical to the theory of phase equilibrium in nonreactive mixtures (Ung and Doherty, 1995*c*).

Equation 10.38 is readily derived from Eq. 10.32 using the definition of the transformed compositions. It also has a simple graphical interpretation shown in Fig. 10.20 (Frey and Stichlmair, 1999). This figure shows a series of tie-lines connecting the liquid and vapor equilibrium curves, together with a set of stoichiometric lines. Vapors in reaction equilibrium always lie on the vapor curve; liquids in reaction equilibrium always lie on the liquid curve. We can move along the phase and reaction equilibrium curves by imagining a series of steps involving first a separation step followed by an equilibrium reaction step.

Consider that we have a saturated vapor at point 1 which is in phase equilibrium with a liquid at point 2 (because the vapor is on the combined phase and reaction equilibrium vapor curve, a phase equilibrium calculation will produce a liquid which is on the liquid reaction equilibrium curve and the phases will be in combined phase and reaction equilibrium). Take the liquid at point 2 and totally vaporize it to its dew point. This vapor will react to achieve reaction equilibrium, and will do so by moving

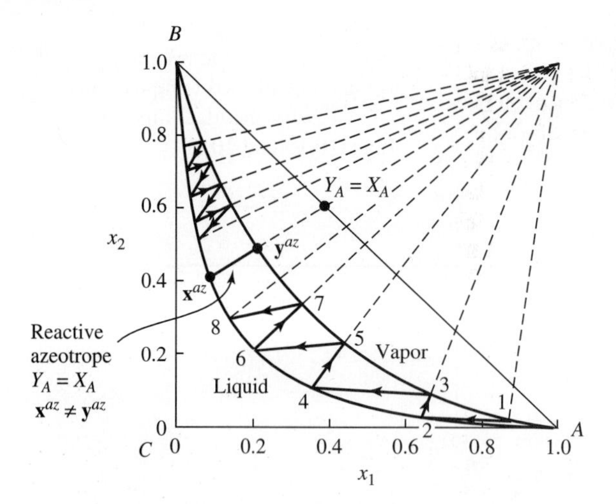

FIGURE 10.20
Graphical interpretation of a reactive azeotrope.

along the stoichiometric line through point 2.[15] This gives a vapor at point 3. A phase equilibrium calculation produces a liquid at point 4, and the process continues until a point is reached where the tie-line is collinear with the stoichiometric line. At this point it is not possible to move any further by these steps; we have reached the maximum-boiling reactive azeotrope. Given the interpretation of X_A and Y_A in Fig. 10.18, the azeotrope corresponds to the condition $X_A = Y_A$.

In order to implement this procedure we need to have tie-lines, but these are easy to get as the tangents to residue curves. We also need the stoichiometric lines, and these are easy to get by locating the reaction pole point and drawing straight lines from it. The final thing needed is the location of the reaction equilibrium curve, which is easy to locate at large and small values of K_{eq}. The position of the reaction equilibrium curve at intermediate values of K_{eq} is easy to estimate. *A reactive azeotrope occurs when a residue curve crosses the reaction equilibrium curve tangent to a stoichiometric line.* This construction is shown in Fig. 10.21 and is due to Frey and Stichlmair (1999); see Exercise 11.

The MTBE phase diagram (Fig. 10.17), can also be projected along stoichiometric lines and represented in transformed compositions, giving Fig. 10.22*a*. The corresponding Y_1-X_1 diagram is shown in Fig. 10.22*b*. These figures present the phase behavior in a much simpler way that is easier to interpret than the diagrams in mole fraction coordinates. For example, we see that an intermediate-boiling reactive azeotrope containing all three components occurs close to $X_1 = 0.5$. There is no ternary azeotrope in the nonreactive mixture. Moreover, both binary azeotropes in the nonreactive mixture have been eliminated from the equilibrium reactive mixture. The existence of the azeotrope is sensitive to the value of K_{eq} used in the calculations. Larger values of K_{eq} cause two reactive azeotropes to occur (one is a maximum-boiling, the other is minimum-boiling), whereas smaller values cause a single minimum-boiling reactive azeotrope to occur. The intermediate-boiling azeotrope shown in Fig. 10.22(*a*) occurs at a special value of K_{eq} when the maximum

---

[15] The vaporization raises the temperature, and the new equilibrium composition is at point 3.

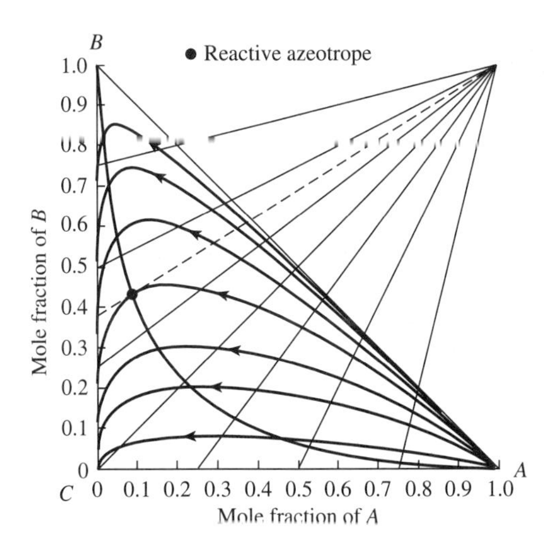

FIGURE 10.21
Graphical interpretation of a reactive azeotrope using the residue curve map.

and minimum azeotropes merge together; by coincidence this happens at a value of K_{eq} close to the value reported by Colombo et al. (1983). These results have been reported by Okasinski and Doherty (1997) using continuation methods. They can also be found using the graphical procedure of Frey and Stichlmair (1999); see Exercise 12.

An important implication of the behavior shown in Fig. 10.22*b* for distillation is the following. Feeds rich in isobutene give MTBE as the bottoms product and excess isobutene in the distillate. In contrast, feeds rich in methanol produce MTBE in the distillate with excess methanol in the bottoms.

Summary of the method

For a single chemical reaction of the form

$$\sum_{i=1}^{c} \nu_i A_i = 0 \tag{10.39}$$

We can define $c - 1$ transformed compositions in each phase

$$X_i = \frac{x_i - \frac{\nu_i}{\nu_k} x_k}{1 - \frac{\nu_T}{\nu_k} x_k} \qquad i = 1, \ldots, c \qquad i \neq k \tag{10.40}$$

$$Y_i = \frac{y_i - \frac{\nu_i}{\nu_k} y_k}{1 - \frac{\nu_T}{\nu_k} y_k} \qquad i = 1, \ldots, c \qquad i \neq k \tag{10.41}$$

where subscript k represents a reference component[16], equivalent to x_C in the transformed compositions developed for the A, B, C chemistry. A formal derivation of

[16]The reference component should be chosen so that the denominator in these transformed compositions is never zero. When $\nu_T = 0$ any component can be chosen, when $\nu_T > 0$ a reactant should be chosen, and when $\nu_T < 0$ a product should be chosen. It is no accident that x_C appears as the reference component in the A, B, C chemistry, Eqs. 10.34 to 10.37.

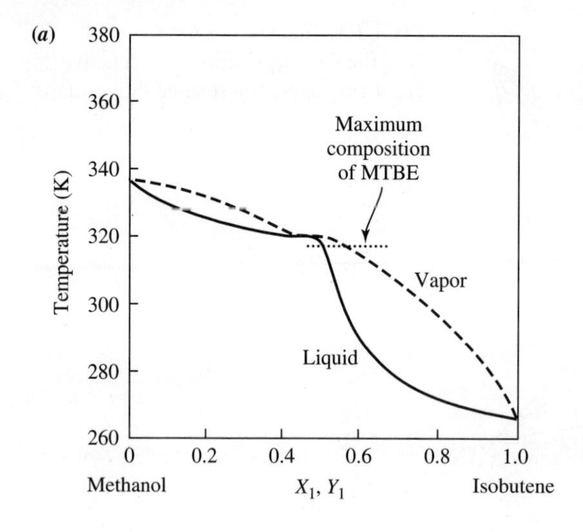

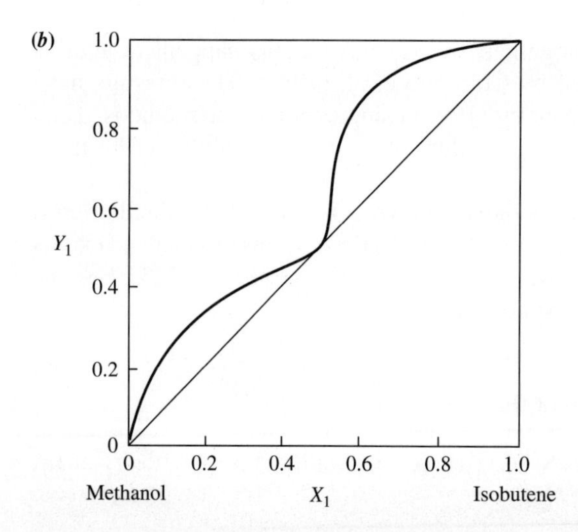

FIGURE 10.22
(*a*) Phase diagram of temperature versus transformed compositions (X_1, Y_1) for the MTBE mixture in Fig. 10.17. (*b*) Y_1-X_1 diagram. The component numbering is isobutene (1), methanol (2), MTBE (3).

these transforms is given by Ung and Doherty (1995*d*) and Ung and Doherty (1995*c*) from two different viewpoints. The transformed compositions have the following two properties:

$$\sum_{\substack{i=1 \\ i \neq k}}^{c} X_i = \sum_{\substack{i=1 \\ i \neq k}}^{c} Y_i = 1 \tag{10.42}$$

and

$$X_i(\varepsilon = 0) = X_i(\varepsilon) \tag{10.43}$$

where ε is the extent of reaction. Both properties are proved in Ung and Doherty (1995*d*). They also make intuitive sense because the transforms represent the fractional amounts of starting materials needed to produce a final composition after reaction has taken place (hence Eq. 10.42), and the transforms represent stoichiometric lines along which the given fractions of starting materials give rise to all allowable product compositions by chemical reaction (hence Eq. 10.43).

The conditions for a reactive azeotrope are given by

$$\frac{y_1 - x_1}{\nu_1 - \nu_T x_1} = \frac{y_i - x_i}{\nu_i - \nu_T x_i} \qquad i = 2, \ldots, c - 1 \tag{10.44}$$

which can be written as

$$Y_i = X_i \qquad i = 1, \ldots, c - 2 \tag{10.45}$$

Residue Curve Maps

The equations for simple distillation in the limit of reaction equilibrium are (Barbosa and Doherty, 1988*d*; Ung and Doherty, 1995*a*).

$$\frac{dX_i}{d\tau} = X_i - Y_i \qquad i = 1, \ldots, c - 1 \;\; i \neq k \tag{10.46}$$

where X_i and Y_i are related by phase and reaction equilibrium. The residue curves correspond to the paths followed by the liquid during evaporation of mixtures that are constrained to lie on the chemical equilibrium surface. The singular points of these equations occur when

$$Y_i = X_i \qquad i = 1, \ldots, c - 2 \tag{10.47}$$

or, equivalently, when Eq. 10.44 is satisfied. This happens at all of the following places:

1. Reactive azeotropes.
2. Pure components that lie on the reaction equilibrium curve or surface, e.g., for the $A + B \rightleftharpoons C$ chemistry these components are A and B, but not C.
3. Nonreactive azeotropes that survive the chemistry, i.e., nonreactive azeotropes that lie on the reaction equilibrium surface. For example, in the chemistry acetic acid + methanol $\rightleftharpoons$ methyl acetate + water, this includes the azeotrope between methanol and methyl acetate but not the azeotrope between methyl acetate and water.

Systems with one degree of freedom

The residue curve map for the ideal mixture in Example 10.4 is generated by solving one equation of the form Eq. 10.46, and the results are shown in Fig. 10.23. There is only one residue curve at chemical equilibrium, which is the chemical equilibrium curve itself. The maximum-boiling reactive azeotrope is a stable node that attracts the residue curve on either side. The residue curves and the reactive azeotrope could also be shown in transformed composition coordinates by projecting them along the stoichiometric lines onto the A-B edge.

At a smaller value of K_{eq}, the residue curve no longer approaches a maximum-boiling reactive azeotrope. Instead it starts at pure A (the low-boiling unstable node) and ends at pure B (the stable node); the azeotrope has disappeared (see Fig. 10.24).

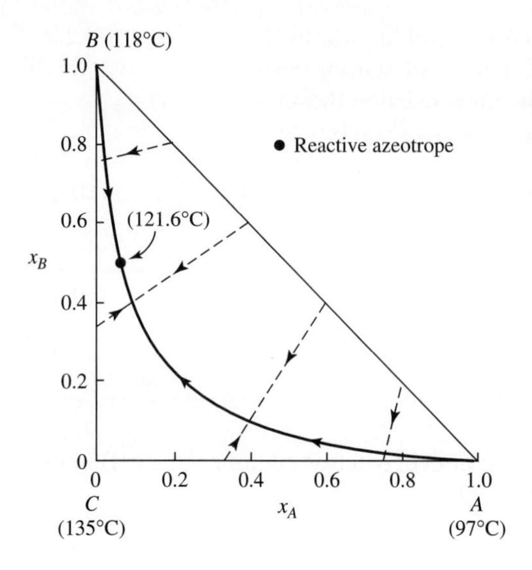

FIGURE 10.23
Residue curve map for the ideal mixture in Example 10.4 ($\Delta G^0 = -8.314$ kJ/mol, $K_{eq} \sim 14$). The solid line is the residue curve, the dashed lines are the stoichiometric lines of constant X_i.

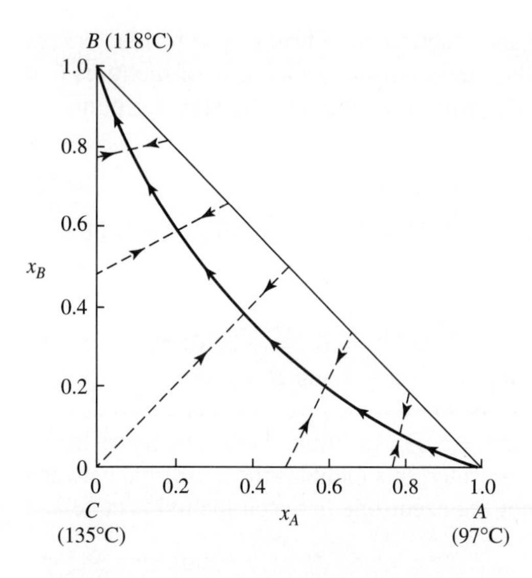

FIGURE 10.24
Residue curve map for the ideal mixture in Example 10.4 ($\Delta G^0 = -0.8314$ kJ/mol, $K_{eq} \sim 1.3$). The solid line is the residue curve; the dashed lines are the stoichiometric lines of constant X_i.

The residue curve map for the MTBE chemistry can also be computed from a single equation of the form Eq. 10.46, and the result is shown in Fig. 10.25.

Systems with two degrees of freedom

For systems with two degrees of freedom, there are two independent transformed compositions and two equations of the form Eq. 10.46. Solutions for the residue curves are most conveniently shown on two-dimensional figures with axes X_1 and X_2.

We begin by revisiting the MTBE chemistry, but now with n-butane present as an inert component (DeGarmo et al., 1992). For an isobaric mixture the phase rule gives

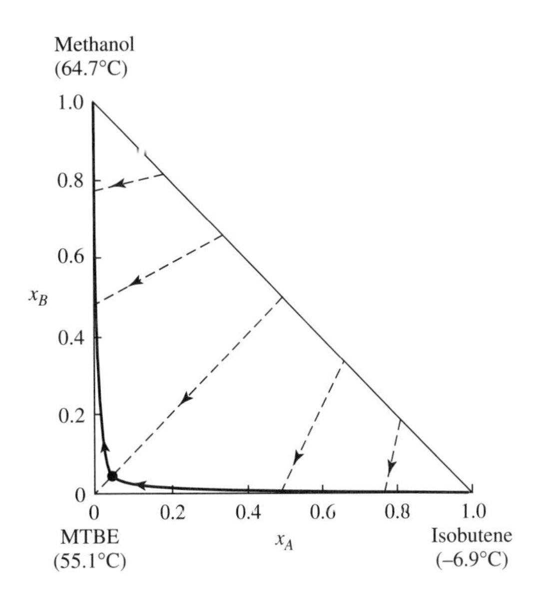

FIGURE 10.25
Residue curve map for the MTBE system in Example 10.5. The solid line is the residue curve; the dashed lines are the stoichiometric lines of constant X_i.

$\mathcal{F} = c - R - 1 = 4 - 1 - 1 = 2$. We have no choice for the reference component other than MTBE, which leads to the following transformed composition variables:

$$X_1 = \frac{x_1 + x_3}{1 + x_3} \tag{10.48}$$

$$X_2 = \frac{x_2 + x_3}{1 + x_3} \tag{10.49}$$

$$X_4 = \frac{x_4}{1 + x_3} \tag{10.50}$$

where the components are (1)isobutene, (2)methanol, (3)MTBE, (4)n-butane. Choosing X_1 and X_2 as the independent variables produces a triangular transformed composition space with isobutene, methanol, and n-butane at the three vertices. The residue curve map is shown in Fig. 10.26 at a pressure of 1 atm (Ung and Doherty, 1995*a*). Pure MTBE is not part of the solution space for the same reason that it is not on the reaction equilibrium curve in Fig. 10.25 or on the transformed composition line in Fig. 10.22. The left side of the triangle represents the binary nonreactive mixture of n-butane + methanol, and the base represents the binary nonreactive mixture of n-butane + isobutene. The hypotenuse represents the reactive ternary mixture isobutene + methanol $\rightleftharpoons$ MTBE. The interior of the triangle represents a four-component mixture at chemical equilibrium. The system has two azeotropes (two solutions of Eq. 10.47); one is the nonreactive azeotrope between methanol and n-butane; the other is the reactive azeotrope between isobutene + methanol + MTBE.[17] The mixture has two unstable nodes (one at the isobutene vertex, the other at the methanol + n-butane azeotrope), one stable node (methanol vertex), one saddle (in the n-butane corner), and a reactive azeotrope of mixed stability. This is a complex residue curve map that leads to some interesting reactive distillation configurations (Ung and Doherty, 1995*b*); see Exercise 13.

[17]This is the same azeotrope as in Fig. 10.22.

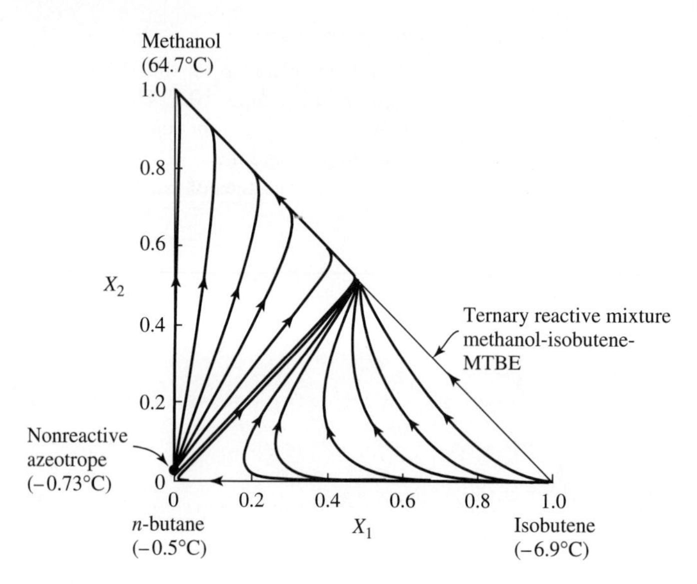

FIGURE 10.26
Residue curve map for the MTBE chemistry with *n*-butane as inert. $P = 1$ atm.

The next system with two degrees of freedom is the esterification of acetic acid with methanol to produce methyl acetate and water

$$\underset{\text{(HOAc)}}{CH_3COOH} + \underset{\text{(MeOH)}}{CH_3OH} \rightleftharpoons \underset{\text{(MeOAc)}}{CH_3COOCH_3} + H_2O \tag{10.51}$$

This reaction is carried out homogeneously with mineral acid catalysts (typically sulfuric acid), or heterogeneously over acidic ion exchange resins (e.g., Amberlyst 15). For this chemistry, $\nu_T = 0$, so we can choose any component as the reference for defining the transformed compositions. It is convenient to pick methyl acetate for the reference component together with acetic acid (A) and methanol (B) for the independent transformed compositions

$$X_A = x_{\text{HOAc}} + x_{\text{MeOAc}} \qquad X_B = x_{\text{MeOH}} + x_{\text{MeOAc}} \tag{10.52}$$

These variables have a simple physical interpretation; they represent the fraction of acetic acid and methanol in a feed mixture consisting of only those components that are required to achieve an equilibrium mixture of given composition. The residue curve map is defined on a square; see Fig. 10.27*a*.

The four corners of the square represent the four pure components;[18] the four edges represent the four binary nonreactive mixtures consisting of one reactant and one product, and the interior of the square represents a four-component mixture in phase and reaction equilibrium. The map has one azeotrope, which is the nonreactive

[18]For this chemistry, all four pure components are on the reaction equilibrium surface because they will not spontaneously react on their own. This is analogous to isobutene and methanol in the MTBE chemistry.

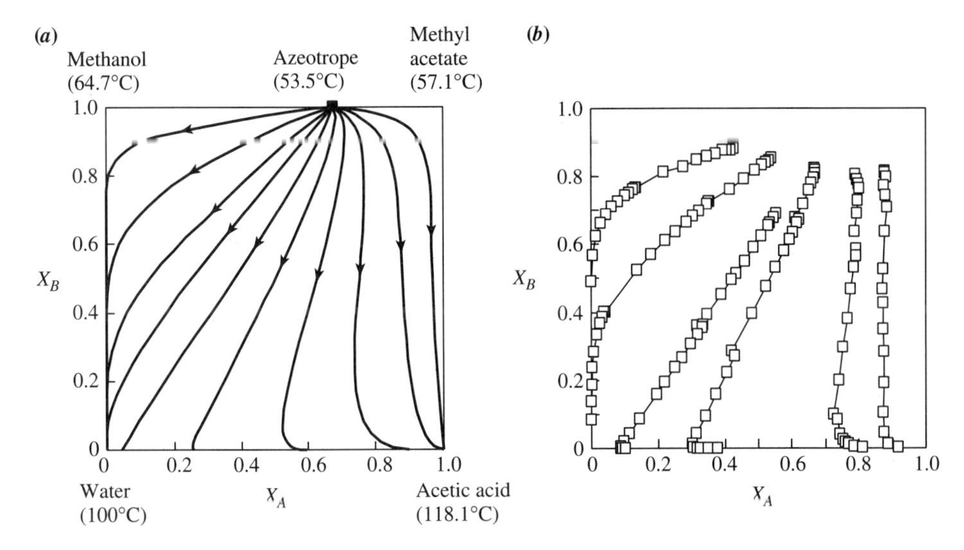

FIGURE 10.27
Residue curve maps for methyl acetate chemistry at $P = 1$ atm. (*a*) Calculated by Barbosa and Doherty (1988*a*). (*b*) Measured by Song et al. (1998).

azeotrope between methanol and methyl acetate. The azeotrope between methyl acetate and water reacts into a four-component equilibrium mixture when catalyst is added.

The residue curve map has a rather simple structure, consisting of one unstable node, one stable node, and three saddles. Figure 10.27*b* shows the residue curves measured in a series of simple distillation experiments. The experiments were designed to be close to phase and reaction equilibrium by using a large amount of catalyst and a low vapor rate, as discussed by Song et al. (1998).

The last example is another esterification of acetic acid, this time with isopropanol to make isopropyl acetate and water.

$$\underset{\text{(HOAc)}}{\mathrm{CH_3COOH}} + \underset{\text{(IPOH)}}{\mathrm{CH_3CH(OH)CH_3}} \rightleftharpoons \underset{\text{(IPOAc)}}{\mathrm{CH_3COOCH(CH_3)_2}} + \mathrm{H_2O} \quad (10.53)$$

This reaction is also carried out over acid catalysts. For this chemistry we pick isopropyl acetate for the reference component together with acetic acid (A) and methanol (B) for the independent transformed compositions

$$X_A = x_{\text{HOAc}} + x_{\text{IPOAc}} \qquad X_B = x_{\text{MeOH}} + x_{\text{IPOAc}} \quad (10.54)$$

The residue curve map is again defined on a square (Fig. 10.28). The map has two binary azeotropes, which are the the nonreactive azeotropes between isopropanol and isopropyl acetate, and isopropanol and water.[19] The most interesting feature in this map is the four-component minimum-boiling reactive azeotrope. The open circles

[19] As soon as the reaction is initiated by adding catalyst, the ternary azeotrope between isopropanol, isopropyl acetate, and water as well as the binary azeotrope between isopropyl acetate and water react into four-component mixtures.

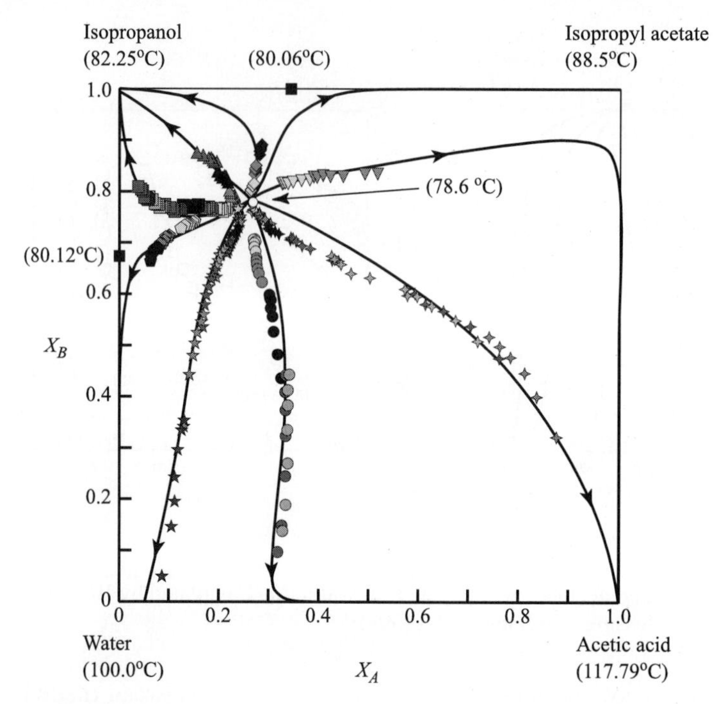

FIGURE 10.28
Residue curve map for isopropyl acetate chemistry at $P = 1$ atm. Solid lines are predicted from a model; the points are measured from simple distillation experiments (Song et al., 1997).

indicate the position of the reactive azeotrope as closely as it could be determined from a simple distillation experiment. This single run consists of five data points taken over a span of ~ 4 hours, during which time the compositions remained nearly constant and the boiling temperature varied by < 0.1°C. The measured molar liquid composition of the reactive azeotrope is 5.4% acetic acid, 56.5% isopropanol, 21.4% isopropyl acetate, and 16.7% water at a temperature of 78.6°C, with a corresponding vapor composition of 0.0% acetic acid, 49.1% isopropanol, 27.0% isopropyl acetate, and 23.9% water. The corresponding transformed compositions are $X_A = 0.268$, $X_B = 0.779$, $Y_A = 0.270$, $Y_B = 0.761$ (Song et al., 1997).

Topological rules for residue curve maps defined on composition squares have been devised by Doherty (1990).

Alternatives and Design Targets

When the design equations for equilibrium reactive distillation are written in terms of transformed compositions, they have a structure almost identical to the design equations for nonreactive distillation. This means that the methods already developed in earlier chapters can be applied to equilibrium reactive distillation with minor modification. This makes it possible to develop distillation systems using residue curve maps, calculate feasible product regions (by tracking pinches), minimum reflux (by pinch alignment), number of stages (by boundary value or initial value design methods), etc. One of the most useful results is that the lever rule applies to transformed compositions in spite of the fact that it does not apply to mole fractions for reactive distillation (Barbosa and Doherty, 1988*a*; Barbosa and Doherty, 1988*b*; Espinosa et al., 1995*a*; Espinosa et al., 1995*b*; Espinosa et al., 1999; Bessling et al., 1997*a*; Bessling et al., 1997*b*; Bessling, 2000).

The McCabe-Thiele design method can be developed for ternary equilibrium reactive distillation. Figure 10.29 shows its application to the MTBE chemistry. Isobutene-rich feeds give MTBE as a bottoms product and excess isobutene as a distillate (Fig. 10.29*a*), whereas methanol-rich feeds give MTBE as a distillate and excess methanol as a bottoms product. This unusual behavior is caused by the presence of the intermediate-boiling reactive azeotrope.

The feasible product composition regions for equilibrium reactive distillation to make methyl acetate can be calculated by extending the methods in Chaps. 4 and 5.

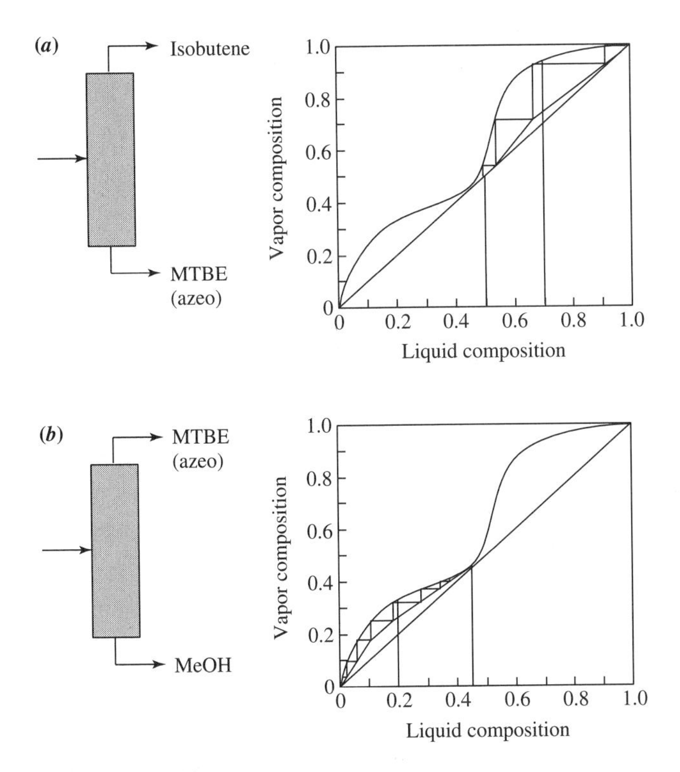

FIGURE 10.29
McCabe–Thiele diagrams for the equilibrium reactive distillation to make MTBE from isobutene and methanol at $P = 1$ atm. (*a*) Isobutene-rich feed. (*b*) Methanol-rich feed.

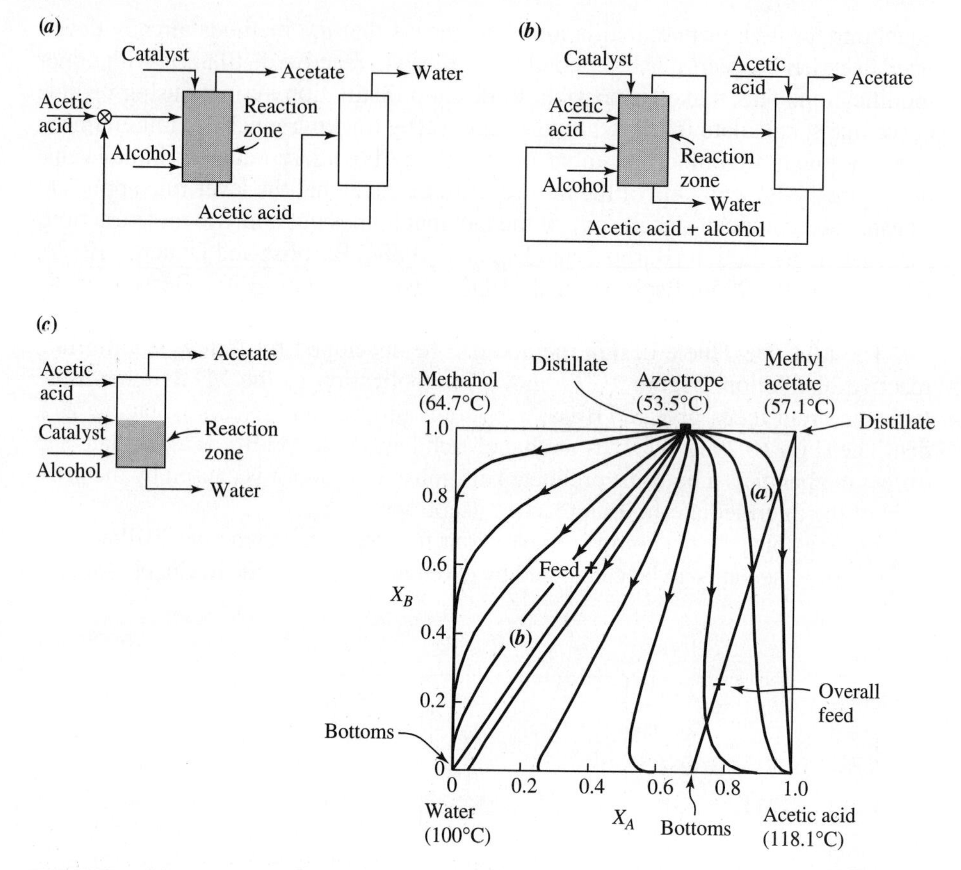

FIGURE 10.30
Process alternatives for making methyl acetate by reactive distillation. Shaded sections are reactive.

From this and the residue curve map in Fig. 10.27 we anticipate at least two process alternatives, shown in Fig. 10.30.

First alternative

In this alternative (marked *a* on the figure) we attempt to take methyl acetate as a distillate from the reactive column, and a mixture of acetic acid and water as bottoms (the overall feed composition must lie on the lever rule line for the column, as shown, and has excess acetic acid). This should be possible using a two-feed column similar to extractive distillation. From the residue curve map we expect the rectifying section composition profile to lie on the methyl acetate-acetic acid edge, the stripping profile to lie on the acetic acid-water edge and then turn at a saddle into the interior of the diagram, and the middle section profile to connect them together (similar to extractive distillation). A second separation task (which may be more than one unit, such as solvent extraction followed by distillation) separates the acetic acid from the water, as shown in flowsheet *a*.

Second alternative

In this alternative (marked *b* on the figure) we take the methyl acetate-methanol azeotrope (unstable node) as distillate from the reactive column and water as bottoms. The azeotrope is broken in a second column using, for example, extractive distillation. Our first choice of extractive agent is to try one of the components already in the process. Acetic acid is a feasible extractive agent (i.e., the residue curve map for methyl acetate, methanol, and acetic acid is the extractive map). The bottoms stream from the extractive column contains methanol and acetic acid. Since both are reactants, this stream is recycled to the reactive column, as shown in flowsheet *b*. We then attempt to match the internal flows in each of the columns, and use a 1:1 molar feed ratio of acetic acid to methanol in an attempt to combine the columns in flowsheet *b*, giving a single hybrid device shown later in flowsheet *c*.

These are some of the alternatives we can generate. Designs for the reactive columns in flowsheets *a* and *b* are shown later in Fig. 10.39.

10.5 KINETICALLY CONTROLLED REACTIVE DISTILLATION

Feasibility and Alternatives

Reactive flash cascades

In a nonreactive distillation column, the product purities can be estimated by a flash cascade arrangement as shown in Fig. 10.31. This sort of cocurrent cascade is discussed briefly in Sec. 2.6 as motivation for the countercurrent cascade, and continuous columns. Residue curve maps for nonreactive systems are used in earlier chapters to assess feasibility, but the cascade model turns out to be more convenient for reactive distillations, in our opinion. For nonreactive systems, the countercurrent and cocurrent cascade models differ mainly in the recovery of key components, which is much higher in the countercurrent cascade than in the cocurrent (Henley and Seader, 1981, Chap. 7.6). The countercurrent cascade is more complex to analyze, because the stages are coupled to one another, while the cocurrent flows are only in the downstream direction. In a feasibility analysis, the product compositions are sought, and it is convenient to use the simpler cocurrent flash cascades to study the feasibility of reactive distillation in continuous columns (Chadda et al., 2001).

In Fig. 10.31, imagine that each stage is a two-phase flash with a well-mixed liquid phase where the chemical reaction occurs. For simplicity we begin with a single reaction, an equimolar chemistry, a saturated liquid feed, and steady-state conditions. The cascade is isobaric, so the temperature changes from stage to stage according to the boiling point of the mixtures. In the rectifying cascade, vapor from each stage is partially condensed and fed to the next unit. This cascade is similar to the rectifying section of a continuous distillation column but without any liquid recycle. The opposite is done for the stripping cascade shown in the bottom half of Fig. 10.31. The liquid stream from each flash device is partially vaporized and

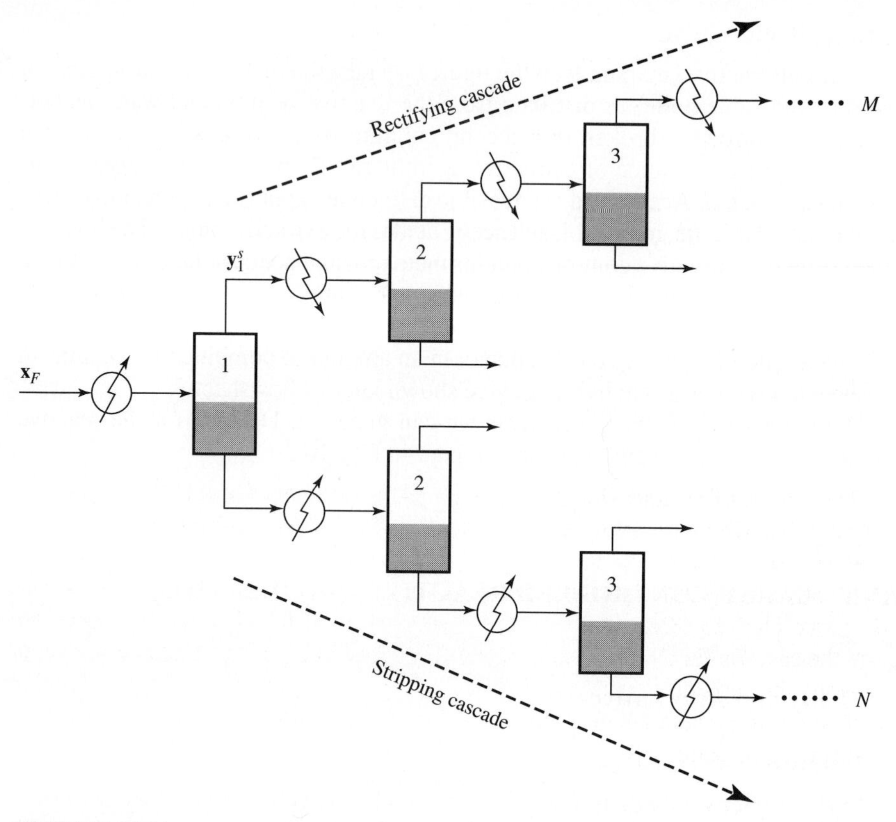

FIGURE 10.31
Cocurrent flash cascades with chemical reaction. The top half is the rectifying cascade and the bottom half is the stripping cascade.

sent as feed to the next unit in the series. The vapor compositions in the rectifying section and the liquid compositions in the stripping section are tracked to estimate the feasible product compositions for a continuous, countercurrent reactive distillation. As you might expect, this model is closely related to the residue curve maps, but in a more complex way than for nonreactive distillation.

The overall mass balance for the jth stage in the stripping cascade (Fig. 10.32) is

$$L_{j-1} = V_j + L_j \qquad (j = 1, \ldots, N) \tag{10.55}$$

The material balance for the ith component is

$$L_{j-1}x_{i,j-1} = V_j y_{i,j} + L_j x_{i,j} - \nu_i r(\mathbf{x}_j) H_j \qquad (i = 1, \ldots, c-1) \quad (j = 1, \ldots, N) \tag{10.56}$$

where r is the reaction rate given by Eq. 10.2. Eliminating L_j from Eq. 10.56 leads to

$$x_{i,j-1} - x_{i,j} = \phi_j(y_{i,j} - x_{i,j}) - \nu_i Da_j \mathcal{R}(\mathbf{x}_j) \qquad (i = 1, \ldots, c-1) \quad (j = 1, \ldots, N) \tag{10.57}$$

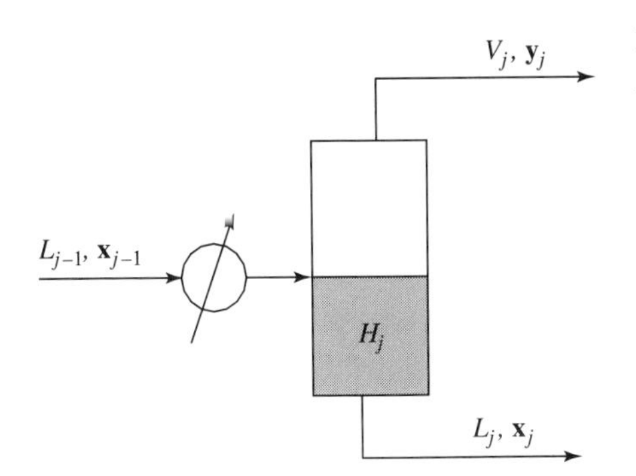

FIGURE 10.32
A schematic of the jth flash in the stripping cascade. L and V are the molar flows of liquid and vapor: H_j is the molar holdup in the stage.

where $\phi_j = \frac{V_j}{L_{j-1}}$ is the fraction of feed vaporized in the jth flash unit and $Da_j = \frac{H_j/L_{j-1}}{1/r_{0,j}}$.

The values of ϕ_j and Da_j are independent parameters for each flash unit. We study the case where the same fraction of feed is vaporized in each stage $\phi_1 = \phi_2 = \cdots = \phi_j = \cdots = \phi_N \equiv \phi$ and where each flash stage has the same residence time relative to the reaction time $Da_1 = Da_2 = \cdots = Da_j = \cdots = Da_N \equiv Da$. These two choices mean that the vapor rate and the liquid holdup both decrease along the cascade for a fixed feed flowrate. This implies a policy of decreasing vapor rate along the cascade similar to a decreasing vapor rate policy in simple distillation which keeps the instantaneous value of Da approximately constant. The model for the stripping cascade becomes

$$x_{i,j-1} = \phi y_{i,j} + (1-\phi)x_{i,j} - \nu_i Da\mathcal{R}(\mathbf{x}_j) \qquad (i = 1, \ldots, c-1) \qquad (j = 1, 2, \ldots, N) \tag{10.58}$$

where $\mathbf{x}_0 = \mathbf{x}_F$.[20] Equation 10.58 can be solved recursively for given values of ϕ and Da, starting with the initial condition $\mathbf{x}_0 = \mathbf{x}_F$. The solution is a trajectory of liquid compositions for the stripping cascade.

The analogous model for the rectifying cascade is

$$y_{i,j-1} = \phi y_{i,j} + (1-\phi)x_{i,j} - \nu_i Da\mathcal{R}(\mathbf{x}_j) \qquad (i = 1, \ldots, c-1) \qquad (j = 2, 3, \ldots, M) \tag{10.59}$$

where $\mathbf{y}_1 = \mathbf{y}_1^s$, and $\mathbf{y}_1^s$ is the vapor stream composition from the first flash device of the stripping cascade shown in Fig. 10.31. The solution to Eq. 10.59 is a trajectory of vapor phase compositions along the rectifying cascade.

For a given feed composition, Eqs. 10.58 and 10.59 are solved recursively for $N, M \to \infty$, until there is no change in successive iterates, i.e., until a stable fixed point is reached. The solutions depend on ϕ and Da, but the fixed points depend on a single quantity involving both parameters (see Eqs. 10.60 and 10.61 below). We choose $\phi = 0.5$ to find feasible product compositions under the following hypothesis.

[20] Equation 10.58 is a nonlinear, autonomous, implicit discrete dynamical system of the form $\mathbf{f}(\mathbf{x}_{j+1}, \mathbf{x}_j, \mathbf{p}) = 0$, where $\mathbf{x}$ is the vector of states and $\mathbf{p}$ is the vector of parameters. Such systems have known mathematical properties for fixed points, stability, etc. (Mira, 1987; Julka, 1995).

Hypothesis

The trajectories of the flash cascades lie in the feasible product regions for continuous reactive distillation.

Note that:

1. The flash trajectories do not provide the entire feasible product regions but generate a subset of the feasible compositions.
2. Selecting an iterate on the stripping cascade trajectory as a potential bottoms and an iterate on the rectifying cascade trajectory as a potential distillate does not imply that these products can be simultaneously obtained from a reactive distillation column, because these compositions may not satisfy the overall mass balance for the column. However, when the flash trajectories are used in conjunction with the *lever rule* for a continuous reactive column, feasible splits for continuous reactive distillation can be quickly predicted.

Fixed points of the flash cascades

Equations 10.58 and 10.59 have fixed points for $j \to \infty$. These points determine fundamental limits to the trajectories, and thus to the feasible product compositions, just as the singular points in residue curve maps determine feasible products in nonreactive distillation.

At a fixed point, successive liquid and vapor mole fractions reach constant values, and these compositions are in phase equilibrium with each other. The fixed points $\hat{\mathbf{x}}$ for the stripping cascade (from Eq. 10.58) are solutions of

$$(\hat{x}_i - \hat{y}_i) + \nu_i \left(\frac{Da}{\phi}\right) \mathcal{R}(\hat{\mathbf{x}}) = 0 \qquad (i = 1, \ldots, c-1) \qquad (10.60)$$

For the rectifying cascade, the fixed points $\hat{\mathbf{y}}$ (from Eq. 10.59) are solutions of

$$(\hat{x}_i - \hat{y}_i) - \nu_i \left(\frac{Da}{1-\phi}\right) \mathcal{R}(\hat{\mathbf{x}}) = 0 \qquad (i = 1, \ldots, c-1) \qquad (10.61)$$

Equation 10.61 has the same fixed points as Eq. 10.59 except that their stability is reversed.[21]

The results for the fixed point behavior of the flash cascades from Eqs. 10.60 and 10.61 can be organized into limiting cases as follows.

1. At $Da = 0$, the fixed point criteria for both the rectifying and stripping cascades reduce to

$$\hat{x}_i - \hat{y}_i = 0 \qquad (i = 1, \ldots, c-1) \qquad (10.62)$$

Equation 10.62 is identical to the fixed point criteria for simple distillation and also to that of a continuous column at total reflux and total reboil (see Chap. 5). In those cases, the fixed points are at all of the pure components and azeotropes. Since there is a symmetry in the rectifying and stripping maps for $Da = 0$, we can find the fixed points for *both* the rectifying and stripping cascades from the single Eq. 10.62. This recovers the criterion for fixed points in the well known limit of no reaction, as expected.

[21]This is an arbitrary choice, simply so that the stability agrees with approaches in earlier chapters for simple distillation without reaction, e.g., so that unstable nodes are distillates.

2. As $Da \to \infty$, the limit of chemical reaction equilibrium is approached and the fixed point criteria reduce to

$$(\nu_i - \nu_T x_i)\mathcal{R}(\hat{\mathbf{x}}) = 0 \qquad (10.63)$$

Equation 10.63 implies either that fixed points are at pole points or that they lie on the reaction equilibrium surface. Like all fixed points, there must also be a simultaneous vapor–liquid equilibrium between $\hat{\mathbf{x}}$ and $\hat{\mathbf{y}}$. The combined reaction and phase equilibrium can be written as (see Sec. 10.4)

$$\hat{X}_i - \hat{Y}_i = 0 \quad (i = 1, \ldots, c-2) \qquad (10.64)$$

Solutions of Eq. 10.64 are fixed points for a simple reactive distillation in the limit of chemical equilibrium and also for a continuous reactive distillation at total reflux and total reboil. As in the nonreactive case, the fixed point criteria for the rectifying and stripping cascades are the same, i.e., Eq. 10.64.

3. For $0 < Da < \infty$, solutions of Eqs. 10.60 and 10.61 give the fixed points for the stripping and rectifying cascades, respectively. Unlike the limits $Da = 0$ and $Da \to \infty$, there can be *different* fixed points for the rectifying and stripping cascades. Solutions of Eq. 10.60 at $\phi \to 1$ are the fixed points for simple reactive distillation, which were studied by Venimadhavan et al. (1999*a*). These fixed points provide information about the potential *bottoms* products, but not about the distillates. *Therefore, the distillate composition from a continuous column for $0 < Da < \infty$ cannot be inferred from a knowledge of fixed points of simple reactive distillation alone.* The fixed points for the rectifying section correspond instead to fixed points in simple *condensation* with reaction (Chadda, 2000). Some of the fixed points in the rectifying cascade will also be solutions for the stripping cascade, but others may be different. That is, condensation followed by chemical reaction does not necessarily result in the same liquid composition as reaction with vaporization, except under conditions of chemical equilibrium. Therefore, we estimate potential distillates from fixed points of the rectifying flash cascade, Eq. 10.61.

Generalized Cascade Model

Most of the development above presumes an equimolar chemical reaction, where the total number of moles is conserved, $\nu_T = 0$. If this is not the case, it is convenient to use parameters defined in terms of mass, in order to define a vapor mass fraction ϕ_m and Da_m in place of ϕ and Da. ϕ_m is bounded between 0 and 1 and independent of conversion. The resulting framework for the stripping cascade is derived by Nisoli et al. (1997), with the results below.

For the stripping cascade we replace Eq. 10.58 with

$$\begin{aligned}(x_{i,j} - x_{i,j-1}) = \phi_m \frac{M(\mathbf{x}_{j-1})}{M(\mathbf{y}_j)}(x_{i,j} - y_{i,j}) \\ + (\nu_i - \nu_T x_{i,j}) Da_m \frac{M(\mathbf{x}_{j-1})}{M(\mathbf{x}_j)}\mathcal{R}(\mathbf{x}_j) \qquad (i = 1, \ldots, c-1) \\ (j = 1, 2, \ldots, N) \qquad (10.65)\end{aligned}$$

where $\mathbf{x}_0 = \mathbf{x}_F$.

The corresponding model for the rectifying cascade, in place of Eq. 10.59, is

$$(x_{i,j} - y_{i,j-1}) = \phi_m \frac{M(\mathbf{y}_{j-1})}{M(\mathbf{y}_j)}(x_{i,j} - y_{i,j}) + (\nu_i - \nu_T x_{i,j}) Da_m \frac{M(\mathbf{y}_{j-1})}{M(\mathbf{x}_j)} \mathcal{R}(\mathbf{x}_j) \quad (i = 1, \ldots, c-1) \quad (j = 2, 3, \ldots, M) \tag{10.66}$$

where $\mathbf{y}_1 = \mathbf{y}_1^s$. $M(\mathbf{x})$ is the average molecular weight of the liquid.

$$M(\mathbf{x}) = \sum_{i=1}^{c} M_i x_i \tag{10.67}$$

A similar definition applies for the vapor $M(\mathbf{y})$.

For the stripping cascade, in place of Eq. 10.60, the fixed points in Eq. 10.65 are found from

$$\frac{M(\hat{\mathbf{x}})}{M(\hat{\mathbf{y}})}(\hat{x}_i - \hat{y}_i) + (\nu_i - \nu_T \hat{x}_i)\left(\frac{Da_m}{\phi_m}\right)\mathcal{R}(\hat{\mathbf{x}}) = 0 \quad (i = 1, \ldots, c-1) \tag{10.68}$$

For the rectifying cascade, in place of Eq. 10.61, the fixed points in Eq. 10.66 are found from

$$\frac{M(\hat{\mathbf{x}})}{M(\hat{\mathbf{y}})}(\hat{x}_i - \hat{y}_i) - (\nu_i - \nu_T \hat{x}_i)\left(\frac{Da_m}{1-\phi_m}\right)\mathcal{R}(\hat{\mathbf{x}}) = 0 \quad (i = 1, \ldots, c-1) \tag{10.69}$$

We are interested in investigating the fixed point branches of the flash cascade model for $0 \le Da \le \infty$ at $\phi = 0.5$. A systematic approach is by a bifurcation analysis of the solutions $\hat{\mathbf{x}}(Da)$ for Eqs. 10.60 and 10.61. The starting points for the analysis are the solutions at $Da = 0$, e.g., using the homotopy continuation method of Fidkowski et al. (1993). This relies on using the pseudo-arc-length continuation method in AUTO (Keller, 1977; Doedel, 1981). We use the mixture boiling point $T(\hat{\mathbf{x}}(Da))$ to represent the solution.

For determining the most important products, the unstable node branches for the rectifying cascade and the stable node branches from the stripping cascade are used to make a composite "feasibility diagram." We use the following rule for selecting feasible products for single-feed, fully reactive columns.

Rule for feasible products

Unstable node branches in the feasibility diagram represent potential distillates while the stable node branches represent the potential bottoms from a continuous reactive distillation column.

To illustrate the approach, we consider two systems used earlier in this chapter. In these, it is convenient to represent the results using $D = \frac{Da}{1+Da}$.

EXAMPLE 10.6. Bifurcations in Example 10.1. Figure 10.33 shows the fixed points in stripping and rectifying cascades from Eqs. 10.68 and 10.69, for the constant volatility mixture in Example 10.1. The unstable nodes from the rectifying cascade and the stable node from the stripping cascade make up the composite feasibility diagram in the lower portion of the figure. Above $D = 0.4$, $(Da = 0.67)$, two unstable nodes are present, corresponding to pure A and to pure B. One or the other of these is a possible distillate product, depending on the feed composition. The stable node branch is always a feasible

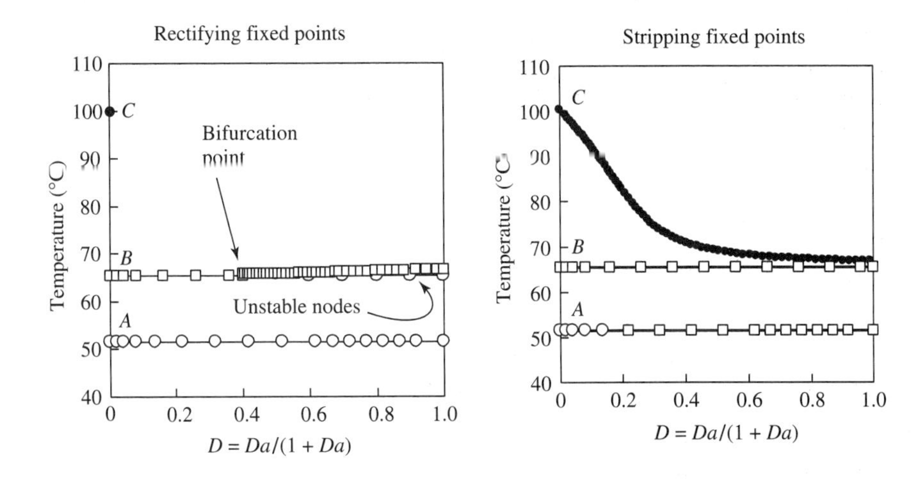

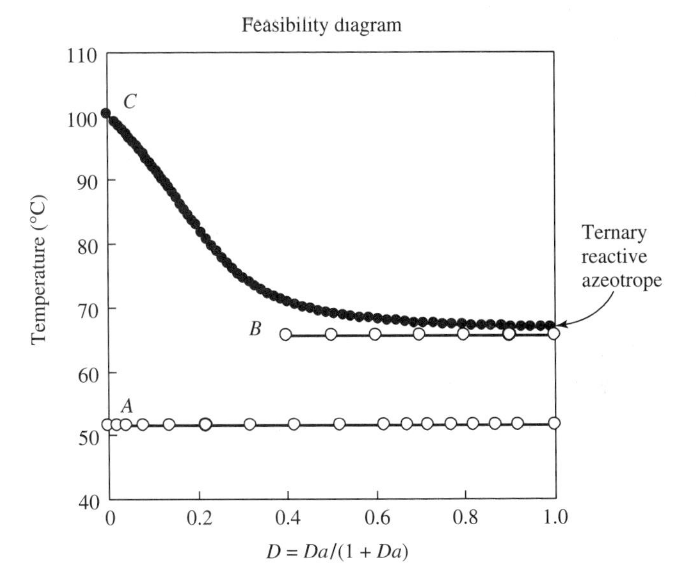

FIGURE 10.33
Fixed points for the rectifying (top left) and stripping (top right) cascades. The filled circles denote stable node branches, open circles denote unstable node branches, and the open squares denote saddle branches. The stable node branch in the stripping map and the unstable node branches in the rectifying map give the composite feasibility diagram (bottom).

bottoms product, while C is not. Based on this diagram, it is not surprising that the fully reactive columns in Example 10.1 did not give pure C as a bottoms product.

EXAMPLE 10.7. Isopropyl Acetate. We also consider the isopropyl acetate system discussed in Sec. 10.4. Figure 10.34 shows the fixed point branches for the rectifying and stripping cascades. The left edge of the diagrams $Da = 0$, $(D = 0)$ represents the limit of no reaction. Here, there is a minimum-boiling ternary azeotrope containing isopropanol,

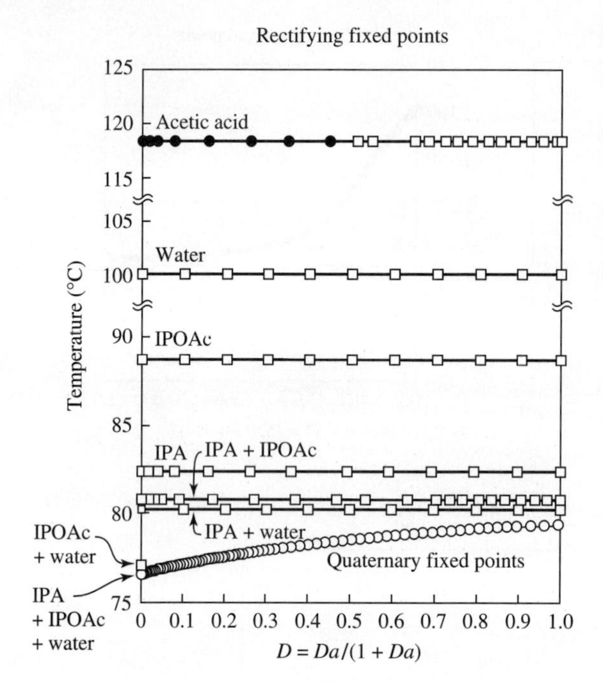

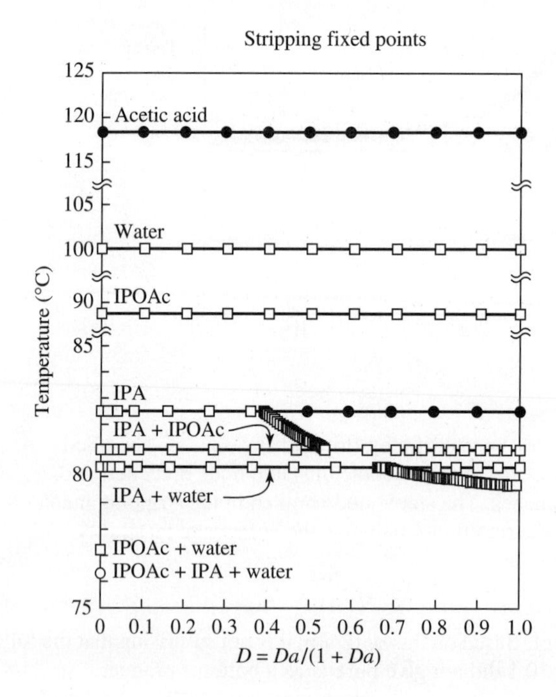

FIGURE 10.34
Bifurcation diagrams of the rectifying cascade (top), stripping cascade (bottom), and the composite feasibility diagram (continued on next page) describing isopropyl acetate at 1 atm pressure. The filled circles denote stable node branches, open circles denote unstable node branches, and the open squares denote saddle branches.

isopropyl acetate, and water (an unstable node). There are also six intermediate boiling fixed points (all saddles), and acetic acid is the heaviest species (a stable node). Starting from these initial conditions, fixed point branches are tracked for both the stripping and rectifying cascades, using Eqs. 10.60 and 10.61, respectively.

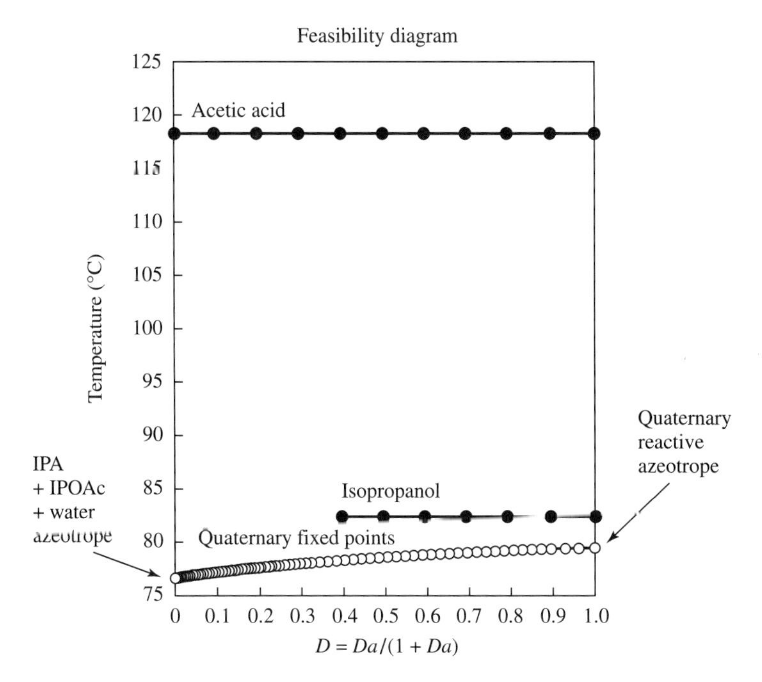

FIGURE 10.34 ***(Continued)***
Bifurcation diagrams of the rectifying cascade (top of last page), stripping cascade (bottom of last page), and the composite feasibility diagram (above) describing isopropyl acetate at 1 atm pressure. The filled circles denote stable node branches, open circles denote unstable node branches, and the open squares denote saddle branches.

The unstable node branches in the rectifying cascade and the stable node branches in the stripping cascade are shown in the feasibility diagram (Fig. 10.34). For $0 \leq D \leq 0.395$, $(0 \leq Da \leq 0.653)$, it is possible to obtain acetic acid as bottoms from a continuous reactive distillation. However, for $0.395 \leq D \leq 1$, $(0.653 < Da < \infty)$ either isopropanol or acetic acid can be obtained as the bottoms product depending on the feed composition. The potential distillates are all quaternary mixtures, due to the unstable node from the rectifying cascade, which is present at all values of Da. Thus, different structures are feasible for different ranges of the Damköhler number.

Note that any given structure may or may not be feasible as the reaction rate or residence time is changed, so that the feasibility of a given separation may depend on production rates, catalyst levels, and liquid holdup.

Hybrid cascades

It is a simple matter to explore feasible products for hybrid columns, by including nonreactive flash stages in the cascade models for both the stripping and rectifying sections. These building blocks can be used to generate alternatives, like the one shown at the end of Example 10.1.

For instance, the reactive stripping and rectifying cascade profiles for $Da =$ 0.277 are shown in Fig. 10.35. The fixed point in the rectifying cascade is at pure A, which is a feasible distillate. In the stripping profile, the stable node in Eq. 10.65

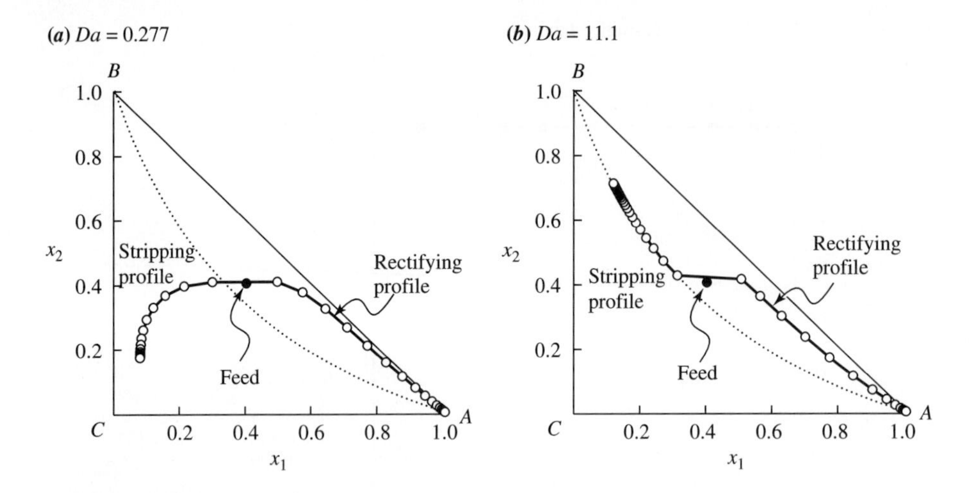

FIGURE 10.35
Stripping and rectifying profiles for the reactive cascade, for the system in Example 10.1 at (*a*) $Da = 0.277$ and (*b*) $Da = 11.1$. The feed composition is •.

is at the point $x_A \approx 0.08$, and $x_B \approx 0.17$. At a larger value of Da, the stripping profile moves closer to the reaction equilibrium curve, and the node is very close to the reactive azeotrope, (see Fig. 10.35(*b*)). The stripping nodes compare with the stable nodes in the residue curve maps in Fig. 10.8 and to the results in Figs. 10.9 and 10.33.

From the feasibility diagram in Fig. 10.33, we can also see that C is a stable node at $Da = 0$. Therefore, pure C is a potential bottoms product from a single-feed, hybrid flash cascade that has a *nonreactive* stripping section that follows a reactive section. The non-reactive section begins at or near the stable node in the reactive section. This leads to designs similar to the hybrid design found for Example 10.1, but with an improved conversion and much lower energy use, e.g., Fig. 10.36.

Design Targets for Kinetically Controlled Columns

The amount of catalyst or holdup and the temperature (controlled by the pressure) for reactive distillation is decided by economic trade-offs. To estimate the corresponding optimal values of Da, the distribution of catalyst and holdup throughout a column, as well as the traditional variables, such as reflux ratio, etc., design methods are necessary.

Geometric design methods that account for finite rates ($Da < \infty$) have been fully developed for two- and three-component reactive mixtures.

For binary reactive mixtures, e.g., $A \rightleftharpoons B$, difference-point methods (Lee et al., 2000*b*) have been developed for deciding the placement of catalyst in a column, the extent of reaction on each stage, the minimum reflux, the number of stages at finite reflux, etc. This is a generalization of the McCabe-Thiele method for systems with

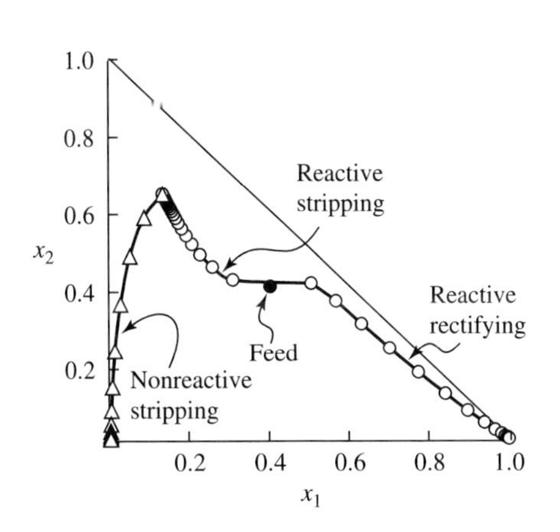

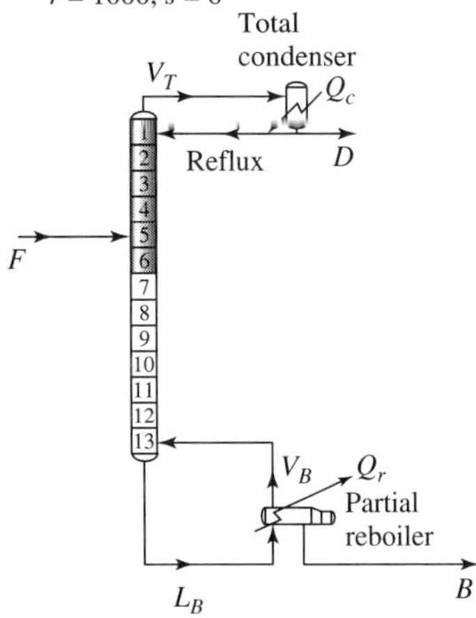

FIGURE 10.36
An improved hybrid design for Example 10.1. The feed composition is •, (*a*) shows the hybrid flash cascade profiles and (*b*) summarizes the column design.

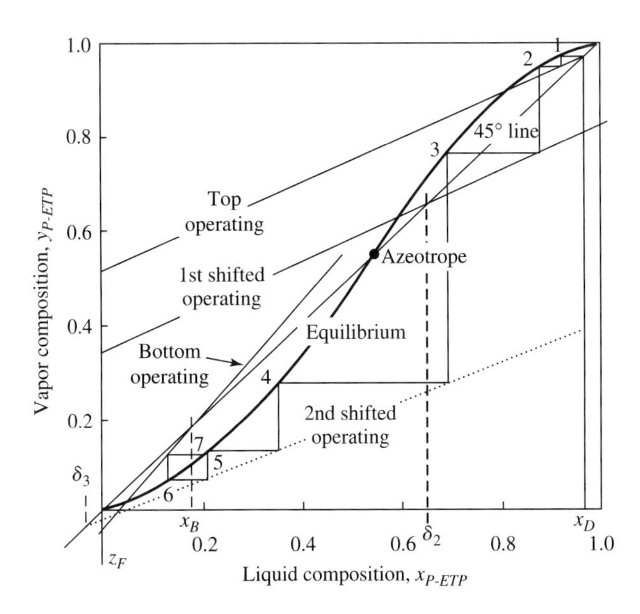

FIGURE 10.37
Reactive distillation for the isomerization of 2-phenyl ethanol to *p*-ethylphenol at 1 atm pressure. High-purity products and high conversion are obtained with seven stages and the proper placement of extent of reaction, even though the mixture has a binary nonreactive azeotrope (Lee et al., 2000*a*, Fig. 2).

chemical reaction. The key new idea in this method is to adjust the extent of reaction on each stage to develop more effective designs. Figure 10.37 shows an example from Lee et al. (2000*a*), where the approach leads to a design that effectively breaks a binary azeotrope by proper placement of reaction.

For ternary mixtures, feasible product compositions can be found prior to design using fixed point methods (Chadda et al., 2000). A surprising result is that the fixed

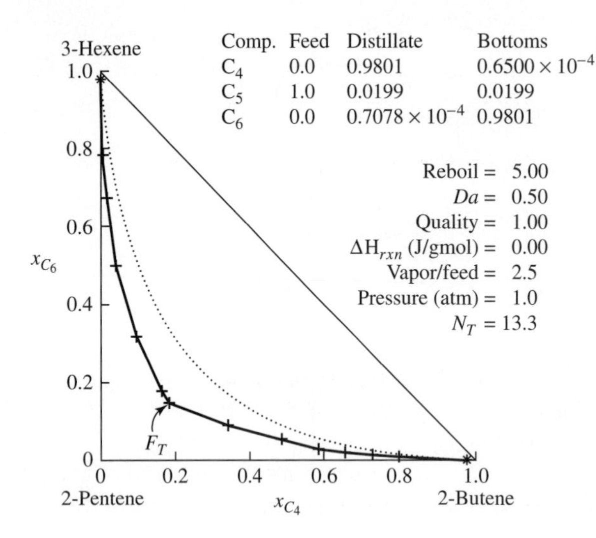

FIGURE 10.38
Reactive distillation for the olefin metathesis $2C_5H_{10} \rightleftharpoons C_4H_8 + C_6H_{12}$ at 1 atm pressure. Column design at $Da = 0.5$ (per stage), $s = 5$, gives 13.3 reactive stages and high-purity products.

points in column profiles always lie on the chemical equilibrium curve, even though $Da < \infty$. For finding the number of stages, reflux, etc., Buzad and Doherty (1994), Buzad and Doherty (1995), Okasinski and Doherty (1998), and Melles et al. (2000) have developed methods similar to the boundary value design procedure (Chaps. 4 and 5), modified to include generation or consumption of species on each stage. Figure 10.38 shows an example for olefin metathesis.

For mixtures with more components, equation-based methods for simulation are useful, though not a satisfactory substitute for a complete design method. For instance, it is difficult to find the minimum reflux using simulation. It is useful, though, to find the equilibrium design and then study successive simulations as Da is decreased in small increments (Chen et al., 2000). Results from this method are discussed in the next section.

Another possibility is to rely on mathematical programming to find solutions, either fully or partially optimized. This typically requires great care in model formulation and solution, at least for a general method, since feasibility constraints and model sensitivities can pose major difficulties in model convergence. For example, Ciric and Gu (1994) used an MINLP approach to find equipment sizes and feed addition policies for ethylene glycol synthesis via reactive distillation. Cardoso et al. (2000) used a simulated annealing approach for the same application. At the moment, these methods have the upper hand for finding designs of realistic complexity, though they do not provide much insight compared to geometric or difference-point methods.

10.6
CASE STUDY: METHYL ACETATE SYNTHESIS

As an example of the methods described above, we consider the reactive distillation of acetic acid with methanol to produce methyl acetate by the reaction

$$CH_3COOH + CH_3OH \rightleftharpoons CH_3COOCH_3 + H_2O$$

A pseudo-homogeneous rate model is

$$r = k_f \left(a_{HOAc} a_{MeOH} - \frac{a_{MeOAc} a_{H_2O}}{K_{eq}} \right)$$

The reaction equilibrium constant (Song et al., 1998) and rate constant are

$$K_{eq} = 2.32 \exp(782.98/T)$$

$$k_f = 9.732 \times 10^8 \exp(-6{,}287.7/T)\, \mathrm{h}^{-1}$$

where T is in Kelvin. The rate constant was obtained by fitting the pseudo-homogeneous rate equation to predictions from a more complex heterogeneous rate model (Song et al., 1998). The normal boiling point of MeOAc is chosen as the reference temperature for the calculation of k_0, giving a value of 5.1937 h^{-1}. The heat of reaction is −3.0165 kJ/mol, slightly exothermic.

The liquid phase activity coefficients are well represented by the Wilson equation with the parameters listed in Table 10.4 (from Song et al., 1998 with minor corrections).

Equilibrium Design

First, we consider equilibrium designs using the methods and transformed compositions developed by Barbosa and Doherty (1988*c*) and Ung and Doherty (1995*d*) and discussed in Sec. 10.4. The transformed compositions are

$$X_A = x_{HOAc} + x_{MeOAc}$$

and

$$X_B = x_{MeOH} + x_{MeOAc}$$

where X_A is the fractional molar composition of acetate groups and X_B is the fractional molar composition of alcohol groups in the mixture.

TABLE 10.4
Phase equilibrium parameters for methyl acetate reactive distillation

Vapor pressures P^{sat} in Pa from $\ln P^{sat} = A + \frac{B}{T+C}$, with T in K

	HOAc	MeOH	MeOAc	H_2O	DME
A	22.1001	23.4999	21.1520	23.2256	21.2303
B	−3,654.62	−3,643.3136	−2,662.78	−3,835.18	−2,164.85
C	−45.392	−33.434	−53.460	−45.343	−25.344

Binary interaction parameters A_{ij} [cal/gmol] for the Wilson equation as described in Chapter 2. Dimerization of acetic acid in the vapor phase is also included.

	HOAc	MeOH	MeOAc	H_2O	DME
HOAc	0	2,535.2019	1,123.1444	237.5248	−96.7798
MeOH	−547.5248	0	813.1843	107.3832	900.9358
MeOAc	−696.5031	−31.1932	0	645.7225	−17.2412
H_2O	658.0266	469.5509	1,918.232	0	703.3566
DME	96.7797	−418.6490	−21.2317	522.2653	0

Figure 10.39 shows composition profiles for three equilibrium reactive column designs. The first design, shown by the open diamonds, is similar to the example presented by Barbosa and Doherty (1988*c*). This column produces methyl acetate as a distillate but uses an excess of acetic acid and therefore has a bottoms stream of water and acetic acid. The second design, shown by the open squares, is a conceptual starting point for the Eastman column (Agreda and Partin, 1984). The third design is an equilibrium version of the Eastman column, where acetic acid is fed in at the top of the column and methanol at the bottom. We find a nonreactive section separating methyl acetate and acetic acid. This can be seen from the rectifying profile in Fig. 10.39, which is confined to the nonreactive edge between methyl acetate and acetic acid.

Figure 10.40 shows the effect of reflux ratio on this column design. The side figures show multiple possible middle trajectories from the bottom section of the column at 1 atm. The top and bottom boxes in part (*b*) show what happens when the reflux ratio is too low and too high, respectively. The middle box shows the best middle profile to minimize the total number of trays in the column. A similar effect has been shown experimentally by Bessling et al. (1998). Increasing the pressure increases the minimum number of trays required and reduces the range of feasible reflux ratios. For a column with a fixed number of stages, the product purities can decrease as the reflux ratio (or reboil ratio) is either increased or decreased from its design value. Similar results are found in other distillations, both reactive and nonreactive, that show both a minimum and maximum reflux (Knapp and Doherty, 1994; Okasinski and Doherty, 1998).

For the final equilibrium design, we specify a methyl acetate mole fraction of 0.985 in the distillate, and a water mole fraction of 0.985 in the bottoms. The

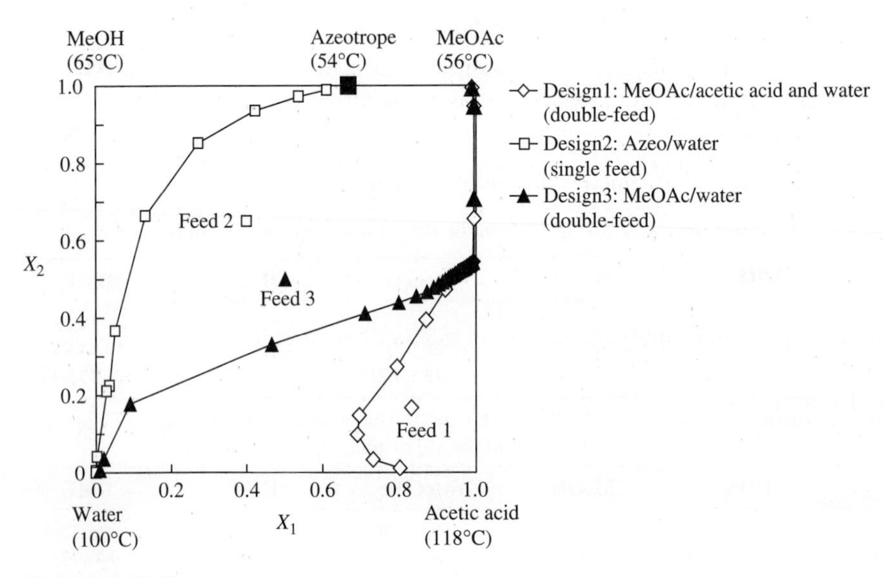

FIGURE 10.39
Equilibrium designs.

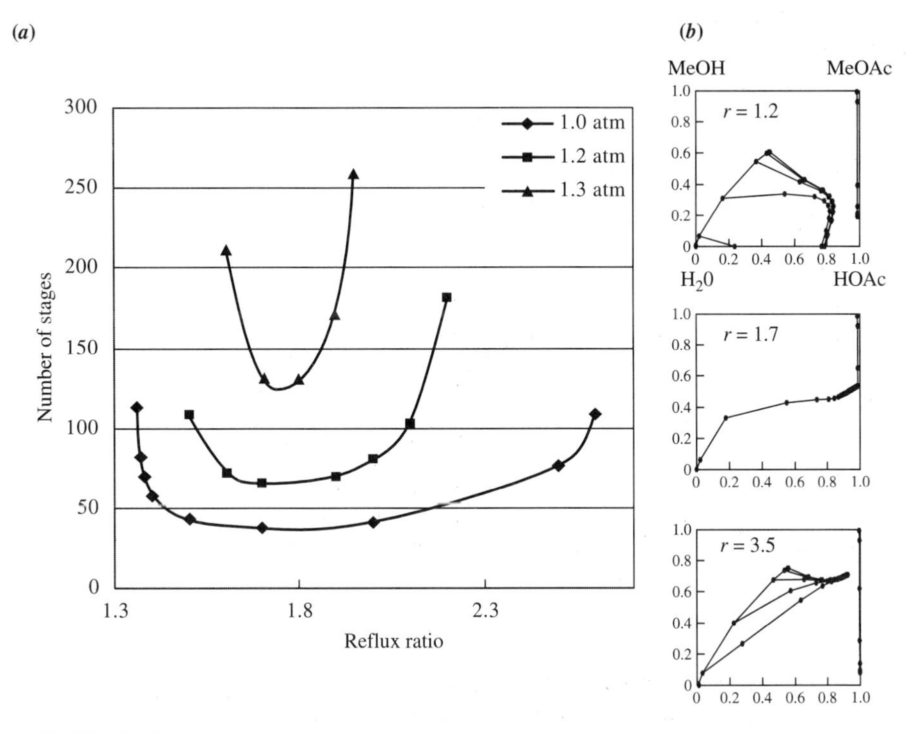

FIGURE 10.40
Effect of reflux ratio on equilibrium design. Part (*a*) shows the number of trays required as a function of reflux ratio for three pressures. Part (*b*) shows the column trajectories for three different values of the reflux ratio at a pressure of 1 atm.

reflux ratio that corresponds to the minimum number of reactive stages is $r = 1.7$, which is chosen as the design value. The corresponding reboil ratio is $s = 2.7$. The equilibrium design results are summarized in Fig. 10.41. The feed rates are based on the published production rate of 400 million lb of MeOAc/year ($\sim$ 280.0 kmol/h) (Agreda et al., 1990).

Kinetic Effects

A model for column simulation including kinetics can be formulated as a system of stage-to-stage ordinary differential equations, which are integrated to steady state. This formulation is a stable and robust method for finding steady-state solutions (Chen et al., 2000). We refer to two configurations of this model, with and without heat effects. The heat effects model considers heat of reaction and variable heats of vaporization.

Before trying to design away from the equilibrium limit, we calculate the feasibility diagram for the stripping and rectifying cascades, as described in Sec. 10.5. It shows no bifurcations, so we anticipate that a column designed in the equilibrium limit can remain feasible for finite Da. The results are shown in Fig. 10.42. The

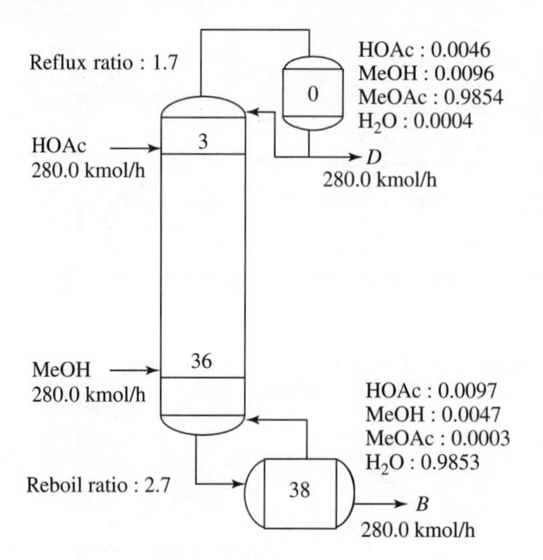

FIGURE 10.41
Equilibrium design results at 1 atm pressure. Compositions are reported in mole fractions.

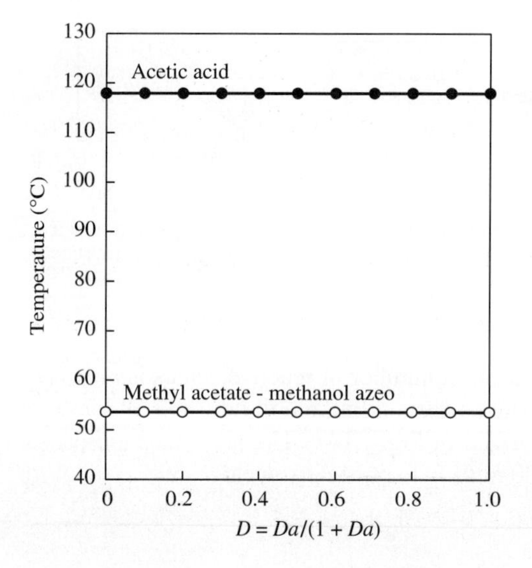

FIGURE 10.42
Feasibility diagram for methyl acetate synthesis at 1 atm pressure.

equilibrium design and a simulation at $Da = 100$ are compared in Fig. 10.43. This shows that the kinetic model approaches the equilibrium design for high Da.

For large Da, simulations with increasing or decreasing reboil ratio, and a constant reflux ratio of 1.7, show two different branches of steady-state solutions. The resulting column profiles for $s = 2.7$ are shown in Fig. 10.44. In simulations, the initial estimates for composition and temperature profiles decide whether a steady-state simulation will converge to a high-conversion or a low-conversion solution. This effect was first reported by Bessling et al. (1998), who also confirmed the result experimentally. The implications of these multiple steady states for column design and operation are quite significant.

The conversion of acetic acid in the column and the average volumetric holdups on each stage are shown in Fig. 10.45. The average molar holdup on each stage is

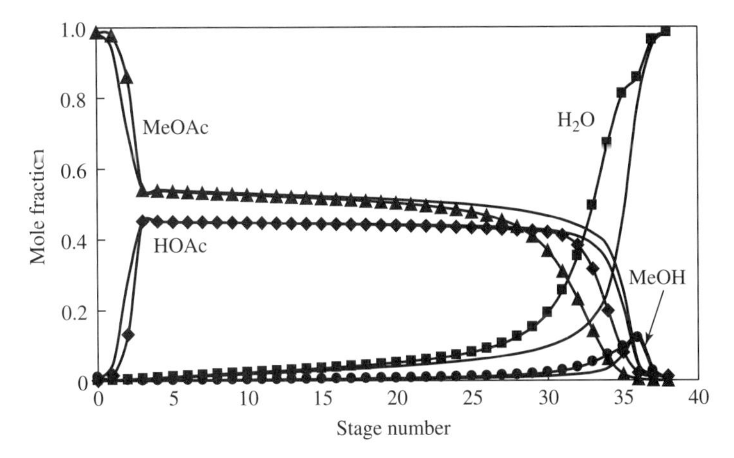

FIGURE 10.43
Methyl acetate column. Comparison of column profiles from the simulation at $Da = 100$ (symbols) and equilibrium design (solid lines).

converted to a volume holdup. Figure 10.45 shows that the column at $Da \approx 20$ is close to a realistic operation because the purities of both products are close to the design specifications, and the average volume holdup on each stage ($\sim$3 m^3) for the production rate of 400 million lb of MeOAc/year is reasonable. Therefore, we start with Da of 20 to do further design and simulation.

We summarize several steps described in more detail by Huss et al. (1999). In these steps, the top section of the column is nonreactive, because in the design we find that this section does not perform much reaction. This also reflects the conceptual diagram of the Eastman column (Agreda et al., 1990). Since constant volume holdups are more practical, we repeat the simulation with constant volume rather than constant molar holdups. Using the configuration from Fig. 10.46, the simulation without heat effects shows that this column produces a distillate containing 98.42 mol % MeOAc and a bottoms containing 97.11 mol % H_2O, which does not reach the design specifications. (In this case, we have specified constant volumes on the reactive stages and then calculated the molar holdups to find a value of $Da = 17.5$.) By adding three reactive stages above the MeOH feed and two below, and increasing the reboil ratio from 2.7 to 2.73, the simulation results show both products meeting or exceeding the design specification (Fig. 10.47). Heat effects have no significant impact on the product purities.

Figure 10.48 shows the effects of reflux ratio on product purities from simulations (including heat effects) of the column described in Fig. 10.47. For these, $D = 281.547$ kmol/h (this implies D/B is constant). The purities in both products exceed the design specification at $r = 1.9$ for reactive stage holdups of 3 m^3. The purities and conversions show a maximum as the reflux ratio changes, and these persist over a wide range of Da. As expected, higher conversions are obtained for high Da, but the gain in maximum conversion is small above $Da \approx 22$. As in the equilibrium design, decreasing or increasing r from its optimal value decreases the product purity.

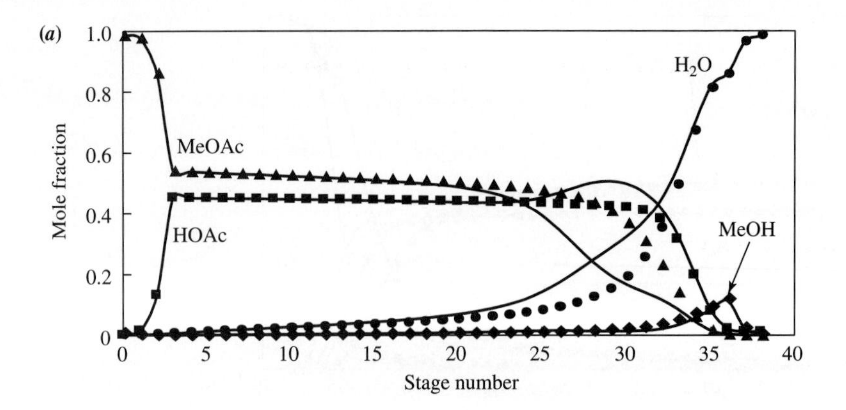

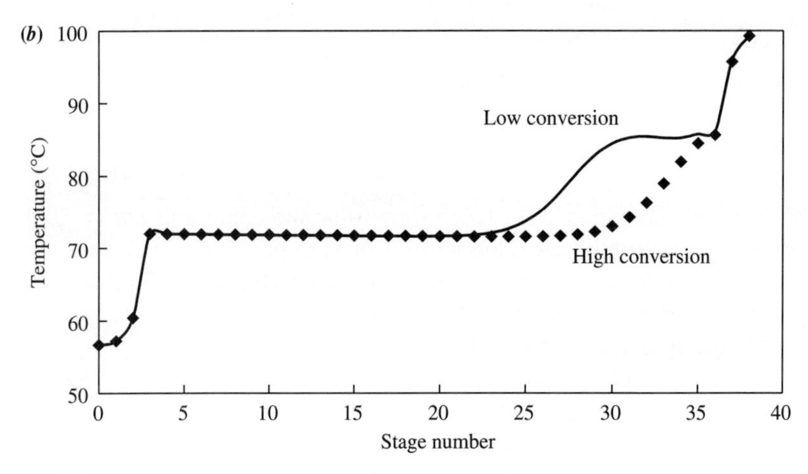

FIGURE 10.44
Multiple steady states in column profiles at $r = 1.7$, $s = 2.7$, $Da = 100$. (*a*) Composition profiles. (*b*) Temperature profiles.

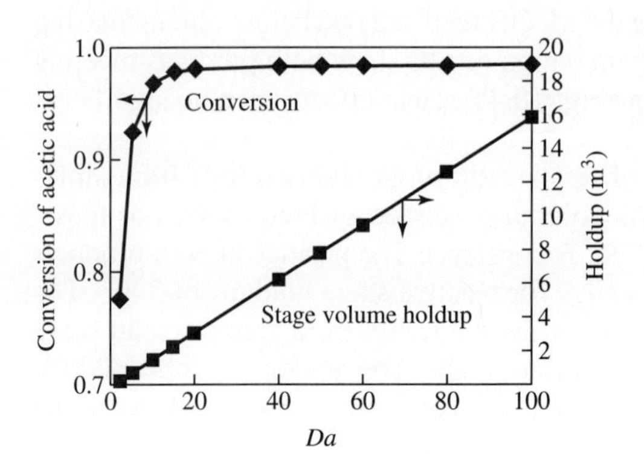

FIGURE 10.45
Conversion of acetic acid and average stage volume holdup for different values of *Da*.

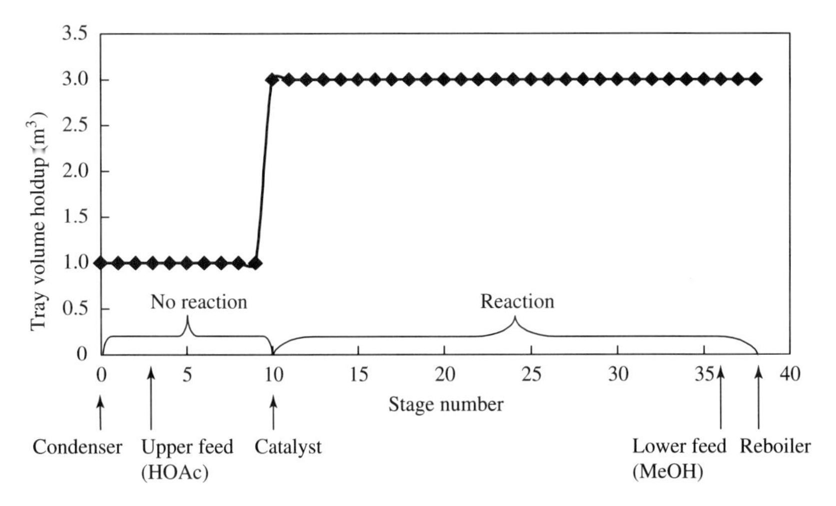

FIGURE 10.46
Volume holdup distribution and reactive status throughout the column based on 400 million lb MeOAc/year (280.0 kgmol MeOAc/h).

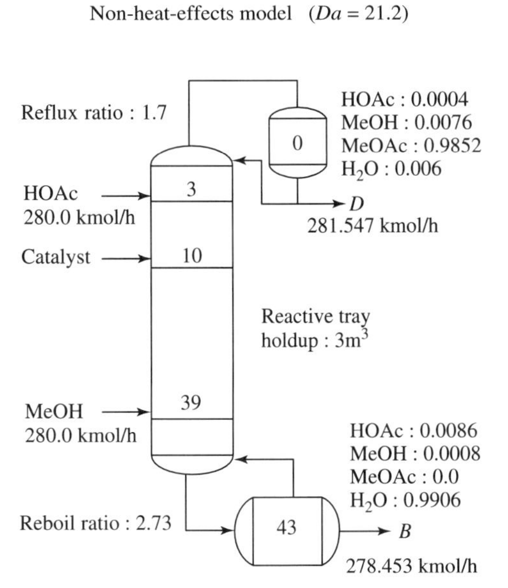

FIGURE 10.47
Summary of simulations without heat effects; $Da = 21.2$. The compositions are reported as mole fractions.

Figure 10.49 summarizes the final design, including heat effects. The profiles for the column in Fig. 10.47 and the column in Fig. 10.49 are compared in Fig. 10.50.

Effects of a Side Reaction

Many systems of interest in reactive distillation have multiple chemical reactions. Often, the side reactions are undesirable, and they may have much larger reaction

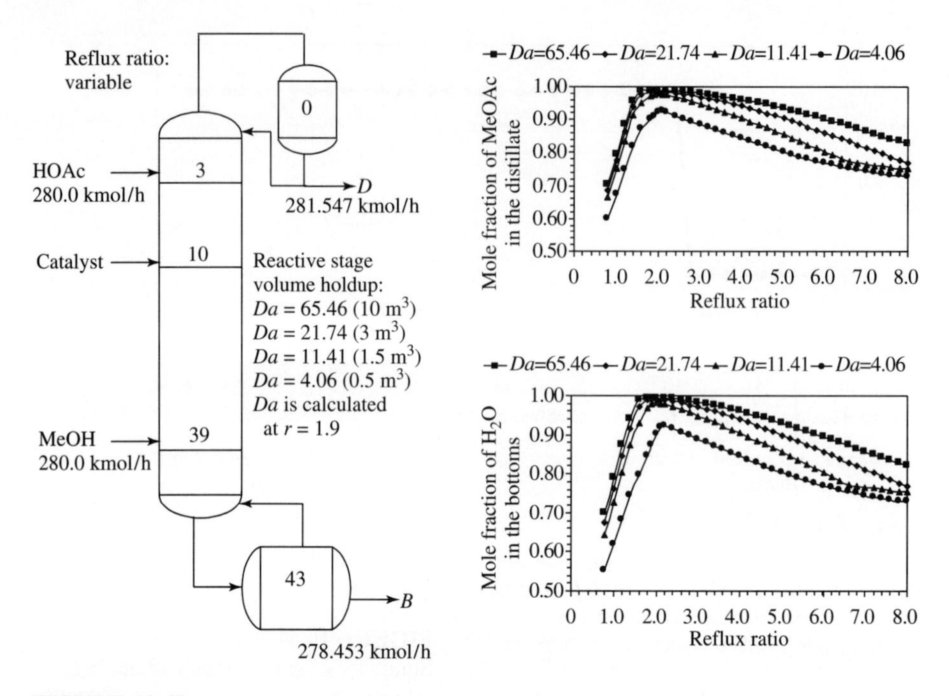

FIGURE 10.48
Influence of reflux ratio on the compositions of MeOAc and H_2O in the distillate and bottoms, respectively, for different values of Da. The simulations include heat effects.

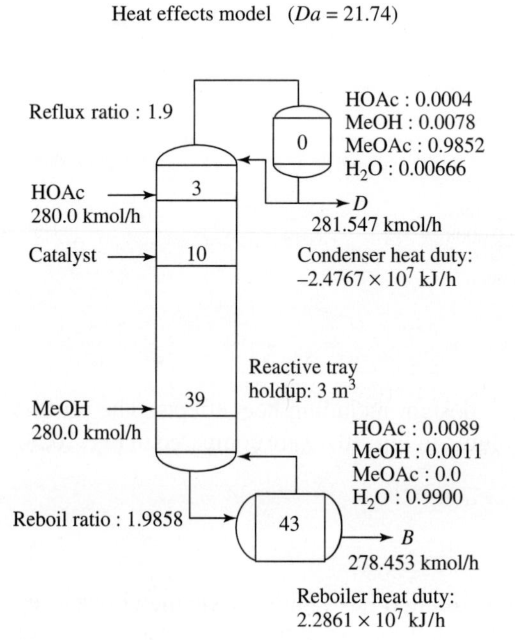

FIGURE 10.49
Final design for methyl acetate reactive distillation. Compositions are reported as mole fractions.

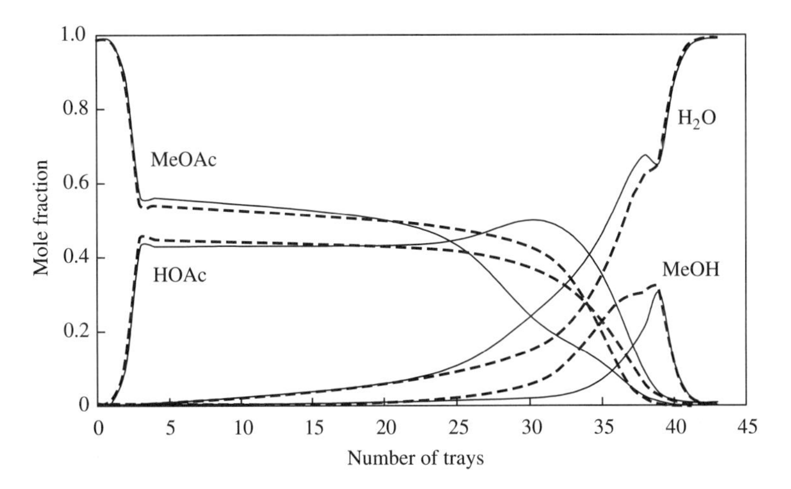

FIGURE 10.50
Comparison of column profiles. Dashed line: profile of column in Fig. 10.48; 44 stages; simulation without heat effects. Solid line: profile of column in Fig. 10.51; 44 stages; simulation including heat effects.

equilibrium constants, so that a design in the limit of $Da \to \infty$ does not give useful results when the side reaction is included. A strategy for this case is to base the equilibrium design on the main (desired) reaction and then to study the effect of a side reaction on the selectivity and product purity using simulation.

We consider a side reaction of methanol dehydration to dimethyl ether (DME) and water.

$$2\text{MeOH} \rightleftharpoons \text{DME} + \text{H}_2\text{O}$$

The rate expression is

$$r = k_f \left(a_{\text{MeOH}}^2 - \frac{a_{\text{DME}} a_{\text{H}_2\text{O}}}{K_{eq}} \right)$$

The reaction equilibrium constant (Song et al., 1998) and rate constant are

$$K_{eq} = 2.145 \exp(1{,}239.8/T)$$

and

$$k_f = 7.602 \times 10^9 \exp(-10{,}654/T) \text{ h}^{-1}$$

where T is in Kelvin. The normal boiling point of MeOAc is chosen as the reference temperature for the calculation of r_0 (5.1937 h^{-1}) using the rate constant of the main reaction.

For the same specifications and column configuration described in Fig. 10.49, the simulation results with the heat effects show no significant DME creation in the column at 1 atm. The side reaction may be more significant operating at higher pressures, but higher pressure requires more trays to achieve the same products;

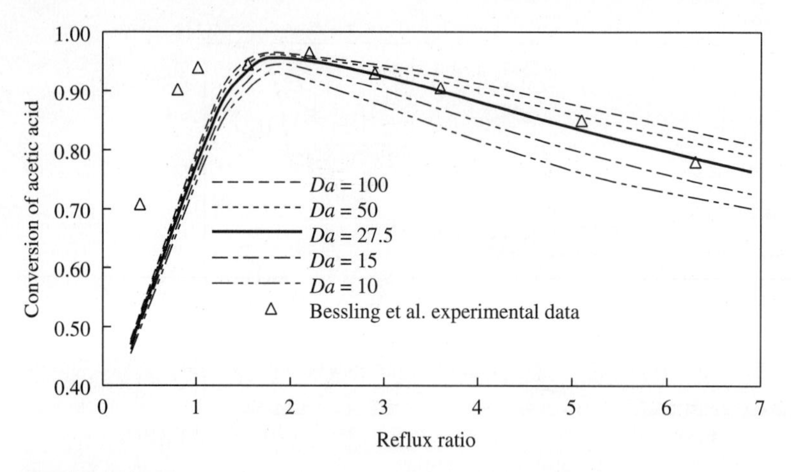

FIGURE 10.51
Influence of reflux ratio on the conversion of acetic acid for different *Da*. The total number of stages is 27, including a total condenser (0) and an equilibrium reactive reboiler (26); acetic acid is fed on stage 7; methanol is fed on stage 20; stages 19 through 7 are reactive; the molar feed ratio MeOH/HOAc $= 1:1$; $D/B = 1.0$; $P = 1$ atm. Simulation includes heat effects.

see Fig. 10.40 for the equilibrium design and Chadda (2001) for a non-equilibrium design.

Comparison with Experiment

The detailed column configuration for a simulation is taken from the test column of Bessling et al. (1998, Fig. 9). The comparison of our simulation results using our physical property and rate models with experimental data from Bessling et al. (1998) is shown in Fig. 10.51. This result shows clearly that,

> Effects of chemical reaction kinetics only become significant at higher reflux ratios in combination with a decreasing residence time of the reactants (Bessling et al., 1998).

10.7
CLOSING REMARKS AND OPEN QUESTIONS

Conceptual design is essentially complete when good alternatives are identified. The most fruitful approach for further evaluation of alternatives combines high-fidelity models, incorporating hydrodynamics and mass transfer with new hardware designs and tests. The hydrodynamics and mass transfer models demand new data because conditions are frequently outside the bounds of known correlations for distillation design. For example, large liquid holdups and multiphase flow over supported heterogeneous catalysts are major issues (Moritz et al., 1999).

Simulation methods have matured significantly in the last decade, and columns of realistic complexity can be simulated, including the effects of tray hydraulics, and mass transfer between liquid and vapor and between fluid and solid catalyst, etc. (Taylor and Krishna, 2000). Many of these models also exhibit multiple steady states. Recent studies show that some of the calculated multiple steady states are not found in models that incorporate realistic constraints. However, it is clear that some multiple steady states do occur for chemistries with small heat of reaction (e.g., methyl acetate synthesis), with large heat of reaction (e.g., ethylene glycol synthesis), and with intermediate heat of reaction (e.g., MTBE synthesis). Experiments confirm these predictions for fuel ethers (Mohl et al., 1999).

Along with greater detail in modeling, many studies have focused on new hardware designs. This has led to novel new designs for packings and supports to accommodate heterogeneous catalysts, which improves contacting for multiphase reacting mixtures.

Operability analysis normally includes multiple simulation steps at increasing levels of detail. A key issue at this step is to check the robustness of the design, i.e., the ability to maintain product purities and conversion in a desired range in the face of disturbances in production rate, feed composition, and other connections to the environment. Process control remains an important issue, and there are two main considerations. The first is to ensure that disturbances in operating parameters do not change the feasible split. This is intimately related to the design choices and the availability of sufficient data on rates of reaction and mass transfer (Taylor and Krishna, 2000). The second is to understand the multiplicity and stability of solutions which have been predicted and, in some cases, measured.

Despite progress in modeling, there are few results to guide practical decisions on operating strategies, control system design, instrumentation, etc. For instance, little work has been done to ensure that the right measurements are available to detect this multiplicity and to avoid sending the column to a lower conversion steady state. In fact, very few of the papers in the data from Fig. 10.1 focus on operability or control.

There are significant opportunities for further study of reactive distillation. From our viewpoint (Malone and Doherty, 2000), the most important problems are as follows.

Experiments

Sometimes the best way to answer a design question is to do an experiment; furthermore, sometimes the best way to plan an experiment is to develop a design model.

Some experiments for reactive distillation must support feasibility studies, e.g., phase and reaction equilibrium, adsorption equilibria for heterogeneous catalysts, reaction rates, and selectivity information. A critical decision at this point is the choice of a rapid and useful batch experiment to support the design development. For instance, batch kinetic studies in closed systems often result in selectivity losses far in excess of those possible with by-product removal in reactive distillation (Gadewar et al., 2000). A batch reactive distillation is more informative on this aspect, though not preferred for building a database of kinetic parameters.

Pilot-scale reactive distillation columns are expensive to construct and maintain. However, with the current state of the art, it is inconceivable that any one would build a new reactive distillation process without a pilot test. This should focus on mass transfer and hydrodynamics. A combination of simulation and experiment design leading to a small number of pilot tests is the sensible current approach. A challenge for the future is to develop simpler experimental validation procedures without the need for pilot plants.

Phase equilibrium

For catalyzed reactions, VLE models can be assembled from the constituent binary pair interactions in a similar way as for nonreactive mixtures. The resulting models seem to give quite good results for esterifiation and etherification reactions, ethylene glycol synthesis, etc. When they do not, it is typically due to a shortage of data, and the models can be improved as new information for the individual binary pairs becomes available. However, self-catalyzed reactions such as formaldehyde chemistry, chlorination chemistries, etc., are much more difficult to model because the individual binary pairs cannot be measured separately since the reaction spontaneously occurs. Although significant progress has been made on VLE models for formaldehyde chemistry in recent years (Albert et al., 1999), little is known about other self-catalyzed reactions, several of which have good potential for reactive distillation technology. One promising avenue that could provide a general framework for modeling these kinds of mixtures is the reaction-ensemble Monte Carlo simulation methods (Lisal et al., 2000, and earlier work; Johnson et al., 1994).

Reaction rates and catalysis

Heterogeneous catalysts give more flexibility for devising hybrid reactive distillation columns at the expense of mass transfer limitations.

A significant advantage can be gained in reactive distillation with catalysts that are more active at the relatively low temperatures where distillation is effective. Many of the reactive distillations currently carried out have relatively long reaction times, on the order of an hour, so that significant gains in productivity with more effective catalysts can be expected. These catalysts should probably be different from those in conventional processes in many cases, since the temperatures, pressures, and flow conditions are different. However, that is not the case today, and a fruitful area of research should be the development of catalysts specifically tailored for use in the multiphase conditions inherent in reactive distillations.

Along with catalyst development, the influence of side reactions on selectivity, for both homogeneous and catalytic reactions, needs further study. The conditions used in a traditional chemistry and catalyst development effort may promote very different side reactions, or extent of those reactions, than will prevail in a reactive distillation. Little is known about experimental protocols to systematically and reliably address this.

Equipment design

For effective mass transfer with heterogeneous catalysts, a number of novel supports have been devised, and numerous patents and technologies are available. It is often unclear which of these is a better choice, and general guidelines would be very useful. These should also address the important issue of catalyst deactivation through fouling or poisoning because regeneration or replacement is significantly more difficult than in conventional reactors.

Hydrodynamics and its interactions with the reaction rate, e.g., getting effective phase disengagement, good mixing, and sufficient holdup for the intended extents of reaction, cannot typically be done by modeling alone. Experiments can be done more effectively with some insights from models, but many open questions remain, e.g., how to decide the scale for pilot experiments and how to reliably scale up the results.

Alternatives and conceptual design

At the moment, there are relatively few general methods available for generating all feasible alternatives, but some systematic studies are published. This is a very hard problem to solve and it is reasonable to expect that methods will be slower in coming.

A promising approach is the modular representation framework being developed by Pistikopoulos and co-workers (Ismail et al., 1999). This approach combines the feasibility step and the sequencing step in a single algorithmic approach. The idea is to replace column sections with inequalities that feasible sections must obey. This sidesteps the need to specify the many internal design variables for each section (number of stages, liquid and vapor rates, stage holdups, catalyst concentrations, etc.), thereby simplifying the problem formulation at the conceptual design stage. This approach is especially appealing for complex systems like reactive distillation, where so many variables influence the behavior of each column section. A drawback is that the resulting feasible sequences may be difficult to evaluate and rank because relatively little is known about those very design variables! Nevertheless, the methodology has already achieved success by inventing known reactive distillation system structures for methyl acetate production as well as novel structures for the production of ethyl acetate.

It is possible to imagine that the geometric feasibility methods could be linked to a synthesis methodology (in a similar way that residue curve maps have been for nonreactive distillations), but so far not much has developed along these lines.

Energy management has received surprisingly little systematic study on aspects specifically for reactive distillation, although traditional methods for distillation, heat, and power integration are useful. For exothermic reactions, one might expect to capture the heat released by reaction to drive the separation. This may be possible, but the reaction must take place at temperatures higher than the boiling point of the bottoms product for complete integration. Furthermore, the reaction requirements for cooling or heating may not correspond to the preferred reflux or reboil requirements needed for the separation.

Heuristics for the use of reactive distillation would also be useful for the early stages of conceptual design. Some early work does address general guidelines, e.g., Xu et al. (1985), and it may be possible to expand these substantially using the methods and tools developed in the last decade.

TABLE 10.5
Selected systems studied for reactive distillation

A	*B*	Reference
	$A \rightleftharpoons B$	
n-Butane	Isobutane	Lebas et al. (1999)
1,4-Dichloro-2-butene	1,2-Dichloro-3-butene	Lee et al. (2000*b*)
Cyclohexanone oxime	Epsilon-caprolactam	Lee et al. (2000*b*)
	$2A \rightleftharpoons B$	
Cyclopentadiene	Dicyclopentadiene	Robinson and Gilliland (1950)
Isobutyraldehyde	Isobutylisobutyrate	Lee et al. (2000*b*)

A	*B*	*C*	Reference
	$A + B \rightleftharpoons C$		
Adipic acid	Hexamethylenediamine	Salt	Grosser et al. (1987)
Butadiene	Sulfur dioxide	Butadiene sulfone	AIChE (1970)
Ethylene oxide	Water	Ethylene glycol	Corrigan and Miller (1968)
Isobutene	Methanol	MTBE	Smith (1984)
Isobutene	Ethanol	ETBE	Frey et al. (1999)
Isoamylene	Methanol	TAME	Ward (1993)
Isoamylene	Water	tert-Amyl alcohol	Gonzalez and Fair (1997)
Propylene	Isopropanol	Diisopropyl ether	Marker et al. (1998)
Monoethanolamine	Ethylene oxide	Diethanolamine	DiGuilio and McKinney (2000)
Benzene	Propylene	Cumene	Smith (1993)
Aniline	Hydrogen	Cyclohexylamine	Hearn and Nemphos (1997)
	$A + 2B \rightleftharpoons C$		
Glyoxal	Acetaldehyde	bis(Cyclic acetal)	Sharma (1995)

TABLE 10.5 *(Continued)*

Selected systems studied for reactive distillation

A	*B*	*C*	Reference
	$A \rightleftharpoons B + C$		
Tertiary butyl alcohol	Isobutylene	Water	Knifton et al. (1998)
Dimethyl xylylene dicarbamate	Xylylene diisocyanate	Methanol	Okawa et al. (1993)
1-Butene	Ethylene	3-Hexene	Jung et al. (1987)
	$A + B \rightleftharpoons 2C$		
Benzene	Xylene	Toluene	Holmgren et al. (1999)
Acetic anhydride	Water	Acetic acid	Costa and Canepa (1969)

A	*B*	*C*	*D*	Reference
		$A + B \rightleftharpoons C + D$		
Acetic acid	Methanol	Methyl acetate	Water	Corrigan and Ferris (1969); Agreda and Partin (1984)
Acetic acid	Ethanol	Ethyl acetate	Water	Komatsu and Holland (1984)
Acetic acid	Isopropanol	Isopropyl acetate	Water	Song et al. (1997)
Acetic acid	Butanol	Butyl acetate	Water	Leyes and Othmer (1945); Hanika et al. (1999)
Acetic acid	Vinyl stearate	Stearic acid	Vinyl acetate	Geelen and Wijffels (1965)
1,2-Propanediol	Acetaldehyde	2,4-Dimethyl-1,3-dioxolane	Water	Broekhuis et al. (1994)
Toluene	Chlorine	Benzyl chloride	Hydrochloric acid	Xu and Dudukovic (1999)
2,3-Butanediol	Formaldehyde	4,5-Dimethyl-1,3-dioxolane	Water	Broekhuis et al. (1994)
Methyl benzoate	Benzyl alcohol	Benzyl benzoate	Methanol	Tang et al. (1984)

TABLE 10.5 *(Continued)*
Selected systems studied for reactive distillation

A	*B*	*C*	*D*	Reference
		$A + B \rightleftharpoons C + D$		
Acrylic acid	Ethanol	Ethyl acrylate	Water	Jelinek and Hlavacek (1976)
Acrylic acid	Butanol	Butyl acrylate	Water	Jelinek and Hlavacek (1976)
Butanol	Monobutyl phthalate	Dibutyl phthalate	Water	Berman et al. (1948)
Ethanol	Lactic acid	Ethyl lactate	Water	Keyes (1932)
Butyl acetate	Ethanol	Ethyl acetate	Butanol	Davies and Jeffreys (1973)
Formic acid	Ethanol	Ethyl formate	Water	Rhim et al. (1985)
m-Xylene	Sodium *p*-xylene	Sodium *m*-xylene	*p*-Xylene	Terrill et al. (1985)
Pentenoyl chloride	Water	Pentenoic acid	Hydrogen chloride	Murphree and Ozer (1996)
Ethanol	Ammonia	Ethylamine	Water	Nemphos and Hearn (1997)
Glycerol	mono-para-Nitrobenzoic acid	mono-para-Nitrobenzoate	Water	Baker et al. (1976)
Dimethyl carbonate	Phenol	Phenyl methyl carbonate	Methanol	Rivetti et al. (1998)
Benzene	Nitric acid	Nitrobenzene	Water	Thelen et al. (1977)
Methanol	3-Pentenoic acid	Methyl-3-pentenoate	Water	Sherman et al. (1996)
		$A + 2B \rightleftharpoons C + D$		
Formaldehyde	Methanol	Methylal	Water	Kolah et al. (1996)
		$A + 2B \rightleftharpoons C + 2D$		
2,3-Butylene glycol	Acetic acid	2,3-Butylene glycol diacetate	Water	Schniepp et al. (1945)
		$A \rightleftharpoons B, B \rightleftharpoons C, B \rightleftharpoons D$		
m-Xylene	*o*-Xylene	*p*-Xylene	Ethylbenzene	Castier et al. (1989)
		$A + B \rightleftharpoons C, A + C \rightleftharpoons D$		
Propylene	Benzene	Cumene	Diisopropylbenzene	Kaeding and Holland (1988); Shoemaker and Jones (1987)
Ethanol	Formaldehyde	Hemiacetal	Acetal	Sharma (1995)

TABLE 10.5 *(Continued)*
Selected systems studied for reactive distillation

A	B	C	D	Reference
$A + B \rightleftharpoons C, 2C \rightleftharpoons B + D$				
Ethylene	Water	Ethanol	Diethylether	Eguchi et al. (1986)
Formaldehyde	Methanol	Hemi-formal	Higher hemi-formal	Hasse et al. (1990), Maurer (1986)
Formaldehyde	Water	Methylene glycol	Polyoxymethylene	Hasse et al. (1990), Maurer (1986)

A	B	C	D	E	Reference
$A + B \rightleftharpoons C, C + B \rightleftharpoons D, 2B \rightleftharpoons E$					
Isobutane	Isobutene	Trimethyl-pentane	Dodecane	Dimethyl-hexene	Albright et al. (1988), Lee and Harriott (1977), Cupit et al. (1961)
$2A \rightleftharpoons B + C, B + D \rightleftharpoons E$					
Acetone	Mesityl oxide	Water	Hydrogen	MIBK	Lawson and Nkosi (1999)
$A + B \rightleftharpoons C + D, 2B \rightleftharpoons E$					
Methanol	Acetic acid	Methyl acetate	Water	Acetic acid dimer	Agreda et al. (1990), Barbosa and Doherty (1988*a*)
$A + B \rightleftharpoons C + D, 2A \rightleftharpoons E + D$					
Ethanol	Acetic acid	Ethyl acetate	Water	Diethyl ether	Castier et al. (1989), George et al. (197[illegible]), Sanderson and Chien (1973), Hawes and Kabel (1968)

TABLE 10.5 *(Continued)*
Selected systems studied for reactive distillation

A	*B*	*C*	*D*	*E*	Reference
		$A + B \rightleftharpoons E, A + C \rightleftharpoons D$			
Formaldehyde	Methanol	Water	Methylene glycol	Hemi-formal	Hasse et al. (1990), Maurer (1986)
		$A + B \rightleftharpoons C, C + D \rightleftharpoons A + E$			
Methanol Cyclohexene	Carbon monoxide Formic acid	Methyl formate Cyclohexyl formate	Water Water	Formic acid Cyclohexanol	Lee et al. (2000*b*) Sharma (1995)
		$A + B \rightleftharpoons C, C \rightleftharpoons D + E$			
Acetic anhydride	Acetaldehyde	Ethylene diacetate	Acetic acid	Vinyl acetate	Zoeller et al. (1998)
		$A + B \rightleftharpoons C + D, B + C \rightleftharpoons D + E$			
Dichlorosilane di-tert-Butylbenzene	Hydrogen chloride *m*-Xylene	Trichlorosilane tert-Butylbenzene	Hydrogen tert-Butyl-*m*-xylene	Silicon tetrachloride Benzene	Ung and Doherty (1995*d*) Ung and Doherty (1995*b*)
		$A + B \rightleftharpoons C + D, 2C \rightleftharpoons A + E$			
Dimethyl carbonate	Phenol	Phenyl Methyl Carbonate	Methanol	Diphenyl Carbonate	Oyevaar et al. (2000)

10.8 EXERCISES

1. What assumptions are needed to obtain the reaction rate model used in Example 10.1 from the general expression in Equation 10.2? What is the corresponding expression for the reaction $A + B \rightleftharpoons 2C$?
2. Prepare a simulation, e.g., using Hysys or Aspen+, or another simulator, for Example 10.1. The effects of Da can be captured by changing the rate constant or by changing the liquid holdup on the stages. Using mass balances, show that for high-purity splits and high recoveries, it is desirable to adjust the reflux and reboil ratios so that $D/B \approx 2\frac{1-x}{x}$, where x is the conversion of A. How does this change the simulation results from those shown in Fig. 10.2?
3. If you solved the previous exercise, modify the simulation to reproduce the results for the hybrid column shown in Fig. 10.3. If the feed is moved down to stage 10, how do the results change? If the value of the reboil ratio or Da is gradually lowered, how do the results change? Starting from a solution at a low value of s or Da, do you reach the same (or similar) solution as in Fig. 10.3? Discuss.
4. How do the results for Example 10.1 change if the feed is a 50/50 mixture of A and B containing no C? If the distillate purity is relaxed and an impure stream is recycled to the feed, will this reduce the stage or energy requirements?
5. Derive a model for simple condensation with chemical reaction. Assume that the vapor phase condenses into a pool of liquid (where the reaction occurs) with constant molar holdup. The condensed vapor is removed continuously from the liquid reservoir.
6. Consider the reactions $A + 2B \rightleftharpoons C$ and $C \rightleftharpoons A + 2B$ in the limit as $K_{eq} \to 0$. Choose a set of initial compositions, e.g., similar to those in Fig. 10.13, and show the paths followed by the reaction up to the final equilibrium composition. Derive the stoichiometric lines and identify the reaction pole point.
7. Consider the reactions $A + B \rightleftharpoons 2C$ and $2C \rightleftharpoons A + B$ in the limit as $K_{eq} \to 0$. Choose a set of initial compositions, e.g., similar to those in Fig. 10.13, and show the paths followed by the reaction up to the final equilibrium composition. Derive the stoichiometric lines and identify the reaction pole point.
8. Derive Eq. 10.31 for the maximum equilibrium liquid composition of C in an ideal reacting mixture $A + C \rightleftharpoons C$ with a constant value of K_{eq}.
9. Explain why ordinary azeotropes (i.e., nonreactive azeotropes, where $x_i = y_i$) do not occur in binary or ternary reactive mixtures. Explain why these types of azeotropes may occur in a four-component mixture undergoing the chemistry $A + B \rightleftharpoons C + D$.
10. Consider an ideal liquid–vapor mixture with constant relative volatilities and a constant value for K_{eq}. Reactive azeotropes occur when the conditions for azeotropy and for phase and reaction equilibrium are satisfied simultaneously

$$\frac{y_1 - x_1}{\nu_1 - \nu_T x_1} = \frac{y_i - x_i}{\nu_i - \nu_T x_i} \qquad i = 2, \ldots, c \tag{10.70}$$

$$y_i = \frac{x_i \alpha_{in}}{\sum_{j=1}^{c} x_j \alpha_{jn}} \qquad i = 1, \ldots, c \tag{10.71}$$

$$K_{eq} = \prod_{i=1}^{c} x_i^{\nu_i} \tag{10.72}$$

where n is any component used to define the relative volatilities.

For the reaction $A + B \rightleftharpoons C$ show that reactive azeotropes can only occur when A and B are either both lighter or both heavier than C.

11. Consider an ideal vapor-liquid mixture with the reaction $A + B \rightleftharpoons C$. Use the graphical method of Frey and Stichlmair (see Fig. 10.21) to show:
    a. That reactive azeotropes never occur when C boils between A and B.
    b. The position of the reactive azeotrope for the case (A light, B intermediate, and C heavy) when K_{eq} is large, when K_{eq} is small, and at various values of K_{eq} in between.
12. Using a residue curve map for MTBE chemistry (isobutene + methanol $\rightleftharpoons$ MTBE) at 1 atm pressure, locate the reaction pole point and draw the stoichiometric lines. Determine whether reactive azeotropes occur at any value of K_{eq}.
13. Develop some reactive distillation column configurations for making pure MTBE from mixtures consisting of isobutene, n-butane, and methanol. Base your process synthesis on the equilibrium reactive residue curve map in Fig. 10.26 and the nonreactive map in Fig. 10.4.
14. The equilibrium residue curve map (Wasylkiewicz and Ung, 2000) and a feasibility diagram (see Sec. 10.5) for butyl acetate synthesis is shown in Fig. 10.52. Devise a process flowsheet for the production of butyl acetate. What value of Da would you choose for the conceptual design?
15. Oyevaar et al. (2000) describe a process to make phenyl-methyl carbonate from phenol and dimethyl carbonate using reactive distillation technology (see the first column of Fig. 1 in U.S. Patent 6,093,842). Make your best estimate of the equilibrium residue curve map(s) for this chemistry, and use them to explain the column described in the patent.
16. Bessling et al. (1997*a*) show that reactive distillation is very suitable for the production of methyl formate from formic acid and methanol. Develop the necessary residue curve maps and use them to invent a reactive distillation column to make methyl formate.

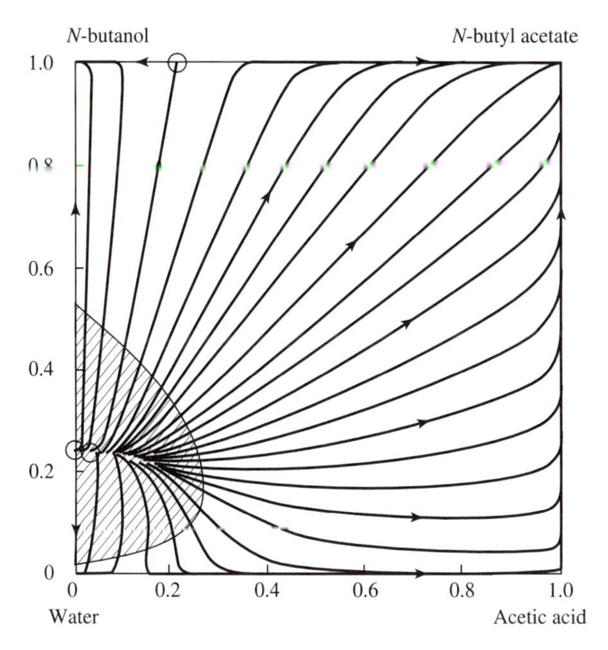

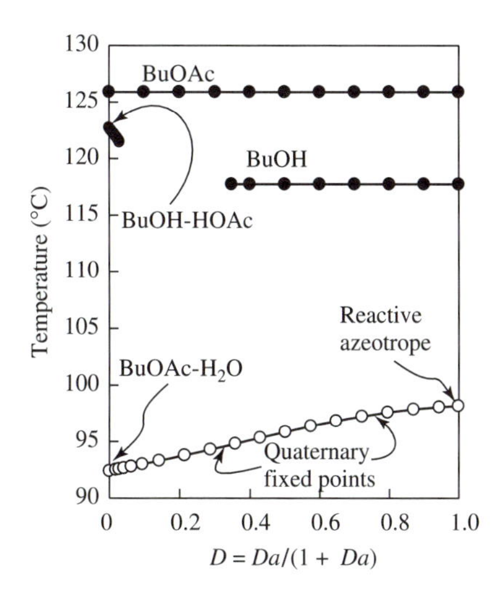

FIGURE 10.52
Equilibrium residue curve map (top) and feasibility diagram (bottom) for butyl acetate at 1 atm pressure.

References

Agreda, V. H., L. R. Partin, and W. H. Heise, "High Purity Methyl Acetate via Reactive Distillation," *Chem. Engng. Prog.,* **86**(2), 40–46 (1990).

Agreda, V. H., and L. R. Partin, Reactive Distillation Process for the Production of Methyl Acetate (1984). U.S. Patent 4,435,595 assigned to Eastman Kodak Company.

AIChE, *Student Contest Problem.* New York (1970).

Albert, M., B. C. Garcia, C. Kreiter, and G. Maurer, "Vapor-Liquid and Chemical Equilibria of Formaldehyde-Water Mixtures," *AIChE J.,* **45,** 3019 (1999).

Albright, L. F., M. A. Spalding, J. A. Nowinski, R. M. Ybarra, and R. E. Eckert, "Alkylation of Isobutane with C_4 Olefins. 1. First-step Reactions Using Sulfuric Acid Catalyst," *Ind. Eng. Chem. Research,* **27,** 391–397 (1988).

Baker, J. A., A. Gray, and J. F. Benford, Esterification of Nitrobenzoic Acids (1976). U.S. Patent 3,948,972.

Barbosa, D., and M. F. Doherty, "Theory of Phase Diagrams and Azeotropic Conditions for Two-Phase Reactive Systems," *Proc. Roy. Soc. London, A,* **413,** 443–458 (1987).

Barbosa, D., and M. F. Doherty, "Design and Minimum Reflux Calculations for Single-Feed Multicomponent Reactive Distillation Columns," *Chem. Engng. Sci.,* **43,** 1523–1537 (1988*a*).

Barbosa, D., and M. F. Doherty, "Design and Minimum Reflux Calculations for Double-Feed Multicomponent Reactive Distillation Columns," *Chem. Engng. Sci.,* **43,** 2377–2389 (1988*b*).

Barbosa, D., and M. F. Doherty, "Influence of Equilibrium Chemical Reactions on Vapor-Liquid Phase Diagrams," *Chem. Engng. Sci.,* **43,** 529–540 (1988*c*).

Barbosa, D., and M. F. Doherty, "Simple Distillation of Homogeneous Reactive Mixtures," *Chem. Engng. Sci.,* **43,** 541–555 (1988*b*).

Berman, S., H. Isbenjian, A. Stedoff, and D. F. Othmer, "Esterification-Continuous Production of Dibutyl Phthalate in a Distillation Column," *Ind. Eng. Chem.,* **40,** 2139–2148 (1948).

Bessling, B., Experiental and Computational Screening Methods for Reactive Distillation, in Malone, M. F., and J. A. Trainham, editors, *Foundations of Computer-Aided Process Design,* number 323 in *AIChE Symp. Ser.,* pp. 385–388, New York, AIChE (2000).

Bessling, B., J.-M. Loning, A. Ohligschlager, G. Schembecker, and K. Sundmacher, "Investigations on the Synthesis of Methyl Acetate in a Heterogeneous Reactive Distillation Process," *Chem. Eng. Tech.,* **21,** 393–400 (1998).

Bessling, B., G. Schembecker, and K. Simmrock, "Design of Processes with Reactive Distillation Line Diagrams," *Ind. Eng. Chem. Research,* **36,** 3032–3042 (1997*a*).

Bessling, B., G. Schembecker, and K. Simmrock, "Design of Reactive Distillation Processes with Reactive and Nonreactive Distillation Zones," *IChemE Symp. Ser.* No. 142, 675–683 (1997*b*).

Block, U., "Performance of Continuous Reactions with Superimposed Distillation," *Ger. Chem. Eng.,* **1,** 79–82 (1978).

Broekhuis, R. R., S. Lynn, and C. J. King, "Recovery of Propylene Glycol from Dilute Aqueous Solutions via Reversible Reaction with Aldehydes," *Ind. Eng. Chem. Research,* **33,** 3230–3237 (1994).

Buzad, G., and M. F. Doherty, "Design of Three-Component Kinetically Controlled Reactive Distillation Columns Using Fixed-Point Methods," *Chem. Engng. Sci.,* **49,** 1947–1963 (1994).

Buzad, G., and M. F. Doherty, "New Tools for the Design of Kinetically Controlled Reactive Distillation Columns for Ternary Mixtures," *Computers Chem. Engng.,* **19,** 395–408 (1995).

Cardoso, M. F., R. L. Salcedo, S. F. de Azevedo, and D. Barbosa, "Optimization of Reactive Distillation Processes with Simulated Annealing," *Chem. Engng. Sci.,* in press (2000).

Castier, M., P. Rasmussen, and A. Fredenslund, "Calculation of Simultaneous Chemical and Phase Equilibria in Nonideal Systems," *Chem. Engng. Sci.,* **44,** 237–248 (1989).

Chadda, N., M. F. Malone, and M. F. Doherty, "Effect of Chemical Kinetics on Feasible Splits for Reactive Distillation," *AIChE J,* in press (2001).

Chadda, N., *Feasibility of Kinetically Controlled Reactive Distillation,* Ph.D. thesis, University of Massachusetts, Amherst MA (2001).

Chadda, N., M. F. Malone, and M. F. Doherty, "Feasible Products for Kinetically Controlled Reactive Distillation of Ternary Mixtures," *AIChE J.,* **46,** 923–936 (2000).

Chen, F., R. S. Huss, M. F. Malone, and M. F. Doherty, "Computer-Aided Tools for the Design of Reactive Distillation Systems," *Computers Chem. Engng.,* **24,** in press (2000).

Ciric, A. R., and D. Gu, "Synthesis of Nonequilibrium Reactive Distillation Processes by MINLP Optimization," *AIChE J.,* **40,** 1479–1487 (1994).

Cleary, W., and M. F. Doherty, "Separation of Closely Boiling Mixtures by Reactive Distillation. 2. Experiments.," *Ind. Eng. Chem. Process Design Dev.,* **24,** 1071–1073 (1985).

Colombo, F., L. Cori, L. Dalloro, and P. Delogu, "Equilibrium Constant for the Methyl *tert*-Butyl Ether Liquid-Phase Synthesis by Use of UNIFAC," *Ind. Eng. Chem. Fundam.,* **22,** 219–223 (1983).

Corrigan, T. E., and W. R. Ferris, "A Development Study of Methanol Acetic Acid Esterification," *Can. J. Chem. Eng.,* **47,** 334–335 (1969).

Corrigan, T. E., and J. H. Miller, "Effect of Distillation on a Chemical Reaction," *Ind. Eng. Chem. Process Design Dev.,* **7,** 383–384 (1968).

Costa, P., and B. Canepa, "Sul Calcolo Di Colonna Per La Distillazione Di Miscele Reagenti," *Quad. Ing. Chim. Ital.,* **5,** 113–121 (1969).

Cupit, C. R., J. E. Gwyn, and E. C. Jernigan, "Special Report Catalytic Alkylation," *Petrol/Chem. Engr.,* **33,** 202–215 (1961).

Damköhler, G., "Strömungs und Wärmeübergangsprobleme in Chemischer Technik und Forschung," *Chem. Ing. Tech.,* **12,** 469–480 (1939).

Davies, B., and G. V. Jeffreys, "The Continuous Trans-esterification of Ethyl Alcohol and Butyl Acetate in a Sieve Plate Column. Part III: Trans-esterification in a Six Plate Sieve Plate Column," *Trans. Inst. Chem. Engrs.,* **51,** 275–280 (1973).

DeGarmo, J. L., V. N. Parulekar, and V. Pinjala, "Consider Reactive Distillation," *Chem. Engng. Prog.,* **88**(3), 43–50 (1992).

DiGuilio, R. M., and M. W. McKinney, Selective Production of Diethanolamine (2000). U.S. Patent 6,075,168.

Doedel, E. J., AUTO: A Program for the Automatic Bifurcation Analysis of Autonomous Systems, in *Proc. 10th Manitoba Conf. on Num. Math. and Comp., Univ. of Maitoba, Winnipeg, Canada,* pp. 265–284 (1981). See http://indy.cs.concordia.ca/auto/ for the current version, Auto97, as of July 9, 2000.

Doherty, M. F., "A Topological Theory of Phase Diagrams for Multiphase Reacting Mixtures," *Proc. R. Soc. Lond. A,* **430,** 669–678 (1990).

Eguchi, K., T. Tokiai, Y. Kimura, and H. Arai, "High Pressure Catalytic Hydration of Olefins over Various Proton-exchanged Zeolites," *Chem. Lett.,* **4,** 567–570 (1986).

Espinosa, J., P. A. Aguirre, and G. A. Perez, "Product Composition Regions of Single-Feed Reactive Distillation Columns: Mixtures Containing Inerts," *Ind. Eng. Chem. Research,* **34,** 853–861 (1995*a*).

Espinosa, J., P. A. Aguirre, and G. A. Perez, "Some Aspects in the Design of Multicomponent Reactive Distillation Columns Including Non-Reactive Species," *Ind. Eng. Chem. Research,* **34,** 469–484 (1995*b*).

Espinosa, J., P. A. Aguirre, T. Frey, and J. Stichlmair, "Analysis of Finishing Reactive Distillation Columns," *Ind. Eng. Chem. Research,* **38,** 187–196 (1999).

Fidkowski, Z. T., M. F. Malone, and M. F. Doherty, "Computing Azeotropes in Multicomponent Mixtures," *Computers Chem. Engng.,* **17,** 1141–1155 (1993).

Frey, S. J., S. P. Davis, S. L. Krupa, and P. R. Cottrell, Process for Producing Ethyl Tertiary Butyl Ether by Catalytic Distillation (1999). U.S. Patent 5,990,361.

Frey, T., and J. Stichlmair, "Reactive Azeotropes in Kinetically Controlled Reactive Distillation," *Chem. Eng. Res. Des. (Trans. IChemE, Part A),* **77,** 613–618 (1999).

Gadewar, S., M. F. Malone, and M. F. Doherty, "Selectivity Targets for Batch Reactive Distillation," *Ind. Eng. Chem. Research,* **39,** 1565–1575 (2000).

Gautam, R., and W. D. Seider, "Computation of Phase and Chemical Equilibrium—2. Phase-Splitting," *AIChE J.,* **25,** 999–1006 (1979).

Geelen, H., and J. B. Wijffels, The Use of a Distillation Column as a Chemical Reactor, in *3rd European Symposium on Chemical Reaction Engineering,* pp. 125–133, Oxford, England. Pergamon Press (1965).

George, B., L. P. Brown, C. H. Farmer, P. Buthod, and F. S. Manning, "Computation of Multicomponent, Multiphase Equilibrium," *Ind. Eng. Chem. Process Design Dev.,* **15,** 372–377 (1976).

Gonzalez, C. J., and J. R. Fair, "Preparation of Tertiary Amyl Alcohol in a Reactive Distillation Column. 1. Reaction Kinetics, Chemical Equilibrium, and Mass Transfer Issues," *Ind. Eng. Chem. Research,* **36,** 3833 (1997).

Grosser, J. H., M. F. Doherty, and M. F. Malone, "Modeling of Reactive Distillation Systems," *Ind. Eng. Chem. Research,* **26,** 939–989 (1987).

Hanika, J., J. Kolena, and Q. Smejka, "Butylacetate via Reactive Distillation-Modelling and Experiment," *Chem. Engng. Sci.,* **54,** 5205 (1999).

Harding, S. T., and C. A. Floudas, "Locating All Heterogeneous and Reactive Azeotropes in Multicomponent Mixtures," *Ind. Eng. Chem. Research,* **39,** 1565–1595 (2000).

Hasse, H., I. Hahnentien, and G. Maurer, "Revised Vapor-liquid Equilibrium Model for Multicomponent Formaldehyde Mixtures," *AIChE J.,* **36,** 1807–1814 (1990).

Hauan, S., and K. Lien, "A Phenomena-Based Design Approach to Reactive Distillation," *Chem. Engng. Res. Design,* **76,** 396–407 (1998).

Hawes, R. W., and R. L. Kabel, "Thermodynamic Equilibrium in the Vapor Phase Esterification of Acetic Acid with Ethanol," *AIChE J.,* **14,** 209–216 (1968).

Hearn, D., and S. P. Nemphos, Process for the Production of Cyclohexyl Amine (1997). U.S. Patent 5,599,997.

Henley, E. J., and J. D. Seader, *Equilibrium-Stage Separation Operations in Chemical Engineering.* Wiley, New York (1981).

Holmgren, J. S., D. B. Galloway, L. B. Galperin, and W. R. Willis, Selective Aromatics Disproportionation/Transalkylation (1999). U.S. Patent 6,008,423.

Holve, W. A., "Theoretical Analysis of Changes in Relative Volatility via Reversible Metalation Reactions and the Application to Fractionation," *Ind. Eng. Chem. Fundam.,* **16,** 56–60 (1977).

Hougen, O. A., and K. M. Watson, *Chemical Process Principles, Part Three: Kinetics and Catalysis.* Wiley, New York (1947).

Hugo, P., "Die Berechnung des chemischen Umsatzes von Mehrkomoponenten–Gasgemischen an porösen Katalysatoren. Teil I.," *Chem. Engng. Sci.,* **20,** 187–194 (1965).

Huss, R. S., F. Chen, M. F. Malone, and M. F. Doherty, "Computer-Aided Tools for the Design of Reactive Distillation Systems," *Computers Chemical Engrg.*, S955–S962 (1999).

Ismail, S. R., E. N. Pistikopoulos, and K. P. Papalexandri, "Modular Representation Synthesis Framework for Homogeneous Azeotropic Distillation," *AIChE J.,* **45,** 1701–1720 (1999).

Jelinek, J., and V. Hlavacek, "Steady State Countercurrent Equilibrium Stage Separation with Chemical Reaction by Relaxation Method," *Chem. Eng. Comm.,* **2,** 79–85 (1976).

Johnson, J. K., A. Z. Panagiotopoulos, and K. Gubbins, "Reactive Canonical Monte Carlo," *Molecular Physics,* **81,** 717–733 (1994).

Julka, V., *A Geometric Theory of Multicomponent Distillation,* Ph.D. thesis, University of Massachusetts, Amherst MA (1995).

Jung, C. W., P. E. Garrou, and G. R. Strickler, Disproportionation of Alkenes (1987). U.S. Patent 4,709,115.

Kaeding, W. W., and R. E. Holland, "Shape-Selective Reactions with Zeolite Catalysts," *J. Catal.,* **109,** 212–216 (1988).

Keller, H. B., Numerical Solution of Bifurcation and Nonlinear Eigenvalue Problems, in Rabinowitz, P. H., editor, *Applications of Bifurcation Theory,* pp. 359–384. Academic Press, New York (1977).

Keyes, D. B., "Esterification Processes and Equipment," *Ind. Eng. Chem.,* **24,** 1096–1103 (1932).

Knapp, J. P., and M. F. Doherty, "Minimum Entrainer Flows for Exractive Distillation. A Bifurcation Theoretic Approach," *AIChE J.,* **40,** 243–268 (1994).

Knifton, J. F., J. R. Sanderson, and M. E. Stockton, Use of Reactive Distillation in the Dehydration of Tertiary Butyl Alcohol (1998). U.S. Patent 5,811,620.

Kolah, A. K., S. M. Mahajani, and M. M. Sharma, "Acetalization of Formaldehyde with Methanol in Batch and Continuous Reactive Distillation Columns," *Ind. Eng. Chem. Research,* **35,** 3707 (1996).

Komatsu, H., and C. D. Holland, "A New Method of Convergence for Solving Reacting Distillation Problems," *J. Chem. Eng. Japan,* **10,** 292–297 (1984).

Lawson, K. H., and B. Nkosi, Production of MIBK Using Catalytic Distillation Technology (1999). U.S. Patent 6,008,416.

Lebas, E., S. Jullian, C. Travers, P. B. Capron, J. Joly, and M. Thery, Paraffin Isomerisation Process Using Reactive Distillation (1999). U.S. Patent 5,948,948.

Lee, J. W., S. Hauan, and A. W. Westerberg, "Circumventing an Azeotrope in Reactive Distillation," *Ind. Eng. Chem. Research,* **39,** 1061–1063 (2000*a*).

Lee, J. W., S. Hauan, and A. W. Westerberg, "Graphical Methods for Reaction Distribution in a Reactive Distllation Column," *AIChE J.,* **46,** 1218–1233 (2000*b*).

Lee, L., and P. Harriott, "The Kinetics of Isobutane Alkylation in Sulfuric Acid," *Ind. Eng. Chem. Process Design Dev.,* **16,** 282–287 (1977).

Lee, L.-S., and M.-Z. Kuo, "Phase and Reaction Equilibria of the Isopropanol-Acetic Acid-Isopropyl Acetate-Water System at 760 mm Hg," *Fluid Phase Equilib.,* **123,** 147–165 (1996).

Leyes, C. E., and D. F. Othmer, "Continuous Esterification of Butanol and Acetic Acid, Kinetic and Distillation Consideration," *Trans. Amer. Inst. Chem. Engrs.,* **41,** 157–196 (1945).

Lin, C. C., and L. A. Segel, *Mathematics Applied to Deterministic Problems in the Natural Sciences.* Macmillan, New York (1974).

Lisal, M., W. R. Smith, and I. Nezbeda, "Molecular Simulation of Multicomponent Reaction and Phase Equilibria in MTBE Ternary System," *AIChE J.,* **46,** 866–875 (2000).

Malone, M. F., and M. F. Doherty, "Reactive Distillation," *Ind. Eng. Chem. Research,* **39,** in press (2000).

Marker, T. L., G. A. Funk, P. T. Barger, and H. U. Hammershaimb, Two-Stage Process for Producing Diisopropyl Ether Using Catalytic Distillation (1998). U.S. Patent 5,744,645.

Maurer, G., "Vapor-Liquid Equilibrium for Formaldehyde and Water-Containing Multicomponent Mixtures," *AIChE J.,* **32,** 932–948 (1986).

Mayer, H. H., and O. Wörz, "Selection of Reactors for Reactions with Superposed Distillation," *Ger. Chem. Eng.,* **3,** 252–257 (1980).

McDonald, C. M., and C. A. Floudas, "GLOPEQ: A New Computational Tool for the Phase and Chemical Equilibrium Problem," *Computers Chem. Engng.,* **21,** 1–23 (1997).

Melles, S., J. Grievink, and S. M. Schrans, "Optimisation of the Conceptual Design of Reactive Distillation Columns," *Chem. Engng. Sci.,* **55,** 2089–2097 (2000).

Mira, C., *Chaotic Dynamics.* World Scientific, Singapore (1987).

Mohl, K.-D., A. Kienle, E.-D. Gilles, P. Rapmund, K. Sundmacher, and U. Hoffmann, "Steady-State Multiplicities in Reactive Distillation Columns for the Production of Fuel Ethers MTBE and TAME: Theoretical Analysis and Experiental Verification," *Chem. Eng. J.,* **72,** 1029–1043 (1999).

Moritz, P., B. Bessling, and G. Schembecker, "Fluiddynamische Betrachtung von Katalysatortraegernd bei der Reaktivdestillation (Fluid Dynamic Considerations about Catalyst Packing in Reactive Distillation)," *Chemie Ingenieur Technik,* **71**(1+2), 1479–1487 (1999).

Murphree, B., and R. Ozer, Preparation of Pentenoic Acid (1996). U.S. Patent 5,536,873.

Nemphos, S. P., and D. Hearn, Amination Process (1997). U.S. Patent 5,679,862.

Nisoli, A., M. F. Malone, and M. F. Doherty, "Attainable Regions for Reaction with Separation," *AIChE J.,* **43,** 374–387 (1997).

Okasinski, M. J., and M. F. Doherty, "Thermodynamic Behavior of Reactive Azeotropes," *AIChE J.,* **43,** 2227–2238 (1997).

Okasinski, M. J., and M. F. Doherty, "Design Method for Kinetically Controlled Staged Reactive Distillation Columns," *Ind. Eng. Chem. Research,* **37,** 2821–2834 (1998).

Okawa, T., Y. Sato, H. Igarashi, and S. Suzuki, Process for Producing Xylylene Diisocyanate (1993). U.S. Patent 5,196,572.

Oyevaar, M. H., B. W. To, M. F. Doherty, and M. F. Malone, Process for Continuous Production of Carbonate Esters (2000). U.S. Patent 6,093,842.

Rev, E., "Reactive Distillation and Kinetic Azeotropy," *Ind. Eng. Chem. Research,* **33,** 2174–2179 (1994).

Rhim, J. K., S. Y. Bae, and H. T. Lee, "Isothermal Vapor-Liquid Equilibrium Accompanied by Esterification. Ethanol-Formic Acid System." *Int. Chem. Eng.,* **25,** 551–557 (1985).

Rivetti, F., R. Paludetto, and U. Romano, Continuous Process for the Preparation of Phenyl Methyl Carbonate (1998). U.S. Patent 5,705,673.

Robinson, C. S., and E. R. Gilliland, *Elements of Fractional Distillation.* McGraw-Hill, 4th ed. (1950).

Sanderson, R. V., and H. H. Y. Chien, "Simultaneous Chemical and Phase Equilibrium Calculation," *Ind. Eng. Chem. Process Design Dev.,* **12,** 81–85 (1973).

Schniepp, L. E., J. W. Dunning, and E. C. Lathrop, "Continuous Process for Acetylation of 2, 3-Butylene Glycol," *Ind. Eng. Chem.,* **37,** 872–877 (1945).

Sharma, M. M., "Some Novel Aspects of Cationic Ion-Exchange Resins as Catalysts," *Reat. Funct. Polym.,* **26,** 3–23 (1995).

Sherman, S. R., C. A. Eckert, and L. S. Scott, "Modeling Multicomponent Equilibria from Binary Equilibrum Data for Reacting Systems," *Chem. Eng. Proc.,* **35,** 363 (1996).

Shoemaker, J. D., and E. M. Jones, Jr., "Cumene by Catalytic Distillation," *Hydrocarbon Processing.,* **66,** 57–58 (1987).

Siirola, J. J., An Industrial Perspective on Process Synthesis, in Biegler, L. T., and M. F. Doherty, editors, *Foundations of Computer-Aided Process Design,* number 304 in AIChE Symp. Ser. pp. 222–233, New York. AIChE (1995).

Smith, L. A., Catalytic Distillation Structure (1984). U.S. Patent 4,443,559.

Smith, L. A., Method for the Alkylation of Organic Aromatic Compounds (1993). U.S. Patent 4,435,595.

Smith, W. R., and R. W. Missen, *Chemical Reaction Equilibrium Analysis: Theory and Algorithms.* Wiley, New York (1982).

Song, W., R. Huss, M. F. Doherty, and M. F. Malone, "Discovery of a Reactive Azeotrope," *Nature,* **388,** 561–563 (1997).

Song, W., G. Venimadhavan, J. M. Manning, M. F. Malone, and M. F. Doherty, "Measurement of Residue Curve Maps and Heterogeneous Kinetics in Methyl Acetate Synthesis," *Ind. Eng. Chem. Research,* **37,** 1917–1928 (1998).

Sundmacher, K., L. K. Rihko and U. Hoffmann, "Classification of Reactive Distillation Processes by Dimensionless Numbers," *Chem. Engr. Commun.,* **127,** 151–167 (1994).

Syed, F. H., C. Egleston, and R. Datta, "*tert*-Amyl Methyl Ether (TAME). Thermodynamic Analysis of Reaction Equilibria in the Liquid Phase," *J. Chem. Eng. Data,* **45,** 319–323 (2000).

Tang, S., S. Li, J. Han, and B. Chen, "Study of the Process of Reactive Distillation for Preparing Benzyl Benzoate," *Chemical Reaction Engineering and Technology,* **39,** 11 (1984).

Taylor, R., and R. Krishna, "Modelling Reactive Distillation," *Chem. Engng. Sci.,* **55,** 5183–5229 (2000).

Terrill, D. L., L. F. Sylvestre, and M. F. Doherty, "Separation of Closely Boiling Mixtures by Reactive Distillation. 1. Theory.," *Ind. Eng. Chem. Process Design Dev.,* **24,** 1062–1071 (1985).

Thelen, B., W. Auge, and K.-W. Thiem, Process for the Production of Aromatic Mononitro Compounds (1977). U.S. Patent 4,064,147.

Thiel, C., K. Sundmacher, and U. Hoffmann, "Residue Curve Maps for Heterogeneously Catalyzed Reactive Distillation of Fuel Ethers MTBE and TAME," *Chem. Engng. Sci.,* **52,** 993–1005 (1997).

Ung, S., and M. F. Doherty, "Calculation of Residue Curve Maps for Mixtures with Multiple Equilibrium Chemical Reactions," *Ind. Eng. Chem. Research,* **34,** 3195–3202 (1995*a*).

Ung, S., and M. F. Doherty, "Synthesis of Reactive Distillation Systems with Multiple Equilibrium Chemical Reactions," *Ind. Eng. Chem. Research,* **34,** 2555–2565 (1995*b*).

Ung, S., and M. F. Doherty, "Theory of Phase Equilibria in Multireaction Systems," *Chem. Engng. Sci.,* **50,** 3201–3216 (1995*c*).

Ung, S., and M. F. Doherty, "Vapor-Liquid Phase Equilibrium in Systems with Multiple Chemical Reactions," *Chem. Engng. Sci.,* **50,** 23–48 (1995*d*).

van Zeggeren, F., and S. H. Storey, *The Computation of Chemical Equilibria.* Cambridge University Press, London (1970).

Venimadhavan, G., G. Buzad, M. F. Doherty, and M. F. Malone, "Effect of Kinetics on Residue Curve Maps for Reactive Distillation," *AIChE J.,* **40,** 1814–1824 (1994).

Venimadhavan, G., M. F. Malone, and M. F. Doherty, "Bifurcation Study of Kinetic Effects in Reactive Distillation," *AIChE J.,* **45,** 546–556 (1999*a*).

Venimadhavan, G., M. F. Malone, and M. F. Doherty, "Novel Distillate Policy for Batch Reactive Distillation with Application to the Production of Butyl Acetate," *Ind. Eng. Chem. Research,* **38,** 714–722 (1999*b*).

Ward, D. J., Etherification of Isoamylenes by Catalytic Distillation (1993). U.S. Patent 5,196,612.

Wasylkiewicz, S. K., and S. Ung, "Global Phase Stability Analysis for Heterogeneous Reactive Mixtures and Calculation of Reactive Liquid-Liquid and Vapor-Liquid-Liquid Equilibria," *Fluid Phase Equilibria,* in press (2000).

Xu, X., B. Zhu, and H. Chen, "Reactive Distillation," *Shiyou Huagong,* **14,** 480–486 (1985).

Xu, Z., and M. P. Dudukovic, "Modeling and Simulation of Semi-batch Photo Reactive Distillation," *Chem. Engng. Sci.,* **54,** 1397–1403 (1999).

Zoeller, J. R., D. W. Lane, E. H. Cwirko, D. W. Fuller, Jr., and S. D. Barnicki, Process for Generating Vinyl Carboxylate Esters (1998). U.S. Patent 5,821,384.

A

Heat Effects

In earlier chapters, we frequently assumed that the liquid and vapor flows in each section of the column are constant. This is true when the column is adiabatic and the heat of vaporization is a constant. Because heats of vaporization are typically much larger than heats of mixing and also larger than the differences among the pure component heats of vaporization, a solution with constant molar flows is usually a good starting point for conceptual design. For some mixtures we need to account for variable internal flows, so we describe a more general approach after identifying cases where the added complexity is necessary. Often, the most important factor in modeling heat effects is good data and a model for the enthalpies of the components in the mixture. Information on pure component heats of vaporization is much easier to obtain than a good description for heats of mixing, so we will consider the importance of differences in the the pure component heats of vaporization as a first step in improving the fidelity of the CMO model.

A.1 APPLICABILITY OF CONSTANT MOLAR OVERFLOW

General Conditions

Constant molar overflow design equations rely on the constancy of L_n/V_{n-1} and D/V_{n-1} in the rectifying section and of L_{n+1}/V_n and B/V_n in the stripping section of a column. These conditions are met if and only if V is constant within each column section (L is constant in this case, by material balance). As a result, L/V is constant in each section of the column and the material balance relating vapor and liquid phase mole fractions is linear.

A general expression for the difference in vapor flow between two successive stages in either section of the column can be obtained from an energy balance around stage n.

$$V_{n-1}h_{n-1}^V + L_{n+1}h_{n+1}^L = V_n h_n^V + L_n h_n^L \tag{A.1}$$

and the overall material balance around stage n is $L_{n+1} - L_n = V_n - V_{n-1}$. These two equations can be combined to find the difference in vapor flowrates between successive stages as

$$V_n - V_{n-1} = \frac{L_n(h_{n+1}^L - h_n^L) + V_{n-1}(h_{n-1}^V - h_n^V)}{h_n^V - h_{n+1}^L} \tag{A.2}$$

Equation A.2 shows that V is constant if h^L and h^V are constant, i.e., independent of composition. This implies that the saturated liquid and vapor enthalpy lines are horizontal and linear, as shown in Fig. A.1. This demands that the reference enthalpies of the saturated pure liquids are equal; we generally choose them to be zero in this book. It follows that the latent heat of the mixture is independent of composition. That is, $\lambda(x) = \lambda_l = \lambda_h = \lambda$, where λ_i is the latent heat of pure i and $\lambda(x)$ is the latent heat of the mixture at composition x. In general, the enthalpy difference between coexisting equilibrium phases $h^V(y) - h^L(x)$ is not the same as the latent heats $\lambda(y)$ or $\lambda(x)$. However, for the special case shown in Fig. A.1, these quantities are all equal and constant.

A more general condition also follows from Eq. A.2, namely, if

$$\frac{h_{n-1}^V - h_n^V}{h_n^L - h_{n+1}^L} = \frac{L_n}{V_{n-1}} \tag{A.3}$$

then V will be constant. Since L is constant if V is constant, Equation A.3 becomes

$$\frac{h_{n-1}^V - h_n^V}{h_n^L - h_{n+1}^L} = \frac{L}{V} \tag{A.4}$$

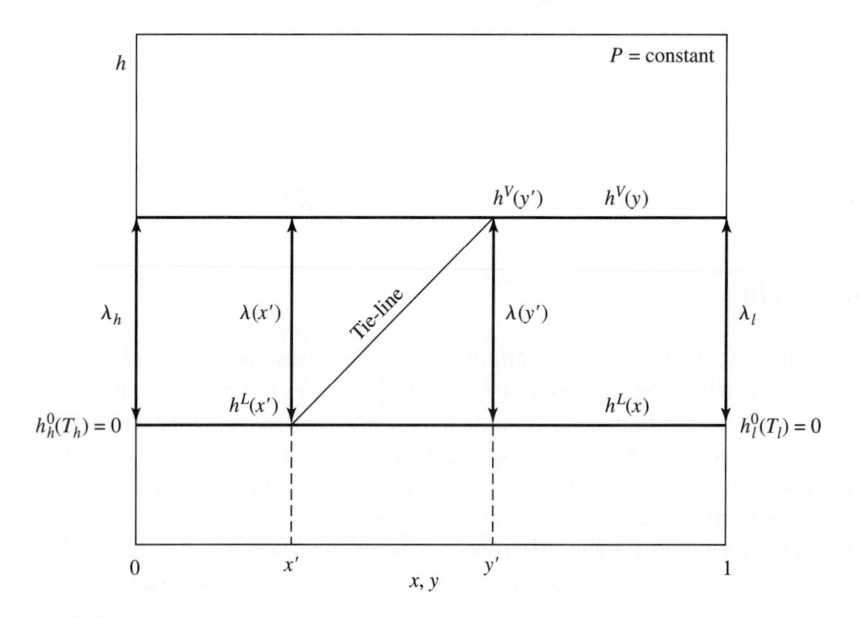

FIGURE A.1
Enthalpy-composition diagram for constant molar overflow; equal reference enthalpies.

A mass balance around plate n, with L and V constant, gives

$$\frac{L}{V} = \frac{y_{n-1} - y_n}{x_n - x_{n+1}} \tag{A.5}$$

Combining Eqs. A.4 and A.5 leads to

$$\frac{h_{n-1}^V - h_n^V}{y_{n-1} - y_n} = \frac{h_n^L - h_{n+1}^L}{x_n - x_{n+1}} \tag{A.6}$$

If Eq. A.6 holds for all n, i.e., for all possible values of the plate compositions, then V and therefore L will be constant. Equation A.6 requires the slopes of the lines segments shown in Fig. A.2 to be equal for all feasible combinations of the tie-lines. The only way that this can be satisfied is if the saturated molar enthalpy curves are *parallel straight lines*. Such a mixture also has the property of a constant latent heat, but this does not imply constant $h^L(x)$, constant $h^V(y)$, or constant $h^V(y) - h^L(x)$, as can be seen in Fig. A.2. Note that the reference enthalpies of the two pure saturated liquids are unequal in Fig. A.2, but otherwise the figure is identical to Fig. A.1. Since the reference enthalpies are arbitrary, it is gratifying to see that the conditions required for constant molar overflow do not depend on them.

We see that constant molar overflow requires that

1. The latent heat must be the same for each pure component, $\lambda_l = \lambda_h$.
2. The h^L vs. x and h^V vs. y lines should be straight.

These conditions insure that the saturated molar enthalpy curves are parallel straight lines, i.e., the latent heat of the mixture is independent of composition.

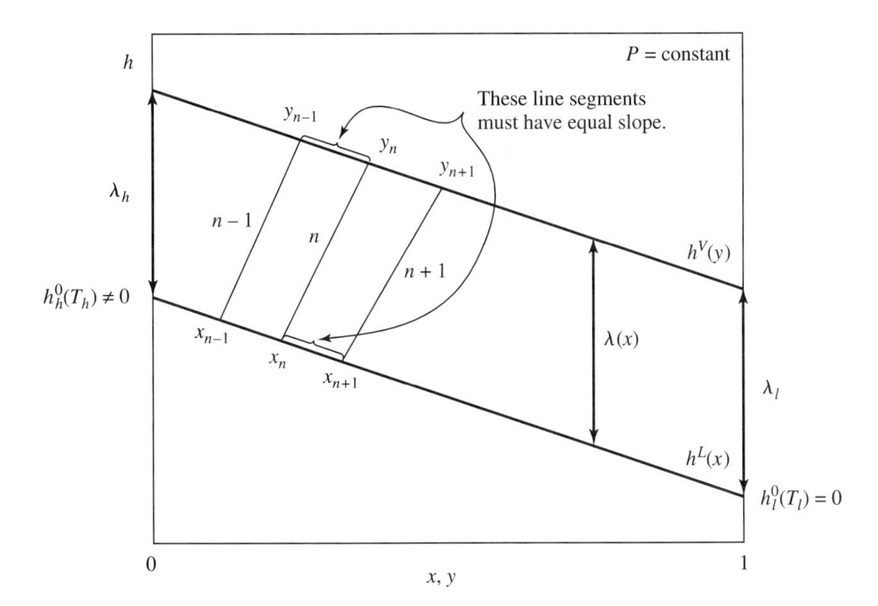

FIGURE A.2
Enthalpy-composition diagram for constant molar overflow; unequal reference enthalpies.

The first requirement is rarely satisfied exactly, even for pure components which form ideal mixtures. For example, data in Reid et al. (1977) show that the latent heat of vaporization for pure substances often encountered in distillation varies from 16,000 to 56,000 J/mol. In fact, the latent heat of vaporization at normal boiling conditions is a monotonically increasing function of the normal boiling temperature for many substances. This is reflected in Kistiakowsky's rule (Reid et al., 1977, p. 215).

$$\lambda_i = 8.75 T_i + R T_i \ln T_i \tag{A.7}$$

where λ_i is the latent heat of component i at the normal boiling point in cal/mol, T_i is the normal boiling point of component i in Kelvin, and $R = 1.987$ cal/mol K. Therefore, a large normal boiling point difference between the components undergoing distillation usually implies unequal latent heats.

The second requirement for CMO is not normally satisfied exactly, even when the mixture is ideal. The enthalpy of a liquid mixture is usually written as

$$h^L(T, \mathbf{x}) = h^{\text{ideal},L}(T, \mathbf{x}) + h^E(T, \mathbf{x}) \tag{A.8}$$

where

$$h^{\text{ideal},L}(T, \mathbf{x}) = \sum_{i=1}^{c} h_i^{0,L}(T) x_i \tag{A.9}$$

and $\mathbf{x}$ is the vector of mole fractions for the c components in the mixture. The saturated liquid enthalpy curve is given by Eq. A.8 with the understanding that T and $\mathbf{x}$ are related through the phase equilibrium equations. For binary mixtures, Eq. A.8 gives

$$h^{L,\text{sat}}(x) = c_{p_l}^L (T - T_l) x + c_{p_h}^L (T - T_h)(1 - x) + h^E \tag{A.10}$$

where T_l and T_h are the boiling points at the pressure of interest for the light and heavy components, respectively; T is the saturation temperature of the mixture (which depends on the composition x); and $c_{p_l}^L$ and $c_{p_h}^L$ are the average molar heat capacities of the two pure components over the temperature range of interest. This is typically T_l to T_h but may be a wider interval if azeotropes are present.

Ideal Binary Mixtures

For ideal mixtures, $h^E = 0$ and Eq. A.10 reduces to

$$h^{L,\text{sat}}(x) = c_{p_l}^L (T - T_l) x + c_{p_h}^L (T - T_h)(1 - x) \tag{A.11}$$

with T and x related implicitly by the phase equilibrium equation

$$P = P_l^{\text{sat}}(T) x + P_h^{\text{sat}}(T)(1 - x) \tag{A.12}$$

Note that $h^{L,\text{sat}}(x)$ can be nonlinear even for ideal phase equilibrium relations. The saturated enthalpy-composition diagram for a mixture of benzene and toluene at 1 atm is only mildly nonlinear, as shown in Fig. A.3. The diagram was calculated using Eqs. A.11 and A.12 together with the corresponding equations for the saturated vapor enthalpy curve

$$h^{V,\text{sat}}(y) = \left[c_{p_l}^V (T - T_l) + \lambda_l\right] y + \left[c_{p_h}^V (T - T_h) + \lambda_h\right](1 - y) \tag{A.13}$$

$$y = \frac{P_l^{\text{sat}} x}{P} \tag{A.14}$$

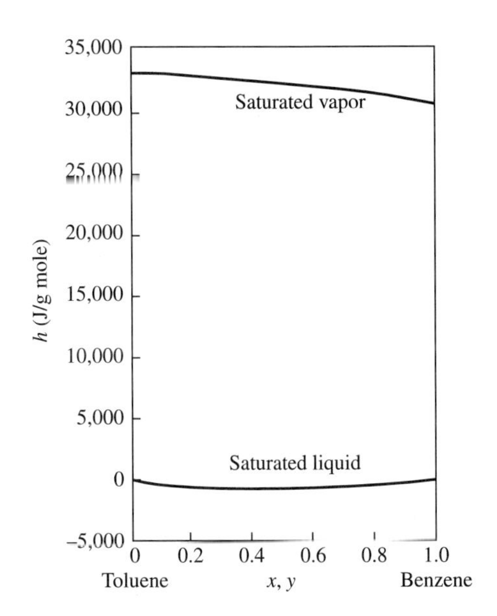

FIGURE A.3
Saturated enthalpy-composition diagram for benzene and toluene at 1 atm.

It should be emphasized that the independent variables in these calculations are taken to be P, which is held constant at 1 atm, and x. As x is varied, T and y are calculated; once these three variables are known, the enthalpies are calculated.

As expected, the saturated enthalpy lines are not quite parallel since $\lambda_l = 30{,}760$ J/mol and $\lambda_h = 33{,}180$ J/mol. However, the deviation of these curves from linearity is relatively slight, and this is typical of the behavior of many ideal mixtures when the difference in boiling points is not too large. In such cases, the curvature of the enthalpy-composition lines can generally be neglected in comparison to the effects arising from inequality of the molar latent heats of the pure components. This can be accounted for by the *Peters* method described below.

Nonideal Mixtures

For nonideal binary mixtures, $h^{L,\text{sat}}$ is given by Eq. A.10, with T and x related by the phase equilibrium expression

$$P = P_l^{\text{sat}} \gamma_l(T, x)x + P_h^{\text{sat}} \gamma_h(T, x)(1 - x) \tag{A.15}$$

In this case, a variety of shapes for the liquid enthalpy-composition diagram are possible. In the examples that follow, regular solution activity coefficient models have been used for the liquid phase, e.g., the Margules or Van Laar equations, with temperature-independent parameters. For such models, the excess entropy of the mixture s^E is zero and therefore[1]

$$h^E = g^E = RT\ x \ln \gamma_l + RT\ (1 - x) \ln \gamma_h \tag{A.16}$$

[1]Data for the enthalpy of boiling mixtures is often scarce, and it is not unusual to have a model that describes the compositions and temperatures accurately but gives much larger errors in the enthalpy.

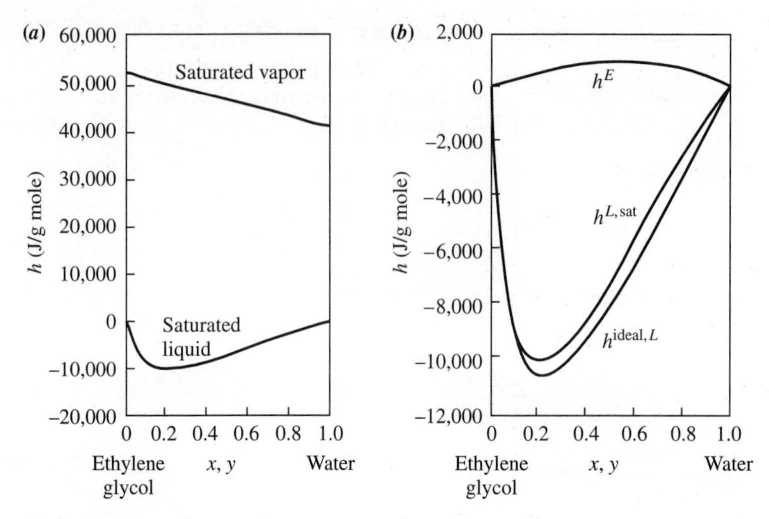

FIGURE A.4
Saturated enthalpy-composition diagram for water and ethylene glycol at 1 atm.

At low pressures, the saturated vapor enthalpy is usually adequately estimated from Eq. A.13; i.e., the excess enthalpy for the vapor phase can be taken as zero. For a given P and x, we calculate T and y from Eqs. A.15 and A.17

$$y = \frac{P_l^{sat}\gamma_l x}{P} \tag{A.17}$$

and the saturated vapor and liquid enthalpy curves can be computed.

The results of such a calculation for a mixture of water(l) and ethylene glycol(h) are shown in Fig. A.4*a* at a constant pressure of 1 atm. The liquid phase activity coefficients were calculated from the two-parameter Margules model. A notable feature of this diagram is the rather significant curvature in the saturated liquid enthalpy line. This, together with the unequal latent heats of the pure components, causes a significant variation in the enthalpy of vaporization of the mixture as its composition is varied. Water-rich mixtures have a latent heat of approximately 40,000 J/mol, while ethylene glycol-rich mixtures can have a value as much as 50% greater, at 60,000 J/mol.

Rather surprisingly, the curvature in the saturated liquid enthalpy-composition line is due almost entirely to the curvature in $h^{ideal,L}$, since the excess enthalpy is very small under saturation conditions. This is shown on an expanded scale in Fig. A.4*b*. Notice that there is a slight cancellation between the ideal and the excess enthalpy contributions, which causes $h^{L,sat}$ to lie between these curves. This cancellation effect can be quite pronounced in other mixtures, even to the extent of causing $h^{L,sat}$ to be nearly constant over the entire composition range. A classic example of such behavior is found in the mixture of acetone(l) and water(h), shown in Fig. A.5.

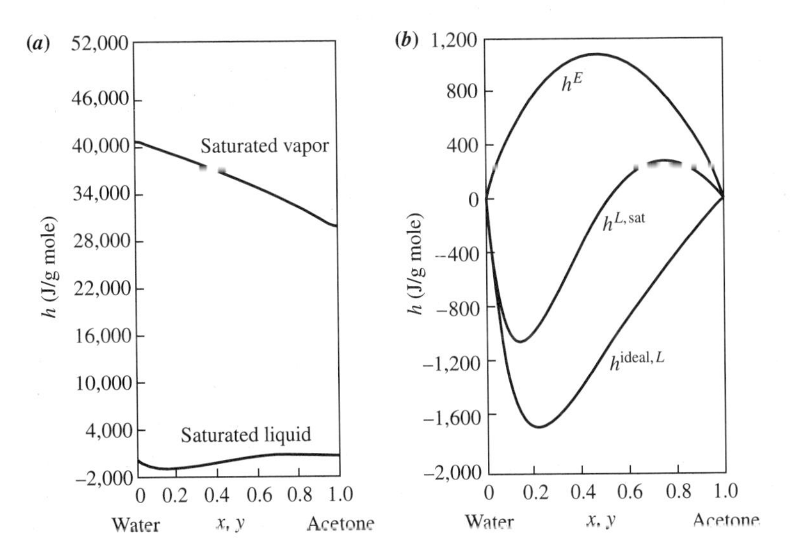

FIGURE A.5
Saturated enthalpy-composition diagram for acetone and water at 1 atm.

A.2
THE PETERS METHOD FOR BINARY MIXTURES

When heat effects are not negligible, the internal liquid and vapor flows vary from stage to stage. Thus, the design equations developed earlier for constant molar overflow need to be augmented by incorporating the energy balances, as given by Eqs. 3.23 and 3.24. Surprisingly, there is an elegant analytical solution to this problem for the important special case of linear enthalpy-composition relations with unequal latent heats. The ideas began with the original work of Peters (Peters, 1922), and additional aspects of the method were developed later (Fisher, 1963; and Bitter, 1983).

Solution Using Mole Fraction Variables

We consider enthalpy-composition diagrams of the form shown in Fig. A.6. The saturated liquid enthalpy is taken as zero and the saturated vapor enthalpy h^V is taken as a linear function of y

$$h^V = (\lambda_l - \lambda_h)y + \lambda_h \tag{A.18}$$

which we rewrite in the form

$$h^V = \lambda_h(1 - \epsilon y) \tag{A.19}$$

where $\epsilon \equiv (\lambda_h - \lambda_l)/\lambda_h$ is the relative difference between the latent heats of the two pure components. This lies in the interval $-\lambda_l/\lambda_h < \epsilon < 1$, and $\epsilon = 0$ corresponds to constant molar overflow. Since the latent heat of the higher-boiling component is

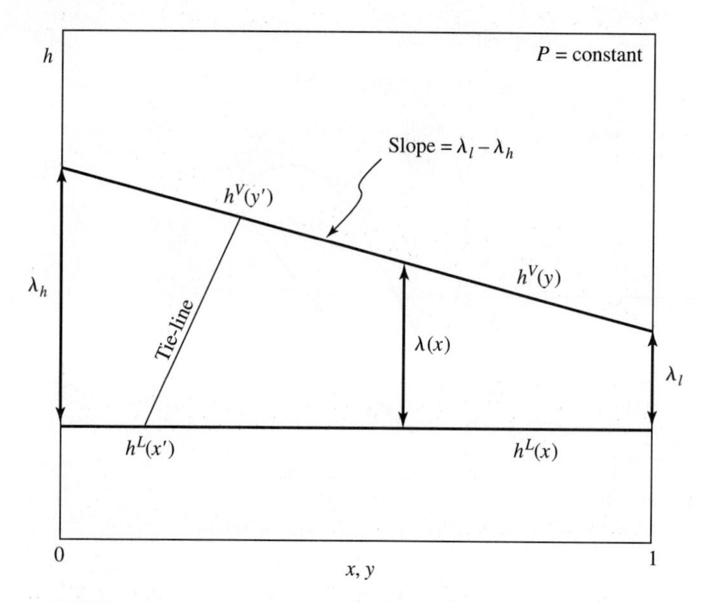

FIGURE A.6
Saturated enthalpy-composition diagram for the Peters method.

frequently greater than that of the lower boiler, ϵ is often, but not always, positive, as shown in Table A.1.

If we denote the *internal reflux ratio* on each stage as $r_n = L_n/D$, the mass balance for the rectifying section can be written

$$y_{n-1} = \frac{r_n}{r_n + 1} x_n + \frac{z_D}{r_n + 1} \tag{A.20}$$

We are interested in cases where r_n varies from stage to stage. Its value at the top stage is equal to the *external reflux ratio* $r \equiv r_{N+1} = L_{N+1}/D$, which is normally set as one of the design variables.

Equations A.20 can be solved along with the VLE if r_n can be expressed in terms of quantities such as r, ϵ, z_D, and x_n. This requires that we solve the energy balance from stage to stage. In the rectifying section, we can write the energy balance around the top of the column including the condenser and any stage n as

$$(L_n + D)h_{n-1}^V = L_n h_n^L + Dh_D + Q_C \tag{A.21}$$

For the condenser, $n = N + 1$ and this becomes

$$(L_{N+1} + D)h_N^V = L_{N+1}h_{N+1}^L + Dh_D + Q_C \tag{A.22}$$

Subtracting these two equations, using the definition of r_n, and taking the liquid enthalpies as zero, we find

$$r_n + 1 = \frac{1 - \epsilon y_N}{1 - \epsilon y_{n-1}}(r + 1) \tag{A.23}$$

This gives the internal reflux ratio for each stage as a function of the vapor composition entering stage n, the composition of the vapor leaving the top of the column,

TABLE A.1
Some latent heats and their relative differences at the normal boiling point of the pure substance (Reid et al., 1977, Appendix A). These binary mixtures have enthalpy-composition diagrams that are nearly linear

Light Component (normal b.p., °C)	Heavy Component (normal b.p., °C)	λ_l (J/mol)	λ_h (J/mol)	ϵ
Acetone (56.1)	Water (100)	29,120	40,660	0.284
Benzene (80.1)	Toluene (110.6)	30,760	33,180	0.073
i–Butane (−11.8)	i–Octane (99.3)	21,300	31,010	0.313
Hexane (68.7)	p–Xylene (138.3)	28,850	35,980	0.198
Methyl acetate (56.9)	Methanol (64.6)	30,120	35,260	0.146
i–Octane (99.3)	n-Dodecane (216.4)	31,010	43,640	0.289
Pentane (36.0)	Hexane (68.7)	25,770	28,850	0.107
Toluene (110.6)	Diphenyl (255.3)	33,180	45,610	0.273
Water (100.0)	Acetic acid (118.0)	40,660	23,680	−0.717
Water (100.0)	Butyric acid (163.2)	40,660	42,000	0.033
Water (100.0)	1-4 Dioxane (101.3)	40,660	36,358	−0.106
Water (100.0)	Ethylenediamine (117.2)	40,660	41,840	0.028
Water (100.0)	Ethylene glycol (197.2)	40,660	52,509	0.226
Water (100.0)	Isopropylamine (117.2)	40,660	27,196	−0.495

and the external reflux ratio. It is also useful to eliminate the vapor compositions in favor of liquid compositions using the mass balance from Eq. A.20 in Eq. A.23 to find

$$r_n = \frac{1 - \epsilon x_{N+1}}{1 - \epsilon x_n} r \tag{A.24}$$

Equation A.24 can be used along with Eq. A.20 to obtain a *nonlinear* operating line for the top of the column.

For the stripping section, the internal reboil ratio is $s_n \equiv V_n/B$, while $s_0 = V_0/B \equiv s$ is the external reboil ratio. The mass balance is

$$y_n = \frac{s_n + 1}{s_n} x_{n+1} - \frac{z_B}{s_n} \tag{A.25}$$

Following a procedure for a solution of the energy balances similar to that for the top of the column, we find

$$s_n = \frac{1 - \epsilon y_0}{1 - \epsilon y_n} s \quad \text{and} \quad s_n + 1 = \frac{1 - \epsilon x_1}{1 - \epsilon x_{n+1}} (s + 1) \tag{A.26}$$

Equation A.26 can be combined with Eq. A.25 to find a nonlinear bottom operating line.

The overall energy balance constrains r, s, and the feed quality, according to Eq. 3.22. Using the assumptions above for the enthalpies, the result is

$$\frac{D}{B} = \frac{s(1 - \epsilon y_0) + (1 - q)(1 - \epsilon z_F)}{(r + 1)(1 - \epsilon y_N) - (1 - q)(1 - \epsilon z_F)} \tag{A.27}$$

The physical interpretation of q is described in the comments following Eq. 3.31. In this equation, y_0 is the mole fraction of light component in the vapor entering the column from the reboiler and y_N is the vapor composition leaving the top of the column. Depending on the type of condenser and reboiler, a dew point or bubble point calculation may be needed to find these values. As expected, Eq. A.27 reduces to Eq. 3.35 when $\epsilon = 0$.

In contrast to the constant molar overflow case, the q-line is *nonlinear,* and the result is

$$\hat{y} = \frac{q(1 - \epsilon z_F)\hat{x} - (1 - \epsilon\hat{x})z_F}{(1 - \epsilon\hat{x})(q - 1 - \epsilon q z_F) - \epsilon q\hat{x}(1 - \epsilon z_F)} \tag{A.28}$$

This is a sufficient basis for a graphical solution like the McCabe-Thiele method but with operating lines that are curved on account of the variation of enthalpy, and thus liquid and vapor rates, from stage to stage.

Figure A.7 shows a comparison of a constant molar overflow construction with the results found using the nonlinear operating lines and q-line developed above. More stages are required for the same external reflux ratio and the minimum reflux ratio calculated using the Peters method will be larger than that found using the constant molar overflow assumption for this mixture. Different conclusions are found for $\epsilon < 0$.

For binary mixtures, the stage-to-stage solution of the general energy balances can also be done graphically instead of analytically as was done above. This is the *Ponchon-Savarit* construction and accounts for curvature in both the liquid and vapor enthalpy-composition relationships. The details of this approach can be found elsewhere (King, 1980; Treybal, 1980; Wankat, 1988).

The Peters Transforms

Analytical solutions can also be developed for linear enthalpy-composition relations using quite a remarkable transform, originally introduced by Peters for binary mix-

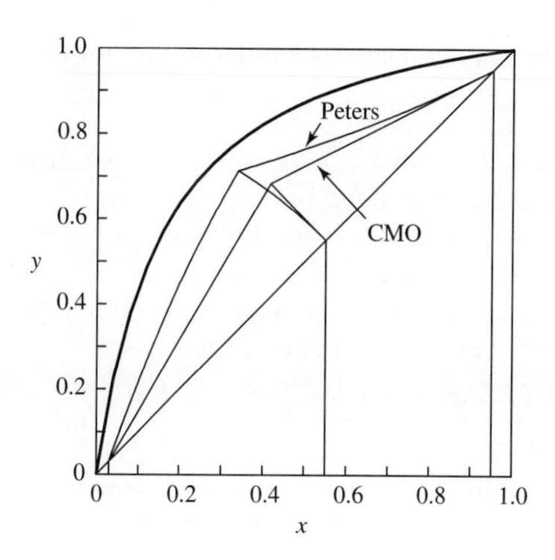

FIGURE A.7
Comparison of the Peters and the constant molar overflow constructions for the (hypothetical) case $\alpha = 7$, $\tilde{q} = 0.5$, $x_F = 0.55$, $x_D = 0.95$, $x_B = 0.03$, and $r = 1$, corresponding to $s_{\text{CMO}} = 1.45$. The reboil ratio is calculated from Eq. A.27 for $\epsilon = 0.5$, which gives $s = 0.723$. The corresponding value of q is 0.58.

tures (Peters, 1922).[2] This converts the nonlinear operating relationships above to linear forms in a new set of variables. In that case the methods developed for the constant molar overflow case such as Underwood's equations, Smoker's equation, and the McCabe-Thiele construction, can be applied. Furthermore, the method can be extended to cases with more components. The original work of Peters treated only binary mixtures, and we illustrate this case before developing a more general approach in the next section.

The transformed composition variables for binary mixtures are

$$u = \frac{\lambda_l x_l}{\lambda_l x_l + \lambda_h x_h} \qquad v = \frac{\lambda_l y_l}{\lambda_l y_l + \lambda_h y_h} \qquad w = \frac{\lambda_l z_l}{\lambda_l z_l + \lambda_h z_h} \tag{A.29}$$

and these expressions may be rearranged into the more convenient forms

$$u = \frac{1-\epsilon}{1-\epsilon x} x \qquad v = \frac{1-\epsilon}{1-\epsilon y} y \qquad w = \frac{1-\epsilon}{1-\epsilon z} z \tag{A.30}$$

Note that these transformed variables range from zero to unity and that $u = x, v = y$, and $w = z$ when $\epsilon = 0$.

For mixtures with constant relative volatility, the VLE expression in Eq. 2.31 remains *unchanged* under the action of the above transforms, i.e.,

$$v = \frac{\alpha u}{1 + (\alpha - 1)u} \tag{A.31}$$

Application of the transforms to the operating lines, Eqs. A.20 and A.25, and the q-line, Eq. A.28, gives

$$v_{n-1} = \frac{r'}{r'+1} u_n + \frac{u_D}{r'+1} \tag{A.32}$$

$$v_n = \frac{s'+1}{s'} u_{n+1} - \frac{u_B}{s'} \tag{A.33}$$

and

$$\hat{v} = \frac{q}{q-1} \hat{u} - \frac{w_F}{q-1} \tag{A.34}$$

where $w_F = (1-\epsilon)z_F/(1-\epsilon z_F)$ and the effective reflux and reboil ratios are

$$r' = \left[\frac{1-\epsilon x_{N+1}}{1-\epsilon z_D}\right] r \qquad s' = \left[\frac{1-\epsilon y_0}{1-\epsilon z_B}\right] s \tag{A.35}$$

These equations are identical in form to those developed in Chap. 3, except that the transformed composition variables u, v, and w are used in place of the mole fractions. *Therefore, all of the results developed for constant molar overflow can be applied to cases with heat effects when the appropriate transformed variables are employed.* Note that Eq. A.27 must be used to relate r and s, rather than Eq. 3.35, and that the VLE must also be transformed.

[2]This is sometimes described in terms of a fictitious molecular weight for the mixture, but it is hard to see the utility of this viewpoint, especially for mixtures with more than two components.

Aside on the feed quality

The thermodynamic state of the feed is known if its composition, temperature, and pressure are specified. It is common to characterize the feed in terms of its *quality* q, and we use the definition in Eq. 3.32.

Unfortunately, the fraction of liquid in a vapor-liquid mixture $\tilde{q}$ is often confused with q as defined in Chap. 3. It is best to think of q as a dimensionless enthalpy; our scale is with respect to the heat of vaporization of the feed, as given in Eq. 3.32. When the constant molar overflow description is accurate, q is given by the simpler expression in Eq. 3.33. In this special case, and for two-phase feeds, $\tilde{q}$ is equal to q. However, when constant molar overflow is not a good approximation, $\tilde{q} \neq q$, even though these quantities are often close in numerical value. Using the enthalpy-composition relationship as in the Peters method, we can find an explicit relationship between $\tilde{q}$ and q for a two-phase feed

$$\tilde{q} = 1 - (1 - q)\frac{\boldsymbol{\lambda} \cdot \mathbf{z_F}}{\boldsymbol{\lambda} \cdot \mathbf{y_F}} \tag{A.36}$$

where $\mathbf{z_F}$ is the overall feed composition and $\mathbf{y_F}$ is the composition of the vapor portion of the feed. For other cases, it is simple to find q using Eq. 3.32 after a separate calculation of the enthalpy for the feed. You should probably forget about relating $\tilde{q}$ and q for such cases, although $\tilde{q}$ is a decent substitute for q for two phase feeds in the absence of the enthalpy calculations.

EXAMPLE A.1. We reconsider Example 3.1, but now with heat effects taken into account. From Table A.1 we see that $\epsilon = 0.198$ for the hexane–*p*-xylene mixture. The information provided in the original description of the example was: $\tilde{q} = 0.5, z_F = 0.55, x_D = 0.95, x_B = 0.03$, and $\alpha = 7.0$. For $\tilde{q} = 0.5$, $x_F = 0.327$, $y_F = 0.773$, and Eq. A.36 gives $q = 0.53$.

- We begin by applying the transforms, Eqs. A.30, to obtain $w_F = 0.495, u_D = 0.938, u_B = 0.024$.
- To calculate $r_{\min}$ we use Eqs. 3.54 and 3.55 written in terms of w_F, u_D, u_p, v_p, and q. Equation 3.54 gives $u_p = 0.282$, Eq. A.31 produces $v_p = 0.734$, and Eq. 3.55 gives $r_{\min} = 0.451$.
- The minimum number of stages required to achieve the separation can be calculated from the transformed Fenske equation, i.e., Eq. 3.48, where $S = \frac{u_D}{1-u_D}\frac{1-u_B}{u_B}$. However, it is easy to show that S is invariant under the action of the transform; thus $N_{\min}$ is not affected by the presence of heat effects. This occurs because the nonlinear operating lines approach the 45° line as r and s become infinite. Thus, $N_{\min} = 3.3$ as before.
- The actual number of stages in the column is calculated at $r = 1.2r_{\min} = 0.578$. From Eq. A.27 we find $s = 0.716$.
- We use Smoker's method (Exercise 3.8) in the transformed variables to find the number of stages in the stripping section. The intersection of the operating lines in the transformed variables occurs at $\hat{u} = 0.290, \hat{v} = 0.700$. Therefore, $U_0 = u_B - \bar{u} = 0.297, U_n = \hat{u} - \bar{u} = -0.031$. The intermediate variables $a = 1 + (\alpha - 1)\bar{u}$ and $\bar{\alpha} = \frac{\alpha}{ma^2}$ take the values $a = 2.926, \bar{\alpha} = 0.326$. Applying Smoker's equation 3.73 between U_0 and U_n gives 4.0 theoretical stages. Thus, there will be three stages in the bottom of the column, plus a partial reboiler.
- In the rectifying section, $m = 0.366, b = 0.594, \bar{u} = 0.233$, and $\bar{v} = 0.680$. The points $\hat{u}$ and $\hat{v}$ are unchanged; therefore, $U_0 = \hat{u} - \bar{u} = 0.057, U_n = u_D - \bar{u} = 0.706$.

The intermediate variables are $a = 2.395$ and $\bar{\alpha}$=3.332. Applying Smoker's equation between U_0 and U_n gives 3.22 theoretical stages. This is the number of stages in the rectifying section of the column since a total condenser is used.

Comparing this design with the one developed in Sec. 3.2, we see that the minimum reflux ratio is more than 20% larger and the total number of theoretical stages is 14% larger than is predicted by ignoring heat effects.

It is quite easy to understand these trends by looking at the shape of the nonlinear operating lines in the Peters model. In fact, at the expense of some algebra, we can show some general results. For given values of r, q, z_F, x_D, and x_B, it follows that:

- The operating lines in the Peters method lie *above* the constant molar overflow lines when $\epsilon > 0$ and $0 \le q \le 1$. A typical set of curves are shown in Fig. A.7.
- The constant molar overflow analysis *underestimates* $r_{\min}$ when $\epsilon > 0$ and *overestimates* $r_{\min}$ when $\epsilon < 0$. As a demonstration of this, Fig. A.7 shows the two sets of operating lines for the p-xylene-hexane example problem.
- The constant molar overflow analysis *underestimates* the number of theoretical plates required to achieve a given separation.
- The constant molar overflow analysis *overestimates* the vapor rate in the bottom of the column, i.e., overestimates s. For the reasons disussed in Chap. 6, it follows that the constant molar overflow analysis *overestimates* the column diameter, the utilities requirements, and the heat exchanger sizes for the reboiler and condenser.
- As a final note, we mention that these arguments apply also to mixtures that cannot be described with a constant volatility, provided that the enthalpy-composition lines are straight.

All systems with positive values of ϵ and $0 \le q \le 1$ will exhibit similar behavior. When $\epsilon > 0$ and $q > 1$, i.e., subcooled liquid feeds, these results need not necessarily hold—although for values of q less than about 2, they typically do.

For the special case of saturated liquid feeds and high-purity distillates, i.e., $q = 1$ and $z_D \approx 1$, it is possible to obtain a simple equation that gives a good estimate of the difference in $r_{\min}$ between the Peters (P) method and the constant molar overflow (CMO) method for constant α mixtures. In this case, Eq. 3.57 simplifies to

$$r_{\min,\text{CMO}} = \frac{1}{\alpha - 1}\frac{1}{x_F} \tag{A.37}$$

The corresponding equation for the Peters method is

$$r_{\min,P} = \frac{1}{\alpha - 1}\frac{1}{u_F} \tag{A.38}$$

The discrepancy between the methods can be represented by the ratio

$$\frac{r_{\min,P}}{r_{\min,\text{CMO}}} = \frac{x_F}{u_F} = \frac{1 - \epsilon x_F}{1 - \epsilon} \tag{A.39}$$

When $\epsilon > 0$, the constant molar overflow approximation *underestimates* the minimum reflux ratio, while for $\epsilon < 0$, the constant molar overflow approximation *overestimates* the minimum reflux ratio. In either case the ratio is independent of α and increases as x_F decreases or as $|\epsilon|$ increases, i.e., as heat effects become more pronounced.

A.3 THE PETERS METHOD FOR MULTICOMPONENT MIXTURES

We now consider heat effects for mixtures with more than two components, but where the enthalpy of both the liquid and the vapor phase can be approximated as a linear function of the composition. In such cases, the enthalpy of the liquid can be taken as zero by a judicious choice of the reference enthalpy for the pure components. This gives

$$h^L = 0 \tag{A.40}$$

$$h^V = \boldsymbol{\lambda} \cdot \mathbf{y} \tag{A.41}$$

where $\boldsymbol{\lambda}$ and $\mathbf{y}$ are vectors of c pure-component heats of vaporization and mole fractions, respectively. The dot product of two vectors is indicated by the "•" symbol. The operating relations can be written most conveniently as

$$(r_n + 1)\mathbf{y}_{n-1} = r_n \mathbf{x}_n + \mathbf{z}_D \tag{A.42}$$

$$s_n \mathbf{y}_n = (s_n + 1)\mathbf{x}_{n+1} - \mathbf{z}_B \tag{A.43}$$

where the reboiler has $n = 0$ and the condenser $n = N + 1$.

In the rectifying section, the energy balances for a general stage n as well as for the condenser (without specializing to any particular case of saturated liquid or vapor products) are

$$(L_n + D)h^V_{n-1} = L_n h^L_n + D h_D + Q_C \tag{A.44}$$

$$(L_{N+1} + D)h^V_N = L_{N+1} h^L_{N+1} + D h_D + Q_C \tag{A.45}$$

Subtracting these, using the definition $r_n \equiv L_n/D$, and taking the liquid enthalpies as zero, we find

$$\frac{r_n + 1}{r + 1} = \frac{\boldsymbol{\lambda} \cdot \mathbf{y}_N}{\boldsymbol{\lambda} \cdot \mathbf{y}_{n-1}} \tag{A.46}$$

It is also convenient to have r_n as a function of $\mathbf{x}_n$, and this can be found by eliminating the vapor phase compositions from Eq. A.46 using Eq. A.42, which gives

$$\frac{r_n}{r} = \frac{\boldsymbol{\lambda} \cdot \mathbf{x}_{N+1}}{\boldsymbol{\lambda} \cdot \mathbf{x}_n} \tag{A.47}$$

Substitution of Eqs. A.46 and A.47 into A.42 and rearranging gives an expression for the top operating relation

$$(r' + 1)\mathbf{v}_{n-1} = r'\mathbf{u}_n + \mathbf{w}_D \tag{A.48}$$

where the transformed composition variables are

$$u_i = \frac{\lambda_i x_i}{\boldsymbol{\lambda} \cdot \mathbf{x}} \tag{A.49}$$

$$v_i = \frac{\lambda_i y_i}{\boldsymbol{\lambda} \cdot \mathbf{y}} \tag{A.50}$$

$$w_i = \frac{\lambda_i z_i}{\boldsymbol{\lambda} \cdot \mathbf{z}} \tag{A.51}$$

and the effective reflux ratio is

$$r' \equiv r\left[\frac{\boldsymbol{\lambda}\cdot\mathbf{x}_{N+1}}{\boldsymbol{\lambda}\cdot\mathbf{z}_D}\right] \tag{A.52}$$

For a total condenser, $\mathbf{x}_{N+1} = \mathbf{z}_D$ so $r' = r$.

For the stripping section, a similar approach gives

$$(V_n + B)h^L_{n+1} = V_n h^V_n + Bh_B - Q_R \tag{A.53}$$

$$(V_0 + B)h^L_1 = V_0 h^V_0 + Bh_B - Q_R \tag{A.54}$$

from which it follows that

$$\frac{s_n}{s} = \frac{\boldsymbol{\lambda}\cdot\mathbf{y}_0}{\boldsymbol{\lambda}\cdot\mathbf{y}_n} \tag{A.55}$$

and

$$\frac{s_n + 1}{s + 1} = \frac{\boldsymbol{\lambda}\cdot\mathbf{x}_1}{\boldsymbol{\lambda}\cdot\mathbf{x}_{n+1}} \tag{A.56}$$

Equations A.55 and A.56, along with a mass balance around the reboiler to eliminate $\mathbf{y}_0$ in favor of $\mathbf{x}_1$ and $\mathbf{z}_B$, to find the bottom operating relation

$$s'\mathbf{v}_n = (s' + 1)\mathbf{u}_{n+1} - \mathbf{w}_B \tag{A.57}$$

from Eq. A.43. The effective reboil ratio is found from

$$s' \equiv s\left[\frac{\boldsymbol{\lambda}\cdot\mathbf{y}_0}{\boldsymbol{\lambda}\cdot\mathbf{z}_B}\right] \tag{A.58}$$

and for a total reboiler, $\mathbf{y}_0 = \mathbf{z}_B$ so that $s' = s$.

The overall energy balance, analogous to Eq. A.27, is

$$\frac{D}{B} = \frac{s'(\boldsymbol{\lambda}\cdot\mathbf{z}_B) + (1-q)(\boldsymbol{\lambda}\cdot\mathbf{z}_F)}{(r'+1)(\boldsymbol{\lambda}\cdot\mathbf{z}_D) - (1-q)(\boldsymbol{\lambda}\cdot\mathbf{z}_F)} \tag{A.59}$$

Thus, all of the results for multicomponent CMO models obtained previously can also be used directly in terms of the transformed variables.

A.4 EXERCISES

1. When heat effects are negligible, the slopes of the operating lines are constant and equal to L/V. What is the geometrical interpretation of L/V when $\epsilon \neq 0$?
2. When $\epsilon > 0$, does the internal reflux ratio increase or decrease from the top stage to the feed stage in the rectifying section? How does s_n vary in the stripping section? What would be typical percentage changes in r_n and s_n between the ends and the middle of the column? Draw typical profiles for the change in the liquid and vapor flowrates over the entire height of the column when $\epsilon > 0$ and $q = 1,\ q = 0$.
3. Derive the expression $r_n = [(1 - \epsilon x_D)/(1 - \epsilon x_n)]\, r$.
4. Show that the VLE expression for constant α systems, $y = \alpha x/[1 + (\alpha - 1)x]$ remains unchanged under the actions of Peters' transforms, Eq. A.30.

5. Show that the nonlinear q-line in the Peters method, i.e., Eq. A.28, reduces to the constant molar overflow form under the action of Peters' transforms.
6. Show that the nonlinear top and bottom operating lines in the Peters method reduce to the constant molar overflow forms under the action of Peters' transforms.
7. Solve Exercise 3.19, accounting for heat effects. The enthalpy-composition lines for the saturated liquid and vapor phases are almost linear and can be represented by the expressions

$$h^L = 0 \qquad h^V = \lambda_h(1 - \epsilon y)$$

where $\lambda_h = 35{,}260$ J/mol and $\epsilon = 0.146$ (see Table A.1). Now solve the problem by combining Peters' method with the VLE expression Eq. 3.65, taking $\alpha' = \alpha_B = 2.6$. Compare your answers with the constant molar overflow design obtained for the case $\epsilon = 0$.

References

Bitter, R., "Comments on: 'Analytical Form of the Ponchon-Savarit Method for Systems with Straight Enthalpy-Composition Phase Lines'," *Ind. Eng. Chem. Process Design Dev.,* **22,** 684 (1983).

Fisher, G. T., "Modification of the McCabe-Thiele Method for Systems of Unequal Heats of Vaporization," *Ind. Eng. Chem. Process Design Dev.,* **2,** 284 (1963).

King, C. J., *Separation Processes.* McGraw-Hill, New York, 2d ed. (1980).

Peters, W. A., Jr., "The Efficiency and Capacity of Fractionating Columns," *Ind. Eng. Chem.,* **14,** 476 (1922).

Reid, R. C., J. M. Prausnitz, and T. K. Sherwood, *The Properties of Gases and Liquids.* McGraw-Hill, New York, 3d ed. (1977).

Treybal, R. E., *Mass Transfer Operations.* McGraw-Hill, New York, 3d ed. (1980).

Wankat, P. C., *Equilibrium Staged Separations.* Prentice-Hall, Englewood Cliffs, NJ (1988).

B

Implicit Functions

Tangent pinches are discussed in Sec. 4.6, where they are analyzed as bifurcations in the solutions of Eqs. 3.60 (written in the more general form of Eqs. 4.81) for pinch compositions. In this appendix, the conditions applied in Sec. 4.6 to find those solutions are described as a general result from the implicit function theorem.

B.1 THE IMPLICIT FUNCTION THEOREM FOR A SINGLE EQUATION

Suppose we are given an equation of the form

$$F(x, \lambda) = 0 \tag{B.1}$$

where x is the unknown variable and λ is a parameter with specified values. Then, for each value of λ there may be one or more real values of x which satisfy Eq. B.1. It is also possible that there are no real values which do so. If I is an interval on the range of values of λ defined by $\lambda_0 - \epsilon_\lambda < \lambda < \lambda_0 + \epsilon_\lambda$, such that for each value of λ on I there is exactly one value of x satisfying Eq. B.1, then we say that $F(x, \lambda) = 0$ defines x as a function of λ *implicitly* on I. Denoting this function by f, we have $x = f(\lambda)$, where $f(\lambda)$ satisfies $F(f(\lambda), \lambda) = 0$ for λ on I.

An implicit function theorem is one that determines conditions under which a relation like Eq. B.1 defines x as a function of λ. The solution is usually a local one because we normally have to restrict the size of the interval on the x axis as well as the interval on the λ axis. For example, Fig. B.1 shows a graph of a relation such as Eq. B.1. We see that F defines x as a unique function of λ in a region about P. However, F defines x as a local function of λ in a region about R. That is to say, unless we restrict our attention to values of x in the interval $x_R - \epsilon_x < x < x_R + \epsilon_x$, then the relation $F(x, \lambda) = 0$ does not produce a unique solution for x for values of λ in the interval $\lambda_R - \epsilon_\lambda < \lambda < \lambda_R + \epsilon_\lambda$. At the point Q, F does not define x as a function of λ, since it is not possible to obtain a unique value of x satisfying $F(x, \lambda) = 0$ for each value of λ in the interval $\lambda_Q - \epsilon_\lambda < \lambda < \lambda_Q + \epsilon_\lambda$. This is

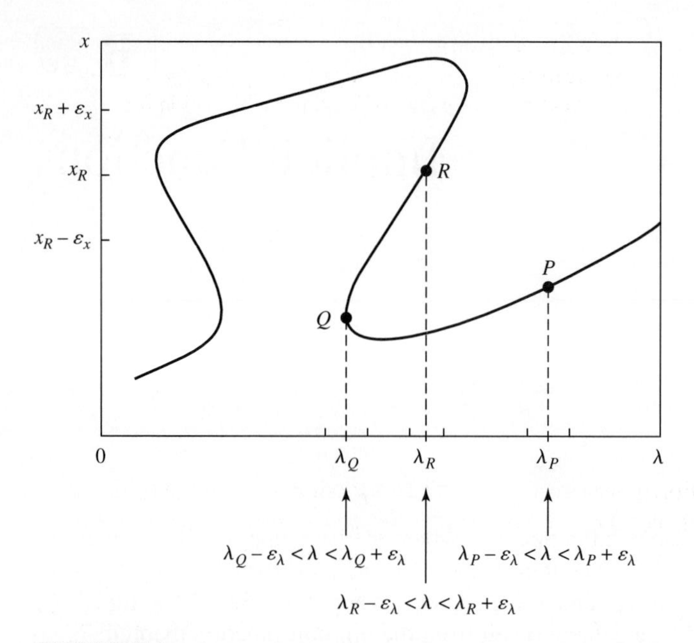

FIGURE B.1
Graph of a relation such as Eq. B.1

all stated more precisely in the following theorem, which gives sufficient conditions for the existence of an implicit function.

The Implicit Function Theorem

Suppose that $F(x, \lambda)$, $F_x(x, \lambda)(\equiv (\frac{\partial F}{\partial x}(x, \lambda))_\lambda)$ and $F_\lambda(x, \lambda)(\equiv (\frac{\partial F}{\partial \lambda}(x, \lambda))_x)$ are continuous in an open set of points A in the x, λ plane containing the point (x^*, λ^*), and suppose that $F(x^*, \lambda^*) = 0$ and $F_x(x^*, \lambda^*) \neq 0$. Then

1. There are positive numbers ϵ_x and ϵ_λ which determine a rectangle R contained in A given by

$$R = \left\{(x, \lambda) : |x - x^*| < \epsilon_x, |\lambda - \lambda^*| < \epsilon_\lambda\right\}$$

such that for *each* λ in $I = \{\lambda : |\lambda - \lambda^*| < \epsilon_\lambda\}$ there is a *unique* number x in the interval $J = \{x : |x - x^*| < \epsilon_x\}$ which satisfies the equation $F(x, \lambda) = 0$. The totality of the points (x, λ) forms a function f whose domain contains I and whose range is in J.
2. The function f and its derivative $df/d\lambda$ are continuous on I.

The proof of this theorem, and corollaries of it, e.g., the inverse function theorem, are given in Protter and Morrey (1977, pp. 332–337).

A bifurcation point (x_0, λ_0) in the relation $F(x, \lambda) = 0$ is a point at which multiple values for x start to satisfy the relation for each value of λ. The graph of points x and λ which satisfy the relation $F(x, \lambda) = 0$ is called a bifurcation diagram.

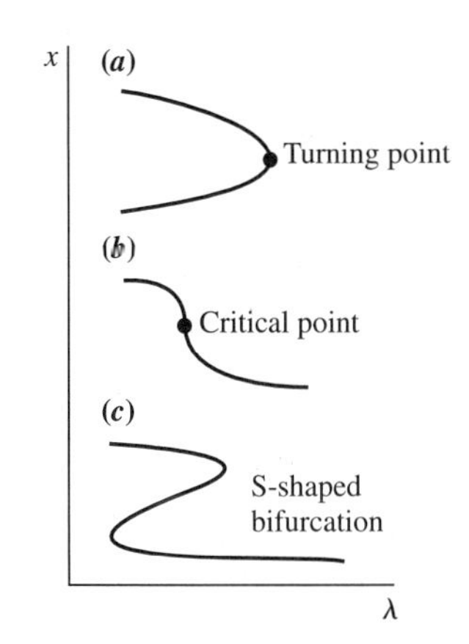

FIGURE B.2
Bifurcation diagram showing a turning point (*a*), a critical point (*b*), and an S-shaped bifurcation (*c*).

The purpose of bifurcation theory is to classify the most common types of bifurcation diagrams, and to provide conditions, in terms of the derivatives of F, that enable us to distinguish between them.

We conclude from the implicit function theorem that $F_x(x_0, \lambda_0) = 0$ is a necessary condition for a solution (x_0, λ_0), of $F(x, \lambda) = 0$ to be a bifurcation point, for otherwise we could solve uniquely for x as a smooth function of λ. A point (x_0, λ_0) at which $F(x_0, \lambda_0) = F_x(x_0, \lambda_0) = 0$ is called a *singularity*. Note that a singularity need not be a bifurcation point, as demonstrated by the function

$$x^3 - \lambda + 2 = 0 \tag{B.2}$$

in which we will take λ to be a positive parameter. The point $x_0 = 0$, $\lambda_0 = 2$ satisfies $F(x_0, \lambda_0) = F_x(x_0, \lambda_0) = 0$, yet the singularity is not a bifurcation point, since the relation may be solved explicitly and uniquely for x, i.e., $x = (\lambda - 2)^{\frac{1}{3}}$.

The simplest kind of bifurcation occurs when the bifurcation diagram exhibits a turning point,[1] as shown in Fig. B.2*a*. It is easy to show that sufficient conditions for a singularity to be a turning point are $F_{xx}(x_0, \lambda_0) \neq 0$, $F_\lambda(x_0, \lambda_0) \neq 0$. These come about because the values of x and λ which satisfy $F(x, \lambda) = 0$ all lie on a level curve of the function $F(x, \lambda)$. Since

$$dF = F_x dx + F_\lambda d\lambda \tag{B.3}$$

those special values of x and λ which lie on a level curve must satisfy

$$F_x dx + F_\lambda d\lambda = 0 \tag{B.4}$$

or

$$\frac{d\lambda}{dx} = -\frac{F_x}{F_\lambda} \tag{B.5}$$

Differentiating this expression with respect to x gives

$$\frac{d^2\lambda}{dx^2} = -\left(\frac{F_\lambda F_{xx} - F_x F_{\lambda x}}{F_\lambda^2}\right) \tag{B.6}$$

[1]Such points are also called *folds* and *limit points* in the literature.

Sufficient conditions for a turning point to occur in the bifurcation diagram are $\frac{d\lambda}{dx} = 0$, $\frac{d^2\lambda}{dx^2} \neq 0$. A singularity is defined by $F(x_0, \lambda_0) = F_x(x_0, \lambda_0) = 0$; thus we see that sufficient conditions for a singularity to be a turning point are $F_{xx}(x_0, \lambda_0) \neq 0$, $F_\lambda(x_0, \lambda_0) \neq 0$. Therefore, if a point (x_0, λ_0) satisfies the conditions $F(x_0, \lambda_0) = F_x(x_0, \lambda_0) = 0$, $F_{xx}(x_0, \lambda_0) \neq 0$, $F_\lambda(x_0, \lambda_0) \neq 0$, then the implicit function theorem fails to guarantee the existence of a unique solution for x, and the resulting singularity is a bifurcation point of the turning point type.

Another simple singularity is the *critical point,*[2] which is shown in Fig. B.2*b*. Following on from our previous logic, sufficient conditions for such a point are $F(x_0, \lambda_0) = F_x(x_0, \lambda_0) = 0$, $F_{xx}(x_0, \lambda_0) = 0$, $F_{xxx}(x_0, \lambda_0) \neq 0$, $F_\lambda(x_0, \lambda_0) \neq 0$. These conditions imply that a stationary inflection occurs at the point (x_0, λ_0) in the bifurcation diagram, i.e., $\frac{d\lambda}{dx}(x_0, \lambda_0) = 0$, $\frac{d^2\lambda}{dx^2}(x_0, \lambda_0) = 0$, $\frac{d^3\lambda}{dx^3}(x_0, \lambda_0) = -\frac{F_{xxx}(x_0,\lambda_0)}{F_\lambda(x_0,\lambda_0)} \neq 0$. Although the implicit function theorem fails to guarantee the existence of a unique solution for x at the singularity, a unique solution does nevertheless exist, as can be seen from the shape of the bifurcation diagram. Note that the singularity at the point $x_0 = 0$, $\lambda_0 = 2$ in relation B.2 is a critical point. (The reader should check that the conditions are satisfied and sketch the graph in order to reinforce the theory.)

In many problems of interest to chemical engineers, turning points occur in pairs. This gives rise to the S-shaped bifurcation diagram shown in Fig. B.2*c*. If the problem contains a second parameter, say α, in addition to λ, then we can draw a family of bifurcation diagrams parameterized by α. A typical family of S-shaped bifurcation diagrams is shown in Fig. B.3*a*. As the value of α is changed, the S in the curves normally gets shallower until it eventually disappears altogether at a critical point. Further change in α produces curves which no longer exhibit multiple solutions for x. The family of curves shown in Fig. B.3*a* can also be represented as a folded surface called the *cusp bifurcation surface,* as shown in Fig. B.3*b*. The set of points (x, λ, α) that define this surface all satisfy the relation

$$F(x; \lambda, \alpha) = 0 \tag{B.7}$$

Fixing the value of α, and following our earlier logic, we find that a set of points (x_0, λ_0, α) which satisfy the following equations[3]

$$\begin{aligned} F(x_0; \lambda_0, \alpha) = F_x(x_0; \lambda_0, \alpha) &= 0 \\ F_{xx}(x_0; \lambda_0, \alpha) &\neq 0 \\ F_\lambda(x_0; \lambda_0, \alpha) &\neq 0 \end{aligned} \tag{B.8}$$

define a locus of turning points in the x-λ graphs. Similarly, a point $(x^c, \lambda^c, \alpha^c)$ which satisfies the equations

$$\begin{aligned} F(x^c; \lambda^c, \alpha^c) = F_x(x^c; \lambda^c, \alpha^c) = F_{xx}(x^c; \lambda^c, \alpha^c) &= 0 \\ F_{xxx}(x^c; \lambda^c, \alpha^c) &\neq 0 \\ F_\lambda(x^c; \lambda^c, \alpha^c) &\neq 0 \end{aligned} \tag{B.9}$$

is a critical point.

[2]Sometimes called a *hysteresis point*.

[3]The symbol $F_x(x_0; \lambda_0, \alpha)$ now means the partial derivative of F with respect to x at constant λ and α, evaluated at the point (x_0, λ_0, α). Similar interpretations apply to the other partial derivatives.

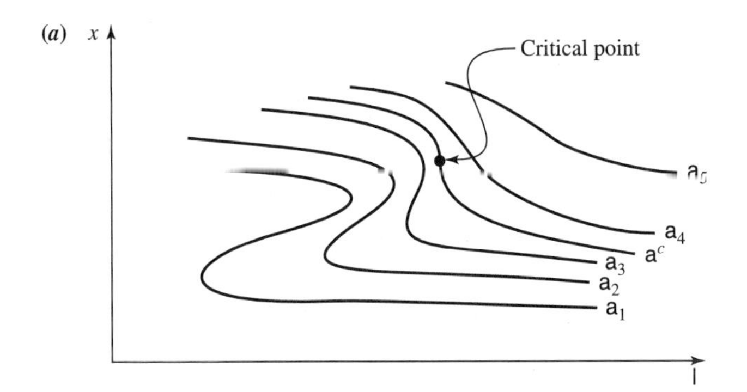

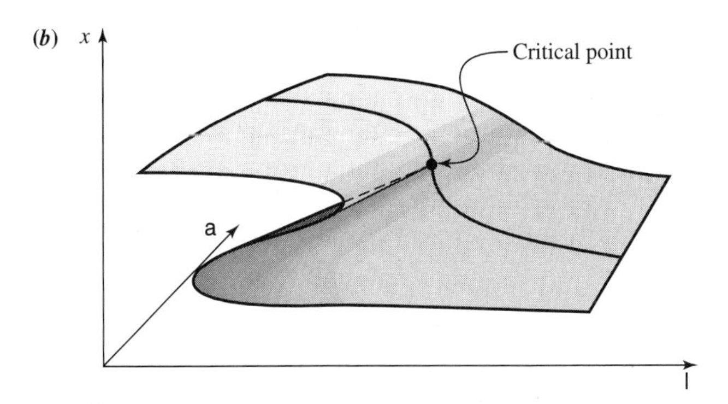

FIGURE B.3
Typical family of S-shaped bifurcation diagrams (*a*), parameterized by α and the *cusp bifurcation surface* (*b*).

In two-parameter problems other types of bifurcations are common, e.g., the *pitchfork*, which occurs when a turning point on one branch of solutions of Eq. B.7 meets a critical point on another branch of solutions. A much more general theory than that presented here is now in an advanced state of development but is best left to the specialist reader [see, e.g., Golubitsky and Schaeffer (1985); Seydel (1994)].

EXAMPLE B.1. One of the most familiar examples of the cusp bifurcation surface is provided by the Van der Waals equation of state for a pure fluid

$$P = \frac{RT}{v-b} - \frac{a}{v^2} \tag{B.10}$$

where a and b are positive constants. A typical family of isotherms on a P-v phase diagram that satisfy this equation are shown schematically in Fig. B.4. There is a clear similarity between this figure and those shown in Fig. B.3, once we make the associations $v \to x$, $P \to \lambda$, $T \to \alpha$. So, for example, we can find the coordinates of the thermodynamic critical point in the fluid by detecting the presence of a critical point in the cusp bifurcation surface generated by Eq. B.10. We write

$$F(v; P, T) = P - \frac{RT}{v-b} + \frac{a}{v^2} = 0 \tag{B.11}$$

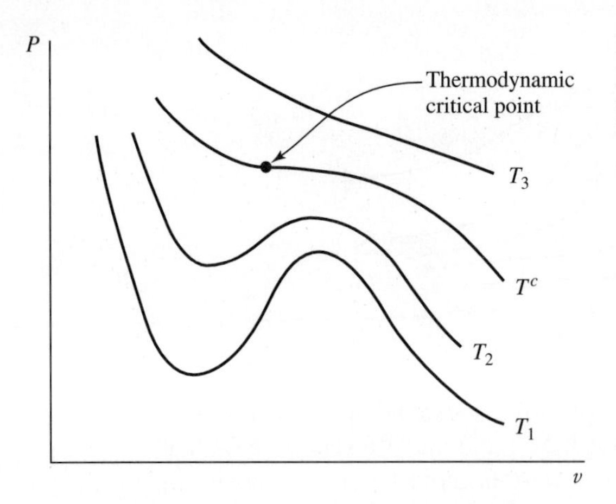

FIGURE B.4
Family of isotherms on a P-v phase diagram.

and now check the sufficiency conditions for the presence of a critical point, i.e.,

$$F(v^c; P^c, T^c) = F_v(v^c; P^c, T^c) = F_{vv}(v^c; P^c, T^c) = 0 \tag{B.12}$$

$$F_{vvv}(v^c; P^c, T^c) \neq 0, F_P(v^c; P^c, T^c) \neq 0 \tag{B.13}$$

Equations B.12 can be solved explicitly for v^c; P^c, T^c, to give $v^c = 3b$, $P^c = a/27b^2$, $T^c = 8a/27Rb$, and at this point the conditions given by Eqs. B.13 are also satisfied since

$$F_{vvv}(v^c; P^c, T^c) = 1.23456\frac{a}{b^5} \neq 0$$

$$F_P(v^c; P^c, T^c) = 1 \neq 0$$

These results are in agreement with those obtained using the more conventional approach of classical thermodynamics.

B.2 THE IMPLICIT FUNCTION THEOREM FOR SYSTEMS OF EQUATIONS

We now wish to extend our methods to systems of equations of the form

$$F_i(x_1, x_2, \ldots, x_n; \lambda_1, \lambda_2, \ldots, \lambda_k) = 0 \qquad i = 1, 2, \ldots, n \tag{B.14}$$

which we will represent in the more compact notation

$$\mathbf{F}(\mathbf{x}; \boldsymbol{\lambda}) = \mathbf{0} \tag{B.15}$$

where $\mathbf{x} = (x_1, x_2, \ldots, x_n)$, $\boldsymbol{\lambda} = (\lambda_1, \lambda_2, \ldots, \lambda_k)$ and $\mathbf{F} = (F_1, F_2, \ldots, F_n)$. Note that the number of equations is equal to the number of unknowns.

The following theorem gives a sufficient condition which guarantees that the system of equations B.15 may be solved locally for a unique value of $\mathbf{x}$ as a function of the parameters $\boldsymbol{\lambda}$. Before stating the theorem, we introduce one additional piece of notation. The symbol $\mathbf{J_x}(\mathbf{x}^*, \boldsymbol{\lambda}^*)$ stands for the $n \times n$ Jacobian matrix of first partial derivatives of F_i with respect to x_j, evaluated at the point $(\mathbf{x}^*, \boldsymbol{\lambda}^*)$.

Thus,

$$\mathbf{J}_{\mathbf{x}}(\mathbf{x}^*, \boldsymbol{\lambda}^*) = \begin{bmatrix} \frac{\partial F_1}{\partial x_1} & \cdots & \frac{\partial F_1}{\partial x_n} \\ \vdots & \frac{\partial F_i}{\partial x_j} & \vdots \\ \frac{\partial F_n}{\partial x_1} & \cdots & \frac{\partial F_n}{\partial x_n} \end{bmatrix}_{\mathbf{x}=\mathbf{x}^*, \boldsymbol{\lambda}=\boldsymbol{\lambda}^*}$$

The Implicit Function Theorem

Suppose that $\mathbf{F}(\mathbf{x}^*, \boldsymbol{\lambda}^*) = 0$ and that $\det \mathbf{J}_x(\mathbf{x}^*, \boldsymbol{\lambda}^*) \neq 0$. Then there exists a neighborhood U of $\mathbf{x}^*$ and V of $\boldsymbol{\lambda}^*$ such that for every $\boldsymbol{\lambda} \in V$, Eq. B.15 has a unique solution $\mathbf{x} = \mathbf{f}(\boldsymbol{\lambda})$ in U. Moreover, if $\mathbf{F}$ is s times differentiable, so is $\mathbf{f}$.

A more precise statement and proof of this theorem is given in Protter and Morrey (1977, pp. 339–349).

As for one-dimensional systems, we conclude from the implicit function theorem that $\det \mathbf{J}_x(\mathbf{x}^*, \boldsymbol{\lambda}^*) = 0$ is a necessary condition for a solution $(\mathbf{x}^*, \boldsymbol{\lambda}^*)$ of $\mathbf{F}(\mathbf{x}, \boldsymbol{\lambda}) = \mathbf{0}$ to be a bifurcation point, for otherwise we could solve uniquely for $\mathbf{x}$ as a smooth function of $\boldsymbol{\lambda}$. We can also develop sufficiency conditions for the existence of various types of singularities in terms of the higher-order derivatives of $F_i(\mathbf{x}, \boldsymbol{\lambda})$, $i = 1, 2, \ldots, n$. However, this is a complex subject best left as a reading exercise for the specialist [see, e.g., Golubitsky et al. (1988)].

As a final comment, it is worth noting that if the function $\mathbf{F}$ does not depend on any parameters, so that Eq. B.15 becomes

$$\mathbf{F}(\mathbf{x}) = \mathbf{0} \tag{B.16}$$

then if $\det \mathbf{J}_{\mathbf{x}}(\mathbf{x}_i^*) \neq 0$, where $\mathbf{F}(\mathbf{x}_i^*) = \mathbf{0}$, the implicit function theorem guarantees that each solution $\mathbf{x}_i^*$ is *locally unique*. That is, solutions of the equation are *isolated*.

References

Golubitsky, M., and D. G. Schaeffer, *Singularities and Groups in Bifurcation Theory,* vol. 1. Springer Verlag, New York (1985).

Golubitsky, M., I. Stewart, and D. G. Schaeffer, *Singularities and Groups in Bifurcation Theory,* vol. 2. Springer Verlag, New York (1988).

Protter, M. H., and C. B. Morrey, Jr., *A First Course and Real Analysis.* Springer Verlag, New York (1977).

Seydel, R., *Practical Bifurcation and Stability Analysis: From Equilibrium to Chaos.* Springer Verlag, New York (1994).

C

Azeotropy and the Gibbs–Konovalov Conditions

In this appendix we present the main ideas behind the characterization of azeotropes and vapor-liquid phase diagrams for multicomponent mixtures that do not react and which have a single liquid phase; see Chap. 5. Chapter 8 considers multiple liquid phases, and Chap. 10 treats reacting mixtures.

C.1 AZEOTROPY

Consider a closed system in which a multicomponent, homogeneous, nonreactive liquid mixture is being vaporized at constant pressure (or temperature, but not both), as shown schematically in Fig. C.1. A material balance on component i gives

$$\frac{d(Lx_i)}{dt} + \frac{d(Vy_i)}{dt} = 0 \qquad i = 1, 2, \ldots, c-1 \tag{C.1}$$

where L and V are the liquid and vapor molar holdups, respectively. The overall material balance for the system is

$$\frac{dL}{dt} + \frac{dV}{dt} = 0 \tag{C.2}$$

If we expand the derivatives in Eq. C.1 and use Eq. C.2 to eliminate the term dV/dt, we can write the component balance for the closed system as

$$L\frac{dx_i}{dt} + V\frac{dy_i}{dt} + \frac{dL}{dt}(x_i - y_i) = 0 \qquad i = 1, 2, \ldots, c-1 \tag{C.3}$$

During an azeotropic transformation, $dx_i/dt = dy_i/dt = 0$ and $dL/dt \neq 0$. Therefore, according to Eq. C.3, a *necessary* condition which must hold during the course of an azeotropic transformation in a closed system is

$$x_i = y_i \qquad i = 1, 2, \ldots, c-1 \tag{C.4}$$

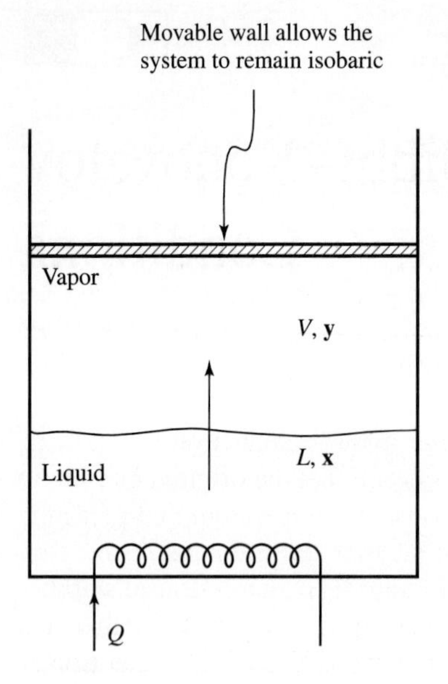

FIGURE C.1
Schematic representation of equilibrium vaporization in an isobaric closed system.

Equation C.4 also applies to the last component c, but this is not an independent condition and is thus omitted.

We now show that when Eq. C.4 holds at each instant during a vaporization or condensation in a closed system, then it follows that both $dx_i/dt = 0$ and $dy_i/dt = 0$. To show this, it is convenient to write the material balances, Eqs. C.1 and C.2, in the equivalent, integrated form

$$Lx_i + Vy_i = N_{i0} \qquad i = 1, 2, \ldots, c - 1 \tag{C.5}$$

$$L + V = N \tag{C.6}$$

where N_{i0} is the total, constant amount of component i present in the closed system (in moles) and N is the total number of moles in the system (also a constant). Putting $x_i = y_i$ in Eq. C.5, and using Eq. C.6, gives

$$Nx_i = N_{i0} \qquad i = 1, 2, \ldots, c - 1 \tag{C.7}$$

Differentiating this equation with respect to time shows that

$$\frac{dx_i}{dt} = 0 \qquad i = 1, 2, \ldots, c - 1 \tag{C.8}$$

That is, the composition of the liquid is constant during the course of the transformation. From Eq. C.7, the value of this constant is N_{i0}/N. By a similar argument we find $dy_i/dt = 0$. Thus, Eq. C.4 is the *necessary and sufficient condition* for an azeotropic transformation to occur in homogeneous, nonreactive multicomponent mixtures.

C.2
MATERIAL STABILITY

The second law of thermodynamics provides criteria to distinguish those processes which may occur spontaneously in a closed system from those which are impossible. In a closed system held at constant temperature and pressure, any process which increases the total Gibbs free energy of the system is thermodynamically impossible. Processes which reduce the total Gibbs free energy, or keep it constant,[1] are thermodynamically feasible. These conditions can be used to determine whether a fluid mixture at a given overall composition, and fixed temperature and pressure, will split into two (or more) phases.

If an arbitrary microscopic fluctuation brings about a reduction in the total Gibbs free energy of the mixture, then we can be certain that the original phase will spontaneously split into two or more phases at constant temperature and pressure. The original phase is called *materially unstable*.[2] Other terms for the same concept include: diffusional stability (Prigogine and Defay, 1967) and intrinsic stability (Modell and Reid, 1983; Tester and Modell, 1997), and such phases can be detected using local analysis based upon Taylor series expansions of the Gibbs free energy function. A materially unstable phase will initially split into two (or more) nonequilibrium phases which continuously change composition until they are ultimately in phase equilibrium. The time scale and size scale for the composition changes during the nonequilibrium transient is not a matter for thermodynamics. These effects can usually be predicted by either one or a combination of several convective diffusion theories which are derived from the various hypothesized mechanisms for the dynamics of phase transitions.[3]

If a mixture is stable with respect to microscopic fluctuations, then it is called *materially stable*. Such mixtures may or may not be stable with respect to macroscopic fluctuations. A materially stable mixture which is unstable with respect to macroscopic fluctuations is said to be *metastable*. Metastable mixtures will survive as a homogeneous phase provided heterogeneous nucleation sites are absent. Such mixtures exist in nature and in the laboratory, e.g., see Reid (1978*a*, *b*, *c*) for a discussion of superheat limit temperatures in pure fluids and mixtures. We refer to a materially stable mixture which is also stable with respect to macroscopic fluctuations as being *absolutely stable*. Such mixtures must exist as a homogeneous phase.

A simple convexity criterion can be developed which makes it easy to distinguish between materially stable and materially unstable mixtures. It is harder, but of

[1] All thermodynamically reversible processes and some irreversible processes maintain the total Gibbs free energy of the system constant. An example of such an irreversible process is the equilibrium azeotropic vaporization of a liquid in a closed system.

[2] This is the terminology used by Rowlinson (1969, p. 143).

[3] Two popular mechanisms for the dynamics of phase transitions are spinodal decomposition (Cahn and Hilliard, 1958; Cahn, 1965; Kwei and Wang, 1978) and nucleation and growth (Christian, 1965).

great practical importance, to distinguish between metastable and absolutely stable mixtures (Michelsen, 1982).

For microscopic fluctuations, the change in the total Gibbs free energy of the closed system can be expressed in the form (Modell and Reid, 1983, Chap. 9).

$$\delta G = k\delta \mathbf{x}^T \mathbf{g} \delta \mathbf{x} + \text{higher-order terms} \tag{C.9}$$

where k is a positive quantity, $\delta \mathbf{x}$ is the vector of $c - 1$ independent composition fluctuations about the overall composition of the original mixture, and $\mathbf{g}$ is a $(c - 1) \times (c - 1)$ square symmetric matrix consisting of the second partial derivatives of the molar Gibbs free energy g with respect to the mole fractions, evaluated at the fixed T and P of the system and at the overall composition of the original mixture; thus

$$\mathbf{g} = \begin{bmatrix} & \vdots & \\ \cdots & \left(\frac{\partial^2 g}{\partial x_i \partial x_j}\right)_{T,P,\mathbf{x}'} & \cdots \\ & \vdots & \end{bmatrix} \tag{C.10}$$

If δG is positive for all possible microscopic fluctuations, then the original mixture is materially stable. On the basis of this information alone, it is not possible to tell whether the phase is metastable or absolutely stable. If δG is negative definite or indefinite, then the original mixture is materially unstable, which guarantees that the mixture will split into two or more phases.

A necessary and sufficient condition for the quadratic form representation of δG to be positive for all microscopic fluctuations is that the matrix $\mathbf{g}$ be positive definite. This, in turn, can be tested by Sylvester's theorem (Mirskey, 1961), which states that a necessary and sufficient condition for a matrix to be positive definite is that the symmetric part have strictly positive principal minors, e.g.,

$$g_{11} > 0; \quad \det \begin{bmatrix} g_{11} & g_{12} \\ g_{21} & g_{22} \end{bmatrix} > 0\,; \quad \ldots \quad ; \det \mathbf{g} > 0 \tag{C.11}$$

where $g_{ij} = \left(\frac{\partial^2 g}{\partial x_i \partial x_j}\right)_{T,P,\mathbf{x}'}$.

The limit of material stability for a homogeneous mixture occurs along the locus of points where $\mathbf{g}$ just ceases to be positive definite. This locus of points is called the *spinodal surface*, and following the arguments of Mirskey (1961) this occurs where the determinant of $\mathbf{g}$ becomes zero. Thus, the spinodal surface is defined by

$$\det \mathbf{g} = 0 \tag{C.12}$$

For binary mixtures, the criterion of material stability reduces to

$$\left(\frac{\partial^2 g}{\partial {x_1}^2}\right)_{T,P} > 0 \tag{C.13}$$

A plot of g vs. x_1 will reveal whether or not the mixture exhibits a miscibility gap, as shown in Fig. C.2. If the curve is concave, the mixture is unstable and cannot exist as a homogeneous phase. The mixture is materially stable wherever the curve is convex. On the same plot, the binodal curve can be located by drawing the locus of

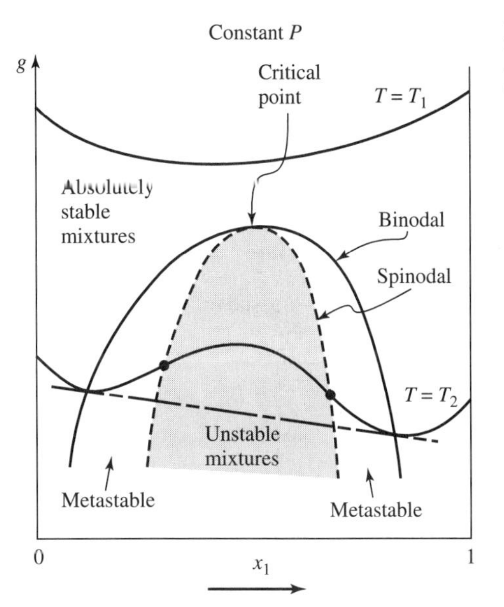

FIGURE C.2
Relationship between the molar Gibbs free energy function $g(T, P, x_1)$ and the various types of stability regions for a binary mixture.

common tangents for the g-x_1 isotherms. The spinodal curve is the locus of inflection points on the g-x_1 isotherms, and the region in between the spinodal and binodal is the metastable region. Homogeneous liquids can exist in this region but only under carefully controlled conditions.

C.3 THE GIBBS–KONOVALOV CONDITIONS FOR HOMOGENEOUS AZEOTROPES

The various possible shapes of isobaric binary T-x, y phase diagrams are easy to visualize and widely discussed in the literature.[4] However, isobaric equilibrium T-$\mathbf{x}, \mathbf{y}$ surfaces for mixtures containing three or more components are less easy to visualize and exhibit a greater variety of shapes than binary phase diagrams. In addition to maxima and minima, new features such as saddles, ridges and valleys are quite commonplace on the equilibrium surfaces for multicomponent mixtures. The resulting phase diagrams can be very complicated, and they frequently contain many azeotropes. In fact, there are examples of practical interest where every subsystem in the mixture contains an azeotrope; e.g., at pressures above 5 atmospheres, the mixture ethanol + water + acetone exhibits one ternary azeotrope and three binary azeotropes (one for each binary subsystem).

[4]See, for example, Smith and VanNess (1996, Chap. 12) and Rowlinson (1969, Chap. 6) for a general discussion of binary phase diagrams. Malesiński (1965, Chap. VI) describes the types of binary phase diagrams predicted by regular solution theory. A vast amount of experimental binary phase equilibrium data is available in the *DECHEMA Vapor-Liquid Equilibrium Data Collection.*

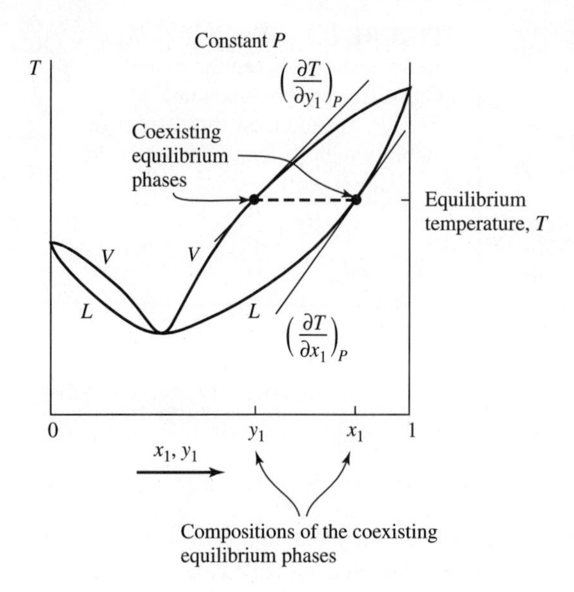

FIGURE C.3
Schematic representation of an isobaric binary phase diagram showing the tangents to the equilibrium curves for coexisting equilibrium phases.

It would be extremely helpful if thermodynamics provided explicit expressions for T as a function of P and $\mathbf{x}$, and T as a function of P and $\mathbf{y}$, thereby explicitly characterizing the bubble point and dew point surfaces, respectively. It is only possible to obtain such explicit expressions under highly idealized limiting conditions, such as those described in Sec. 2.3. Therefore, if we are to successfully develop fundamental relationships between the two equilibrium surfaces we need to find another approach. This is provided by the theory of equilibrium displacements, which gives explicit expressions for the partial derivatives along the equilibrium surfaces.[5] A surprisingly large amount of useful information can be obtained from this technique.

For binary liquid-vapor mixtures under isobaric conditions, the tangents $\left(\frac{\partial T}{\partial x_1}\right)_P$ and $\left(\frac{\partial T}{\partial y_1}\right)_P$ to the bubble point and dew point curves, respectively, for coexisting equilibrium phases are represented schematically in Fig. C.3. The theory of equilibrium displacements shows that these derivatives can be expressed as

$$\left(\frac{\partial T}{\partial x_1}\right)_P = \frac{T\left(\frac{\partial^2 g^L}{\partial x_1{}^2}\right)_{T,P}(x_1 - y_1)}{y_1 \Delta h_1 + y_2 \Delta h_2} \tag{C.14}$$

and

$$\left(\frac{\partial T}{\partial y_1}\right)_P = \frac{T\left(\frac{\partial^2 g^V}{\partial y_1{}^2}\right)_{T,P}(x_1 - y_1)}{x_1 \Delta h_1 + x_2 \Delta h_2} \tag{C.15}$$

where Δh_i is the difference between the partial molar enthalpy of component i in the vapor and the partial molar enthalpy of component i in the liquid for the coexisting equilibrium phases, i.e., $\Delta h_i = h_i{}^V(T, P, y_1) - h_i{}^L(T, P, x_1)$. The quantities x_1 and

[5] A quite comprehensive treatment of this subject is given in Malesiński (1965, Chap. V). Additional material can be found in Rowlinson (1969, Chap. 6) and Prigogine and Defay (1967).

y_1 are the compositions of the coexisting equilibrium phases at the fixed pressure P and equilibrium temperature T. The functions g^L and g^V represent the molar Gibbs free energy functions of the liquid and vapor phases, respectively. They can be expressed in the form

$$g^L(T, P, x_1) = x_1 \mu_1{}^L(T, P, x_1) + (1 - x_1)\mu_2{}^L(T, P, x_1) \qquad \text{(C.16)}$$

$$g^V(T, P, y_1) = y_1 \mu_1{}^V(T, P, y_1) + (1 - y_1)\mu_2{}^V(T, P, y_1) \qquad \text{(C.17)}$$

where the chemical potentials can be represented either by an equation of state or by a solution model (Smith and VanNess, 1996, Chaps. 10–13; Prausnitz et al., 1998).

Disregarding the critical region, the quantities Δh_1 and Δh_2 are positive. The absolute temperature, appearing explicitly on the right-hand side of Eqs. C.14 and C.15, is positive, and for materially stable phases, the second derivatives of the Gibbs free energy are also positive. Thus, it follows that

$$\left(\frac{\partial T}{\partial x_1}\right)_P = 0 \Leftrightarrow \left(\frac{\partial T}{\partial y_1}\right)_P = 0 \Leftrightarrow x_1 = y_1 \qquad \text{(C.18)}$$

These conditions are known as the Gibbs–Konovalov conditions for binary homogeneous azeotropes, and any one implies the other two. As discussed earlier, when the compositions of both phases are identical, the mixture exhibits a homogeneous azeotrope. Clearly, the dew point and bubble point curves touch at such a point by virtue of the fact that $x_1 = y_1$ and the equilibrium temperature is the same in each phase. What the Gibbs–Konovalov conditions show is that when the mixture is azeotropic, the curves touch with a common tangent of zero slope.[6] The reverse of this is also true; if the equilibrium curves exhibit a stationary point, then the coexisting phases must have identical composition. As discussed by Malesiński (1965, Chap. V), similar results hold for isothermal phase diagrams. Differentiating Eqs. C.14 and C.15 with respect to x_1 and y_1, respectively, leads to simple thermodynamic inequalities that distinguish between maximum-boiling, minimum-boiling, and inflecting binary azeotropes (Malesiński, 1965, pp. 60–63).

Equations C.14 and C.15 may be generalized to multicomponent mixtures and written in the form (Malesiński, 1965, Chap. V)

$$\nabla_x T = \beta_1 \mathbf{g}^L\,(\mathbf{x} - \mathbf{y}) \qquad \text{(C.19)}$$

and

$$\nabla_y T = \beta_2 \mathbf{g}^V\,(\mathbf{x} - \mathbf{y}) \qquad \text{(C.20)}$$

where $\nabla_x T$ is the vector of derivatives $\left(\frac{\partial T}{\partial x_j}\right)_{P,\mathbf{x}'}$, $j = 1, 2, \ldots, c-1$ along the bubble point surface and $\nabla_y T$ is the corresponding vector of derivatives $\left(\frac{\partial T}{\partial y_j}\right)_{P,\mathbf{y}'}$, $j = 1, 2, \ldots, c-1$ along the dew point surface. The quantities β_1 and β_2 are the scalar functions $T/\Sigma_{i=1}^{c} y_i \Delta h_i$ and $T/\Sigma_{i=1}^{c} x_i \Delta h_i$, respectively. These functions are strictly positive away from the critical region. The matrices $\mathbf{g}^L$ and $\mathbf{g}^V$ are each $(c-1) \times (c-1)$ square symmetric matrices with elements $\left(\frac{\partial^2 g^L}{\partial x_i \partial x_j}\right)_{T,P,\mathbf{x}'}$ and $\left(\frac{\partial^2 g^V}{\partial y_i \partial y_j}\right)_{T,P,\mathbf{y}'}$, respectively.

[6] In some sources, homogeneous binary azeotropic phase diagrams are drawn with a cusp in one of the curves at the azeotropic point. This is incorrect.

For materially stable liquid and vapor phases, $\mathbf{g}^L$ and $\mathbf{g}^V$ are positive definite (hence nonsingular). Therefore,

$$\nabla_x T = \mathbf{0} \Leftrightarrow \nabla_y T = \mathbf{0} \Leftrightarrow \mathbf{x} = \mathbf{y} \tag{C.21}$$

Hence, the Gibbs–Konovalov conditions extend to multicomponent mixtures.

It is a straightforward application of the Gibbs phase rule to show that the equality of temperature, pressure, chemical potential, and composition across phases results in a thermodynamic state with one degree of freedom, regardless of the number of components present. Therefore, at constant pressure (or temperature), homogeneous azeotropes occur at *isolated* points in the thermodynamic state space.[7] This result, coupled with the Gibbs–Konovalov conditions, shows that homogeneous azeotropes occur when the isobaric (or isothermal) dew point and bubble point surfaces touch with a common tangent of zero slope, and that this necessarily happens at isolated points.

Thus, it is not possible to have a line of points where $x_i = y_i,\ \forall i = 1, 2, \ldots, c-1$, or where $\left(\frac{\partial T}{\partial x_i}\right)_{P,\mathbf{x}'} = 0,\ \forall i = 1, 2, \ldots, c-1$, or where $\left(\frac{\partial T}{\partial y_i}\right)_{P,\mathbf{y}'} = 0,\ \forall i = 1, 2, \ldots, c-1$, in an isobaric (or isothermal) phase diagram. However, it is possible to find separate families of lines where (*a*) $x_i = y_i$ for each component separately, (*b*) $\left(\frac{\partial T}{\partial x_i}\right)_{P,\mathbf{x}'} = 0$ for each component separately, (*c*) $\left(\frac{\partial T}{\partial y_i}\right)_{P,\mathbf{y}'} = 0$ for each component separately, and (*d*) $\alpha_{ij} = 1$ for each pair of components separately. These lines correspond to a variety of topographical features of the boiling temperature surfaces [a more complete discussion of these features is given by Swietoslawski (1963) and Van Dongen and Doherty (1984)].

Using the laws and methods of classical thermodynamics, we have succeeded in characterizing the main features of the boiling temperature surfaces for isobaric homogeneous multicomponent azeotropic mixtures. We conclude this section with three isobaric ternary phase diagrams, which exhibit increasingly complex features as different third components are added to the binary azeotropic mixture of acetone and methanol. At atmospheric pressure, acetone boils at 56.1°C, methanol at 64.6°C, and the mixture has a minimum-boiling azeotrope at approximately 80 mole% acetone which boils at 55.4°C (Gmehling and Onken, 1977, p. 85). Ethanol boils at 78.4°C and forms no additional azeotropes when mixed with methanol and acetone (Gmehling and Onken, 1977, pp. 60, 328, 617). A schematic representation of the resulting ternary phase diagram is shown in Fig. C.4. The most prominent feature on this phase diagram is a valley which runs from the pure ethanol vertex to the acetone-methanol face. The liquid and vapor boiling temperature surfaces do not touch at any point along the length of the valley, except at the endpoints. Figure C.5 shows the effect of replacing ethanol with dichloromethane. This schematic diagram was constructed from the data given by Ewell and Welch (1945) and from the slightly

[7] This result is sometimes misinterpreted as meaning that the azeotrope occurs at a *unique point* under isobaric or isothermal conditions. This is false, since it is well known that a system of independent nonlinear algebraic equations with zero degrees of freedom may generate multiple isolated solutions. Moreover, multiple azeotropy has been observed experimentally and can be successfully modeled with a variety of representations of the liquid phase nonidealities (Malesiński, 1965, Chap. 1; Gaw and Swinton, 1966; Doherty and Perkins, 1978).

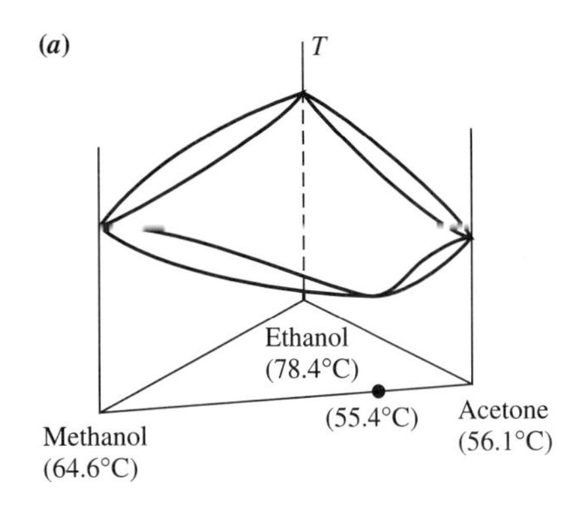

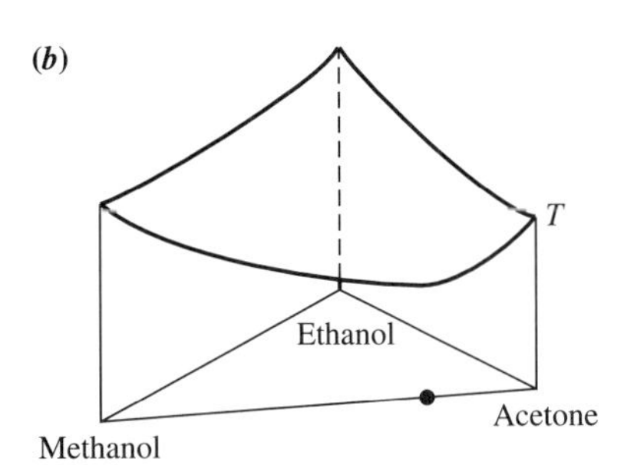

FIGURE C.4
Schematic representation of the isobaric phase diagram for the ternary mixture, acetone + methanol + ethanol at 1 atm pressure. Fig. C.4*a* shows the three binary faces, Fig. C.4*b* shows the liquid boiling temperature surface.

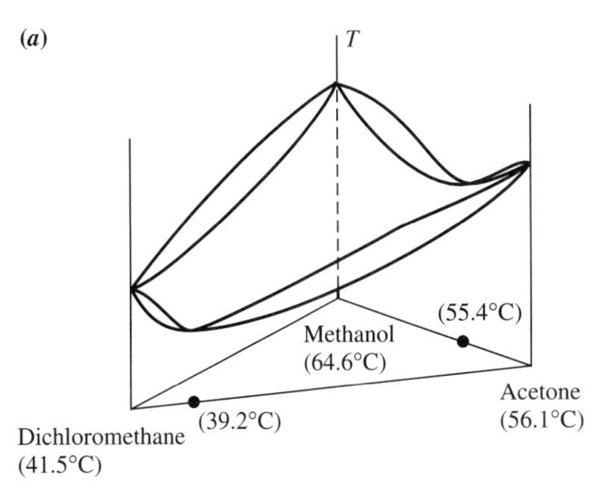

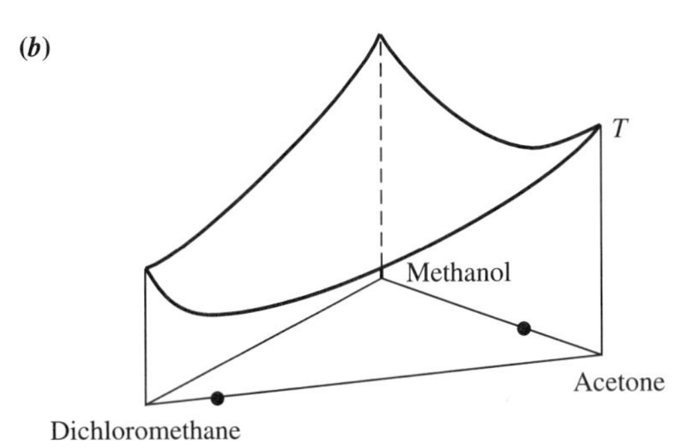

FIGURE C.5
Schematic representation of the isobaric phase diagram for the ternary mixture, acetone + methanol + dichloromethane at 1 atm pressure. Figure C.5*a* shows the three binary faces; Fig. C.5*b* shows the liquid boiling temperature surface.

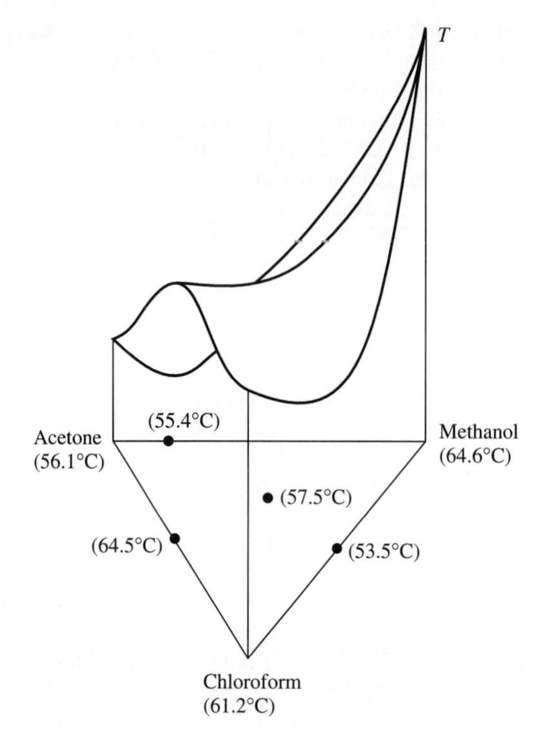

FIGURE C.6
Schematic representation of the isobaric liquid boiling temperature surface for the ternary mixture, acetone + methanol + chloroform at 1 atm pressure.

lower pressure data (1 bar instead of 1 atm) given in Gmehling and Onken (1977, p. 24). There is now a second minimum-boiling binary azeotrope, between dichloromethane and acetone, which boils at 39.2°C. The resulting phase diagram shows a valley which now runs between the acetone-methanol face and the dichloromethane-acetone face. Chloroform makes the mixture even more complex, as shown in Fig. C.6. In addition to the binary azeotrope between acetone and methanol, there is now a minimum-boiling azeotrope between chloroform and methanol, a maximum-boiling azeotrope between acetone and chloroform, and a ternary saddle azeotrope (see Gmehling and Onken, 1977, pp. 19, 609–611; Gmehling et al., 1979, p. 106). Using this mixture, Ewell and Welch (1945) provided the first experimental demonstration of the existence of ternary saddle azeotropes.[8]

[8]In 1955, Wilson et al. (1955) reported the existence of saddle azeotropes in ternary mixtures of aniline and water with benzene, toluene, and xylene. Since then, many more saddle azeotropes have been discovered, and they are now considered commonplace.

References

Cahn, J. W., "Phase Separation by Spinodal Decomposition in Isotropic Systems," *J. Chem. Phys.,* **42,** 93 (1965).

Cahn, J. W., and J. E. Hilliard, "Free Energy of a Nonuniform System. I. Interfacial Free Energy," *J. Chem. Phys.,* **28,** 258 (1958).

Christian, J. W., *The Theory of Transformations in Metals and Alloys.* Pergamon, Oxford UK (1965).

Doherty, M. F., and J. D. Perkins, "On the Dynamics of Distillation Processes—II. The Simple Distillation of Model Solutions," *Chem. Engng. Sci.,* **33,** 569–578 (1978).

Ewell, R. H., and L. M. Welch, "Rectification in Ternary Systems Containing Binary Azeotropes," *Ind. Eng. Chem.,* **37,** 1224–1231 (1945).

Gaw, W. J., and F. L. Swinton, "Occurrence of a Double Azeotrope in the Binary System Hexafluorobenzene-Benzene," *Nature,* **212,** 283 (1966).

Gmehling, J., and U. Onken, *Vapour-Liquid Equilibrium Data Collection,* vol. 1/2a, Organic Hydroxy Compounds: Alcohols, of *Chemistry Data Series.* DECHEMA, Frankfurt/Main (1977).

Gmehling, J., U. Onken, and W. Arlt, *Vapour-Liquid Equilibrium Data Collection,* vol. 1/3+4, Aldehydes and Ketones, Ethers, of *Chemistry Data Series.* DECHEMA, Frankfurt/Main (1979).

Kwei, T. K., and T. T. Wang, Phase Separation Behavior of Polymer-Polymer Mixtures, in Paul, D. R., and S. Newman, editors, *Polymer Blends,* vol. 1. Academic Press, New York (1978).

Malesiński, W., *Azeotropy and Other Theoretical Problems of Vapour-Liquid Equilibrium.* Interscience, London (1965).

Michelsen, M. L., "The Isothermal Flash Problem. Part I. Stability," *Fluid Phase Equilibria,* **9,** 1–19 (1982).

Mirskey, L., *An Introduction to Linear Algebra.* Oxford University Press, London (1961).

Modell, M., and R. C. Reid, *Thermodynamics and Its Applications.* Prentice-Hall, Englewood Cliffs, NJ, 2d ed. (1983).

Prausnitz, J. M., R. N. Lichtenthaler, and E. G. deAzevdeo, *Molecular Thermodynamics of Fluid Phase Equilibria.* Prentice-Hall, Englewood Cliffs, NJ, 3d ed. (1998).

Prigogine, I., and R. Defay, *Chemical Thermodynamics.* Longmans, Green and Co., London, 4th ed. (1967).

Reid, R. C., "Superheated Liquids. A Laboratory Curiosity and, Possibly, An Industrial Curse, Part 1.," *Chem. Eng. Educ.,* Spring, 60 (1978*a*).

Reid, R. C., "Superheated Liquids. A Laboratory Curiosity and, Possibly, An Industrial Curse, Part 2.," *Chem. Eng. Educ.,* Summer, 108 (1978*b*).

Reid, R. C., "Superheated Liquids. A Laboratory Curiosity and, Possibly, An Industrial Curse. Part 3.," *Chem. Eng. Educ.*, Fall, 194 (1978*c*).

Rowlinson, J. S., *Liquids and Liquid Mixtures.* Butterworths, London, 2d ed. (1969).

Smith, J. M., and H. C. VanNess, *Introduction to Chemical Engineering Thermodynamics.* McGraw-Hill, New York, 5th ed. (1966).

Swietoslawski, W., *Azeotropy and Polyazeotropy.* Pergamon Press, New York (1963).

Tester, J. W., and M. Modell, *Thermodynamics and Its Applications.* Prentice-Hall PTR, Upper Saddle River, NJ, 3d ed. (1997). Revised edition of Modell and Reid (1983).

Van Dongen, D. B., and M. F. Dohery, "On the Dynamics of Distillation Processes—V. The Topology of the Boiling Temperature Surface and Its Relation to Azeotropic Distillation," *Chem. Engng. Sci.*, **39,** 883 (1984).

Wilson, R. Q., W. H. Mink, H. P. Munger, and J. W. Clegg, "Dehydration of Hydrazine by Azeotropic Distillation," *AIChE J.*, **1,** 220 (1955).

Credits

Chapter 1

Figure 3 from "Process Synthesis" by J. J. Siirola in *ADVANCES IN CHEMICAL ENGINEERING*, Volume **23,** Copyright ©1996 by Academic Press, reproduced by permission of the publisher. All rights of reproduction in any form reserved.

Figure 5 from "High Purity Methyl Acetate via Reactive Distillation," V. H. Agreda, L. R. Partin and W. H. Heise, *Chemical Engineering Progress,* **86**(2), 40-46 (1990). Reprinted with permission from Eastman Chemical Company, Copyright ©2000 Eastman Chemical Company. All rights reserved.

Chapter 4

Figure 10 reprinted with permission from "Estimation of the Number of Theoretical Plates as a Function of the Reflux Ratio" E. R. Gilliland, *Industrial & Engineering Chemistry,* **32,** 1220-1223, (1940). Copyright ©1940 American Chemical Society.

Figure 15a from "Feasibility of Separations for Distillation of Nonideal Ternary Mixtures," Z. T. Fidkowski, M. F. Malone and M. F. Doherty, *AIChE Journal,* **39,** 1303-1321 (1993). Reproduced with permission of the American Institute of Chemical Engineers. Copyright ©1993 AIChE. All rights reserved.

Figures 16, 17, 18, 19, 24, 25, 26, 27, 28, 41, 42 and 43 from "Nonideal Multicomponent Distillation: Use of Bifurcation Theory for Design," Z. T. Fidkowski, M. F. Malone and M. F. Doherty, *AIChE Journal,* **37,** 1761-1779 (1991). Reproduced with permission of the American Institute of Chemical Engineers. Copyright ©1991 AIChE. All rights reserved.

Figures 20, 22 and 23 reprinted with permission from "Design and Synthesis of Homogeneous Azeotropic Distillations: 2. Minimum Reflux Calculations for Nonideal and Azeotropic Columns," S. F. Levy, D. B. Van Dongen, and M. F. Doherty, *Industrial & Engineering Chemistry Fundamentals,* **24,** 463-474 (1985). Copyright ©1985 American Chemical Society.

Figures 30, 31, 32, 33 and 34 reprinted from Chemical Engineering Science, **45,** V. Julka and M. F. Doherty, "Geometric Behavior and Minimum Flows for Nonideal Multicomponent Distillation," 1801-1822, copyright ©1990, with permission from Elsevier Science.

Figures 35, 36 and 37 reprinted from Chemical Engineering Science, **48,** V. Julka and M. F. Doherty, "Geometric Nonlinear Analysis of Multicomponent Nonideal Distillation: A Simple Computer-Aided Design Procedure," 1367-1391, copyright ©1993, with permission from Elsevier Science.

Chapter 5

Figure 7b from "Consistency Test of Ternary Azeotropic Data by Use of Simple Distillation," Y. Yamakita, and H. Matsuyama, *Journal of Chemical Engineering of Japan,* **16,** 145-146 (1983). Reproduced with permission of the Society of Chemical Engineers, Japan. Copyright ©1983 SCEJ.

Figures 8a and 8c reprinted from *Chemical Engineering Science,* **39,** D. B. Van Dongen, and M. F. Doherty, "On the Dynamics of Distillation Processes. V. The Topology of the Boiling Temperature Surface and its Relation to Azeotropic Distillation," 883-892, copyright ©1984, with permission from Elsevier Science.

Figures 8b and 14 reprinted with permission from "Design and Synthesis of Homogeneous Azeotropic Distillations: 1. Problem Formulation for a Single Column," D. B. Van Dongen, and M. F. Doherty, *Industrial Engineering Chemistry Fundamentals,* **24,** 454-462 (1985). Copyright ©1985 American Chemical Society.

Figure 9 from "Isobaric Liquid-Vapor Equilibrium in a Ternary System with an Azeotrope of the Saddlepoint Type," I. N. Bushmakin and I. N. Kish, *J. Appl. Chem. USSR (Engl. Trans.),* **30,** 205-215 (1957). Reproduced with permission of Plenum Publishers. Copyright ©1957 Plenum Publishers. All rights reserved.

Figure 10b reprinted from Chemical Engineering Science, **43,** D. Barbosa and M. F. Doherty, "The Simple Distillation of Homogeneous Reactive Mixtures," 541-550, copyright ©1988, with permission from Elsevier Science.

Figures 11, 12, 18b, 19, 20, 21, 24, 26, 37 and 38 from "Feasibility of Separations for Distillation of Nonideal Ternary Mixtures," Z. T. Fidkowski, M. F. Malone and M. F. Doherty, *AIChE Journal,* **39,** 1303-1321 (1993). Reproduced with permission of the American Institute of Chemical Engineers. Copyright ©1993 AIChE. All rights reserved.

Figure 13a reprinted with permission from "Three Component Distillation at Total Reflux," K. W. Free and H. P. Hutchison, p. 231-237 in *Proceedings of the International Symposium on Distillation,* Brighton, England. Copyright ©(1960) The Institution of Chemical Engineers.

Figure 13b from "Distribution of Liquid Composition along the Height of Columns of Various Types," N. V. Lutugina, O. F. Kovalichev, L. P. Shandalova and I. V. Antipina, *Theoretical Foundations of Chemical Engineering,* **7**(2), 234-237 (1974). Reproduced with permission of Plenum Publishers. Copyright ©1974 Plenum Publishers. All rights reserved.

Figure 15 reprinted with permission from "Distillation Lines for Multicomponent Separation in Packed Columns: Theory and Comparison with Experiment," S. Pelkonen, R. Kaesemann, and A. Gorak, *Industrial & Engineering Chemistry Research,* **36,** 5392-5398 (1997). Copyright ©1997 American Chemical Society.

Figure 16 reprinted from *Chemical Engineering Science,* **40,** M. F. Doherty, "The Presynthesis Problem for Homogeneous Azeotropic Distillation has a Unique Explicit Solution," 1885-1889, copyright ©1985, with permission from Elsevier Science.

Figures 35 and 57 reprinted with permission from "Automatic Screening of Entrainers for Homogeneous Azeotropic Distillation," E. R. Foucher, M. F. Malone and M. F. Doherty, *Industrial Engineering Chemistry Research,* **30,** 762-772 (1991). Copyright ©1991 American Chemical Society.

Figures 36, 39, 40, 41 and 42 from "Minimum Entrainer Flows for Extractive Distillation: A Bifurcation Theoretic Approach," J. P. Knapp and M. F. Doherty, *AIChE Journal,* **40,** 243-268 (1994). Reproduced with permission of the American Institute of Chemical Engineers. Copyright ©1994 AIChE. All rights reserved.

Figure 44 reprinted with permission from "A New Pressure-Swing Distillation Process for Separating Homogeneous Azeotropic Mixtures," J. P. Knapp and M. F. Doherty, *Industrial Engineering Chemistry Research,* **31,** 346-357 (1992). Copyright ©1992 American Chemical Society.

Figure 50 reprinted with permission from "Vapor-Liquid Equilibrium in Perfluorobenzene-Benzene-Methylcyclohexane System," J. C. Wade and Z. L. Tayler Jr., *Journal of Chemical Engineering Data,* **18,** 424 (1973). Copyright ©1973 American Chemical Society.

Figure 51 reprinted with permission from "Isobaric Vapor-Liquid Equilibria for 2-Methoxy-2-methylpropane + Ethanol + Octane and Constituent Binary Systems at 101.3 kPa," T. Hiaki, K. Tatsuhana, T. Tsuji, and M. Hongo, *Journal of Chemical & Engineering Data,* **44,** 323-327 (1999). Copyright ©1999 American Chemical Society.

Figure 52 reprinted with permission from "Isobaric Vapor-Liquid Equilibria for the Ternary System Acetone-Ethyl Acetate-Ethanol," R. Carta, S. Dernini, and P. Sanna, *Journal of Chemical & Engineering Data,* **29,** 463-466 (1984). Copyright ©1984 American Chemical Society.

Figures 53 and 54 reprinted with permission from "Design and Synthesis of Homogeneous Azeotropic Distillation. 3. The Sequencing of Columns for Azeotropic and Extractive Distillations," M. F. Doherty and G. A. Caldarola, *Industrial & Engineering Chemistry Fundamentals,* **24,** 424-485 (1985). Copyright ©1985 American Chemical Society.

Chapter 6

Figure 1 from "Distillation," J. R. Fair, Chapter 5 in *Handbook of Separation Process Technology,* 1st Edition, R. W. Rousseau (ed) Wiley, New York. Copyright ©1987 Reprinted by permission of John Wiley & Sons, Inc.

Figure 3 reproduced with permission from *Chemical Engineering Handbook,* R. Perry (ed), 6th edition. Copyright ©1984 the McGraw-Hill Companies.

Chapter 7

Figure 4 reprinted with permission from "A Geometric Design Method for Side-Stream Distillation Columns," R. E. Rooks, M. F. Malone and M. F. Doherty, *Industrial & Engineering Chemistry Research,* **35,** 3653-3664 (1996). Copyright ©1996 American Chemical Society.

Figures 17 and 19 reprinted with permission from "Synthesis of Distillation Column Configurations for a Multicomponent Separation," R. Agrawal, *Industrial Engineering Chemistry Research,* **35,** 1059-1071 (1996). Copyright ©1996 American Chemical Society.

Figure 20 reproduced with permission from "Improving Efficiency of Distillation with New Thermally Coupled Configurations of Columns," R. Agrawal and Z. T. Fidkowski, *AIChE Symp. Series,* **96,** 381-384 (2000). Copyright ©2000 Air Products and Chemicals, Inc.

Figures 21, 22, 23, 24, 25, 26, 27 and 29 from "Structure of Distillation Regions for Multicomponent Azeotropic Mixtures," R. E. Rooks, M. F. Doherty, and M. F. Malone, *AIChE Journal,* **44,** 1382-1391 (1998). Reproduced with permission of the American Institute of Chemical Engineers. Copyright ©1998 AIChE. All rights reserved.

Figures 32 and 33 from "A Simple Synthesis Method Based on Utility Bounding for Heat-Integrated Distillation Sequences," M. J. Andrecovich and A. W. Westerberg, *AIChE Journal,* **31,** 363-375 (1985). Reproduced with permission of the American Institute of Chemical Engineers. Copyright ©1985 AIChE. All rights reserved.

Chapter 8

Figures 1, 2, 3 and 4 reprinted from *Chemical Engineering Science,* **45,** H. N. Pham and M. F. Doherty, "Design and Synthesis of Heterogeneous Azeotropic Distillations. I. Heterogeneous Phase Diagrams," 1823-1826, copyright ©1990, with permission from Elsevier Science.

Figures 5, 8, 12, 13 and 20 from "Design/Optimization of Ternary Heterogeneous Azeotropic Distillation Sequences," P. J. Ryan and M. F. Doherty, *AIChE Journal,* **35,** 1592-1601 (1989). Reproduced with permission of the American Institute of Chemical Engineers. Copyright ©1989 AIChE. All rights reserved.

Figure 7 reprinted from *Chemical Engineering Science,* **45,** H. N. Pham and M. F. Doherty, "Design and Synthesis of Heterogeneous Azeotropic Distillations. II. Residue Curve Maps," 1837-1843, copyright ©1990, with permission from Elsevier Science.

Figure 11 reprinted from *Chemical Engineering Science,* **45,** H. N. Pham and M. F. Doherty, "Design and Synthesis of Heterogeneous Azeotropic Distillations. III. Column Sequences," 1845-1854, copyright ©1990, with permission from Elsevier Science.

Chapter 9

Figures 4 and 10 reprinted from *Chemical Engineering Science,* **45,** C. Bernot, M. F. Doherty and M. F. Malone, "Patterns of Composition Change in Multicomponent Batch Distillation," 1207-1221, copyright ©1990, with permission from Elsevier Science.

Figure 6 reprinted from *Chemical Engineering Science,* **40,** D. B. Van Dongen and M. F. Doherty, "On the Dynamics of Distillation Processes. VI. Batch Distillation," 2087-2093, copyright ©1985, with permission from Elsevier Science.

Figure 8 from "Feasibility of Separations for Distillation of Nonideal Ternary Mixtures," Z. T. Fidkowski, M. F. Malone and M. F. Doherty, *AIChE Journal,* **39,** 1303-1321 (1993). Reproduced with permission of the American Institute of Chemical Engineers. Copyright ©1993 AIChE. All rights reserved.

Figures 9, 11 and 31 reprinted with permission from "Rectification in Ternary Systems Containing Binary Azeotropes," R. H. Ewell and L. M. Welch, *Industrial & Engineering Chemistry,* **37,** 1224-1231 (1945). Copyright ©1945 American Chemical Society.

Figures 13, 15, 16, 17, 18, 19, 20, 26, 27, 28, 29 and 30 reprinted from *Chemical Engineering Science,* **45,** C. Bernot, M. F. Doherty and M. F. Malone, "Feasibility and Separation Sequencing in Multicomponent Batch Distillation," 1311-1326, copyright ©1991, with permission from Elsevier Science.

Figures 21b and 22 reprinted with permission from "Design and Operating Targets for Nonideal Multicomponent Batch Distillation," C. Bernot, M. F. Doherty and M. F. Malone, *Industrial & Engineering Chemistry Research,* **32,** 293-301 (1993). Copyright ©1993 American Chemical Society.

Chapter 10

Figures 4 and 6 reprinted with permission from "Synthesis of Reactive Distillation Systems with Multiple Equilibrium Chemical Reactions," S. Ung and M. F. Doherty, *Industrial & Engineering Chemistry Research,* **34,** 2555-2565 (1995). Copyright ©1995 American Chemical Society.

Figure 11 reprinted with permission from "Equilibrium Constant for the Methyl tert-Butyl Ether Liquid-Phase Synthesis by Use of UNIFAC," F. Colombo, L. Cori, L. Dalloro, and P. Delogu, *Industrial & Engineering Chemistry Fundamentals,* **22,** 219-223 (1983). Copyright ©1983 American Chemical Society.

Figures 15, 16 and 17 reprinted from *Chemical Engineering Science,* **43,** D. Barbosa and M. F. Doherty, "The Influence of Equilibrium Chemical Reactions on Vapor-Liquid Phase Diagrams," 529-540, copyright ©1988, with permission from Elsevier Science.

Figure 19 reprinted with permission from "New Set of Composition Variables for the Representation of Reactive Phase Diagrams," D. Barbosa and M. F. Doherty, *Proc. R. Soc. London Ser. A,* **413,** 459-464 (1989). Copyright ©1989 The Royal Society.

Figure 22 reprinted from *Chemical Engineering Science,* **50,** S. Ung and M. F. Doherty, "Vapor-Liquid Phase Equilibrium in Systems with Multiple Chemical Reactions," 23-48, copyright ©1995, with permission from Elsevier Science.

Figures 23, 24 and 25 reprinted from *Chemical Engineering Science,* **43,** D. Barbosa and M. F. Doherty, "Simple Distillation of Homogeneous Reactive Mixtures," 541-550, copyright ©1988, with permission from Elsevier Science.

Figure 26 reprinted with permission from "Calculation of Residue Curve Maps for Mixtures with Multiple Equilibrium Chemical Reactions," S. Ung and M. F. Doherty, *Industrial & Engineering Chemistry Research,* **34,** 3195-3202 (1995). Copyright ©1995 American Chemical Society.

Figure 27a reprinted from *Chemical Engineering Science,* **43,** D. Barbosa and M. F. Doherty, "Design and Minimum-Reflux Calculations for Single Feed Multicomponent Reactive Distillation Columns," 1523-1537, copyright ©1988, with permission from Elsevier Science.

Figure 27b reprinted with permission from "Measurement of Residue Curve Maps and Heterogeneous Kinetics in Methyl Acetate Synthesis," W. Song, G. Venimadhavan, J. M. Manning, M. F. Malone and M. F. Doherty, *Industrial & Engineering Chemistry Research,* **37,** 1917-1928 (1998). Copyright ©1998 American Chemical Society.

Figure 28 reprinted by permission from *Nature* (http://www.nature.com) "Discovery of a Reactive Azeotrope," W. Song, R. S. Huss, M. F. Doherty and M. F. Malone, **388,** 561-563, copyright ©1997 Macmillan Magazines, Ltd.

Figure 30 reprinted from *Chemical Engineering Science,* **43,** D. Barbosa and M. F. Doherty, "Design and Minimum-Reflux Calculations for Double-Feed Multicomponent Reactive Distillation Columns," 2377-2389, copyright ©1988, with permission from Elsevier Science.

Figure 37 reprinted with permission from "Circumventing an Azeotrope in Reactive Distillation," J. W. Lee, S. Hauan and A.W. Westerberg, *Industrial & Engineering Chemistry Research,* **39,** 1061-1063 (2000). Copyright ©2000 American Chemical Society.

Figure 38 reprinted with permission from "Design Method for Kinetically Controlled Staged Reactive Distillation Columns," M. J. Okasinski and M. F. Doherty, *Industrial & Engineering Chemistry Research,* **37,** 2821-2834 (1998). Copyright ©1998 American Chemical Society.

Figures 39 and 40 from "Computer-Aided Tools for the Design of Reactive Distillation Systems," R.S. Huss, F. Chen, M. F. Malone and M. F. Doherty, *Computers and Chemical. Engineering Supplement* (1999) S955-S962. Reproduced with permission of Pergamon Press. Copyright ©1999 Elsevier Science Ltd.

Appendix A

Figures 3, 4 and 5 reprinted with permission from "Design and Synthesis of Homogeneous Azeotropic Distillations. 5. Columns with Nonnegligible Heat Effects," J. R. Knight and M. F. Doherty, *Industrial & Engineering Chemistry Fundamentals,* **25,** 279-289 (1986). Copyright ©1986 American Chemical Society.

Index